Vincenzo Dovì and Antonella Battaglini (Eds.)

Energy Policy and Climate Change

MDPI

This book is a reprint of the Special Issue that appeared in the online, open access journal, *Energies* (ISSN 1996-1073) in 2015 (available at: http://www.mdpi.com/journal/energies/special_issues/energy-policy-climate-change).

Guest Editors
Vincenzo Dovì
University of Genoa
Italy

Antonella Battaglini
Potsdam Institute for Climate Impact Research (PIK)
Germany

Editorial Office
MDPI AG
Klybeckstrasse 64
Basel, Switzerland

Publisher
Shu-Kun Lin

Senior Assistant Editor
Guoping (Terry) Zhang

1. Edition 2016

MDPI • Basel • Beijing • Wuhan • Barcelona

ISBN 978-3-03842-157-3 (Hbk)
ISBN 978-3-03842-158-0 (PDF)

Table of Contents

Chapter 2: Corporate Policies and Investment Decisions

Chapter 3: Public Policy Issues

Chapter 4: Global Phenomena and Global Governance

List of Contributors

Claudio A. Agostini: School of Government, Universidad Adolfo Ibañez, Avenida Diagonal Las Torres 2640, Peñalolén, Santiago 7941169, Chile.

Sharifah Rafidah Wan Alwi: Process Systems Engineering Centre (PROSPECT), Faculty of Chemical Engineering, Universiti Teknologi Malaysia, 81310 UTM Johor Bahru, Johor, Malaysia.

Reynir Smari Atlason: Department of Industrial Engineering, Mechanical Engineering and Computer Science, Centre for Productivity, Performance and Processes, University of Iceland, Hjardarhagi 6, 107 Reykjavik, Iceland.

Jake Badger: Technical University of Denmark, Wind Energy Department, Frederiksborgvej 399, Roskilde 4000, Denmark.

Antonella Battaglini: Potsdam Institute for Climate Impact Research (PIK), Research Domain IV—Transdisciplinary Concepts & Methods, Potsdam Institute for Climate Impact Research (PIK), Telegrafenberg A31, Potsdam 14473, Germany.

Carlos Benavides: Department of Electrical Engineering, Energy Center, Faculty of Physical and Mathematical Sciences, University of Chile, Santiago 8370451, Chile.

Andreas Beneking: Potsdam Institute for Climate Impact Research (PIK), Research Domain IV—Transdisciplinary Concepts & Methods, Telegrafenberg A31, 14473 Potsdam, Germany.

Luís Bernardes: EMEPC, Rua Costa Pinto 165, 2770-047 Paço de Arcos, Portugal.

Filipe Brandão: EMEPC, Rua Costa Pinto 165, 2770-047 Paço de Arcos, Portugal.

Júlio Carneiro: Departamento de Geociências, Escola de Ciências e Tecnologia, Instituto de Investigação e Formação Avançada, Instituto de Ciências da Terra, Universidade de Évora, Évora, Portugal.

Ana Castillo-Martinez: Department of Computer Sciences, Polytechnic School, University of Alcala, Madrid-Barcelona Road, Km 33.6, Alcala de Henares 28871, Spain.

Andrzej Ceglarz: Potsdam Institute for Climate Impact Research (PIK), Research Domain IV—Transdisciplinary Concepts & Methods, Telegrafenberg A31, 14473 Potsdam, Germany.

Hong Chang: Key Laboratory of Advanced Control and Optimization for Chemical Processes, East China University of Science and Technology, Shanghai 200240, China.

Kew Hong Chew: Process Systems Engineering Centre (PROSPECT), Faculty of Chemical Engineering, Universiti Teknologi Malaysia, 81310 UTM Johor Bahru, Johor, Malaysia.

Soo Cho: Korea Institute of Energy Research, Daejeon 305-343, Korea.

Young-Hum Cho: School of Architecture, Yeungnam University, Gyeongsan 712-749, Korea.

Bo-Eun Choi: Department of Architectural Engineering, Graduate School of Yeungnam University, Gyeongsan 712-749, Korea.

Xuankai Deng: School of Resource and Environmental Sciences, Wuhan University, Wuhan 430079, China.

Alessandro Di Bella: Technical University of Denmark, Wind Energy Department, Frederiksborgvej 399, Roskilde 4000, Denmark.

Manuel Diaz: Department of Electrical Engineering, Energy Center, Faculty of Physical and Mathematical Sciences, University of Chile, Santiago 8370451, Chile.

Vincenzo Dovì: Department of Chemistry and Industrial Chemistry, University of Genoa, Via Dodecaneso 31, Genova 16146, Italy.

Antanas Dumbrauskas: Institute of Water Resources Engineering, Aleksandras Stulginskis University, 10 Universiteto str., Akademija, Kaunas District LT-53361, Lithuania.

Saskia Ellenbeck: Potsdam Institute for Climate Impact Research (PIK), Research Domain IV—Transdisciplinary Concepts & Methods, Telegrafenberg A31, 14473 Potsdam, Germany.

Rosario Ferrara: Department of Law, University of Turin, Lungo Dora Siena, 100 A 10153 Torino, Italy.

Elisha R. Frederiks: CSIRO Adaptive Social and Economic Systems, Ecosciences Precinct, 41 Boggo Road, Dutton Park, QLD 4102, Australia.

Rodrigo Fuentes: Institute of Economics, Pontifical Catholic University of Chile, Santiago 7820436, Chile.

Gonzalo García: Instituto de Economía, Pontificia Universidad Católica de Chile, Santiago 7820436, Chile.

Jose M. Gomez-Pulido: Department of Computer Sciences, Polytechnic School, University of Alcala, Madrid-Barcelona Road, Km 33.6, Alcala de Henares 28871, Spain.

Luis Gonzales: Latin American Center for Economic and Social Policy, Pontifical Catholic University of Chile, Santiago 8331010, Chile.

Alberto Gutierrez-Escolar: Department of Computer Sciences, Polytechnic School, University of Alcala, Madrid-Barcelona Road, Km 33.6, Alcala de Henares 28871, Spain.

Jose-Maria Gutierrez-Martinez: Department of Computer Sciences, Polytechnic School, University of Alcala, Madrid-Barcelona Road, Km 33.6, Alcala de Henares 28871, Spain.

Fanghua Hao: School of Environment, State Key Laboratory of Water Environment Simulation, Beijing Normal University, No. 9 Xinjiekouwai Street, Haidian District, Beijing 100875, China.

Charlotte Bay Hasager: Technical University of Denmark, Wind Energy Department, Frederiksborgvej 399, Roskilde 4000, Denmark.

Yujun He: Department of Electronic & Communication Engineering, North China Electric Power University, Baoding 071003, Hebei, China.

Elizabeth V. Hobman: CSIRO Adaptive Social and Economic Systems, Ecosciences Precinct, 41 Boggo Road, Dutton Park, QLD 4102, Australia.

Romain Husson: Collecte Localisation Satellites, Avenue La Pérouse, Bâtiment le Ponant, Plouzané 29280, France.

Martin Jänicke: Environmental Research Center, Free University Berlin and Institute of Advanced Sustainability Studies (IASS), Potsdam, Patschkauer Weg 51, 14195 Berlin, Germany.

Jae-Hun Jo: Department of Architectural Engineering, Inha University, Incheon 151-402, Korea.

Egidijus Kasiulis: Institute of Water Resources Engineering, Aleksandras Stulginskis University, 10 Universiteto str., Akademija, Kaunas District LT-53361, Lithuania.

Seok-Hyun Kim: Department of Architectural Engineering, Graduate School of Yeungnam University, Gyeongsan 712-749, Korea.

Dagmar Kiyar: Wuppertal Institute for Climate, Environment and Energy, Döppersberg 19, 42103 Wuppertal, Germany.

Jiří Jaromír Klemeš: Centre for Process Integration and Intensification—CPI2, Research Institute of Chemical and Process Engineering—MŰKKI, Faculty of Information Technology, University of Pannonia, Egyetem u. 10, Veszprém H-8200, Hungary.

Markus Klimscheffskij: Grexel Systems Oy, Lautatarhankatu 6, FI-00580 Helsinki, Finland.

Nadejda Komendantova: International Institute for Applied Systems Analysis (IIASA), Schlossplatz 1, Laxenburg A-2361, Austria; Climate Policy Group, Institute for Environmental Decisions (ETH), Zurich 8092, Switzerland.

José M. Laínez-Aguirre: Department of Chemical Engineering, Universitat Politècnica de Catalunya, Av. Diagonal 647, PG-2, 08028 Barcelona, Spain; Current Address: Department of Industrial and Systems Engineering, University at Buffalo, Amherst, NY 14260, USA.

Marko Lehtovaara: Grexel Systems Oy, Lautatarhankatu 6, FI-00580 Helsinki, Finland.

Diane Lescot: Observatoire des énergies renouvelables, 146 rue de l'Université, 75007 Paris, France.

Wei Li: Department of Economics and Management, North China Electric Power University, No. 619 Yonghua Street, Baoding 071003, Hebei, China.

Yanfang Liu: School of Resource and Environmental Sciences, Wuhan University, Wuhan 430079, China.

Pedro Madureira: EMEPC, Rua Costa Pinto 165, 2770-047 Paço de Arcos, Portugal.

Zainuddin Abdul Manan: Process Systems Engineering Centre (PROSPECT), Faculty of Chemical Engineering, Universiti Teknologi Malaysia, 81310 UTM Johor Bahru, Johor, Malaysia.

Jose-Amelio Medina-Merodio: Department of Computer Sciences, Polytechnic School, University of Alcala, Madrid-Barcelona Road, Km 33.6, Alcala de Henares 28871, Spain.

Shahriyar Nasirov: Facultad de Ingeniería y Ciencias, Universidad Adolfo Ibáñez, Avenida Diagonal Las Torres 2640, Peñalolén, Santiago 7941169, Chile.

Zim Nwokora: Law School, University of Melbourne, Carlton, Victoria 3053, Australia.

Gudmundur Valur Oddsson: Department of Industrial Engineering, Mechanical Engineering and Computer Science, Centre for Productivity, Performance and Processes, University of Iceland, Hjardarhagi 6, 107 Reykjavik, Iceland.

Wei Ouyang: School of Environment, State Key Laboratory of Water Environment Simulation, Beijing Normal University, No. 9 Xinjiekouwai Street, Haidian District, Beijing 100875, China.

Jenny Palm: Department of Thematic Studies—Technology and Social Change, Linköping University, Linköping SE-581 83, Sweden.

Rodrigo Palma-Behnke: Department of Electrical Engineering, Energy Center, Faculty of Physical and Mathematical Sciences, University of Chile, Santiago 8370451, Chile.

Enrica Papa: Environmental and Spatial Management, Faculty of Engineering and Architecture, Ghent University, Vrijdagmarkt 10-301, 9000 Ghent, Belgium.

Alfredo Peña: Technical University of Denmark, Wind Energy Department, Frederiksborgvej 399, Roskilde 4000, Denmark.

Mar Pérez-Fortes: Department of Chemical Engineering, Universitat Politècnica de Catalunya, Av. Diagonal 647, PG-2, 08028 Barcelona, Spain; Current Address: European Commission, Joint Research Centre, Institute for Energy and Transport, P.O. Box 2, 1755 ZG Petten, The Netherlands.

Luis Puigjaner: Department of Chemical Engineering, Universitat Politècnica de Catalunya, Av. Diagonal 647, PG-2, 08028 Barcelona, Spain.

Petras Punys: Institute of Water Resources Engineering, Aleksandras Stulginskis University, 10 Universiteto str., Akademija, Kaunas District LT-53361, Lithuania.

Claudia Raimundo: IT Power Consulting Ltd., St. Brandon's House 29 Great George Street, Bristol BS1 5QT, UK.

Catalina Ravizza: Institute of Economics, Pontifical Catholic University of Chile, Santiago 7820436, Chile.

Gerard Reid: Alexa Capital, 17 Old Court Place, London W8 4PL, UK.

Andrea Pietro Reverberi: Department of Chemistry and Industrial Chemistry (DCCI), University of Genova, Via Dodecaneso 31, Genova 16146, Italy.

Cristina Roque: EMEPC, Rua Costa Pinto 165, 2770-047 Paço de Arcos, Portugal.

David Schillebeeckx: Institute of Physics, Carl von Ossietzky University, Ammerländer Heerstraße 136, 26129 Oldenburg, Germany.

Peter Schmidt: Potsdam Institute for Climate Impact Research (PIK), Research Domain IV—Transdisciplinary Concepts & Methods, Telegrafenberg A31, 14473 Potsdam, Germany.

Christopher A. Scott: School of Geography & Development; Udall Center for Studies in Public Policy, University of Arizona,Tucson, AZ 85719, USA.

Dominik Seebach: Oeko-Institut e.V., Merzhauser Strasse 173, PO Box 17 71, 79017 Freiburg, Germany.

Joshua D. Shackman: College of Business Administration, Trident University International, Cypress, CA 90630, USA.

Pengfei Sheng: School of Economics, Henan University, North Part of Jinming Street, Jinming District, Kaifeng 475004, China.

Kyung-Ju Shin: Department of Architectural Engineering, Graduate School of Yeungnam University, Gyeongsan 712-749, Korea.

Linas Šilinis: Institute of Water Resources Engineering, Aleksandras Stulginskis University, 10 Universiteto str., Akademija, Kaunas District LT-53361, Lithuania.

Carlos Silva: Facultad de Ingeniería y Ciencias, Universidad Adolfo Ibáñez, Avenida Diagonal Las Torres 2640, Peñalolén, Santiago 7941169, Chile.

Zlatko Stapic: Faculty of Organization and Informatics, University of Zagreb, Pavlinska 2, Varazdin 42000, Croatia.

Karen Stenner: Behavioural Economics Team, Behavioural & Social Sciences Group, Adaptive Social & Economic Systems Program, CSIRO (Commonwealth Scientific and Industrial Research Organisation), Dutton Park, Queensland 4102, Australia.

Zachary P. Sugg: School of Geography & Development, University of Arizona, ENR2 Building, 1064 E. Lowell St., P.O. Box 210137, Tucson, AZ 85721, USA.

Wei Sun: School of Economics and Management, North China Electric Power University, Baoding 071003, Hebei, China.

Shuang Sun: Department of Economics and Management, North China Electric Power University, No. 619 Yonghua Street, Baoding 071003, Hebei, China.

Patrik Thollander: Department of Management and Engineering, Division of Energy Systems, Linköping University, Linköping SE-581 83, Sweden; Department of Building, Energy and Environment Engineering, University of Gävle, Gävle SE-801 76, Sweden.

Christof Timpe: Oeko-Institut e.V., Merzhauser Strasse 173, PO Box 17 71, 79017 Freiburg, Germany.

Angela Tschernutter: Energie-Control Austria, Rudolfsplatz 13a, 1010 Vienna, Austria.

Runar Unnthorsson: Department of Industrial Engineering, Mechanical Engineering and Computer Science, Centre for Productivity, Performance and Processes, University of Iceland, Hjardarhagi 6, 107 Reykjavik, Iceland.

Samuel Van Ackere: Environmental and Spatial Management, Faculty of Engineering and Architecture, Ghent University, Vrijdagmarkt 10-301, 9000 Ghent, Belgium.

Thierry Van Craenenbroeck: Vlaamse regulator van de elektriciteits-en gasmarkt, 1000 Brussels, Belgium.

Greet Van Eetvelde: Environmental and Spatial Management, Faculty of Engineering and Architecture, Ghent University, Vrijdagmarkt 10-301, 9000 Ghent, Belgium.

Karel Van Wyngene: Power-Link, Ghent University, Wetenschapspark 1, 8400 Ostend, Belgium.

Lieven Vandevelde: Power-Link, Ghent University, Wetenschapspark 1, 8400 Ostend, Belgium.

Pauline Vincent: Collecte Localisation Satellites, Avenue La Pérouse, Bâtiment le Ponant, Plouzané 29280, France.

Marco Vocciante: Department of Civil, Chemical and Environmental Engineering, University of Genoa, Genoa 16145, Italy.

Patrick J. H. Volker: Technical University of Denmark, Wind Energy Department, Frederiksborgvej 399, Roskilde 4000, Denmark.

Gitana Vyčienė: Institute of Water Resources Engineering, Aleksandras Stulginskis University, 10 Universiteto str., Akademija, Kaunas District LT-53361, Lithuania.

Wenyan Wang: School of Environment, State Key Laboratory of Water Environment Simulation, Beijing Normal University, No. 9 Xinjiekouwai Street, Haidian District, Beijing 100875, China.

Bettina B. F. Wittneben: Environmental Change Institute, School of Geography and the Environment, University of Oxford, Oxford OX1 3QY, UK; Pentland Centre for Sustainability in Business, Lancaster University, Lancaster LA1 4YX, UK.

Gerard Wynn: GWG Energy, 78 Belle Vue Road, Salisbury SP1 3YD, UK.

Jun Yang: School of Economics and Business Administration, Chongqing University, Shanzheng Street 174, Shapingba District, Chongqing 400044, China.

Yanhua Yu: Institute for Interdisciplinary Research, Jianghan University, Wuhan 430056, China.

About the Guest Editors

Vincenzo Dovì graduated in Chemical Engineering and obtained a second Doctor's Degree in Physics. He worked with Tecnimont (Milan, Italy), EMBL (Heidelberg, Germany) and CEA (Fontenay-aux-Roses, France). He is presently Professor of Theory of Industrial Process Development at the University of Genova. He has been President of the European Section of the International Environmetrics Society and Scientific Advisor to the Italian Embassy in Berlin.He participated in several Interministerial Conferences as a Member of the Italian Delegation and Sherpa in one G8 Science Meeting (Carnegie Meeting). He has served as a reviewer in the European Framework Programmes and the EuropeAid Programme. He has been on the Editorial Board of the *Environmetrics* journal and is presently on the Editorial Board of the *Energies* journal. He is the author of over one hundred articles in the areas of his competence, as well as of five books (in Russian) on energy efficiency issues.

Antonella Battaglini is Senior Scientist at the Potsdam Institute for Climate Impact Research (www.pik-potsdam.de) where she leads the work of the SuperSmart Grid Team. The team investigates if it is possible and desirable to move to a 100% renewable electricity sector, and how societal processes and forces influence and define energy policies and investments decisions. Antonella is also the CEO of the Renewables Grid Initiative www.renewables-grid.eu), an organisation dealing with solutions for the energy transition. She is an expert member of the World Economic Forum's Global Agenda Council on the "Future of Electricity". In 2015 she was nominated as one of the Tällberg Foundation's five global leaders. She has authored dozens of articles in the areas of her competence.

Preface

Energy Policy and Climate Change: A Multidisciplinary Approach to a Global Problem

Vincenzo Dovì and Antonella Battaglini

Reprinted from *Energies*. Cite as: Dovì, V.; Battaglini, A. Energy Policy and Climate Change: A Multidisciplinary Approach to a Global Problem. *Energies* **2015**, *8*(12), 13473-13480.

1. Introduction

In the period between the end of the Second World War and the oil crises of 1973 and 1979, the most critical issues in the energy debate were the impending depletion of non-renewable resources and the level of pollution that the environment is able to sustain.

At the time, large investments in nuclear energy technologies were the answer to the growing energy needs, whereas local pollution was reduced by dilution in a wider environment. By the end of the seventies, it was clear that these policies and approaches were largely inadequate.

The Three Mile Island accident—the first of a series of catastrophic events—reduced the diffusion of traditional nuclear reactors around the world. Moreover, also as a consequence, the development of breeder reactors was abandoned in most countries.

Similarly, the attempt to reduce air and water pollution by using higher chimneys or by diluting emissions into larger river basins and lakes, while partially mitigating local problems, was giving rise to severe global issues, including the depletion of the ozone layer, the onset of acid rains and the eutrophication of coastal waters, among others.

Sustainability became the new paradigm that both the academic world and public opinion were ready to embrace. By that time, scientists all over the world were starting to raise concerns about climate change and its impacts at the global level. Research demonstrated the correlation between carbon dioxide emissions and increasing global mean temperatures. Over time, the need to reduce fossil fuel emissions to prevent serious climate change grew together with scientific evidence. Since then, innovation in energy generation, distribution and consumption has become inextricably tied up with climate change research.

Technology was called upon to improve energy efficiency by providing better building construction practices and improved industrial design procedures, to foster energy conservation by developing alternative, less energy-intensive products and to speed up the transition to a carbon-free energy environment by building up renewable energy sources.

Similarly, science has been committed to developing ever more refined models designed to predict the consequences of increasing carbon dioxide levels and other greenhouse gases in the atmosphere.

However, it soon became clear that science and technology alone could not provide the solutions for averting or partially alleviating the negative consequences of global warming and the conflicting objectives of economic growth, affordable prices and sustainability.

Indeed, an interdisciplinary approach including the natural and social sciences as well as economics and investment theory proved crucial to the development and implementation of mitigation and adaptation strategies.

What we are experiencing today is a real revolution that is transforming the global energy sector; it impacts all aspects on how we generate, distribute and consume energy. This revolution is not just a technological process; it is first and foremost, a societal and political process in which technology is one determining factor.

Thus, the role of different stakeholders, as well as societal needs and wants, are to be duly taken into account for the transition to be timely and effective. International agreements, as well as small and large-scale infrastructure projects to support the transition phase, all need to be based on a clear vision of the future and require laborious dealings and negotiation skills.

Finally, a strong juridical framework must be put in place at both the national and supranational levels to provide direction and reduce the risks for investors in implementing innovative solutions.

A multi-disciplinary approach is at the core of any serious action for sustainable development: any decision-making on energy issues should consistently pursue the parallel objectives of preventing serious climate change and protecting the natural environment. A partial or incoherent strategy could lead to ineffective and unsatisfactory results.

The goal of this Special Issue is to bring together a wide range of disciplines and technical fields under the same roof of a thematic area, which can be broadly defined as Energy Policy. Indeed, Energy Policy is called upon to solve some of the most pressing problems that our society is confronted with, *i.e.*, the prevention of dramatic climate change, the disruption of food supply chains, the onset of worldwide water stress and the collapse of political institutions.

As Guest Editors of this Special Issue, we were able to ascertain that this interdisciplinary position on fundamental issues of Energy Policy is now widespread among scientists and scholars, who are well aware that a wider and more embracing paradigm should be established for the Energy Policy Research.

The positive echo from the scientific community, reflected by the number and the quality of the contributions submitted, proves the significance of this general approach and the necessity of examining energy issues from a more general perspective, which should increasingly become part of the general public discourse beyond the strict framework of academic debates. Society at large needs to be actively involved in developing, suggesting and supporting possible and sustainable pathways, which are relevant at both the national and international levels. This will give decision makers the confidence and sense of urgency to develop and implement necessary policies. This is the main message being conveyed by this Special Issue.

Indeed, the articles submitted, as well as the ones finally selected to be published, cover a broad range of thematic areas: science and technology, economics, corporate finance, law, public policy, social analysis, national and international issues. As a matter of fact, several manuscripts straddle different subjects. We regard them as a sign of vitality in academic research and as a further confirmation of the necessary interdisciplinary character of innovative Energy Policy Research.

In the next section, we provide a brief review of the papers published, roughly classifying them according to the previously outlined thematic areas.

2. A Short Review of the Contributions in This Issue

A number of articles can be grouped under the broad heading of Science and Technology.

As far as energy efficiency is concerned, two articles address optimal building construction practices and adequate industrial design procedures respectively.

Kim *et al.* [1] show how the optimal design of shading installations can substantially affect the heating and cooling load for horizontal devices and venetian blinds in office buildings. The significant potential reduction of up to 13% of energy consumption that might be reached in Korea reflects the importance of this type of analysis in the construction sector.

In the article by Chew *et al.* [2], process integration is applied to the important case of non-negligible pressure drops between plant units. Indeed, process integration, based on pinch analysis, has become one of the most powerful techniques for efficient process design. Greeted as a veritable breakthrough in the eighties, pinch analysis has been successively extended to include the energy optimization of whole industrial sites and the energy interactions between industrial processes and the surrounding territory.

The role of technology in the development of renewable energy sources is analyzed in four articles dedicated to wind energy, energy from biomass and small hydropower.

While wind energy technology is approaching a level of maturity allowing it to be considered a well-established and economically viable source of energy, its very diffusion gives rise to new problems that the scientific community is called upon to solve.

Thus, the influence of the wind farm wake on the power production of neighboring wind farms is analyzed by Hasager *et al.* [3] using advanced methods based on satellite synthetic aperture radar observations. Furthermore, as shown by Van Ackere *et al.* [4], the inclusion of small and medium turbines in distributed energy systems, though reducing the overall impact of the spatial footprint, requires more detailed knowledge of the spatial distribution of the average wind speed.

As for the use of biomass in large centralized systems, Puigjaner *et al.* [5] show how an innovative adjustable design platform, conceptually derived from Process Systems Engineering, can be used for the optimal establishment of biomass-based supply chains, which constitute an essential requirement for the introduction of co-combustion/co-gasification projects in the current electricity production scenario.

Punys *et al.* [6] explore the potential negative impact of small hydropower systems on environmental aquatic systems due to sudden changes in the flow rate, especially during start-up and shutdown phases or as a consequence of changes in the electrical load. They provide an effective design method which includes the type of turbines employed and evaluate the largest ramping rate that does not strand or isolate fish populations.

Basic science and advanced engineering methods are employed by Bernardes *et al.* [7] to estimate the areas of profitable exploitation of gas hydrate fields in sub-seabed sediments. In addition to tapping methane hydrates, the technique might provide a suitable storage for a large amount of carbon dioxide sequestered in power plants and other industrial processes.

Contributions dedicated to corporate policies and investment decisions on energy issues include five articles.

In particular, Atlason *et al.* [8] show, on the basis of data related to Iceland's geothermal sector, how the establishment of a critical mass of local experts in the maintenance of innovative infrastructure can give rise to stable clusters based on knowledge and be beneficial to the further development of the sectors involved.

In the article by Thollander and Palm [9], the introduction of operational measures into the energy management system of production processes is recommended as an effective means of overcoming the barriers which reduce the propensity to the adoption of cost-effective measures. In particular, the authors recommend embedding additional models into the frequently used input-output analysis for improving the overall efficiency of energy management systems in the industry. Allowing for periodic adjustments would further enhance their reliability.

Similarly, Nasirov *et al.* [10] examine the barriers to the deployment of renewable energy technologies in Chile, despite the favorable political and economic climate. Using surveys and interviews, they identify these barriers as being caused by grid limitations, excessive red tape, scarce credit and uncertain regulations for the lease of physical commodities. The authors provide recommendations for overcoming these difficulties in the Chilean system.

Kiyar and Wittneben [11] analyze the portfolio management strategies of the four large electricity providers in Germany and come to the conclusion that the fossil fuel divestment campaign launched in 2012 (and presently including over two hundred institutions) is not at the root of their corporate decisions on reducing fossil fuel assets. Rather, they seem to be the result of a compromise brought about by dropping prices, the current composition of stockholder equity, political regulations, and the economic and social importance of lignite extraction in Germany.

Ellenbeck *et al.* [12] question the widespread reliance on the sole market design for secure electricity supplies through the full development of capacity markets at national or regional level. Using an interdisciplinary approach based on economic analysis and sociological insight, they theorize that the market as a social institution is capable of shaping the investment behavior of its participants. The creation of a European Energy Union is proposed as a possible tool for positively influencing the four behavioral determinants identified by the authors.

A number of contributions can be classified as evaluations of policy responses to public problems or as proposals for the modification of existing regulations on energy issues. In other words, they fall under the broad definition of Public Policy.

Two articles consider the effectiveness of CO_2 mitigation measures adopted or recommended in China. Deng et al. [13] assess the consequences of the CO_2 reduction plan contained in the 12th five-year plan (2011–2015) in each of the 30 Chinese provinces by analyzing temporal and spatial variations of emissions. Taking into account the considerable differences encountered in the geographic distribution (mainly between coastal and internal regions), they suggest different targets for the geographical areas with a view of a more stringent future reduction of CO_2 emissions at the national level. Li et al. [14] use a general equilibrium model to forecast the consequences of the introduction of two mitigating measures by the year 2050. In particular, a carbon tax is shown to generate a strong reduction of emissions in the short term, but a decline of economic growth and innovation in the energy sector. On the other hand, the expansion of the emission trading scheme would lead not only to a significant emissions reduction but also to an increased deployment of renewable energy sources. A combination of the two measures is shown to be the optimal strategy.

Two more articles analyze energy issues in China and, using the resulting outcomes, recommend the adoption of suitable public policies. Sheng et al. [15] investigate energy efficiency in China using estimated shadow prices of energy and comparing them with market prices for the evaluation of the level of energy utilization. They show that although considerably higher since 1998, shadow prices are below the market value in eighteen provinces, which implies an inefficient use of energy. A reallocation of inputs is recommended to increase overall energy efficiency. Wang et al. [16] analyze the policies promoting the national industry which produces densified biomass solid fuel from agricultural and forestry residues in China. They recommend the use of a framework based on the supply chain steps to eliminate fluctuations in the availability of biomass feedstocks and consequently enhance the development of this industrial sector.

Benavides et al. [17] examine the implications of the present 5 US$/tCO₂e carbon tax in Chile and the possible consequences of its increase up to 50 US$/tCO₂e, using simulation models. They predict both an emission reduction and a price increase, which depend parametrically on factors beyond the control of policy-makers, including fossil fuel prices and non-conventional renewable energy investment costs. Alternative measures are considered and the resulting emission reductions evaluated.

In their article, Gutierrez-Escolar et al. [18] show that the regulations adopted by the Spanish government for the limitation of the impact of street lighting on the energy consumption of Spanish municipalities have missed the nationally set targets. By analyzing the performances of the various components which make up street lighting, the authors identify some aspects which should be considered for an efficient use of energy.

Reliable forecasts for the future development of national energy markets play a vital role in the correct formulation of energy policies by governments and other decision makers' bodies.

Reid and Wynn [19] investigate the solar power market in the United Kingdom and estimate that large-scale solar farms, as well as commercial and residential rooftop installations, will be able to compete in the market without subsidies in the next decade thanks to expected cost reductions due to technological innovation. The additional grid integration costs required by solar power variability could then be weighed against the advantages of using a secure, environmentally friendly and carbon-free energy source.

The article by Sun *et al.* [20] has a strong methodological content and includes a new algorithm for the forecasting of fossil fuel consumption in the generation of electricity at the national level. Comparing the predictions of their model with the actual data of the Chinese market, they show the high reliability of their approach.

The analysis of global phenomena and global governance institutions constitutes the basis of any consistent energy policy capable of tackling the problems to be solved at an international level.

Scott and Sugg [21] examine, using national and global data, the strong interactions between water scarcity caused by energy generation in several large and diversified economies and, conversely, limitations to energy development due to water shortage (especially in arid areas and small island states). They estimate that a virtuous water-energy-climate cycle is possible and that it should be fostered through institutional arrangements. To this purpose, they suggest a number of coupled energy-water policies.

Jänicke's [22] attempt to ascertain mechanisms for the global diffusion of climate-friendly technologies leads to the identification of three interactive, partly overlapping processes based on virtuous policy-market cycles, the emulation of best practices from domestic markets and pioneer countries and the integration of vertical and horizontal dynamics in multi-level systems of governance. Even if the focus is on the European system of multi-level climate governance, general recommendations are addressed to policy makers.

A strong juridical framework is a prerequisite for any efficient and consistent energy policy. On the other hand, creating the necessary consensus for the establishment of international institutions with the appropriate authority to produce binding rulings becomes more and more arduous with increasing extension of the territorial jurisdiction.

Two articles consider the role of European legislation and the conditions for its enforcement.

Considering a number of European directives and regulations, Ferrara [23] shows that the progress of smart cities in Europe is critically dependent on the further development of energy efficiency and renewable energy sources. However, he finds that for the current pledges for an efficient development of smart cities to be transformed into binding rules at national level, a clear legal framework should be established.

Klimscheffskij *et al.* [24] analyze the reliability of the guarantees of origin of sold electricity by considering the residual mix resulting from the combination of implicit disclosure systems and explicit tracking procedures. They identify a number of double-counting errors related to implicit electricity disclosure and suggest the use of a new residual mix calculation methodology developed in RE-DISS project.

Energy-related decisions have a considerable impact on a number of societal issues, including a direct influence on peoples' lifestyles, as well as negative externalities. Indeed, energy can contribute to the solution of many severe social problems, but it can also aggravate some of them, reinforcing the necessity of involving all stakeholders in the decision-making procedure on energy strategies.

This Special Issue contains three articles on energy-related societal issues.

Stenner and Nwokora [25] question the general validity of Inglehart's post-materialism theory even in affluent countries. Using a large cross-sectoral data set, they argue that particular interests, rather than broad value systems, determine global trends in environmental attitudes. Accordingly, the environment's future champions are expected to be the citizens of the developing nations most at risk of real material harm from climate change and environmental degradation.

Similarly, Frederiks *et al.* [26] regard the variables frequently employed to explain household energy usage as inconclusive. Instead, they suggest that a multitude of factors should be considered to avoid inconsistent interpretations, which might lead to questionable conclusions. Understanding the nature and the mechanisms of these factors might provide useful guidance to policymakers.

Komendantova *et al.* [27] examine the concerns of local stakeholders affected by the construction of electricity transmission infrastructure. Since several thousands of kilometers of new lines have to be constructed and upgraded at the European level to accommodate growing volumes of intermittent renewable electricity, the dialogue between transmission system operators and non-governmental organizations, as exemplified by the BestGrid approach, is crucial. The stakeholders' concerns disclosed in four pilot projects constitute the main subject of the article. The authors present an analysis of the reasons underlying them and evaluate the BestGrid process on the basis of the attitudes of the stakeholders involved in the four projects.

3. Conclusions

It has been argued by the authors of "Limits to Growth" [28] in the 30-Year Update of their book that *"humanity is in overshoot"*, *i.e.*, it has exceeded the limits, beyond which any further evolution can lead to collapse, unless *"the damage from overshoot [is] reduced through wise global policy, changes in technology and institutions, political goals, and personal aspirations"*. Sustainability in energy generation and consumption is one of these limits.

Over the following ten years, more and more evidence has been accumulated suggesting this program is more pressing than ever. The task of energy policy research is to identify means and tools of this *"wise global policy"*. As reflected by the content of this Special Issue, ingenuity and commitment among members of the scientific community have proven to be equal to the task: the political institutions are now called upon to urgently put the resulting knowledge into practice.

Acknowledgments: The authors are grateful to the MDPI Publisher for the invitation to act as guest editors of this special issue and are indebted to the editorial staff of *"Energies"* for the kind co-operation, patience and committed engagement. We would like to give special thanks to the Italian Embassy in Berlin for hosting a panel discussion on the main outcomes of this issue. The panel was composed of the two academic editors, Vincenzo Dovì and Antonella Battaglini, Luis Puigjaner and Rosario Ferrara. In particular, we are indebted to the Director of the Italian Institute of Culture in Berlin, Aldo Venturelli, for including the panel presentation in the events calendar of the Institute, and to Verena Vittur for the management of the organization.

Author Contributions: The authors contributed equally to this work.

Conflicts of Interest: The authors declare no conflict of interest.

References

1. Kim, S.H.; Shin, K.J.; Choi, B.E.; Jo, J.H.; Cho, S.; Cho, Y.H. A Study on the Variation of Heating and Cooling Load According to the Use of Horizontal Shading and Venetian Blinds in Office Buildings in Korea. *Energies* **2015**, *8*, 1487–1504.
2. Chew, K.H.; Klemeš, J.J.; Wan Alwi, S.R.; Zainuddin Abdul Manan, Z.A.; Reverberi, A.P. Total Site Heat Integration Considering Pressure Drop. *Energies* **2015**, *8*, 1114–1137.
3. Hasager, C.B.; Vincent, P.; Badger, J.; Badger, M.; di Bella, A.; Peña, A.; Husson, R.; Volker, P. Using satellite SAR to characterize the wind flow around offshore wind farms. *Energies* **2015**, *8*, 5413–5439.
4. Van Ackere, S.; van Eetvelde, G.; Schillebeeckx, D.; van Wyngene, K.; Vandevelde, L. Wind resource mapping using landscape roughness and spatial interpolation methods. *Energies* **2015**, *8*, 8682–8703.
5. Puigjaner, L.; Pérez-Fortes, M.; Laínez-Aguirre, J.M. Towards a Carbon-neutral Energy Sector: Opportunities and Challenges of Coordinated Bioenergy Supply Chains—A PSE Approach. *Energies* **2015**, *8*, 5613–5660.
6. Punys, P.; Dumbrauskas, A.; Kasiulis, E.; Vyčienė, G.; Šilinis, L. Flow Regime Changes: From Impounding a Temperate Lowland River to Small Hydropower Operations. *Energies* **2015**, *8*, 7478–7501.
7. Bernardes, L.M.; Carneiro, J.; Madureira, P.; Brandão, F.; Roque, C. Definition of priority study areas for coupling CO_2 storage and CH4 gas hydrates recovery in the Portuguese offshore area. *Energies* **2015**, *8*, 10276–10292.
8. Atlason, R.S.; Oddsson, G.V.; Unnthorsson, R. Theorizing for Maintenance Management Improvements: Using Case Studies from the Icelandic Geothermal Sector. *Energies* **2015**, *8*, 4943–4962.
9. Thollander, P.; Palm, J. Industrial energy management decision making for improved energy efficiency—Strategic system perspectives and situated action in combination. *Energies* **2015**, *8*, 5694–5703.
10. Nasirov, S.; Silva, C.; Agostini, C.A. Investors' perspectives on barriers to renewables deployment in Chile. *Energies* **2015**, *8*, 3794–3814.
11. Kiyar, D.; Wittneben, B.B.F. Carbon as Investment Risk—The Influence of Fossil Fuel Divestment on Decision Making at Germany's Main Power Providers. *Energies* **2015**, *8*, 9620–9639.
12. Ellenbeck, S.; Battaglini, A.; Beneking, A.; Ceglarz, A.; Schmidt, P. Security of supply in European electricity markets—Determinants of investment decisions and the European Energy Union. *Energies* **2015**, *8*, 5198–5216.

13. Deng, X.; Yu, Y.; Liu, Y. Temporal and Spatial Variations Provincial CO_2 Emissions and Assessment of a Reduction Plan in China from 2005 to 2015. *Energies* **2015**, *8*, 4549–4571.

14. Li, W.; Li, H.; Sun, S. China's Low-Carbon Scenario Analysis of CO_2 Mitigation Measures towards 2050 Using Hybrid AIM/CGE Model. *Energies* **2015**, *8*, 3529–3555.

15. Sheng, P.; Yang, J.; Shackman, J.D. Energy's Shadow Price and Energy Efficiency in China: A Non-Parametric Input Distance Function Analysis. *Energies* **2015**, *8*, 1975–1989.

16. Wang, W.; Ouyang, W.; Hao, F. A Supply-Chain Analysis Framework for Assessing Densified Biomass Solid Fuel Utilization Policies in China. *Energies* **2015**, *8*, 7122–7139.

17. Benavides, C.; Gonzales, L.; Diaz, M.; Rodrigo Fuentes, R.; García, G.; Palma-Behnke, R.; Catalina Ravizza, C. The impact of a Carbon Tax in the Chilean Electricity Generation Sector. *Energies* **2015**, *8*, 2674–2700.

18. Gutierrez-Escolar, A.; Castillo-Martinez, A.; Gomez-Pulido, J.M.; Gutierrez-Martinez, J.M.; Stapic, Z.; Medina Merodio, J.A. A Study to Improve the Quality of Street Lighting in Spain. *Energies* **2015**, *8*, 976–994.

19. Reid, G.; Wynn, G. The Future of Solar in the United Kingdom. *Energies* **2015**, *8*, 7818–7832.

20. Sun, W.; He, Y.; Chang, H. Forecasting Fossil Fuel Energy Consumption for Power Generation Using QHSA-Based LSSVM Model. *Energies* **2015**, *8*, 939–959.

21. Scott, C.A.; Sugg, Z.P. Global energy development and climate-induced water scarcity—Physical limits, sectoral constraints, and policy imperatives. *Energies* **2015**, *8*, 8211–8225.

22. Jänicke, M. Horizontal and Vertical Reinforcement in Global Climate Governance. *Energies* **2015**, *8*, 5782–5799.

23. Ferrara, R. The Smart City and the Green Economy in Europe: A Critical Approach. *Energies* **2015**, *8*, 4724–4734.

24. Klimscheffskij, M.; Van Craenenbroeck, T.; Lehtovaara, M.; Lescot, D.; Tschernutter, A.; Raimundo, C.; Seebach, D.; Timpe, C. Residual Mix Calculation at the Heart of Reliable Electricity Disclosure in Europe—Case Study on the Effect of the RE-DISS Project. *Energies* **2015**, *8*, 4667–4696.

25. Stenner, K.; Nwokora, Z. Current and Future Friends of the Earth: Assessing Cross-National Theories of Environmental Attitudes. *Energies* **2015**, *8*, 4899–4919.

26. Frederiks, E.; Stenner, K.; Hobman, E. The socio-demographic and psychological predictors of residential energy consumption: A comprehensive review. *Energies* **2015**, *8*, 573–609.

27. Komendantova, N.; Vocciante, M.; Battaglini, A. The BestGrid process: Going beyond the existing practices of stakeholder involvement in electricity transmission projects? *Energies* **2015**, *8*, 9407–9433.

28. Meadows, D.; Randers, J.; Meadows, D. *Limits to Growth: The 30-Year Update*; Chelsea Green Publishing: White River Junction, VT, USA, 2004; pp. xiii-xv.

Chapter 1:
Technology

A Study on the Variation of Heating and Cooling Load According to the Use of Horizontal Shading and Venetian Blinds in Office Buildings in Korea

Seok-Hyun Kim, Kyung-Ju Shin, Bo-Eun Choi, Jae-Hun Jo, Soo Cho and Young-Hum Cho

Abstract: The construction industry has made considerable energy-saving efforts in buildings, and studies of energy-savings are ongoing. Shading is used to control the solar radiation transferred through windows. Many studies have examined the position and type of shading in different countries, but few have investigated the effects of shading installation in Korea. In this study, the case of the shading installation according to the standard of Korea, and variations of the heating and cooling load in the unit area on the performance of the windows were examined. This study compared the variations of the heating and cooling load in the case of horizontal shading and the changing position of venetian blinds. This study confirmed that horizontal shading longer than the standard length in Korea saved a maximum of 13% energy consumption. This study confirmed the point of change of energy consumption by the Solar Heat Gain Coefficient (SHGC) variations. The exterior venetian blinds and those between glazing were unaffected by the SHGC. On the other hand, in the case of a south façade, the interior venetian blinds resulted in 24% higher energy consumption than the installation of horizontal shading in case of Window to Wall Ratio (WWR): 80%, U-value: 2.1 and SHGC: 0.4.

Reprinted from *Energies.* Cite as: Kim, S.-H.; Shin, K.-J.; Choi, B.-E.; Jo, J.-H.; Cho, S.; Cho, Y.-H. A Study on the Variation of Heating and Cooling Load According to the Use of Horizontal Shading and Venetian Blinds in Office Buildings in Korea. *Energies* **2015**, *8*, 1487-1504.

1. Introduction

The Window to Wall Ratio (WWR) of buildings has increased through the modern advances in architectural design. As the window area increases, the performance of windows becomes more important. Most efforts to improve this performance have been based on thermal insulation and air-tightness, when actually the cooling load is affected by the solar radiation transmitted through windows in buildings.

The thermal insulation performance and air-tightness performance help reduce the energy consumption in buildings. In addition, energy labeling, which is regulated in Korea, was proposed to grade the performance of thermal insulation (U-value, $W/m^2 \cdot K$) and air-tightness (flow rate, $m^3/h \cdot m^2$). As a result, the performance of windows has been improved by the manufacturers. On the other hand, solar heat gain increases the cooling load in summer and decreases the heating load in winter. This must be controlled appropriately. The Solar Heat Gain Coefficient (SHGC) and shading were used to consider solar control. Studies of the SHGC and shading are ongoing, and shading has been studied according to the material and method of installation to determine the most efficient shading effect. Despite this, few studies have examined energy consumption through a comprehensive examination of the window orientation and climate.

A range of studies have examined energy consumption in terms of windows. The complex application of window elements was confirmed [1], and the correlation between energy consumption and the impact of energy consumption was analyzed by simulation. On the other hand, they only confirmed energy consumption according to the variation of the window performance. Therefore, the present study confirmed the variation of the heating and cooling load by a similar review and a simulation of base modeling. In addition, the effects of heating and cooling energy consumption and lighting were confirmed by simulation of office buildings [2]. This study confirmed the variation of energy consumption according to the window performance. A study of the energy performance confirmed the correlation with glazing and windows [3]. The elements of this study were the U-value and G-value. Another study considered the elements of the windows for energy consumption and optical comfort [4]. The present study confirmed the importance of the WWR. A study of the change in energy consumption by the influence of WWR was published [5]. On the other hand, these studies did not consider the correlation between the elements of window performance and shading. Studies of a double-skin façade considered various shading factors. The intermediate space of the glass-skin is the installation position of shading, as the shading device is installed in this space. This study confirmed the effects of the shading device on the interior environment [6,7]. In addition, a study of the double skin façade design parameters confirmed that the design process changes according to the variations of the blind position and reflectance [8]. Studies of the external horizontal shading confirmed the variation of solar radiation incidence caused by six types of shading in tropical climates [9]. One study of the envelope design confirmed the effect of the various sunshields in Taiwan [10]. For sustainable design guidelines, a design parameter affecting the energy performance of shading was proposed, and its contribution of shading and reflective surfaces on the cooling load was confirmed [11]. Nevertheless, more studies of the variation of energy consumption according to the shading properties are needed.

The Korean government has provided guidelines and regulations for window installation. Shading is also defined by regulations in Korea. The guidelines and regulations for reducing the energy consumption in buildings are not perfect, and any study related to windows must adhere to the guidelines and regulations of the Korean government. The Building Energy Conservation in Korea defined shading as blocking solar radiation, and identified exterior shading, interior shading and between the glass shading as parameters [12]. In addition, the Energy Performance Indicator (EPI) checks the exterior shading and admits just the auto controlled interior shading. Green building certification is the proposed standard for decreasing the level of greenhouse gas emissions [13]. This standard proposes the minimum length of horizontal shading to decrease glare and provide environmental improvements. The minimum length of horizontal shading was determined using Equation (1):

$$P = \frac{H}{\tan A} \tag{1}$$

where P is the length of horizontal shading, H is the horizontal length to shading from the bottom of the window and A is the meridian altitude in summer (90-latitude + 23.5). Figure 1 presents the length of horizontal shading.

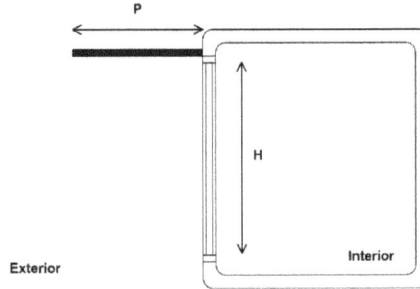

Figure 1. Length of horizontal shading.

"The Window Design Guidelines for Energy-saving of Buildings" were published the Ministry of Land, Transport and Maritime Affairs [14]. The purpose of these guidelines is to allow a variety of designs to consider the energy performance in building design. These guidelines confirmed the impact of the window design of office buildings on the energy consumption of buildings and proposed the orientation, WWR and types of windows in each region. In addition, these guidelines can be used to calculate the energy savings. The guidelines proposed the exterior shading for energy saving but it was just one case of a length of 600 mm. Table 1 lists the regulation and guidelines for windows in Korea.

Table 1. Regulation and guidelines for windows in Korea.

Title	Section	Contents
Building Energy Conservation Design Standards	Design performance	WWR/Orientation
	Window performance	Heat transmission coefficient (U-Value)/Air-tightness
	Shading	Position: Exterior/Between Glazing/Interior
Green Building Certification	Design performance	WWR
	Window performance	Heat transmission coefficient (U-Value)/Air-tightness
	Shading	Proposed minimum length (P) of shading
The Window Design Guideline for Energy-saving of Buildings	Design performance	WWR/Orientation
	Window performance	Heat transmission coefficient (U-Value)/Air-tightness/SHGC
	Shading	Optional Exterior shading (600 mm)

The aim of this study was to confirm the energy saving effect of shading installation. Because the remodeling or new buildings in Korea needs to install shading using the renewed Korean regulations, this study confirmed the energy saving effects and proposed basic research guidelines for shading design. In addition, this study referenced the regulations and guidelines in Korea. This study also confirmed the horizontal shading installation effect. The variation of heating and cooling load was confirmed by changing the length of horizontal shading. The results were compared according to the variation of SHGC. In addition, this study confirmed the correlation between the horizontal shading and the various venetian blind types through the variations of the heating and cooling load. This study confirmed the energy saving ratio of the various venetian blinds than the horizontal shading.

2. Standard Building Modeling and Simulation Condition

Standard buildings are needed to confirm the variation of energy consumption of a building according to the variation of shading. The standard building was not defined in Korea but this study referenced the standard building in previous research results. The standard building in this study references "The Window Design Guideline for Energy-saving of Buildings" and uses "unit area" as defined in the guideline. This method revealed the energy demand and the best way according to the variation of the window elements at each orientation. The unit space in the standard buildings was selected; the size of this unit space was 6 m × 4.5 m × 2.7 m. This size is the result of research that considered the average commercial building by an analysis of various buildings. The gap between the columns of the building was 6 m, the depth to the considered environment of light was 4.5 m, and the height of the room was 2.7 m [2]. Figure 2 presents a model of the standard building.

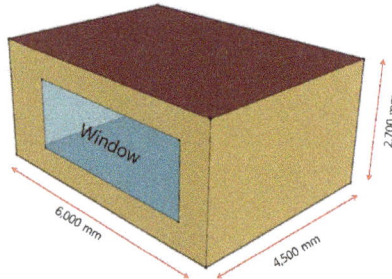

Figure 2. Schematic diagram of standard building modeling.

This study used the simulation tool, COMFEN 4.1, by the Lawrence Berkeley National Laboratory (LBNL, Berkeley, CA, USA) and simulated standard modeling. This tool is a façade design tool based on the Energy Plus engine and provides a systematic evaluation of various elevations. The Energy Plus engine is normally used to confirm the heating and cooling load in buildings. Many studies confirmed the accuracy of the algorithm for the daylighting analysis method [15,16]. COMFEN 4.1 can model the fenestration façade according to the number of windows, size, location, glazing, frame, and outside shading. The façade can select the daylight controls and has the option of orientating the buildings. The annual energy consumption (heating, cooling, fan, and lighting) and peak energy were analyzed by comparing the charts [17]. Table 2 lists the simulation conditions and Figure 3 presents the schedule.

Table 2. Simulation conditions.

Section	Contents
Heating, Ventilation, Air Conditioning	Packaged Single Zone
Temperature Set point	Cooling: 24 °C, Heating: 21 °C
Lighting/Equipment Load	16 W/m^2/10 W/m^2
People	3 people
U-Value of façade wall	0.4 W/m^2·K
Simulation period	Annual

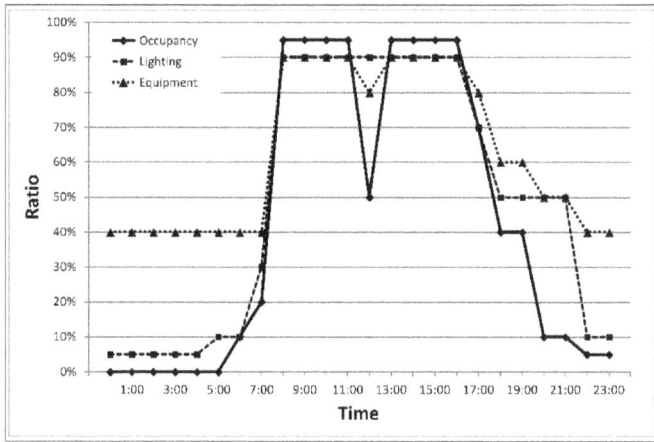

Figure 3. Simulation schedule.

3. Analysis of the Heating and Cooling Load with Horizontal Shading

3.1. Variation of Heating and Cooling Load with the Installation of Horizontal Shading

This study examined the heating and cooling load of buildings according to the horizontal exterior shading installation. The annual heating and cooling load of a building with non-shading and horizontal exterior shading were compared. The performance of the window was the level of regulation. The SHGC, WWR and shading length were not proposed standards in the regulation, therefore, this study referenced previous studies [2]. Table 3 lists the simulation parameters. Figure 4 shows the window area and position according to each WWR. This study used the weather data for Seoul in Korea. Seoul is located in the Northern Hemisphere, altitude 37.3° and longitude 127°. The weather data was provided from the Korean Solar Energy Society [18]. In addition, the data from previous research was used. The length of horizontal shading was calculated using Equation (1), which used latitude 37.3° for Seoul.

8

Table 3. Simulation parameter.

Section		Contents			
Orientation		East/West/South/North			
WWR		20%/40%/60%/80%			
	–	Type 1	Type 2	Type 3	Type 4
Window Type	U-Value (W/m²·K)	1.8	1.8	2.1	2.1
	SHGC	0.4	0.6	0.4	0.6
Shading	WWR	20%	40%	60%	80%
	P = Length (m)	0.35	0.5	0.55	0.6

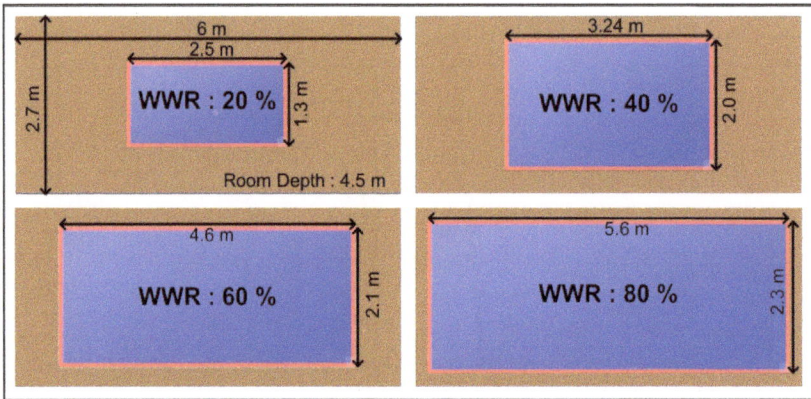

Figure 4. Simulation modeling by WWR.

This study confirmed the simulation result of the south façade according to the horizontal shading installation. In the case of WWR 20%, the energy consumption of non-shading was 79–87 kWh/m²·y and the energy consumption by horizontal shading installation was 78–80 kWh/m²·y. In the case of 40%, the energy consumption of non-shading was 93–111 kWh/m²·y and the energy consumption by horizontal shading installation was 86–96 kWh/m²·y. In the case of 60%, the energy consumption of non-shading was 106–137 kWh/m²·y and the energy consumption by horizontal shading installation was 94–112 kWh/m²·y. Finally, in the case of 80%, the energy consumption of non-shading was 122–165 kWh/m²·y and the energy consumption by horizontal shading installation was 104–129 kWh/m²·y. From the results, this study confirmed an energy saving of 9% by the installation of horizontal shading in the case of WWR = 40%. The energy saving was 22% by the installation of horizontal shading in the case of WWR = 80%. Figure 5 shows the variation of the heating and cooling load by shading installation on the south façade.

This study confirmed the saving ratio for all orientation façades using type 4 glazing, and installation of horizontal shading. The length of horizontal shading was regulation level (0.6 m). In the case of the south façade, the energy consumption savings increased steadily with increasing WWR. The maximum saving was 22% at WWR = 80%. In case of the west and east façades, this study confirmed the maximum 10% energy saving at WWR = 80%. On the other hand, in the case of the north façade, there were no energy consumption savings at WWR < 40%. In addition, in the case of WWR = 60% and 80%, the energy consumption saving was only 1%. This means that to reduce energy consumption, the installation of horizontal shading on the south façade is recommended. Figure 6 shows the energy consumption saving ratio according to the WWR and orientation.

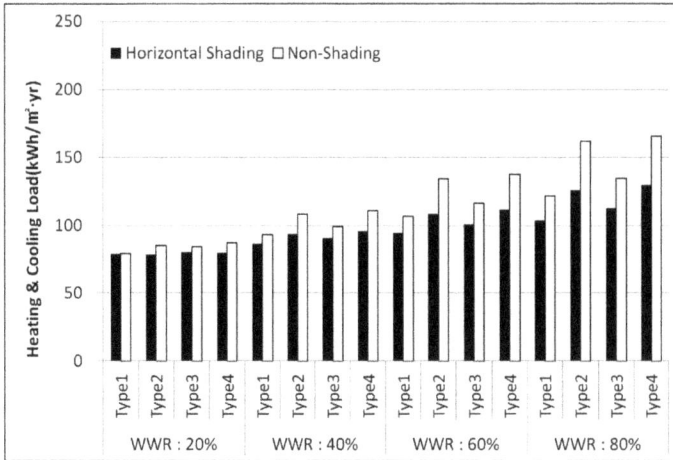

Figure 5. Variation of the heating & cooling load according to the shading installation (south façade).

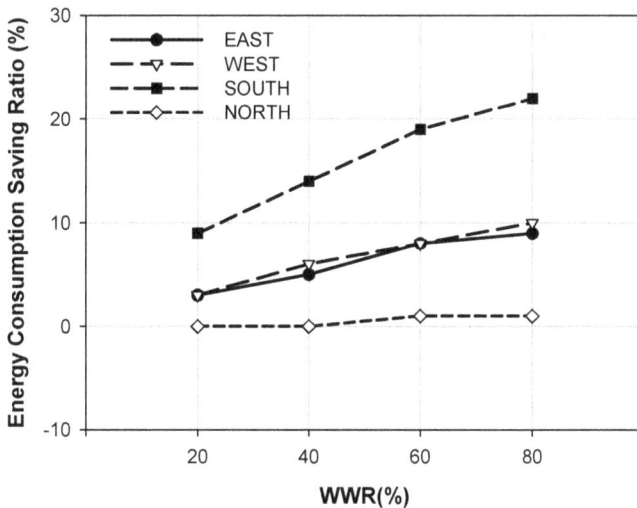

Figure 6. Energy consumption saving ratio by WWR and orientation.

3.2. Variation of the Heating and Cooling Load by the Length of Horizontal Shading

The regulations of Korea propose a minimum length (*P*) of horizontal shading. The length of horizontal shading in the exterior was calculated using Equation (1). The length was affected by the window height (*H*) and the latitude of Seoul, but the regulation does not explain the reference of the length of shading. This study confirmed the variation of the heating and cooling load by changing the length of horizontal shading, as listed in Table 4. The length of horizontal shading was changed to 50% from 200%, WWR = 60%, the performance of window was U-value = 2.1 (W/m^2·K) and SHGC = 0.6. Figure 7 shows the energy consumption ratio variation of the heating and cooling load by the orientation and length.

Table 4. Simulation parameter and case.

Section			Contents			
Orientation			East/West/South/North			
WWR			60% (*W*:5 m, *H*:1.94 m)			
Window Performance	U-Value (W/m^2·K)		2.1			
	SHGC		0.6			
Shading	CASE	*P* (m)	*H*/tan*A* = 1.94/tan(90 − 37.3 + 23.5) = 0.48 ≈ 0.5			
		Ratio (%)	50	100	150	200
		Length (m)	0.25	0.5	0.75	1

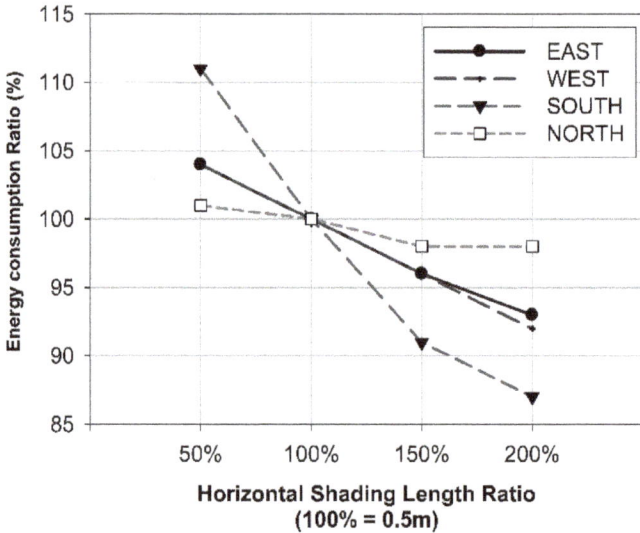

Figure 7. Ratio variation of the heating and cooling load with shading length.

The results of WWR = 60% and various lengths of shading confirmed the heating and cooling load dependence on the orientation and length of shading. The energy consumption ratio is all the orientation façades based on the installation of horizontal shading of 0.5 m. In the case of the east façade, the energy consumption was 150 kWh/m^2·y after the installation of horizontal shading. That length of shading was 0.5 m, which is the level of the regulation. The energy consumption was 139 kWh/m^2·y when the length was 200% (1 m). The energy consumption savings in that case is 7%. In the case of the west façade, the energy consumption was 142 kWh/m^2·y with a length of

100% (0.5 m). The energy consumption was 154 kWh/m²·y at a length of 200% (1 m). That case saved 8% energy consumption. In the case of the south façade, the energy consumption was 112 kWh/m²·y at a length of 100% (0.5 m). The energy consumption was 97 kWh/m²·y for a length of 200% (1 m). That case saved 13% of energy consumption. In the case of the north façade, the energy consumption was 111 kWh/m²·y at a length of 100% (0.5 m). The energy consumption was 109 kWh/m²·y for a length of 200% (1 m). That case saved 2% of energy consumption. This means that the length of shading needs to be longer than the proposed regulation length to improve the energy savings. When the length of shading was less than 100% (0.5 m), the energy consumption ratio increased in all cases. This study confirmed that the heating and cooling load decreased with a length longer than the length of the regulation. The South façade showed the most effective energy savings for an extended shading length.

3.3. Analysis of the Heating and Cooling Load Variation with the SHGC and Shading

From the amount of heat gain from the window, the SHGC of the window affects the heating and cooling load. This study confirmed the variation of the heating and cooling loads according to various SHGC values and horizontal shading. In addition, this study found a correlation between the SHGC and shading. For the analysis, the WWR was fixed at 60%, the U-value of the glazing was 2.1 W/m²·K and the length of horizontal shading was 0.5 m. Those conditions are the same as those listed in Table 3.

From the results of the simulation, in the case of the east façade, the use of the regulation horizontal shading length and changing SHGC from 0.2 to 0.75 resulted in an increase in energy consumption from 109 to 150 kWh/m²·y. In the case of the west façade, the energy consumption was increased from a 107 to 149 kWh/m²·y after changing the SHGC from 0.2 to 0.75. A 37% (east) 39% (west) increase in the heating and cooling load was observed by increasing SHGC. Figure 8 shows the variation of heating and cooling load with SHGC. This means that horizontal shading does not affect the energy consumption in the case of the east and west façades. In the case of the south façade, however, the energy consumption was not increased by changing the SHGC from 0.2 to 0.75. From 0.2 to 0.4 SHGC, the energy consumption showed a tendency to increase from 97 to 109 kWh/m²·y, which is approximately a 12% increase. On the other hand, from 0.4 to 0.75 SHGC, the energy consumption showed a tendency to decrease from 109 to 105 kWh/m²·y, which is approximately a 3% decrease. This shows that a high SHGC and horizontal shading resulted in less energy consumption than at a low SHGC. This study confirmed the correlation between the horizontal shading and SHGC. In the case of the north façade, energy consumption was increased from 103 to 125 kWh/m²·y by changing the SHGC from 0.2 to 0.4. On the other hand, from 0.4 to 0.75 SHGC, the energy consumption changed only slightly, 124–125 kWh/m²·y. At SHGC > 0.4, in the case of the north façade, the horizontal shading had a slight effect on the energy consumption.

12

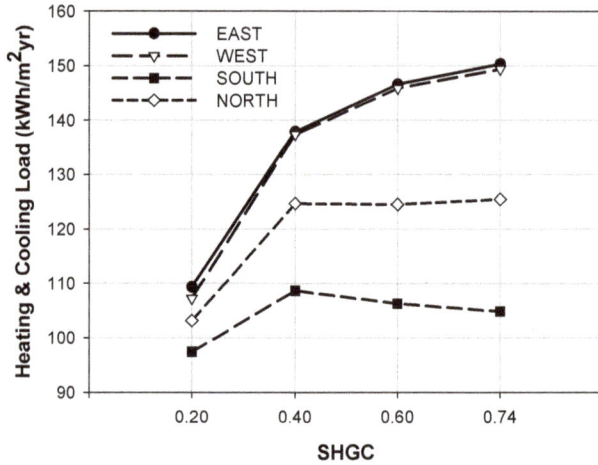

Figure 8. Variation of heating and cooling load with the SHGC.

4. Analysis of the Heating and Cooling Load According to the Type of Venetian Blind

The Building Energy Conservation in Korea has proposed the position of shading. The guidelines for the venetian blind position vary. To confirm the variation of the heating and cooling load with venetian blinds, this study compared the case of horizontal shading with the case of the various venetian blind positions. The angle of the venetian blind was 45°. This study used the venetian blind types of the interior position/between glazing position/exterior position. Table 5 lists the simulation case and venetian blind type. Figure 9 shows the type of shading. In the case of WWR 60% and 80%, and SHGC 0.4 and 0.6 of the window, this study confirmed the heating and cooling load of a standard building. Table 6 lists the results of the simulation.

Table 5. Simulation parameter and details of various shading.

Section			Contents			
Orientation			East/West/South/North			
WWR						
Window Performance	U-Value		2.1			
	SHGC		0.4/0.6			
CASE	Horizontal Shading	WWR (%)	60		80	
		Length (m)	0.55		0.6	
	Venetian Blind	Position	Angle (°)	Width of slat (mm)	Spacing (mm)	Thickness (mm)
		Exterior	45	76.96	70.1	1.02
		Between glazing		4	7.62	
		Interior		25.4	20.07	

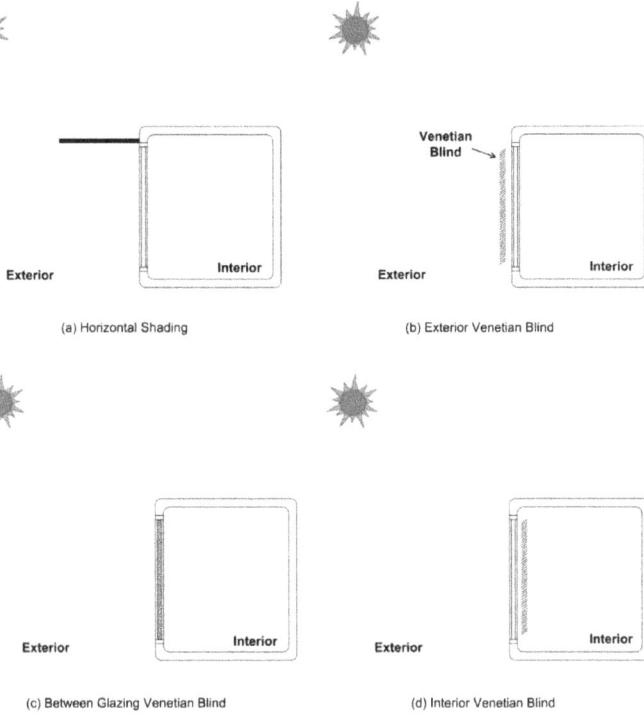

Figure 9. Type of shading and venetian blinds.

Table 6. Heating and cooling load variation with horizontal shading and venetian blinds.

WWR	Window Performance	Shading Type	Position	Period	Heating and Cooling load (kWh/m²)			
					East	West	South	North
60%	U-Value: 2.1 (W/m²·K) SHGC: 0.4	Horizontal shading	Exterior	May–Oct	57.5	58.2	40.7	59.5
				Nov–Apr	74.3	76.2	60.0	48.8
		Venetian blind	Exterior	May–Oct	55.7	55.5	50.1	57.9
				Nov–Apr	49.5	49.5	48.6	44.4
			Between glazing	May–Oct	51.2	51.7	45.4	54.7
				Nov–Apr	59.9	58.9	60.7	54.1
			Interior	May–Oct	58.6	60.1	47.0	57.9
				Nov–Apr	83.4	85.1	72.3	50.4
	U-value: 2.1 (W/m²·K) SHGC: 0.6	Horizontal shading	Exterior	May–Oct	57.9	59.6	41.6	55.9
				Nov–Apr	91.8	94.6	69.9	54.8
		Venetian blind	Exterior	May–Oct	54.4	54.2	47.7	56.8
				Nov–Apr	52.2	52.6	51.4	45.7
			Between glazing	May–Oct	51.5	52.0	45.6	55.0
				Nov–Apr	59.6	58.7	60.4	53.9
			Interior	May–Oct	58.3	60.3	47.2	56.4
				Nov–Apr	89.2	91.4	76.7	54.0

14

Table 6. *Cont.*

WWR	Window Performance	Shading Type	Position	Period	Heating and Cooling load (kWh/m²)			
					East	West	South	North
80%	U-value: 2.1 (W/m²·K) SHGC: 0.4	Horizontal shading	Exterior	May–Oct	66.1	67.5	46.6	68.8
				Nov–Apr	85.7	88.2	66.0	51.7
		Venetian blind	Exterior	May–Oct	64.0	63.8	57.1	66.9
				Nov–Apr	51.7	52.2	50.7	44.4
			Between glazing	May–Oct	58.0	58.9	51.0	62.4
				Nov–Apr	65.1	64.4	66.4	58.3
			Interior	May–Oct	68.6	71.1	56.3	67.1
				Nov–Apr	98.2	100.4	82.9	54.3
	U-value: 2.1 (W/m²·K) SHGC: 0.6	Horizontal shading	Exterior	May–Oct	68.1	70.9	50.3	64.1
				Nov–Apr	109.3	113.2	79.1	59.1
		Venetian blind	Exterior	May–Oct	62.1	62.0	53.9	65.3
				Nov–Apr	55.3	56.2	54.4	46.1
			Between glazing	May–Oct	58.4	59.2	51.4	62.8
				Nov–Apr	64.7	64.0	66.1	58.0
			Interior	May–Oct	68.9	72.0	57.7	65.1
				Nov–Apr	105.8	109.2	88.8	58.3

In case of WWR = 60% and SHGC = 0.4, the case of south façade, the energy consumption was decreased by 2% using the exterior venetian blind installation compared to horizontal shading installation. The energy consumption was increased 5% and 18% by between glazing and interior venetian blind installation, respectively, compared to horizontal shading installation. In the case of the east façade, the energy consumption was decreased 20% and 16% by exterior venetian blinds and between glazing venetian blind installation, respectively, compared to horizontal shading installation. The energy consumption was increased 8% by interior venetian blind installation compared to horizontal shading installation. In addition, in the case of the west façade, the energy consumption was decreased 22% and 18% by exterior venetian blinds and between glazing venetian blind installation, respectively, compared to horizontal shading installation. The energy consumption was increased 8% by interior venetian blind installation compared to horizontal shading installation. In the case of the north façade, the energy consumption was decreased 6% by exterior venetian blind installation compared to horizontal shading installation. The energy consumption was increased 1% by between glazing venetian blind installation compared to horizontal shading installation. In the case of interior venetian blind installation, the energy consumption was the same as case of horizontal shading installation.

At WWR = 60% and SHGC = 0.6, in the case of the south façade, the energy consumption was decreased 11% and 5% by exterior venetian blind and between glazing venetian blind installation compared to horizontal shading installation. The energy consumption was increased 11% by interior venetian blind installation compared to horizontal shading installation. In the case of the east façade, the energy consumption was decreased 29%, 26% and 1% by exterior venetian blind, between glazing venetian blind and interior venetian blind installation, respectively, compared to horizontal shading installation. In addition, in the case of the west façade, the energy consumption was decreased 31%, 28% and 2% by exterior venetian blind, between glazing venetian blind and interior venetian blind installation, respectively, compared to horizontal shading installation. In the case of the north façade, the energy consumption was decreased 7% and 2% by exterior venetian blind and between glazing venetian blind installation, respectively, compared to horizontal shading installation. In the case of interior venetian blind installation, the energy consumption was the same as that of horizontal shading installation.

At WWR = 80% and SHGC = 0.4, in the case of the south façade, the energy consumption was decreased 4% by exterior venetian blind installation compared to the case of horizontal shading installation. The energy consumption was increased 4% and 24% by between glazing and interior venetian blind installation, respectively, compared to horizontal shading installation. In the case of the east façade, the energy consumption was decreased 24% and 19% by exterior venetian blind and between glazing venetian blind installation, respectively, compared to horizontal shading installation. The energy consumption was increased 10% by interior venetian blind installation compared to horizontal shading installation. In the case of the west façade, the energy consumption was decreased 26% and 21% by exterior venetian blind and between glazing venetian blind installation, respectively, compared to horizontal shading installation. The energy consumption was increased 10% by interior venetian blind installation compared to horizontal shading installation. In the case of the north façade, the energy consumption was decreased 8% by exterior venetian blind installation compared to horizontal shading installation. In the case of between glazing venetian blind installation, the energy consumption was the same as that of horizontal shading installation. On the other hand, the energy consumption was increased 1% by interior venetian blind installation compared to horizontal shading installation.

At WWR = 80% and SHGC = 0.6, in the case of the south façade, the energy consumption was decreased 16% and 9% by exterior venetian blind and between glazing venetian blind installation, respectively, compared to horizontal shading installation. The energy consumption was increased 13% by interior venetian blind installation compared to horizontal shading installation. In the case of the east façade, the energy consumption was decreased 34%, 31% and 2% by exterior venetian blind, between glazing venetian blind and interior venetian blind installation, respectively, compared to horizontal shading installation. In the case of the west façade, the energy consumption was decreased 36%, 33% and 2% by exterior venetian blind, between glazing venetian blind and interior venetian blind installation, respectively, compared to horizontal shading installation. In the case of the north façade, the energy consumption was decreased 10% and 2% by exterior venetian blind and between glazing venetian blind installation, respectively compared to horizontal shading installation. In the case of interior venetian blind installation, the energy consumption was the same as that of horizontal shading installation. Figures 10–13 shows the saving ratio of heating and cooling energy consumption according to the type of shading. The result of the simulation showed generally low energy consumption in the case of the south façade. In the case of the north façade, the simulation showed little change. At WWR = 80% and SHGC = 0.6, the heating energy consumption was increased 11%–170% by exterior venetian blinds compared to horizontal shading. At WWR = 60% and SHGC = 0.4, the heating energy consumption was increased 8%–161% by exterior venetian blinds compared to horizontal shading. That reason for this is that venetian blinds tended to block more solar radiation in the winter season than horizontal shading.

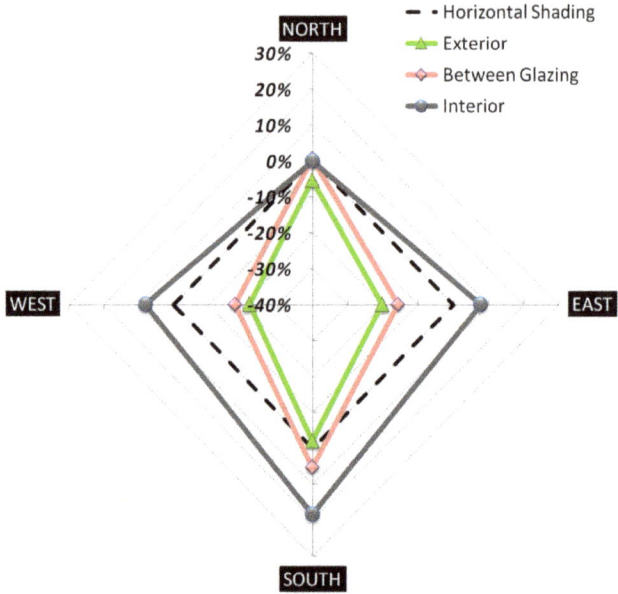

Figure 10. Saving ratio of heating and cooling load according to the type of shading (WWR 60%, U-value: 2.1, SHGC: 0.4).

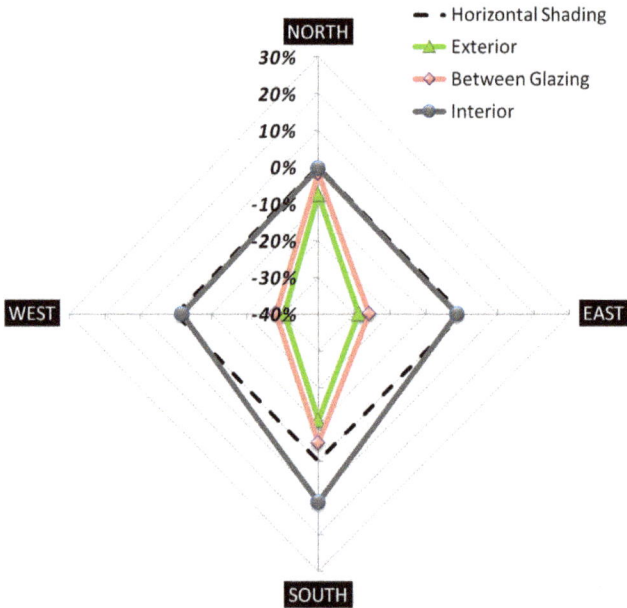

Figure 11. Saving ratio of heating and cooling load according to the type of shading (WWR 60%, U-value: 2.1, SHGC: 0.6).

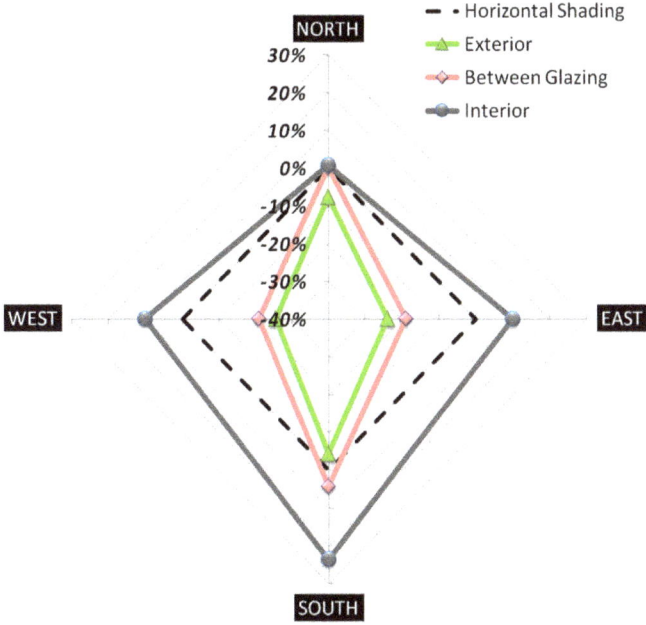

Figure 12. Saving ratio of the heating and cooling load according to the type of shading (WWR 80%, U-value: 2.1, SHGC: 0.4).

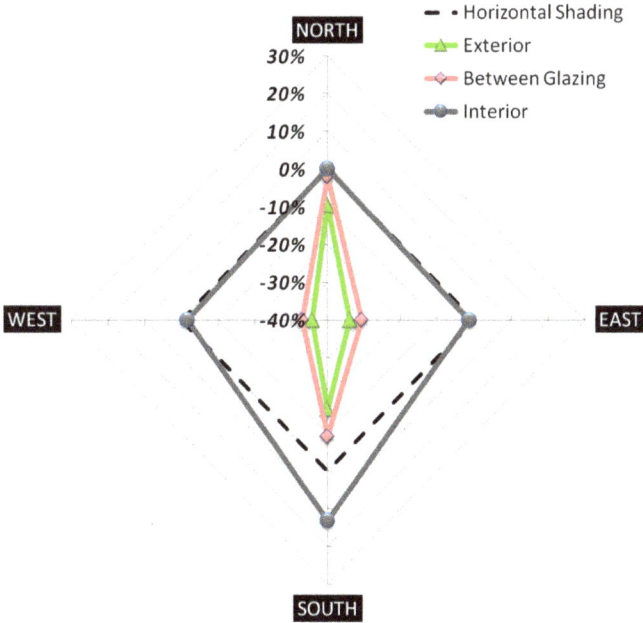

Figure 13. Saving ratio of the heating and cooling load according to the type of shading (WWR 80%, U-value: 2.1, SHGC: 0.6).

In the case of the south façade, the interior venetian blind resulted in higher heating and cooling energy consumption because the heating load was increased by blocked solar radiation in winter. At WWR = 80%, the exterior venetian blind saved 4%–36% of the heating and cooling load than the horizontal shading. This means that the exterior venetian blind was more efficient at saving energy than horizontal shading in a large WWR. In the case of the east and west façades, the exterior venetian blinds and between glazing venetian blinds improved energy consumption compared to horizontal shading. The maximum saving ratio was 36% at WWR = 80%, exterior venetian blind and west façade. In the case of the east and west façades, the exterior venetian blind and between glazing venetian blind is more efficient than horizontal shading. Shading the north façade had little or no effect on heating and cooling load regardless of the type of shading.

5. Conclusions

This study confirmed the annual heating and cooling load of a standard building by simulation modeling. The results of this study are as follows:

(1) In the case of horizontal shading installation, the heating and cooling load was lower than the case of no shading. The decrease in the heating and cooling load was most efficient in the south façade. Horizontal shading installation reduced the heating and cooling load by a maximum of 22% in the case of type 4 and WWR = 80%.

(2) Extended horizontal shading (1 m) compared to the regulation level (0.5 m) reduced the heating and cooling load. In the case of extending the length of horizontal shading, the south façade showed a maximum 13% decrease in heating and cooling load.

(3) This study confirmed the variation of heating and cooling load by the variation of the SHGC. In case of the south and north façades, the heating and cooling load was decreased at SHGC = 0.4. This means that the high SHGC and horizontal shading is more efficient in energy savings than the case of a low SHGC and horizontal shading.

(4) A comparison of horizontal shading with various venetian blind types showed that in the case of the west façade, the energy consumption was decreased 36% by exterior venetian blind installation compared to the case of horizontal shading installation.

The exterior venetian blinds and between glazing were unaffected by the SHGC. The exterior venetian blind was more effective in reducing the heating and cooling load than horizontal shading. In addition, the heating and cooling load of the interior venetian blind was higher than that of horizontal shading. From the results of this study, our next study will develop indicators and guidelines for the shading choice. These indicators or the guidelines for the shading design will be proposed to a designer.

Acknowledgments

This work was supported by the 2014 Yeungnam University Research Grant (214A380044).

Author Contributions

All authors contributed to this work. Seok-Hyun Kim performed the result analysis of simulation and wrote the major part of this article. Kyung-Ju Shin and Bo-Eun Choi conducted the energy simulation. Jae-Hun Jo proposed the case of the energy simulation and conducted data analysis. Soo Cho performed the result discussion and gave technical support. Young-Hum Cho was responsible for this article and gave conceptual advice.

Conflicts of Interest

The authors declare no conflict of interest.

References

1. Kim, S.H.; Kim, S.S.; Kim, K.W.; Cho, Y.H. Analysis of the energy consumption of window elements through simulation. *Korea J. Archit. Instit. Spring Conf.* **2013**, *33*, 261–262. (In Korean)
2. Kim, S.H.; Kim, S.S.; Kim, K.W.; Cho, Y.H. A study on the proposes of energy analysis indicator by the window elements of office buildings in Korea. *Energy Build.* **2014**, *73*, 153–165.
3. Nielsen, T.R.; Duer, K.; Svendsen, S. Energy performance of glazings and windows. *Solar Energy* **2000**, *69*, 137–143.
4. Carlos, E.O.; Myriam, B.C.A.; Evert, J.L.; Jan, L.M.H. Considerations on design optimization criteria for windows providing low energy consumption and high visual comfort. *Appl. Energy* **2012**, *95*, 238–245.
5. Perssona, M.L.; Roosa, A.; Wall, M. Influence of window size on the energy balance of low energy houses. *Energy Build.* **2006**, *38*, 181–188.
6. Kim, B.S.; Kim, K.H. A study on thermal environment and the design methods to save energy in small glass-skin commercial buildings. *J. Asian Archit. Build. Eng.* **2004**, *3*, 115–123.
7. Leigh, S.B.; Bae, J.I.; Ryu, Y.H. A study on cooling energy savings potential in high-rise residential complex using cross ventilated double skin façade. *J. Asian Archit. Build. Eng.* **2004**, *3*, 275–282.
8. Seok, H.T.; Jo, J.H.; Kim, K.W. Establishing the design process of double-skin façade elements through design parameter analysis. *J. Asian Archit. Build. Eng.* **2009**, *8*, 251–258.
9. Ossen, D.R.; Hohd, H.A.; Mardros, N.H. Optimum overhang geometry for building energy saving in tropical climates. *J. Asian Archit. Build. Eng.* **2005**, *4*, 563–570.
10. Lai, C.M.; Wang, Y.H. Energy-saving potential of building envelope designs in residential houses in Taiwan. *Energies* **2011**, *4*, 2061–2076.
11. Kang, H.J.; Rhee, E.K. Development of a sustainable design guideline for a school building in the early design stage. *J. Asian Archit. Build. Eng.* **2014**, *13*, 467–474.
12. Ministry of Land, Transport and Maritime Affairs. Building Energy Conservation Design Standard. Available online: http://www.molit.go.kr/ (accessed on 9 February 2015). (In Korean)
13. Ministry of Land, Transport and Maritime Affairs. Ministry of Environment, Green Building Certification Criteria. Available online: http://www.molit.go.kr/ (accessed on 9 February 2015). (In Korean)

14. Ministry of Land, Transport and Maritime Affairs. The Window Design Guide-line for Energy-saving of Buildings. Available online: http://www.molit.go.kr/ (accessed on 9 February 2015). (In Korean)
15. Yoon, Y.B.; Jeong, W.R.; Lee, K.H. Window material daylighting performance assessment algorithm: Comparing radiosity and split-flux methods. *Energies* **2014**, *7*, 2362–2376.
16. Yoon, Y.B.; Manandhar, R.; Lee, K.H. Comparative study of two daylighting analysis methods with regard to window orientation and interior wall reflectance. *Energies* **2014**, *7*, 5825–5846.
17. Lawrence Berkeley National Laboratory, COMFEN. Available online: http://lbl.gov/ (accessed on 9 February 2015).
18. The Korean Solar Energy Solar Energy Society. Available online: http://kses.re.kr/ (accessed on 9 February 2015). (In Korean)

Total Site Heat Integration Considering Pressure Drops

Kew Hong Chew, Jiří Jaromír Klemeš, Sharifah Rafidah Wan Alwi, Zainuddin Abdul Manan and Andrea Pietro Reverberi

Abstract: Pressure drop is an important consideration in Total Site Heat Integration (TSHI). This is due to the typically large distances between the different plants and the flow across plant elevations and equipment, including heat exchangers. Failure to consider pressure drop during utility targeting and heat exchanger network (HEN) synthesis may, at best, lead to optimistic energy targets, and at worst, an inoperable system if the pumps or compressors cannot overcome the actual pressure drop. Most studies have addressed the pressure drop factor in terms of pumping cost, forbidden matches or allowable pressure drop constraints in the optimisation of HEN. This study looks at the implication of pressure drop in the context of a Total Site. The graphical Pinch-based TSHI methodology is extended to consider the pressure drop factor during the minimum energy requirement (MER) targeting stage. The improved methodology provides a more realistic estimation of the MER targets and valuable insights for the implementation of the TSHI design. In the case study, when pressure drop in the steam distribution networks is considered, the heating and cooling duties increase by 14.5% and 4.5%.

Reprinted from *Energies*. Cite as: Chew, K.H.; Klemeš, J.J.; Wan Alwi, S.R.; Manan, Z.A.; Reverberi, A.P. Total Site Heat Integration Considering Pressure Drops. *Energies* **2015**, *8*, 1114-1137.

1. Introduction

Pressure drop is an important factor to consider during a Heat Integration (HI) system design [1]. It is especially so with Total Site Heat Integration (TSHI) when distances between the different plants are large and the heat exchangers are often installed at different elevations within a plant. Pressure drop is mainly due to frictional losses as the fluids flow through pipes and fittings as well as pressure losses across the heat exchangers. When the fluids are liquid phase, there is additional pressure loss due to elevation changes. Failure to include the pressure drop factor in the early stages of design can lead to serious problems at the later stages. Exclusion of pressure drops when targeting minimum energy requirement (MER) may lead to too optimistic energy targets resulting in undersizing of central utilities systems. Neglecting pressure drops at the heat exchanger network (HEN) synthesis stage may render a proposed design infeasible if the actual pressure drops are higher than that what is allowable by the pumps or compressors. The need to replace the pumps or compressors may outweigh the savings from Heat Integration.

Most studies on pressure drop issues are associated with the retrofitting or synthesis of HEN for a single process. Mathematical Programming (MP)-based methodologies were mostly used to address the impact of pressure drop in the optimisation of HEN. Polley *et al.* [2] introduced the concept of pressure drop targeting in HEN retrofits where the pressure drop is correlated to the heat exchange area and heat transfer coefficient. The allowable pressure drop is used as an objective to optimise the heat exchange area. Ciric and Floudas [3] addressed the pressure drop issue based on the distances between heat exchangers and used a piping cost factor to minimise HEN modification costs. Ahmad and Hui [4] considered the pressure drop issue, in terms of distance between processes, by grouping the processes into "areas of integrity" and incorporated the impact in the methodology in the form of forbidden matches. Sorsak and Kranvanja [5] extended the Mixed Integer Non-linear Programming (MINLP) model of Yee and Grossman [6] to optimise the pressure drop and heat transfer coefficient. The pressure drop across the heat exchangers, both tube and shell sides, were estimated and considered in terms of pumping costs. Nie and Zhu [7] considered pressure drops in HEN retrofits by first estimating the pressure drop limits and then tackled the pressure drop constraints by optimising the area allocation, shell arrangement and use of heat transfer enhancement option. Panjeshahi and Tahouni [8] proposed a procedure whereby the pressure drop is considered together with the possibility of pump/compressor replacement when optimising area and utility costs. Soltani and Shafiei [9] introduced a new procedure which uses a genetic algorithm along with linear programming to retrofit HEN, including pressure drops. Stream pressure drop is correlated to area and heat transfer coefficient and the allowable pressure drops are introduced as constraints in the network optimisation.

Few studies have addressed pressure drops in the MER targeting stage. Zhu and Nie [10] considered the pressure drop aspect simultaneously with area and utility costs during the targeting and design stages. The pressure drop estimated for the heat exchanger is used to determine the optimum minimum approach temperature (ΔT_{min}) along with area and utility cost in the targeting stage. Inclusion of pressure drop (for heat exchangers only) in the proposed MP model led to different network structures and costs. Chew et al. [11] highlighted the significance of considering distribution piping pressure drop on steam generation from a Site Source. In the case study, the amount of steam recovered from the Site Source is significantly reduced when steam has to be generated at a higher pressure level to overcome the pressure drop in the pipes. Without considering pressure drops, the estimated utility targets maybe too optimistic and would result in undersizing of central steam generation systems. Liew et al. [12] extended the numerical algorithms, Total Site Problem Table Algorithm and Total Site Utility Distribution, to consider pressure drops and heat losses in steam pipes. The utility targets are based on a steam level which is at higher pressure (i.e., to overcome the pressure drop) and superheated (i.e., at a sufficient degree of superheat such that after heat loss the steam will reach the user at saturated conditions).

The studies so far have addressed the pressure drop factor in the optimisation of HEN in terms of pumping costs (based on distance or heat exchanger pressure drops), allowable pressure drop as constraints or objectives, or forbidden matches. The consideration of pressure drops in MER targeting has been at the heat exchanger (ΔT_{min}) or due to distance (steam pipes). None had looked at the pressure drop implications in a Total Site (TS) context which would encompass distance, equipment and utility distribution systems. Moreover, the MP-based methods provide few design insights required by designers [1]. In this paper, the graphical pinch-based TSHI methodology is extended to consider the pressure drop factor during the MER targeting stage. The methodology provides a more realistic estimation of MER targets and better understanding of the TSHI design for implementation later.

2. Pressure Drop Factor in TSHI

In the established TSHI methodology, the utility targeting are based on temperatures and heat loads. The overall heat surplus (Source) and deficit (Sink) of the processes in a TS are represented by the Total Site Profile (TSP). The potential utility generation from the source and heating requirement of the Sink are shown by the Site Utility Composite Curves (SUCC) which are then used to set the targets for site heating and cooling utilities requirements [13]. The steam utilities are generated (from Site Source) and utilised (at Site Sink) at the same temperatures, see Figure 1a,b.

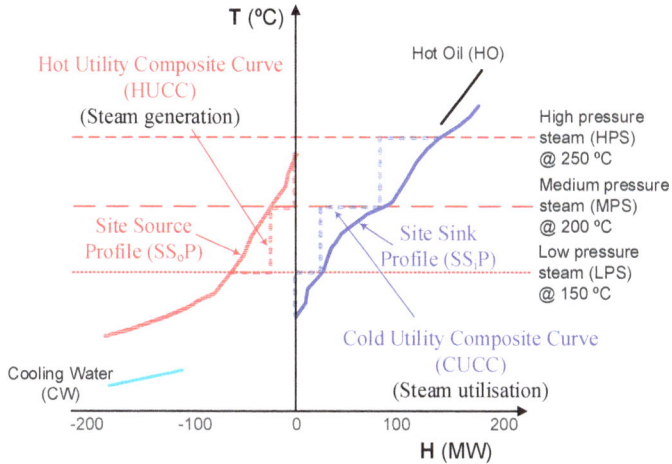

(a) TSP

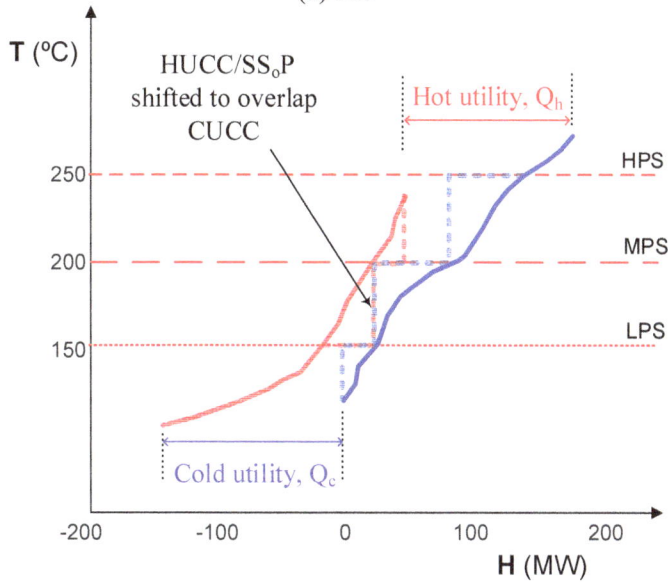

(b) Utilities targeting

Figure 1. *Cont.*

24

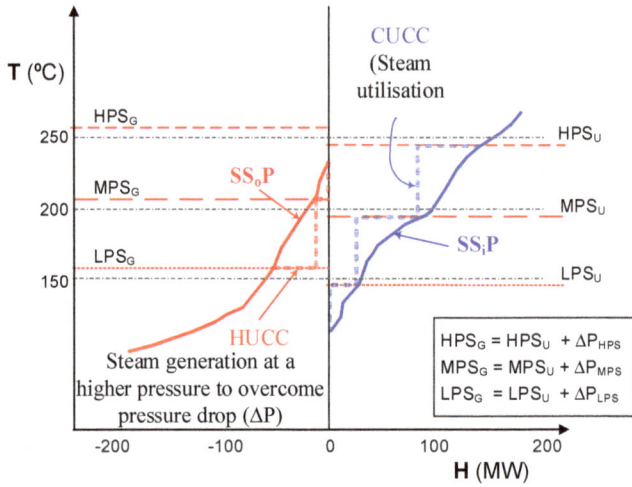

(c) TSHI considering pressure drop

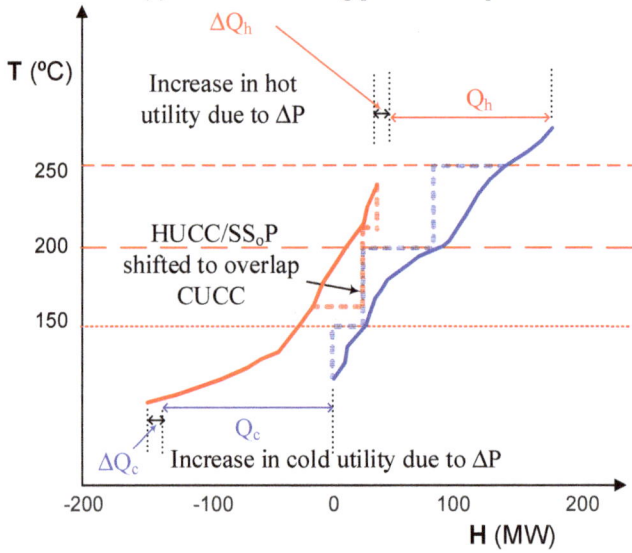

(d) Utilities targeting considering pressure drop

Figure 1. TSHI utilities targeting (adapted from Klemeš *et al.* [13]).

In a TS, the utilities are distributed by an array of headers, sub-headers and pipes. Figure 2 gives a flow schematic of a TS comprising four plants with hot oil (HO), high pressure steam (HPS), medium pressure steam (MPS), low pressure steam (LPS) and cooling water (CW) utilities.

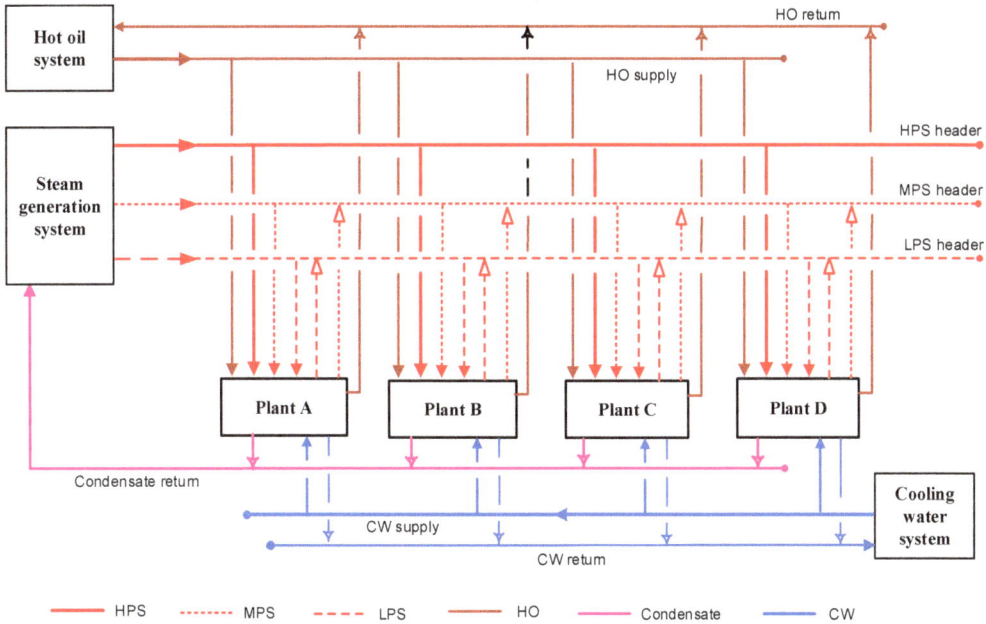

Figure 2. Schematic of a typical utilities distribution system at a TS with HO, HPS/MPS/LPS and cooling water.

2.1. Steams: HPS, MPS, LPS

The main headers take supply from the boilers and various steam generators which recover heat from the Site Source. Steam is then distributed to the various plants via sub-headers and distribution pipes. The main header operates at a sufficiently high pressure to supply steam to the furthest steam users. Figure 3 gives the process flow diagram of a typical steam generation and distribution system. Because of pressure drops in the headers, pipes and equipment, steam will be generated at a higher pressure and used at a lower pressure. The pressures and pressure drops of the steam distribution system are summarised in Figure 4. As saturated steam temperature is a function of its pressure, the difference in pressures between generation and usage can be represented by the difference in temperatures for generation and usage as shown in Figure 1c,d. As shown, consideration of the pressure drop factor will increase the heating (ΔQ_h) and cooling (ΔQ_c) utilities. In addition, the discharge head of the boiler feed water (BFW) pumps will have to be specified accordingly and the information used as input in the cost optimisation exercise.

26

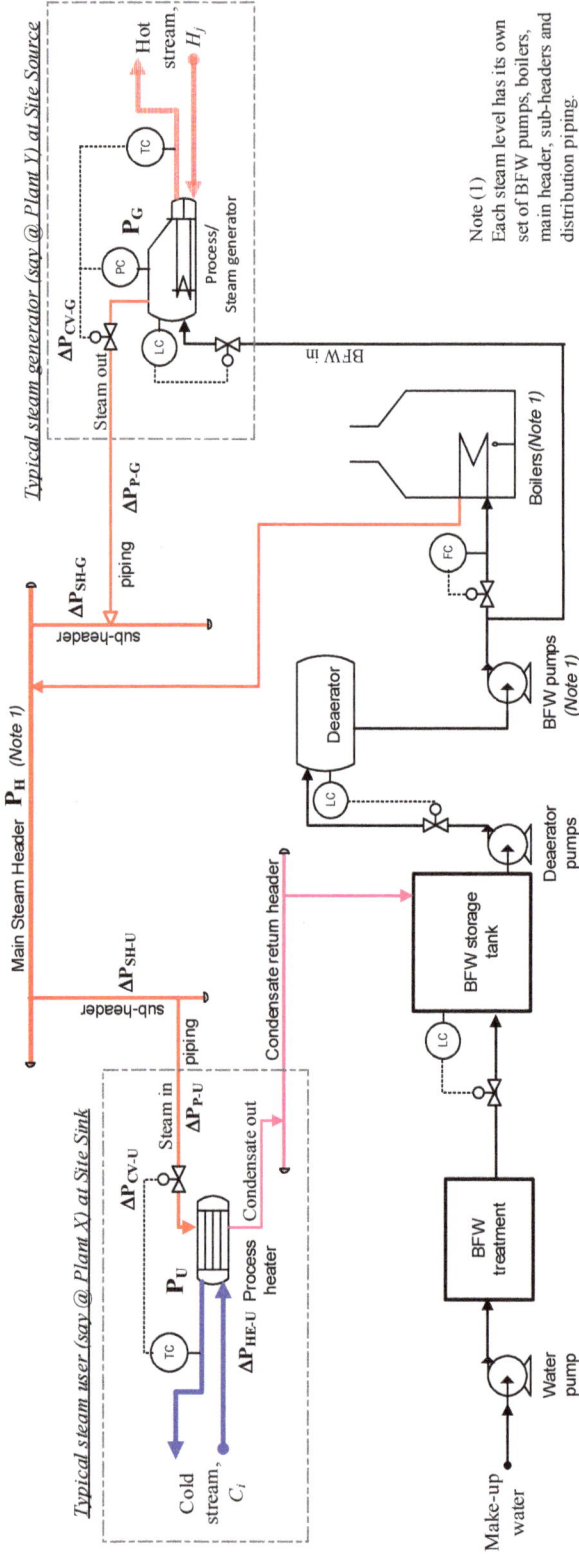

Figure 3. Process flow diagram—a typical steam generation, distribution and utilisation system at a TS.

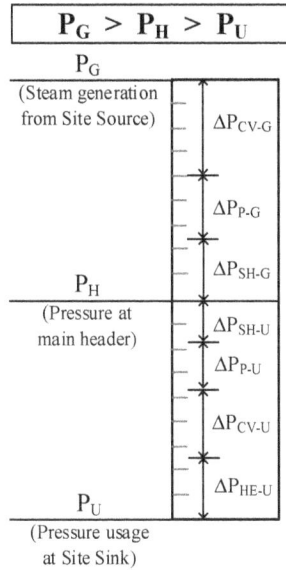

$$P_G > P_H > P_U$$

P_G
(Steam generation
from Site Source)

ΔP_{CV-G}

ΔP_{P-G}

ΔP_{SH-G}

P_H
(Pressure at
main header)

ΔP_{SH-U}

ΔP_{P-U}

ΔP_{CV-U}

ΔP_{HE-U}

P_U
(Pressure usage
at Site Sink)

Figure 4. Pressures and pressure drops at a steam distribution system.

2.2. Cooling Water

Figure 5 is the flow diagram of a typical CW distribution system. The CW pumps deliver CW to the various users, *i.e.*, the process coolers, on the TS via the supply header, sub-headers and distribution piping. Warm CW exiting the coolers is routed to the cooling towers via the return header. For liquids like water, the temperature is not affected by its flow pressure. As long as there is no phase change, the pressure drop does not affect the MER targeting. However, the CW pumps have to be specified for a sufficient discharge head so as to overcome the pressure drop in the distribution system to ensure adequate volumes of the utility are delivered to the users as required. For liquid utilities such as CW, the elevation pressure drop due to liquid column static head (above the pump) is important and has to be considered. The required pump discharge head can then be used as an input parameter in the cost optimisation exercise.

2.3. Hot Oil

Like CW, pressure drops do not affect MER targeting. Pressure drops due to elevation have to be included when estimating the discharge head for the HO circulation pumps and used as an input to the cost optimisation exercise. As with the liquid utilities, the impact of pressure drop on the process streams are seen in the penalty of additional pumping or compression costs so long as there is no phase changes, *i.e.*, liquid remains as liquid and gas stays as gas in the pipes. The impact of pressure drop on TSHI is summarised in Table 1.

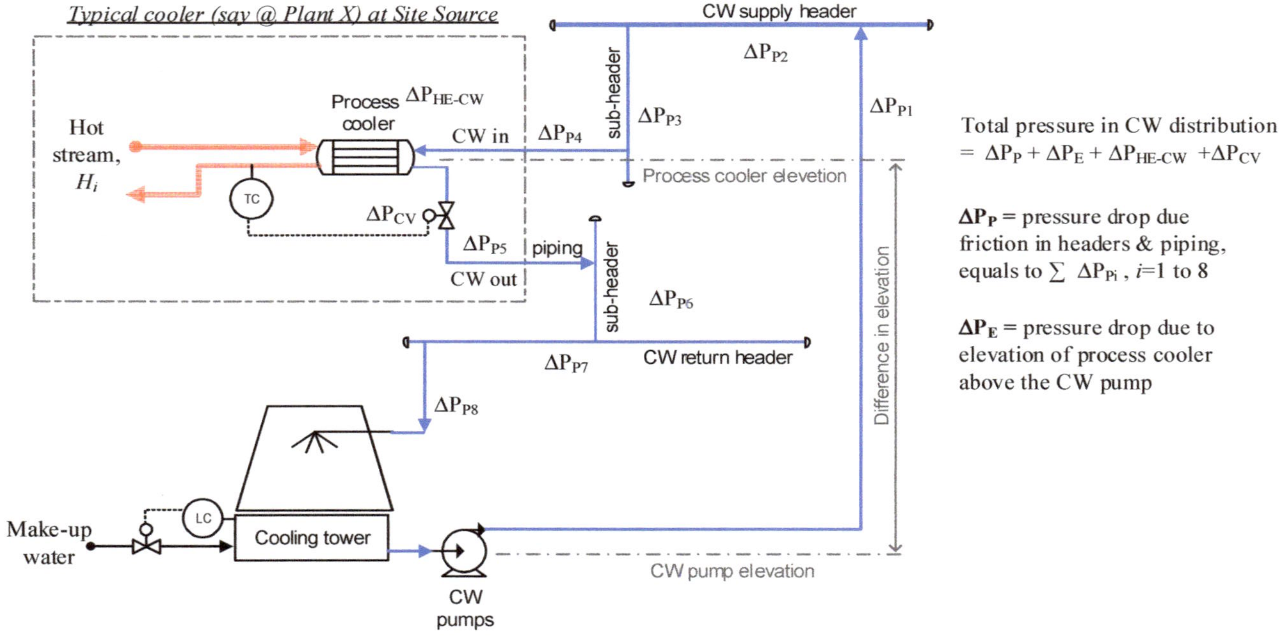

28

Figure 5. Process flow diagram—cooling water distribution system.

Table 1. Impact of pressure drop on TS.

Fluid	MER Targeting	Cost Optimisation
Steam e.g., HPS, MPS, LPS	Increase ΔQ_h and ΔQ_c	Higher BFW pump capital and pumping costs
Liquid utilities (e.g., CW, HO, *etc.*)	No impact	Higher utility circulation pump capital and pumping costs
Process—liquids [a]	No impact	Higher pump capital and pumping costs
Process—gas [b]	No impact	Higher compressor capital and compressing costs

[a] Assume no phase change, liquid remains as liquid in the pipes; [b] assume no phase change, gas stays as gas in the pipes.

3. Pressure Drop Estimates

3.1. Steam Distribution System

Figure 3 is a process flow diagram of a steam generation and distribution system. The main steam header takes supply from the boilers and process/steam generators which recover heat from the Site Source. At the process/steam generator, a pressure control valve regulates the pressure at the heat exchanger ensuring that steam is generated at a sufficient pressure for delivery to the main header via the sub-header. The pressure drops between the process/steam generator and the main header, ΔP_{G-S}:

$$\Delta P_{G-S} = \Delta P_{CV} + \Delta P_P + \Delta P_{SH} + \Delta P_H \tag{1}$$

The pressure drops between the process/steam user and the main header, ΔP_{S-U}:

$$\Delta P_{S-U} = \Delta P_H + \Delta P_{SH} + \Delta P_P + \Delta P_{CV} + \Delta P_{HE} \tag{2}$$

where ΔP_{CV} is the pressure drop across the control valve, ΔP_{HE} is the pressure drop across the heat exchanger and ΔP_P, ΔP_{SH}, ΔP_H are the frictional pressure drops in the distribution pipe, sub-header and header.

The steam is assumed to be saturated and dry throughout the distribution network. Any condensate dropouts due to heat losses from the insulated pipe to the ambient and/or due to the Joule-Thompson effects of pressure drops are removed by steam traps located at strategic locations [14]. Heat loss from a steam distribution system occurs in several ways. In addition to the heat loss from the insulated pipes to the ambient a majority of the heat loss is through leaks in steam pipes, condensate return lines as well as steam traps. It is more appropriate to account for steam losses (which have to be made up by extra steam generation) as a percentage of steam consumption than to use a degree of superheating in the steam temperature as proposed by [12] to account for heat losses.

The frictional pressure drop in steam lines can be calculated using the Babcock equation [15]:

$$\Delta P_f = 2489 \left\{ \frac{d + 3.6}{d^6} \right\} \frac{W^2 L}{\rho} \tag{3}$$

where W is the mass flow (kg/h), L is the pipe length (m), ρ is the single phase density (kg/m^3) and d is the pipe internal diameter (mm).

Alternatively, a steam line sizing nomograph, see Appendix 1, can be used for quick estimate of steam line pressure drops [14]. Commercial software such as Pipe module, which estimate pressure drop and heat loss in pipes, in the Aspen-HYSYS process simulator can also be used [16].

3.2. Cooling Water Distribution System

In Figure 5, pressure drop ΔP_{CW} at the CW distribution system, for a process/CW cooler, can be described as:

$$\Delta P_{CW} = \Delta P_P + \Delta P_E + \Delta P_{HE} + \Delta P_{CV} \tag{4}$$

where ΔP_P is the frictional pressure drop, ΔP_E is the elevation pressure drop, ΔP_{HE} and ΔP_{CV} are as described before. Equation (4) can generally be used for other liquid phase utilities such as HO, *etc.*

3.3. Frictional Pressure Drop in Liquid and Gas Lines, ΔP_P

Fluid flow always results in energy losses due to friction. The frictional losses will be have to be overcome by additional head required on the pump. The pressure drop due to friction can be estimated by the well-known Darcy-Weisbach equation [17]:

$$\Delta P_f = 0.5 \, \rho \, f_m \, L \, V^2 / d \tag{5}$$

where ρ is the density (kg/m^3), L is the length (m), V is the velocity (m/s) and d is the internal diameter of pipe (mm). f_m is the Moody friction factor, which depends on the Reynolds number (Re) and ε, the absolute roughness of the pipe for turbulent flow, typical of fluids flow in plant. Appendix 2 gives the values of ε and f_m for different pipe materials. These values are the iterative solution of the Colebrook correlation [17]:

$$\frac{1}{\sqrt{f_m}} = -2 \log_{10} \{ \frac{\varepsilon}{3.7 \, d} + \frac{2.51}{Re \, \sqrt{f_m}} \} \tag{6}$$

Equation (5) can be directly applied for liquid lines.

To estimate pressure drop in gas lines within plant or battery limits, the Darcy-Weisbach formula can be written in a simple form, assuming that the pressure drop through the line is less than 10% of the line pressure [17]. Pressure drop per 100 m of equivalent pipe length can be written as:

$$\Delta P_{100} = \frac{W^2}{\rho} \{ \frac{62\,530\,(10^2)f}{d^5} \} \tag{7}$$

where W is the mass flow (kg/h), ρ is the single phase density (kg/m^3), f is the friction factor and d is the pipe internal diameter (mm).

3.4. Elevation Pressure Drop for Liquid Lines, ΔP_E

For liquid lines, the pressure drop due to static head of liquid column above the utility circulation pump need to be included. The elevation pressure drop has to be calculated separately using the following equation which is based on Bernoulli's Theorem:

$$\Delta P_E = 0.00981 \, \rho_l \, Z_E \tag{8}$$

where ρ_l is the liquid density (kg/m^3) and Z_E is the elevation of the heat exchanger above the utility circulation pump centre line (m).

3.5. Pressure Drop across Heat Exchanger, ΔP_{HE}

During conceptual design, the type of heat exchanger or detailed geometry of the heat exchanger are often not available. Typical values of pressure drop based on company's guidelines or designer's experience can be used. Alternatively, the heat exchangers pressure drop can be estimated using established equations with some explicit assumptions on the heat exchanger geometries, for e.g., shell and tube heat exchangers: number of passes, tube diameter, tube length, tube pitch, tube configurations, baffle cuts, etc. [18].

3.6. Pressure Drop across Control Valve, ΔP_{CV}

The pressure drop across a control valve can be estimated if the characteristics of the control valve, C_v, is known. A larger pressure drop will increase pumping costs while a smaller pressure drop will increase valve costs. During the conceptual stage, when the details of the valves are not known, the usual rule of thumb is to use an allowable pressure drop of 10%–15% of total pressure drop, or 70 kPa, whichever is greater [19].

4. Methodology

The proposed methodology is presented in Figure 6 and described as follows.

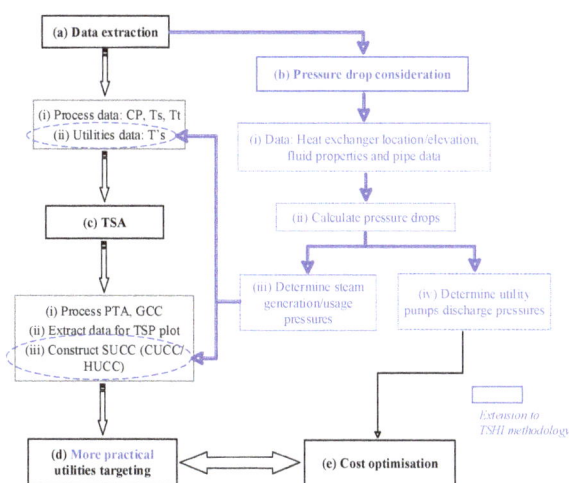

Figure 6. Algorithm to consider pressure drops in TSHI.

(a) Data extraction—extract stream and utilities data, *i.e.*, heat capacities (CP) and temperatures.

(b) To consider the pressure drop factor in TSHI:

(i) Information on the location of the heat exchangers, fluid properties and pipe data are required in order to estimate the pressure drops. Location and elevation of the heat exchangers can be obtained from the site plot plan, individual plant layout and elevation drawings. Fluid properties such as mass flow and density can be extracted from the heat and mass balances. Pipe data required are the internal diameter and roughness factor. For each plant on site, determine the header, sub-header and pipe lengths based on the process/utility heat exchangers located furthest from the reference point and the process/utility heat exchanger at the highest elevation.

(ii) The pressure drops can be estimated using the equations given in Section 3. Alternatively, pressure drops can be based on the typical ΔP per unit length for pipes, control valves and heat exchangers available from company guidelines or designer's experiences.

(iii) Determine steam generation/usage pressure and corresponding steam saturation temperatures. Referring to Figure 3 again, the steam usage pressure, P_U, is the steam pressure at the steam/process heater, furthest from the utility reference point:

$$P_U = \Delta P_{S-U} + P_H \tag{9}$$

where, P_H is the header pressure.

The steam generation pressure, P_G, is the steam pressure at the process/steam generator furthest from the utility reference point

$$P_G = P_H + \Delta P_{G-S} \tag{10}$$

The steam saturation temperatures at P_U and P_G can be obtained from the steam tables.

(iv) Determine utility pumps discharge pressure.

Referring to Figure 5, the CW pump discharge pressure reads as:

$$P_{PUMP} = P_{DES} + \Delta P_{CW} \tag{11}$$

where subscript DES denotes destination, at the process/CW cooler furthest from the CW pumps.

(c) Carry out TS analysis:

(i) Prepare the TSP from individual process PTA and GCC [13]. The utility usage and generation are directly interpolated on the TS-PTA at the respective utilities temperatures [20]. An example of the TS-PTA is given in Table 3.

(ii) A graphical representation of the SUCC can be obtained from the TS-PTA, see Figure 8.

(d) Utilities targeting—Steam is generated and used at different temperatures due to the pressure drops in the steam distribution network. The TS energy targets are determined using the pinch-based graphical and algebraic method [20].

(e) Pressure drops determined for liquid utility systems can be used as an input to the cost optimisation in terms of higher pumping cost and the constraints in allowable ΔP.

5. Illustrative Examples

The TS consists of four plants A, B, C and D with hot oil (HO), HPS, MPS, LPS and cooling water (CW) utility systems as depicted in Figure 2. A simplified plot plan and elevation drawing is given in Figure 7.

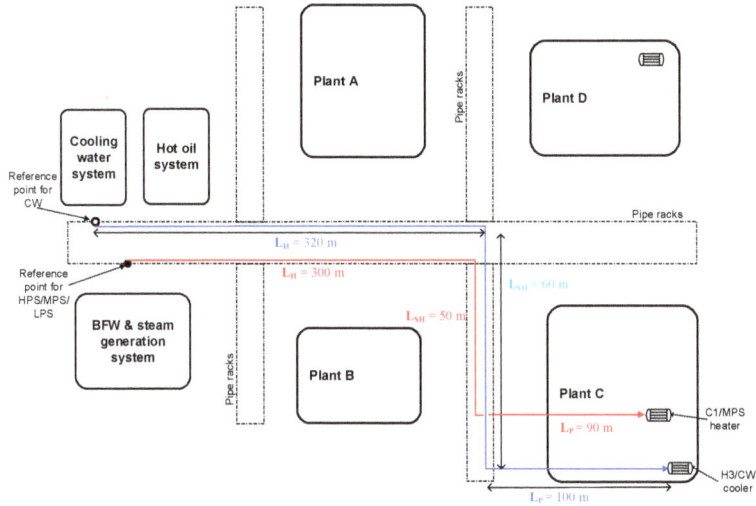

(a) TS plot plan

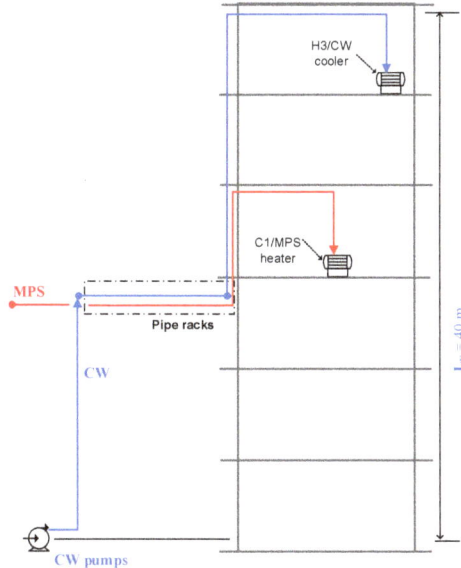

(b) Plant C elevation view

Figure 7. Simplified plot plan and elevation drawing for the TS.

For LPS, $P_{U\text{-LPS}}$ is governed by stream C1/LPS heater at Plant D, located furthest from the main LPS header, while $P_{G\text{-LPS}}$ is governed by stream H1/LPS steam generator at the same Plant D. For

MPS, $P_{U\text{-MPS}}$ is governed by stream C1/MPS heater at Plant C located furthest from the main MPS header while $P_{G\text{-MPS}}$ is governed by stream H1/MPS steam generator at the same Plant C. For HPS, $P_{U\text{-HPS}}$ is governed by stream C1/HPS heater at Plant C, located furthest from the main HPS header. There is no HPS steam generation on site.

A summary of the stream data, layout and elevation information for the estimation of pressure drops is given in Table 2. Figure 8 shows the TSP and SUCC of the TS. The results of the pressure drops estimation and the corresponding steam generation and usage temperatures for the steams and CW distribution networks are summarised in Table 3. Table 4 gives the modified TS-PTA by which the utilities usage and generation are interpolated from the Site Sink and Site Source PTA. The revised SUCC, with consideration for pressure drops, are superimposed on Figure 8.

Table 2. Summary of input data for TS analysis.

Process	Stream	CP (MW/°C)	T_s (°C)	T_t (°C)	L_H (m)	L_{SH} (m)	L_P (m)	E (m)	Heat exch. Furthest from Utility Reference Point
A	H1	0.35	260	225					
	H2	1.15	260	195					
	H3	0.50	195	130					
	C1	1.25	240	255					
	C2	0.65	175	260					
	C3	0.20	155	205					
B	H1	0.36	260	175					
	H2	0.60	260	115					
	H3	0.75	175	95					
	C1	1.10	175	255					
	C2	0.20	110	175					
	C3	0.89	95	155					
C	H1	0.62	225	155	300	50	90		H1/MPS
	H2	0.32	195	95					
	H3	1.00	130	85	320	60	100	40	H3/CW
	C1	0.60	110	240	300	50	85		C1/HPS, C1/MPS
	C2	0.40	155	240					
	C3	0.70	110	175					
D	H1	0.41	130	85	300	50	90		H1/LPS
	H2	0.10	110	80					
	H3	0.15	95	70					
	C1	0.20	90	140	300	50	80		C1/LPS
	C2	0.50	60	110					
	C3	0.40	50	100					
Utility	T_s (°C)		T_r (°C)						
HO	300		260		$\Delta T_{\text{min-pp}}$ is		20		°C
HPS	250		-		$\Delta T_{\text{min-pu}}$ is		15		°C
MPS	200		-						
LPS	150		-						
CW	25		45						

[1] Piping lengths are only extracted for those heat exchangers furthest from the utilities reference point, *i.e.*, which govern the steam generation and usage levels and utility circulation pump sizing. Only steam and CW are considered.

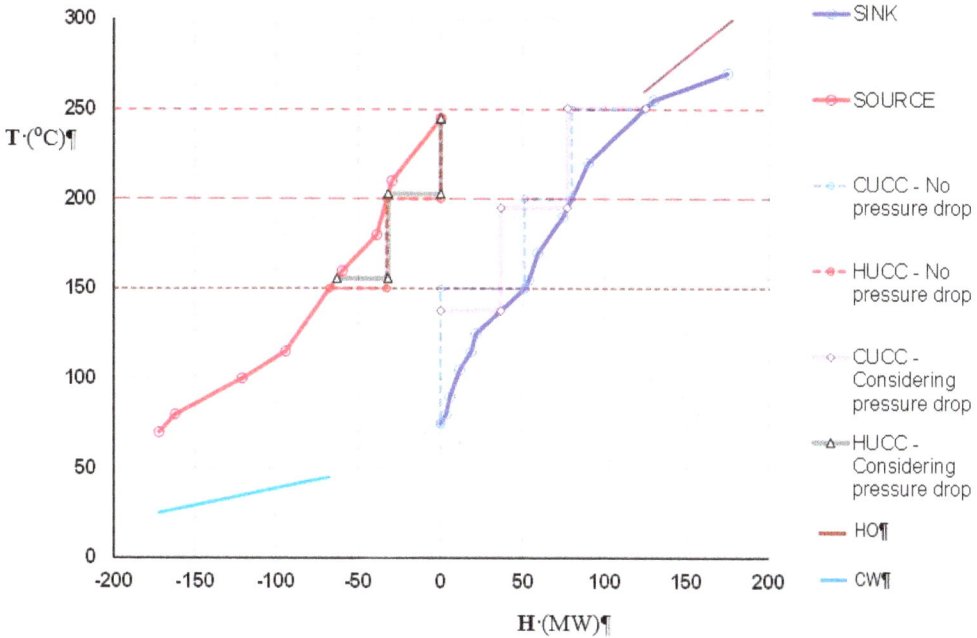

Figure 8. Simplified plot plan and elevation drawing for the TS.

Due to pressure drops in the headers, sub-headers, piping and across control valves and heat exchangers, steams have to be generated at a higher value than their usage. A comparison of utility targeting with and without consideration of pressure drop is given in Table 5. The impact of pressure drop is more notable for steam at low pressure, due to its higher volumetric flow. From Table 3, the differences in steam usage and generation temperatures are 18.1 °C and 7.7 °C for LPS and MPS. From Table 5, the overall heating utilities increases by 14.7 MW (14.5%) and the cooling utility increases by 4.7 MW or (4.5%) when pressure drop is taken into consideration. The HPS requirement increases by 3.4 MW, the MPS usage increases by 11.9 MW while LPS usage reduces by 14.4 MW. Excluding pressure drop could lead to over estimation of the amount of steam that can be raised at the Site Source for HPS and MPS leading to the undersizing of central HPS and MPS generation capacities.

Table 3. Summary of pressure drops estimation for steam and CW distribution networks.

Operating parameter			LPS Usage	LPS Generation	MPS Usage	MPS Generation	HPS Usage	CW
Header pressure		kPag	375	375	1453	1453	3831	-
Header temp.		°C	150	150	200	200	250	-
Pressure drops [1]:								
	ΔP_H	kPa	33	33	33	33	33	-
	ΔP_{SH}	kPa	5.5	5.5	5.5	5.5	5.5	-
	ΔP_P	kPa	10.8	9.6	9	9.6	9	-
	ΔP_{CV}	kPa	35	35	50	50	50	70
	ΔP_{HE}	kPa	50	-	50	-	50	100
	ΔP_f	kPa	-	-	-	-	-	218 [2]
	ΔP_E	kPa	-	-	-	-	-	392.4
Total pressure drops		kPa	134	83.1	148	98.1	148	780.4
Pressure @ user	P_U	kPag	240.7	-	1305.5	-	3683.5	-
Temperature @ user	T_U	°C	137.8	-	195.1	-	247	
Pressure @ generation	P_G	kPag	-	458.1	-	1551	-	
Temperature @ generation	T_G	°C	-	155.9	-	202.8	-	
ΔT between usage and generation		°C	18.1		7.7			
Pressure @ CW pump		kPag						980.4 [3]

[1] A frictional pressure drop of 0.11 kPa/m has been assumed for the headers and sub-headers and 0.12 kPa/m for the piping; [2] Total frictional pressure drops at supply/return headers, sub-headers and piping; [3] Based a destination pressure, i.e., pressure at the H3/CW cooler within Process C, of 200 kPag.

Table 4. TS-PTA for Site Source and Site Sink with utilities usage and generation.

(a) Site Sink PTA

T** (°C)	ΔT (°C)	Process CP				ΣCP (MW/°C)	ΔH (MW)	Cascade H (MW)	H(1) (MW)	Utility Usage H (MW)
		A (MW/°C)	B (MW/°C)	C (MW/°C)	D (MW/°C)					
75						0	0	0		
80	5				0.65	0.65	3.3	3.3		
90	10				0.24	0.24	2.4	5.7		
105	15				0.39	0.39	5.9	11.5		
115	10				0.69	0.69	6.9	18.4		
125	10				0.29	0.29	2.9	21.3		
137.8								36.4	36.4	LPS = 36.4
150	25			0.98	0.20	1.18	29.5	50.8		
155	5			0.36	0.20	0.56	2.8	53.9		
170	15			0.36		0.36	5.4	59.0		
190	20			0.76		0.76	15.2	74.2		
195.1									76.9	MPS = (76.9 – 36.4) = 40.5
220	30		0.14	0.38		0.52	15.6	89.8		
250.7									124.8	HPS = (124.8 – 76.9) = 47.9
255	35		0.14	1.00		1.14	39.9	129.7		
270	15	1.90	1.10			3.00	45.0	174.7		
275	5	0.65				0.65	3.3	178.0		

(1) Interpolate at the steam temperatures.

Table 4. *Cont.*

(b) Site Source PTA

T** (°C)	ΔT (°C)	Process CP A (MW/°C)	B (MW/°C)	C (MW/°C)	D (MW/°C)	ΣCP (MW/°C)	ΔH (MW)	Cascade H (MW)	Utility Usage H^(1) (MW)	H (MW)
245						0	0	0		
210	35	0.85				0.85	29.8	29.8		
202.8									31.9	MPS = 31.9
180	30	0.30				0.30	9.0	38.8		
160	20	0.30	0.76			1.06	21.2	60.0		
155.9									63.1	LPS = (63.1 − 31.9) =31.2
115	45	0.50	0.26			0.76	34.2	94.2		
100	15		0.46	1.32		1.78	26.7	120.9		
80	20		0.75	1.32		2.07	41.4	162.3		
70	10				1.00	1.00	10.0	172.3		

T^{**} Double shifted temperature for TSP plot and TS-PTA, °C; (1) Interpolate at the steam temperatures.

Table 5. Impact of pressure drop on TS.

Utilities	Base Case (No Pressure Drops) Usage (MW)	Generation (MW)	Nett (MW)	Case 1 (with Pressure Drops) Usage (MW)	Generation (MW)	Nett (MW)	(Case 1)—(Base Case) Usage (MW)	Generation (MW)	Nett (MW)
HO	54.0	-	54.0	53.2	-	53.2	−0.8	-	−0.8
HPS	44.6	-	44.6	48.0	-	48.0	+3.4	-	+3.4
MPS	28.6	32.8	−4.2 (1)	40.5	31.9	8.6	+11.9	−0.9	+8.6
LPS	50.8	34.8	11.8 (1)	36.4	31.2	5.2	−14.4	−3.8	−6.8
Total heating	-	-	100.4			115.0			+14.6
CW	-	-	104.7			109.2			+4.7

(1) Excess MPS generated is used for LPS heating. Nett LPS heating is (50.8 − 34.8 − 4.2) = 11.8 MW.

The pressure drop in liquid utilities does not affect TSHI MER targeting, however it should be considered and used as an input parameter when evaluating the TSHI options for economic evaluation. Exclusion of pressure drops will lead to undersizing of pumps or compressors leading to infeasible design solutions, and expensive re-design at the detailed design stage.

6. Conclusions

A systematic methodology that considers pressure drops in TSHI utility targeting has been developed. The case study proved that ignoring pressure drops in TSHI design led to optimistic MER targets and resulted in undersizing of external steam generation capacity. While pressure drops of liquid utilities such as water do not affect MER targeting, the pressure drop information should be incorporated in the economic evaluation of TSHI options. Pressure drops due to pipe friction, elevation changes and pressure drops across control valve and heat exchangers all need to be accounted for. Incorporation of pressure drops leads to closer to real life MER targeting and design. The proposed methodology can benefit from the visualisation advantages of the graphical method and from the precision of the numerical method and should be of the benefit to both industry and academia [21].

Acknowledgments

The authors gratefully acknowledge the financial supports from the Universiti Teknologi Malaysia (UTM) Research University Grant under Vote No. Q.J130000.2509.07H35 and the EC FP7 project ENER/FP7/296003/EFENIS "Efficient Energy Integrated Solutions for Manufacturing Industries—EFENIS". The support from the Hungarian project Társadalmi Megújulás Operatív Program "TÁMOP-4.2.2.A-11/1/KONV-2012-0072—Design and optimisation of modernisation and efficient operation of energy supply and utilisation systems using renewable energy sources and ICTs" significantly contributed to the completion of this analysis.

Author Contributions

Jiří Jaromír Klemeš, Sharifah Rafidah Wan Alwi, Zainuddin Abdul Manan and Andrea Pietro Reverberi conceived the idea of research, provide guidance and supervision; Kew Hong Chew implemented the research, performed the analysis and wrote the paper. All authors have contributed significantly to this work.

Appendix 1. Steam Line Sizing Chart—Pressure Drop (Spirax Sarco, 2014).

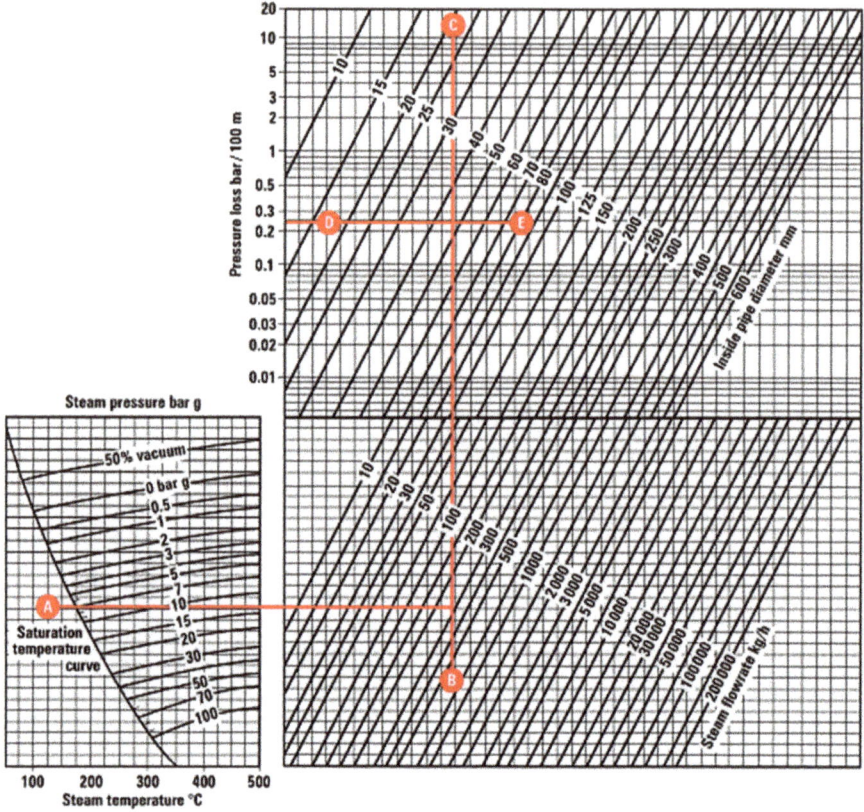

- Select the steam pressure at the saturated steam line (7 barg), A;
- From A, draw a horizontal line to the steam flowrate (286 kg/h) and mark B;
- From B, draw a vertical line to the top of nomograph, C;
- Draw a horizontal line from 0.24 bar/100 m (allowable DP) on the pressure loss scale (DE);
- Point which BC crosses DE will indicate the pipe size required.

Appendix 2. Relative Roughness of Pipe Materials and Friction Factors for Complete Turbulence (GPSA, 1998).

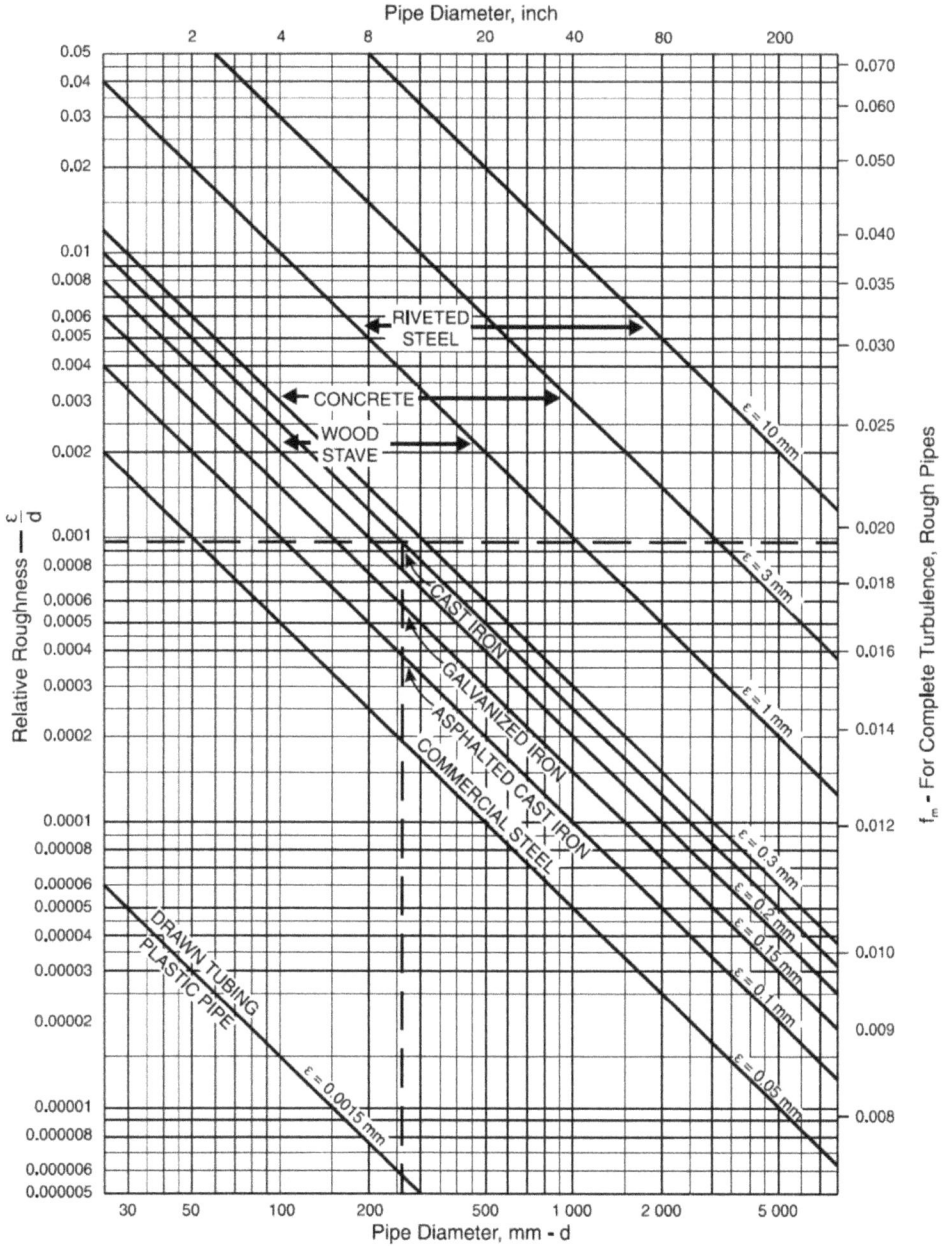

42

Abbreviations

BFW	Boiler feed water
CUCC	Cold Utility Composite Curve
CV	Control valve
CW	Cooling water
FC	Flow control
GCC	Grand Composite Curve
HE	Heat exchanger
HEN	Heat exchanger network
HI	Heat Integration
HO	Hot oil
HPS	High pressure steam
HUCC	Hot Utility Composite Curve
LC	Level control
LPS	Low pressure steam
MINLP	Mixed integer non-linear programming
MP	Mathematical Programming
MPS	Medium pressure steam
MER	Minimum energy requirement
PC	Pressure control
PTA	Problem Table Analysis
SUCC	Site utility Composite Curves
SS_iP	Site Sink Profile
SS_oP	Site Source Profile
TC	Temperature control
TS	Total Site
TS-PTA	Total Site Problem Table Analysis
TSHI	Total Site Heat Integration
TSP	Total Site Profile
VHPS	Very high pressure steam

Nomenclature

CP	Heat capacity flowrate, MW/°C
d	Pipe internal diameter, mm
ε	Pipe roughness factor, m
Q_c	Cooling utilities heat flowrate, MW
Q_h	Heating utilities heat flowrate, MW
H	Process heat flowrate, MW
ΔP	Pressure drop, kPa
ΔT_{min}	Minimum approach temperature, °C

$\Delta T_{\text{min-pp}}$	Minimum approach temperature between process and process, °C
$\Delta T_{\text{min-pu}}$	Minimum approach temperature between process and utility, °C
f_m	Moody friction factor
L	Length, m
P	Pressure, kPag
Re	Reynolds number
ρ	Density, kg/m^3
T	Temperature, °C
T^*	Shifted temperature for process PTA, °C
T^{**}	Double shifted temperature for TSP plot and TS-PTA, °C
V	Velocity, m/s
W	Mass flow, kg/h
Z	Elevation, m

Subscripts

CV	Control valve
CW	Cooling water
DES	Destination
E	Elevation
f	Friction
G	Generation
H	Header
HE	Heat exchanger
l	Liquids
P	Piping
r	Return
S	Steam
s	Supply
SH	Sub-header
t	Target
U	Utilisation

Conflicts of Interest

The authors declare no conflict of interest.

References

1. Klemeš, J.J.; Varbanov, P.S.; Kravanja, Z. Recent developments in process integration. *Chem. Eng. Res. Des.* **2013**, *91*, 2037–2053.
2. Polley, G.T.; Panjehshahi, M.H.; Jegede, F.O. Pressure drop considerations in the retrofit of heat exchanger networks. *Chem. Eng. Res. Des.* **1990**, *68*, 211–220.

3. Ciric, A.M.; Floudas, C.A. A mixed integer nonlinear programming model for retrofitting heat-exchanger networks. *Ind. Eng. Chem. Res.* **1990**, *29*, 239–251.

4. Hui, C.W.; Ahmad, S. Minimum cost heat recovery between separate plant regions. *Comput. Chem. Eng.* **1994**, *18*, 711–728.

5. Sorsak, A.; Kranvanja, Z. Simultaneous MINLP synthesis of heat and power integrated heat exchanger network. *Comput. Chem. Eng. Suppl.* **1999**, S143–S147.

6. Yee, T.F.; Grossman, I.E. Simultaneous optimisation models for heat integration—II. Heat exchanger network synthesis. *Comput. Chem. Eng.* **1990**, *14*, 1165–1184.

7. Nie, X.R.; Zhu, X.X. Heat exchanger network retrofit considering pressure drop and heat transfer enhancement. *AIChE J.* **1999**, *45*, 1239–1254.

8. Panjaeshahi, M.H.; Tahouni, N. Pressure optimisation in debottlenecking of heat exchanger networks. *Energy* **2008**, *33*, 942–951.

9. Soltani, H.; Shafiei, S. Heat exchanger networks retrofit with considering pressure drop by coupling genetic algorithm with LP (linear programming) and ILP (integer linear programming) methods. *Energy* **2011**, *36*, 2381–2391.

10. Zhu, X.X.; Nie, X.R. Pressure drop considerations for heat exchanger network grassroots design. *Comput. Chem. Eng.* **2002**, *26*, 1661–1676.

11. Chew, K.H.; Klemeš, J.J.; Wan Alwi, S.R.; Abdul Manan, Z. Industrial implementation issues of Total Site Heat Integration. *Appl. Therm. Eng.* **2013**, *61*, 17–25.

12. Liew, P.Y.; Wan Alwi, S.R.; Klemeš, J.J. Total Site Heat Integration targeting algorithm incorporating plant layout issues. *Comput. Aided Chem. Eng.* **2014**, *33*, 1801–1806.

13. Klemeš, J.J.; Dhole, V.R.; Raissi, K.; Perry, S.J.; Puigjaner, L. Targeting and design methodology for reduction of fuel, power and CO_2 on total sites. *Appl. Therm. Eng.* **1997**, *17*, 993–1003.

14. Spirax Sarco, Resources. Available online: spiraxsarco.com/resources/steam-engineering-tutorials/steam-distribution/pipes-and-pipe-sizing.asp (accessed on 23 September 2014).

15. *Flow of Fluids through Valves, Fittings and Pipe*; Technical Paper No. 410; Crane Co.: Chicago, IL, USA, 1988.

16. Aspen-HYSYS (Version 2006) (computer software). AspenTech: Burlington, MA, USA, 2014.

17. Gas Processing Supplier Association (GPSA). Fluid flow and piping. In *GPSA Engineering Data Book*, 11th ed.; GPSA: Tulsa, OK, USA, 1998; Volume 17, pp. 1–28.

18. Smith, R. *Chemical Process—Design and Integration*; McGraw-Hill: New York, NY, USA, 2005.

19. Choudhury, A.A.S.; Nwaoha, C.; Vishwasrao, S.V. In *Process Plant Equipment: Operation, Control and Reliability*; Holloway, M.D., Nwaoha, C.N., Onyewuenyi, O.A., Eds.; Wiley: New York, NY, USA, 2012; pp. 9–15.

20. Chew, K.H.; Klemeš, J.J.; Wan Alwi, S.R.; Manan, Z.A. Process Modifications to Maximise Energy Savings in Total Site Heat Integration. *Appl. Therm. Eng.* **2014**, doi:10.1016/j.applthermaleng.2014.04.044.

21. Klemeš, J.J. Industry-academia partnership. *Clean Technol. Environ. Policy* **2013**, *15*, 861–862.

Using Satellite SAR to Characterize the Wind Flow around Offshore Wind Farms

Charlotte Bay Hasager, Pauline Vincent, Jake Badger, Merete Badger, Alessandro Di Bella, Alfredo Peña, Romain Husson and Patrick J. H. Volker

Abstract: Offshore wind farm cluster effects between neighboring wind farms increase rapidly with the large-scale deployment of offshore wind turbines. The wind farm wakes observed from Synthetic Aperture Radar (SAR) are sometimes visible and atmospheric and wake models are here shown to convincingly reproduce the observed very long wind farm wakes. The present study mainly focuses on wind farm wake climatology based on Envisat ASAR. The available SAR data archive covering the large offshore wind farms at Horns Rev has been used for geo-located wind farm wake studies. However, the results are difficult to interpret due to mainly three issues: the limited number of samples per wind directional sector, the coastal wind speed gradient, and oceanic bathymetry effects in the SAR retrievals. A new methodology is developed and presented. This method overcomes effectively the first issue and in most cases, but not always, the second. In the new method all wind field maps are rotated such that the wind is always coming from the same relative direction. By applying the new method to the SAR wind maps, mesoscale and microscale model wake aggregated wind-fields results are compared. The SAR-based findings strongly support the model results at Horns Rev 1.

Reprinted from *Energies.* Cite as: Hasager, C.B.; Vincent, P.; Badger, J.; Badger, M.; Di Bella, A.; Peña, A.; Husson, R.; Volker, P.J.H. Using Satellite SAR to Characterize the Wind Flow around Offshore Wind Farms. *Energies* **2015**, *8*, 5413-5439.

1. Introduction

In the Northern European Seas offshore wind farms are planned as clusters. The wind farm wake from one wind farm thus has the potential to influence the power production at neighboring wind farms. The expected wake loss due to wind farm cluster effects is investigated in the present study. The wind farm wake cluster effects are strongly dependent upon the atmospheric conditions. These vary spatially and temporally. One remote sensing method for observing ocean surface winds is satellite Synthetic Aperture Radar (SAR). The advantage of SAR is that a large area is observed and several wind farms are covered. The derived wind map from SAR provides a snapshot of the wind field during a few seconds at the time of acquisition.

In the satellite SAR data archives covering the North Sea, thousands of wind turbines are visible as white dots in calm conditions. For low wind speed the backscatter signal over the ocean is low and the images appear dark while hard targets such as wind turbines and ships provide high backscatter and the objects appear very bright. During windy conditions wind farm wakes are sometimes visible as dark elongated areas downwind of a wind farm while the surrounding seas appear brighter. This is a result of the differences in wind speed with reduced winds downwind of large operating wind farms. The intensity of backscatter of microwave electromagnetic radiation from the ocean surface is a non-linear function of the wind speed over the ocean. The physical relationship is due to the capillary and short gravity waves formed at the ocean surface by the wind. For higher wind speeds the backscatter is higher.

Previous wind farm wake studies based on SAR from ERS-1/-2, Envisat, RADARSAT-1/-2, TerraSAR-X and airborne SAR show great variability in wind farm wakes [1–4]. This reflects the natural variability in atmospheric conditions at the micro- and mesoscale. Wind farm wakes are often not clearly visible in the SAR archive data. This may be explained by wind turbines out of operation or presence of oceanic features, e.g., bathymetry, currents, surfactants. However, the great variability in the wind field is most likely a major cause.

In order to show that wind farm wakes are detectable from SAR, we present in this study one case based on RADARSAT-2 ScanSAR Wide. This scene is a good example where ideal conditions for wake analysis occur and the coverage is just right for capturing 10 large offshore wind farms located in the southern North Sea. We compare the instantaneous SAR-based wind farm wakes to micro- and mesoscale wake model results.

For the rest of our SAR wind archive, we wish to find out if wakes can be detected even if they are not so clearly visible. We first use a simple method which has some disadvantages. Next we apply the aggregated method to overcome some of these disadvantages.

This is the first time that a suitable number of SAR scenes covering several large operating winds farms have become available [5]. The present study focuses on the wind farm wake climatology using many overlapping SAR scenes. The wide-swath-mode (WSM) products from the Advanced SAR (ASAR) on-board Envisat are selected. This data source is sampled routinely so there are many more samples but with less spatial detail (original resolution 150 m) than those used in previous studies (of the order 5 to 30 m spatial resolution) where the data are rare and infrequent and only sampled upon request [1–4]. It is questionable whether wake effects can be detected from Envisat ASAR WSM data as we cannot always visually see the wind farm wakes. However, we hypothesize that the combination of many satellite samples will show an aggregated effect of the wind farm wakes on the mean wind climate in the area. The SAR-based wind farm wake climatology results therefore can be used to validate wake model results. The wind farm wake climatology can be modelled by micro- and mesoscale models but perfect agreement cannot be expected between SAR and wake model results. This is due to the different nature of data with SAR based on the sea surface while wake models operate around wind turbine hub-height.

The main topic of the study is on the potential of using SAR for characterization of wind flow around offshore wind farms. Our mission is to find out how to best utilize SAR for wake mapping. Three modes of investigation are considered: (1) Wind flow observed at 10 offshore farms with wind farm wakes concurrent in one SAR scene; (2) Wind flow observed at two wind farms concurrent and wind farm wakes average value based on 7 to 30 SAR scenes; (3) Wind flow observed at one wind farm at a time and the aggregated wind farm wake based on 100 to 800 SAR scenes. Basic information on the three modes of investigation is listed in Table 1. The advantages and limitations of each investigation mode in regard to wind farm wake model comparison are provided. Selected wake model results are presented as demonstration for each of the modes. Presenting the three modes in such an order clearly shows the evolution of wake studies using SAR.

Table 1. SAR source, number of wind farms covered in the method, number of SAR scenes used in each method for wake identification based on no averaging, geo-located SAR wind field averaging and rotated SAR wind field averaging. The spatial resolution of the wind fields are given. The Section in this paper where each mode is presented is also indicated.

Satellite Data	Number of Wind Farms	Number of SAR Scenes	Averaging	Analysis Type	Resolution (km)	Section
RADARSAT-2	10	1	None	Qualitative	1	3
Envisat ASAR	2	7–30	Geo-located	Quantitative	1	4
Envisat ASAR	1	100–800	Rotated	Quantitative	1	5

The structure of the paper includes in Section 2 a description of the study site, satellite data and the two wake models used. In Section 3 the results from the case study based on RADARSAT-2 and the results from two wake models are presented and discussed. Section 4 presents the Envisat SAR-based wind farm wake climatology based on simple averaging of the wind fields at Horns Rev 1 and 2 wind farms and comparison to results from one wake model. Section 5 gives introduction to the new methodology developed in which the wind field maps are rotated such that the wind is always coming from the same relative direction. The Envisat SAR-based results from the new methodology as well as wake model results from two models are presented and Envisat SAR-based results from four other wind farms are presented. In Section 6 is the discussion of results. Conclusions are given in Section 7.

2. Study Site, Satellite SAR and Wake Modelling

2.1. Study Site

Wind flow around the wind farms in the southern part of the North Sea is investigated. The wind farms studied are listed in Table 2 and the location of most of the wind farms is shown in Figure 1. Those not shown in Figure 1 are the Alpha ventus wind farm located in the German North Sea and Horns Rev 1 and 2 located in the Danish North Sea. The information in Table 2 includes the year of start of operation and key data on the wind turbines and area covered.

RS-2 20130430 17:41:53 UTC SAR intensity image

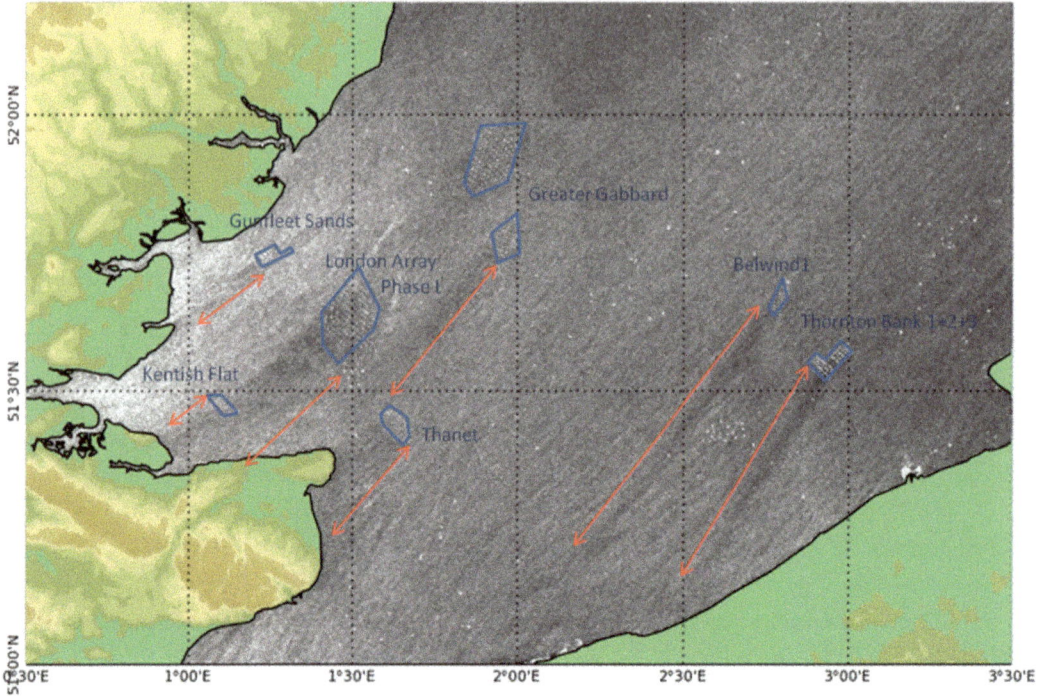

Figure 1. RADARSAT-2 intensity map of the southern North Sea observed 30 April 2013 at 17:41 UTC. The blue lines outline wind farms and the red arrows the wind farm wake.

Table 2. Wind farm info: Country, start year of operation, approximated latitude and longitude, number and size of turbines, wind park capacity and area covered.

Wind Farm	Nationality	Year	Latitude (°)	Longitude (°)	Number of Turbines	Turbine Size (MW)	Park (MW)	Area (km²)
Alpha ventus	Germany	2009	54.010	6.606	12	5	60	4
Belwind 1	Belgium	2010	51.670	2.802	55	3	165	13
Greater Gabbard	United Kingdom	2012	51.883	1.935	140	3.6	504	146
Gunfleet Sands 1 + 21	United Kingdom	2010	51.730	1.229	48	3.6	172.8	16
Horns Rev 1	Denmark	2002	55.486	7.840	80	2.0	160	21
Horns Rev 2	Denmark	2009	55.600	7.582	91	2.3	209.3	33
Kentish Flats	United Kingdom	2005	51.460	1.093	30	3	90	10
London Array Phase 1	United Kingdom	2012	51.626	1.495	175	3.6	630	100
Thanet	United Kingdom	2010	51.430	1.633	100	3	300	35
Thornton Bank 1	Belgium	2009	51.544	2.938	6	6	30	1
Thornton Bank 2	Belgium	2012	51.556	2.969	30	6.15	184.5	12
Thornton Bank 3	Belgium	2013	51.540	2.921	18	6.15	110.7	7

2.2. Satellite SAR

SAR data from RADARSAT-2 and Envisat ASAR WSM are used. From RADARSAT-2 only one scene is investigated. It is ScanSAR Wide in VV polarization. The wind field retrieval requires input information about the wind direction. From the RADARSAT-2 image the wake direction has been estimated as 40° and using this input for wind direction, wind speed has been retrieved using CMOD-IFR2 [6]. It is the equivalent neutral wind (ENW) at the height 10 m. The calculated wind speed is presented in Figure 2a. The original ScanSAR Wide product has spatial resolution 100 m. The spatial resolution is reduced to approx. 1 km in connection with the processing of wind fields. This is performed to eliminate effects of random noise and long-period waves.

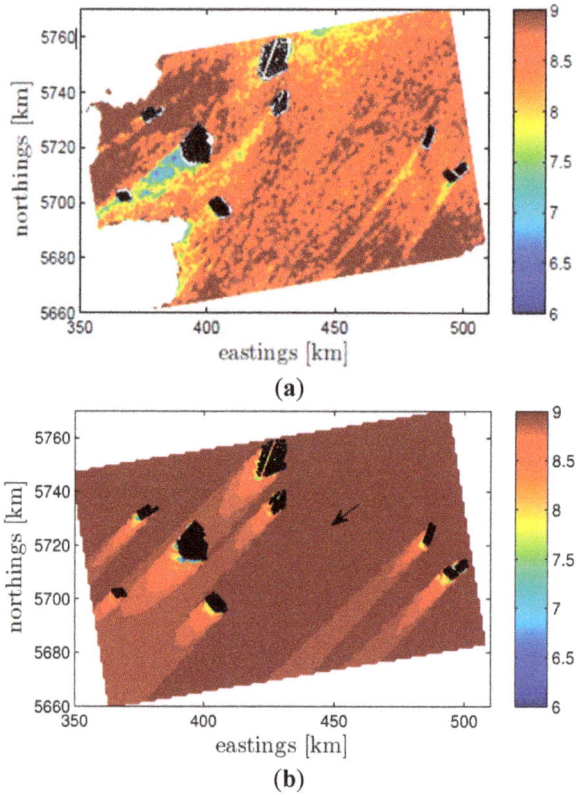

Figure 2. (a) Satellite 10-m SAR wind retrieval observed 30 April 2013 at 17:41 UTC and **(b)** modified PARK wake results at 70 m for the wind farms in the UK and Belgium. The wind direction used for the modeling is indicated with the black arrow.

The Envisat ASAR data were processed to wind fields as part of the project NORSEWInD [5]. The wind field retrieval gives the ENW at the height 10 m. For processing of large image archives, it is desirable to use wind direction information from an atmospheric model. In this case the wind directions were obtained from the European Centre for Medium-Range Weather Forecasts (ECMWF) model and interpolated spatially to match the higher resolution of the satellite data.

Further details about the SAR-wind processing chain, which was setup by Collecte Localisation Satellites (CLS), are given in [5]. The original WSM product has spatial resolution 150 m. The spatial resolution is to approx. 1 km in the wind field.

2.3. Wake Modelling with PARK and WRF

A modified version of the PARK wake model [7], also implemented in the Wind Atlas Analysis and Application Program (WAsP) [8], is here used for wake calculations. The main difference between this modified version and that in WAsP is that the former does not take into account the effects of the "ground reflecting back wakes" and so it only takes into account the shading rotors both directly upstream and sideways. The PARK wake model is based on the wake deficit suggested by [9], who derived a mass-conservation-like equation for the velocity immediately before a turbine u_2, which is affected by a wake:

$$u_2 = u_1 \left[1 - \frac{a}{\left(1 + \frac{k_w\, x}{r_r}\right)^2} \right] \tag{1}$$

where u_1 is the upstream wind speed, a the induction factor which is a function of the thrust coefficient (C_t), k_w the wake decay coefficient, x the downstream distance and r_r the turbine's rotor radius. The square of the total wake deficit is estimated as the sum of the square of all contributing wake deficits. We implemented the model in a Matlab script. This allows us to compute wake deficits at any given point. The wake model can be compared to satellite derived wind maps which contains information over a large area. We use $k_w = 0.03$ for the wake computations.

The Weather Research and Forecasting (WRF) mesoscale model [10] is also used for wake modelling. The advantage of WRF is that the dynamic synoptic flow is considered. The computational cost is much higher than that for the PARK model though.

Mesoscale models have been developed to simulate the atmosphere flow over areas on the order of hundreds of kilometers. Due to their low horizontal resolution unresolved processes, such as turbulence and turbine induced wakes have to be parametrized. In common wind farm parametrisations [11–16] the local turbine interaction is not accounted for, instead the wind speed reduction within the wind farm is obtained from the interaction between the turbine containing grid-cells. The Explicit Wake Parametrisation (EWP) is used for the parametrization of wind farms [16]. In this approach a grid-cell averaged deceleration is applied, which accounts for the unresolved wake expansion with the turbine containing grid-cell. Turbulence Kinetic Energy (TKE) is provided by the Planetary Boundary Layer (PBL) scheme from a changed vertical shear in horizontal velocity in the wake. The EWP scheme is independent of the PBL scheme, although, a second order scheme is recommended.

We use WRF V3.4 with the selected mesoscale model physics parametrizations: PBL [17] (MYNN 2.5), convection [18] (Domain I and II), micro-physics [19], long-wave radiation [20], shortwave radiation [21], land-surface [22] and Nudging of U and V in the outer domain (outside PBL). The number of grid cells in the innermost domain were 427 times 304 in the x and y direction, with a 2 km grid-spacing. We used [16] for the wind farm parametrization.

The model outer domain is driven by ERA-Interim reanalysis data [23] and two nests are inside. The horizontal resolution for the three domains is 18 km, 6 km and 2 km, respectively. The inner nest is run twice, without and with the wind farm parametrization. The number of vertical layers is set to 60. The second mass level is at around 12 m above sea level and it is used for the comparison to the satellite images.

3. Case Study Based on RADARSAT-2

The case study is based on the RADARSAT-2 scene from 30th April 2013 at 17:41 UTC (selected from around 30 images with visible wakes). Figure 1 shows the backscatter intensity map. The wind is from the northeast and the map shows elongated long dark areas downwind of most of the wind farms. These are the wind farm wakes. The approximate extent of the individual wind farm wakes is outlined in the image. The longest is at Belwind around 55 km long while at Thornton Bank it is 45 km, London Array 15 km and Thanet 14 km. At Kentish Flat the wind farm wake is only 10 km long but it is probably passing over the coast and inland in the UK. This cannot be mapped from SAR. It should be noticed that all wakes are very straight and with similar direction. In the intensity map the wind turbines can be seen as small regularly spaced white dots while numerous ships can be noted in irregular spatial pattern. Some large ships show higher backscatter than the turbines.

The retrieved wind speed map is shown in Figure 2a. The wind speed in the northern part of the map is slightly lower than in the southern part. Yet the synoptic flow appears to be fairly homogenous across the entire area. Coastal speed up is seen particular near the UK and Belgium coastlines. The wind speed varies around 8.5–9.5 $m \cdot s^{-1}$ in areas not affected by wind farms while the wind farm wake regions show lower wind speed around 7–8 $m \cdot s^{-1}$ dependent upon location. The wake at London Array is very wide and it appears to influence Kentish Flat at this time. The wake at London Array has a large wake deficit with much lower wind speeds in the wake than in the upwind free stream region. Wake meandering is not pronounced.

The case illustrates a rather unique situation. Firstly because we observe wakes in the satellite image for all wind farms distributed in a large area of the North Sea (all farms in the area show clear speed deficits). Secondly the wind speed and wind the direction do not seem to largely change over such an extended area. Therefore we are able to simulate with the PARK wake model all wind farms at the same time (assuming the same background inflow conditions for all of them). The background wind speed is about 9 $m \cdot s^{-1}$ and direction 40°. We use these two values at 70 m as inflow conditions for the wake modeling.

Figure 2a shows the SAR wind retrieval at 10 m, where most of the variability seems to come from the wake deficits downstream the wind farms, and the wake model results at 70 m in Figure 2b. In this case, we do not extrapolate the satellite background conditions up to 70 m or extrapolate downwards the model results to 10 m as we assume the same wind speed at around hub height when performing the wake simulations. The comparison is only qualitative.

Interestingly, the speed deficits seem to be rather well reproduced by the wake model, extending in most cases nearly as long as the wakes observed in the SAR image.

The WRF wake model with the EWP wind farm scheme is also used for simulation. We include only London Array, Greater Gabbard, Thanet, Belwind1 and Thornton Bank which are the largest wind farms. The domain is rotated around 10° at the wind farm location. The simulation is from the 24 April to 1 May 2013. The velocity deficit at 10 m at 30 April 2013 at 18:00 UTC is shown in Figure 3. We have chosen to plot the velocity deficit since due to the gradients in the background velocity the wake is not visible in the velocity field from the wind farm simulation.

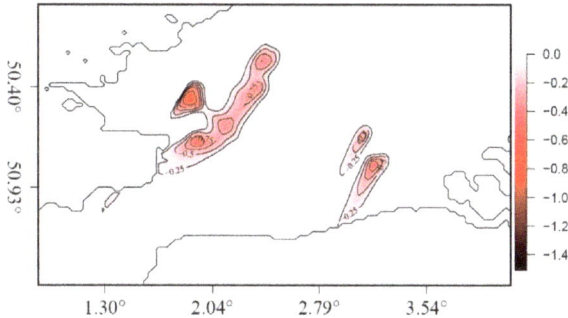

Figure 3. WRF wake model results on velocity deficit in $m \cdot s^{-1}$ at 10 m AMSL at 30 April 2013 at 18:00 UTC at the wind farms London Array, Greater Gabbard, Thanet, Belwind 1 and Thornton Bank.

The WRF modelled wakes at the UK wind farms are oriented slightly more towards the eastern direction than the satellite wakes. The orientations differ by around 10°. For the Belgian wind farms we find that the wakes are well aligned in the SAR and WRF results. Regarding the wake extension behind the wind farms we find for the London Array short wakes both in SAR and WRF while the wakes at the Thanet and Greater Gabbard wind farms are considerably longer both in SAR and WRF. However WRF shows even longer wakes than SAR for Greater Gabbard and Thanet. The extension of wakes at the Belgian wind farms compare well in SAR and WRF.

From WRF it is found that the synoptic conditions two hours before 18:00 UTC show intensified pressure gradients, leading to increased wind speeds near the English coast from 4 $m \cdot s^{-1}$ to higher winds in the order of 10 $m \cdot s^{-1}$. The Greater Gabbard and Thanet wind farms experienced high wind speeds for two hours at 18:00 UTC. The wind speeds started to increase at the London Array only shortly before 18:00 UTC. It might be that the increasing model wind speeds are for some hours out of phase, which would explain the longer wakes behind Greater Gabbard and Thanet wind farms in WRF compared to SAR.

In summary, the wind farm wakes from several wind farms are visually compared between SAR and WRF simulations of the velocity deficit obtained without wind farm and with wind farms using the EWP scheme. The wind farm wake directions and extension of wake are found to compare well despite that mesoscale features, such as that resulting from unsteady flow conditions, are noted in the wind farm wakes in the WRF simulation. We cannot expect the WRF simulations to match the observed velocity and wind direction in SAR satellite data perfectly. The PARK model results do not include unsteady flow but even so the PARK model results have overall good agreement to SAR. This can in part be attributed to the rather unique atmospheric conditions at the time of this SAR acquisition. The results for single events only can be used qualitatively. For a quantitative comparison statistics over longer periods are needed.

4. Wind Farm Wake Climatology Geo-Located Wind Maps

The case in Section 3 was selected based on clear visual observation of wind farm wakes at several wind farms within one satellite SAR image. However, we would like to study the behavior of the wind farm wake in a climatological fashion and investigate whether this can be performed using our Envisat ASAR WSM data set. We select to study the Horns Rev 1 and 2 wind farms for which we have 356 SAR scenes in total for the period of dual wind farm operation from September 2009 to the termination of the Envisat mission in March 2012.

As the wake behavior is highly dependent on the inflow wind conditions, in particular the wind direction, we perform the study for 12 directional sectors based on the ECMWF model wind direction used to retrieve the SAR winds. The SAR scenes are first binned according to the wind directions extracted for a single point near the two wind farms. The data set is then filtered such that only scenes with wind speeds in the range 4–14 m·s^{-1}at the same point are included (total of 241). This is the range where wind farm wakes are expected to be most detectable. At lower wind speeds, the turbines are not operating and at higher wind speeds, wind penetration through the wind farms is expected. For each directional bin, we extract the inflow conditions from a point upstream of the wind farms for every SAR scene in the bin. The reference points are located on two circles circumscribing the wind farms Horns Rev 1 and Horns Rev 2 with radii of 7.5 km and 10 km, respectively, as shown in Figure 4. To get representative inflow conditions, the satellite winds are extracted within a radius of 10 km for Horns Rev 2 and a radius of 7.5 km for Horns Rev 1 wind farm.

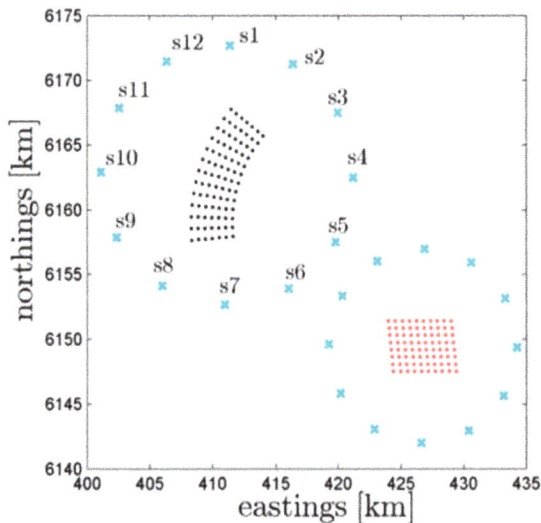

Figure 4. Horns Rev 1 and 2 wind farms in red and black markers, respectively. The locations of the points where the inflow conditions are extracted per sector (cyan markers) is also illustrated.

We perform simulations using the modified PARK model at all positions on the satellite grid and for all the inflow conditions per sector. The point around the wind farms where we extract the inflow conditions is selected based on the sector analyzed, e.g., we use the point north of Horns Rev 2 when performing simulations for Sector 1. In the following subsections we perform qualitative comparisons of the 10 m wind speed SAR retrievals with the results of the wake model per sector at the same height. Both results (wind speed maps) show the average wind speed per sector. We choose to show results for Sector 2 and 3 in Figure 5 because the coastal wind speed gradient and wind farm wakes can be seen in these results even though relatively few data are available. Table 3 shows the number of samples per sector. The inflow wind speed for Sector 2 and Sector 3 is 8.52 and 8.25 $m \cdot s^{-1}$ in average with a standard deviation of 3.50 and 2.55 $m \cdot s^{-1}$, respectively (these are the values at 70 m height). This means that simulations are performed for a rather wide range of wind speeds. These two examples have several overlapping images and several features can be noted such as the coastal wind speed gradient and wind farm wake.

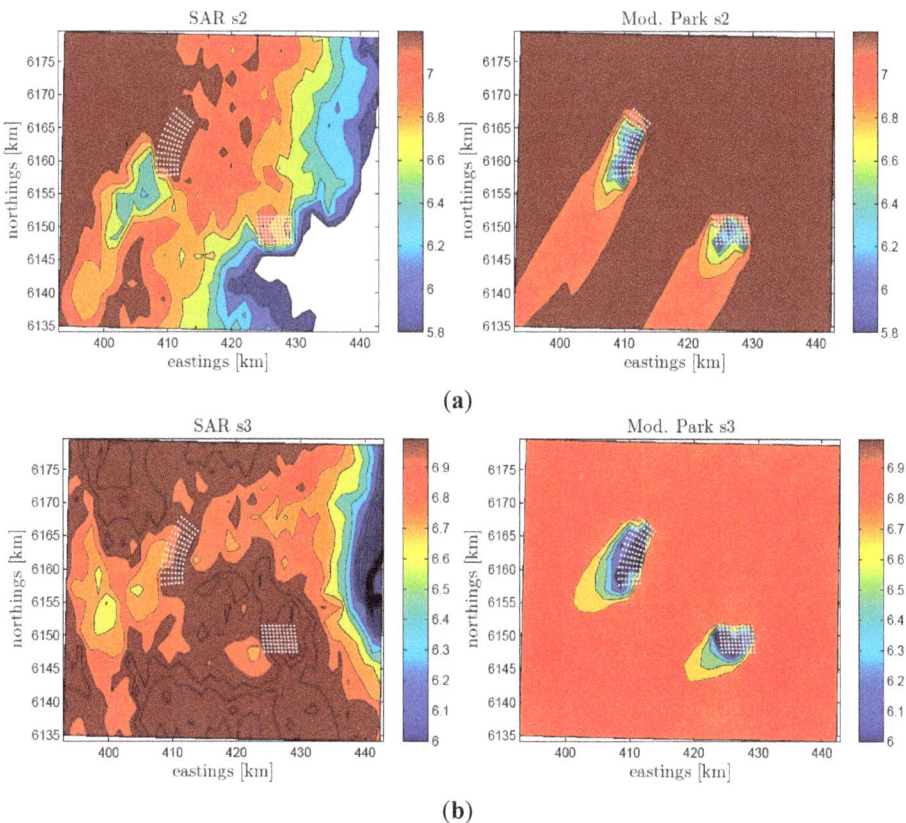

(a)

(b)

Figure 5. Average wind speed based on satellite SAR (**left**) and modified PARK wake model results (**right**) at 10 m height for the Horns Rev wind farm area for winds from Sector 2 (**a**) and Sector 3 (**b**). The color bar indicates wind speed in $m \cdot s^{-1}$.

Table 3. Number of Envisat ASAR samples available per wind directional bin at Horns Rev.

Sector	1	2	3	4	5	6	7	8	9	10	11	12	Total
Samples	13	7	12	16	24	21	22	28	22	30	20	26	241

For Sector 2 there are seven available SAR images for the analysis and the average wind speed is 7.2 m·s^{-1}. It is surprising we do not observe wakes from the Horns Rev 1 wind farm (Figure 5a). This might be simply because of the high horizontal wind speed gradient approaching the coast, which is located east of Horns Rev 1. There is a clear wake spreading towards the southwest direction of the Horns Rev 2 wind farm. The PARK model results show clear wakes spreading southwest of both wind farms.

For Sector 3 twelve SAR images are available for the analysis (west of Horns Rev 2 the number is reduced to 10) with an average wind speed of 7.0 m·s^{-1}. This case shows agreement in terms of the location of the areas where wakes are observed in both the SAR and the PARK wake model at both wind farms (Figure 5b).

The number of samples per sector varies from 7 to 30 within the 12 sectors. The overall agreement between SAR and the wake model is variable. For some sectors (1 and 7) the bathymetry effect at Horns Rev appears to be particularly strong as previously noted by [1]. This results in the lack of wake effects in the mean wind speed maps from SAR due to the interaction of bathymetry and currents, which sometimes leaves a detectable "imprint" at the sea surface. This effect is most visible when winds blow directly from the north or south at Horns Rev (not shown here).

Another reason for the difficulties to systematically observe wakes of offshore wind farms from satellite-derived wind products are inhomogeneous flow. Although the ocean surface is rather homogenous, e.g., when compared to the land surface, the effects of the horizontal wind variability diminish those of the wakes. In the particular case of the Horns Rev area, there is a systematic wind speed gradient near the coast also obstructing the observation of wakes, particularly for easterly and westerly winds. These effects are not taken into account in the PARK modeling. The coastal gradient in wind speed is noticeable in the SAR images in Figure 5.

Finally it can be noted that the distribution of wind maps into direction sectors is performed with some uncertainty. The model wind directions used to drive the SAR wind speed retrieval are not always accurate. The accuracy of the wind direction input could be improved through implementation of higher-resolution regional model simulations, e.g., from WRF. Another option is to detect the wind direction directly from wake signatures whenever they are visible in the images. The distribution of satellite scenes into the 12 sectors is based on information extracted at a single point. Local turning of the wind is possible but not accounted for in the analysis. Each directional bin is 30° wide thus the peak wake directions are expected to vary within this and it will diffuse the observed aggregated wake features. Due to the nature of the SAR images (specifically its number) and due to other phenomena causing spatial variability in the wind speed (like coastal gradients and mesoscale phenomena), it seems not suitable to perform the SAR wake analysis per sectors this wide.

5. Wind Farm Wake Climatology Based on Rotation of Wind Maps

In this section a new approach to analyze SAR-derived wind farm wakes in a climatological way is presented. The method aligns (rotates) all SAR wind field samples such that the wind farm wakes are overlapping before the wake deficit is calculated. This increases the number of samples considerably compared to the method presented in Section 3. Furthermore the 30° wind direction bins used previously give diffuse results whereas in the new method wind directions alignment at 1° resolution is used. The new method is based on extracting wind speeds along points inscribed by circles centered on the wind farm under analysis. While the method was developed for analysis of wakes in SAR scenes, here it is also applied to the WRF simulations as a way to validate the mesoscale simulated wakes.

5.1. Description of the Method

For each SAR scene (or WRF simulated wind field) the wind speeds as a function of compass direction θ are extracted along 3 concentric circles centered on the wind farm. The radii depend on the wind farm in question, and are given in Table 3.

Figure 6 gives an example of the circles centered on the Horns Rev 2 wind farm. The wind fieldsare based on SAR data (1 km). It can be difficult to determine a wake by eye. For each SAR scene the wind speeds along these 3 circles are extracted and stored as U_i (θ_j), where U_i is the wind speed for circle i where $i = 1, 2, 3$ and θ_j is the compass direction relative to the center of the wind farm. θ_j steps through values from 0 to 359° with a 1 degree increment. The number of scenes used for the analysis depends on the wind farm under examination. The number is given in Table 4. The WRF model is run for all SAR scenes from Horns Rev 1 and 2, and results are extracted in a similar way from the WRF simulation results as for the SAR wind fields. The SAR results are valid at 10 m AMSL while WRF model results are available at 14 m AMSL.

First the sum of wind speeds is calculated:

$$S_i^N(\theta_j) = \frac{1}{N_{scene}} \sum_{k=1}^{N_{scene}} (U_i(\theta_j))_k \tag{2}$$

where k is the scene number, and N_{scene} is the total number of scenes. Figure 7a shows S^N plotted against θ, for the Horns Rev 1 wind farm. We see from this plot how the mean wind speed depends on θ. This can be explained in terms of the gradient of the mean wind in the vicinity of the coastline. We call this the coastal gradient. Similar results of S^N based on WRF are shown in Figure 7b. The coastal gradient in wind speed at Horns Rev 1 shows lowest values around 80°–110° (east), where the inscribed circles are closest to the coastline, and highest around 250° (west), where the inscribed circles are furthest from the coastline both in SAR and WRF. SAR shows a direction closer to 80° while WRF shows a direction closer to 110°. It is expected that there is an east-west gradient at Horns Rev as reported in [24]. For the eastern sector SAR shows higher wind speed values at the inner radius (6 km) and progressively lower values at outer radii (10 and 13 km) (nearer to the coastline). For the western sector SAR shows slightly higher wind speed at outer radii (further from the coastline).

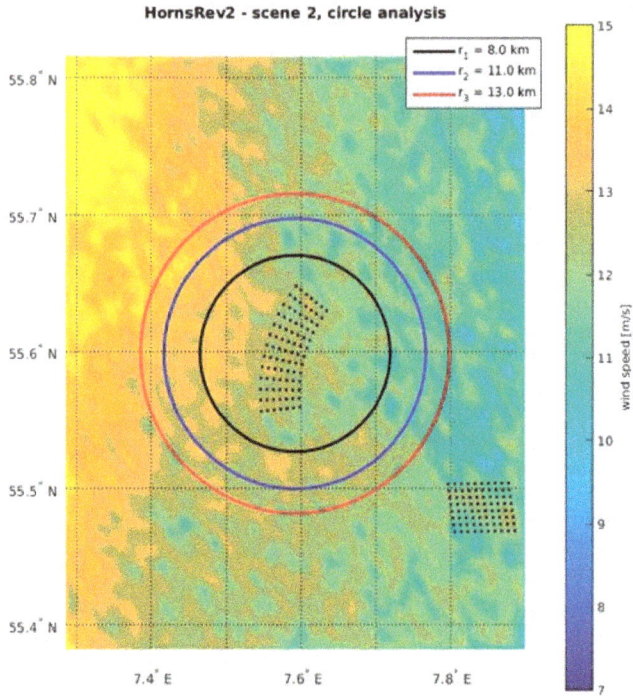

Figure 6. Example of the circles centered on Horn Rev 2 wind farm. The radii are 8, 11 and 3 km.

Table 4. The radii of the three concentric circles for the different wind farms and the number of Envisat ASAR scenes used for the analysis.

Wind Farm	r_1 (km)	r_2 (km)	r_3 (km)	N_{scenes}
Alpha ventus	5	10	15	245
Belwind1	6	11	15	97
Gunfleet Sands 1 + 2	4	5	6	153
Horns Rev 1	6	10	13	835
Horns Rev 2	8	12	15	303
Thanet	7	9	11	128

WRF shows a similar pattern as SAR for the western sector but shows a reverse order in the wind speed at the eastern sector at different radii. This most likely is due to the simulated wake effects of Horns Rev 2 influencing the results at the 13 km radius around 260°–350°. This is supported by examining the results from using WRF without simulating the wind farms, shown in Figure 8. In this plot a very much cleaner signature of the coastal gradient is seen. The next step is to rotate the direction frame of reference for each SAR scene by using each scene's reported wind direction, θ_k, to give a new direction reference, φ. In the new direction reference frame for each scene $\varphi = 0°$ is aligned in the upwind direction and thus one may expect that the wake direction is in the region of $\varphi = 180°$. Now we can determine the wind speeds on the inscribed circles as a function of φ_j instead of θ_j by using:

$$\varphi_j = \theta_j - \theta_k \tag{3}$$

The sum of wind speeds for all scenes is now calculated with respect to the new direction frame, *i.e.*:

$$S_i^R(\varphi_j) = \frac{1}{N_{\text{scene}}} \sum_{k=1}^{N_{\text{scene}}} (U_i(\varphi_j))_k \tag{4}$$

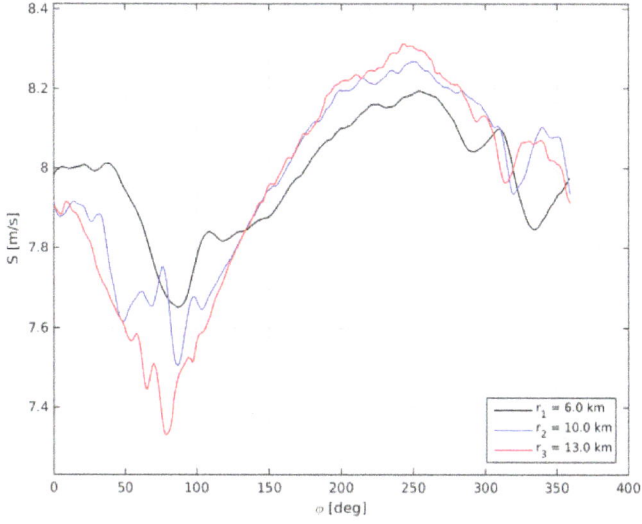

(a)

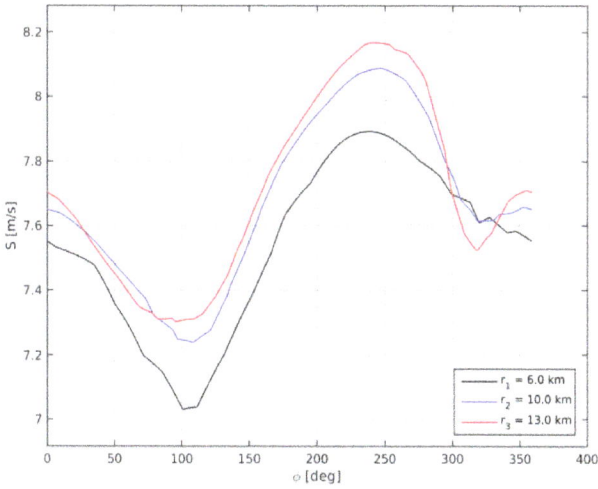

(b)

Figure 7. Horns Rev 1 wind speed summations without rotation (mean wind speed gradient) (*S*) based on SAR (**a**) and WRF (**b**).

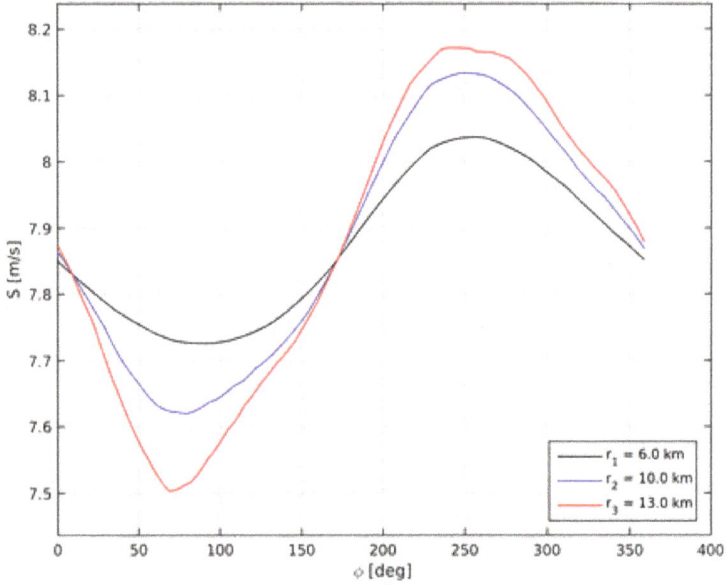

Figure 8. Horns Rev 1 wind speed summations without rotation (*S*) based on WRF without wind farms, so showing only the coastal wind speed gradient.

Figure 9a shows S^R plotted against φ, for the Horns Rev 1 wind farm. The corresponding results of S^R based on WRF simulations are shown in Figure 9b. Please note that the differences between the top and bottom panels are due to the individual rotation of each scene prior to averaging and not the result of a single rotation by one angle. The similarity in form between SAR and WRF for the rotated maps (S_{rot}) is very good with low wind speeds showing at all radii at around φ = 180°. The inner radius shows more wake effect than outer radii. The SAR derived results are less smooth than those from WRF because the SAR scenes capture variability at smaller scales, due to the heterogeneity of the wind field, than is modelled by WRF.

To further reveal the wind farm wake from the heterogeneous wind field around the wind farm a method to calculate a wake wind speed deficit is employed. It is based on calculating a local perturbation of the wind speed on each SAR scene based on the side lobe wind speeds. The side lobe wind speeds are used at the directions φ + Δφ$_i$ and φ − Δφ$_i$. For the smallest radius Δφ$_1$ = 90°, this means that the side lobe wind speed is from the left and right of the wind farm, at a distance of r_1 from the farm center. For the other radii, the side lobes have the same distance, r_1, from the line aligned with the wind direction and passing through the center of the wind farm, thus:

$$\Delta\varphi_i = \arcsin\left(\frac{r_1}{r_i}\right) \qquad (5)$$

The wake wind speed deficit is defined by:

$$U_i^P = U_i(\varphi_j) - \frac{1}{2}\left(U_i(\varphi_j + \Delta\varphi_i) + U_i(\varphi_j - \Delta\varphi_i)\right) \qquad (6)$$

and the wake wind speed deficit summation is:

$$S_i^D(\varphi_j) = \frac{1}{N_{scene}} \sum_{k=1}^{N_{scene}} (U_i^D(\varphi_j))_k \qquad (7)$$

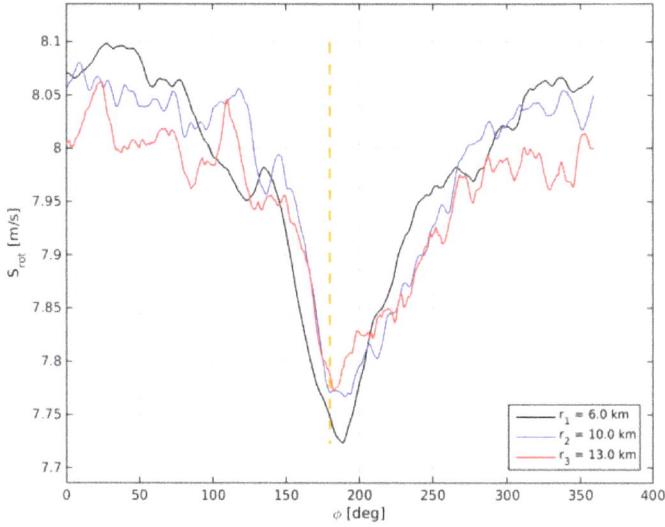

(a)

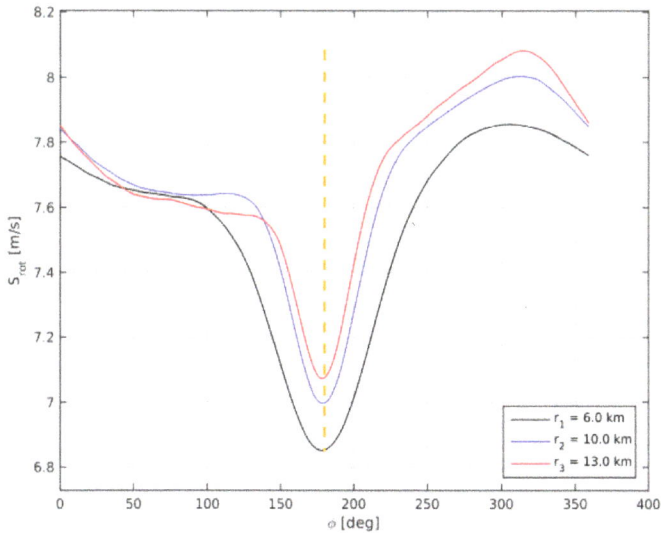

(b)

Figure 9. Horns Rev 1 wind speed summations with rotation (S_{rot}) based on SAR (**a**) and WRF (**b**).

Figure 10 shows S^D plotted *versus* φ for the Horn Rev 1 wind farm for SAR and WRF. In Figure 10 the wake wind speed deficit results based on SAR wind fields and WRF simulations both show the deepest wake at the inner radius and gradual recovery at the outer radii. Both SAR and WRF results

62

show a speed up along the sides of the wake. This shows most clearly at the inner radius but is also noted at the outer radii. The SAR results on wake deficit compares well to the WRF results at Horns Rev 1, however the magnitude of the SAR derived wake is weaker compared to the WRF wakes. It should be noted that the WRF simulations here are one embodiment of WRF simulations and that broader variability in WRF-generated wakes would be generated by other choices of PBL schemes, vertical resolution and approach for representing the wind farm effect.

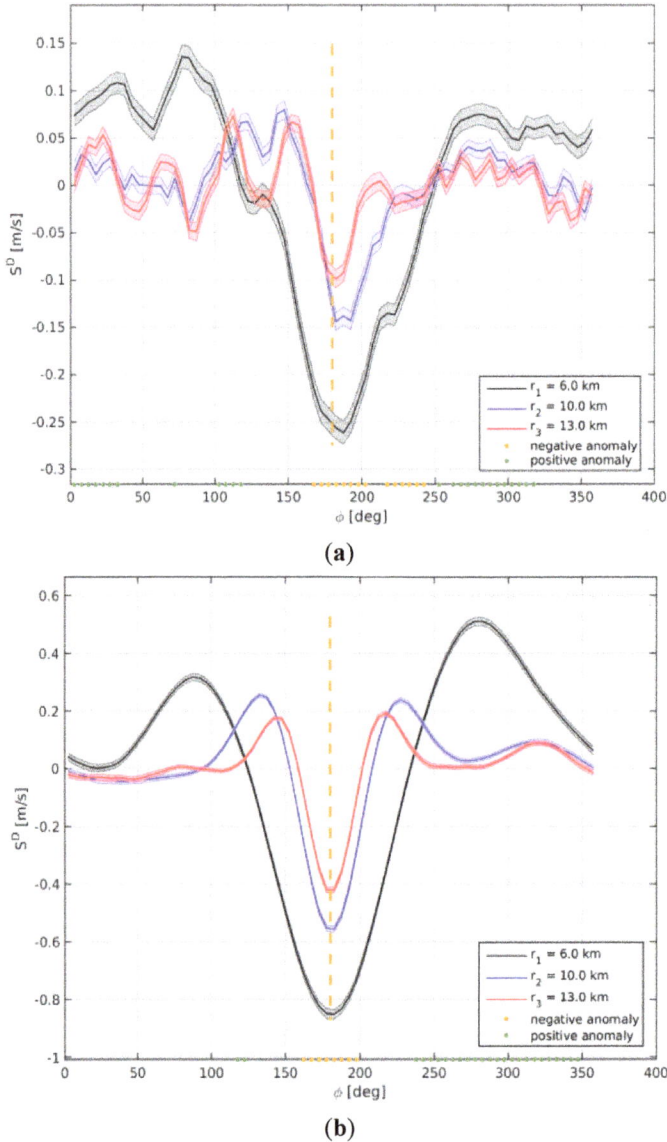

(a)

(b)

Figure 10. Horns Rev 1 wake wind speed deficit (S^D) based on SAR (**a**) and WRF (**b**). The shaded areas in SAR indicate the standard error.

5.2. PARK Model Results

The PARK model is used to simulate 855 cases (a few more cases than used above but the results are expected to be comparable) at Horns Rev 1. The SAR wind time series at 10 m is used as input, the winds are extrapolated to 70 m and the wake is modeled at that height and finally the winds are extrapolated downwards to 10 m again. The extrapolation is done using the logarithmic wind profile assuming a constant roughness length of 0.0002 m. The results are presented at 10 m. Three wake decay coefficients are used. The three wake decay coefficients are: 0.03, 0.04 and 0.05. The wake decay coefficient 0.04 is often used offshore while the lower and higher values are used in case of more stable or unstable cases. The results are rotated and averaged and the results are shown in Figure 11. The coastal gradient is not accounted for in the PARK model results. In case the coastal gradient should be added in the PARK model this could either be from SAR or from WRF, but it has not been attempted in the current study. The results are comparable to the wake wind speed deficit (S^D) results.

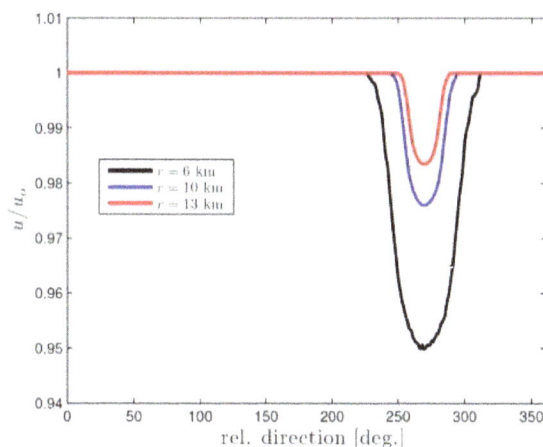

(a)

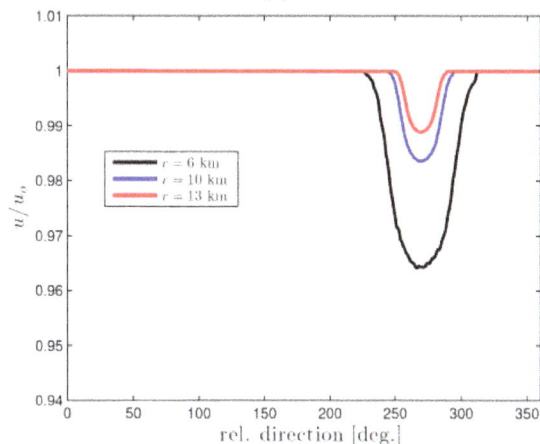

(b)

Figure 11. *Cont.*

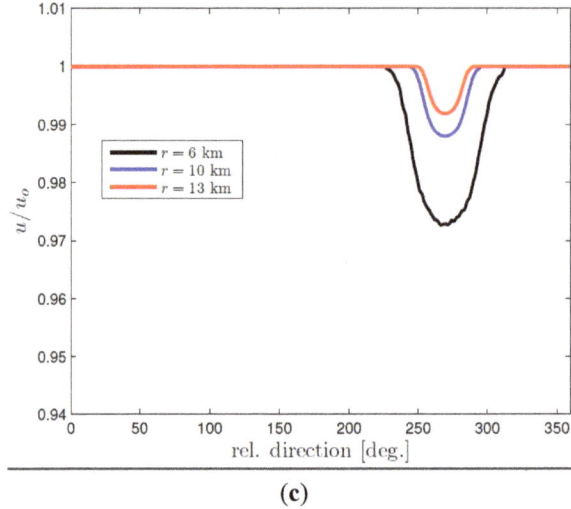

(c)

Figure 11. Wake results from the modified PARK model at Horns Rev 1 using wake decay coefficients (**a**) 0.03; (**b**) 0.04 and (**c**) 0.05.

Figure 11 shows the systematic variation in wake with deeper wakes at the inner radius and progressively weaker wakes further from the wind farm. The wake decay coefficient of 0.03 gives much deeper wakes than for the higher wake decay coefficients, in particular for the inner radius. The shape of the wake compares well to Figure 10 from SAR and WRF.

5.3. Horns Rev 2 Results

Horns Rev 2 is located further offshore than Horns Rev 1 thus similar directional but lower wind speed gradients are expected. The mean wind speed gradient results based on SAR and WRF for Horns Rev 2 are presented in Figure 12a,b.The coastal wind speed gradient observations in SAR at Horns Rev 2 (Figure 12a) show a very peaked and significant minimum around 110° corresponding to the direction of Horns Rev 1. The feature (drop of around 0.2 m·s^{-1}) is observed at all radii (8, 11 and 13 km) and is most pronounced at the outer radius. This narrow fine-scale feature is only fully observed in SAR. SAR resolves features at smaller spatial scales than the WRF simulations presented here. This minimum value might be related to the wind farm wake from Horns Rev 1. Interestingly WRF shows a broad minimum with a shift in direction between radii from 100° at the 8 km radius to 110° at the 10 km radius and 120° at the 13 km radius (Figure 12b). Thus the WRF simulation may in fact here capture a blend of coastal gradient and wind farm wake from Horns Rev 1. At the western sector SAR shows a peaked maximum around 250° and similar wind speeds at all three radii while WRF shows flatter maximum and slightly higher winds at outer radii.

Figure 12c,d shows the rotated maps (S_{rot}) for Horns Rev 2. For SAR a minimum around 180° at 8 km radius is observed while at 10 km and 13 km the minima are around 250° and 110°, respectively. Only at 8 km do the WRF simulations agree with the SAR observations. The three radii at Horns Rev 2 are each located 2 km further from the wind farm center than the results for Horns Rev 1. This was necessary because the Horns Rev 2 wind farm is larger than the Horns Rev 1 wind farm. This however means that the wind alignment between inflow conditions and the deepest wind farm wake potentially deviate relatively more at Horns Rev 2 than Horns Rev 1.

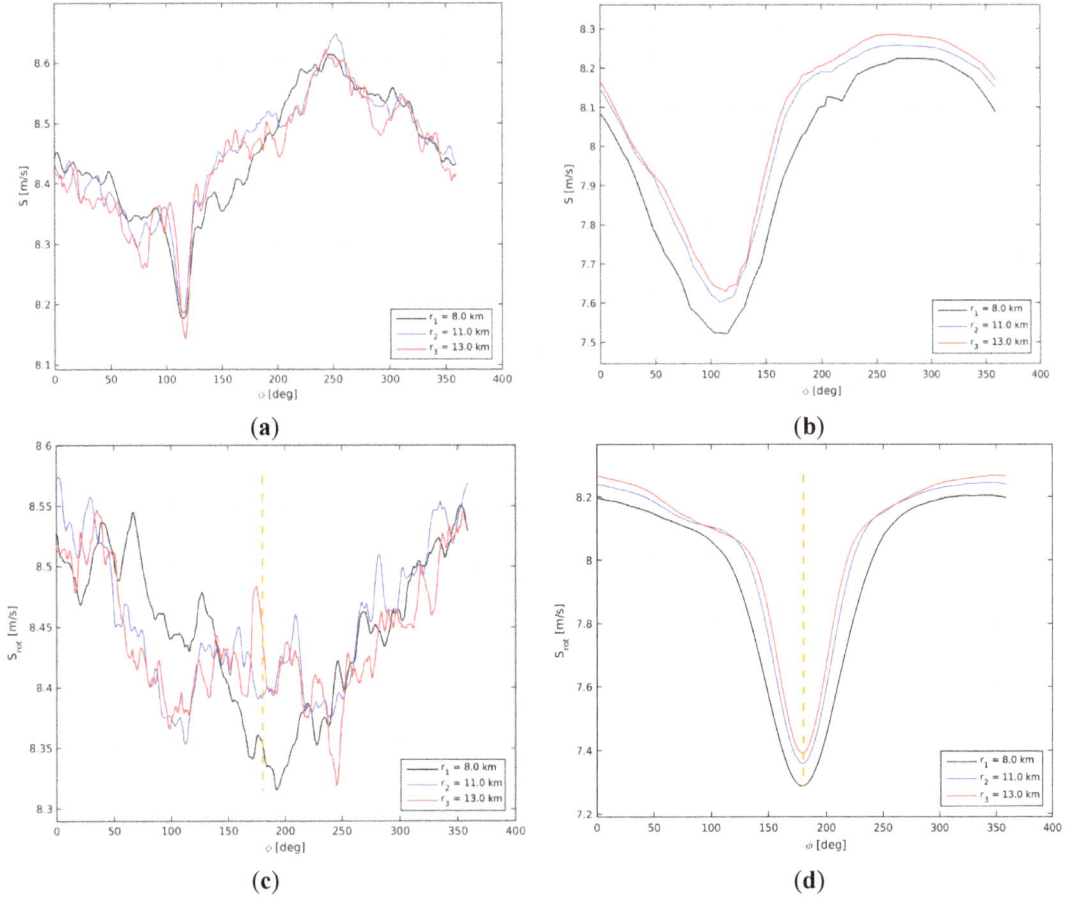

Figure 12. Results for Horns Rev 2. (**a**) wind speed summations without rotation (mean wind speed gradient) (S) based on SAR; (**b**) WRF; (**c**) wind speed summations with rotation (S_{rot}) based on SAR; (**d**) WRF.

Figure 13 shows the wake wind speed deficit (S^D) results from SAR and WRF. The results at the inner radius compare well even though the SAR results show a broader wake than WRF. Speed up in the side lobe winds are noticed both in SAR and WRF at the inner radii and no residual wind speed gradient is noted. At the middle and outer radii WRF shows gradual decrease in the wake winds speed deficit and speed up at the sides while the SAR results are difficult to interpret. In SAR the minimum wake wind speed deficit is not observed around 180° but around 100° and 250°. The analysis appears not to work so well in this case, in part due to the significant minimum around 110° in Figure 12a. This feature may possibly be the wake feature of Horn Rev 1, which acts to contaminate the analysis, as this feature can be as strong as the Horn Rev 2 wake itself.

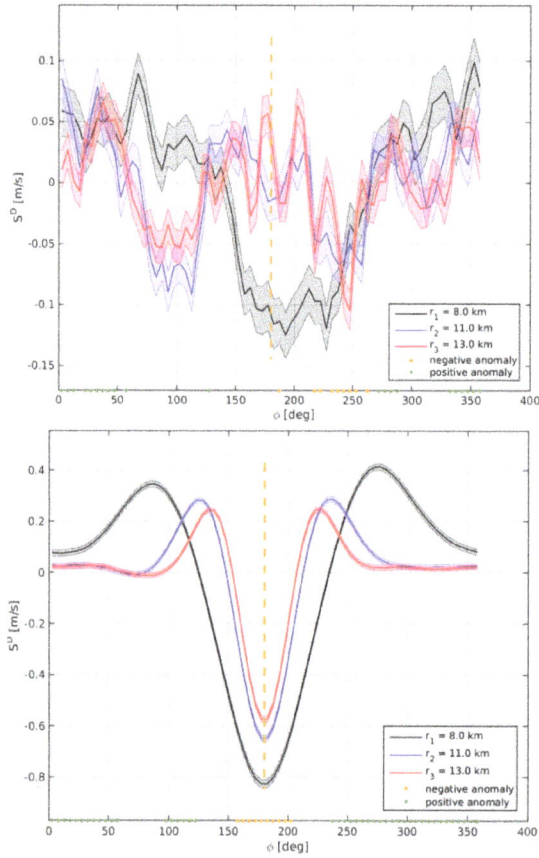

Figure 13. Similar to Figure 10 but for Horns Rev 2.

5.4. SAR-Based Results for Six Wind Farms

Based on the available Envisat ASAR wind field archive we find it interesting to compare the observed aggregated wind farm wakes at four other wind farms in the southern North Sea using the new methodology of rotation of the wind maps. The results of the wake wind speed deficit (S^D) are shown in Figure 14 together with the results from Horns Rev 1 and Horns Rev 2 already discussed.

It is the results for the average of the three radii (see Table 4) for each wind farm that is shown. Figure 14 shows results for six wind farms. It is noted that Gunfleet Sands 1 + 2 show the deepest wake wind speed deficit. The side lobe speed-up effects are clear. A residual wind speed gradient is not noted. For the wind speed summation (S) (not shown) there is a weaker signature of a climatological wind speed gradient across the wind farm compared to other wind farms, this may be because the radii used are smaller.

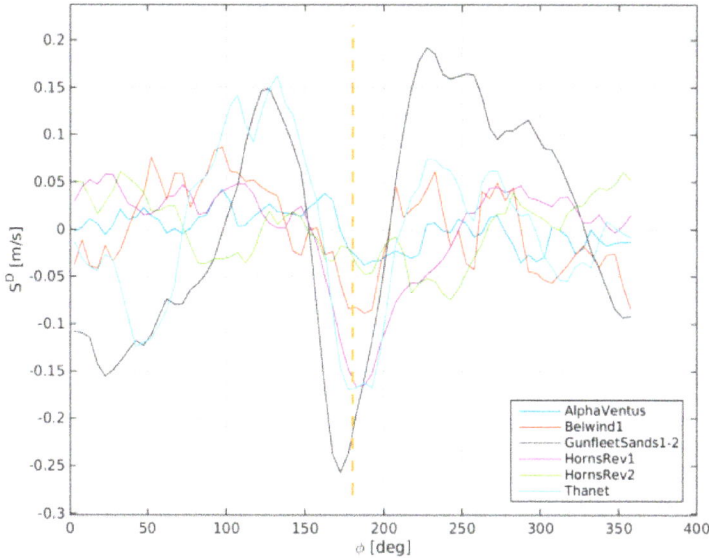

Figure 14. SAR-based wind farm wake observed at six wind farms, Alpha ventus, Belwind 1, Gunfleet Sands 1 + 2, Horns Rev 1, Horns Rev 2 and Thanet, showing the wake wind speed deficit (S^D).

At the Thanet wind farm, the wake wind speed deficit results show very clear wake at around 180° and side lobe speed-up effects. A residual of the coastal wind speed gradient is noted. There is a feature at 50° with lower winds that would need further investigation.

At the Belwind 1 wind farm the wake wind speed deficit is clear but not as pronounced as for the Gunfleet Sands 1 + 2, Thanet and Horns Rev 1 results. Weak side lobe effects are observed at Belwind 1. A residual of coastal wind speed gradient is not noted in the wake wind speed deficit despite that strong coastal gradient mean wind speed gradient is found in the coastal gradient plot (S) (not shown).

Finally, at the Alpha ventus wind farm, the smallest wind farm in terms of installed capacity and area, there is observed wake wind speed deficit at around 180°. The wind farm wake is not as pronounced as for the larger wind farms investigated. This is expected due to the size of the wind farm. The coastal gradient is not observed in the result but is noted in the coastal wind speed gradient (S) (not shown).

6. Discussion

The very long wind farm wakes observed in the RADARSAT-2 scene have in qualitative terms successfully been modeled both by the PARK and WRF model at several wind farms in the southern North Sea. The comparison is qualitative due the different nature of data. The SAR-based results are near-instantaneous observations of the sea surface while wake model results are time-averaged results withbest representation of the conditions at around hub-height. Thus we focus on the apparent wind farm wake direction and the length of the wakesin this comparison instead of the wake deficit at any given location. The SAR image has the advantage of clear visible wake features. Thus the retrieved wind field can be used to evaluate the wake model results in qualitative terms.

We are interested in developing more robust SAR-based wind farm wake data representation for the evaluation of wake models. Therefore the climatology of wind farm wake is necessary. The Envisat ASAR data of wind fields enable us to study the wind farm wake at Horns Rev 1 and 2 with the data set divided into 12 wind directional bins. The results are compared to the PARK model. Occasionally good agreement is found but due to strong coastal wind speed gradients, bathymetry effects and too few samples firm conclusions cannot be drawn.

The new method, in which the wind field maps are rotated, overcomes two of the main issues when trying to isolate the wind farm effect of the wake on the wind fields: the low number of samples and the coastal gradient. The first advantage is that the inflow wind is aligned (rotated) with 1° bins instead of 30° bins. This gives more certainty that the deepest wind farm wake are overlapping in the aggregated results. With the inflow wind speed used to normalize the winds in the wake, the wake are clearly seen but at the same time a residual of the coastal gradient is often noted, e.g., at Horns Rev 1. The clearest wind farm wake results are typically obtained using the side lobe winds for normalization, the wake wind speed deficit method (S^D). This is not too surprising as the coastal wind speed gradients at most wind farms are significant and the circles used around the large offshore wind farms need to be at some distance. Therefore any inhomogeneity in the flow, most importantly the coastal gradient, but in fact also meandering and other atmospheric features gain importance. Also in [2] the nearby parallel transects winds along the wind farm wake were optimal for normalization, rather than the inflow winds upwind of the wind farm.

The number of samples at Horns Rev 2 is 303 while at Horns Rev 1 it is 835. So the lower number of samples at Horns Rev 2 could be one reason for the lesser clarity in data at this site when compared to Horns Rev 1. Also the influence of the Horns Rev 1 wind farm wake may hinder full interpretation at Horns Rev 2 in particular at radii far from the Horns Rev 2 wind farm. Finally it should be mentioned that the Horns Rev 1 wind farm has a geometric shape (turbine lay-out) more convenient for the proposed new methodology of analysis than that of the Horns Rev 2 wind farm. We assume that all wind turbines are in operation at all times a characteristic which may not be fulfilled.

The SAR-based aggregated wind farm wake data compare well both to the WRF simulations and the PARK model results. It is the first time that assessment of the wind farm wake climatology has been attempted based on SAR (to our knowledge) and the results are promising. The main importance of the establishment of SAR-based wake wind speed aggregated results is for the validation of wind farm wake models in the far-field wake region where other observations are extremely limited. In the future more wind farms will operate offshore thus cluster-scale wind farm wake therefore become an even more important focus area. It is suggested to continue this type of research using new SAR data from the Sentinel-1 mission.

7. Conclusions

The case study based on a RADARSAR-2 scene is a unique situation with fairly homogeneous flow across the southern North Sea. The observed wind farm wakes are visible in the SAR scene and thus appealing for demonstration. Both WRF and PARK reproduce the observed very long wind farm wakes convincingly regarding their direction and extent. SAR archive renders possible climatology studies.

The available Envisat ASAR data archive at Horns Rev is the most comprehensive. It has therefore been used for geo-located wind farm wake climatology studies. However the results are only occasionally clear for interpretation due to the limited number of samples per 30° sectors, the coastal wind speed gradient and oceanic bathymetry effects in SAR at Horns Rev.

The key results are based on a new methodology of rotating wind maps. By applying the new methodology to SAR-based wind fields, mesoscale model WRF and microscale model PARK results comparable aggregated wind farm wake results are obtained. The SAR-based findings strongly support the model results at Horns Rev 1. The new methodology increases the number of samples, aligns the wind direction of inflow much more accurately (1° bins) and in most cases but not always overcome the coastal wind gradient. The most convincing results are obtained for the wind wake deficit results in which the side lobe winds are used for normalization.

Acknowledgments

Support from the European Energy Research Alliance-Design Tools for Offshore wind farm Clusters (EERA DTOC) project FP7-ENERGY-2011-1/n 282797 and satellite images from RADARSAT-2 from Data and Products © MacDonald, Dettewiler and Associates Ltd and Envisat ASAR data from the European Space Agency are acknowledged. We are thankful to the Northern Seas Wind Index Database (NORSEWInD) project for the Envisat ASAR wind field archive processed by Alexis Mouche.

Author Contributions

Charlotte Bay Hasager coordinated the main theme of this paper and wrote the manuscript. Pauline Vincent and Romain Husson processed the RADARSAT-2 scene, identified the wind farm wakes and retrieved the wind speed. Jake Badger and Alessandro Di Bella developed the new methodology of rotating the wind maps for aggregated wake results and Alessandro Di Bella programmed and applied the methodology on Envisat ASAR wind fields and WRF model results and produced the graphics. Merete Badger prepared the Envisat ASAR wind field archive and extracted all information for the wake research. Alfredo Peña programmed the PARK model in Matlab and produced results and graphics. Patrick Volker set up the WRF model and produced the results used for wake comparison. All authors discussed the research results and commented on the manuscript. All the authors read and approved the final manuscript.

Conflicts of Interest

The authors declare no conflict of interest.

References

1. Christiansen, M.B.; Hasager, C.B. Wake effects of large offshore wind farms identified from satellite SAR. *Remote Sens. Environ.* **2005**, *98*, 251–268.
2. Christiansen, M.B.; Hasager, C.B. Using airborne and satellite SAR for wake mapping offshore. *Wind Energy* **2006**, *9*, 437–455.
3. Li, X.; Lehner, S. Observation of TerraSAR-X for Studies on Offshore Wind Turbine Wake in near and Far Fields. *IEEE* **2013**, *5*, 1757–1768.
4. Hasager, C.B.; Vincent, P.; Husson, R.; Mouche, A.; Badger, M.; Peña, A.; Volker, P.; Badger, J.; Di Bella, A.; Palomares, A.; *et al.* Comparing satellite SAR and wind farm wake models. *J. Phys. Conf. Ser.* **2015**, *625*, in press.
5. Hasager, C.B.; Mouche, A.; Badger, M.; Bingöl, F.; Karagali, I.; Driessenaar, T.; Stoffelen, A.; Peña, A.; Longépé, N. Offshore wind climatology based on synergetic use of Envisat ASAR, ASCAT and QuikSCAT. *Remote Sens. Environ.* **2015**, *156*, 247–263.
6. Quilfen, Y.; Chapron, B.; Elfouhaily, T.; Katsaros, K.; Tournadre, J. Observation of tropical cyclones by high-resolution scatterometry. *J. Geophys. Res.* **1998**, *103*, 7767–7786.
7. Katic, I.; Højstrup, J.; Jensen, N.O. A simple model for cluster efficiency. In Proceedings of the European Wind Energy Association Conference & Exhibition, Rome, Italy, 7–9 October 1986.
8. Mortensen, N.G.; Heathfield, D.N.; Myllerup, L.; Landberg, L.; Rathmann, O. *Getting Started with WAsP 9*; Tech. Rep. Risø-I-2571(EN); Risø National Laboratory: Roskilde, Denmark, 2007.
9. Jensen, N.O. *A Note on Wind Generator Interaction*; Tech. Rep. Risø-M-2411(EN); Risø National Laboratory: Roskilde, Denmark, 1983.
10. Skamarock, W.C.; Klemp, J.B.; Dudhia, J.; Gill, D.O.; Barker, D.M.; Duda. M.; Huang, X.Y.; Wang, W.; Powers, J.G. A description of the advanced research WRF version 3. *Tech. Rep.* **2008**, doi:10.5065/D68S4MVH.
11. Adams, A.S.; Keith, D.W. A wind farm parametrization for WRF. Available online: http://www2.mmm.ucar.edu/wrf/users/workshops/WS2007/abstracts/5-5_Adams.pdf (accessed on 2 June 2015).
12. Baidya Roy, S. Simulating impacts of wind farms on local hydrometeorology. *J. Wind Eng. Ind. Aerodyn.* **2011**, *99*, 491–498.
13. Blahak, U.; Goretzki, B.; Meis, J. A simple parametrisation of drag forces induced by large wind farms for numerical weather prediction models. In Proceedings of the European Wind Energy Conference & Exhibition 2010 (EWEC), Warsaw, Poland, 20–23 April 2010.
14. Jacobson, M.Z.; Archer, C.L. Saturation wind power potential and its implications for wind energy. *Proc. Natl. Acad. Sci. USA* **2012**, *109*, 15679–15684.
15. Fitch, A.; Olson, J.; Lundquist, J.; Dudhia, J.; Gupta, A.; Michalakes, J.; Barstad, I. Local and mesoscale impacts of wind farms as parameterized in a mesoscale NWP model. *Mon. Weather Rev.* **2012**, *140*, 3017–3038.
16. Volker, P.J.H.; Badger, J.; Hahmann, A.H.; Ott, S. The Explicit Wake Parametrisation V1.0: A wind farm parametrisation in the mesoscale model WRF. *GMDD* **2015**, *8*, 3481–3522.

17. Nakanishi, M.; Niino, H. Development of an improved turbulence closure model for the atmospheric boundary layer. *J. Meteorol. Soc. Jpn.* **2009**, *87*, 895–912.

18. Kain, J.S. The Kain-Fritsch convective parameterization: An update. *J. Appl. Meteorol. Climatol.* **2004**, *43*, 170–181.

19. Thompson, G.; Field, P.R.; Rasmussen, M.; Hall, W.D. Explicit forecasts of winter precipitation using an improved bulk micro- physics scheme. Part II: Implementation of a new snow parameterization. *Mon. Weather Rev.* **2008**, *136*, 5095–5115.

20. Mlaver, E.J.; Taubman, S.J.; Brown, P.D.; Iacono, M.J.; Clough, S.A. Radiative transfer for inhomogeneous atmosphere: RRTM, a validated corrected-k model for the long wave. *J. Geophys. Res.* **1997**, *102*, 16663–16682.

21. Dudhia, J. Numerical study of convection observed during the wind monsoon experiment using a mesoscale two-dimensional model. *J. Atmo. Sci.* **1989**, *46*, 3077–3107.

22. Chen, F.; Dudhia, J. Coupling an advanced land surface-hydrology model with the Penn State-NCAR MM5 modeling system. Part I: Model implementation and sensitivity. *Mon. Weather Rev.* **2001**, *129*, 569–585.

23. Uppala, S.M.; Kallberg, P.W.; Simmons, A.J.; Andrae, U.; Bechtold, V.; Fiorino, M.; Gibson, J.K.; Haseler, J.; Hernandez, A.; Kelly, G.A.; *et al.* The ERA-40 re-analysis. *Quart. J. R. Meteorol. Soc.* **2005**, *131*, doi:10.1256/qj.04.176.

24. Barthelmie, R.J.; Badger, J.; Pryor, S.C.; Hasager, C.B.; Christiansen, M.B.; Jørgensen, B.H. Offshore coastal wind speed gradients: Issues for the design and development of large offshore windfarms. *Wind Eng.* **2007**, *31*, 369–382.

Wind Resource Mapping Using Landscape Roughness and Spatial Interpolation Methods

Samuel Van Ackere, Greet Van Eetvelde, David Schillebeeckx, Enrica Papa, Karel Van Wyngene and Lieven Vandevelde

Abstract: Energy saving, reduction of greenhouse gasses and increased use of renewables are key policies to achieve the European 2020 targets. In particular, distributed renewable energy sources, integrated with spatial planning, require novel methods to optimise supply and demand. In contrast with large scale wind turbines, small and medium wind turbines (SMWTs) have a less extensive impact on the use of space and the power system, nevertheless, a significant spatial footprint is still present and the need for good spatial planning is a necessity. To optimise the location of SMWTs, detailed knowledge of the spatial distribution of the average wind speed is essential, hence, in this article, wind measurements and roughness maps were used to create a reliable annual mean wind speed map of Flanders at 10 m above the Earth's surface. Via roughness transformation, the surface wind speed measurements were converted into meso- and macroscale wind data. The data were further processed by using seven different spatial interpolation methods in order to develop regional wind resource maps. Based on statistical analysis, it was found that the transformation into mesoscale wind, in combination with Simple Kriging, was the most adequate method to create reliable maps for decision-making on optimal production sites for SMWTs in Flanders (Belgium).

Reprinted from *Energies*. Cite as: Van Ackere, S.; Van Eetvelde, G.; Schillebeeckx, D.; Papa, E.; van Wyngene, K.; Vandevelde, L. Wind Resource Mapping Using Landscape Roughness and Spatial Interpolation Methods. *Energies* **2015**, *8*, 8682-8703.

1. Introduction

Next to energy savings and reduction of emissions, an increased share of renewables in the European energy mix is a key priority of the Energy Union [1]. With a target of 20% by 2020 and 27% by 2030, Europe has set ambitious goals for renewable energy, requiring a broad mix of clean technologies, both large and small scale, to take a share.

Over time, technical research and innovation projects on distributed renewable energy sources (DRES)—such as small and medium wind turbines (SMWTs)—have been a primary focal area of interest. However, wind energy generation is difficult to manage because of the irregular nature of wind flows. Further, the current transition in energy demand and supply also encompasses many aspects, such as the resource availability evaluation, the compliance with environmental and legal constraints, and many more technical aspects. In this complex context, understanding the spatial distribution of the long-term average wind speed is essential for decision-making, particularly in regards to the siting of wind turbines. Hence, the current transition in distributed energy demand and supply prompts a new area of research: spatial energy planning. Further, by combining technical and spatial wind research and integrating it with regulatory, economic and social constraints, a new interdisciplinary research and innovation area is unfolded with a high valorisation potential for energy prosumers on a local scale.

Understanding the spatial distribution of the long-term mean wind speed is essential for decision-making, particularly in regards to the siting of wind turbines. However, there is often a lack of measurements to enable accurate wind speed mapping. Despite the long evolution of wind mapping and method development for assessing wind as a resource, along with increasing computational capabilities, a single general method for creating predicative wind maps does not exist. Indeed, a reliable approach depends on a number of factors that are context-related: the size of the analysed area, the required resolution of the results, the climatic and topographical characteristics of the analysed area, the density of the available meteorological measurements, *etc.* [2].

In regions like Flanders (Belgium), an area of 13,522 km^2 with *ca.* 6.4 million inhabitants and a high potential in terms of wind power generation, efficient energy planning based on renewables is a complex task. In fact, the region is characterised by a composite topography, a compound of land covers and dispersed buildings. The open space is no longer a monofunctional agricultural production area but, rather, a complex structure of fragments with varying densities and functions [3]. Marked by a dense matrix of meteorological stations, this region is challenging for identifying optimal SMWT locations.

Next to meteorological data, basic wind speed measurements are equally available at various heights, covering the entire Flemish region. As shown in Figure 1, a primary wind study for Belgium was performed in 1984 by Hirsch. Although an interesting effort, it provides insu□cient insight in local wind availabilities to enable detailed siting for SMWTs. In 2014, a roughness map was generated for the Flemish region by converting land cover categories into sequences of roughness length [4].

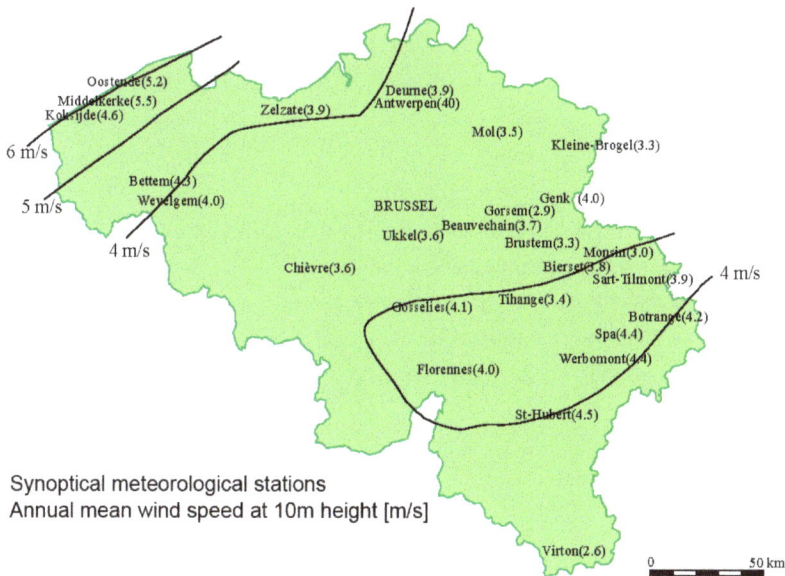

Synoptical meteorological stations
Annual mean wind speed at 10m height [m/s]

Figure 1. Annual mean wind speed map of Belgium [5].

This article starts from the results of the latter study and develops a detailed low-height wind speed map, providing a useful tool for the identification of optimal locations for SMWTs in Flanders. The research aim is not to develop new methodologies, although some are described in Section 3, but to analyse data by applying already existing methods and producing an updated wind map, which is valuable for deployment of micro-wind energy.

Another advantage of the proposed methodology is that it uses open source data and software. Compared to the method we propose here, more sophisticated models (*i.e.*, Wasp or windPRO) and data assimilation techniques have been developed in literature, but they are not affordable for use by small municipalities.

In Section 2, the wind speed measurements for Flanders and the roughness map are presented, providing the geo-database used in this study. Section 3 describes two types of exposure corrections and introduces the seven interpolation methods assessed in this study. The results are discussed and mapped in Section 4, and presents the conclusions of the selected methodologies for wind resource mapping. In order to demonstrate how the Annual Energy Production (AEP) can be calculated for a specific small or medium wind turbine, an AEP map is created in Section 5 for a 10 kW 3-blade, upwind, horizontal axis wind turbine. A Rayleigh distribution, which is identical to a Weibull distribution with shape factor 2, is used as the reference wind speed frequency distribution.

2. Data collection

2.1. Wind speed Measurements

This work is based on the wind data recorded in a number of Flemish meteorological stations spread over the region. The study used recent observed data since both the wind climate and the environment have changed in the past decades. The collected data, location of meteorological stations, relative recording dates and local wind speed measurements used in this study are summarised in Table 1. The geographic location of the meteorological stations is visualised in Figure 2.

Daily wind speed observations obtained from the National Climatic Data Center (NCDC) of the US National Oceanic and Atmospheric Administration (NOAA) [6] were collected for all available stations in Belgium, with the addition of some frontier mast data from the Netherlands and France. All stations are equipped with an anemometer at the height of 10 m, hence, the observed wind speed is the so-called "surface wind speed", which is further averaged over a calendar year so as to rule out seasonal bias. Data validation is performed by using the more accurate and precise dataset created by the Royal Meteorological Institute (RMI), which refers to a smaller group of 18 stations selected from the original number.

The data from NCDC include the average wind speed at 10 m for France, the Netherlands and Belgium. The data are validated through comparison with the corresponding dataset from RMI (for Flanders). Apart from being rounded to one decimal place, both sets are identical, therefore the extensive open-source database of NCDC is selected for producing regional wind maps in this study.

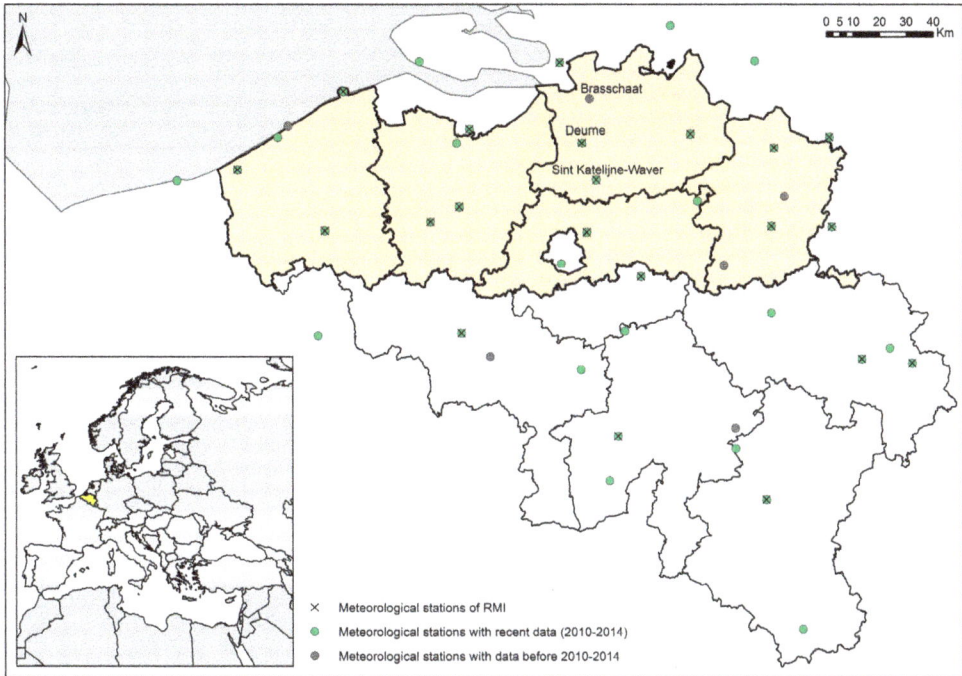

Figure 2. Location of the measurement stations used.

Since recent studies suggest that climate change is affecting the prevailing wind profiles, 40-year-old observations are considered inadequate for future wind modelling [7]. Likewise, the landscape in Flanders has significantly changed over the past decades due to the development of built-up areas [8]. Therefore, meteorological stations with recent wind data (2010–2014) were selected and the annual mean wind speed was calculated based on five years of measurements, with the exception of Sint-Katelijne-Waver (only 2013–2014 available). Even with a reduced number of recent observations as recorded in Table 1, it is observed that there is a decreasing trend of the annual mean wind speed over the last five years.

2.1.1. Roughness Map Flanders

To account for the different surfaces in Flanders, a roughness map, developed in 2014 [4], was used. The map uses the roughness length of a land mark as indicator, defined by [9]. In this case, a resolution of 250 by 250 meters is presented (see Figure 3).

Table 1. Summary of the measurement stations (height 10 m). The recent data is (2010–2014) indicated in bold letter type [6].

Station	Roughness [m]	Latitude [°]	Longitude [°]	Begin Date	End Date	Mean Wind Speed [m/s]	Mean Wind Speed (2010–2014) [m/s]
Beauvechain	0.03	50.758	4.768	1/01/1973	42.004	3.88	3.70
Beitem	0.469	50.900	3.116	1/02/2008	42.035	3.69	3.67
Brasschaat	0.14	51.333	4.500	1/02/1973	31/01/2006	3.26	
Brussels NATL	0.037	50.902	4.485	1/01/1973	31/12/2014	3.99	3.62
Brussels South	0.2	50.459	4.453	1/01/1973	31/12/2014	3.96	4.00
Buzenol	0.6	49.616	5.583	26/10/2009	25/10/2014	2.76	2.74
Casteau/Heli	0.8	50.500	3.980	1/01/2011	31/12/2014	2.18	
Chievres	0.1	50.575	3.831	1/01/1973	31/12/2014	3.73	3.75
Deurne	0.896	51.189	4.460	1/01/1973	31/12/2014	3.55	3.58
Diepenbeek	0.08	50.916	5.450	1/01/2010	31/12/2014	2.92	2.92
Dourbes	0.6	50.100	4.600	1/01/2010	31/12/2014	2.52	2.52
Elsenborn	0.6	50.466	6.183	1/01/1987	31/12/2014	3.11	3.12
Ernage	0.1	50.583	4.683	1/01/2008	31/12/2014	4.06	4.04
Florennes	0.15	50.243	4.645	1/01/1973	31/12/2014	3.75	3.69
Genk/Zwartberg	0.676	51.012	5.522	7/01/1973	6/01/2004	3.60	
Gent/Industrie	0.021	51.187	3.799	1/01/1985	31/12/2014	3.31	3.32
Humain	0.4	50.200	5.250	1/03/2010	28/02/2015	3.69	3.66
Kleine Brogel	0.054	51.168	5.470	1/01/1973	31/12/2014	3.01	3.01
Koksijde	0.06	51.090	2.652	1/01/1973	31/12/2014	4.68	4.57
Liege	0.15	50.637	5.443	1/01/1973	31/12/2014	4.07	4.11
Melle	0.2	50.983	3.816	1/01/2010	31/12/2014	3.42	3.42
Mont-Rigi	0.2	50.516	6.066	16/01/2008	15/01/2015	3.83	3.74
Oostende	0.64	51.198	2.862	1/01/1973	31/12/2014	5.22	4.75
Oostende (Pier)	0.98	51.235	2.914	1/01/1973	31/12/2005	6.91	
Retie	0.118	51.216	5.033	26/10/2009	25/10/2014	2.64	2.63
Saint Hubert Mil	0.2	50.035	5.404	1/01/1973	31/12/2014	3.88	3.29
Schffen	0.03	51.000	5.066	2/01/1973	1/01/2015	3.93	3.21
Semmerzake	0.231	50.933	3.666	1/01/1973	31/12/2014	3.70	3.26
Sinsin	0.3	50.266	5.250	20/09/1984	19/09/1995	3.49	
Sint Katelijne-waver	0.278	51.070	4.535	1/10/2012	30/09/2014	3.02	3.05
Sint Truiden	0.03	50.791	5.201	1/01/1973	31/12/1991	3.62	
Spa/La Sauveniere	0.1	50.483	5.916	1/01/1974	31/12/2014	3.87	3.74
Uccle	0.621	50.800	4.350	1/01/1973	31/12/2014	3.48	3.44
Zeebrugge	0.001	51.350	3.200	26/10/2009	25/10/2014	6.05	6.02
Dunkerque	0.01	51.050	2.333	2/01/1973	1/01/2015	6.20	5.26
Lesquin	0.1	50.561	3.089	1/01/1973	31/12/2014	4.37	4.09
Eindhoven	0.1	51.450	5.374	1/01/1973	31/12/2014	3.94	3.64
Ell AWS	0.15	51.200	5.766	1/01/2002	31/12/2014	3.53	3.46

Table 1. *Cont.*

Station	Roughness [m]	Latitude [°]	Longitude [°]	Begin Date	End Date	Mean Wind Speed [m/s]	Mean Wind Speed (2010–2014) [m/s]
Gilze Rijen	0.05	51.567	4.931	1/01/1973	31/12/2014	3.81	3.53
Maastricht	0.05	50.911	5.770	1/01/1973	31/12/2014	4.25	4.06
Vlissingen	0.25	51.450	3.600	1/01/1973	31/12/2014	6.07	6.10
Westdorpe	0.25	51.233	3.866	1/01/1995	31/12/2014	4.02	4.00
Woensdrecht	0.3	51.449	4.342	1/01/1996	31/12/2014	3.45	3.48

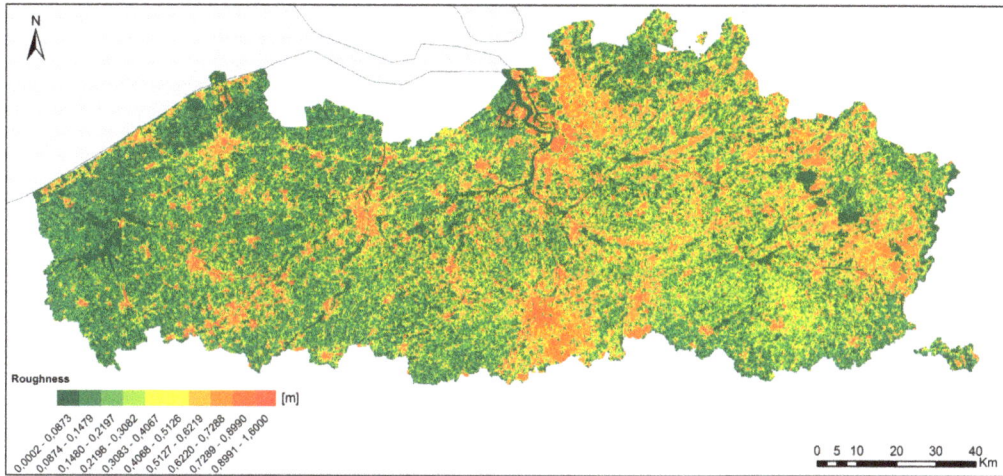

Figure 3. Roughness map of Flanders [4].

Whilst the roughness length (z_0) is not a physical length, it can be considered as a length-scale representing the roughness of the surface: for example, forests have a much larger roughness length than open sea areas. At a low height above the ground, or the surface layer, the roughness of a terrain affects the turbulence intensity as well as the vertical wind pattern and, by consequence, the wind speed. The roughness map was constructed by the Flemish Institute for Technological Research (VITO) based on the CoORdination of INformation on the Environment (CORINE) Land Cover 2000 data set [9]. In this project, the National Geographic Institute (NGI) constructed the national land cover map using high resolution (Landsat Thematic Mapper) satellite images [10]. The detailed map enables correction of the observations for local sheltering and topography.

3. Methodology

In this section, the statistical interpretation process of the wind time series is explained. In detail, the section describes how seven different interpolation methods are tested and assessed in order to select the most performant way to generate a wind resource map of Flanders.

3.1. The PBL Two Layer Model

Over the last two decades, several studies have been carried out with the aim of creating an adequate statistical model for describing the wind speed frequency distribution. One of the more recent studies developed for the Netherlands [11] used the Planetary Boundary Layer (PBL) two-layer model. This two-layer transformation model from Wieringa [12] was further developed by Verkaik [13–17] and more recently by Wever and Groen [18]. The methodology is generally accepted and recommended [19–22], however, the report admits to not having explored the potential benefits of using Kriging (see Section 3.2.2), as detailed in Section 3.3 of [11].

In the research carried out by Wieringa in the 70s and 80s [12,23], wind speed variations on a resolution of 250 by 250 meters are caused mainly by differences in atmospheric stability and surface roughness. At a certain so-called blending height [12] these variations become negligible compared to the average speed, yielding a spatially homogeneous dataset suitable for interpolation. A roughness correction is applied to the observations by using the measured surface wind speed to calculate the regional wind speed that is representative for a larger area by using the roughness length of the meteorological station. After completion of the spatial interpolation, the regional winds are used to calculate the wind speed at 10 m by using the inverse roughness correction and by using the roughness map of Flanders [24]. Finally, according to this methodology, two different regional wind speeds are used for interpolation. This "roughness blending height" is set to be $z_b = 60$ m [23]. The macrowind speed is measured at the top of the PBL.

3.1.1. Mesowind

At the blending height z_b defined above, land covers and local obstacles have a minimal influence on the wind speed. This height is set to be $z_b = 60$ m. The observed surface wind speed, U_s, can be used to calculate the mesowind speed, U_{meso}, by assuming a logarithmic wind profile [24]:

$$U_{meso} = U_s \frac{\ln(\frac{z_b}{z_{0s}})}{\ln(\frac{z_s}{z_{0s}})} \tag{1}$$

with z_{0s} as the roughness length at the meteorological station site and z_s as the anemometer height, equal to 10 m for all stations in this study. For all stations in Flanders, Wallonia, France and the Netherlands, the roughness length z_{0s} was estimated from a terrain description and by using data based on satellite images of the sites [25].

It is shown that the mesoscale wind climate is spatially more homogeneous than the surface wind [11], hence, it is better suited for interpolation. The interpolated U_{meso} values in Flanders are then reconverted to the surface wind speed at 10 m, U_{10m}, by using [24]:

$$U_{10m} = U_{meso} \frac{\ln(\frac{10}{z_0})}{\ln(\frac{z_b}{z_0})} \tag{2}$$

with z_0 as the roughness length at each 250 m pixel from the roughness map.

3.1.2. Macrowind

The use of macrowind for interpolation purposes is described by Wieringa and is further used to create a gridded wind speed map of the Netherlands [11]. In this method, two layers are defined. In the lower layer, the surface layer, Monin-Obukhov theory is used [26,27]. In this theory, the logarithmic wind speed profile is used to express the increase in wind speed U in the lower layer [28]:

$$U = \frac{u^*}{\kappa} \ln(\frac{z}{z_{0s}}) \tag{3}$$

by using the local roughness length at the site z_{0s} and the Von Kármán constant $\kappa = 0.4$ [29,30]. u^* is the friction velocity and is constant with height over homogeneous terrain, which makes it possible to calculate u^* at the meteorological stations.

Geostrophic drag relations apply in the second higher layer, the planetary boundary layer (PBL). In the PBL, the wind speed increases further and in addition the wind direction veers (turns clockwise) such that a second wind speed component perpendicular to the surface wind speed (V_{macro}) is formed [31]:

$$\frac{\kappa(U - U_{macro})}{u^*} = \left[\ln\left(\frac{z\,f}{u^*}\right) + A\right] \tag{4}$$

$$V_{macro} = B\frac{u^*}{\kappa} \tag{5}$$

with the Coriolis parameter $f = 1.129 \times 10^{-4}$ at 51°N [32]. The stability parameters A and B are equal to 1.9 and 4.5 respectively, as is generally accepted in literature [17] when assuming neutral stability [33]. The vertical extrapolation methods rely on the neutral stability assumption; although neutral conditions characterised by log-profiles are common in general, stable and unstable conditions with non-log vertical profiles occur often as well [34].

The wind at this PBL is called the macrowind, S_{macro}, and varies on a larger scale than the mesowind [12]. The macrowind S_{macro} consists of two components: U_{macro} is parallel to the surface wind and V_{macro} is perpendicular to U_{macro}. Matching the two layers at the mesolevel according to Equations (3) and (4) leads to [17]:

$$U_{macro} = \frac{u^*}{\kappa}\left[\ln\left(\frac{u^*}{f\,z_{0s}}\right) - A\right] \tag{6}$$

The PBL ranges from a few hundred meters to a few kilometres above the surface of the Earth [35], the height of the top of the PBL is given by [17]:

$$h = \frac{u^*}{f e^A} \tag{7}$$

Both components U_{macro} and V_{macro} and the root of the squared sum (macrowind speed, S_{macro}) are interpolated by using Simple Kriging separately onto the 250 m resolution grid of the regional surface roughness map. Such obtained S_{macro} values are cross-checked with the values calculated from the interpolated U_{macro} and V_{macro}, yielding differences that are negligibly small.

After spatial interpolation of the S_{macro} values at different location points, S_{macro} is used to calculate the surface wind speed at these locations, by using the inverse process. First the friction velocity u^* needs to be calculated, by using Equations (5) and (6), in order to calculate the surface wind speed U_{10m} with Equation (3) using z_0 instead of z_{0s} (see Figure 4b).

3.2. Spatial Interpolation Methods

In order to obtain an accurate picture of the Flemish wind potential, in addition to the roughness map and meteorological observations, the wind speed was estimated at un-recorded sites via spatial interpolation of the measured data.

Various interpolation techniques are available, of which seven are commonly used for generating lacking data in meteorological variables (rainfall, solar radiation, sunshine, temperature, *etc.*). In [36] an overview of climatological studies using different interpolation methods is presented. For wind speed, spatial interpolation is commonly used [11,37–42].

In this study, seven interpolation methods are tested: Inverse Distance Weighting (IDW), Global Polynomial Interpolation (GPI), Local Polynomial Interpolation (LPI), Radial Basis Functions (RBF), Ordinary Kriging (OK), Universal Kriging (UK) and Simple Kriging (SK).

Prior to the interpolation process, the wind data at the different meteorological stations are used to calculate the wind speed at the blending height in order to reduce the wind speed variations and to obtain a spatially homogeneous dataset suitable for interpolation. The two different blending height methods, mesoscale wind (see Section 3.1.1) and macroscale wind (see Section 3.1.2), are used and compared to accomplish this exposure correction. The maps are evaluated by using Leave-One-Out-Cross-Validation (LOOCV) where one data point is discarded from the sample and the remaining observations are used to estimate the missing value [43]. A comparison between the observed and predicted wind speeds then leads to statistical values on which the quality of the methods can be validated. All methods are applied to create wind speed maps that are further analysed in Section 3.3. Based on this analysis, the most appropriate spatial interpolation technique for wind resource mapping was selected.

Upon correction of the observed wind speed for the influence of the land cover and local obstacles, a spatial interpolation is required to construct a gridded wind speed map. In this section, the basic principles of the interpolation techniques used in the study are explained. All interpolations are performed by using the Geostatistical Analyst from the geographic information system ArcGIS 10.1.

In general, interpolation methods are either denoted as deterministic or as geostatistical. Deterministic interpolation techniques use the configuration of sample points to create a surface defined by a mathematical function, while geostatistical techniques make use of the statistical properties of sample data to create a surface.

3.2.1. Deterministic methods

Inverse Distance Weighted (IDW)

Inverse Distance Weighted (IDW) is one of the most simple interpolation methods. It is based on the assumption that the influence of each sample point is reduced with distance. Every predicted value is calculated by a linear combination of the surrounding measured values within a search neighbourhood, multiplied by a weight that is proportional to the inverse of the distance. Therefore, the closest values will have a larger influence on the estimated values than sample points that are located farther away. IDW is a so-called exact method, meaning that the surface passes through all measured sample points. The estimated value at location s_0, $\hat{Z}_{s0}$ can be determined from [44,45]:

$$\hat{Z}_{s0} = \frac{\sum_{i=1}^{N} Z_i \, d_i^{-p}}{\sum_{i=1}^{N} d_i^{-p}} \tag{8}$$

with Z_i as the sample values, N as the total number of sample values, d_i as the distance between the sample point and the estimated point and p as the inverse distance weighting power (IDP). The IDP factor determines the rate at which the influence of the sample point decreases with distance [40,46]. In this study, IDP values ranging from 1 to 5 are tested and the minimum and maximum number of points are set to 10 and 15 respectively.

Global Polynomial Interpolation (GPI)

Global Polynomial Interpolation (GPI) fits a polynomial function on all sample points by using a least-squares regression fit in order to create a surface. The degree of the polynomial can be adjusted so the surface can describe a physical process. A first-order global polynomial fits a flat plane through the sample points, while going to higher order polynomials will allow for bends, such that valleys and peaks can be represented by the surface [46]. In this study, a first-order global polynomial is used.

Local Polynomial Interpolation (LPI)

Local Polynomial Interpolation (LPI) creates a surface by combining many different polynomials, all fit for smaller (overlapping) neighbourhoods, in contrast to GPI, which fits a polynomial function over the entire data set. Therefore, LPI is able to better account for more short-range variations. Again the order of the polynomial function can be chosen and similar. As for GPI, the coefficients of the polynomials are found using the least-squares method [46,47]. For GPI, first-order polynomials are selected for this interpolation method.

Radial Basic Functions (RBF)

Radial Basic Functions (RBF) or spline interpolation tries to minimise the curvature of a basis function in order to create a smooth surface that goes through all the measured points. Therefore, like IDW, RBF is an exact interpolator. However, in contrast to IDW, RBF is able to predict values above or below the measured maximum or minimum value, respectively. RBF can be seen as fitting a rubber membrane through the sample points while still keeping the surface as smooth as possible. RBF is appropriate for slowly varying surface values but is less suitable when the sample data are subject to measurement errors [46,48]. Here the choice was made to use the "completely regularised spline" as basis function.

3.2.2. Geostatistical Methods

Geostatistical techniques are predominantly found within the Kriging family. Similar to IDW, Kriging uses linear interpolation of the neighbouring measured points to estimate the unsampled points. However, with Kriging, both the distance and the degree of variation between the measured data points are taken into account. For the latter, a variogram is required, which indicates the rate at which the values change with distance. This is obtained by calculating the semi-variance from the sample data. The expression to predict the unmeasured data is similar to IDW but the weights, λ_i, are calculated differently:

$$\hat{Z}_{s0} = \sum_{i=1}^{N} \lambda_i Z_i \tag{9}$$

The weight λ_i depends on the distance to the estimated value, a trend model fitted through the measured data and an auto-correlation as a function of distance. For Kriging methods, the variable of interest, Z, can be broken down into a deterministic trend, μ, and an error term, ε:

$$Z(s) = \mu(s) + \varepsilon(s) \tag{10}$$

with s denoting the location. The way $\mu(s)$ is modelled depends on the Kriging method that is used. The error term is estimated by using the variogram and by assuming spatial autocorrelation.

Kriging is most appropriate when there is a spatially correlated distance or directional bias in the data [40,46,48]. Three types of Kriging methods are used in this study—Ordinary Kriging, Universal Kriging and Simple Kriging:

Ordinary Kriging (OK) is the most widespread Kriging method. In OK, the trend in Equation (10) is assumed to be an unknown constant $\mu(s) = \mu$ over a local subset [46,49].

Universal Kriging (UK) models the trend $\mu(s)$ as a deterministic function. The function is subtracted from the measured data to obtain random errors, $\varepsilon(s)$. The autocorrelation is then calculated from these errors. Later, the deterministic function is added back to the model that was fitted on the random errors to get the predicted data [36,46]. In this study, a first-order trend model is used.

Simple Kriging (SK) uses the trend as a known constant and therefore the errors are also known exactly. Hence, the expected mean of the residuals equals 0 and all variation is statistical [46,50].

3.3. Validation

In order to evaluate and compare the different interpolation methods, LOOCV is used [43]. As detailed above (see Section 3.2), one observation is temporarily removed from the measured data set, upon which the wind speed at that site is estimated with the remaining measurements. This procedure is done one at a time for all observations in Flanders. Next, the estimated wind speeds are compared with the observed wind speed initially discarded from the data set. The following test statistics are used in this study:

1. Mean Error (ME) indicates the degree of bias. A negative value signifies an underestimation while a positive ME means that the predictions are an overestimation of the real values:

$$ME = \frac{1}{N}\sum_{i=1}^{N}(\hat{z}(s_i) - z(s_i)) \tag{11}$$

2. Mean Absolute Percentage Error (MAPE) is a simple measure of accuracy:

$$MAPE = \frac{100}{N}\sum_{i=1}^{N}(\frac{|\hat{z}(s_i) - z(s_i)|}{z(s_i)}) \tag{12}$$

3. Root mean square error (RMSE) represents the standard deviation and is sensitive to outliers:

$$RMSE = \sqrt{\frac{1}{N}\sum_{i=1}^{N}(|\hat{z}(s_i) - z(s_i)|^2)} \tag{13}$$

4. R^2 indicates how well the predicted data match the observations:

$$R^2 = 1 - \frac{\sum_{i=1}^{N}(\hat{z}(s_i) - z(s_i))^2}{\sum_{i=1}^{N}(\bar{z}(s_i) - z(s_i))^2} \tag{14}$$

with N representing the number of observations in Flanders, $z(s_i)$ as the observed values, $\hat{z}(s_i)$ as the predicted values and $\bar{z}(s_i)$ as the mean observed value.

4. Results and Discussion

4.1. Exposure Correction

The measured surface wind speeds are used to calculate either mesowinds, U_{meso}, or macrowinds, S_{macro}, in order to correct for the influence of the terrain or local obstacles. These regional wind speeds are both interpolated by using SK before calculating the surface wind speeds at 10 m by using the roughness map. All available data summarised in Table 1 are used to create both maps. In Figure 4, the two surface wind maps resulting from the two methods are shown. Table 2 gives the statistics of the prediction errors.

From Table 2 it is clear that the methods produce very different results. The statistical values in Table 2 are all largely in favour of the U_{meso} method. For the S_{macro} method, the values indicate a larger error in the prediction map in comparison with the U_{meso} method. The poor results obtained here by using S_{macro} are in direct contrast to good results presented in [11], where the same method was applied. Two possible reasons for this difference are given. The first is that for this method the roughness has a very large influence. In this study the roughness length at the stations is obtained by using satellite images of the sites and updated pictures of the surrounding areas, identifying the relative land use. In a second step, the land uses derived were assigned their relative roughness through the use of roughness tables available in literature [25]. However, in [11] the roughness lengths at the masts are determined by analysing the wind gust ratio. This method is more accurate and less dependent on the exact mast location. Another reason is that the relationship between the mesoscale wind and the macrowind is based on the PBL similarity theory, which assumes a homogeneous PBL with neutral stability. However, in coastal areas horizontal temperature gradients are present and the method is unlikely to be applicable [12]. This may explain the failure of this method since the wind climate in Flanders is heavily determined by the presence of the sea. On the other hand, the U_{meso} exposure correction model gives much better results. It is expected that the results will even ameliorate when only recent observations recorded in Table 1 are used. Therefore, it was decided to use the recent data, together with the U_{meso} method, for the evaluation of the interpolation methods.

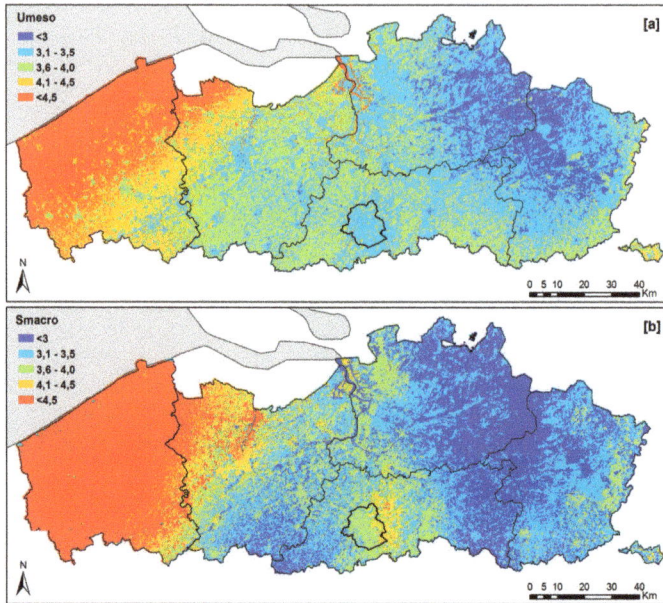

Figure 4. Yearly mean surface wind speed generated using the mesowind method and the macrowind method (**a**) Mesoscale wind interpolation. (**b**) Macroscale wind interpolation.

Table 2. Statistical details of the measurement errors for the U_{meso} and S_{macro} exposure correction methods.

Method	ME [m/s]	MAPE [%]	RMSE [m/s]	R^2
U_{meso}	−0.069	13.82	0.596	0.68
S_{macro}	0.035	19.42	0.945	0.56

4.2. Spatial Interpolation Methods Comparison

The comparison of different interpolation methods is here presented, showing differences in the yearly mean surface wind speed, with a direct consequence on turbine sitting. The annual mean wind speed maps generated with the different interpolation methods are all shown in Figure 5. The evaluation of the results is again done by comparing the LOOCV validation statistics and is summarised in Table 3.

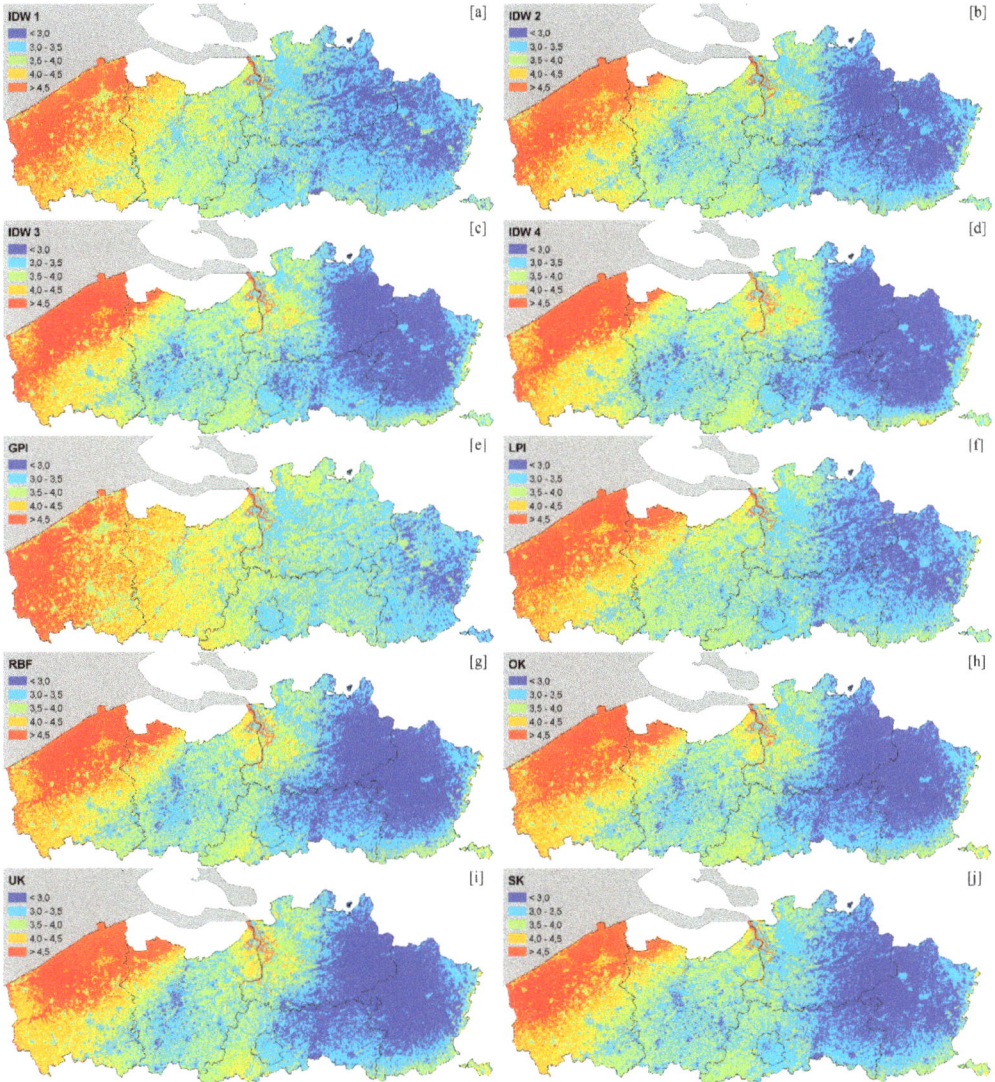

Figure 5. Yearly mean surface wind speed generated using different spatial interpolation methods. (**a**) Inverse Distance Weighted IDP = 1. (**b**) Inverse Distance Weighted IDP = 2. (**c**) Inverse Distance Weighted IDP = 3. (**d**) Inverse Distance Weighted IDP = 4. (**e**) Global Polynomial Interpolation. (**f**) Local Polynomial Interpolation. (**g**) Radial Basic Function. (**h**) Ordinary Kriging. (**i**) Universal Kriging. (**j**) Simple Kriging.

When comparing the statistics from SK where only recent data are used (Table 3) with the SK results where all data are used (Table 2), it is clear that the use of recent, overlapping data is most effective.

The results from IDW are largely dependent on the IDP value. The IDP = 3 is the power factor with the best overall performance. The GPI produces the most inaccurate results, with the highest

RMSE and MAPE values and a large over-prediction. The LPI method also has a large negative ME value but the other statistical values are somewhat better. This is due to the fact that the mesoscale is smaller than the area of Flanders and hence GPI is not applicable here.

It can be seen that the Kriging methods and the RBF method yield very good results: they all have an RMSE value of about 0.48 m/s. OK and UK give similar results, with UK being slightly better than OK. This is caused by the fact that in Flanders a linear trend of the wind speed can be assumed, with higher values near the sea and decreasing values farther inland; this is a trend that can be incorporated in the UK method. Still, the RBF and SK methods rendered an even smaller MAPE value of 0.5%.

Table 3. LOOCV statistical values for the different spatial interpolation methods.

Method	ME [m/s]	MAPE [%]	RMSE [m/s]	R^2
IDW 1	−0.143	14.58	0.577	0.56
IDW 2	−0.127	13.06	0.520	0.64
IDW 3	−0.121	12.10	0.509	0.67
IDW 4	−0.125	11.93	0.521	0.68
IDW 5	−0.132	12.29	0.540	0.68
GPI	−0.258	17.43	0.688	0.43
LPI	−0.257	13.15	0.504	0.77
RBF	−0.125	10.88	0.479	0.74
SK	−0.030	10.82	0.484	0.67
OK	−0.133	11.37	0.487	0.72
UK	−0.144	11.38	0.477	0.72

It is difficult to prefer one of the latter two methods over the other without additional verification measurements since the RBF method has a better R^2-value but with SK, the ME value is closer to 0. Nevertheless, the under-prediction of the wind speed when using the RBF method is undesirable since it leads to an underestimation of the power production by a potential wind turbine. When comparing the maps (g) and (j) from Figure 5, the largest difference between the two interpolation methods is situated around Antwerp, where RBF predicts higher wind speeds than SK. This is due to the high wind speed recorded in Deurne (Antwerp city district) which has a higher influence on the RBF interpolation as it is an exact method. Hence, the predicted surface is forced through the measured points and thus the SK method is considered more robust and more suited to deal with wind speed measurement uncertainties.

As the station of Brasschaat (Table 1) in the Antwerp region only has recorded data up to 2006, it is not used for the construction of the wind speed maps but can be applied for evaluation of both maps. For Brasschaat, the RBF and SK methods predicted a mean wind speed of 3.40 m/s and 3.21 m/s, respectively. Knowing that the average measured wind speed from 1973 to 2006 in Brasschaat is 3.26 m/s, the prediction error equals to 4.20% for RBF and 1.36% for SK. Hence, it is demonstrated that SK has the best performance for this observation point, with RBF acting as a valuable alternative method since a small margin of error is equally observed.

The above reasoning leads to the conclusion that the SK spatial interpolation method is slightly more realistic than RBF for the interpolation of the annual mean wind speed in Flanders. With the method described above, it is possible to create wind resource maps on different heights (see Figure 6).

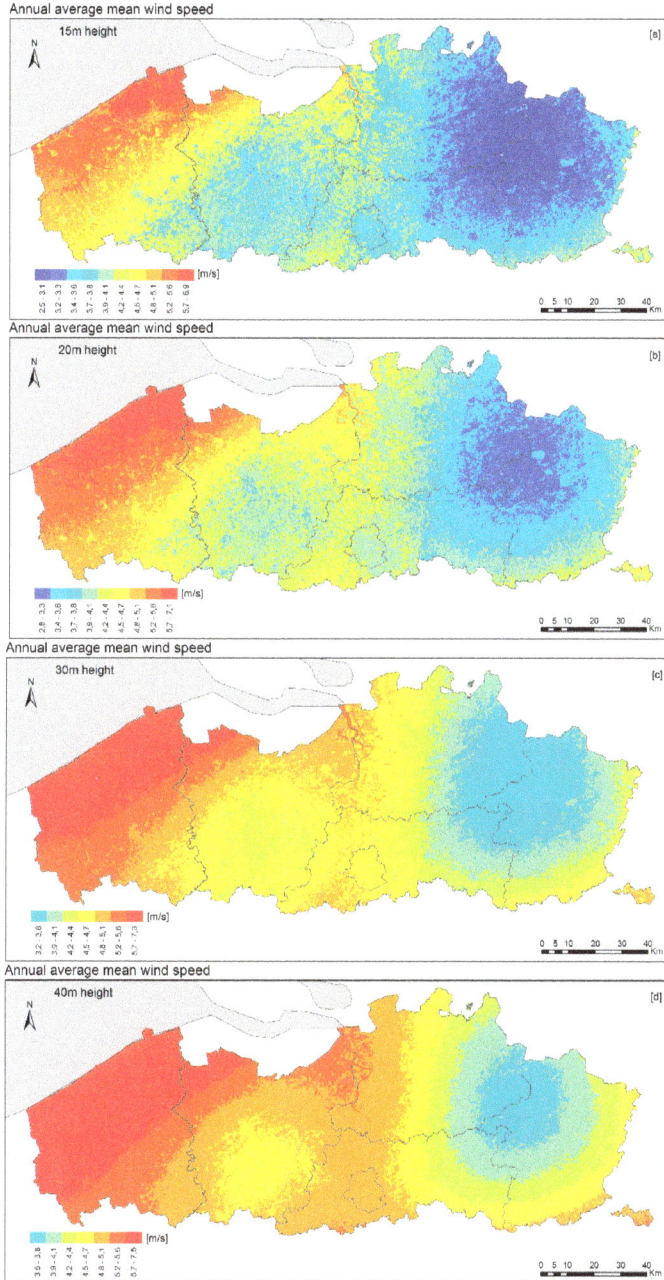

Figure 6. Annual mean wind speed, (**a**) 15 m, (**b**) 20 m, (**c**) 30 m, (**d**) 40 m.

5. Energy Resource Mapping

Mean wind speed is not a representative value for energy production assessment by itself. Accordingly, in this section we use a Rayleigh distribution as an indicator to estimate energy yield, which is a Weibull distribution with a shape factor of 2 [51–54].

The International Standard IEC 61400-12-1 [55] describes methods for determining the power performance of electricity producing horizontal axis wind turbines. The annual energy production (AEP) curve, described in this standard, allows the estimation of the annual production at different reference wind speed frequency distributions, assuming 100 % availability. A Rayleigh distribution, which is identical to a Weibull distribution with a shape factor of 2, is used as the reference wind speed frequency distribution. Starting from the wind speed map of Flanders, energy resource maps could be developed for different wind turbines, based on their AEP curve (see Figure 7). In order to demonstrate how the AEP can be calculated, in this article an AEP map is created for a 10 kW 3-blade, upwind, horizontal axis wind turbine.

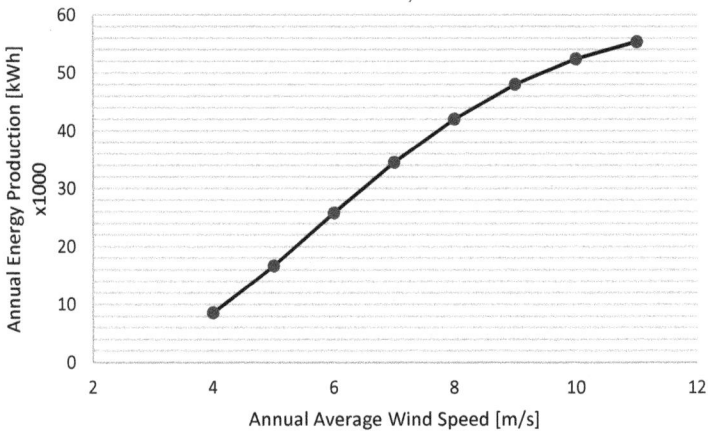

Figure 7. Estimated Annual Energy Production for a 10 kW 3-blade, upwind, horizontal axis wind turbine (reference air density: 1.225 kg/m^3).

The horizontal axis wind turbine in this example has a swept area of 40.7 m^2. The above-described small wind turbine (SWT) is certified by the Small Wind Certification Council to be in conformance with the American Wind Energy Association (AWEA) Small Wind Turbine Performance and Safety Standard (AWEA Standard 9.1–2009) [56].

The equation of the best-fit curve needs to be calculated. With this, it becomes possible to calculate the estimation of the AEP in Flanders for this wind turbine (See Figure 8). This is the equation for the best fit curve:

$$AEP = -1 \times 10^{-5} x^6 - 0.0011x^5 + 0.0696x^4 - 1.3928x^3 + 12.477x^2 - 42.413x + 51.124 \qquad (15)$$

with x representing the annual average wind speed (m/s), and taking into account the cut-in wind speed of 2.2 m/s.

In Figure 9 the estimated simple payback period is visualised for the above-described SWT. Hence it is possible to predict an area with reasonable payback times for the wind turbine. In Figure 9 only red to orange areas yield payback times that are commercially acceptable; the blue area is ruled out for the concerned SWT at a height of 15 m.

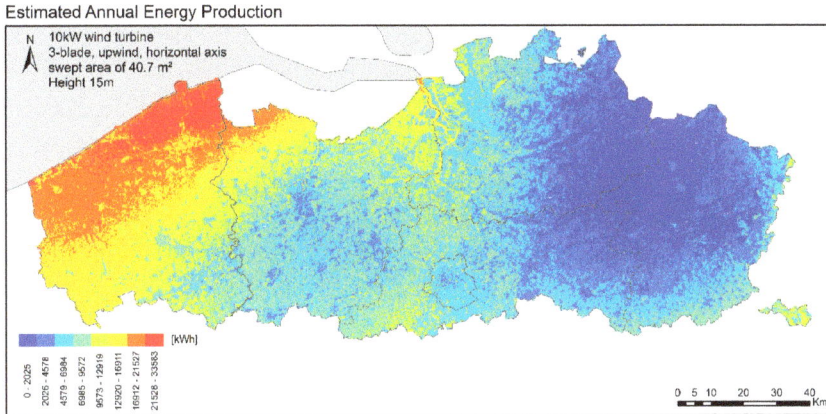

Figure 8. Estimated annual energy production (kWh) for a 10 kW 3-blade, upwind, horizontal axis wind turbine at 15 m height.

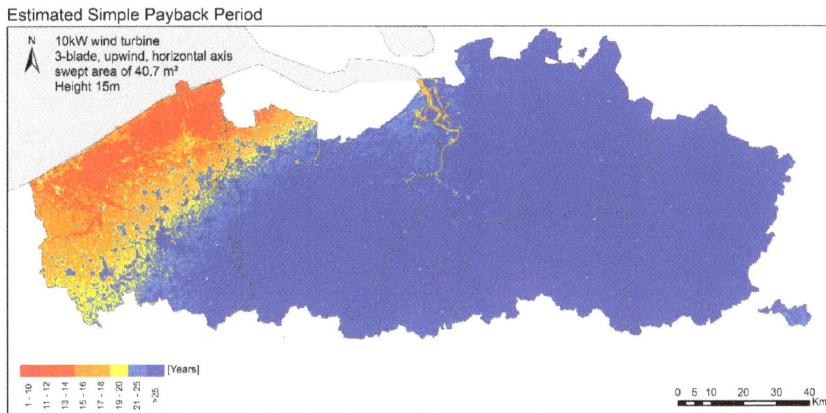

Figure 9. Estimated simple payback period (years).

6. Conclusions

The present study has produced a reliable wind speed map of Flanders based on measurement data and roughness maps, and likewise has provided insight on spatial interpolation methods. The study demonstrated how local wind conditions, and thus the local wind energy generation potential, can be calculated by modelling available wind measurements.

The method used is based on a traditional wind mapping methodology but adds an integrated spatial interpolation and transformation model to create reliable location-specific wind resource maps.

By applying the model to Flanders, it was observed that transformation of the surface wind to the mesoscale level yielded better results for wind resource mapping than the macroscale level. Likewise, the comparison of seven different spatial interpolation methods led to the observation that geostatistical and RBF methods outperformed IDW and Polynomial interpolation methods.

In contrast to the findings of [11,37,38,47], the robust Simple Kriging interpolation method was found to produce the best results for developing regional wind resource maps since it has the lowest MAPE, a very low RMSE of 0.48 m/s and a negligible bias (see Table 3).

As an overall conclusion, based on statistical analysis, it was found that the transformation of surface wind measurements into mesoscale wind data in combination with Simple Kriging interpolation is the most adequate method to create reliable wind resource maps that enable the selection of optimal production sites for SMWTs in Flanders.

A limitation of the study is that an average wind speed map alone is not sufficient for wind energy applications. Accordingly, further steps for research might include additional information, such as seasonal maps and statistics on diurnal variability, to improve the energy map applications.

Another open issue is the transferability of our results, and to what extent this application for Flanders can be used as a reference for other implementations. Further steps for research should analyse whether phenomena described in the study are general characteristics under the practical applications.

Acknowledgments

This study is part of the *Windkracht 13* project that seeks to open the market for SMWTs in Flanders [57]. *Windkracht 13* is a demonstration and dissemination project in the frame of the New Industrial Policy of the Flemish government. The project studies current barriers by performing a focused LESTS (Legal, Economic, Spatial, Technical, Social) analysis [58]. Based on this pentagonal mapping, recommendations are made for lowering the thresholds to install SMWT in Flanders.

Author Contributions

Samuel Van Ackere conceived and designed the research; Samuel Van Ackere and David Schillebeeckx performed the experiments, and analysed the data; Samuel Van Ackere wrote the paper; Enrica Papa, Greet Van Eetvelde, Lieven Vandevelde and Karel Van Wyngene helped during the editing, rewriting and review process.

Conflicts of Interest

The authors declare no conflict of interest.

References

1. Giacomarra, M.; Bono, F. European Union commitment towards RES market penetration: From the first legislative acts to the publication of the recent guidelines on State aid 2014/2020. *Renew. Sustain. Energy Rev.* **2015**, *47*, 218–232.

2. Hanslian, D.; Hošek, J. Combining the VAS 3D interpolation method and Wind Atlas methodology to produce a high-resolution wind resource map for the Czech Republic. *Renew. Energy* **2015**, *77*, 291–299.

3. Tempels, B.; Pisman, A. Open Ruimte in Verstedelijkt Vlaanderen: Een Vergelijkende Studie Naar Vier Onderschatte Ruimtegebruiken. *Ruimte Maatsch.* **2013**, *5*, 33–58.

4. VMM. *Luchtkwaliteit in het Vlaamse Gewest—Jaarverslag Immissiemeetnetten*; Vlaamse Milieumaatschappij: Erembodegem, Belgium, 2014. (In Dutch)

5. Ministerie van de Vlaamse Gemeenschap, afdeling Natuurlijke Rijkdommen en Energie. *Windenergie Winstgevend*; Ministerie van de Vlaamse Gemeenschap, afdeling Natuurlijke Rijkdommen en Energie: Brussel, Belgium, 1998; p. 16. Available online: http://stro.vub.ac.be/wind/windenergie_winstgevend.pdf (accessed on 12 August 2015). (In Dutch)

6. National Oceanic & Atmospheric Administration (NOAA). U.S. Daily Observational Data. NOAA: Washington, DC, USA. Available online: http://gis.ncdc.noaa.gov/map/viewer/#app=cdo (accessed on 12 August 2015).

7. Pryor, S.; Barthelmie, R. Climate change impacts on wind energy: A review. *Renew. Sustain. Energy Rev.* **2010**, *14*, 430–437.

8. Antrop, M. Landscape change and the urbanization process in Europe. *Landsc. Urban Plan.* **2004**, *67*, 9–26.

9. Silva, J.; Ribeiro, C.; Guedes, R. Roughness length classification of Corine Land Cover classes. In Proceedings of the European Wind Energy Conference, Milan, Italy, 7–10 May 2007; pp. 1–10.

10. NGI. De eenheid beeldverwerking van het NGI. Available online: http://www.ngi.be/Common/articles/CA_Td/artikel_td.htm (accessed on 12 August 2015). (In Dutch)

11. Stepek, A.; Wijnant, I.L. *Interpolating Wind Speed Normals from the Sparse Dutch Network to a High Resolution Grid Using Local Roughness from Land Use Maps*; Koninklijk Nederlands Meteorologisch Instituut: De Bilt, The Netherlands, 2011.

12. Wieringa, J. Roughness-dependent geographical interpolation of surface wind speed averages. *Quart. J. R. Meteorol. Soc.* **1986**, *112*, 867–889.

13. Verkaik, J. Evaluation of two gustiness models for exposure correction calculations. *J. Appl. Meteorol.* **2000**, *39*, 1613–1626.

14. Verkaik, J. *A Method for the Geographical Interpolation of Wind Speed over Heterogeneous Terrain*; Koninklijk Nederlands Meteorologisch Instituut: De Bilt, The Netherlands, 2001.

15. Verkaik, J.; Smits, A. Interpretation and estimation of the local wind climate. In Proceedings of the 3rd European & African Conference on Wind Engineering, Eindhoven, The Netherlands, 2–6 July 2001; pp. 2–6.

16. Verkaik, J.W. *On Wind and Roughness over Land*; Wageningen Universiteit: Wageningen, The Netherlands, 2006.

17. Verkaik, J.W. *Windmodellering in het KNMI-hydra project—Opties en Knelpunten*; Koninklijk Nederlands Meteorologisch Instituut: De Bilt, The Netherlands, 2000.

18. Wever, N.; Groen, G. *Improving Potential Wind for Extreme Wind Statistics*; Koninklijk Nederlands Meteorologisch Instituut: De Bilt, The Netherlands, 2009.

19. Lopes, A.S.; Palma, J.M.L.M.; Piomelli, U. On the Determination of Effective Aerodynamic Roughness of Surfaces with Vegetation Patches. *Bound.-Layer Meteorol.* **2015**, *156*, 113–130.

20. Lorente-Plazas, R.; Montávez, J.P.; Jimenez, P.A.; Jerez, S.; Gómez-Navarro, J.J.; García-Valero, J.A.; Jimenez-Guerrero, P. Characterization of surface winds over the Iberian Peninsula. *Int. J. Climatol.* **2014**, *35*, 1007–1026.

21. Baas, P.; Bosveld, F.C.; Burgers, G. The impact of atmospheric stability on the near-surface wind over sea in storm conditions. *Wind Energy* **2015**, doi:10.1002/we.1825.

22. Kreibich, H.; Bubeck, P.; Kunz, M.; Mahlke, H.; Parolai, S.; Khazai, B.; Daniell, J.; Lakes, T.; Schröter, K. A review of multiple natural hazards and risks in Germany. *Nat. Hazards* **2014**, *74*, 2279–2304.

23. Wieringa, J. An objective exposure correction method for average wind speeds measured at a sheltered location. *Quart. J. R. Meteorol. Soc.* **1976**, *102*, 241–253.

24. Manwell, J.F.; McGowan, J.G.; Rogers, A.L. *Wind Energy Explained: Theory, Design and Application*; John Wiley & Sons: Hoboken, NJ, USA, 2010.

25. Wieringa, J. Representative roughness parameters for homogeneous terrain. *Bound.-Layer Meteorol.* **1993**, *63*, 323–363.

26. Obukhov, A. Turbulence in an atmosphere with a non-uniform temperature. *Bound.-Layer Meteorol.* **1971**, *2*, 7–29.

27. Businger, J.; Yaglom, A. Introduction to Obukhov's paper on 'turbulence in an atmosphere with a non-uniform temperature'. *Bound.-Layer Meteorol.* **1971**, *2*, 3–6.

28. Tennekes, H. The logarithmic wind profile. *J. Atmos. Sci.* **1973**, *30*, 234–238.

29. Högström, U. Von Karman's constant in atmospheric boundary layer flow: Reevaluated. *J. Atmos. Sci.* **1985**, *42*, 263–270.

30. Frenzen, P.; Vogel, C.A. A further note "on the magnitude and apparent range of variation of the von karman constant". *Bound.-Layer Meteorol.* **1995**, *75*, 315–317.

31. Garratt, J.R.; Hess, G.D.; Physick, W.L.; Bougeault, P. The atmospheric boundary layer—Advances in knowledge and application. *Bound.-Layer Meteorol.* **1996**, *78*, 9–37.

32. Tieleman, H.W. Strong wind observations in the atmospheric surface layer. *J. Wind Eng. Ind. Aerodyn.* **2008**, *96*, 41–77.

33. Arya, S. Suggested revisions to certain boundary layer parameterization schemes used in atmospheric circulation models. *Mon. Weather Rev.* **1977**, *105*, 215–227.

34. Newman, J.F.; Klein, P.M. The impacts of atmospheric stability on the accuracy of wind speed extrapolation methods. *Resources* **2014**, *3*, 81–105.

35. Duda, J.D. A Modelling Study of PBL Heights, 2010. Available online: http://www.meteor.iastate.edu/~jdduda/portfolio/605_paper.pdf (accessed on 12 August 2015).

36. Apaydin, H.; Sonmez, F.K.; Yildirim, Y.E. Spatial interpolation techniques for climate data in the GAP region in Turkey. *Clim. Res.* **2004**, *28*, 31–40.

37. Cellura, M.; Cirrincione, G.; Marvuglia, A.; Miraoui, A. Wind speed spatial estimation for energy planning in Sicily: Introduction and statistical analysis. *Renew. Energy* **2008**, *33*, 1237–1250.

38. Chinta, S. A Comparison of Spatial Interpolation Methods in Wind Speed Estimation across Anantapur District, Andhra Pradesh. *J. Earth Sci. Res.* **2014**, *2*, 48–54.

39. Luo, W.; Taylor, M.; Parker, S. A comparison of spatial interpolation methods to estimate continuous wind speed surfaces using irregularly distributed data from England and Wales. *Int. J. Climatol.* **2008**, *28*, 947–959.

40. Rehman, S.U.; Uddin Qazi, M.; Siddiqui, I.; Shah, N.H.S. Comparing Geostatistical and Non-geostatistical Techniques for the Estimation of Wind Potential in Un-sampled Area of Sindh, Pakistan. *Eur. Acad. Res.* **2013**, *I*, 1770–1792.

41. Wei, Y. Spatial Variation and Interpolation of Wind Speed Statistics and Its Implication in Design Wind Load. Postdoctoral Thesis, The University of Western Ontario, London, ON, Canada, 2013.

42. Dobesch, H.; Dumolard, P.; Dyras, I. *Spatial Interpolation for Climate Data: The Use of GIS in Climatology and Meteorology*; John Wiley & Sons: Hoboken, NJ, USA, 2013.

43. Michaelsen, J. Cross-validation in statistical climate forecast models. *J. Clim. Appl. Meteorol.* **1987**, *26*, 1589–1600.

44. Keckler, D. *The Surfer Manual*; Golden Software, Inc.: Golden, CO, USA, 1995.

45. Song, J.; DePinto, J.V. A GIS-based Data Query System. In Proceedings of the International Association for Great Lakes Research (IAGLR) Conference, Windsor, ON, Canada, 10–14 January 1995.

46. Johnston, K.; ver Hoef, J.M.; Krivoruchko, K.; Lucas, N. *Using ArcGIS Geostatistical Analyst*; ESRI: Redlands, CA, USA, 2001; p. 300.

47. Ali, S.M.; Mahdi, A.S.; Shaban, A.H. Wind Speed Estimation for Iraq using several Spatial Interpolation Methods. *Br. J. Sci.* **2012**, *7*, 48–55.

48. Lang, C.-Y.; de Mesnard, L. Types of Interpolation Methods. Available online: http://www.gisresources.com/types-interpolation-methods_3/ (accessed on 3 July 2015).

49. Shi, G. *Chapter 8—Kriging Data Mining and Knowledge Discovery for Geoscientists*; Elsevier: Amsterdam, The Netherlands, 2014; pp. 238–274.

50. Eberly, S.; Swall, J.; Holland, D.; Cox, B.; Baldridge, E. *Developing Spatially Interpolated Surfaces and Estimating Uncertainty*; United States Environmental Protection Agency: Washington, DC, USA, 2004.

51. Alodat, M.T.; Anagreh, Y.N. Durations distribution of Rayleigh process with application to wind turbines. *J. Wind Eng. Ind. Aerodyn.* **2011**, *99*, 651–657.

52. Olaofe, Z.O.; Folly, K.A. Statistical Analysis of Wind Resources at Darling for Energy Production. *Int. J. Renew. Energy Res.* **2012**, *2*, 250–261.

53. Olaofe, Z.O.; Folly, K.A. Wind energy analysis based on turbine and developed site power curves: A case-study of Darling city. *Renew. Energy* **2013**, *53*, 306–318.

54. Ahmmad, M.R. Statistical Analysis of the Wind Resources at the Importance for Energy Production in Bangladesh. *Int. J. U- & E-Service Sci. Technol.* **2014**, *7*, 127–136.

55. IEC. International Standard IEC 61400-12-1. Available online: ftp://ftp.ee.polyu.edu.hk/wclo/Ext/OAP/IEC61400part12_1_WindMeasurement.pdf (accessed on 12 August 2015).

56. *AWEA Small Wind Turbine Performance and Safety Standard*; American Wind Energy Association: Washington, DC, USA, 2009.

57. Van Ackere, S.; van Wyngene, K. Windkracht 13. Available online: http://www.windkracht13.be/ (accessed on 3 July 2015).

58. Van Eetvelde, G.; van Zwam, B.; Maes, T.; Vollaard, P.; De Vries, I.; Tavernier, P.; D'Hooge, E.; Geenens, D.; Verdonck, L.; Leynse, L. *Praktijkboek duurzaam bedrijventerreinmanagement*; POM West-Vlaanderen: Sint-Andries, Belgium, 2008; p. 117. (In Dutch)

Towards a Carbon-Neutral Energy Sector: Opportunities and Challenges of Coordinated Bioenergy Supply Chains-A PSE Approach

Luis Puigjaner, Mar Pérez-Fortes and José M. Laínez-Aguirre

Abstract: The electricity generation sector needs to reduce its environmental impact and dependence on fossil fuel, mainly from coal. Biomass is one of the most promising future options to produce electricity, given its potential contribution to climate change mitigation. Even though biomass is an old source of energy, it is not yet a well-established commodity. The use of biomass in large centralised systems requires the establishment of delivery channels to provide the desired feedstock with the necessary attributes, at the right time and place. In terms of time to deployment and cost of the solution, co-combustion/co-gasification of biomass and coal are presented as transition and short-medium term alternatives towards a carbon-neutral energy sector. Hence, there is a need to assess an effective introduction of co-combustion/co-gasification projects in the current electricity production share. The purpose of this work is to review recent steps in Process Systems Engineering towards bringing into reality individualised and ad-hoc solutions, by building a common but adjustable design platform to tailored approaches of biomass-based supply chains. Current solutions and the latest developments are presented and future needs under study are also identified.

Reprinted from *Energies*. Cite as: Puigjaner, L.; Pérez-Fortes, M.; Laínez-Aguirre, J.M. Towards a Carbon-Neutral Energy Sector: Opportunities and Challenges of Coordinated Bioenergy Supply Chains-A PSE Approach. *Energies* **2015**, *8*, 5613-5660.

1. Introduction

A greener and more sustainable society needs renewable energy under all its forms, higher efficiency systems and a change of habits. Oil, chemicals and related industries, are nowadays evolving considerably due to market demands, unprecedented globalisation and the arising limitations from environmental concerns and security. Moreover, sustainable considerations combined with tools such as stakeholder analysis, key performance indicators (KPIs) and life cycle assessment (LCA) approach may cover the supply chain from cradle-to-the grave, being powerful approaches in a pre-design step [1].

The energy sector is moving towards a new energy paradigm, which favours more efficient conversion processes (due to more scarce and expensive fossil fuels), renewable sources and micro-generation (*i.e.*, smart grids), through tailor-made approaches, adapted to the needs and resources of each area. Decentralisation ideally involves more population participation and supply security. There will not be a unique technology or renewable source massive implementation, but a combination of various conversion technologies to meet the energy demand [2]. The alternatives to centralised and conventional sources of energy should be sustainable in the time, which implies a responsible resource exploitation, by balancing source availability with electricity demand, and therefore with the capacity of the plant.

Biomass can play an important role in both centralised (large scale) and decentralised or distributed (small scale) energy systems. Each scale evolves into different challenges in the use of biomass. As immediate solutions, where technology is already well developed, biomass at large scale can be co-used properly with fossil fuels. Biomass systems at small scale are appropriate for residential uses and rural electrification in emerging countries [3].

Bioenergy or energy from biomass is a promising contributor in the upcoming energy mix. In order to become a key actor, technological, economic, environmental and social aspects need to be advantageous if compared to conventional fuels. As one of the main points to improve, biomass needs to be densified to increase its calorific value, while easing its transportation and stabilising moisture and dry matter contents. That is the reason why biomass pre-treatment becomes crucial for the development of sustainable supply chains.

The use of coal can be reduced if appropriately mixed with biomass. Around the world, requirements for energy and electricity are largely met by fossil fuels, and coal is widely selected as it is a secure, low-cost and high energy density source; coal resources are abundant and broadly distributed geographically [4]. Coal is also relatively easy to mine, ship, and store. It is expected to contribute significantly in the future energy needs in many nations, especially in fast-developing countries such as China and India [5]. These qualities make coal-fuelled power plants important electricity price stabilisers and reliable power producers, especially in electricity systems with price-volatility or intermittently available resources.

In this context, it is worth noticing that coal demand had an average growth rate of 3.3% per year, between 2010 and 2013, and is expected to reach nine billion tonnes per year by 2019. According to the International Energy Agency (IEA) Executive Director, Maria van der Hoeven, although the contribution that coal makes to energy security and access to energy is undeniable, coal use in its current form is unsustainable, which makes the deployment of carbon capture and sequestration a priority [6].

On the other hand, according to the Energy International Agency (EIA), the global energy demand is set to grow by 37% by 2040 [7,8]. As example, the European Union has established a target of 20% share of renewable energy out of the total European energy consumption by 2020 [9]. The U.S. in its Energy Independence and Security Act (EISA) of 2007 states that advanced biofuels shall supply at least 21 billion gallons of U.S. motor fuels by 2022 [10]. In this context, biomass exploitation becomes into a need.

1.1. Challenges in the Bioenergy Sector

The development of a successful bioenergy sector in developed and developing nations, through centralised and more decentralised systems, will make a useful long-term contribution to diversity, security and self-sufficiency of energy supply [11]. Current challenges in the worldwide energy sector reflect three main issues: natural sources diminution, climate change and technology development. Within this context, bioenergy is one of the most appreciated options to mitigate Greenhouse Gases (GHG) emissions by replacing conventional sources in vehicles fuel and in electric power generation, certainly by adequately exploiting biomass resources and the multiple technology options [12].

Bioenergy challenges are classified into two main blocks: energy generation and biomass as a source. Figure 1 shows the major topics to be addressed by these blocks: (i) energy generation deals with the different biomass conversion routes, for fuels or electricity production, to be brought into market status; and (ii) biomass as a source, faces controversies like land use, while it is also concerned with globalisation and global markets.

Bioenergy inter-linkages

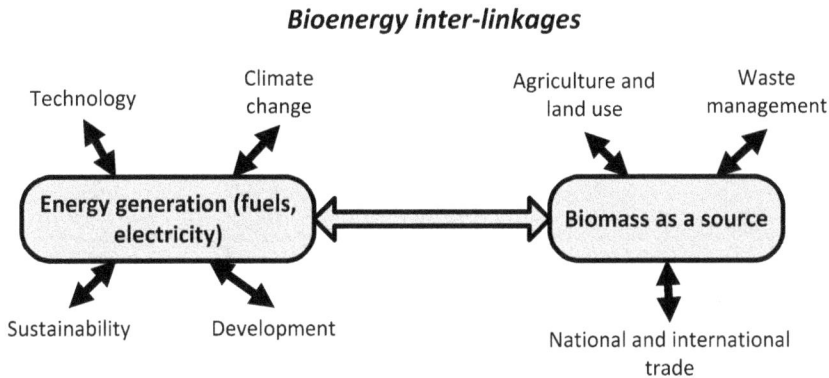

Figure 1. Bioenergy and concerned actuation areas.

In relation to the first block, global trends are promoting the utilisation of renewable sources as alternatives to fossil fuels to mitigate climate change and to alleviate the peak oil effects. In poorer areas, it is equally important to promote energy access through renewable sources at affordable prices [13]. As [14] and [15] point out, the current energy models are biased towards industrialised countries because these are usually developed by experts and/or organisations that live and have been educated in industrialised countries. Accordingly, not only economic factors should be revised, but also types of demands, capability for operational and maintenance tasks or accessibility to the grid, to come up with more versatile models capable of looking into each project through its own reality, context and particularities.

Concerning the second block, biomass as energy source is coupled with two important economy sectors: agriculture and waste management. Agriculture can be used to produce food, feed, fuel and fibre (the so-called "4Fs") creating a certain controversy and competitiveness for the land use, and therefore, for water use. As last instance, land as a resource is a protector of ecosystem systems, deals with the pressure of population growth, life styles variations and climate change consequences [16]. Residues management is interlinked with other markets: they can be used as raw materials, as feed or as fertilizer, or in other industries that treat them to be further used in other processes. This leads into a complex competitive trade, where prices are set by the demand [12]. Biomass markets are changing from exclusively national to international markets: globalisation makes accessible a broad range of globally dispersed potential suppliers and consumers. In order to develop a stable market with biomass as a commodity, supply and demand should be secured, in a sustainable way, while meeting the appropriate technical standards [17]. As the works of

Janssen *et al.* [18] and Madjera [19] point out, developing countries such as those in Africa, have the potential to become significant producers and exporters of raw biomass while supplying their basic needs.

Overall, biomass can provide a larger energy share than the one that provides nowadays. For that to become a reality, technological, economic and social barriers have to be overcome [20]. As a result, efforts are concentrated on developing integrated frameworks to support the decision-making process. This is further described in the current paper, which is principally focused on gasification and combustion technologies.

The Scale of the Problem

Biomass as energy source, in comparison with fossil fuels, has a lower calorific value as well as intrinsic characteristics that derive into technological limitations. That is the reason why 100% biomass to energy projects typically employ small scale conversion systems. Moreover, they tend to be placed close to the biomass generation source as well as close to the biomass demand points, to avoid high logistic and network infrastructure constraints [21–23]. According to [12], large gasification systems are from 10 MW_{th}, and small gasification systems cover the range from few MW_{th} to less than 100 kW_{th}. In terms of electricity and in accordance with [24], small scale gasification plants enclose plants with a power up to 200 kW_e. These ranges lead to significant differences in terms of land use for the plant infrastructure, investment, operation and maintenances costs and evidently, plant dimensions (as example, the ELCOGAS integrated gasification combined cycle (IGCC) power plant uses a land extension of 480,000 m^2, while a small scale gasification plant can occupy around 25 m^2, as it is the case of the real scale pilot plant built in our laboratory at the Universitat Politècnica de Catalunya). *Centralised energy systems* (CES) are defined here as large power plants that inject electricity to the grid and transport the raw material or energy source to the plant; *decentralised* or *distributed energy systems* (DES) entail localised electricity generation near the demand points and near the biomass production places. There exists no agreement in the literature about the definition of distributed generation; nevertheless it is usually perceived as small scale electricity generation [25]. The literature overviews from [26] and [25], point out that the term refer to: (i) stand alone or autonomous applications; (ii) stand-by sources that supply power during grid outages; (iii) co-generation (or waste heat recovery) installations with power injection to the grid (if the DES has a higher power production than the local demand); (iv) DES that support the grid by decreasing power losses and improving the system voltage profile and (v) to energy systems connected directly to the grid that sell the electricity produced. This work uses the term DES as stand alone applications, with co-generation possibilities. See in Figure 2 an overview of centralised *vs.* distributed systems.

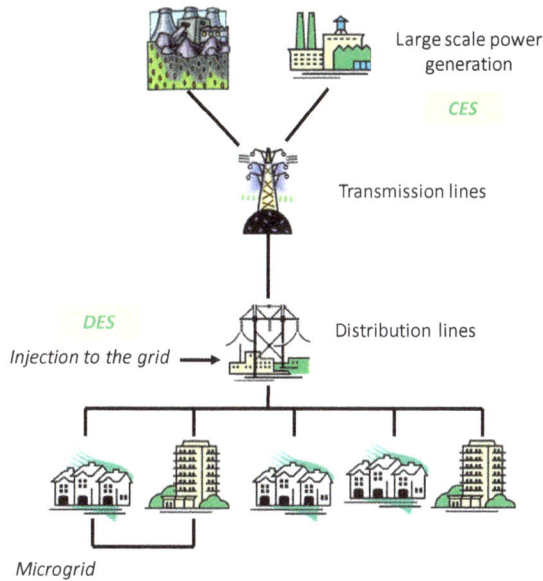

Figure 2. Conventional centralised and decentralised based systems.

The supply chains of decentralised and centralised systems are studied in this paper, for two well differentiated concepts: rural/urban areas in developing/developed countries. This terminology does not have a well extended norm of usage; one possible definition for rural area uses a threshold of 150 inhabitants/km^2, including countryside, towns and small cities. Other definitions take into account towns and municipalities outside the urban centres, with population of 10,000 or more; or population living outside regions with major urban settlements of 50,000 or more people, dividing the areas into "metropolitan adjacent" or "not adjacent" categories [27]. Urban areas include a central city and the surrounding dense areas that have together a population of 50,000 or more, encompassing a minimum of 2500 people, the minimum of which (1500 people) residing outside institutional group quarters, according to the United States Census Bureau, [28]. The developed-developing countries division is more controversial, since it is difficult to assess the standards of living for worldwide countries. For instance, the World Bank (WB) classifies the countries according to their gross national income (GNI) for year 2013, being developing countries those ones with a GNI lower than US$4125 [29]. The World Energy Assessment from the United Nations Development Program [30] (UNDP), the United Nations Department of Economic and Social Affairs, and the World Energy Council, (2004) uses the term industrialised country to refer to high-income countries that belong to the Organisation for Economic Co-operation and Development (OECD). In this way, developed countries are also called industrialised. In this review, both terms, industrialised and developed are used.

Energy chains should be developed according to the context of each country/project and taking into account economic, environmental and social issues. Consequently, even if the technology to be implemented is the same in developed and developing countries (*i.e.*, gasification), the specific power to produce, and the energy chain itself (distance to raw materials, to existing grids, *etc.*)

should be characterised according to features such as the sector financing, the existence of a grid, the grid distribution losses, the demand, *etc.* [14,31]. Gasification principles for large and small scale gasification are the same, but the type of reactor as well as the final syngas or producer gas composition and uses are generally different. This work describes further three plant layouts, which have been chosen because of the current challenges in process design and in supply chain management: large scale gasification of biomass-coal blends, represented by IGCC power plants, with carbon capture and storage (CCS) technology (IGCC-CCS), small scale biomass gasification which considers the produced gas usage in a gas engine (BG-GE), and co-combustion in large scale power plants, *i.e.*, in retrofitted pulverised coal (PC) power plants. The next two sections describe: (i) the main characteristics of biomass and the range of available technologies and (ii) the techniques used in a pre-design stage, to approach a biomass-to-energy problem.

1.2. Biomass as a Resource

Biomass is defined as "all the organic matter contained in plant and animal based products (including organic wastes) that can be captured and used as a source of stored chemical energy" [11]. Biomass can be classified into three large categories according to its origin [11,12]: primary, secondary and tertiary biomass. See in Figure 3 this well extended biomass classification, detailing sources and raw materials.

According to Sim [11], biomass contributes significantly to the world's primary energy supply, with 45 EJ/yr utilised in traditional and modern uses of biomass. Inside this number, the traditional use of biomass is estimated in 38 EJ/yr: it is the first energy source in developing countries (involving a 20%–35% of their national primary energy demand). The traditional use of biomass includes cooking and heating in a non-sustainable and inefficient way, through direct firing. As Silveira [32] points out, "biomass is the fuel of the rural poor in developing countries". There is no global information about the biomass market size; nevertheless it is assumed that the non-conventional use of the biomass is around 29 EJ/yr. The most relevant properties of biomass as energy carrier or chemical feedstock in thermochemical conversion processes (described in point 1.2.1) are, according to Rubiera *et al.* [33]: proximate and ultimate analyses, moisture content, lower and higher heating values (LHV and HHV), heats of formation, ashes content, biochemical composition (hemicellulose, cellulose, lignin and extractives), bulk density and grindability.

In Mathews [34] is stated that the world is in a transition, from an economy fuelled by carbon from the past ("petro-economy"), to an economy fuelled by biomass, which is created through photosynthesis ("bio-economy"). According to Rosillo-Calle *et al.* [17] and Mathews [34], bioenergy is extensively considered as carbon neutral, since the carbon emitted replaces the carbon absorbed during the crop growth. Nonetheless, each specific situation should be treated separately, and a LCA is recommended to calculate a complete carbon balance.

Three situations can be identified in general for fossil and biomass fuels, and are depicted in Figure 4. Carbon positive fuels describe fossil fuels, as they release (net) CO_2 into the atmosphere. Carbon neutral fuels symbolise biomass resources, which absorb CO_2 from the atmosphere and release it again. Nevertheless, in practice, the carbon balance may be positive if fossil fuels are used at some echelon of the supply chain (mainly in biomass production and logistics). Carbon negative fuels represent biomass resources that absorb CO_2 from the atmosphere and release less CO_2 into it, because of directing part of the captured emissions during growing to the soil, as bio-char, or because of CSS use (called bionenergy with CCS, BECCS). Analogously, fossil fuels with CCS aim at a complete carbon neutrality, even if a small fraction of the CO_2 is not captured and hence it is discharged in the atmosphere. This neutral-carbon objective is theoretical, since a complete LCA should be performed to evaluate the trade-off between the emissions captured, and the emissions derived from the utilities consumed to perform this capture.

Figure 3. Biomass sources classification, based on [11,12].

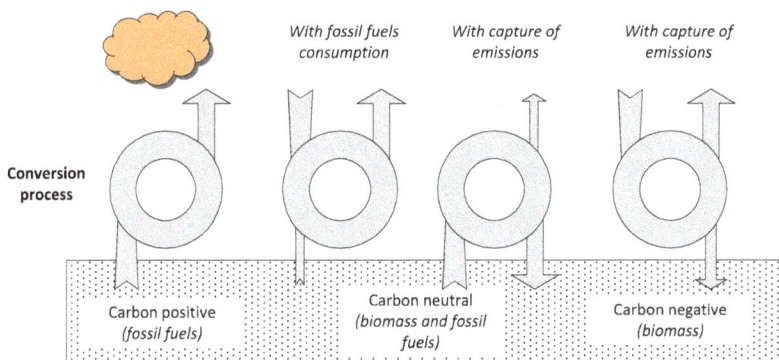

Figure 4. Bio-sources carbon balance, based on Mathews [34].

1.2.1. Available Technologies

An important portfolio of technologies allows for biomass transformation into heat, electricity, co-generation or transport fuels, and chemical feedstock. The most suitable conversion technology for a specific type of biomass depends on the composition, characteristics and amount of the resource, the desired final product, the environmental standards and the economic and project specific conditions [12,35]. Figure 5 shows the different available technologies and products obtained. Thermochemical conversion processes are suitable for low moisture content biomass (less than 50%), while physic-chemical and biological ones are adequate for humid biomass. A biorefinery integrates different technologies to produce heat, electricity, fuels and chemicals, at the same facility.

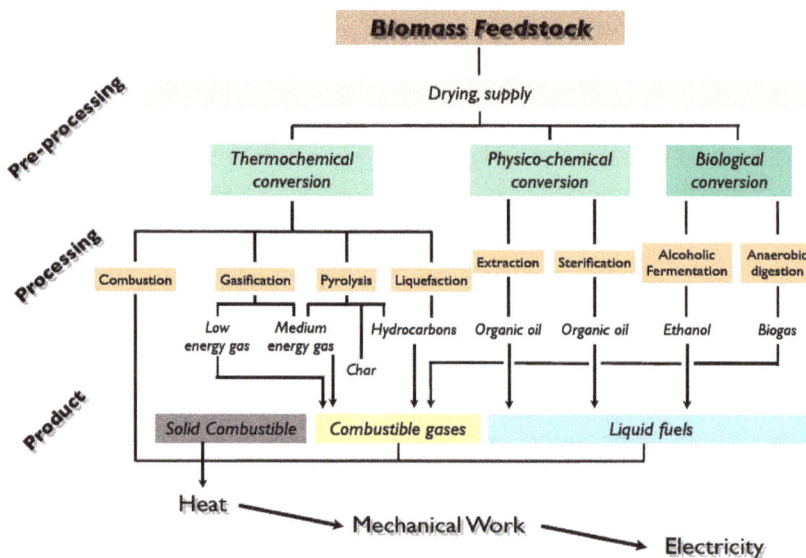

Figure 5. Outline of the main biomass conversion processes.

1.2.2. Biomass Trade

The sub-Saharan Africa (SSA) has the lowest electrification rate worldwide and still relies on the traditional use of wood for cooking and heating; 587 million people, 75% of the population, had no access to electricity in 2009, a number that is believed to rise up to 652 million by 2030 [36,37]. SSA agricultural waste potential for energy purposes is estimated at 136 PJ per year, while the forest residues in west and central Africa can reach 95 PJ per year [36]. Identification of best strategies for modern wood technologies implementation and efficient small scale techniques are underway. As aforementioned this type of technologies are necessary to enable developing countries to become significant producers and exporters of raw biomass while supplying their basic needs. Along this lines, the European Union is a significant pellets consumer and importer: international imports have grown from 56 PJ in 2000 to 300 PJ in 2010 [38]. According to

EURELECTRIC, the solid-biomass provided by external suppliers may increase up to 1650 PJ by 2020. This is an example of how a global context can set the basis for a clear offer-demand opportunity: there is true momentum for solid-biomass market (with pellets as its main representative) in Europe, and SSA, together with other areas in America and Canada, has a potential that is nowadays underexploited.

In the medium and short term, the use of waste, which entails disposal problems, is a continuous source of organic matter for power production. Due to its distributed nature, biomass is appropriate for decentralised power generation in local areas, with certain centralisation for cost optimisation purposes [39]. The pellets industry should not only use "high quality" woody resources as raw material, but also the wide range of available organic residues. Combinations of grass and woody materials, biomass with coal pellets, are under research as raw materials for pellets [40].

Strategies for rural electrification and pellets market development must use criteria other than economic: sustainability, combining economic, environmental and social measures. The study of the whole bio-based supply chain at the design and planning stage is essential to propose long-term projects. Multi-objective optimisation tools to support decision-making (see next sections) in the field of renewable energy are being developed, used and improved. Solid-biomass (and specifically, pellets) is evolving towards an established renewable commodity internationally commercialised. Two main concerns must be solved in the context of SSA as a potential exporter: (i) resources evaluation and (ii) rural areas electrification.

1.3. Decision Making

Different tools and methodologies can be utilised to take decisions and diminish their associated risk. Multiple criteria decision analyses (MCDA) comprise the methods for process optimisation that contemplate multiple objectives. MCDA is applied to two types of systems: process and supply chain (SC) systems. In order to apply this decision analysis, the system is previously modelled or simulated. The following subsections explain the basis of this methodology.

1.3.1. Conceptual Design

The term "conceptual design" can be understood as the product design cycle phase where the basic solution is established through the formulation of abstract ideas with approximate concrete representations. Moreover, those ideas are evaluated with different selected criteria. This stage starts with "high-level" requirements descriptions and continues with "high-level" solutions descriptions. By the end of the conceptual design phase, a decision must be taken [41]. According to Douglas [42], a more chemical process conceptual design implies to find the best process flowsheet (selection of process units and connections among them) and estimate the optimal operating conditions. It is often referred to as a preliminary design stage. In a more abstract level, and extrapolating both previous definitions, conceptual design can be applied not only to processes development, but also to the development of the whole SC.

1.3.2. Process Modelling

The aim of process design is to specify the most economic and effective practical procedures to transform raw materials into a new product, to manufacture an existing product by new means or to bring about some designated material transformation to a commercial scale, so as to satisfy a market need [43]. The classical design procedure is seen as an iterative procedure to estimate in advance the resource implications. To reduce uncertainty in the decision-making (dimensions, materials, type of units, *etc.*), the use of process simulation is a convenient strategy. Process simulation is understood as the use of computer software to construct mathematical models of process components which provide an accurate representation of the whole chemical process. The simulation aims at understanding the process behaviour during regular plant operation. Depending on the degree of model accuracy ("granularity"), the precision of process design cost estimates varies within a wide range. This work deals with preliminary design, where the precision of the cost has a margin of 10%–25% but it represents only the 0.4%–0.8% of the total project cost [42]. The design level includes the optimisation approach to identify the best design according to selected custom criteria. Optimal process design assesses the performance of a process according to economic, technical, thermodynamic and/or environmental indicators. Process modelling coupled with LCA incorporates the environmental aspects, which can guide the process design towards a wider analysis, rather than the plant as a unique entity [44].

1.3.3. Process Systems Engineering (PSE) Approach: the Concept of Superstructure

PSE is considered an interdisciplinary field in chemical engineering that generally deals with how complex engineering projects should be designed and managed. More specifically, it uses computational techniques for mathematical modelling and simulation, process design, process control and process optimisation. In this section, features of this discipline are identified and the focus that vertebrates this *Review* is justified.

PSE has earned an important place for a wide range of chemical engineering activities. A basic requirement for the application of the techniques offered by this relatively young discipline is based on the notion of a model. As a requirement, the model represents relevant properties (structural and behavioural) of the system under study. The essential feature of a model (with respect to PSE) is that it can be formally evaluated to make statements about a system. This feature allows the use of digital computers, which have become an essential tool for many tasks in systems engineering process now coined as computer aided process engineering. Models should be considered valuable for engineering in general and particularly for decision-making processes, as they are not only data but embody a wealth of knowledge about the process studied and can be used to generate information on the same. The models allow virtual experiments through simulation and/or optimisation processes that would be expensive or even infeasible to implement them differently [45].

Modelling activities consider a variety of elements of chemical engineering at different levels of complexity [46]. Model-based studies cover a range from the design of molecules [47,48] at one end of a scale of size, as well as studies of the SC between different plants or even sites in the other end [49]. The relevant time scales ranging from microseconds to months or even years, respectively. Between these two extremes, the most common models used today represent thermodynamic phases, individual unit operations or a complete chemical process (see Sections 2 and 3). In addition to modelling physical processes, models of operating modes (Section 3) are also of interest for simulation and optimisation applications (Sections 4 and 5).

The modelling work process is also important with respect to developing supporting tools for model development because any software tool must focus on the work processes it is intended to support. Several steps including documentation, conceptual modelling, model implementation and model application have been considered in this field study [50]. More recently, models are becoming part of a flexible design framework called *modelling superstructure* that facilitates process conceptual design, synthesis, simulation and optimisation. According to Biegler *et al.* [51], the superstructure is able to compile feasible options for topological changes of a determined flowsheet, embracing equipment combinations that affect the final results or products and by-products characteristics. The superstructure representation involves the appearance of units that develop the same role in the flowsheet. Therefore, if using process simulators, those options can be considered by adding splitters and mixers according to the process layout. Mathematical programming is the usual representation for model implementation in a specific numerical application. It includes a way to represent and generate process superstructures, as well as all the elements required to formulate complete superstructure optimisation models using an entirely modular approach and standard processing unit models. The models for all superstructure elements (*i.e.*, processing units and connectivity elements) are created from detailed simulation models. Specifically, in the approach proposed by Biegler *et al.* [51] the process synthesis problem is formulated as a mathematical programming problem. The whole superstructure, which is understood as the ensemble of all feasible flowsheets, of all possible combinations of equipment, raw material and products is programmed as a Mixed Integer Non-Linear Problem (MINLP). Integer (binary) variables are related to the presence or not of given equipment in the solution while real variables represent equipment parameters such as temperatures, pressures or flowrates. It is worth noting that the complexity of the problem posed in this rigorous way may lead to intractable situations in terms of computational time. Instead, one important method of solving these kinds of problems is the use of meta-models or *surrogate models*, which are specially suited for sequential modular simulations. This is the approach followed in Section 4, where a specific application of the superstructure to a bio-based co-gasification process [52] is presented. Mathematical programming as solution methodology for designing and planning the whole bioenergy SC [53] is contemplated in Section 5.

1.3.4. Multiple-Criteria Decision Analyses (MCDA)

Decision analysis refers to the methodological process of identifying, modelling, assessing and determining a suitable way of action for a given decision problem. This usually presents multiple and conflicting criteria to evaluate alternatives. It is then necessary to make compromises or trade-offs regarding the results of the different possible choices. In MCDA context, the term objective is used to designate a direction that should be followed to "improve", as perceived by the decision maker. In contrast, the concept goal is a specific target of an objective, attained by the best choice.

If the criteria of the decision maker is not specific or concise (no prioritisation of the objective functions), instead of providing one specific solution, a set of feasible solutions may be possible, the so-called Pareto optimal solutions. These are also called the *Pareto Frontier* [54].

From the PSE perspective, modelling of IGCC together with CCS, abridged (IGCC-CCS), and biomass gasification (BG) coupled with a gas engine (GE), abbreviate (BG-GE), represent the aspects of interest to gain knowledge about the system's performance in terms of thermodynamics, mass and energy flows; while IGCC-CCS and BG-GE supply chains modelling enable the investigation of possible alternatives for SC management.

1.4. Scope and Objectives

The bioenergy sector should deal with environmental, social and economic issues and adopt decisions that take into account biomass intrinsic characteristics, availability and population demand.

The main objective of this survey is to contribute to the bioenergy sector by studying the co-combustion and co-gasification of biomass using advanced process modelling techniques, and incorporating specific PSE strategies, from different perspectives. This work distinguishes between centralised (large scale) and decentralised (small scale) power generation layouts in different contexts. Representative and current case studies have been selected in this work to exemplify the utility of design methods and supply chain optimisation when tackling bioenergy problems. This general aim can be divided into three more specific objectives:

- To assess the effective introduction of co-combustion projects in the current electricity production share, preferably by using biomass waste. Special consideration is given to biomass intrinsic heterogeneity.
- To develop a PSE approach for IGCC-CCS modelling and optimisation and propose working conditions guidelines in co-gasification and co-production of H_2 and electricity in IGCC-CCS plants.
- To apply existing models and tools in SC management to two bio-based supply chains differing in scale and social/economic contexts, and propose sustainable networks.

2. Co-Combustion of Biomass and Coal

The technologies outlined in Section 1.2.1 include large and small scale typical applications. Large scale systems to produce power and heat by means of a gas contemplate biogas production through anaerobic digestion, combustion; or flue gas production through combustion; and syngas generation through gasification. Combustion and gasification are the two possibilities treated here. They offer five alternatives for biomass usage: combustion, co-combustion or co-firing, gasification, co-gasification and gasification for co-firing [12]. From the efficiency point of view, GHG emissions reduction and solution immediacy in centralised energy systems, and the co-firing and co-gasification options are studied. These two options have in common the range of power produced (hundreds of MW) and the profitability of already existing installations originally design to operate with 100% fossil fuels. Typical 100% biomass combustion plants are around 20–50 MW$_e$ and 100% biomass gasification plants are in the range of 10 MW$_e$ [55]. Co-firing and co-gasification permit the usage of local biomass sources, being of special interest the organic wastes management area. CO_2, sulphur and nitrogen emissions reduction are direct benefits from the coal fraction substitution.

2.1. Process Description

Co-firing can be defined as the simultaneous combustion of two or more fuels in the same combustion plant [56] using biomass along with a fossil fuel [57], coal in this case. This biomass application is the cheapest one if compared with other biomass uses and other renewable sources: it can cost from 2 to 5 times less than other bioenergy alternative [56]. The study by Gómez *et al.* [58] reports a range for specific investment costs that varies from 100 €/kW to 880 €/kW, with kW of thermal power contained in the flowrate of biomass used. These estimates depend on the type of coal power plant and the selected co-firing system. According to Faaij [12], co-firing is the largest conversion technology of biomass that is growing in the European countries. It offers clear advantages. These are mainly high efficiency in energy terms (due to the already existing economies of scale in the thermal power plants) and low investment costs by appropriately matching the biomass quality, the co-firing option and the coal percentage substitution.

The coal-fired power plants can be of different types according to the reactor used: fluidised bed boilers, PC boilers and grate-fired boilers. The *biomass quality* should mimic as far as possible the main properties of a fossil fuel. These are low moisture content (MC), optimal grindability to be pulverised and high bulk and energetic densities. Biomass has in fact all these drawbacks: high MC, due to its fibrous nature it is hard to be reduced into powder, and low bulk and energetic densities. Those are the reasons why the quality of biomass should be improved to optimise its transport, its handling and its processing. Pre-treated biomass is needed to further develop the supply and use of it in CES. Pellets, torrefied biomass, torrefied pellets (TOP) and bio-oil (pyrolysis oil) are the state-of-the-art options currently proposed as "enhanced biomass" [59] (with torrefaction and fast pyrolysis being in pre-commercial stage) and used in co-firing, as well as in co-gasification. From the cheapest to the most expensive technique, *the biomass co-firing processes* include: blending biomass with coal or co-milling, separate injection and parallel co-firing in separate boilers that are connected to a common steam turbine (ST). Indirect co-firing also contemplates advanced techniques such as gasification to burn the syngas, thus syngas co-firing [60–62].

2.2. Biomass Pre-Treatments

The pre-treatment echelon in a bio-based supply chain is the bottleneck of biomass as a fuel if compared with other organic fuels. Different available pre-treatment characteristics are briefly described as follows.

- *Torrefaction.* This is a thermal step at relatively low temperature (225–300 °C, depending on the type of biomass) performed at atmospheric pressure in an inert atmosphere. The heating rate is low, approximately 50 °C/min [63]. The final solid product is a uniform solid with lower MC and higher calorific value than the raw material. By-products are a condensable liquid and a non-condensable gas. It is a ratter new technique applied to biomass, but there has been already seen benefits on the bulk density and grindability, since the needed energy for milling is notably reduced [59,64]. During torrefaction the biomass achieves a 0% of humidity. Nevertheless, after the process, it can capture some environmental moisture. This capture of humidity is limited since torrefied biomass has a hydrophobic nature. Moreover, torrefaction limits biological degradation [65]. In the paper by Couhert *et al.* [66] several pilot analysis are conducted with torrefied woods in an entrained flow gasification reactor. It is concluded that the syngas produced has a better quality than the syngas produced by gasifying wood. Moreover, char from torrefied raw material is less reactive with steam. Also as example, the work by Deng *et al.* [67] evaluates the performance of torrefied agricultural residues in co-gasification with coal. They prove that torrefied biomass can be grinded to lower diameters, being for all its properties more similar to coal than untreated biomass.

- *Pelletisation.* Mass densification and homogenisation, implying a higher bulk density, easing the handling, transport and storage of the biomass as fuel due to the uniform size, the high density and the low MC of the pellet [68]. This pre-treatment also limits biological degradation; nevertheless some drawbacks are moisture uptake and mechanical resistance *versus* crushing and dust formation. A pellet is a cylinder of 6–8 mm diameter, and pelletisation implies drying, milling, conditioning, shaping and cooling. Usually, lignin acts as binding agent, not being necessary to add any external additive. Nowadays, pellets market is growing, being the pellet the usual standard biomass shape used to commercialise biomass as fuel. The most usual raw material to be transformed is wood [69]. Pelletisation of torrefied biomass is described in [59,68], calling the product TOP pellets. The combination of both processes can overcome the main pellet and torrefied biomass disadvantages: biomass is completely dry after torrefaction, its humidity uptake is limited and biological degradation is practically completely inhibited; on the other hand, torrefied biomass has a relatively low energy density. Even more, the storage of TOP pellets can be simplified. As a result, the TOP pellets production process consumes less energy than conventional pelletisation. Torrefaction gas can be used for raw material drying at the beginning of the process. In contrast to conventional pellets, the TOP pellets can be produced from a wide variety of feedstock (sawdust, willow, larch, grass, demolition wood, straw) yielding similar physical properties.

- *Fast pyrolysis.* Pyrolysis can be defined as the thermal decomposition of biomass in the absence of oxygen, in a range of temperatures of 400–800 °C. It produces gas, liquid and char, with variable proportions according to the pyrolysis method, the biomass type and the reaction parameters [59]. The pyrolysis methods comprise slow pyrolysis and fast pyrolysis. This last takes into account a high heating rate, taking place at 450–550 °C. This option allows obtaining a liquid product called pyrolysis oil or bio-oil, consisting of 70% oxygenated organics and 30% water (on a mass basis) [68]. The proportion of water can cause corrosion problems. The pyrolysis can provide a cheaper transport and handling, due to the liquid state of the bio-oil. This oil can be used as a transport fuel, even directly in a diesel engine. The works by Wu *et al.* [70] and Abdullah *et al.* [71] use bioslurry from mallee biomass as fuel to be transported. The bioslurry is formed by combining bio-oil and biochar (which is the solid product that results from the fast pyrolysis) into the bio-oil: the biochar is milled into fine particles (due to its favourable grindability) and suspended into the bio-oil. In that way, the LHV of the bioslurry profits energy concentration of char, enhancing the efficiency of the process. The works by Uslu *et al.* [59] and Magalhaes *et al.* [72] evaluate the use of pre-treatment technologies in a bio-based SC. In the first paper, those are seen as alternatives to promote international trade: among the considered options (which are the same than the technologies described in this section), TOP pellets are the preferred selection, while pyrolysis has as main drawback from the economic point of view. In the second one, three different pre-treatment technologies are evaluated to select the most profitable and the most environmental friendly option (by means of CO_2 emissions) in a biomass-to-liquid SC. The final numbers favour the case study with rotating cone reactor to perform a fast pyrolysis, *vs.* fast pyrolysis in a fluidised bed, torrefaction, and torrefaction combined with pelletisation. Large scale *vs.* small scale plants issue is also presented by means of different scenarios evaluation. It seems that transportation costs are not crucial. Efficiencies increase and cheaper biomasses would enhance the financial pattern.

- *Biomass storage*. Even if it is not a pre-treatment, it can change biomass properties, such as MC, LHV and dry matter content, mainly due to degradation (microbiological) processes. A critical parameter here is the temperature of the pile. In order to avoid as much as possible biomass degradation, biomass stored should be homogeneous and with low MC (usually under 20% on a mass basis) [73]. The work by Rentizelas *et al.* [74] points out that biomass waste can be a seasonal fuel. Therefore, storage is crucial to provide the adequate supply in each period. A bio-based SC should be a multi-source SC that pays special attention on transportation costs, embedding storage as a potential part of the process. The type of storage used depends on the type of biomass to be stored. Especially during summer or in tropical climates, open air storage is used to dry the biomass. Different possibilities exist: open or closed storage. For pre-treated biomass, usually a closed storage such as in silos or bunkers is used. Storage of liquids is done in tanks. Those last types have less, or non-influence on biomass characteristics.

The different pre-treatments imply changes on MC, dry matter, LHV and the bulk density of the biomass. All pre-treatments enhance transportation costs. Examples of enterprises that commercialise those state-of-the-art processes are for instance Dynamotive Inc. or BTG group for the fast pyrolysis technology, and Topell BV, for the torrefaction. Note that those technologies are still on their development or pilot phase. All the supply chain steps, except storage, need energy (electricity or liquid/solid fuel such as diesel or biomass itself) to be run.

2.3. Co-Combustion Process Characterisation

The paper by Damen and Faaij [75] describes the main energy loss sources due to biomass co-firing: increase of electricity internal use caused by the higher milling, drying, *etc.* needs, lower boiler efficiency and boiler de-rating as a consequence of the air consumption increase. In van Loo and Koppejan [76], it is commented that the impact of co-firing on the thermal efficiency of the boiler depends on the co-firing ratio and the MC of the biomass. Nonetheless, a range of 3% to 5% substitution has a very small effect on the efficiency. The studies by Perry and Rosillo-Calle [61] and Berndes *et al.* [56] point out that it is important to limit the biomass share due to problems of corrosion, slagging and fouling. The paper by Baxter [77] also mentions the formation of striated flows, fly ash utilisation and fuel conversion. These are the reasons why the co-firing rates are limited. According to Chiaramonti *et al.* [78], co-firing bio-oil does not have any major technical problem, being the economic issue its main drawback. A successful operation, after a proper boiler modification is reported for a 5% substitution of coal on thermal basis. Faaij [12] reports low co-firing rates, up to 10% of substitution on thermal basis, with no important consequences to the boiler. The work by Damen and Faaij [75] reports that the energy penalty is not significant when co-firing up to 7% of biomass on a weight basis. In the range 7% to 20% on a weight basis, the net energy penalty (that takes into account the impact of biomass transportation too) is reduced. Nevertheless, an energy efficiency loss of 3% should be assumed for the overall combustion plant. Different experiences in UK demonstrate that biomass can be effectively co-fired up to 20% in weight basis, even if some technical problems have been encountered (namely corrosion, and no space for biomass storage before burning) [61]. Nonetheless, due to the large range of biomass and coals types, every blend can have its own optimal characteristics. Thermogravimetric Analysis (TGA) is used to determine blends characteristics at a laboratory level [79,80].The paper by Gómez *et al.* [81] uses as a reference value, in coal and biomass co-firing plants in Spain, a 10% on thermal basis, since higher fractions can decrease the boiler efficiency and cause corrosion problems. It also states that less than a 5% on thermal basis does not imply a valuable change. The affordable shares according to Berndes *et al.* [56] are, for mid-term solutions and no technical penalisations, 15% for fluidised bed boilers and 10% for PC boilers and grate-fired boilers, on energy basis.

Usually, a conventional coal-fired power plant can have an energy efficiency on a LHV basis between 30% and 45% [78], depending on the technology and the antiquity of the plant. The study by Van Den Broek *et al.* [82] reports a range from 39% to 44% for a co-firing PC power plant, remarking the fact that only large scale plants can be counted in this range (up from 100 MW). Van Den Broek *et al.* [82] show the difference in efficiency terms related to the biomass combusted alone in small scale plants and biomass co-combusted in larger plants. These values differ in seven points, from 30% to 37%, respectively. The value of the efficiency reduction in a conventional power plant is reported by a quadratic equation that depends on the biomass percentage in the blend in the study by Tillman [60]. This value is relatively small (up to 1.9 points) in the range 5% to 20% on a mass basis. The work by Gómez *et al.* [58] uses a value of 38% as co-firing efficiency.

The energy efficiency in the work by [83] increases. It reports an experimental work that puts into relevance the synergetic effect between coal and biomass. Not only the efficiency is enhanced, but also NO_x, SO_x and CO_2 are reduced. Similar results related to NOx emissions and CO_2 emissions are reported by Kalisz *et al.* [84] and Munir *et al.* [85], respectively. In the conclusions by Damen and Faaij [75] and Perry and Rosillo-Calle [61] it is manifested the fact that the emissions from production, conversion and transport of biomass, as well as land use change and displacement, should be taken into account. The final balance shows net avoided GHG emissions.

2.4. CCS

Several works can be found in the field of CCS applied to power plants. Desideri and Paolucci [86] is one of the first works developed in the carbon capture topic concerning modelling. They reproduce, in Aspen Plus, a carbon capture technology in post-combustion configuration for conventional power plants. Their approach contemplates an exhaustive description of the system, model validation with literature data, whole plant performance evaluation and cost analysis. The developed approach allows for optimisation when changing input conditions. It is concluded that 90% CO_2 emissions can be reduced using this methodology, but capital costs are significant and penalise the final cost of electricity (COE). The work by Hamelinck and Faaij [87] is based on biomass gasification for methanol, hydrogen and electricity production. This last is produced taking advantage of the remaining gases after methanol or hydrogen production units. Those products have a relatively low LHV if compared with fossil fuels, but they offer the possibility of being self-sustained in electricity consumption through the proposed configuration. The considered process involves pre-treatment, gasification, gas cleaning, reforming of higher hydrocarbons, a shift step to obtain a proper H_2/CO ratio and the final gas separation for H_2 production or methanol synthesis and purification. The software used is again Aspen Plus. The main purpose of the work is to identify biomass to methanol and H_2 conversion key points that may drive to higher efficiencies at lower costs. The study by Kanniche and Bouallou [88] investigates an IGCC power plant with CCS technology in pre-combustion configuration, fuelled with coal. The authors perform an evaluation of scenarios considering different physical and chemical solvents, contrasting them by means of technical and economic parameters. Aspen Plus is again the chosen simulation tool. They aim at being as much conservative as possible, then avoiding big modifications to an already existing IGCC power plant. Consequently, the existing operating conditions without CCS technology are conserved as much as possible. The work carried out demonstrates that physical processes, concretely Selexol and Rectisol, and activated amines have lower thermal consumption (mainly in the desorption column) than other options. Capturing CO_2 leads to 24% efficiency reduction, penalising the power produced. Therefore, CCS technology should be included carefully integrated in the already existing power plant.

The article by Descamps *et al.* [89] describes a Rectisol process (with methanol as solvent) for CO_2 abatement in a pre-combustion configuration for an IGCC power plant. Before the absorption process, a CO_2 removal process should be placed. In this case, this process counts with three WGS reactors to obtain a high CO conversion rate. The necessary steam is obtained from the integration with the CC. The performed sensitivity analyses demonstrate that CO conversion depends on the amount of used water, concretely in a way that the H_2O/CO ratio of 1 in the first reactor optimises the conversion. The final conversion achieved is around 92% on a molar basis. The CO_2 absorption rate varies between 77% and 98% on a molar basis. Higher rates imply a slight increase of GT power production and a slight decrease of ST power production. The work by Chen and Rubin [90] develops an integrated platform to evaluate CCS costs and performance for IGCC power plants. Their base case considers a Selexol system for CO_2 separation. All the rest of units that constitute the plant are based on commercial components. The WGS step has two stages (one for syngas steam consumption and the other one for external steam supply), and the Selexol unit includes two stages, one for sulphur and another one for carbon removal. It is observed that a redesign of the heat integration system of the plant should be done because of the addition of new units. A probabilistic uncertainty analysis is also performed and shows that most of the uncertainty in costs estimation comes from the plant itself rather than from the carbon capture system. Design optimisation is studied by Biagini *et al.* [91]. The authors in this case consider different biomass conversion processes to produce H_2: gasification and combustion, with pre and post-combustion configurations, at small scale. Sensitivity analyses are performed taking into account the most influencing parameters: the amount of air and steam added to the gasifier and the MC of the biomass.

CCS is applicable to point emission sources [92]. CO_2 is considered one of the most important GHG. Many current industrial processes, not only for energy production, release CO_2; *i.e.*, refineries, iron and steel industries, oil and gas extraction, cement production, paper and mills, *etc.* Moreover, virtually, all industries produce (directly or indirectly) CO_2 emissions, mainly due to their electricity consumption. CCS aim at liquefying the CO_2 stream before its release to the environment, and transport it to a final geological storage. In order to implement such a solution, it is necessary to have an integrated approach considering the whole supply chain. It means that a CCS process in a factory, performed with existing and well proved technology in the field of gas purification, should be directly linked with the localisation of a possible geological reservoir or used for the captured CO_2. It also has to be considered the different CO_2 transportation network possibilities, by pipelines or by boats (similar to the ones used in natural gas). For combustion power plants, the implementation of a post-combustion carbon capture technology penalises the global efficiency of the plant in around 7%, calculated based on a LHV basis [93].

Besides the technical and logistic aspects, also the public acceptance is important to be considered together with the requirements on legal developments which altogether have a key role to implement CCS as a part of the climate change solution. For example, to decide the obligatory nature of the CO_2 capture measure and the purity of the CO_2 to be injected, the subject has to be extensively discussed and assessed from legal and technological points of view [93].

There exist three types of carbon capture techniques: oxy-fuel combustion, pre-combustion and post-combustion. The first one can be applied in combustion plants, where the reaction takes place with pure oxygen instead of air. Therefore, the CO_2 from combustion is almost pure and easy to separate. This option is beyond the scope of the superstructure described before, so it will not be further described. Post-combustion can be installed in combustion and gasification plants, and separates the CO_2 from the flue gas resulting from the combustion. Pre-combustion can be installed in gasification plants, and separates CO_2 before syngas combustion (before the gas turbine—GT).

The partial pressure of the CO_2 in the gas mixture is a key parameter that represents the CO_2 concentration and that is directly related to the CO_2 capture efficiency. The higher the CO_2 partial pressure is, the easier to separate the CO_2. In gasification power plants, the difference between syngas CO_2 partial pressure before and after the GT is complex to assess. In general, higher partial pressures are found before the GT. This is due to the fact that even though CO_2 is generated during the combustion step, the flue gas is diluted with N_2 from the air. Moreover, the flue gas is expanded due to the inherent turbine expansion. Therefore, a noticeable difference exists between pre and post-combustion carbon capture techniques [94].

Usually, chemical solvent processes are used for CO_2 partial pressures below 15 bar. Then, physical solvent processes are applicable to gas streams which have higher CO_2 partial pressure and/or a high total pressure. Post-combustion techniques are represented mainly by chemical absorption, in which amines play an important role. The outlet CO_2 stream is treated, compressed and liquefied to be prepared for the transport to its final disposal location. As shown in Figure 6, this is done after the GT combustion, thus, in the case of gasification, after syngas production and use, the flue gas is processed. Figure 7 is further described in Section 3.2.2.

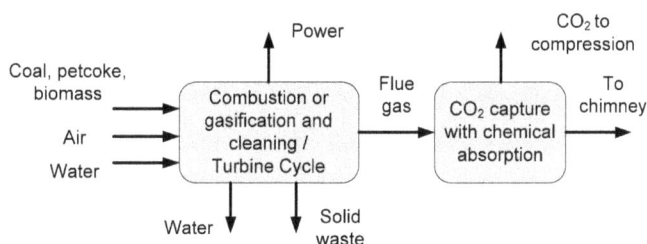

Figure 6. Post-combustion carbon capture configuration.

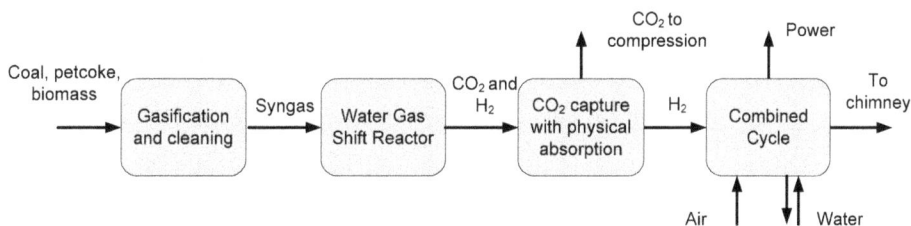

Figure 7. Pre-combustion carbon capture configuration.

3. Co-Gasification of Biomass and Coal

The product resulting from gasification is the synthesis gas, called syngas, and it is a mixture of mainly H_2 and CO, with different proportions of H_2O and CO_2. Usually, the term producer gas is used to describe a syngas with H_2, CO and CH_4, coming from a low temperature gasification. Typically, low temperature gasification uses air as gasifying agent [95]. Thus, producer gas normally has an important fraction of N_2. Flexibility is one of the main characteristics of syngas, since it is not restricted to a single source of fuel; it can be obtained from natural gas, coal, petroleum refinery fractions, biomass and organic wastes. Traditionally, natural gas and petroleum fractions have been the largest syngas sources worldwide, due to the trade-off between costs and availability; however, because of global economic, energetic and environmental contexts, coal and biomass are of growing interest and use. Moreover, syngas is the worldwide most used source of H_2 and CO productions. The proportion of H_2/CO depends on the source and on the syngas generation process and the performing parameters [96]. Two main routes are currently available for syngas generation, both of them traditionally and highly used for H_2 generation from fossil fuels, specifically from natural gas and coal. Syngas from natural gas mainly refers to partial oxidation with oxygen, oxidation with steam or oxidation with steam and oxygen; being the principal steam reforming. Syngas from coal involves gasification. Syngas is referred to as a medium energy gas, ranging from 4 to 18 MJ/m^3 of calorific value, depending on the gasifying agent [35].

Gasification can be defined as a partial combustion of an organic matter; producing as a result a combustible gas. The usual gasification process refers to solid organic matter as feedstock; where gas-solid and gas-phase reactions take place. On the other hand; several applications demonstrate that the concept of gasification is also applicable to liquid and gas feedstock's; being referred to as a "partial oxidation" [97]. Gasification takes place into three main types of reactors that differ between them in the type of bed. They are moving or fixed bed; fluidised bed and entrained bed gasifiers.

The most relevant gasification aspects for process design are (i) type of reactor and (ii) feedstock characteristics. A general gasification picture is given in U.S. Department of Energy's National Energy Technology Laboratory (NETL) [98], which provides gasification data for the year 2007, revealing that the global marketplace has coal as dominant feedstock, and that Sasol Lurgi, General Electric energy and Shell are the main gasifier providers. It is important to mention that China is developing its own technology, as the current project Tianjin IGCC power plant exemplifies. The preferred products from gasification are mainly chemicals (such as fertilisers). They are followed by Fischer-Tropsch (FT) liquids, power and gaseous fuels. Find in Section 3.2 a summary of the different products that can be generated from syngas. Efficiency and CO_2 emissions favour the use of gasification *versus* the use of combustion to take advantage of the heating value of a solid combustible. The main benefit of the former method if compared to the latter is the production of a versatile gas. Focusing on the production of electricity, the syngas allows its use in a combined cycle (also high efficient gas turbines) while a solid combustion can be only used to produce electricity by steam turbines. The efficiency difference can be of around 6 points.

3.1. Biomass and Gasification

Biomass gasification intrinsically produces tars. As defined in Milne *et al.* [99] "tars are the organics produced under thermal or partial oxidation regimes of any organic material and are generally assumed to be largely aromatics". Tars tolerance of gasifier downstream units is a matter of research. It is stated through the experience that tars constitute a problem when the syngas is not simply burnt in a combustor. They may condense before syngas usage: because of their carcinogenic effects they can cause health damage and generate environmental issues due to their disposal [100]. Tars avoidance counts with two methodologies. Firstly, tar formation reduction in the gasifier itself: primary methods include adequate selection of main operating parameters (pressure and temperature), the use of a catalyst and specific design modifications (shape, dimensions, *etc*). Secondly, tars removal from syngas: secondary methods entail hot gas cleaning downstream the gasifier by means of thermal or catalytic tar cracking, as well as wet scrubbing or mechanical methods such as cyclones and filters. The challenge of all the actual small gasification pilot plants that use biomass as a feedstock is to find an adequate gasifier design to produce a syngas free of tars, avoiding the syngas cleaning process before its final application, thus gaining compactness and saving in costs. Nevertheless, nowadays, the most used approach for tars avoiding is gasification with secondary methods.

The formation of carbonaceous materials (char, or particle fines) and that of heavy compounds (tars), as well as the inorganic release (in form of fly ashes or slag), are strictly correlated to the fuel structure and composition. Apparently, due to the low ashes melting point, an entrained bed gasifier looks very attractive to obtain a tar free syngas, with less oxidant consumption. Nevertheless, due to the aggressive behaviour of ashes, a non-slagging process is recommended (except if the biomass is mixed with high amounts of other feeds, such as coal or petcoke). Moreover, entrained bed gasifiers require of small particle diameter, however, there is no effective method for size reduction of fibrous biomass. Fixed beds, with no highly restrictive particles size, are extensively used for small scale gasification of biomass applied successfully in rural areas [101]. According to Mastellone *et al.* [102], among all gasification technologies applied, fluidised beds are the most promising one as a result of their operation flexibility for different oxidants (thus, for different fluidising agents), temperature and residence time ranges. They also allow for catalyst addition. According to Highman and van der Burght [97], low rank coals and biomass are more suitable for fluidised beds owing to their ashes reactivity. Nonetheless, biomass ashes have low melting point and in molten state have an aggressive behaviour with refractory material.

Tars formation and ashes reactivity are the main drawbacks in biomass gasification. The most extended bed for big scale application is the fluidised one, while the most extended bed for small scale is the fixed one. Entrained bed gasifiers are normally used in co-gasification [103].

3.2. Syngas Purification Units and Final Applications

Syngas is an intermediate product for further elaboration of a wide range of end-use products. The term polygeneration may refer to one gasification plant that makes different products; when only two products are manufactured, the term used is co-production. The concept of polygeneration and co-gasification is the essence of the biorefineries, which aim at mimicking the energy efficiency of oil refineries through the production of fuels, power and chemicals from biomass. An integrated biorefinery optimises the biomass use to produce biofuels, bioenergy and biomaterials; the approach includes knowledge from plant genetics, biochemistry, biotechnology, biomass chemistry, separation and process engineering [104].There are four types of biorefineries, being one of them the biosyngas-based refinery. The other types are pyrolysis, hydrothermal and fermentation based [105].

Final syngas application(s) downstream the gasification process, feedstock type and syngas generation conditions (mainly pressure, temperature and oxygen purity) decide the layout of the cleaning processes, which aim at meeting the needed conditions of cleanliness and temperature before the syngas usage [106]. Nevertheless, the train of purification units should work optimally in a wide range of syngas compositions (H_2/CO ratio, sulphur, nitrogen, chlorine and phosphorous) and operating conditions, as derived from the variability in the feedstock [97]. Analogously to the tar removal methods, syngas cleaning units can be divided into two types according to the syngas generation process: during gasification (generally for solid removal) and after gasification (fluid pollutants removal), being called respectively primary and secondary cleaning methods. IGCC-CCS and BG-GE approaches have different needs of syngas cleanliness and temperature, also depending on the size of the system.

3.2.1. Syngas Cleaning

Syngas requirements before its final application mainly include temperature, pressure and *pollutants level conditioning*. Knoef [107] specifies that syngas cooling is required for combustion in gas engines, for filters having a maximum acceptable temperature and for an optimal syngas compression. The pressure level can be reached in the gasification reactor. In turn, gas purity, independently from the scale, ranges from pollutant levels of mg/m^3, passing through ppm, and reaching ppb: the syngas cleaning level is dictated by the flue gas emission requirements and the specific devices conditions to work properly and during long time. Wet and dry, hot and cold cleaning systems have been developed and implemented. The most efficient option in a gasification plant is to determine the pressure in the gasifier itself and try to maintain it until the syngas usage. High temperature can be used downstream heat exchanger integrated with the heat requirements of the plant. In general, final syngas uses require from specific H_2/CO ratios. Acid and basic pollutants should be removed. Gas purity and composition, selectivity and economic issues are of concern when choosing a cleaning method [97,106].

Syngas pollutants mainly include solids, tars, heavy metals, halogens, alkalines, acid and basic species. Some of them are released as by-products. In turn, CO_2 absorption has the purpose of concentration, where H_2 is the desired product. Heterogeneous and homogeneous mixtures require different cleaning methods. In the case of heterogeneous mixtures, *i.e.*, a solid-gas mixture, *mechanical separation methods* such as filtration or water scrubbing are applied to separate the different phases. In contrast, for homogeneous mixtures, *i.e.*, only the gas phase, *diffusion based separation processes* are suitable. Its aim is to convert a feed mixture into two or more products that differ in composition. The most widely used processes in syngas cleaning are *absorption* and *adsorption*. Physical and chemical (reactive) absorption are the type of separation process typically used for syngas purification, where a liquid solvent is used to selectively remove acid and basic species. The absorption process includes a regeneration step where the solvent is cleaned from pollutants and recycled to be used again in the absorber. In general they are formed by two columns (one for absorption and the other for desorption) and a set of a heat exchangers and pumps that transform the solvent back to the absorber conditions. Physical solvents are for example methanol and Dimethyl Ethers of Polyethylene Glycol (DMPEG) that work using common processes called

Rectisol and Selexol, respectively. Water-based chemical solvents are for instance the amines. The MDEA is the most widely used one due to its high selectivity [108]. Adsorption systems are normally formed by a solid bed that adsorbs the selected species. The bed has to be either periodically changed, or regenerated *in situ*. This adsorption- desorption process involves changes in temperature and pressure. For example, the Pressure Swing Adsorption (PSA) cycle operates at a constant T, and at high P for the adsorption, and at low P for desorption. This unit can be used for H_2 concentration and purification.

According to Sharma *et al.* [109]; a gas cleaning process can be operated at three temperature regimes as a consequence of the syngas final application in a gasification plant. Cold (less than 25 °C); warm (less than 300 °C) and hot cleaning (more than 300 °C). Comparatively; all the commercially available processes operate using cold and warm syngas. It means that for gasification plants where the syngas is obtained at high temperature; there exists a considerable loss of energetic and exergetic efficiencies. In addition to that; hot gas cleaning can lower operational costs when final syngas applications need high temperature (for instance H_2 production by steam reforming and WGS; or combined heat and power generation in a FC). The study by Pisa *et al.* [110] is focused on IGCC power plants alternative designs; in desulphurisation processes in particular. The authors evaluate a hot desulphurisation process with ferrite ($ZnFe_2SO_4$). This bed needs oxygen to convert H_2S on the one hand; and steam to provide the humidity for the optimal operation work; on the other. The final result shows that the high steam consumption penalises power production. Therefore; the steam consumption finally penalises the global efficiency of the plant. Absorption processes require temperatures around 200 °C. In contrast; adsorption processes require nearly ambient temperatures. The syngas cooling has several problems inherent to ashes presence; due to their slugging condition at certain temperature ranges. Figure 8 shows an outline of the main applications of syngas; and the different processes to synthesise it.

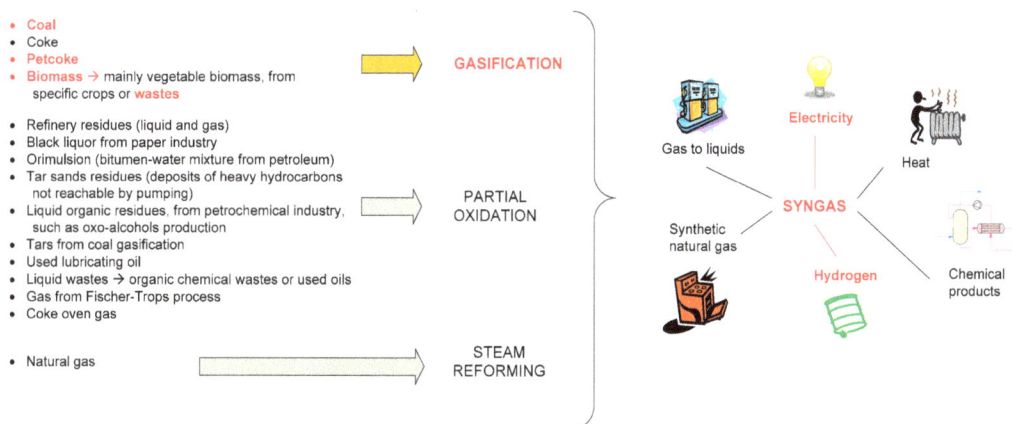

Figure 8. Syngas generation pathways and final products possibilities.

118

3.2.2. CCS

Pre-combustion installation in an IGCC plant, aims at obtaining H_2 as a product. As Figure 7 shows, it requires of a water-gas shift (WGS) reactor. The CO_2 produced is captured and then CO_2 and H_2 are separated. The relatively pure H_2 is then sent to the combined cycle to produce power. And analogously to the post-combustion scenario (Figure 6), CO_2 is sent to a compression system to be liquefied before its transport. A purer H_2 stream can be obtained through a PSA. As a consequence, the objective is to sell the H_2 on the market. This pre-combustion technique counts with a physical solvent that absorbs acid compounds. That is the reason why Huang *et al.* [111] evaluate the same absorption process for both, CO_2 and H_2S abatement by means of process intensification. It is concluded that sulphur penalises the WGS reactor performance.

In general, for oxygen blown gasifiers at high operating pressures and relatively high CO_2 concentrations, the predominant choice is a physical solvent absorption system. According to Metz *et al.* [94], the most extended technology to capture CO_2 before the GT combustion is the Selexol process. It uses dimethyl ether of polyethylene glycol (dimethyl ether of PEG, the key ingredient of Selexol) as solvent, achieving a CO_2 capture efficiency of more than 90%.The optimum pressure for H_2 purification is in the range of 15–30 bars. Finally, the H_2 concentration in the outlet stream of a modern PSA unit usually lies between 80% and 92%. The PSA process is based on the different adsorption behaviour of the molecules. There exists a gap between the extended knowledge of the mentioned processes and their integrated use in gasification or even in combustion plants.

Research in the field of CCS is still in the pre-design or pilot stage. As a consequence, very few full-scale experiences can be found. This is mainly due to the fact that the installation of such a process diminishes the overall efficiency of a power plant. Therefore, the implementation of a carbon capture process should principally obey to environmental reasons.

3.3. Biomass Gasification Conceptual Design

Generation and use of syngas, or producer gas, from biomass in centralised and distributed systems depends essentially on the characteristics of four major components: the percentage of gasified biomass, the type of gasifier, the specific final gas usage and the plant scale. Consequently, even if the raw material and the basics of gasification technology basically remain the same, the resulting plant design will be different in each particular case. Accordingly, the review has been organised around these main components. The conceptual design (also called "preliminary") links the different issues treated on this work. It is the phase between the "laboratory scale" research and the detailed engineering design of the final plant. To this end, the concept of *superstructure* is used. The superstructure built supports process system modelling, process system alternatives and process system optimisation. Then, mathematical programming has been chosen for the representation and optimisation of the whole underlying supply chain. Consequently, the referred papers in the following sections are mainly focused on these methodologies.

The following sections present Section 3 topics at various levels of detail, from the modelling of individual plant's units until the aggregate modelling of the whole integrated supply chain of a

CES. It is worth noting that the distinction between levels (CES and DES) tends to disappear when considering DES, where energy plants operate as "islands", being individually optimised and becoming eventually part of a grid. Consequently, in this latter case, the description is organised following the inverse path: from the most general level to the particular level, to finally identify the challenges in plant operation. See in Figure 9 the outline of Section 3.

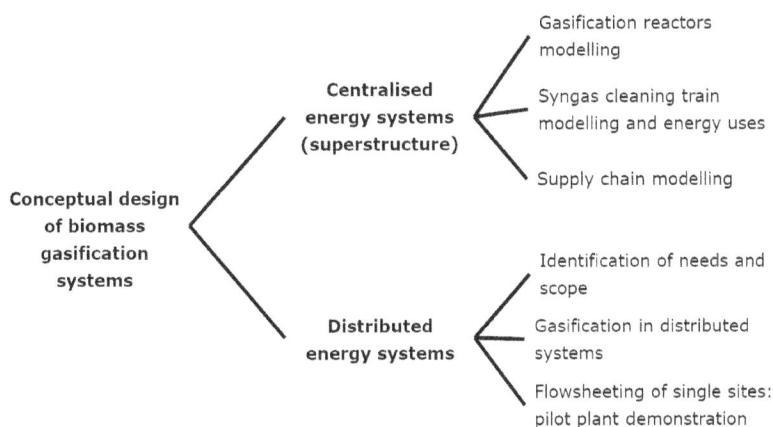

Figure 9. Scheme of the subjects developed in Sections 4 and 5.

3.4. Centralised Energy Systems

Large scale gasification systems normally use entrained or fluidised beds for the production of syngas. Increasing the share of biomass in the energy supply would be associated with the reduction of GHG emissions and the independence from imported and domestic fossil fuels. There exists an interest on the use of biomass and waste material as fuel, therefore, there is much effort devoted in enhancing their conditions for transport, handling and processing. Conceptual modelling should take into account the biomass properties to determine the feasibility in terms of efficiency and most appropriate mixtures of feedstocks and products mix.

Biomass use with coal in combustion and gasification offer five alternatives for biomass usage: combustion, co-combustion or co-firing, gasification, co-gasification and gasification for co-firing [12]. These two options have in common the range of power produced (hundreds of MW) and the profitability of already existing installations originally using 100% fossil fuels. Typical 100% biomass combustion plants are around 20–50 MW$_e$ and 100% biomass gasification plants are in the range of 10 MW$_e$ [55]. Co-firing and co-gasification permit the usage of local biomass sources, being of special interest in the organic wastes management area. CO_2, sulphur and nitrogen emissions reduction are direct benefits from the coal fraction substitution.

3.5. Distributed Energy Systems—Gasification of 100% Biomass

Gasification at small scale utilises fixed beds or fluidised bed gasifiers. Small scale systems are employed to meet the requirements of DES using locally available biomass at or near the point of use. The main characteristics of a DES are sustainability (thus, source sustainability and no need of grid support), high efficiency, demand accomplishment, the consumer implication and fossil fuels independence. There is no unique choice of using biomass for energy demand, but a solution to a specific case study comprises different ranges of scale and different technologies, depending on the available biomass. Rural areas and rural areas from developing countries in particular, require new approaches to optimisation, different from those that have been considered so far, as well as proven and reliable technology.

Rural electrification benefits from biomass residues closest to the treatment plant. The same SC can be depicted for both scales, except that transport is not the main bottleneck in DES. However, a different situation is found when considering the trade of biomass, since the excess of raw material, which is not consumed in the place of production, can be processed to be operated as a raw material for other processes. Moreover, the objectives considered for optimisation in small scale gasification in rural areas, are somewhat different from those considered in a large scale plant. The study by Silva and Nakata [112] remarks that one of the main reasons why renewable energy technologies in modular configuration have not been highly extended in rural areas is the lack of an integrated approach in rural electrification planning. Those integrated approaches should include economic, environmental and social criteria, according to each specific case study context. The paper is focused on a specific case study situated in a remote area in Colombia, evaluating two possible energy access options: electrification and electrification with traditional fuel substitution (cooking purposes), comparing this commitment for diesel and for renewable units. The paper uses goal programming to assess a qualitative response in terms of electricity generation cost ($/kWh), employment generation (jobs/kWh), land use (m^2/kWh per year) in terms of interference with land use for agriculture or habitat conservation due to the plant extension and the needed place for storage, and avoided emissions (kgCO$_2$/kWh). In a previous work from the same authors [113], they use linear programming (LP) to deal with the energy planning model. The considered case study is the same rural region from Colombia. The aim of the authors is to demonstrate that such a rural electrification projects can be financially sustainable, if taken into account the appropriate data concerning reliable geographical location of sources and clients, income levels and energy demand. The mathematical problem deals with an objective function based on the minimisation of subsidised costs. The share of possible technologies takes into account electricity generation with diesel engines, biomass boilers, gasification-gas engines and fast-pyrolysis matched with diesel engines. As a result, the technology that minimises costs is the combustion of biomass. The main drawback found is that at the moment, the performance advantages of gasification and pyrolysis are penalised by the high investment. It leads to a most important conclusion: the proliferation of advanced techniques to take profit from biomass will come with environmental policies that should motivate the implementation of more environmental friendly systems. Kanase-Patil *et al.* [114] also use LP formulation to ensure a reliable integrated renewable energy system, by evaluating COE and costumer interruption costs, and expected energy not supplied. The renewable share of technologies takes into account biomass, solar, hydrological and wind speed. Then, four scenarios are considered to meet with the energy demands in the areas of domestic, agricultural, community and rural industries of an specific area in India, based on combinations of the abovementioned sources. LINGO and HOMER software, which are specific tools for renewable energy mix determination, are used to verify the results. Finally, the system that combines micro-hydrological power, biomass gasification, biogas production, wind and solar photovoltaic is the best one in terms of reliability and cost.

The work by Kanagawa and Nakata [13] is also focused on India, and aims at finding quantitative relations between social and economic development. In this direction the authors evaluate the literacy rate *versus* the electrification rate. In this sense, the paper by Hiremath *et al.* [115] takes into account a high number of state-of-the-art evaluation parameters used for decentralised energy planning. The authors compare goal programming *versus* LP concluding that the first one is the chosen method based on the level of subjectivity. The selected objective functions are cost, system efficiency, petroleum products usage, locally available resources, employment generation, emissions (CO_2, NO_x and SO_x) and reliability on renewable energy systems, subjected to demand and supply constraints. Finally, the results demonstrate that biomass-based systems have the potential to meet with the rural needs, having reliability, promoting local participation, local control and creation of skills. Cherni *et al.* [116] and Brent and Kruger [117] develop, describe and use a multi-criteria decision tool called SURE, that aims at choosing the appropriate energy set of technologies to match the energy demand of a rural area while reducing poverty. The tool combines quantitative and qualitative parameters, and allows for changes on the priorities according to the user criterion. The model analyses the strengths and weaknesses of a community according to five resources: physical, financial, natural, social and human. Then, it tries to find compromise solutions in terms of energy. Behind the software, a local survey should be drawn to state the baseline of a rural community in Colombia, in order to identify the energy needs and the growing tendencies. In Brent and Kruger [117], the authors use experienced individuals in the field of energy and poverty to assess a Delphi research methodology. SURE and the tool developed by the Intermediate Technology Development Group (ITDG) [118] are integrated, and compared with the results from the experts panel. It is put into relevance the fact that technology assessment methods should be further developed to formulate more appropriate implementation strategies. Finally, the paper by Ferrer-Martí *et al.* [119] is an example of a renewable energy source implementation problem, wind, which uses MILP to assess the optimal location of wind generators and the extension of the micro-grid in a specific community from Perú, while minimising the initial investment.

Janssen *et al.* [18] promote the use of African land to produce bioenergy, in a sustainable way. It is stated that it is unfavourable to limit the bioenergy development of Africa, since the country has an important extension of marginal and degraded land that can be suitable for a socio-economic development based on biomass. The study assesses the suitable areas for bioenergy: all regions used for food and with severe water, terrain and soil constraints are not included. Therefore, this use of land should be developed by the appropriate formulation of policies and development plans. Those political issues should deal with rural development, sustainable production, community participation in the projects, modernisation of agricultural policies, creation of standards to guide and facilitate the bioenergy market, avoid fuel-food conflicts and ensure both, food security and bioenergy development. Hamimu [120] is another work that promotes biomass trade from biomass waste from Sub-Saharan countries. Biomass should be used not only for exportation, but also for consumption in the countries themselves, to assure their independence from fossil fuels. This work pays special attention on land tenure: in some countries in Africa, lands cannot be a property of the farmers. Governments should avoid speculation with land. On the contrary, the positive paradigm will count with the partnership between local farmers and foreign investors. To end, the work by Otto [16] distinguishes between the two markets mentioned in the previous paper: biofuels production for exportation and biofuels production for local use (advanced uses of biomass). The emergent business models in the sector, should deal with the link of the two markets.

Overall, LP and goal programming methods do not take into account the allocation problem. Therefore, only the balance between source and demand should be taken into account. Nevertheless, new trends such as biomass sharing between communities and bioenergy trade need to consider the allocation problem. It is observed that there is a lack of systematic energy models that promote international trade; biomass should be promoted in developing areas for exportation and for local use. Moreover, there is also a lack of energy models for rural development that take into account economic, environmental and social issues of the communities.

Gasification at Small Scale

Gasification at small scale is placed in the range of less than 10 MW_{th} and less than 200 kW_e (see Section 1.1.1). Even if it is not a "new" process, research is still needed due to the low commercialisation level achieved by small gasifiers. The first experiences with gasification are from XVIII century in England and France, where coal gasification was used to light the city. Later, at the beginning of the XIX century, "gasworks" using mainly coal and coke, were employed to produce gas for lighting and cooking in some American countries. Then, during the two World Wars, this technology was further used for fuel supply in transport vehicles. At this time, wood gasifiers were used as mobile sources of gas to power cars. Finally, cheap prices of fossil fuels determined the end of a high extended use of gasification [97,101]. During the nineties, small scale biomass gasification was again encouraged by the new restrictive environmental laws and the pressure to be independent from fossil fuels. Nevertheless, small scale gasification has been characterised by a discontinuous technology development, changeable government interests and a pioneering role of research associations and non- governmental organisations (NGO's). Concerning technical aspects, there has been a low deployment of research results but at the same time a progressive development exists guided by the demand, especially on quality producer gas. Investment costs in general are still high [107].

The producer gas generated in a gasifier can be used in one of the applications shown in Figure 10, sorted from the smallest to the largest scale in power terms. Small scale covers till the engine alternative, including the boiler only for heat production. They offer the possibility to produce electricity or the combination of electricity and heat in the same installation, being called co-generation or waste heat profit. More or less restrictive producer gas quality depends on its final application. The less restrictive is the boiler option, while FC's are the most special alternative. According to Lapuerta et al. [121], gasification-gas engine presents more benefit than gasification-GT due to a higher efficiency in terms of electricity generation but also due to the possibility of heat profit for thermal applications.

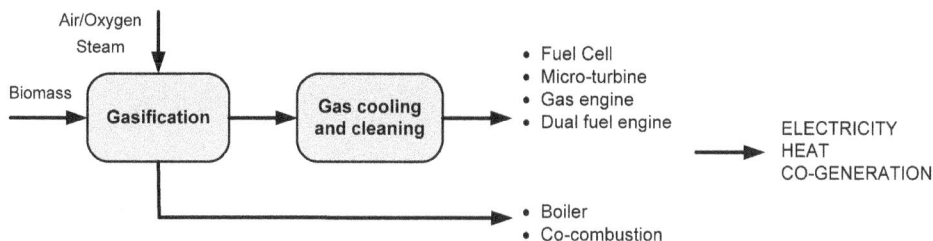

Figure 10. Most extended uses of producer gas from small scale gasification. Based on Bridgwater *et al.* [27] and Karellas *et al.* [122].

The paper by Dornburg and Faaij [123] presents the duality large-small scale biomass gasification as competing alternatives, regarding the trade-off between transport cost, economies of scale and easiness in heat utilisation. From the studied technologies, that comprise heat, power and combined heat and power options through firing and gasification between 0.03–300 MW$_{th}$ input, it is concluded that the relative primary energy consumed improves with the scale, and that gasification is better in energetic performances terms than combustion. It is not the case of economic parameters, in which combustion is better. Husain *et al.* [124] puts this detail into relevance through a case study that reflects the extended practice offering residues. In Malaysia, they profit palm oil mills residues to produce heat and power by means of boiler-turbines installations. This is a clear example of local wastes used to generate inputs for the palm oil industry itself. The authors conclude that the installations have low thermal efficiencies due to the heterogeneity of the residues, as well as that more advanced technologies should be used.

The review by Dong *et al.* [125] states that co-generation alternatives at small scale are the major alternative to traditional systems in energy savings and environmental damage mitigation. Gasification combined with internal combustion (IC) engines, micro-turbines (GT), and/or fuel cells are among the emerging possibilities having higher efficiency than combustion-based cogeneration options. But research is still needed, since efficiencies should be improved. Moreover, fully automatic operated plants are needed at a minimum level of pollutants. The Indian perspective described in Buragohain *et al.* [126] is somehow showing a good picture of the new energy paradigm, in which the emerging country aims at supplying present and future thermal and electrical needs through decentralised generation, concretely through a big use of gasification at small scale, coupled with IC engines, boiler-steam turbines and in bigger scales with CCS. The economic feasibility of the gasification option is analysed in terms of its comparison with the diesel market. Also, the load factor of the plant is a crucial decision parameter to be considered since rural demand is very changeable during the day and small if compared with other contexts. Gasification is a valuable option because of its low expertise requirement and its social effects through jobs creation.

The most important barrier towards the commercial stage of small scale biomass gasifiers are still the high investment cost and the already small amount of expert people in the field. The increase in process efficiency does not seem enough to reach the combustion status. Even if it is not a fully commercial choice, it is possible to depict a wide range of successful and failed gasification case studies to produce power and/or heat.

3.6. Trends and Challenges

The greatest opportunities and challenges come from the not fully commercialised nature of IGCC-CCS systems and projects BG-GE, and the potential of biomass as a resource. The context of these biomass-based options is favourable due to the change of energy paradigm. However, the use of land for energy crops should be carefully evaluated to avoid further problems. In order to contrast strengths and weaknesses, decision tools are needed to evaluate the trade-off. Therefore, the following two sections are focused on the development of decision-making tools for the biomass use at large and small scale, in different contexts in a sustainable way. This is equally useful for biomass co-combustion in power plants, in order to depict a systematic and consistent approach for biomass projects.

4. Bio-Based Superstructure

A *bio-based superstructure*, can be defined as the workspace that facilitates the allocation of individual unit operations and their connectivity, defined as the ensemble of all feasible flowsheets, combinations of equipment, raw materials and products, using biomass as raw material. The main objective is to ease the evaluation of different process configurations to evaluate the trade-off between different criteria (KPI). Figure 11 represents the information workflow of a generic process analysis.

The different flowsheet configurations are evaluated: scenarios approach, or mathematical modelling, (see Bojarski *et al.* [52] for further detail). KPI values can be depicted in Pareto Frontiers for comparison and configurations selection or prioritisation. Our developed evaluation tool utilises Aspen Plus as process simulator and MS Excel to process the KPI values. Particularly, for the co-combustion case study, the superstructure concept is applied to the selection of the most suitable pre-treatments (see next section). In that case, no process simulation has been performed. However, the superstructure concept applied to co-combustion plants would include the evaluation of different flue gas cleaning units and/or carbon capture materials.

The purpose of R&D in the IGCC power plants field is to improve the environmental performance, decrease marginal costs and investment and assure the technology availability/reliability. The idea that IGCC power plants are an opportunity is supported by the fact that nowadays, there are a lot of new projects envisaged around the world, mainly based on coal and located (in order of starting projects) in USA, Canada, China, and Europe. The report by Metz *et al.* [94] shows that Shell, Texaco and E-gas are demonstrating the real and practical interest of the concept. The main used technology is the Selexol capture system in pre-combustion configuration. New IGCC power plants with CO_2 capture technologies are included in the superstructure developed by the authors. Several works can be cited that measure the global performance of large scale gasification plants [127].

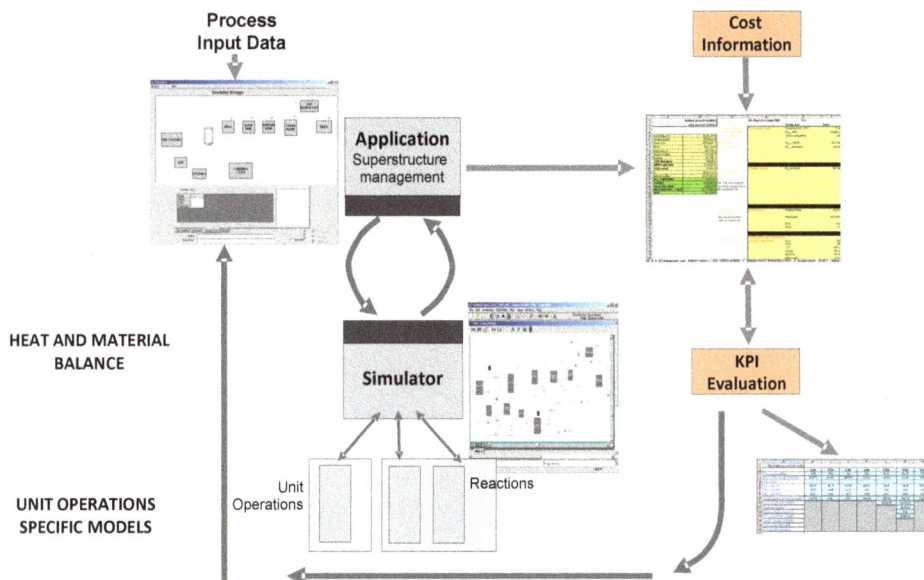

Figure 11. Flowsheet analysis workflow.

The work by Hamelinck and Faaij [87] evaluates technical and economic parameters of gasification plants to produce methanol and hydrogen, taking into account future prospects. Even if they have not developed a superstructure as understood in this work, they also use an Aspen Plus simulation to obtain energy and mass balances of interest for the economic evaluation. When large-scale production is of concern, biomass supply is an important item in operation costs when long distances should be covered. Hydrogen and methanol should be considered as conventional fuels alternatives; nevertheless the main bottleneck lies on the distribution infrastructure, mainly for hydrogen delivery. The work by Chiesa et al. [128] considers the production of hydrogen and electricity from coal; the authors evaluate different scenarios, considering CO_2 venting or CO_2 capture; electricity production with conventional gas turbines, with turbines for burning syngas and H_2, and with steam cycle (thus, pure H_2) as final syngas usages. Process intensification of acid species is also included by removing CO_2 and sulphur acid species in the same unit operation. They propose different analyses considering performance and emissions using simulation of real commercial units. In their economic analysis; performed by Kreutz et al. [129], it is interesting to appreciate that one of the barriers found for a wide H_2 economy is the lack of a cost effective method of storage and the lack of a large interested market on it. Also the CO_2 storage capacity and CO_2 transportation have to be addressed in an efficient way to promote such a solution.

The specific issue of CCS in different plant types is tackled by Rubin et al. [130]. Natural gas combined cycle plant (NGCC), IGCC plant and PC plant are considered. It takes into account different possibilities of final transport and storage of CO_2: geologic, saline storage and enhanced oil recovery (EOR). They found, while comparing coal gasification and combustion with CCS, that costs are very sensitive to the coal quality. Moreover, depending on coal quality, PC plants or IGCC plants are the cheapest options among the three possibilities considered here, being IGCC plants the most penalised by the extra energy consumption from the CCS system. The most relevant contribution by Chen and Rubin [90] is the consideration of uncertainty in the cost of CCS in an IGCC power plant by taking into account coal quality and CCS removal efficiency.

The complete IGCC-CCS superstructure is shown in Figure 12. The diagram assembles all the technical possibilities that an IGCC plant offers. The options considered in our work are in red. Among all the options that a general IGCC plant offers to be optimised, the dashed lines in red indicate the design choices that are taken into account. Raw materials can be from different origins. Pre-treatment options include energy and matter densifications. Feedstock mixture and final syngas usage elections are carried out with MCDA. Note that in Aspen Plus we use stream splitters and mixers to perform the choice of different unit operations executing the same function in the process (see [52,131,132]).

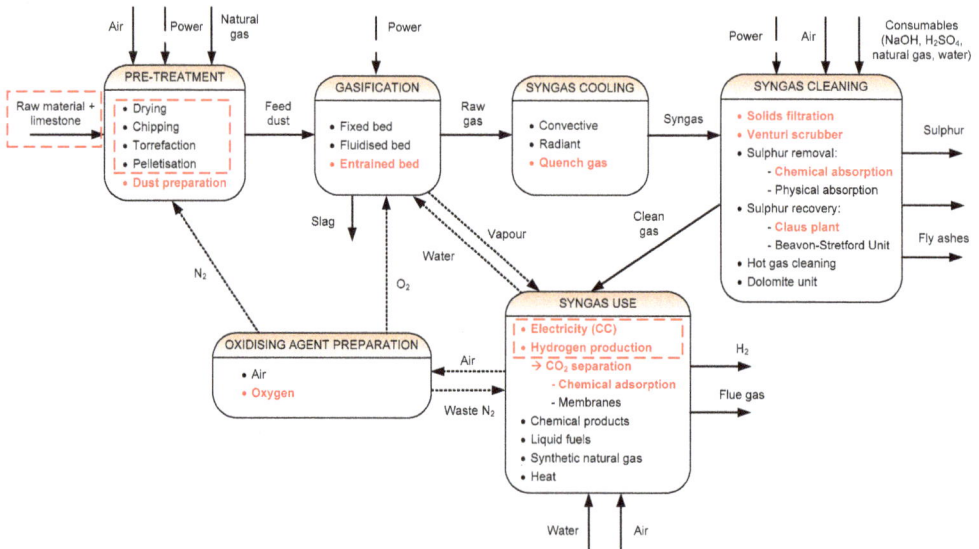

Figure 12. IGCC-CCS superstructure. Dashed lines in red indicate the superstructure options considered. The modelled flowsheet in the process modeller, among the different unit's alternatives, is highlighted in red (for more detail see [131,133]).

Hydrogen separated from the syngas may be used in different ways: (i) sold as a product; (ii) converted in fuel cells (if purified until their standards); or (iii) burnt in a gas turbine, as happens with syngas. Figure 13 depicts the superstructure implemented by the authors (see Bojarski *et al.* [52]) to evaluate these possibilities. Concerning the splitting units used to model the superstructure, separation factors will allow the distribution of total or partial rates among the different technological options. Firstly, the choice whether combined cycle or H_2 needs to be done. Then, the purity of the H_2 in order to be sent to the turbine or to be sold to the market (*i.e.*, the use of PSA), is the variable to select. Co-generation of power and H_2 is one of the possible choices in the superstructure.

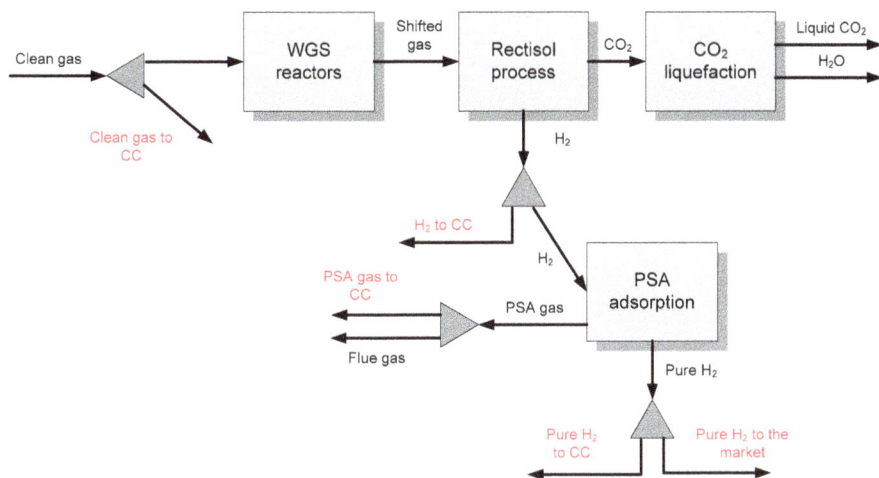

Figure 13. CO_2 capture and H_2 production process superstructure.

5. Bio-Based Supply Chain Modelling

It is recognised that in order to achieve the posed targets in the consumption of renewable energy (see Section 1): (i) efficient networks to sustainable supply the amounts of biomass required; (ii) cost effective technologies to convert biomass and (iii) improved distribution infrastructures to deliver the final product (*i.e.*, energy or fuel) are to be developed [134]. Moreover, the efficient integration of these three elements is equally relevant to achieve these targets. In this context, a supply chain modelling approach can be exploited as a tool that can support decision making towards accomplishing such integration.

The concept of supply chain (SC) refers to the network of interdependent entities (*i.e.*, processing sites, distributors, transporters, warehouses and raw material suppliers) which is the processing and distribution channels of a product from the origin of its raw materials to the final delivery to the customer. Then, supply chain management (SCM) can be defined as the management of material, information and financial flows through a SC that aims at producing and delivering goods or services to consumers [135]. Notice that a SC is comprised by components that may be geographically distributed. One of the main objectives of SCM is to synchronise and coordinate the flows of materials that go through the different processes so that the final product is delivered in the most efficient manner. This is especially important for biomass to energy projects which are highly geographically dependent and whose profitability can be strongly influenced by the location of the different processes and biomass sources. Commonly, biomass production and transportation account for a significant part of the whole bioenergy supply chain cost [136]. Therefore, a tool capable of evaluating the possible trade-offs between the different feedstock sources, each one with specific properties (*i.e.*, humidity and energy density) and the location of processing sites and consumption points is a requisite to develop efficient bioenergy networks.

Typically, a Biomass SC problem considers the possible use of multiple biomass sources from different origins that are geographically distributed, and the subsequent pre-treatment required to homogenise the material in mass and energy terms. These features imply the combination of different moisture contents (MC), dry matters (DM), lower heating values (LHV) and bulk densities (BD). Biomass, with high MC, low BD, low LHV and fibrous nature, may lead to biomass pre-treatment so as to optimise its transport, handling and treatment. Biomass properties can change along the SC.

In general, the major steps that a Biomass SC superstructure may include are (see Figure 14):

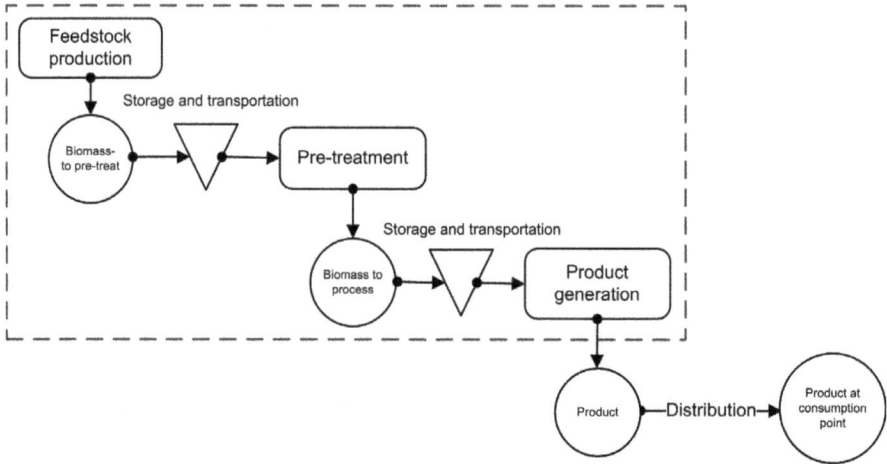

Figure 14. Schematic of a generic Biomass supply chain superstructure.

- *Biomass growing, harvesting and collecting* involve biomass production, by recovering biomass waste or using energy crops. Processes included here are drying, *i.e.*, natural drying in the land field, baling or chipping. Resource seasonality determines the harvesting or collection period. Different seasonal sources mix and storage could mitigate the impact on supply continuity.

- *Biomass pre-treatment* includes all the necessary steps to produce an upgraded fuel. Such fuel is homogeneous, has no impurities and is denser in terms of mass and energy. Pre-treatment allows costs reduction in biomass treatment, storage and transportation. Briquetting, pelletisation, torrefaction, pyrolysis, and combinations of them, are pre-treatment techniques [59]. Research in biomass pre-treatment is crucial for the large scale biomass market.

- *Storage* can be considered throughout the Biomass SC. Biomass can be stored whether before the transportation stage, at the biomass origin, in an intermediate step or at the power station site. This is crucial to create backups against seasonality. A drying phenomenon takes place here, even some dry matter loss may occur [74]. Costs depend on the location and type of storage, *i.e.*, open air, roof covered, air fan, indoor storage. Their selection depends on climate conditions, shape and volume of biomass and time of storage [23].

- *Distribution* can be expensive due to the biomass low energy density. Costs depend on distance, speed, tortuosity, haulier's capacity and amount of biomass to be transported.

- *Biomass treatment* refers to the biomass processing plant to produce the desired product, *i.e.*, biofuels, bioproducts, heat, cool or electricity.

Next, we review some relevant works that have dealt with the Biomass SC problem from a quantitative perspective. The study by van Belle *et al.* [137] presents a qualitative and quantitative study of this first Biomass SC step, considering chipping, central storage of the wood residues and intra-land transport. The harvesting or collection period depends on the seasonality, thus facing with an amount of fuel that can be discontinuous during the year. Hamelinck *et al.* [138] considers an international bioenergy SC taking into account the fact that biomass production and consumption do not need to be in the same region. Biomass compacted to briquettes or pellets can be used to save in distribution costs since they present a higher density. Nevertheless, there exists a trade-off between the distribution cost, the distance and the densification methods. This is important to consider especially when long distances should be covered. They demonstrate that an international bioenergy trade has real potential; however, improvements in terms of prices, policies and social acceptance are necessary. Panichelli and Gnansounou [136] contemplate forest wood residues (FWR) from final cuttings to produce torrified wood that supplies a gasification unit in order to produce electricity. They are able to allocate biomass quantities between predefined combinations of candidate sites to find the best set of locations for the energy units. They fix values for the torrefaction and gasification units, and take into account cost minimisation. A mixed integer linear program (MILP) that determines the optimal sizes and locations of biomass-based methanol plants (biofuel plants) is developed by Leduc *et al.* [139]. The objective function to be optimised is the operating costs and the investment required to establish the Biomass SC. The supply is given by poplar coppice, as energy crop, and the demand is based on gasoline-methanol car blend use. The possible consumer sites are the already existing gas stations in Austria, while methanol production is considered through the use of gasification plants. By-production of heat is also considered as economic revenue, and CO_2 emissions are accounted finally, but not introduced in their model as an environmental objective to be accomplished. The work of Rentizelas *et al.* [74] emphasises the multi-biomass seasonal availability and combines this fact with the biomass storage problem. The stages considered before the conversion plant of the raw material include harvesting and collection, in field handling and transport, storage, loading and unloading, transport, and biomass pre-treatment. This last stage can be included in any of the abovementioned stages, and could precede the transportation stage. Storage can be also located in the biomass origin in an intermediate step or at the power station site. The authors state that one of the main drawbacks of the use of biomass as a source, is its relatively low density and heating value when compared for instance with other fossil sources. On a later work, Rentizelas *et al.* [140], exemplify the fact that a biomass SC can account for multiple sources, as well as for multiple final products production, such as electricity, heat and cooling. They apply the methodology in a specific region of Greece. The results provide with optimal locations and investment details for potential investors. The works by Zamboni *et al.* [141,142] formulate a MILP model to minimise operating costs and GHG emissions of a biofuel SC. The second work performs a multi-objective optimisation by taking into account the whole SC, from biomass cultivation to fuel distribution. The region under study is discretised into a grid of square regions to map the network and to estimate the biomass cultivation potential for each region. Using as starting point the formulation developed in Zamboni *et al.* [141], Akgul *et al.* [143] also minimise costs, while decreasing calculation time including spatial restrictions through a "neighbourhood" flow limitation. Elia *et al.* [144] uses a MILP multi-objective optimisation to propose a sustainable SC for a novel hybrid concept for a plant that produces liquid fuels, *i.e.*, gasoline, diesel and kerosene, using coal, biomass or natural gas as main feedstocks. The authors conclude that the hybrid configuration (fossil fuels—biomass) allows for competitive biofuels prices. Zhang *et al.* [145] adopt a GIS-based method to find the best location for a biofuel plant, based on the distributed nature of the woody biomass sources and the associated transportation costs. The methodology comprises two steps: location of site candidates and selection of the cost-optimal area. The paper by Chiueh *et al.* [146] uses GIS to identify the biomass waste resources and the specific transportation routes for a co-combustion retrofitting problem, evaluating different levels of pre-treatment (torrefaction) centralisation.

Undoubtedly, energy policies are partially driven by environmental considerations, more specifically by the pressure on reducing greenhouse gas (GHG) emissions. Biomass is an energy source that is expected to provide significant reductions of environmental impacts related to GHG emissions when compared to the classical fossil fuels technologies. Therefore, it is relevant the integration of environmental thinking into SCM in order to assess such expected reduced environmental impacts. The aforementioned integration may be achieved through the concept regarded as "Green Supply Chain Management" (GrSCM). This concept considers the environmental interventions associated with the raw materials sourcing and selection, manufacturing process selection, delivery of final product to the consumers as well as end of life management of the product after its useful life [147]. Traditionally, the methodologies devised to assist SC operation and design have focused on finding a solution that maximises a given economic performance indicator while satisfying a set of operational constraints imposed by the manufacturing/processing technology and the topology of the network. In recent years, however, there has been a growing awareness of the importance of including environmental aspects as objectives and not only as constraints associated with the SC decision support [148,149].

The environmental science and engineering community have developed several systematic methodologies for the detailed characterisation of the environmental impacts of chemicals, products, and processes. All of these methodologies have embodied the concepts of life cycle, *i.e.*, they are based on a LCA which is described in a series of ISO documents [150]. The LCA framework includes the entire life cycle of the product, process or activity, encompassing extraction and processing of raw materials; manufacturing, transport and distribution; re-use, maintenance recycling and final disposal. Most importantly, it takes a holistic approach, bringing the environmental impacts into one consistent framework [151]. The idea is to determine process conditions or topology using a multi-criteria optimisation strategy in order to evaluate the trade-off between economic and environmental issues.

As aforementioned, the concept of SC refers to the network of interdependent entities that constitute the processing and distribution channels of a product from the supply of its raw materials to its delivery to the final consumer. Because an LCA study ideally covers a cradle-to-grave approach, it can be clearly seen that LCA fits as a suitable tool for quantitatively assessing the environmental burdens associated with designing and operating a SC.

Cherubini and Stromman [152] review works in the field of LCA of bioenergy systems, concluding that most studies found a reduction in GHG emissions and in fossil fuel energy consumption. Nevertheless, even if the LCA follows a well-established methodology, the selection of the functional unit (FU) changes according to authors criteria (for instance, power produced or energy/mass flowrates of biomass introduced into the system) and makes it difficult the comparison among different works. Damen and Faaij [75] presents a life cycle inventory of a biomass-based SC for co-firing. This work compares a biomass co-use plant with a 100% coal power plant. The net avoided primary energy and GHG emissions are significantly reduced in the biomass co-use plant. Perry and Rosillo-Calle [61] treat exhaustively the subject of CO_2 emissions along the whole SC, considering emissions from production, conversion, distribution and land use displacement caused by the modification of the biomass use. Ayoub et al. [153] present a methodology for designing and evaluating the biomass utilisation networks (so called B-NETs), which are process networks aiming at producing different bio-products, from one or more biomass resources. The idea of this methodology is to provide a framework to create the underlying superstructure that relates the biomass resources to their products via current and possible future available processes, which can be used to develop an optimisation model. Their methodology is applied at a local level and proposes better biomass uses by means of economic parameters (costs) and environmental impacts accounting. The environmental indicator that the authors use accounts for emissions to air, water pollutants and solid wastes. Later on, Van Dyken et al. [154] develop a linear optimisation model for planning the capacity expansions in energy systems where several alternative biomass and technologies are considered simultaneously. The main objective of this work is to present a generic model including different component such as sources, handling, processing, storage and final usage. Heating value, moisture content and bulk density are the key parameters changes that biomass undergoes along the SC. The objectives to be optimised are the operating cost and emissions of the whole SC.

The review by Cambero and Sowlati [155] remarks that the use of biomass has an important potential to substitute fossil fuel, while all three aspects of sustainability (economic, environmental, and societal) have to be considered in the optimisation problem. Tavares et al. [156] use GIS to identify the most suitable locations for municipal solid waste plants, according to socio-economic, technical and environmental criteria. Pérez-Fortes et al. [39] optimise a decentralised biomass problem that uses cassava waste as raw material to power a gasifier—gas engine plant. This problem considers MO optimisation with three criteria: economic, environmental and social. Employment generation is used as social criteria. The developed MILP allows the selection of: (i) optimum flow rates, sites connectivity and capacity of units; and (ii) the best technology to be employed from a superstructure of technological options for biomass processing as shown in Figure 15. The algorithm developed in this work is the starting point of the work of Pérez-Fortes et al. [157]. In this work, the MILP program is adapted to solve a SC problem in Spain that considers biomass waste to fulfil the biomass required to replace a portion of the power in the current coal power plants installed.

From another standpoint, uncertainty has been increasingly considered when modelling biomass SC's: Shastri *et al.* [158] points out the seasonal and distributed origin of biomass in its BioFeed MILP model which is developed to optimise a farm, *i.e.* the biomass collection step. Gebreslassie *et al.* [159] also models a stochastic MILP program to address the optimal design of a biorefinery SC under supply and demand uncertainties, a risk management optimisation approach is presented. In Osmani and Zhang [160], a stochastic MILP model is developed, considering uncertainty in the supply of biomass-to-bioethanol, demand of biofuel, biomass and biofuel prices, with the purpose of determining the location and the efficiency of the biorefineries on the one hand, and the connections and storage sites, until the bioethanol is sold, on the other. Recently, Yilmaz, Balaman and Selim [161] design an anaerobic digestion SC, under cost and environmental criteria optimisation considering uncertainties. In this case, the method employed is a Fuzzy MO-MILP.

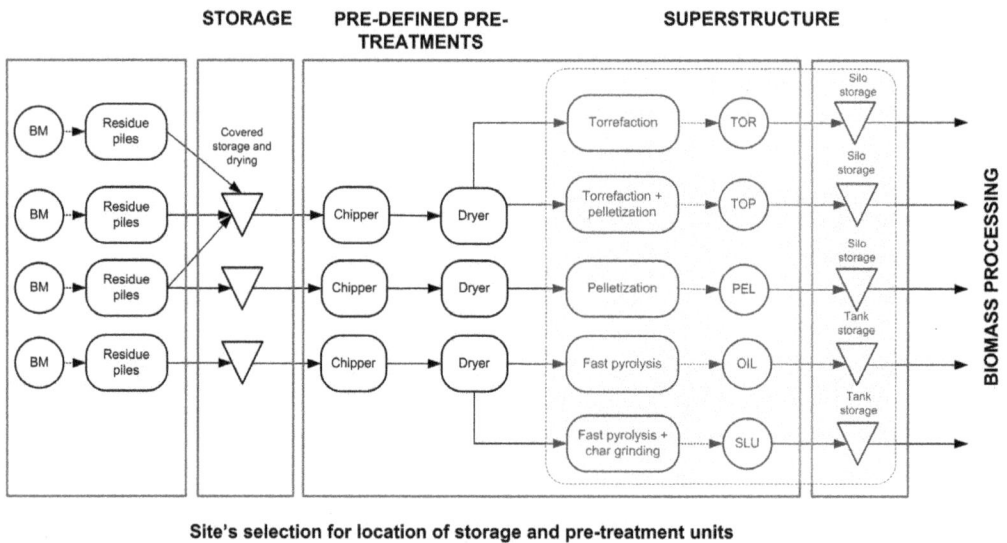

Figure 15. Superstructure utilised for the Biomass SC model presented in Pérez-Fortes *et al.* [157].

High complex and computational demanding programs are resulting from the development of MILPs for the evaluation of biomass related SC's, thus calling for decomposition methods that can attenuate the heavy computational load that is needed for the solution of biomass SC related programs. The article from Osmani and Zhang [160] employs a decomposition based on the Sample Average Approximation method. Gebreslassie *et al.* [159] utilises the Multicut L-shaped method, while the work by Shastri *et al.* [158] uses a Decomposition scheme together with a Distributed Computing approach. More recently, Laínez-Aguirre *et al.* [162] propose a Lagrangian relaxation based approach, the Optimal Condition Decomposition (OCD), to tackle medium size models which require intensive computational power. The methodology is applied to solve the biomass SC case study located in Spain, previously studied in Pérez-Fortes *et al.* [157].

The aforementioned works demonstrate that biomass supply modelling has become a research area of great interest since it provides a tool that supports strategic and tactical decision making which is vital to configure the highly geographical distributed biomass networks. These works point that the necessity of also considering environmental and societal aspects during the evaluation of Biomass SC projects. The main issues that still require further effort and should be overcome are the consideration of seasonality and heterogeneity, in terms of biomass properties and sources. Overall, traditional techniques of mathematical programming for SC modelling have a new challenge in the treatment of biomass properties along the chain. The mathematical models resulting from the Biomass SC are of significant size involving thousands of variables and constraints. Initial efforts have been devoted to develop decomposition techniques in order to make real size problems tractable. However, there are still many challenges in this arena and the application of decomposition techniques is an area that deserves further investigation.

6. Conclusions

This review considers bioenergy from two perspectives and two exploiting technologies: centralised and distributed energy systems, gasification and combustion. The different proposals are evaluated through engineering, economic, environmental and social aspects. Centralised and distributed energy systems place different constraints on the use of biomass. For large scale use, biomass can be adequately co-used with fossil fuels for co-gasification or co-combustion in already existing power plants. Multiple products can be derived taking advantage from the syngas versatility. Polygeneration mimics the energy efficiency of oil refineries through the production of fuels, power and chemicals from biomass. For small scale use, such as residential applications in urban areas or electricity generation in rural areas, biomass waste should be produced locally to avoid transportation costs and biomass degradation. Tailor-made approaches are needed evaluate the characteristics of each possible system and to provide appropriate optimal solutions regarding the available biomass and the energy demand.

This review finishes in Sections 4 and 5 with a systematic and versatile approach for decision making for biomass use at large and small scale, in different contexts. The methodology is based on conceptual design using modelling, simulation and optimisation theories. On this basis, simulation and multiple criteria decision analysis are used to support the decision-making process.

Bioenergy offers decisive advantages in terms of environmental and social impact. Its deployment already straightforwardly supports current energy conversion technologies. Challenges concern the improvement of biomass pre-treatment processes and storage, to meet with the standards for energy generation. Despite all the striking advantages, biomass conversion, combined with carbon capture and storage needs further motivation (such as environmental), to be introduced into the market.

Acknowledgments

The authors are grateful to the CEPIMA members for the support provided. We acknowledge the financial funding received from the Generalitat de Catalunya with the ESF (FI grants).

134

Author Contributions

This work was developed during the PhD thesis of M.P.F. and J.M.L.A. M.P.F. thesis was about conceptual design of bioenergy systems. J.M.L.A. thesis was about mathematical modelling for supply chain optimisation. M.P.F. and J.M.L.A. have contributed to the manuscript. L.P., supervisor of both PhD Theses, has also contributed, coordinated and edited the manuscript.

Conflicts of Interest

This work was performed while author's main affiliation was the Universitat Politècnica de Catalunya. The authors affirm that this study and all related work were carried out without any conflict of interest. Responsibility for the information and views set out in this work lies entirely with the authors.

References

1. Charpentier, J.C. Foreword. In *Syngas from Waste: Emerging Technologies*, 1st ed.; Puigjaner, L., Ed.; Springer-Verlag London Ltd.: London, UK, 2011; p. v.
2. Boullard, F. Kirschen, D.S. Centralised and distributed electricity systems. *Energy Policy* **2008**, *36*, 4504–4508.
3. Pérez-Fortes, M. Conceptual Design of Alternative Energy Systems from Biomass. Ph.D. Thesis, Universitat Politècnica de Catalunya, Barcelona, Spain, 2011.
4. Tchapda, A.H.; Sarma, V.; Pisupati, S.V. A Review of Thermal Co-Conversion of Coal and Biomass/Waste. *Energies* **2014**, *7*, 1098–1148.
5. International Energy Agency (IEA). *21st Century Coal Advanced Technology and Global Energy Solution*; IEA Coal Industry Advisory Board (CIAB): Paris, France, 2013.
6. International Energy Agency (IEA). *Coal: Medium-Term Market Report*; EIA: Paris, France, 2014.
7. U.S. Energy Information Administration (EIA). *Annual Energy Outlook 2014 with Projections to 2040*; EIA: Washington, DC, USA, 2014.
8. International Energy Agency (IEA). *Co-production of Hydrogen and Electricity by Coal Gasification with CO_2 Capture—Updated Economic Analysis*; IEA-GHG: London, UK, 2008.
9. Capros, P.; Mantzos, L.; Tasios, N.; De Vita, A.; Kouvaritakis, N. *EU Trends to 2030: Update 2009*; Directorate-General for Energy: Brussels, Belgium, 2010.
10. *The U.S in Its Energy Independence and Security Act (EISA)*; United States Congress: Washington, DC, USA, 2007.
11. Sims, R. *Bioenergy Options for a Cleaner Environment in Developed and Developing Countries*, 1st ed.; Elsevier: Oxford, UK, 2004; pp. 141–168.
12. Faaij, A. Bio-energy in Europe: Changing technology choices. *Energy Policy* **2006**, *34*, 322–342.
13. Kanagawa, M.; Nakata, T. Assessment of access to electricity and socio-economic impacts in rural areas of developing countries. *Energy Policy* **2008**, *36*, 2016–2029.

14. Urban, F.; Benders, R.M.J.; Moll, H.C. Modelling energy systems for developing countries. *Energy Policy* **2007**, *35*, 3473–3482.

15. Van Ruijven, B.; Urban, F.; Benders, R.; Moll, H.; van der Sluijs, J.; de Vries, B; van Vuuren, D. Modeling energy and development: An evaluation of models and concepts. *World Dev.* **2008**, *36*, 2801–2821.

16. Otto, M. Trade-offs, risks and opportunities linked to bioenergy for sustainable development. In Proceedings of the 17th European Biomass Conference and Exhibition, Hamburg, Germany, 29 June–3 July 2009; pp. 71–77.

17. Rosillo-Calle, F.; de Groot, P.; Hemstock, S.L.; Woods, J. *The Biomass Assessment Handbook. Bioenergy for a Sustainable Environment*; Earthscan: London, UK, 2007.

18. Janssen, R.; Rutz, D.; Helm, P.; Woods, J.; Diaz-Chavez, R. Bioenergy for sustainable development in Africa: Environmental and social aspects. In Proceedings of the 17th European Biomass Conference and Exhibition, Hamburg, Germany, 29 June–3 July 2009; pp. 2422–2430.

19. Madjera, M. The necessity of policies in sub-Saharan Africa (SSA) for the production of energy crops. In Proceedings of the 17th European Biomass Conference and Exhibition, Hamburg, Germany, 29 June–3 July 2009; pp. 2452–2458.

20. Bridgwater, A.V. Renewable fuels and chemicals by thermal processing of biomass. *Chem. Eng. J.* **2003**, *91*, 87–102.

21. Strachan, N.; Dowlatabadi, H. Distributed generation and distribution utilities. *Energy Policy* **2002**, *30*, 649–661.

22. Caputo, A.; Palumbo, M.; Pelagagge, P.; Scacchia, F. Economics of biomass energy utilization in combustion and gasi_cation plants: Effects of logistics variables. *Biomass Bioenergy* **2005**, *28*, 35–51.

23. Gold, S.; Seuring, S. Supply chain and logistics issues of bio-energy production. *J. Clean. Prod.* **2011**, *19*, 32–42.

24. Siemons, R. Identifying a role for biomass gasification in rural electrification in developing countries: The economic perspective. *Biomass Bioenergy* **2001**, *20*, 271–285.

25. Mitra, I.; Degner, T.; Braun, M. Distributed generation and microgrids for small island electrification in developing countries: A review. *Sol. Energy Soc. India* **2008**, *18*, 6–20.

26. Bayod, A.; Mur, J.; Bernal, J.; Domínguez, J. Definitions for distributed generation: A revision. In Proceedings of the International Conference on renewable Energies and Power Quality, Zaragoza, Spain, 16–18 March 2005.

27. Du Plessis, V.; Beshiri, R.; Bollman, R.D. *Definitions of Rural*. Statistics Canada, Agriculture Division: Ottawa, ON, Canada, 2002.

28. United States Census Bureau. Available online: https://www.census.gov/geo/reference/ua/uafaq.html (accessed on 4 April 2015).

29. The World Bank (WB) Database. Available online: http://data.worldbank.org/about/country-and-lending-groups (accessed on 4 April 2015).

30. Goldemberg, J.; Johansson, T.B. World Energy Assessment, 2004 Update. Available online: http://www.undp.org/content/dam/aplaws/publication/en/publications/environment-energy/www-ee-library/sustainable-energy/world-energy-assessment-overview-2004-update/World%20Energy%20Assessment%20Overview-2004%20Update.pdf (accessed on 4 April 2015).

31. Čuček, L.; Varbanov, P.S.; Klemeš, J.J.; Kravanja, Z. Total footprints-based multi-criteria optimisation of regional biomass energy supply chains. *Energy* **2012**, *44*, 135–145.

32. Silveira, S. *Bioenergy—Realizing the Potential*, 1st ed.; Elsevier: Oxford, UK, 2005.

33. Rubiera, F.; Pis, J.J.; Pevida, C. Raw materials selection, preparation and characterization. In *Syngas from Waste: Emerging Technologies*, 1st ed.; Puigjaner, L., Ed.; Springer Verlag: London, UK, 2011; pp. 11–22.

34. Mathews, J. Carbon-negative biofuels. *Energy Policy* **2008**, *36*, 940–945.

35. McKendry, P. Energy production from biomass (part 2): Conversion technologies. *Bioresour. Technol.* **2002**, *83*, 47–54.

36. Dasappa, S. Potential of biomass energy for electricity generation in sub-Saharan Africa. *Energy Sustain. Dev.* **2011**, *15*, 203–213.

37. Welsch, M.; Bazilian, M.; Howells, M.; Divan, D.; Elzing, D.; Strbac G.; Jones, L.; Keane, A.; Gielen, D.; Balijepalli, V.S.K.M.; *et al.* Smart and just grids for sub-Saharan Africa: Exploring options. *Renew. Sustain. Energy Rev.* **2013**, *20*, 336–352.

38. Lamers, P.; Junginger, M.; Hamelinck, C.; Faaij, A. Development in international solid biofuel trade—An analysis of volumes, policies, and market factors. *Renew. Sustain. Energy Rev.* **2012**, *16*, 3176–3199.

39. Pérez-Fortes, M.; Laínez-Aguirre, J.M.; Arranz-Piera, P.; Velo, E.; Puigjaner, L. Design of regional and sustainable bio-based networks for electricity generation using a multi-objective MILP approach. *Energy* **2012**, *44*, 79–95.

40. Obernberger, I.; Theck, G. *The Pellet Handbook: The Production and Thermal Utilization of Pellets*; Earthscan LLC: London, UK, 2010.

41. Wang, L.; Shen, W.; Xie, H.; Neelamkavil, J.; Pardasani, A. Collaborative conceptual design—State of the art and future trends. *Comput. Aided Des.* **2002**, *34*, 981–996.

42. Douglas, J. *Conceptual Design of Chemical Processes*; McGraw-Hill Book Company: New York, NY, USA, 1988.

43. Turton, R.; Bailie, R.; Whiting, W.; Shaeiwitz, J.; Bhattacharyya, D. *Analysis, Synthesis and Design of Chemical Processes*, 4th ed.; Prentice Hall: Upper Saddle River, NJ, USA, 2012.

44. Marquardt, W.; von Wedel, L.; Bayer, B. Perspectives on life cycle process modeling. In *Foundations on Computer-Aided Process Design*; Malone, F., Trainham, J.A., Carnahan, B., Eds.; AIChE Symposium Series 323; American Institute of Chemical Engineers and Computer Aids for Chemical Engineering Education: Breckenbridge, CO, USA, 2000; Volume 96, pp. 192–214.

45. Von Wedel, L.; Marquardt, W. ROME: Modelling frameworks. In *Software Architectures and Tools for Computer Aided Process Engineering*; Braunsweig, B., Gani, R., Eds.; Elsevier Science: Amsterdam, The Netherlands, 2002; Volume 1, pp. 89–125.

46. Pantelides, C.C. New Chalenges and oportunities for process modeling. In *European Symposium on Computer-Aided Process Engineering-11*; Gani, R., Jorgensen, S.B., Eds.; Elsevier: Amsterdam, The Netherlands, 2001; Volume 9, pp. 15–26.

47. Harper, P.M.R.; Gani, R.; Kolar, P.; Ishikawa, T. Computer-Aided Molecular Design with combined molecular modelling and group contribution. *Fluid Phase Equilibria* **1999**, *158–160*, 337–347.

48. Meniai, A.H.; Newsham, D.M.T.; Khalfaoui, B. Solvent design for liquid extraction using calculated molecular interaction parameters. *Chem. Eng. Res. Des.* **1998**, *11*, 942–950.

49. Backs, T.O.; Bosgra, O.; Marquardt, W. Towards intentional dynamics in supply chain conscious process operation. In *Foundations of Computer-Aided Process Operations*; Pekny, J., Blau, G.E., Eds.; AIChE Symposium Series 320; American Institute of Chemical Engineers and Computer Aids for Chemical Engineering Education: Danvers, MA, USA, 1998; Volume 94, p. 5.

50. Foss, B.; Lohman, B.; Marquardt, W. A field study of the industrial modeling process. *J. Process Control* **1998**, *8*,325–337.

51. Biegler, L.T.; Grossmann, I.E.; Westerberg, A.W. *Systematic Methods of Chemical Process Design*, 1st ed.; Prentice-Hall International: Upper Saddle River, NJ, USA, 1997; pp. 1–21.

52. Bojarski, A.; Pérez-Fortes, M.; Nougués, J.M.; Puigjaner, L. Modeling superstructure for conceptual design of syngas generation and treatment. In *Syngas from Waste: Emerging Technologies*, 1st ed.; Puigjaner, L., Ed.; Springer-Verlag London Ltd.: London, UK, 2011; pp. 169–199.

53. Laínez-Aguirre, J.M.; Puigjaner, L. *Advances in Integrated and Sustainable Supply Chain Planning*, 1st ed.; Laínez-Aguirre, J.M., Puigjaner, L., Eds.; Springer-Verlag London Ltd.: London, UK, 2015.

54. Turban, E.; Aronson, J.; Liang, T. *Decision Support Systems and Intelligent Systems*, 7th ed.; Pearson Prentice Hall, Inc.: Upper Saddle River, NJ, USA, 2005.

55. Rodrigues, M.; Water, A.; Faaij, A. Performance evaluation of atmospheric biomass integrated gasifier combined cycle systems under different strategies for the use of low calorific gases. *Energy Convers. Manag.* **2007**, *48*, 1289–1301.

56. Berndes, G.; Hansson, J.; Egeskog, A.; Johnsson, F. Strategies for 2nd generation biofuels in EU—Co-firing to stimulate feedstock supply development and process integration to improve energy efficiency and economic competitiveness. *Biomass Bioenergy* **2010**, *34*, 227–236.

57. Sami, M.; Annamalai, K.; Wooldridge, M. Co-firing of coal and biomass fuel blends. *Progess Energy Combust. Sci.* **2001** *27*, 171–214.

58. Gómez, A.; Rodrigues, M.; Montañés, C.; Dopazo, C.; Fueyo, N. The potential for electricity generation from crop and forestry residues in Spain. *Biomass Bioenergy* **2010**, *34*, 703–719.

59. Uslu, A.; Faaij, A.; Bergman, P. Pre-treatment technologies, and their effect on international bioenergy supply chain logistics. Techno-economic evaluation of torrefaction, fast pyrolysis and pelletisation. *Energy* **2008**, *33*, 1206–1223.

60. Tillman, D. Biomass cofiring: The technology, the experience, the combustion consequences. *Biomass Bioenergy* **2000**, *19*, 365–384.

138

61. Perry, M.; Rosillo-Calle, F. Recent trends and future opportunities in U.K. bioenergy: Maximising biomass penetration in a centralised energy system. *Biomass Bioenergy* **2008**, *32*, 688–701.

62. Basu, P.; Butler, J.; Leon, M. Biomass co-firing options on the emission reduction and electricity generation costs in coal-fired power plants. *Renew. Energy* **2011**, *36*, 282–288.

63. Prins, M.; Ptasinski, K.; Janssen, F. Torrefaction of wood: Part 1. Weight loss kinetics. *J. Anal. Appl. Pyrolysis* **2006**, *77*, 28–34.

64. Prins, M.; Ptasinski, K.; Janssen, F. More efficient biomass gasification via torrefaction. *Energy* **2006**, *31*, 3458–3470.

65. Bergman, P.; Boersma, A.; Zwart, R.; Kiel, J. *Torrefaction for Biomass Co-Firing in Existing Coal-Fired Power Stations*; Energy research Centre of the Netherlands, ECN: Petten, The Netherlands, 2005.

66. Couhert, C.; Salvador, S.; Commandré, J. Impact of torrefaction on syngas production from wood. *Fuel* **2009**, *88*, 2286–2290.

67. Deng, J.; Wang, G.; Kuang, J.; Zhang, Y.; Luo, Y. Pretreatment of agricultural residues for co-gasification via torrefaction. *J. Anal. Appl. Pyrolysis* **2009**, *86*, 331–337.

68. Maciejewska, A.; Veringa, H.; Sanders, J.; Peteves, S.D. *Co-Firing of Biomass with Coal: Constraints and Role of Biomass Pre-Treatment*; Institute for Energy from the European Commission: Petten, The Netherlands, 2006.

69. Sultana, A.; Kumar, A.; Harfield, D. Development of agri-pellet production cost and optimum size. *Bioresour. Technol.* **2010**, *101*, 5609–5621.

70. Wu, H.; Yu, Y.; Yip, K. Bioslurry as a fuel. 1. Viability of a bioslurry-based bioenergy supply chain for mallee biomass in Western Australia. *Energy Fuels* **2010**, *24*, 5652–5659.

71. Abdullah, H.; Mourant, D.; Li, C.; Wu, H. Bioslurry as a fuel. 3. Fuel and rheological properties of bioslurry prepared from the bio-oil and biochar of mallee biomass fast pyrolysis. *Energy Fuels* **2010**, *24*, 5669–5676.

72. Magalhaes, A.; Petrovic, D.; Rodriguez, A.; Putra, Z.; Thielemans, G. Techno-economic assessment of biomass pre-conversion processes as a part of biomass-to-liquids line-up. *Biofuels Bioprod. Biorefining* **2009**, *3*, 584–600.

73. Verbong, G.; Christiaens, W.; Raven, R.; Balkema, A. Strategic niche management in an unstable regime: Biomass gasification in India. *Environ. Sci. Policy* **2010**, *13*, 272–281.

74. Rentizelas, A.; Tatsiopoulos, I.; Tolis, A. An optimization model for multi-biomass tri-generation energy supply. *Biomass Bioenergy* **2009**, *33*, 223–233.

75. Damen, K.; Faaij, A. A greenhouse gas balance of two existing international biomass import chains. The case of residue co-firing in a pulverised coal-fired power plant in The Netherlands. *Mitig. Adapt. Strateg. Glob. Chang.* **2006**, *11*, 1023–1050.

76. Van Loo, S.; Koppejan, J. *The Handbook of Biomass Combustion & Co-Firing. Biomass Energy—Handbooks, Manuals, etc.*; Earthscan: Sterling, VA, USA, 2008.

77. Baxter, L. Biomass-coal co-combustion: Opportunity for affordable renewable energy. *Fuel* **2005**, *84*, 1295–1302.

78. Chiaramonti, D.; Oasmaa, A.; Solantausta, Y. Power generation using fast pyrolysis liquids from biomass. *Renew. Sustain. Energy Rev.* **2007**, *11*, 1056–1086.

79. Vhathvarothai, N.; Ness, J.; Yu, J. An investigation of thermal behaviour of biomass and coal during co-combustion using thermogravimetric analysis (TGA). *Int. J. Energy Res.* **2014**, *38*, 804–812.

80. Parshetti, G.K.; Quek, A.; Betha, R.; Balasubramanian, R. TGA-FTIR investigation of co-combustion characteristics of blends of hydrothermally carbonised oil palma biomass (EFB) and coal. *Fuel Process. Technol.* **2014**, *118*, 228–234.

81. Gómez, A.; Zubizarreta, J.; Rodrigues, M.; Dopazo, C.; Fueyo, N. An estimation of the energy potential of agro-industrial residues in Spain. *Resour. Conserv. Recycl.* **2010**, *54*, 972–984.

82. Van Den Broek, R.; Faaij, A.; Van Wijk, A. Biomass combustion for power generation. *Biomass Bioenergy* **1996**, *11*, 271–281.

83. Nussbaumer, T. Combustion and Co-combustion of Biomass: Fundamentals, Technologies, and Primary Measures for Emission Reduction. *Energy Fuels* **2003**, *17*, 1510–1521.

84. Kalisz, S.; Pronobis, M.; Baxter, D. Co-firing of biomass waste-derived syngas in coal power boiler. *Energy* **2008**, *33*, 1770–1778.

85. Munir, S.; Nimmo, W.; Gibbs, B. The effect of air staged, co-combustion of pulverised coal and biomass blends on NOX emissions and combustion efficiency. *Fuel* **2011**, *90*, 126–135.

86. Desideri, U.; Paolucci, A. Performance modelling of a carbon dioxide removal system for power plants. *Energy Convers. Manag.* **1999**, *40*, 1899–1915.

87. Hamelinck, C.; Faaij, A. Future propects for production of methanol and hydrogen from biomass. *J. Power Sources* **2002**, *111*, 1–22.

88. Kanniche, M.; Bouallou, C. CO_2 capture study in advanced integrated gasification combined cycle. *Appl. Therm. Eng.* **2007**, *27*, 2693–2702.

89. Descamps, C.; Bouallou, C.; Kanniche, M. Efficiency of an integrated gasification combined cycle (IGCC) power plant including CO_2 removal. *Energy* **2008**, *33*, 874–881.

90. Chen, C.; Rubin, E.S. CO_2 control technology effects on IGCC plant performance and cost. *Energy Policy* **2009**, *37*, 915–924.

91. Biagini, E.; Masoni, L.; Tognotti, L. Comparative study of thermo-chemical processes for hydrogen production from biomass fuels. *BioResource Technol.* **2010**, *101*, 6381–6388.

92. Jacobson, M.Z. Review of solutions to global warming, air pollution, and energy security. *Energy Environ. Sci.* **2009**, *2*, 148–173.

93. International Energy Agency, Greenhouse Gas R&D Programme, (IEA-GHG). *Co-Production of Hydrogen and Electricity by Coal Gasification with CO_2 Capture—Updated Economic Analysis*; International Energy Agency (IEA): Cheltenham, UK, 2008.

94. Metz, B.; Davidson, O.; Coninck, H.; Loos, M.; Meyer, L. *Special Report on Carbon Dioxide Capture and Storage*; International Panel on Climate Change (IPCC): New York, NY, USA, 2005.

95. Pérez-Fortes, M.; Bojarski, A. Modeling syngas generation. In *Syngas from Waste: Emerging Technologies*, 1st ed.; Puigjaner, L., Ed.; Springer Verlag: London, UK, 2011; pp. 55–88.

140

96. Wender, I. Reactions of synthesis gas. *Fuel Process. Technol.* **1996**, *48*, 189–297.
97. Highman, C.; van der Burght, M. *Gasification*; Elsevier Science: Amsterdam, The Netherlands, 2003.
98. NETL. *Gasification World Database 2007: Current Industry Status*; DOE, National Energy Technology Laboratory, 2007. Available online: http://www.netl.doe.gov/research/ energy-analysis/publications/details?pub=5c73f5a8-d949-47cc-9ead-28456a48ffc5 (accessed on 28 May 2015).
99. Milne, T.; Evans, R.; Abatzoglou, N. *Biomass Gasifier -tars-: Their Nature, Formation and Conversion*; National Renewable Energy Laboratory (NREL): Springfield, VA, USA, 1998.
100. Van de Kamp, W.; Wild, U.; Zielke, P.; Suomalainen, M. *Tar Measurement Standard for Sampling and Analysis of Tars and Particles in Biomass Gasification Product Gas*; Energy Research Centre of The Netherlands (ECN): Petten, The Netherlands, 2005.
101. Reed, T.; Das, A. *Handbook of Biomass Downdraft Gasifier Engine Systems*; DOE, Solar Energy Research Institute (SERI): Golden, CO, USA, 1988.
102. Mastellone, M.; Zaccariello, L.; Arena, U. Co-gasification of coal, plastic waste and wood in a bubbling fluidized bed reactor. *Fuel* **2010**, 89, 2991–3000.
103. Hernandez, J.; Aranda-Almansa, G.; Serrano, C. Co-gasification of biomass wastes and coal-coke blends in an entrained flow gasifier: An experimental study. *Energy Fuels* **2010**, *24*, 2479–2488.
104. Ragauskas, A.; Williams, C.; Davison, B.; Britovsek, G.; Cairney, J.; Eckert, C.; Frederick, W.; Hallet, J.; Leak, D.; Liotta, C.; *et al*. The path forward for biofuels and biomaterials. *Science* **2006**, *311*, 484–489.
105. Demirbas, M. Biorefineries for biofuel upgrading: A critical review. *Appl. Energy* **2009**, *86*, 151–161.
106. Wang, L.; Weller, C.; Jones, D.; Hanna, M. Contemporary issues in thermal gasification of biomass and its application to electricity and fuel production. *Biomass Bioenergy* **2008**, *32*, 573–581.
107. Knoef, H. *BTG Biomass Gasification*; Biomass Technology Group (BTG): Enschede, The Netherlands, 2008.
108. Reimert, R.; Schaub, G. *Gas Production, Chapter 4: Gas Production from Coal, Wood, and other Solid Feedstocks*, 7th ed.; Ullmann's Encyclopedia of Industrial Chemistry, Wiley-VCH: Weinheim, Germany, 2003.
109. Sharma, S.D.; Park, M.D.D.; Morpeth, L.; Ilyushechkin, A.; McLennan, K.; Harris, D.J.; Thambimuthu, K.V. A critical review of syngas cleaning technologies—Fundamental limitations and practical problems. *Powder Technol.* **2008**, *180*, 115–121.
110. Pisa, J.; Serra de Renobales, L.; Moreno, A.; Valero, A. Evaluación de alternativas en un diseño de planta GICC con gasificador PRENFLO. In Proceedings of the XII Congreso Nacional de Ingeniería Mecánica, Bilbao, Spain, 1997; ISSN:0212-5072, pp. 269–278.
111. Huang, Y.; Rezvain, S.; McIlveen-Wright, D.; Minchener, A.; Hewitt, N. Techno-economic study of CO2 capture and storage in coal fired oxygen-fed entrained flow IGCC power plants. *Fuel Process. Technol.* **2008**, *89*, 916–925.

112. Silva, D.; Nakata, T. Multi-objective assessment of rural electrification in remote areas with poverty considerations. *Energy Policy* **2009**, *37*, 3096–3108.

113. Silva, D.; Nakata, T. Optimization of decentralized energy systems using biomass resources for rural electrification in developing countries. *Energy Policy* **2008**, *37*, 3096–3108.

114. Kanase-Patil, A.; Saini, R.; Sharma, M. Integrated renewable energy systems for off grid rural electrification of remote area. *Renew. Energy* **2010**, *35*, 1342–1349.

115. Hiremath, R.; Kumar, B.; Deepak, P.; Balachandra, P.; Ravindranath, R.; Raghunandan, B. Decentralized energy planning through a case study of a typical village in India. *J. Renew. Sustain. Energy* **2009**, *1*, 043103:1–043103:24.

116. Cherni, J.; Dyner, I.; Henao, F.; Jaramillo, P.; Smith, R.; Font, R. Energy supply for sustainable rural livelihoods. A multi-criteria decision-support system. *Energy Policy* **2007**, *35*, 1493–1504.

117. Brent, A.; Kruger, W. Systems analyses and the sustainable transfer of renewable energy technologies: A focus on remote areas of Africa. *Renew. Energy* **2009**, *34*, 1774–1781.

118. Intermediate Technology Development Group (ITDG). Schumacher Centre for Technology and Development, Bourton Hall, Bourton on Dunsmore, Rugby, Warwickshire CV23 9QZ, United Kingdom. Available online: http://www.itdg.org, http://www.itdg.org.pe/ (accessed on 4 December 2011).

119. Ferrer-Martí, L.; Pastor, R.; Capó, G. Un modelo de ubicación de microaerogeneradores para el diseño de proyectos de electrificación rural con energía eólica. In *Industrial Engineering: A Way for Sustainable Development*, Proceedings of 3rd International Conference on Industrial Engineering and Industrial Management. XIII Congreso de Ingeniería de Organización, Barcelona-Terrassa, Spain, 2–4 September 2009; Companys, R., Coves, A.M., Fernández, V., Lusa, A., Mateo, M., Sallán, J.M., Saez, J., Eds.; pp. 657–666. Available online: http://www.adingor.es/Documentacion/CIO/cio2009/Book%20of%20Full%20Papers%20CIO20 09.pdf (accessed on 4 April 2015).

120. Hamimu, H. The prospects and challenges of biofuel production in developing countries (Tanzania experience). In Proceedings of the 17th European Biomass Conference and Exhibition, Hamburg, Germany, 29 June–3 July 2009; pp. 1755–1759.

121. Lapuerta, M.; Hernández, J.; Pazo, A.; López, J. Gasification and co-gasification of biomass wastes: Effect of the biomass origin and the gasifier operating conditions. *Fuel Process. Technol.* **2008**, *89*, 828–837.

122. Karellas, S.; Karl, J.; Kakaras, E. An innovative biomass gasi_cationprocess and its coupling with microturbine and fuel cells systems. *Energy* **2008**, *33*, 284–291.

123. Dornburg, V.; Faaij, A. Efficiency and economy of wood-fired biomass energy systems in relation to scale regarding heat and power generation using combustion and gasification technologies. *Biomass Bioenergy* **2001**, *21*, 91–108.

124. Husain, Z.; Zainal, Z.; Abdullah, M. Analysis of biomass-residuebased cogeneration system in palm oil mills. *Biomass Bioenergy* **2003**, *24*, 117–124.

125. Dong, L.; Liu, H.; Riffat, S. Development of small-scale and microscale biomass-fuelled CHP systems—A literature review. *Appl. Therm. Eng.* **2009**, *29*, 2119–2126.

126. Buragohain, B.; Mahanta, P.; Moholkar, V. Biomass gasification for decentralized power generation: The Indian perspective. *Renew. Sustain. Energy Rev.* **2010**, *14*, 73–92.

127. Sahraei, M.H.; McCalden, D.; Hughes. R.; Ricardez-Sandoval, L.A. A survey on current advanced IGCC power plant technologies, sensors and control systems. *Fuel* **2014**, *137*, 245–259.

128. Chiesa, P.; Lozza, G.; Mazzocchi, L. Using hydrogen as gas turbine fuel. *J. Eng. Gas Turbines Power* **2005**, *127*, 73–80.

129. Kreutz, T.; Williams, R.; Consonni, S.; Chiesa, P. Co-production of hydrogen, electricity and CO_2 from coal with commercially ready technology. Part B: Economic analysis. *Int. J. Hydrog. Energy* **2005**, *30*, 769–784.

130. Rubin, E.S.; Chen, C.; Rao, A.B. Cost and performance of fossil fuel power plants with CO_2 capture and storage. *Energy Policy* **2007**, *35*, 4444–4454.

131. Pérez-Fortes, M.; Bojarski, A.D.; Puigjaner, L. Advanced simulation environment for clean power production in IGCC plants. *Comput. Chem. Eng.* **2011**, *33*, 1501–1520.

132. Pérez-Fortes, M.; Bojarski, A.D.; Velo, E.; Nougués, J.M.; Puigjaner, L. Conceptual model and evaluation of generated power and emissions in an IGCC plant. *Energy* **2009**, *34*, 1721–1732.

133. Laínez-Aguirre, J.M.; Pérez-Fortes, M.; Bojarski, A.D.; Puigjaner, L. Raw Materials Supply. In *Syngas from Waste: Emerging Technologies*, 1st ed.; Puigjaner, L., Ed.; Springer Verlag: London, UK, 2011; pp. 23–54.

134. U.S. Department of Energy. *Energy Efficiency and Renewable Energy: Biomass Program*; Technology Report; U.S. Department of Energy: Washington, DC, USA, 2010.

135. Tang, C.S. Perspectives in supply chain risk management. *Int. J. Prod. Econ.* **2006**, *103*, 451–488.

136. Panichelli, L.; Gnansounou, E. {GIS}-based approach for defining bioenergy facilities location: A case study in Northen Spain based on marginal delivery costs and resources competition between facilities. *Biomass Bioenergy* **2008**, *32*, 289–300.

137. Van Belle, J.; Temmerman, M.; Schenkel, Y. Three level procurement of forest residues for power plant. *Biomass Bioenergy* **2003**, *24*, 401–409.

138. Hamelinck, C.; Suurs, R.; Faaij, A. International bioenergy transport costs and energy balance. *Biomass Bioenergy* **2005**, *29*, 114–134.

139. Leduc, S.; Schwab, D.; Dotzauer, E.; Schmid, E.; Obersteiner, M. Optimal location of wood gasification plants for methanol production with heat recovery. *Int. J. Energy Res.* **2008**, *32*, 1080–1091.

140. Rentizelas, A.; Tolis, A.; Tatsiopoulos, I. Logistics issues of biomass: The storage problem and the multi-biomass supply chain. *Renew. Sustain. Energy Rev.* **2009**, *13*, 887–894.

141. Zamboni, A.; Bezzo, F.; Shah, N. Spatially explicit static model for the strategic design for future bioethanol production systems. 2. Multi-objective environmental optimization. *Energy Fuels* **2009**, *23*, 5134–5143.

142. Zamboni, A.; Shah, N.; Bezzo, F. Spatially explicit static model for the strategic design for future bioethanol production systems. 1. Cost minimization. *Energy Fuels* **2009**, *23*, 5121–5133.

143. Akgul, O.; Zamboni, A.; Bezzo, F.; Shah, N.; Papageorgiou, G. Optimization-based approaches for bioethanol supply chains. *Ind. Eng. Chem. Res.* **2011**, *50*, 4927–4938.

144. Elia, J.; Baliban, R.; Xiao, X.; Floudas, C. Optimal energy supply network determination and life cycle analysis for hybrid coal, biomass, and natural gas to liquid (CBGTL) plants using carbon-based hydrogen production. *Comput. Chem. Eng.* **2011**, *35*, 1399–1430.

145. Zhang, F.; Johnson, D.; Sutherland, J. A GIS-based method for identifying the optimal location for a facility to convert forest biomass to biofuel. *Biomass Bioenergy* **2011**, *35*, 3951–3961.

146. Chiueh, P.-T.; Lee, K.C.; Syu, F.S.; Lo, S.L. Implications of biomass pretreatment to cost and carbon emissions: Case study of rice straw and pennisetum in Taiwan. *Bioresour. Technol.* **2012**, *108*, 285–294.

147. Srivastava, S.K. Green supply chain management: A state of the art literature review. *Int. J. Manag. Rev.* **2007**, *9*, 53–80.

148. Puigjaner, L.; Guillén, G. Towards an integrated framework for supply chain management in the batch chemical process industry. *Comput. Chem. Eng.* **2008**, *32*, 650–670.

149. Bojarski, A.; Laínez-Aguirre, J.M.; Espuña, A.; Puigjaner, L. Incorporating environmental impacts and regulations in a holistic supply chains modeling: An LCA approach. *Comput. Chem. Eng.* **2009**, *33*, 1747–1759.

150. ISO14040. *Environmental Management—Life Cycle Assessment—Principles and Framework*. International Organization for Standardization (ISO): Geneva, Switzerland, 1997.

151. Guinee, J.; Gorree, M.; Heijungs, R.; Huppes, G.; Kleijn, R.; de Konig, A.; van Oers, L.; Sleeswijk, A.; Suh, S.; de Haes, H.U.; *et al.* *Life Cycle Assessment. An Operational Guide to the ISO Standards. Part 3: Scientific Background*; Ministry of Housing, Spatial Planning and the Environment (VROM) and Centre of Environmental Science—Leiden University (CML): Leiden, The Netherlands, 2001.

152. Cherubini, F.; Stromman, A. Life cycle assessment of bioenergy systems: State of the art and future challenges. *Bioresour. Technol.* **2011**, *102*, 437–451.

153. Ayoub, N.; Seki, H.; Naka, Y. Superstructure-based design and operation for biomass utilization networks. *Comput. Chem. Eng.* **2009**, *33*, 1770–1780.

154. Van Dyken, S.; Bakken, B.H.; Skjelbred, H.I. Linear mixed-integer models for biomass supply chains with transport, storage and processing. *Energy* **2010**, *35*, 1338–1350.

155. Cambero, C.; Sowlati, T. Assessment and optimisation of forest biomass supply chains from economic, social and environmental perspectives—A review of literature. *Renew. Sustain. Energy Rev.* **2014**, *36*, 62–73.

156. Tavares, G.; Zsigraiová, Z.; Semiao, V. Multi-criteria GIS-based sitting of an incineration plant for solid waste. *Waste Manag.* **2011**, *31*, 1960–1972.

157. Pérez-Fortes, M.; Laínez-Aguirre, J.M.; Bojarski, A.D.; Puigjaner, L. Optimisation of pre-treatment selection for the use of woody waste in co-combustion plants. *Chem. Eng. Res. Des.* **2014**, *92*, 1539–1562.

158. Shastri, Y.; Hansen, A.; Rodríguez, L.; Ting, K.C. A novel decomposition and distributed computing approach for the solution of large scale optimisation models. *Comput. Electron. Agric.* **2011**, *76*, 69–79.

159. Gebreslassie, B.H.; Yao, Y.; You, F. Design under uncertainty of hydrocarbon biorefinery supply chains: Multiobjective stochastic programming models, decomposition algorithm, and a comparison between CVaR and downside risk. *AIChE J.* **2012**, *58*, 2155–2179.

160. Osmani, A.; Zhang, J. Economic and environmental optimisation of a large scale sustainable dual feedstock lignocellulosic-based bioethanol supply chain in a stochastic environment. *Appl. Energy* **2014**, *114*, 572–587.

161. Yilmaz Balaman, S.; Selim, H. A fuzzy multiobjective linear programming model for design and management of anaerobic digestion based bioenergy supply chains. *Energy* **2014**, doi:10.1016/j.energy.2014.07.073.

162. Laínez-Aguirre, J.M.; Puigjaner, L. Optimal Condition Decomposition. In *Advances in Integrated and Sustainable Supply Chain Planning*, 1st ed.; Springer-Verlag International Publishing: London, UK, 2015; p. 276.

Flow Regime Changes: From Impounding a Temperate Lowland River to Small Hydropower Operations

Petras Punys, Antanas Dumbrauskas, Egidijus Kasiulis, Gitana Vyčienė and Linas Šilinis

Abstract: This article discusses the environmental issues facing small hydropower plants (SHPs) operating in temperate lowland rivers of Lithuania. The research subjects are two medium head reservoir type hydro schemes considered within a context of the global fleet of SHPs in the country. This research considers general abiotic indicators (flow, level, water retention time in the reservoirs) of the stream that may affect the aquatic systems. The main idea was to test whether the hydrologic regime has been altered by small hydropower dams. The analysis of changes in abiotic indicators is a complex process, including both pre- and post-reservoir construction and post commissioning of the SHPs under operation. Downstream hydrograph (flow and stage) ramping is also an issue for operating SHPs that can result in temporary rapid changes in flow and consequently negatively impact aquatic resources. This ramping has been quantitatively evaluated. To avoid the risk of excessive flow ramping, the types of turbines available were evaluated and the most suitable types for the natural river flow regime were identified. The results of this study are to allow for new hydro schemes or upgrades to use water resources in a more sustainable way.

Reprinted from *Energies*. Cite as: Punys, P.; Dumbrauskas, A.; Kasiulis, E.; Vyčienė, G.; Šilinis, L. Flow Regime Changes: From Impounding a Temperate Lowland River to Small Hydropower Operations. *Energies* **2015**, *8*, 7478-7501.

1. Introduction

Humans often modify rivers to meet specific objectives, including hydropower generation. Flexibility in the timing of water use is a primary reason for regulating rivers. It is accepted that the construction of dams causes hydrologic alterations with consequent changes in ecological conditions. However, the effect of the impact depends on many factors and has to be considered on a case by case basis. For example, irrigation dams have completely different hydrologic alterations than hydropower operations.

The vast majority of studies considering these issues are based on large hydropower plants (HP) or large dams, where impacts are obvious. However, small hydropower plants (SHP) cannot be treated as a scaled down version of the large dams. A literature review has shown that there is a lack of associated studies, especially for low to medium head facilities operating in temperate lowland rivers.

Hydropower plants are divided into different categories according to their installed power (P), which are usually small ($p < 10$ MW), medium (10–100 MW) and large (>100 MW) [1]. However these classification systems may be very different elsewhere. The European standard of SHP plants has been set to the installed capacity ($p < 10$ MW), although many countries adopted a lower SHP limit than indicated in ESHA website [2]. A number of publications indicates that SHP impacts on the environment are not yet clearly revealed; therefore, they receive public support in the EU countries [3–5]. Well designed, monitored and sustainably operating SHPs may have a smaller impact on the ecosystem, if this is the case, they may be regarded as "environmentally friendly" [6,7]. The intensive land use for agriculture plays a more important role in terms of environmental impacts on the water quality of small lowland rivers than deployed SHP dams [8].

In contrast to the outline above, there are studies suggesting much more significant effects of SHPs [9,10]. The results of Kibler and Tullos [11] revealed that the biophysical impact of small hydropower plants in China might exceed even those of large hydropower plants, particularly with regard to habitat security and hydrologic change. However, it should be noted that in China, the small hydro capacity threshold is considered to be below 50 MW [1].

The EU Water Framework Directive (WFD) notes that hydropower has hydromorphological effects on the aquatic environment [12]. Many of these effects can be minimized; however, some effects are irreversible and it is impossible to achieve a good ecological status of the water body in accordance with WFD requirements [13].

Numerous metrics have been developed to represent the magnitude and direction of hydrologic alteration. The typical approach is the Flow Duration Curve (FDC) method, widely used in pre-post impact assessment of multipurpose uses of dams, e.g., hydropower [14–16]. Richter *et al.* [17], Olden and Poff [18] assessed and offered from three dozen to more than a hundred and fifty Indicators of Hydrologic alteration (IHA), which could be used for pre-post evaluations in different flow regime situations [19]. The prevailing opinion is that in most rivers, more than 90% of the flow variation can be explained by several statistically significant criteria [14]. Some hydrological indices are based on the average daily flow rate, which does not reflect irregularities of hydro plant operations within 24 h (e.g., hydropeaking or ramping). Moreover, not all indices can be applied equally to large and small hydropower plants. Any impoundment in spatial and temporal terms affects river flow regimes and global river environments. To represent quantitatively the spatial and temporal intensity of the water mass (inflow or outflow) in the control area, some indicators were proposed [20,21] which can be described using a number of terms e.g., Water Residence, Retention, Exposure, Turnover or Exchange time.

The ratio of storage capacity to mean annual flow has been used previously, for example by Brune [22], as the independent variable from which to predict reservoir trap efficiency. Impounded runoff index (IR) was used for large-scale dam affects to flow regime in California and Spain [23]. It should be noted that it is the dimensionless indicator. For example, an IR of 1 implies an average residence time of 1 year. Similarly to IR, a more detailed index D designated for reservoir hydropower plants with regard to the retention of cumulative inflow for a certain period of time at upstream storage was proposed [24]. According to the water retention time in reservoir or its filling period expressed in hours, HP plants can be divided into 3 groups: (a) run-of-river; (b) pondage; and (c) storage (Table 1). Water retention time (reservoir filling period D in hours) can be determined as follows:

$$D = \frac{V_u}{Q_0}, \left[\frac{m^3}{m^3/h} \right] \qquad (1)$$

where V_u—useful capacity of reservoir, designated for power generation, m^3 and Q_0—inflow into reservoir (annual mean flow), m^3/h.

The smaller the reservoir capacity and the higher the inflow into it, the less regulated natural flow will be assured. It can be accepted that RoR developments with $D \leq 2$ h have a relatively minimal impact on the river flow regime.

The alternative indicator closely related to water retention time is K (Table 2). It is a percentage of the useful reservoir capacity to the volume of annual inflow [16,25]. To distinguish between flowing waters (rivers) and stagnant waters (lakes or reservoirs) K limit was proposed $K = 100$ [26].

Table 1. Classification of hydropower plants according to the retention time of the cumulative inflow into reservoir.

The Mode of Operation or Development Type	Water Retention Time in Reservoir (D)	Comments
Run-of-river (RoR)	$D \leq 2$ h (~0.1 day)	No possibilities to significantly regulate flow
Pondage	$2\,h < D < 400\,h$ (~17 days)	Daily or weekly river flow regulation. Stores water at off-peak times and releases water through turbines at peak times
Storage	$D \geq 400$ h	Long-term impounding of water to meet seasonal and annual fluctuations in water availability. Not typical for an SHP

Note: Compilation is based on [24].

Table 2. Water retention indicator.

Water Retention Indicator K	1	10	100	200	500	1000	2000
Reservoir Useful Capacity, V_u (Percentage of Volume of Annual Inflow)	100	10	1	0.5	0.2	0.1	0.05
Water Retention Time (Days)	365	36.5	3.65	1.82	0.73	0.36	0.18
Water Body State	"Stagnant" (lake)		⟷	"Running" (stream or river)			

Although D and K are meaningfully related, their key estimates in the above tables have not been explicitly based. They only reflect the annual inflow transformation regardless of seasonal or shorter duration flow variations. In addition, RoR developments (small to medium hydro) can be described as follows [27–29]:

- Turbines use the natural flow of the river with very little alteration to the terrain stream channel at the site and little impoundment of the water.
- A type of hydro project that releases water at the same rate as the natural flow of the river (outflow equals inflow).

Run-of-river power projects are sometimes known as "small hydro" projects, though not all are small. References can be found in publications of hydropower plants with reservoir areas of several hundred square kilometres and more that are attributed to RoR projects [30]. For instance, in Austria, most of conventional large hydropower plants are considered RoR plants [31].

A simple sketch of a hydropower plant associated with a small reservoir provides additional explanations on the above mentioned indicators (Figure 1). The operation of some hydropower plants may lead to rapid and frequent changes in flow regimes, usually as a result of turbine start-up or shut-down, or in response to changes in electrical load. This mode of the operation, called hydropeaking or flow ramping, may have adverse impacts on the ecological integrity of streams and rivers. The negative effects of hydropeaking to the downstream river were first documented in Europe at the end of 1930. According to Bain [32], the first environmental impact studies in North America, related to this mode of operation, appeared in 1970s. Currently, there are plenty of investigations demonstrating the biological effects of this phenomenon, including its mitigation measures [33–36]. There is an obvious tendency to note that hydropeaking is mostly attributed to the hydro operations of storage power plants [37]. In run-of-river plants, hydropeaking is only observed following the emergency shut-down of turbines—a very rare occurrence, for instance, in Switzerland [38].

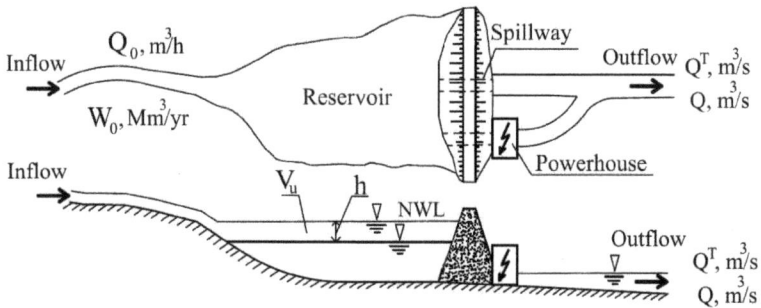

Figure 1. River flow distribution in a hydropower scheme associated with an embankment dam. Notes: Annual inflow into reservoir: Q_0, W_0 (flow and volume, respectively), h—drawdown depth of reservoir corresponding to the useful capacity of reservoir (V_u) and designated for power generation. Outflow: Q^T—turbine flow, Q—river flow (spill, leakage, instream flow). NWL—normal water level.

A decade ago, Baumann and Klaus [39] reviewed more than 200 articles indicating a clear lack of knowledge about the ecological effect of hydropeaking, the same was confirmed [40]. Despite the abundance of publications on the interrelations of hydropeaking and riparian or shallow zone ecology the ability to describe, correlate and quantitatively estimate biological responses on hydropeaking today is still weak [14]. Bain [32] reviewed 43 cases of hydropeaking in US and Europe. The effects ranged in accordance to their significance from moderate to severe alterations of downstream river environments. By analysing the hydropeaking mode practiced on three medium-sized HP ($P = 25$–50 MW), it was found that it was not a significant threat to the river environment and its biota. The restriction of flow ramping rates of hydroelectric turbines located on a river with a medium-sized HP in Canada did not show significant differences in the majority of measured biotic parameters relative to a natural system. However, once ramping restrictions were lifted, some changes in biotic parameters were found [15,41].

The requirement to restore the natural flow regime of hydropower projects shifting from the peak to run-of-river mode is becoming more and more popular in the US [42,43] and Europe, including Lithuania [44]. It is widely believed that the run-of-river operation has to be mandatory to protect aqua biota. However, it reduces the production of electricity per flow unit and at the same decreases incomes [45]. Technical measures have been studied to reduce the effects of hydropeaking [14,37,46–48]. They can be structured into two main groups: Operational (non-structural) and structural (involving any physical construction). Structural measures or the application of river engineering techniques cannot be easily realisable. In Canada and the US, the ramping rate standards (non-structural measures) provide guidance to run-of-river developers and operators for proposed or operating hydropower projects [49].

In Lithuania, more than 1000 small and medium sized reservoirs are currently registered, of which more than 90 are SHPs and only one large conventional power plant, operating at Kaunas on the country's largest river the Nemunas. It should be noted that the RoR operating mode for SHPs is an obligatory requirement in Lithuania.

An examination of the effects that small reservoirs have on the hydrological regime of Lithuanian rivers was carried out [16,50]. It was determined that the large, 100 MW Kaunas HP plant hydropeaking operations resulted in long-term river bed erosion on the Nemunas River [51]. In addition the SHP impacts on the downstream river flow were noted [52]. The findings from these studies indicate that no comprehensive research on the quantification of environmental issues due to rapid HP flow fluctuations on biotic systems in Lithuanian rivers was carried out. One exception was for macro invertebrate composition [53].

The main objective of this research was to investigate and quantify changes in the hydrological regime—abiotic indicators of a small lowland river with pre- and post-impoundments including further construction of small hydropower plants. To achieve this objective, the following tasks were carried out:

(1) Past research focusing on SHPs operating in run-of-river mode on temperate lowland rivers was reviewed;
(2) Historical flow and stage data to quantify the changes in flow regime pre- and post-river impoundment and after SHP construction was collected;
(3) Hydrograph ramping key characteristics using the hourly data of flow/stage downstream power plants to determine the causes was assessed;
(4) Measures to reduce the effects of SHPs by adapting turbines to the river natural flow were proposed.

Hydrologic regime here is understood in a simplified manner, notably, flow and stage fluctuations downstream from the river and storage timing upstream. The following are not in the scope of this paper: Impacts on water quality, riparian vegetation, fluvial geomorphology, and macro invertebrate and fish communities that must be underpinned by a sound hydraulic-hydrological analysis.

150

2. Materials and Methods

The study area is the Susve River Basin located in central Lithuania. There are two reservoirs, Angiriai and Vaitiekunai, which have SHPs installed downstream on the river (Figure 2). The Susve River is the largest tributary of the Nevezis River, which flows into the largest Lithuanian river—the Nemunas. Its length is 130 km, its catchment area is 1165 km^2 and the average annual flow is 6.2 m^3/s [54]. The average slope of the longitudinal river profile is 0.086%, with a gentle gradient upstream and a steeper gradient downstream. According to a number of authors [55], the catchment area over the past five decades has experienced significant land use changes, from deploying the tile drainage that reached 50% of the total catchment area in 1990, to increased woodland area (from 21% in 1960 to 31% in 2010). These changes influenced the hydrological regime, although estimates of the regime changes differ. The Susve River is not abundant in fish resources. The Lithuanian fish index LFI (0 = bad and 1 = very good) for the Susve River is between 0.19 below the Siaulenai gauging station (GS) and 0.42, on average, at its mouth. Sediment transport is not an issue in lowland rivers of the country. In this regard, the life span of reservoirs is long, and their useful capacity can be insignificantly changed during a long time period [16].

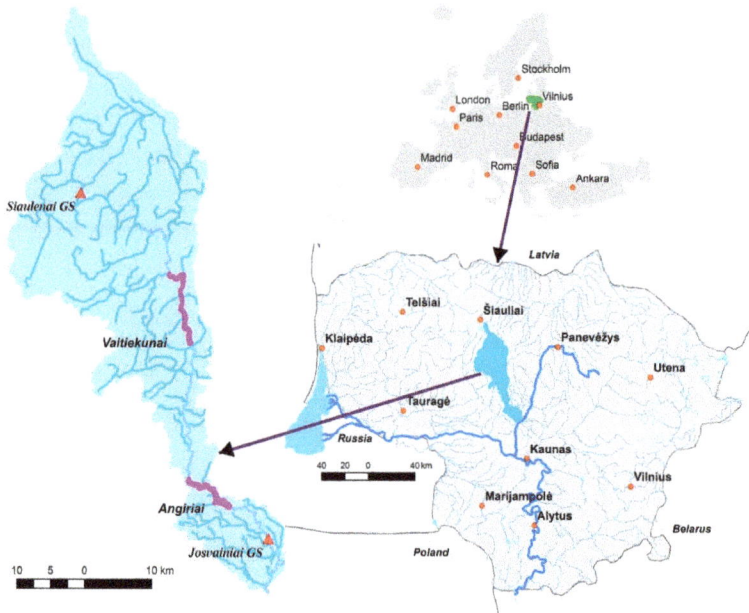

Figure 2. The catchment area of the Susve River with the location of the Angiriai and Vaitiekunai SHP reservoirs (purple color) and the Siaulenai and Josvainiai longstanding gauging stations (GS, red triangles).

Both reservoirs were built 35 years ago as water storage for irrigation at that time. After a couple decades, they were adjusted for hydropower (Table 3).

Table 3. Description of reservoirs and SHPs on the Susve River.

SHP Name	Distance of the Mouth, km	Catchment Area A km²	Mean annual Flow Q_0 m³/s	Year of Construction Reservoir/SHP	Reservoir			SHP				Reservoir Filling Period "D"
					Surface Area km²	Total Volume Mm³	Installed Power MW	Turbine Type and Number	SHP Discharge Q^T m³/s	Head, m	Q^T/Q_e	
Angiriai	25	1050	6.0	1980/1998	2.48	15.6	1.3	Propeller 2	10.2 (5.1 + 5.1)	14.5	15	23
Vaitiekunai	60	799	5.1	1979/2001	1.42	5.0	0.37	Cross-flow 2	5.5 (5.2 + 0.3)	9.7	1.4	15

The hydropower schemes involve an earth-fill dam with an integral and separated intake, and a powerhouse located at the dam toe (Figure 1). They are not traditional diversion schemes.

The Angiriai SHP in terms of the installed capacity is the third largest of the SHPs in the country, while Vaitiekunai SHP falls within the average. In line with the definition of International Commission on Large Dams [56], the dams are considered to be large (reservoir storage >3 Mm³, and dam height >5 m) and SHPs are medium head. Angiriai SHP turbines are propeller-type and consequently flow rate adjustment limits are very narrow. Cross-flow turbines are operating in the Vaitiekunai SHP; they are of varying capacity and flow regulation is very flexible. As a result, the ratio between the turbine minimum discharge Q^T_{min} and instream flow Q_e is 10 times less than in the former plant. The standard flow measurements at Siaulenai and Josvainiai longstanding gauging stations were used (Table 4).

Table 4. Characteristic of the Susve River gauging stations.

Gauging Station	Distance from the Mouth, km	A, km²	Data Series Length	Q_0	$Q_{0\ V-XI}$ $Q_{0\ XII-II}$	Q^{max} Q_0^{max}	Q^{max}/Q_0 Q_0^{max}/Q_0
Josvainiai	14.2	1080	1956–1999	6.2	3	272	47.4
			2003–2014		6.2	32.7	5.2
Siaulenai	108.6	162	1940–1999	1.29	1.29	64.7	50.2
			2000–2014		1.31	6.14	4.75

The observation period of both GS is longer than 50 years; the samples are representative and unbiased and are free from significant diversion or regulation prior to dam construction. In addition, short-term series of recorded water levels at 1 hour time intervals upstream and downstream from the dams (including turbines discharge) of the SHPs were used. The Siaulenai GS can be considered as a reference station—upstream there is no flow regulation, in contrast to the Josvainiai GS. The latter is located downstream of both reservoirs and the flow data integrates both land use impacts and the effects of reservoirs with SHPs.

When evaluating the effect of any impact on the hydrological regime, it is important that the data, which is divided into separate periods, should be assessed in relation to the variations of the long-term data. It is necessary to ensure that the data chosen covers the necessary cycles and determine whether they have downward or upward trends, therefore, a review was made of the Susve at Josvainiai annual flow fluctuations in a light of a long-term historic flow data series of the Nemunas River at Smalininkai (sample exceeds more than 200 years). For this purpose, the most commonly used methods for the assessment are: Cumulative deviations, moving average or trend detection using statistical tests. The Mann-Kendal test was applied for trend detection for the annual average flow at Smalininkai and Josvainiai GS. Flow duration curves (FDC) were used for comparisons of flow regime in three selected periods, and in two subbasins monitored by the Siaulenai and Josvainiai gauging stations. FDC is considered to be not only one of the most popular methods of comparison [19,57], in addition its statistical parameters also enable the formulation of appropriate findings [58]. Periods of construction are excluded from analysed flow series. To obtain a more accurate analysis when comparing two subbasins with different catchment areas, in some cases flow data are presented as water depths in millimetres per day.

The terms hydropeaking and flow (stage) ramping are interchangeable in view of hydrological response. However, no SHP in Lithuania operates under the peaking mode; therefore, the expression "ramping" is more accurate. The notions flow and stage (hydrograph) ramping here are used concurrently.

General hydropeaking indicators were used in the study [14,43,48,59,60]. The results indicate that flow variability statistics should be quantified using subdaily datasets to accurately represent the nature of hydropower operations, especially for daily peaking facilities. For the purposes of this article, "ramping rate" refers to the rate of change in water flow (in cubic meters per second/per hour) or water level range (depth per unit time—cm/h). There can be either up-ramping or down-ramping rates. The first ramping indicator, HR_1 (in the normalised form), represents river flow variations with respect to the river daily average flow due to start-up and shutdown turbines:

$$HR_1 = \frac{Q_{max}^T - Q_{min}^T}{Q_0} \qquad (2)$$

where Q_{max}^T—turbine maximum discharge within 24 h; Q_{min}^T—turbine minimum discharge within 24 h and Q_0—river daily mean discharge (inflow into reservoir).

If turbine flow is steady within a daily period, then the above difference in the numerator does not make sense: The ramping does not occur. The ideal case is when $HR_{1d} = 1$ (inflow to a reservoir equals outflow).

The second indicator, ramping rate (HR_2), describes the gradient in stage (or flow) change and is defined as the change in water level (flow) between two successive records divided by the observation time interval, as defined by:

$$HR_{2,i} = \frac{WL_i - WL_{i-1}}{t_i - t_{i-1}} \qquad (3)$$

where WL_i—water level, cm at time t_i, hour and WL_{i-1} water level at prior time t_{i-1}.

There is no special flow or level ramping restrictions in force for SHPs in Lithuania. Restrictions are imposed only on SHP reservoir water level drawdowns (±10 cm of NWL), which, as observations show, have little effect on ramping characteristics downstream of power plants.

A field experiment was carried out to discover the mechanism of stage ramping at the Angiriai SHP downstream the river [16]. Before starting up two turbines at their full capacity (Q^T = 10.2 m^3/s), inflow into the reservoir and outflow was minimal (only environmental flow—approx. 0.4 m^3/s). It was the end of summer, which is the low flow season; the river channel was extensively vegetated. Water level data logger and gauge staffs (3 in total) were used to monitor water level fluctuations and traveling of a wave along the river reach at 4 locations (last one at some 8 km from SHP) induced by starting and shutting down turbines.

3. Results and Discussion

3.1. Pre- and Post-Impoundment Hydrologic Changes of the Susve River

Moving 11-year averages of long-term data show that the Susve River follows synchronical fluctuations of the Nemunas River and that the Susve flow covers two cycles of long-term fluctuations (Figure 3). According to the Mann-Kendal test, no statistically significant trends are detected for the annual average flow at Smalininkai and Josvainiai GS.

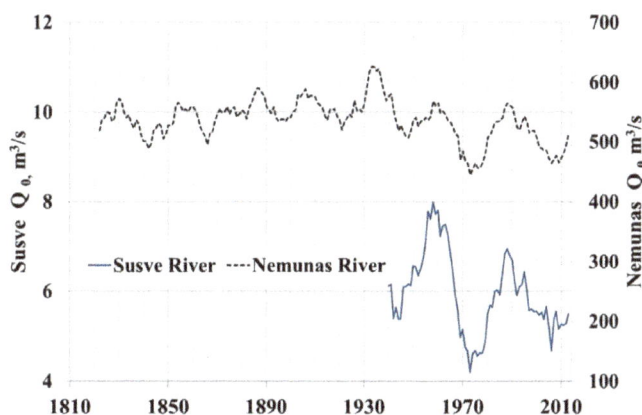

Figure 3. 11-year moving average flow (Q_0) graphs for the Susve River at Josvainiai and the Nemunas River at Smalininkai.

Taking into account long-term fluctuations and knowing the dates of human interventions, the following three periods were selected:

- Free flowing river (1956–1978);
- Regulated river (2 impoundments in place) (1981–1997);
- 2 SHPs deployed (2003–2014).

River flow regime can be clearly observed in daily flow hydrographs (Figure 4), representing each selected period.

154

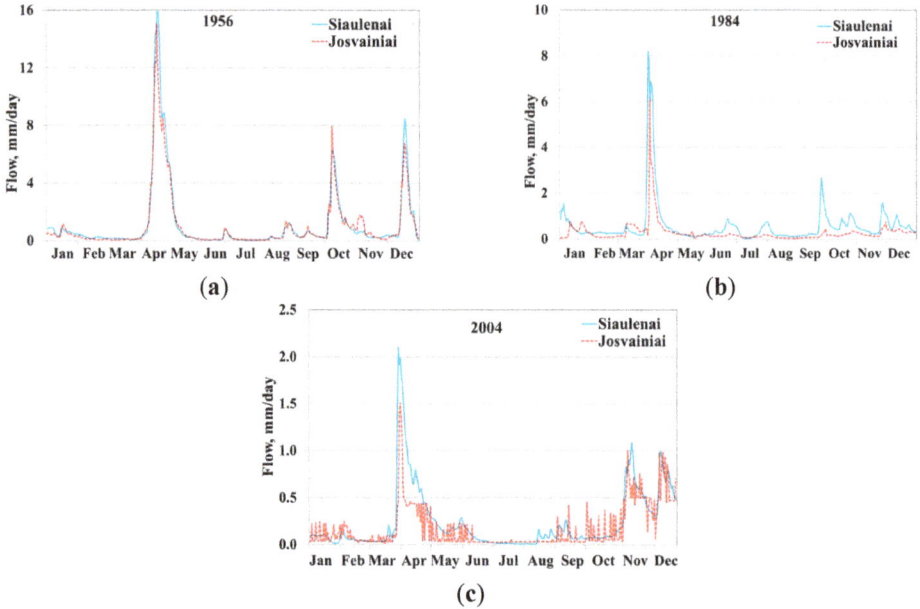

Figure 4. Daily flow of selected years from different periods at Siaulenai (reference station, control river—no impact), and Josvainiai (impact of reservoirs and SHP) gauging stations. (**a**) 1956—pre-impoundment: No reservoirs at either subbasin; (**b**) 1984—post-impoundments: Vaitiekunai and Angiriai reservoirs without SHP; (**c**) 2004—SHPs are operating.

The year 1956 represents the first period showing synchronical flow at both stations and hydrographs almost overlapping with each other. In 1984, decreases of peak flow can be observed at Josvainiai GS due to what is known as the effect of flow transformation in the reservoir. In the third graph (2004), we already see the saw-type hydrograph of the Josvainiai GS, which is completely different from the Siaulenai GS, especially in low flow periods. Figure 4 illustrates evident impacts but does not give us the magnitudes of the impacts. The main parameters of descriptive statistics are presented in Table 5.

Table 5. Statistical parameters of analysed daily flow (mm/day) data.

Period	Gauging Station	Mean	Min	Max	STD	Coefficient of Variation	Skewness	Kurtosis
1956–1978	Siaulenai	0.64	0.0011	24.75	1.30	2.04	7.245	81.01
	Josvainiai	0.45	0.0096	21.76	1.05	2.33	7.528	87.13
1981–1997	Siaulenai	0.68	0.0037	10.72	1.19	1.75	3.846	18.16
	Josvainiai	0.48	0.0004	9.92	0.81	1.67	4.442	27.74
2003–2010	Siaulenai	0.58	0.0085	6.08	0.84	1.46	2.765	9.07
	Josvainiai	0.44	0.0152	6.18	0.65	1.47	2.997	12.67

The high magnitudes of skewness indicate that no one analysed time series follows a normal distribution, although this parameter consistently decreases with each period. That points to the fact of influence of the construction of reservoirs, and the later installation of hydropower plants, on flow regime. Decreasing variation coefficients indicate that the flow regime becomes smoother in last two periods when the reservoirs are constructed and HP are installed. However, general statistics do not explain the impacts on seasonal daily flows.

Comparison of FDC in each period between two gauging stations—Siaulenai and Josvainiai—is presented in Figure 5 and a comparison FDCs of different periods at Josvainiai GS is presented in Figure 6. In the first period (1956–1978) (Figure 5a), the FDCs of two subbasins represent the situation without reservoirs in either subbasin. Curves do not overlap each other because of a different base flow that is slightly higher in the upper subbasin. The second period (1981–1997) (Figure 5b) gives almost overlapped curves in the lower part. Erected reservoirs slightly increase downstream flow in the dry season. This effect lifts up the lower part of Josvainiai FDC. The last period (2004–2014) (Figure 5c) gives the sharpest difference in the lower part of Josvainiai FDC. This can be explained by analysing the saw-shape hydrographs (Figure 4) of the last period. All of the peaks in the hydrographs are due to pulses when SHP turbines are started up or shut down. According to flow magnitudes, these pulses in FDC are moved to the left. At the same time, the rest of the flow magnitudes of the dry period are lower comparing to the Siaulenai GS flow what is the main reason of FDC drop down in interval from 60% to 90%.

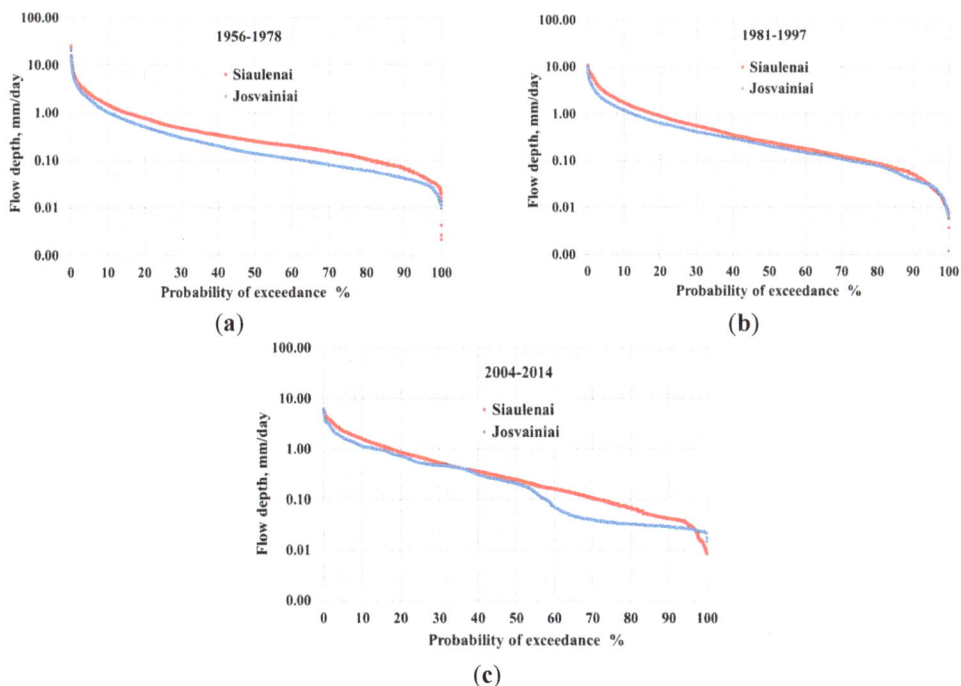

Figure 5. Comparison of FDCs in each period between two gauging stations—Siaulenai and Josvainiai. (a) The first period (1956–1978); (b) The second period (1981–1997); (c) The last period (2004–2014).

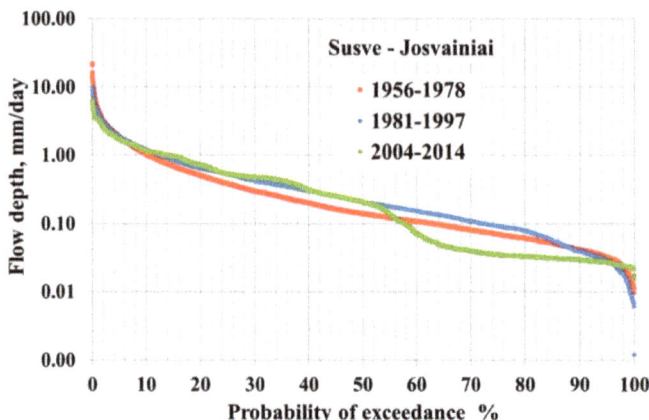

Figure 6. Comparison of FDCs of different periods at Josvainiai GS.

To ensure that there are no impacts of any unknown factors that are upstream or downstream flow on the form of FDC, we compared curves of Josvainiai GS with each other (Figure 6). Here we see a similar situation. The FDC of the first two periods almost overlaps in the lower part and the third one differs greatly. To confirm the obtained visual facts, the FDCs of Josvainiai have been approximated applying Pearson type III distribution (Table 6). Both parameters—bias and RMSE—90 day and 180 day minima indicate the difference of the third period from the first two.

Table 6. Comparison of statistical parameters of approximated FDC by Pearson type III distribution.

Period	FDC Parameters		90 Day Minima		180 Day Minima	
	Mean Q_0 m^3/s	Q_{95} m^3/s	RMSE	Bias m^3/s	RMSE	Bias m^3/s
1956–1978	0.450	0.0319	0.0117	−0.2059	0.0168	−0.2294
1981–1997	0.482	0.0280	0.0137	−0.3518	0.0175	−0.2377
2004–2014	0.446	0.0264	0.0306	−1.6032	0.0834	−2.9760

3.2. River Flow Alterations by Upstream Storage

To date, more than 90 SHP plants are operating in the country. Taking into account the fact that drawdown depth in SHP reservoirs is very limited (±0.1 m of NWL), they cannot be considered as highly dynamic water storage systems. Consequently, SHPs are unlikely to provide peak power because of the limited capacity of upstream storage volume.

Referring to Tables 1 and 2, the indicators D and K were determined for the reservoir useful storage (drawdown depth of 1 m) and power storage (Figure 7). The latter corresponds to the depth $h = 0.2$ m and is exclusively used for the power generation.

Nearly half of the SHPs (42 out 90) can be viewed as run-of-river types, if the reservoir drawdown depth of 0.2 m designed for power generation is considered (water retention time in reservoir $D_{0.2m} < 2$ h). For the SHPs under consideration, Angiriai and Vaitiekunai, this indicator ranges between 23 hours and 15 hours, respectively, and their magnitudes correspond to the pondage mode of operation (Table 1). By contrast, this indicator for the 100 MW Kaunas HP reservoir (A = 63.5 km², h = 0.8 m or ±0.4 m of NWL) operating exclusively at peaking mode is 2 to 3 times bigger ($D_{0.8m}$ = 50 h).

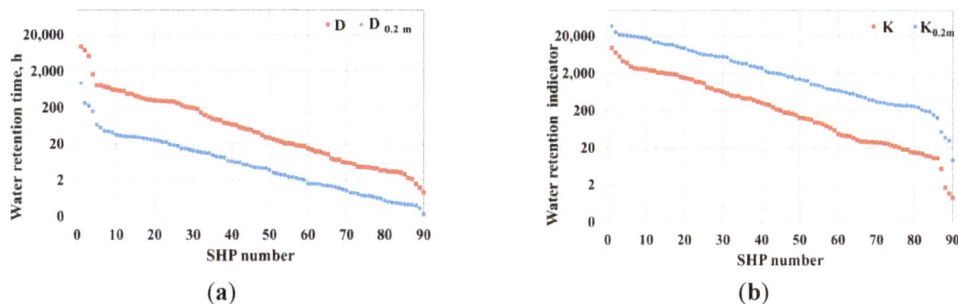

Figure 7. Water retention indicators of SHP reservoirs. D **(a)** and K **(b)** determined for the reservoir useful and power storage. The subscript 0.2 m indicates a drawdown height of 0.2 m in reservoir.

According to the indicator $K_{0.2m}$ almost all reservoirs can be considered as run-of-river reservoirs, with a few exceptions. On an annual scale, water is renewed every day in the Angiriai reservoir and every half-day in the Vaitiekunai reservoir ($K_{0.2m}$ = 381 and 572, respectively). These indicators reflect the operations of reservoirs located on small or medium sized lowland rivers, because the reservoirs do not experience significant seasonal or shorter-term water level fluctuations due to relatively moderate river flow regimes.

3.3. Flow (Stage) Ramping

3.3.1. Mechanics of Ramping

The primary type of turbine operations causing flow ramping downstream in the river reach is turbine start-up and shutdown, not a response to changes in electrical loads in Lithuania. Observations show that ramping occurs at irregular intervals during 24 hours or weekday periods that is different from the hydropeaking mode of operation, where constant flow is mostly observed at weekends. Ramping consequences are most crucial in the lean period of the river flow, especially when the inflow into an SHP reservoir is minimal. The main causes of flow ramping are as follows:

- Operators, believing that they are producing more power are deliberately starting/shutting down turbines frequently during day/night periods for a certain number of hours. During the remaining time, turbines are operating at minimum flow or they can be completely stopped to comply with the prescriptions of instream flow. This mode of operation is inappropriate because energy output will be the same if turbines are operated at a reduced capacity but with stable patterns during 24 hours.
- Turbines are not well adapted to the natural streamflow regime. This means that the design discharge of the turbines is too high, and control of the discharge flowing through the turbines is not flexible. In particular, this is evident for Angiriai SHP, which has a very high design discharge of propeller-type turbines ($Q^T \approx 2\ Q_0$).

158

On the other hand, these artificial downstream flow (stage) fluctuations are due to the presence of an impoundment, and they cannot be completely avoided. Completely different geomorphological properties of the reservoir and the river channel downstream from the plant determine the significance of water level fluctuations. In other words, the upstream storage cross-sectional area is many times larger than that of downstream river (Figure 1). For instance, the drawdown depth of 1 cm in the Vaitiekunai reservoir results in an increase of 23 cm in the downstream water level (Q^T = 0.87 m³/s). Approximately the same amount of drawdown in the Juodeikiai reservoir (the Varduva river) incurs a rise in downstream water levels of 52 cm (Q^T = 2.42 m³/s). These findings were obtained during a low water season (the end of summer) when inflow into the reservoir was extremely minimal and no turbine was able to operate. It is absolutely clear that damping the range of water stage fluctuations is very problematic. Additionally, the requirement to maintain upstream storage water levels as stable as possible (±10 cm of NWL) is not an effective means of regulating the range of downstream stage fluctuations.

3.3.2. Downstream River Stage Fluctuations

The operation mode of Angiriai SHP obviously changes the river's natural flow and stage variations, which were recorded 10 km downstream of the plant (Figure 8). 1971 and 1972 correspond to the years with the reservoir and 2003 and 2005 complemented with SHP. For the latter, a saw-toothed pattern is clearly seen. This hydroelectric facility is a pondage type ($D_{0.2m}$ = 23 h), although its hydrograph ramping has nothing to do with peak power demand.

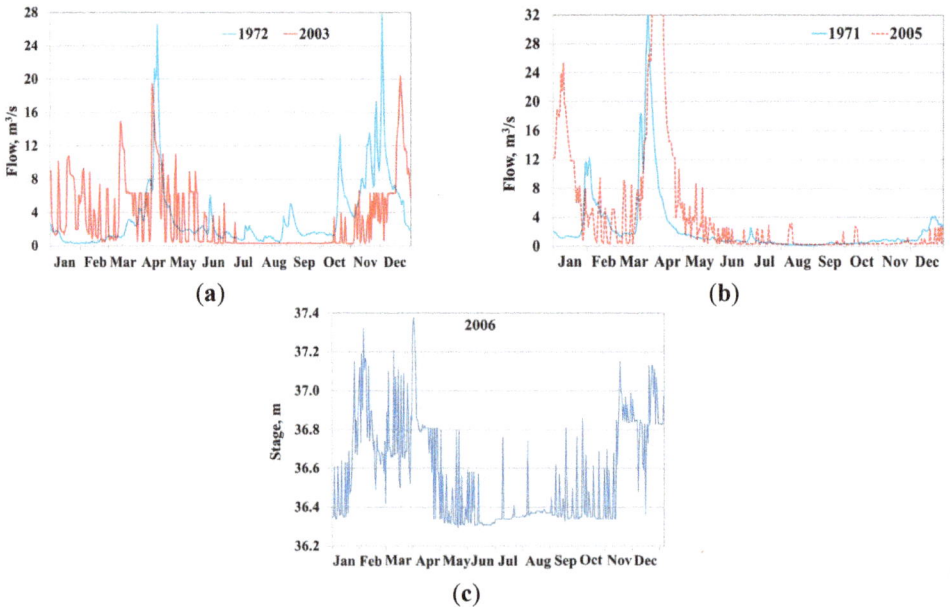

Figure 8. Daily hydrographs (flow and stage) before the construction of Angiriai SHP (1971 (**b**), 1972 (**a**)) and under its operation (2003 (**a**), 2005 (**b**), 2006 (**c**)). Gauging station Susve at Josvainiai.

More obvious are the hourly flow and water level fluctuations downstream. For this SHP, the stage range can reach up to 70 cm within a 24 h period (Figure 9). For the country's SHPs, the stage range at their tailwater varies between 10 and 70 cm, and for large, peaking Kaunas HP it can reach even 150 cm during a 24 h period [16,61]. This is because the turbines are not well adapted to the natural stream flow regime, which means that the design flow of the turbines is too high and the control of the discharge passing through the turbines is not flexible.

The ratio between turbine minimum flow and instream flow (Q^T/Q_e) varies between 15 for Angiriai SHP and 1.4 for Vaitiekunai SHP. For comparison, this ratio in Swiss and Austrian rivers with operating hydropower facilities ranges from 2 to more than 50 [62]. Bain [32] reported extremely high ratios (maximum/minimum flows) for large hydropower facilities of 200 to 1000 (average of 166) in some rivers in the United States, Canada, Finland and France, and these rivers were classified as having moderate to severe alterations of aquatic ecosystems. Up-ramping and down-ramping stage rates (HR_2) for Angiriai SHP are similar (25–30 cm/h), with slightly more intensive rising limbs.

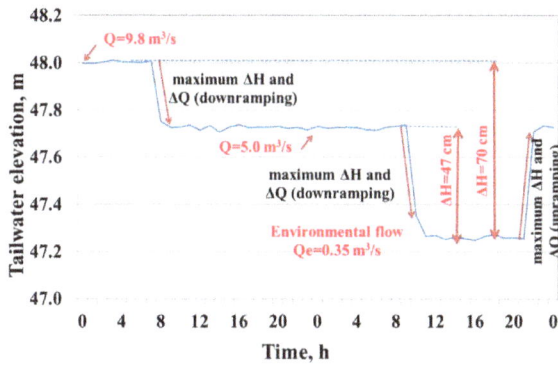

Figure 9. Hourly tailwater level (*H*) fluctuations at Angiriai SHP. At the beginning, 2 turbines ($Q^T = 9.8$ m³/s) are working, then 1 turbine was shut down ($Q^T = 5$ m³/s) and finally both are shut down. Later, 1 turbine was started up again (3–4 March 2009). Inflow into the reservoir $Q_0 = 3.7$ m³/s, $HR_1 = 1.33$.

Open channel hydraulics and hydrological theory demonstrate that hydrograph is modified as water flows downstream a river channel [63–65]. It means that flow ramping attenuates as a function of the distance downstream from the turbines. However, this attenuation (for instance, stage range) over quite a long distance is not impressive, with a lagging time of approximately 4 hours (Figure 10).

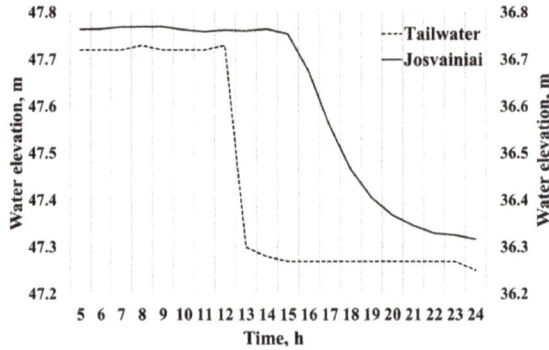

Figure 10. Stage dynamic at the tailwater of Angiriai SHP after shutting down 1 turbine (Q^T = 5 m³/s, HR_1 = 1.22) and 10 km downstream river at the Josvainiai gauging station. 10 May 2013. HR_1 = 1.22.

The stage down-ramping rate at the power plant tailwater is 43 cm/hour, and at a distance of 10 km downstream it decreases to 7 cm/hour and approaches the natural hydrograph recession rate upper limit.

The same path was confirmed by a field experiment to determine the mechanics of travelling artificial flood wave downstream at the start-up and shut down of the turbines (Figure 11).

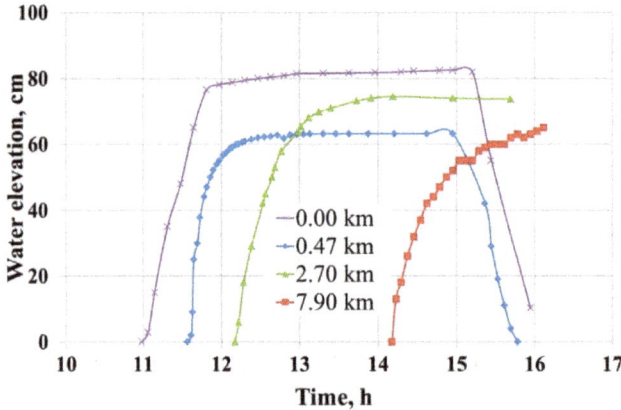

Figure 11. Stage hydrographs at different distances of the release of turbine water at Angiriai SHP. 18 August 2009, on the Susve River. Two turbines were started at a full capacity and shut down after 3 h. (Elevation not to datum).

It must be underlined that this was an unusual mode of turbine operation. After starting the turbines, the velocity in the tailwater was 1 m/s; later, due to significant channel resistance (vegetation, sinuosity, *etc.*) it was reduced to 0.8 m/s. As expected, the maximum stage up-ramping rate (HR_2 = 90 cm/h) was recorded at the turbine water outlet—the tailrace. This is obvious because of a rapid increase in discharged flow by turbines (from 0 to 5.2 m³/s and very swiftly to 10.5 m³/s). At the second transect (0.47 km below the powerhouse), the up-ramping rate (HR_2 = 90 cm/h) did not change, though it progressively lessened flowing downstream from the third and fourth transects (60 and 70 cm/h, respectively). The down-ramping rate was also significant—$HR_2 = 75$ cm/h—at the second transect. Stage range diminished progressively through the river length from 80 cm to 60 cm, but its value was dependent on the channel geometry. This and other recently completed experiments on SHPs confirmed the initial hypothesis that the negative effects of turbine surge waves can be felt a long distance downstream the river. To mitigate this effect, turbine start-up and shutdown times should be extended as much as possible. Hydropower operations must begin and end with the turbine passing the least flow possible.

The analysis of historic flood records of the Susve River (natural river flow and pre-impoundment) reveals that the daily maximum stage rising is 6 cm/h and its recession is two times lower, 3 cm/h. Hourly rates are at least 1.5 times higher, approximately 9 and 5 cm/h, respectively. This bias is due to a smaller interval of measurements. Downstream from the Angiriai SHP, these estimates are exceeded at least 3 times. One important issue must be mentioned, which is that the frequency of natural floods is low compared with moderate artificial flooding caused almost daily by turbine operations.

Down-ramping rates of less than 5 cm/h are naturally observed after flood events in alpine rivers. Gentle flow ramping slower than 10 cm/h drastically decreases the stranding of juvenile fish [14]. An average value of 12–20 cm/h for the down-ramping rate is generally seen as an ecologically acceptable value [61]. Of course, rates are determined by channel geometry and hydraulic parameters, which vary temporally and spatially. Pools, side-channels, and gravel bars attenuate the ramping rate by storing water from higher flows and release this water gradually. Tributary inflow will attenuate the ramping rate and the ramping range.

The highest stage range magnitudes are observed for SHPs with high values of turbine design flow, mostly exceeding $Q_d^T > (1.5...2.0)Q_0$, with unregulated turbines and minimal turbine numbers. However, the latter is seldom valid for the double regulated Kaplan and crossflow turbines, which are able to adapt well to the natural flow regime.

3.4. Assessment of Turbine Types with Regard to Possible Alterations of River Flow

It is convenient to present a river flow regime in the form of a flow duration curve (FDC), which was considered in the previous section. In hydropower engineering, the FDC serves as a basis for determining the number of turbines and the installed capacity. To preserve as possibly as natural river flow pattern and assure a reasonable use of water resources for power generation the area under this curve must be completely used by water discharged from turbines.

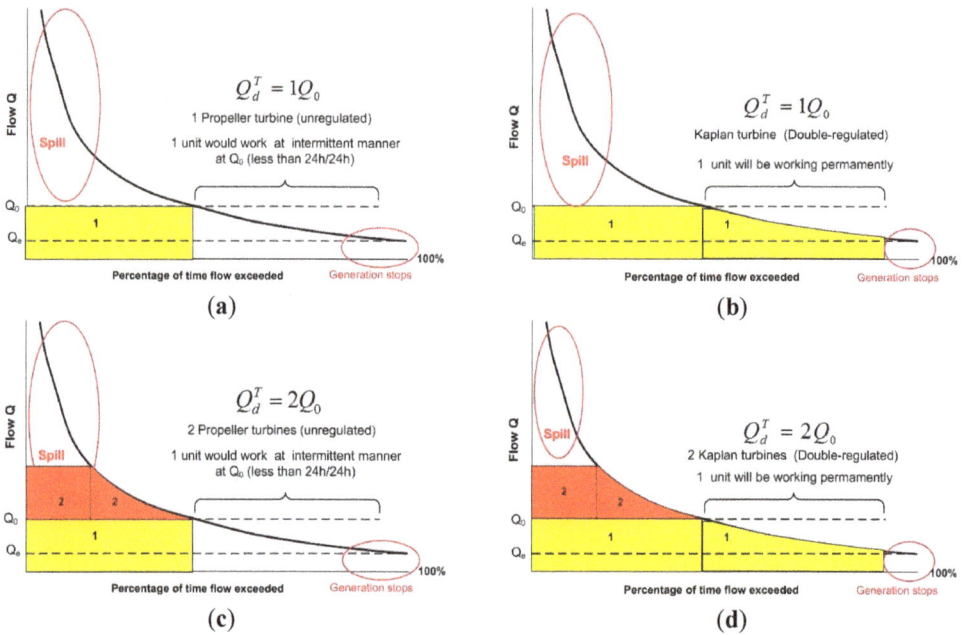

Figure 12. (a) 1 Propeller turbine, $Q_d^T = 1Q_0$; **(b)** 1 Kaplan turbine, $Q_d^T = 1Q_0$; **(c)** Two propeller turbines, $Q_d^T = 2Q_0$; **(d)** Two Kaplan turbines, $Q_d^T = 2Q_0$. Hypothetical FDC and its coverage by the propeller and Kaplan turbines (from 1 to 2 units). Q—daily flow; Q_d^T—design discharge of turbines; Q_0—mean annual flow; Q_e—environmental (instream) flow. Red and yellow indicate turbines working field.

However, for economic and technical considerations, neither the upper part (high water) nor the lower part (low water) of an FDC can be practically used (Figure 12). In practice, hydropower schemes generally have the greatest impacts on mid-range river flows and low flows [66]. A clear difference is seen between different types of turbines (Kaplan and propeller) and their number of "covering" FDC. The bigger the number of turbines and their flexibility to pass a larger range of flow through them, the better the coverage is. It is obvious that the Kaplan turbine (there might be other types as well) outperforms the propeller turbine (unregulated) in terms of using natural flows much larger than its range. The installed capacity of a plant, which is closely related to design discharge, cannot always play a definitive role; the number of turbines and their ability to adjust to a variety of river flows is much more important. As a result, the risk for intermittent modes of operation would be minimised (with some exceptions at the lower end of FDCs) and possible ramping avoided.

Typically, a hydropower developer chooses a design flow for the scheme that allows it to use a good proportion of the higher flows but also to continue to operate down to reasonably low flows so that output can be sustained for as much of the year as possible. In the UK, for low head schemes the common practice has been to use Q_0 flow as the design flow [66]. For SHPs operating in Lithuania, the maximum design flow of turbines exceeds $1.5Q_0$. The more advanced and higher the turbine's technical performance is, the more adaptive it will be to natural stream flow regimes (Table 7). The Kaplan double regulated turbine fits any river flow regime.

From the above, it can be concluded that the risk of emerging flow ramping is much greater in flashy rivers than in rivers with little variation in flow regime.

Table 7. Recommended turbine type taking into account river flow regime.

Turbine Type		Description of Most Suitable River Flow Regime
Kaplan and propeller	Propeller (unregulated)	Only for low variation in flow regime, not suitable for flashy rivers. This means that the low flow has a considerable proportion of the mean flow. The FDC must have a very flat slope. *
	Kaplan (single regulated)	Moderate variation in flow regime
	Kaplan (double regulated)	Any variation in flow regime (any shape of FDC)
Francis		Only for low variation in flow regime. Not suitable for streams with initially steeply sloped FDCs
Cross-flow (Ossberger, Banki-Michell, Cink)		Any variation in flow regime

* A good indication for variation in river flow regimes can be given by the shape of the FDC (for daily discharges), which can be expressed numerically.

4. Conclusions

(1) The basic statistical test allows us to conclude that no significant trends were detected in the long-term hydrological data series and that the employed data represents necessary hydrological cycles.

(2) The FDC derived from long-term records consisting of pre-post impoundment and SHP development does not markedly differ in the first and second period, but differs in the third period, especially the lower part of the curve, this points to possible impact of the Angiriai SHP.

(3) Water retention time (D and K) is a simple and good indicator to evaluate the significance of probable impacts of an impoundment on river flow. Despite this fact, the proposed threshold values need to be based more scientifically.

(4) Downstream river flow (stage) ramping is an environmental issue for large hydro, and small hydropower alike. The operation of small run-of-river power plants that do not necessarily follow energy peak demand could possibly result in this negative phenomenon.

(5) More intensive fluctuations in the downstream river flow and stage can be observed in SHPs with high turbine design flows, unregulated types of turbines and a low number of turbines.

(6) When an SHP is not intended for operations covering peak energy demand, its design turbine flow should not exceed the mean annual river discharge. There are currently a large number of advanced turbines available. The turbines have a wide range of capacities, there are also advances in turbine design (e.g., double regulated or cross-flow). These features will allow for adapting higher values of design flows with a minimum risk of flow ramping.

164

(7) By applying simple turbine operational measures—step-wise turbine start up and shut-down together with varying their number and capacities during 24 h—river flow ramping rates can be substantially alleviated.

(8) Recommended turbine types are most suitable for a particular natural flow regime. However, total avoidance of downstream hydrograph ramping is not possible without applying structural measures (involving physical constructions) for run-of-river projects with impoundments.

Acknowledgments

The authors would like to thank Kelly Kibler of University of Central Florida and the anonymous reviewer for investing their spare time in reviewing this manuscript, for their valuable comments and suggestions in order to help improve this manuscript. Some information for this paper was used from the RESTOR Hydro project (Intelligent Energy-Europe Program, 2012–2015, coordinated by the European Small Hydropower Association) and past and ongoing studies funded by the Lithuanian Environment Agency and the Research Council of Lithuania.

Author Contributions

Petras Punys designed this research and gave the whole guidance; Antanas Dumbrauskas and Gitana Vyčienė largely wrote pre- and post-hydrological changes, Egidijus Kasiulis and Linas Šilinis collected all the data, carried out calculations, result display and analysis. All authors read and approved the final manuscript.

Conflicts of Interest

The authors declare no conflict of interest.

Nomenclature

SHP	small hydropower plant
HP	hydropower plant
Q_0	mean flow
Q^T	turbine discharge
Q^T_d	design flow of turbine
Q_e	environmental (instream) flow
D	water retention time in a reservoir (reservoir filling period)
K	water retention indicator related to water retention time in a reservoir
FDC	mean daily flow duration curve
RoR	run-of-river (HP operation mode)
GS	river flow and stage gauging station
h	drawdown depth of a reservoir needed for power generation
WL, NWL	water level, reservoir normal water level

References

1. Jia, J.; Punys, P.; Ma, J. Hydropower. In *Handbook of Climate Change Mitigation*; Springer Science: New York, NY, USA, 2012; pp. 1357–1401.
2. Elizabeth Stewart Hands and Associates (ESHA). The European Small Hydropower Association. Available online: http://www.esha.be/ (accessed on 6 March 2015).
3. Punys, P.; Pelikan, B. Review of small hydropower in the new member states and candidate countries in the context of the enlarged European Union. *Renew. Sustain. Energy Rev.* **2007**, *11*, 1321–1360.
4. Arcadis. Hydropower Generation in the Context of the EU WFD. Report to EC DG Environment. Project No 11418, 2011. Available online: http://www.arcadis.de/Content/ArcadisDE/docs/projects/11418_WFD_HP_final_110516.pdf (accessed on 10 June 2015).
5. Elizabeth Stewart Hands and Associates (ESHA). Small Hydropower Roadmap. Report. Condensed Research Data for EU-27. The Stream Map Project, 2012. Available online: http://streammap.esha.be/fileadmin/documents/Press_Corner_Publications/SHPRoadmap_FINAL_Public.pdf (accessed on 9 June 2015).
6. Reihan, A.; Loigu, E. Small hydropower in Estonia—Problems and perspectives. In Proceedings of the European Conference on Impacts of Climate Change on Renewable Energy Sources, Reykjavik, Iceland, 5–9 June 2006.
7. Abbasi, T.; Abbasi, S.A. Small hydro and the environmental implications of its extensive utilization. *Renew. Sustain. Energy Rev.* **2011**, *5*, 2134–2143.
8. Vaikasas, S.; Bastiene, N.; Pliuraite, V. Impact of small hydropower plants on physicochemical and biotic environments in flatland riverbeds of Lithuania. *J. Water Secur.* **2015**, *1*, 1–13.
9. Kubecka, J.; Matena, J.; Hartvich, P. Adverse ecological effects of small hydropower stations in the Czech Republic: 1. Bypass plants. *Regul. Rivers Res. Manag.* **1997**, *13*, 101–113.
10. Fu, X.; Tang, T.; Jiang, W.; Li, F.; Wu, N.; Zhou, S.; Cai, Q. Impacts of small hydropower plants on macroinvertebrate communities. *Acta Ecol. Sin.* **2008**, *28*, 45–52.
11. Kibler, K.M.; Tullos, D.D. Cumulative biophysical effects of small and large hydropower development, Nu River, China. *Water Resour. Res.* **2013**, doi:10.1002/wrcr.20243.
12. European Commission. Directive 2000/60/EC of the European Parliament and of the Council of 23 October 2000 establishing a framework for Community action in the field of water policy. *Off. J. Eur. Communities* **2000**, *L327*, 1–73.
13. European Commission. Common Implementation Strategy for the Water Framework Directive. WFD and Hydro-Morphological Pressures Policy Paper. Focus on Hydropower, Navigation and Flood Defense Activities. Recommendations for Better Policy Integration. Available online: http://www.sednet.org/download/Policy_paper_WFD_and_Hydro-morphological_pressures.pdf (accessed on 6 March 2015).
14. Meile, T.; Boillat, J.L.; Schleiss, A.J. Hydropeaking indicators for characterization of the Upper-Rhone, river in Switzerland. *Aquat. Sci.* **2011**, *73*, 171–182.

15. Smokorowski, K.E.; Metcalfe, R.A.; Jones, N.E.; Marty, J.; Niu, S.; Pyrce, R.S. Flow management: Studying ramping rate restrictions. *Hydro Rev.* **2009**, *28*, 68–87.
16. Aplinkos apsaugos agentūra. Aplinkosauginių rekomendacijų hidroelektrinių neigiamam poveikiui sumažinti parengimas (Report to EPA on environmental measures to reduce the negative impact of hydropower plants). Aplinkos apsaugos agentūra: Vilnius, Lietuva, 2010; p. 316. (In Lithuanian)
17. Richter, B.D.; Baumgartner, J.V.; Powell, J.; Braun, D.P. A method for assessing hydrologic alteration within ecosystems. *Conserv. Biol.* **1996**, *10*, 1163–1174.
18. Olden, J.D.; Poff, N.L. Redundancy and the choice of hydrologic indices for characterizing streamflow regimes. *River Res. Appl.* **2003**, *19*, 101–121.
19. Magilligan, F.J.; Nislow, K.H. Changes in hydrologic regime by dams. *Geomorphology* **2005**, *71*, 61–78.
20. Rueda, F.; Moreno-Ostos, E.; Armengol, J. The residence time of river water in reservoirs. *Ecol. Model.* **2006**, *191*, 260–274.
21. Delhez, E.J.M.; de Brye, B.; de Brauwere, A.; Deleersnijder, E. Residence time *vs.* influence time. *J. Mar. Syst.* **2014**, *4*, 185–195.
22. Brune, G.M. The trap efficiency of reservoirs. *Trans. Am. Geophys.* **1953**, *34*, 407–418.
23. Kondolf, G.M.; Batalla, R.J. Hydrological effects of dams and water diversions on rivers of mediterranean-climate regions: Examples from California. In *Catchment Dynamics and River Processes: Mediterranean and Other Climate Regions Developments in Earth Surface Processes*; Batalla, R.J., Ed.; Elsevier: Amsterdam, The Netherland, 2005; pp. 197–211.
24. Unipede-Eurelectric. *Statistical Terminology Employed in the Electricity Supply Industry*; Unipede-Eurelectic: Brussels, Belgium, 1991.
25. Punys, P.; Sabas, G. Small hydropower operations and natural hydrological regime. Case study in Lithuania. In Proceedings of the Hidroenergia 2012: International Congress and Exhibition on Small Hydropower, Wroclaw, Poland, 23–26 May 2012.
26. Hursie, U. Designation of HMWB & GEP. In Proceedings of the Workshop Water Framework Directive and Heavily Modified Water Bodies, Brüssel, Belgium, 12–13 March 2009.
27. Warnick C.C. *Hydropower Engineering*; Prentice-Hall: Englewood Cliffs, NJ, USA, 1984.
28. American Society of Civil Engineers (ASCE). *Civil Engineering Guidelines for Planning and Designing Hydroelectric Developments*; Small Scale Hydro. ASCE: New York, NY, USA, 1989; Volume 4, p. 333.
29. RETScreen® International. RETScreen® Engineering & Cases Textbook. Small Hydro Project Analysis Chapter. Natural Resources Canada, 2004. Available online: http://www.retscreen.net/ (accessed on 11 June 2015).
30. Douglas, T. "Green" hydro power understanding impacts, approvals, and sustainability of run-of-river independent power projects in British Columbia. Watershed Watch Salmon Society. Available online: http://www.watershed-watch.org/publications/files/Run-of-River-long.pdf (accessed on 6 March 2015).
31. World Atlas & Industry Guide, 2013. *The International Journal on Hydropower & Dams*; Aqua Media International Ltd.: Wallington, UK, 2013.

32. Bain, M.B. *Report: Hydropower Operations and Environmental Conservation: St. Marys River, Ontario and Michigan*; International Lake Superior Board of Control: Conrnwall, ON, Canada; Cicinnati, OH, USA, 2007.

33. Hunter, M.A. *Hydropower Flow Fluctuations and Salmonids: A Review of the Biological Effects, Mechanical Causes and Options for Mitigation*; Report No.119; State of Washington, Department of Fisheries: Washington, DC, USA, 1992.

34. Harpman, D.A. Assessing the short-run economic cost of environmental constraints on hydropower operations at Glen Canyon Dam. *Land Econ.* **1999**, *75*, 390–401.

35. Tuhtan, J.A.; Noack, M.; Wieprecht, S. Estimating stranding risk due to hydropeaking for juvenile European grayling considering river morphology. *KSCE J. Civil Eng.* **2012**, *16*, 197–206.

36. Schmutz, S.; Bakken, T.H.; Friedrich, T.; Greimel, F.; Harby, A.; Jungwirth, M. Response of fish communities to hydrological and morphological alterations in hydropeaking rivers of Austria. *River Res. Appl.* **2014**, doi:10.1002/rra.2795.

37. Charmasson, J.; Zinke, P. *Mitigation Measures against hydropeaking Effects. A Literature Review*; Report No. TR A7192-Unrestricted. Stiftelsen for Industriell og Teknisk Forskning (SINTEF): Trondheim, Norway, 2011.

38. Meile, T. Hydropeaking on Watercourses. EAWAG News 61e, November 2006; pp. 28–29. Available online: http://www.eawag.ch/medien/publ/eanews/archiv/news_61/en61e_meile.pdf (accessed on 15 March 2015).

39. Baumann, P.; Klaus, I. *Conséquences Ecologiques des Eclusées. Etude Bibliographique*; Informations concernant la pêche No 75; L'Office Fédéral de l'Environnement, des Forêts et du Paysage (OFEFP): Berne, Switzerland, 2003; p. 116. (In French)

40. Water Framework Directive & Hydropower. Key Conclusions. Common Implementation Strategy Workshop: Berlin, Germany, 4–5 June 2007. Available online: http://www.ecologic-events.de/hydropower/documents/key_conclusions.pdf (accessed on 6 March 2015).

41. Smokorowski, K.E.; Metcalfe, R.A.; Finucan, S.D.; Jones, N.; Marty, J.; Power, M. Ecosystem level assessment of environmentally based flow restrictions for maintaining ecosystem integrity: A comparison of a modified peaking *vs.* unaltered river. *Ecohydrology* **2011**, *4*, 791–806.

42. Jager, H.I.; Bevelhimer, M.S. How run-of-river operation affects hydropower generation and value. *Environ. Manag.* **2007**, *40*, 1004–1015.

43. Haas, N.A.; O'Connor, B.L.; Hayse, J.W.; Bevelhimer, M.S.; Endreny, T.A. Analysis of daily peaking and run-of-river operations with flow variability metrics, considering subdaily to seasonal time scales. *JAWRA J. Am. Water Resour. Assoc.* **2014**, *50*, 1622–1640.

44. Vaikasas, S.; Poškus, V. HE turbinų įjungimo sukeliamo potvynio bangos žemutiniame bjefe tyrimai (The investigation of HP turbines switch impact in river lower reaches). *Vandens Inž. (Water Manag. Eng.)* **2009**, *35*, 103–109. (In Lithuanian)

45. Niu, S.; Insley, M. On the economics of ramping rate restrictions at hydro power plants: Balancing profitability and environmental costs. *Energy Econ.* **2013**, *9*, 39–52.

168

46. Heller, P.; Schleiss, A. Aménagements hydroélectriques fluviaux à buts multiples: Résolution du marnage artificiel et conséquences sur les objectifs écologique, énergétique et social. *Houille Blanch.* **2011**, *6*, 34–41. (In Lithuanian)

47. Ribi, J.; Boillat, J.; Schleiss A. Flow Exchange between a Channel and a Rectangular Embayment Equipped with a Diverting Structure, 2010. Available online: http://infoscience.epfl.ch/record/151681 (accessed on 2 January 2015).

48. Gostner, W.; Lucarelli, C.; Theiner, D.; Kager, A.; Premstaller, G.; Schleiss, A.J. A holistic approach to reduce negative impacts of hydropeaking. In Proceedings of the International Symposium on Dams and Reservoirs under Changing Challenges—79th Annual Meeting of ICOLD—Swiss Committee on Dams, Lucerne, Switzerland, 1 June 2011.

49. Fishers and Oceans Canada. *Flow Ramping Study. Study of Ramping Rates for Hydropower Developments*; Report No. Va103-79/2-1; Fisheries and Oceans Canada, Knight Piésold Consulting: Vancouver, BC, Canada, 2005.

50. Gailiušis, B.; Kriaučiūnienė, J. Runoff changes in the Lithuanian rivers due to construction of water reservoirs. In Proceedings of the Rural development 2009: The 4th International Scientific Conference, Akademija, Kaunas Region, Lithuania, 15–17 October 2009.

51. Ždankus, N.; Vaikasas, S.; Sabas, G. Impact of a hydropower plant on the downstream reach of a river. *J. Environ. Eng. Landsc. Manag.* **2008**, *16*, 128–134.

52. Ždankus, N.; Sabas, G. The influence of anthropogenic factors to Lithuanian rivers flow regime. In Proceedings of the 6th International Conference Environmental Engineering, Rome, Italy, 26–27 May 2005; pp. 515–522.

53. Vaikasas, S.; Palaima, K.; Pliuraite, V. Influence of hydropower dams on the state of macroinvertebrates assemblages in the Virvyte river, Lithuania. *J. Environ. Eng. Landsc. Manag.* **2013**, *21*, 305–315.

54. Gailiušis, B.; Jablonskis, J.; Kovalenkovienė, M. *Lietuvos Upės: Hidrografija ir Nuotėkis Monografija (Lithuanian Rivers: Hydrography and Runoff)*; Lietuvos energetikos institutas: Kaunas, Lietuva, 2001; p. 791. (In Lithuanian)

55. Dumbrauskas, A.; Larsson, R. The Influence of Farming on Water Quality in the Nevėzis Basin. *Environ. Res. Eng. Manag.* **1997**, *2*, 48–55.

56. International Commission on Large Dams (ICOLD). Available online: http://www.icold-cigb.org/ (accessed on 15 March 2015).

57. Ye, S.; Yaeger, M.; Coopersmith, M.; Cheng, L.; Sivapalan, M. Exploring the physical controls of regional patterns of flow duration curves—Part 2: Role of seasonality, the regime curve, and associated process controls. *Hydrol. Earth Syst. Sci.* **2012**, *16*, 4447–4465.

58. The Nature Conservancy. Indicators of Hydrologic Alteration Version 7.1—User's Manual. Available online: https://www.conservationgateway.org/Files/Pages/indicators-hydrologic-altaspx47.aspx (accessed on 6 March 2015).

59. Gore, J.A. *The Restoration of Rivers and Streams: Theories and Experience*; Butterworth/Ann: Arbor, MI, USA, 1985.

60. Sauterleute, J.; Charmasson, J. Characterisation of rapid fluctuations of flow and stage in rivers in consequence of hydropeaking. In Proceedings of the 9th International Symoposium on Ecohydraulics, Viena, Austria, 17–21 September 2012.

61. Sabas, G. Analysis of hydropower plant influence to the river hydrological and hydraulic regimes. In Proceedings of the 6th International Conference Environmental Engineering, Vilnius, Lithuania, 26–27 May 2005.

62. *Kraftwerksbedingter Schwall und Sunk. Eine Standortbestimmung, Im Auftrag des Schweizerischen Wasserwirtschaftsverbands*; ETH Zürich: Lausanne, Switzerland, 2006. (In Lithuanian)

63. Chow, V.T. *Open-Channel Hydraulics*; McGraw-Hill: New York, NY, USA, 1959.

64. *Handbook of Hydrology*; Maidment, D.R., Ed.; McGraw-Hill: New York, NY, USA, 1993.

65. Shaw, E.M.; Beven, K.J.; Chappell, N.A.; Lamb, R. *Hydrology in Practice*, 4th ed.; CRC Press, Taylor & Francis Group: New York, NY, USA, 2010.

66. Environment Agency (EA). The environmental assessment of proposed low head hydro power developments. In *Good Practice Guidelines Annex to the Environment Agency Hydropower Handbook*; EA: Rotherham, UK, August 2009.

Chapter 2:
Corporate Policies and Investment Decisions

Determination of Priority Study Areas for Coupling CO$_2$ Storage and CH$_4$ Gas Hydrates Recovery in the Portuguese Offshore Area

Luís Bernardes, Júlio Carneiro, Pedro Madureira, Filipe Brandão and Cristina Roque

Abstract: Gas hydrates in sub-seabed sediments is an unexploited source of energy with estimated reserves larger than those of conventional oil. One of the methods for recovering methane from gas hydrates involves injection of Carbon Dioxide (CO$_2$), causing the dissociation of methane and storing CO$_2$. The occurrence of gas hydrates offshore Portugal is well known associated to mud volcanoes in the Gulf of Cadiz. This article presents a determination of the areas with conditions for the formation of biogenic gas hydrates in Portugal's mainland geological continental margin and assesses their overlap with CO$_2$ hydrates stability zones defined in previous studies. The gas hydrates stability areas are defined using a transfer function recently published by other authors and takes into account the sedimentation rate, the particulate organic carbon content and the thickness of the gas hydrate stability zone. An equilibrium equation for gas hydrates, function of temperature and pressure, was adjusted using non-linear regression and the maximum stability zone thickness was found to be 798 m. The gas hydrates inventory was conducted in a Geographic Information System (GIS) environment and a full compaction scenario was adopted, with localized vertical flow assumed in the accrecionary wedge where mud volcanoes occur. Four areas where temperature and pressure conditions may exist for formation of gas hydrates were defined at an average of 60 km from Portugal's mainland coastline. Two of those areas coincide with CO$_2$ hydrates stability areas previously defined and should be the subject of further research to evaluate the occurrence of gas hydrate and the possibility of its recovery coupled with CO$_2$ storage in sub-seabed sediments.

Reprinted from *Energies*. Cite as: Bernardes, L.; Carneiro, J.; Madureira, P.; Brandão, F.; Roque, C. Determination of Priority Study Areas for Coupling CO$_2$ Storage and CH$_4$ Gas Hydrates Recovery in the Portuguese Offshore Area. *Energies* **2015**, *8*, 10276-10292.

1. Introduction

Anthropogenic CO$_2$ emissions are pointed as the main driver of global climate change and its negative impacts in nature and society [1]. The energy demand is increasing in many regions of the world and particularly in developing countries, such as China, India and Brazil [2]. Increased energy efficiency and renewable energy sources are essential to reduce CO$_2$ emissions from energy production. However, for the coming decades it is unlikely that renewable energy sources will be able to cope with the rising energy demand, and fossil fuels will remain the main energy source. CO$_2$ Capture and Storage (CCS) is seen as a technology that could bridge the current fossil fuel based society to a low carbon future.

Geological storage of CO$_2$ generally focuses on saline aquifers, depleted oil and gas fields and the use of CO$_2$ for Enhanced Oil Recovery (EOR). Recent years have also seen an increased interest in CO$_2$ storage through *in situ* mineral carbonation in basalts. However, a less studied option available for those countries with a deep geological continental margin at close distance from shore is the storage of CO_2 in sub-seabed sediments in the form of hydrates, as proposed by Koide *et al.* [3] and Li *et al.* [4].

Bernardes *et al.* [5] defined three areas suitable for CO_2 storage in hydrates form in the Portuguese geological continental margin, at a distance of about 70 km from the coast and from the main CO_2 emission clusters in the country. However, that analysis also acknowledged the economic unfeasibility of that option, mainly due to the more than 1100 meters of water column required for the CO_2 hydrates stability in the Portuguese geological continental margin.

However, coupling CO_2 storage with methane recovery from gas hydrates can result in economic benefits. In that process, under the appropriate pressure and temperature conditions, gas hydrates are dissociated by CO_2 injection, releasing methane (CH_4) from the hydrates and replacing it by CO_2 molecules in the hydrate molecular cages.

In 2012, the first *in situ* test of CO_2 injection to dissociate gas hydrates and recover CH_4, was performed. During the field test, 6.11 Million cubic meters (Mm^3) of gas were injected, from which 4.74 Mm^3 of nitrogen (N_2) and 1.37 Mm^3 of CO_2. During the production phase, 24.21 Mm^3 of CH_4 were recovered with 70% and 40% of the previously injected N_2, and CO_2, respectively [6].

Several works based on geophysical, geochemical and geological data [7,8] have identified the occurrence of gas hydrates at more than 100 offshore sites around the world, including in the Gulf of Cadiz, just south from the Algarve coast in Portugal, where several mud volcanoes are known to hold gas hydrates [9–11].

An economically viable way to store CO_2 in seabed sediments, in its hydrate form, on the Portuguese geological continental margin could be to couple it with CH_4 recovering from gas hydrates. However, at present, there is no inventory of methane hydrates occurrences on Portugal's deep offshore.

Within this article, an estimate of the amount of CH_4 hydrates on the Portuguese geological continental margin is presented as a first step towards assessing the possibility of coupling CO_2 storage and CH_4 recovery. This article aims to define areas of possible formation of gas hydrates and their overlap with the CO_2 hydrates potential storage areas defined in Bernardes *et al.* [5] (Figure 1).

2. Background

Natural CH_4 hydrates occur and remain stable in sub-seabed sediments, mud volcanoes and other geological formations, under a specific range of pressure and temperature conditions. Thermodynamic calculations show that CH_4 hydrates are stable at relatively low-pressure (low water depths) conditions at typical seabed temperatures, as long as gas concentration is sufficient. Figure 2 depicts the methane hydrates stability conditions, according to several authors.

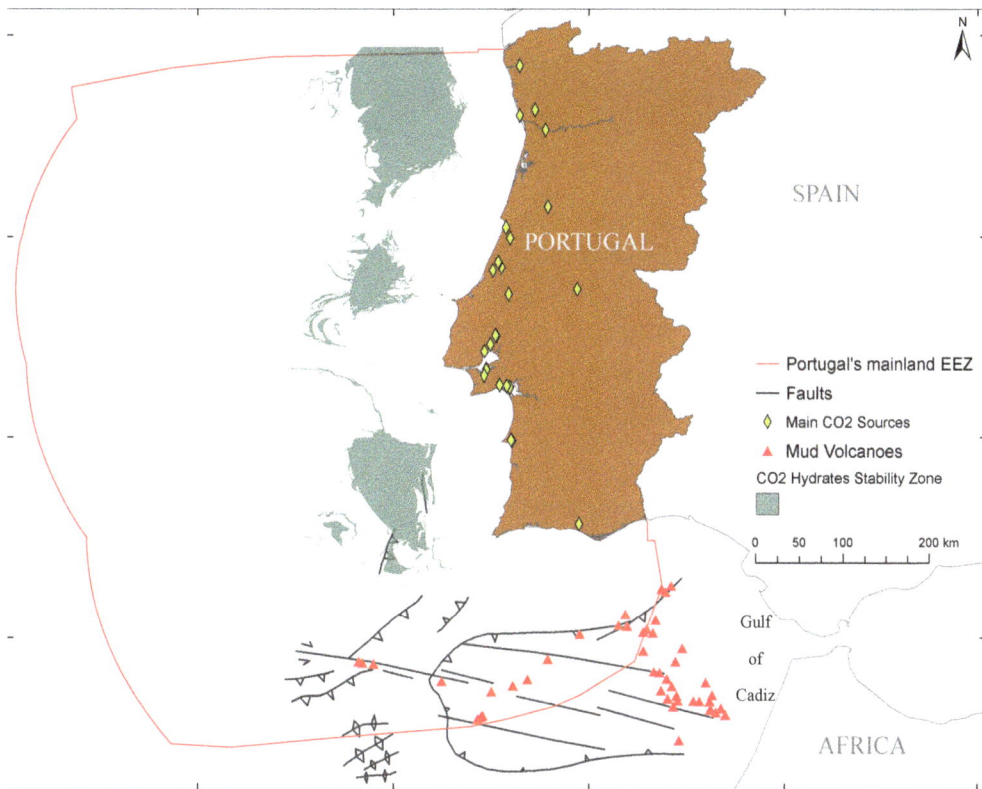

Figure 1. Location of the CO_2 hydrates stability zones defined in Bernardes *et al.* [5].

Hydrates are polyhedral structures that form a cage composed by water molecules capable of retaining non-polar neutral guest gases. The interconnection of water molecules is strong since each molecule can donate and receive two hydrogen bonds. The stability of hydrates is only possible if more than 90% of the total cavities are fully filled by gas [13].

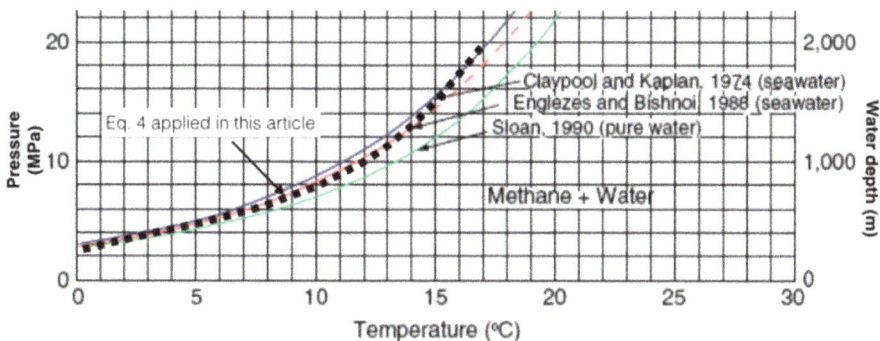

Figure 2. Gas hydrates phase boundary conditions, according to several authors (Adapted from Thakur [12]). Also depicted (dashed black line) the non-linear equation applied in this article.

Due to chemical reactions, the decomposition of organic matter by anaerobic bacteria, at low temperature generates biogenic CH_4 [12]. When Sedimentation rates (S_r) and Particulate Organic Carbon (P_c) are higher than 1 cm/kyr and higher than 1%, respectively, anaerobic bacteria produces CH_4 few cm below the sea floor [12]. Kvenvolden *et al.* [14] defined the range of S_r from 3 to 30 cm/kyr to form biogenic CH_4. Collett [15] and Klauda and Sandler [16] have shown that hydrates do not form if P_c values are below 0.4% to 0.5%.

Hydrates formation is limited by a certain number of variables such as temperature-pressure pairs, organic matter availability, salinity, hydro and geothermal gradients and petrophysical conditions. Methane formation in ocean sediments is created by sulfur reduction and hydrogen sulphide release, from the upper parts of seafloor to hundreds of meters below. At larger depths, methane is produced by catagenesis, where temperature and pressure are the driving parameters [12].

Pore water salinity limits the capacity of the sediments to host gas hydrate and also lowers methane solubility. This leads to reduction of the amount of gas required for hydrates formation. Salinity causes two effects: it decreases the solubility of methane hydrates, but the existence of salts inhibits the formation of hydrate, increasing the dissociation pressure and methane concentration in the solution, although the solubility effect overweighs the effect of an increasing dissociation pressure [12].

Recent studies [17–19], estimate the global inventory of gas hydrates in marine sediments, taking the geological evolution, the sedimentation rate and the Particulate Organic Carbon data [19]. The Gas Hydrate Inventory (GHI) applied transfer functions derived from a numerical model developed by Wallmann *et al.* [18] simulating the gas hydrates formation under normal compaction and full compaction scenarios. The transfer functions, fitted by Piñero *et al.* [19], attempt to estimate the gas hydrates mass based on a reduced number of parameters, namely the sedimentation rate, the particulate organic content, vertical fluid flow and the thickness of the gas hydrates stability zone.

Equation (1) represents the transfer function for estimating the GHI in the full-compaction scenario [19] used to estimate values on Portugal's offshore.

$$GHI = b_1 L_{sz}^{b_2} \left(P_c \frac{b_3}{S_r^{b_4}} \right) \exp[-(b_5 + b_6 \ln[S_r])^2] \tag{1}$$

where L_{sz} is the thickness of hydrate stability zone, P_c is the particulate organic content, S_r is the sedimentation rate, and b_1 to b_5 are the transfer function fit parameters: $b_1 = 0.00285$, $b_2 = 1.681$, $b_3 = 24.4$; $b_4 = 0.99$, $b_5 = -1.44$ and $b_6 = 0.393$ [19].

The thickness of the gas hydrate stability zone (L_{sz}) depends on several parameters: bathymetry, geothermal gradient, temperature, pressure, gas concentration, salinity, porosity and permeability. For analysis at the regional scale, without considering the local variation of petrophysical quantities, the geothermal gradient, temperature/pressure pairs and water salinity are the crucial parameters. Pressure can be assimilated to hydrostatic pressure and readily estimated, while water salinity can be measured. Temperature and its variation with depth (*i.e.*, the geothermal gradient) vary with the geological conditions of the area in study and can be estimated based on bottom-hole temperature measurements in boreholes.

Regarding to geothermal gradient calculations, Piñero et al., [19] compared two models published by Pollack et al. [20] and Hamza et al. [21]. They found Hamza et al. [21] models' to have a better match with Bottom Simulating Reflector (BSR) from Ocean Drilling Program (ODP) sites [22]. The maximum L_{sz} of 900 m was calculated with a thermal conductivity of 1.5 $Wm^{-1} \cdot K^{-1}$.

Leon et al. [23] applied a GIS model to estimate the L_{sz} in the Gulf of Cadiz, and retrieved a maximum thickness of 770 m for biogenic hydrates, with the expected BSR at around 800 m water depth.

Piñero et al. [19] also introduced vertical fluid flow (q) in the gas hydrate inventory, using Equations (2) and (3) for different ranges of vertical fluid flow. The fit parameters are the same as those ones published on [19].

For fluid flow:

$$q \geq 0.001 S_r (2 + \ln[P_c])$$

GHI is estimated from:

$$GHI_q = c_1 L_{sz}^{c_2} c_3 + \left(c_3 + \frac{1}{S_r}\right)(P_c + c_4 q^{c_5}) P_c^{c_6} \tag{2}$$

For fluid flow:

$$q < 0.001 S_r (2 + \ln[P_c])$$

GHI is estimated from:

$$GHI_q = GHI - 10^{-8} c_7 L_{sz}^{c_4} \left(1 + \frac{1}{S_r}\right) q P_c^{c_9} \tag{3}$$

Gas hydrates accumulation is inhibited when fluid flow is:

$$q > 1.3 \times 10^{-8} L_{sz}^2 S_r P_c$$

The transfer function coefficients in Equations (2) and (3) are $c_1 = 0.024$, $c_2 = 1.587$, $c_3 = 0.0224$, $c_4 = 266084$, $c_5 = 2.75$, $c_6 = 0.063$, $c_7 = 0.003$, $c_8 = 4.68$ and $c_9 = 2.31$ [19].

Piñero et al. [19] found in the Full Compaction scenario, hydrates are estimated to occur in areas larger volumes. If vertical fluid flow is also considered, the global amount of GHI is estimated at nearly 550 Giga ton of Carbon (GtC) [19].

The scenario with vertical fluid flow may be appropriate for, at least, part of our study area, namely the Gulf of Cadiz sector included in the Portuguese Economic Exclusive Zone (EEZ) (Figure 1). Geologically, the Gulf of Cadiz is located on a complex transpressive setting nearby the Africa-Eurasia plate boundary [24]. Mud volcanism in the Gulf of Cadiz is triggered by compressional stress related with the Africa plate WNW-ESE trajectory at a rate of about 4.0 mm/yr [5]. It is generally associated to fluid escape, along tectonic structures such as thrust faults, extensional faults, strike-slip faults and diapirs, that opens up pathways to over pressurized fluids, culminating on the formation of mud volcanoes [25,26]. Somoza et al. [27] suggested that shallow fluid upward flow could be originated by the perturbation of gas hydrates rich sediments by the Mediterranean Outflow Water (MOW). In any circumstance vertical fluid flow occurs and the gas hydrate inventory should consider the model given by Equations (2) and (3).

3. Methodology

The methodology implemented aimed to identify areas where biogenic gas hydrates are likely to be formed in the Portuguese geological continental margin and their overlap with areas defined by Bernardes et al. [5] as suitable for CO_2 storage in hydrates form in the subseabed sediments.

Two scenarios were considered based on the Piñero et al. [19] transfer-functions: (1) Assuming full compaction of the sediments at higher depth and (2) Full compaction of the sediments combined with vertical fluid flow on the Gulf of Cadiz. Piñero et al. [19] transfer function for the normal compaction scenario was also applied, but in that case there are no conditions for gas hydrates formation in the Portuguese geological continental margin.

The full compaction scenario favors hydrate formation due to the decrease in porosity and subsequent increase in CH_4 concentration within the pore water. The full compaction scenario also decelerates the loss of methane-rich pore fluids below the GHSZ, due to the increasing difference in burial velocity of pore water and bulk sediment [19]. The overburden load of younger sediments being continuously deposited on seabed drives sediment compaction by inducing an exponential decrease in porosity and water content with the sediment depth [18]. Upward fluid flow happens when sedimentary pressure gradient exceeds the hydrostatic pressure [18]. Full compaction may occur at the base of thick sedimentary deposits and in compressive tectonic settings, although the porosity of sediments usually is not completely eliminated by compaction [18]. Simulations by Wallmann et al. [18] show that GHI is significantly enhanced by Full Compaction scenario.

The methodology (Figure 3) was implemented in a GIS environment gathering data on the relevant features of the deep offshore, although the determination of the gas hydrate stability zone thickness (L_{sz}) was computed using a Fortran code, due to the non-linearity of the equations applied that cannot be easily implemented in a GIS environment.

The GIS includes data on bathymetry, hydrostatic pressure, temperature at the bottom of water column, oil exploration drilling wells, geothermal gradient, sediment thickness, accumulation of particulate organic carbon on the sea floor (P_c) and sedimentation rates (S_r), considered as steady state.

Raster datasets of seabed temperatures, hydrostatic pressure and sediment thickness for the study zone were the same used on Bernardes et al. [5]. The other datasets where built for this study and described in the next section.

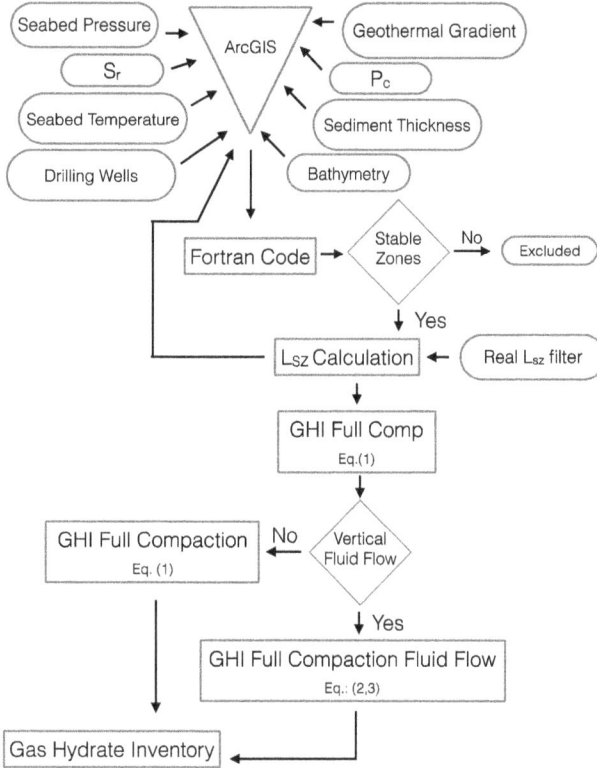

Figure 3. Methodology flow diagram.

4. Results and Discussion

4.1. Thickness of Gas Hydrates Stability Zone (L_{sz})

The determination of the GHSZ thickness (L_{sz}) depends on the thermodynamics of the gas hydrates boundary, namely pressure and temperature, and the inhibitor effect of water salinity. Since the equation that describe the thermodynamic behavior of hydrates is difficult to implement in a GIS, an approximation to the phase boundary computed by the *CSMHYD* code [28] was used, considering a seawater salinity of 36 g/L. Pressure and temperature dissociation pairs were found in *CSMHYD* to which a non-linear regression fitted the following function.

$$\frac{1}{T_{sf} + L_{sz}\delta_g} = \sum_{i=1}^{3} c_i \left[\log(L_{sz}\gamma_w g + P_{sf}\right]^{i-1} \tag{4}$$

where T_{sf} and P_{sf} are the seabed temperature and pressure, respectively, δ_g is the geothermal gradient, γ_w is the sub-seabed formation water salinity (assigned as 3.6% weight), g is gravity and c_i are the non-linear regression factors $c_1 = 3.8 \times 10^{-3}$, $c_2 = -4.09 \times 10^{-4}$ and $c_3 = 8.64 \times 10^{-5}$.

The phase boundary approximation (Equation (4)), is represented in Figure 2 and implemented in a Fortran code, *CO2hydrate*, coupled with the GIS model. The raster datasets of geothermal gradient

(δ_g), seafloor pressure (P_{sf}) and seafloor temperature (T_{sf}) used in *COHydrate* are shown in Figure 4 and were derived by Bernardes *et al.* [5], although the geothermal gradient dataset was updated using data listed in Benazzouz [29].

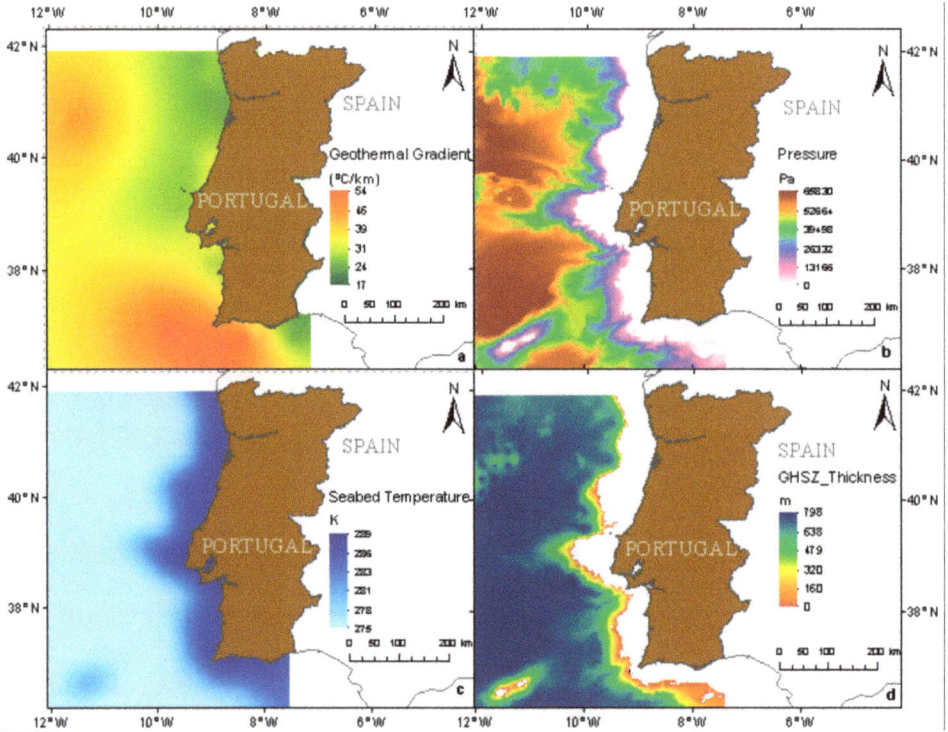

Figure 4. Raster datasets. (**a**) Updated geothermal gradient; (**b**) Hydrostatic pressure; (**c**) Seabed temperature; (**d**) GHSZ thickness (L_{sz}).

The maximum L_{sz} thickness of 798 m (Figure 4d) was found to be in agreement with the values computed by Leon *et al.* [23], and by Peter R. Miles (725 m) as quoted by [29], but contrasts with the 550 m depth estimated by Benazzouz [29].

According to the model implemented, the L_{sz} is highly influenced by the high seawater temperatures in the Portuguese offshore, and it shows that gas hydrates could be stable at depths varying from 600 m to 3000 m. The minimum depth required for GHSZ is 600 m, much shallower than the 1100 m minimum depth for the CO_2 hydrates stability zone computed by Bernardes *et al.* [5]. Thus, for the purpose of this paper, only the GSHZ occurring deeper than 1100 m and overlapping with the CO_2 hydrates stability zones is of interest.

4.2. Sedimentation Rate (S_r)

The sedimentation rate (S_r) raster dataset was built from the interpolation of the available S_r values for the study area. Data was downloaded from the online database PANGEA (Table 1). The dataset is not ideal, with only 62 samples and poor spatial distribution, likely the main limitation for this study (Figure 5a).

Table 1. Sedimentation rate data and source. *Pangea* website (2014).

Lat dd	Long dd	S_r cm/ky	Lat dd	Long dd	S_r cm/ky	Lat dd	Long dd	S_r cm/ky	Reference
36.69	−11.43	0.389	37.70	−9.47	11.620	37.75	−9.73	20.400	
35.35	−10.42	0.969	37.61	−9.28	5.300	37.71	−9.23	7.100	
40.96	−10.72	2.891	37.92	−10.85	7.940	37.70	−9.47	11.620	
36.69	−11.43	0.389	37.72	−10.55	12.040	37.61	−9.28	5.300	Thiede *et al* [30]
35.35	−10.42	0.969	37.70	−9.77	19.000	37.92	−10.85	7.940	
40.96	−10.72	4.167	37.65	−9.53	7.550	37.72	−10.55	12.040	
37.65	−9.53	7.550	37.68	−10.08	16.180	37.70	−9.77	19.000	
37.75	−9.73	20.400	37.71	−9.23	7.100	34.80	−10.25	2.900	
37.77	−10.18	16.000	37.68	−10.08	16.180	-	-	-	Broecker *et al.* [31]
40.58	−9.86	35.600	37.77	−10.18	33.150	-	-	-	Thomson *et al.* [32]
34.89	−7.82	4.864	34.91	−7.58	4.896	34.86	−8.13	3.210	Sarnthein, Michael (2006) [33]
39.04	−10.66	5.917	-	-	-	-	-	-	Voelker *et al.* [34]
38.63	−9.51	0.001	38.63	−9.51	0.001	-	-	-	Alt-Epping, Ulrich [35]
41.23	−9.07	0.240	41.84	−8.99	0.070	41.21	−9.04	0.350	Recent sedimentation
41.81	−9.08	0.170	41.17	−9.03	0.420	41.24	−9.03	0.410	and sedimentary
41.84	−9.11	0.200	41.20	−9.04	0.300	41.32	−9.00	0.170	budgets on the western
41.78	−9.07	0.230	41.85	−9.07	0.140	-	-	-	Iberian shelf

4.3. Particulate Organic Carbon (Pc)

Particulate Organic Carbon (P_c) depends on the total organic carbon flux to the sea floor [19] and the re-mineralization process on the top centimeters of the sediment column [36–38].

The particulate organic content (P_c) raster applied in this study resulted from kriging interpolation of particulate organic concentration values obtained from the PANGEA online database [39] and core samples analysis U1385 and U1391 from the IODP Expedition 339. Figure 5b depicts the resulting raster file, with larger values of P_c of 2.94 wt. % obtained in the Nazaré's canyon area, and smaller values in the South Portugal's EEZ limit area. Again, the limited number of a data points its poor spatial distribution is an important especially in areas distant from the coast. Remineralization processes were not taken into account.

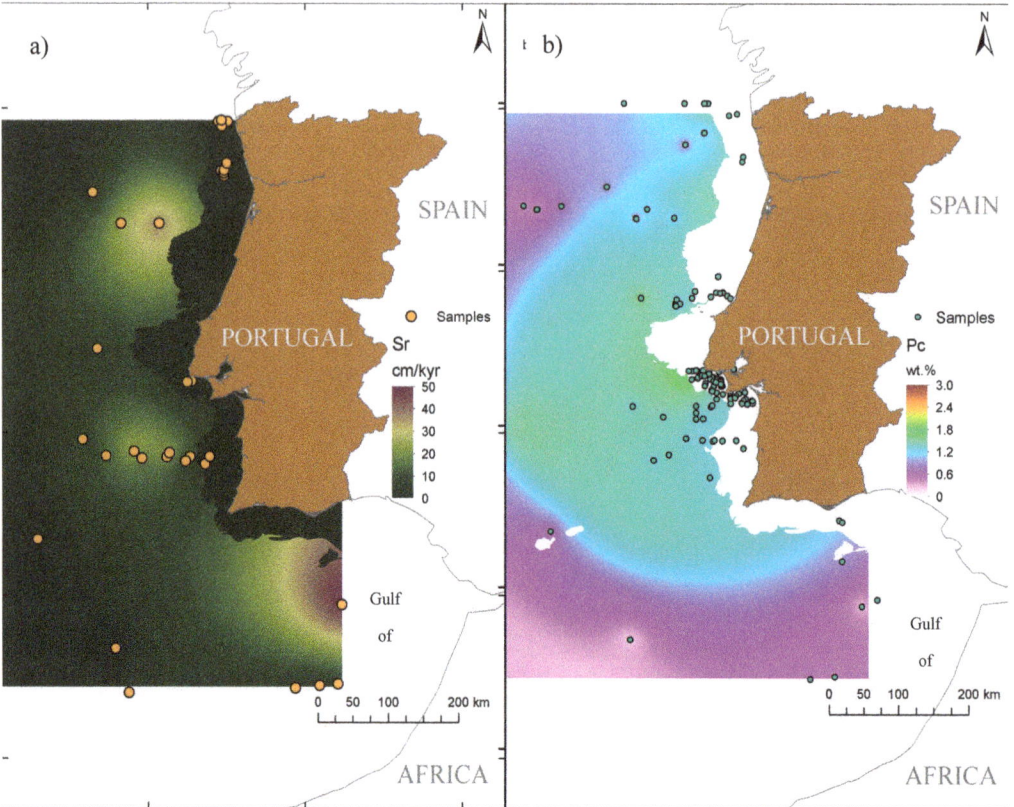

Figure 5. (a) Sedimentation rate raster dataset map; **(b)** Particulate Organic Carbon.

4.4. Gas Hydrates Inventory (GHI)

The gas hydrate stability zone defines the area where P and T conditions are sufficient for gas hydrates to be stable, but according to Piñero *et al.* [19] model, the gas hydrates inventory (GHI) is also a function of the organic content and sedimentation rate. GHI was computed for the Full Compaction scenario (Equations (1)) and the Fluid Flow scenarios (Equations (2) and (3)) and sensitivity analysis was conducted varying by 10% each of the variables in Equations (1) to (3).

4.4.1. Full Compaction Scenario

The full compaction scenario retrieved three areas where the formation of biogenic gas hydrates is possible, given the combination of organic matter content and sedimentation rates (Figure 6). The largest area occurs just offshore at some 70 km West-North-West from the coast. Water depths varies in this region from 700 m to 3000 m. A smaller area occurs offshore from the Sines region, in the Tagus abyssal plain, at some 100 km from the coast. Water column varies from 2100 m to 3500 m in this region. The third area occurs just south from the Algarve, at 50 km from the coast and with maximum water depth of 2600 m.

The maximum value estimated for the GHI in the full compaction scenario is 154 kg/m^2, and taking into account the area and concentration variation, the estimated mass is about 493 MtC.

The estimation of the GHI using the full compaction scenario without vertical fluid flow, rises 17% from the original value when S_r is increased by 10% by initial value, while a 10% increase in the P_c value induces a 25% increase in the amount of the GHI, regarding to the maximum of 154 kg/m^2 (Table 2). For the Full Compaction scenario, an average value of 493 MtC of hydrates increases to 840 Mton for a 10% increase in S_r and of 770 MtC for a 10% increase in P_c. In any case the change in the size of the areas where hydrates can form is small indicating that, despite the uncertainty in the S_r and P_c values, gas hydrates are not likely to be formed in areas other than those depicted in Figure 6.

Table 2. GHI estimates for the several scenarios and sensitivity analysis.

Average GHI (kg·m⁻²)	Parameter Variation	Vertical Fluid Flow with Full Compaction Scenario (cm·yr⁻¹)				Parameter Variation
		$q = 0.005$	$q = 0.01$	$q = 0.015$	$q = 0.02$	
		22.13	72.69	147.65	282.27	
	22.11	22.13	75.7	151.11	286.55	P_c 10% wt.%
		22.13	68.08	137.5	262.19	S_r 10% (cm·kyr⁻¹)
Average GHI (MtC)	493.04	493.04	3590.87	7293.92	13947.23	
	840.49	493.02	3739.71	7465.06	14158.53	P_c 10% wt.%
	769.62	493.04	3362.88	6792.66	12954.94	S_r 10% (cm·kyr⁻¹)

4.4.2. Vertical Fluid Flow and Full Compaction Scenario

The geochemistry of the gas hydrates sampled in the Gulf of Cadiz mud volcanoes indicates a thermogenic origin, and it has been considered that the organic content is too low to allow the formation of gas hydrates with biogenic origin [11]. This is in agreement with the results of our simulations with the Full Compaction scenario (Figure 6a). Nevertheless, Leon et al. [23] do not discard the possibility that, in the deeper mud volcanoes, part of the gas hydrates may have a biogenic origin. The vertical fluid flow scenario was included in our analysis to understand if Pinero et al. [19] model with full compaction and vertical fluid flow could explain gas hydrates observed in the deeper mud volcanoes. Vertical fluid flow was considered only in the accretionary wedge, where the mud volcanoes are also known to occur (Figure 1).

Given the absence of information about the flow rate in the area, values of $q = 0.005$ cm/yr; $q = 0.01$ cm/yr; $q = 0.015$ cm/a and $q = 0.02$ cm/yr were considered. Notice that these values of q were considered over the entire area of the accretionary wedge. This is a simplifying assumption, not only because fluid flow is dependent on the petrophysical properties of sediments, but mainly because it is well known that in the Gulf of Cadiz, vertical fluid flow is associated to tectonic features and mud volcanoes. However, these linear features are nor amenable to be analyzed with a large-scale GIS as applied in this study, and thus fluid flow was distributed uniformly in all the accretionary wedge.

Figure 6b shows the results for the $q = 0.01$ cm/yr scenario. According to the vertical fluid flow model, biogenic gas hydrates can form in the accretionary wedge zone. Even for the lowest flow rate ($q = 0.005$ cm/yr), gas hydrates can form in the accretionary wedge zone. Simulations with other values of q indicate that GHI estimation rises with the fluid flow value (Table 2). Thus, Piñero *et al.* [19] model with vertical fluid flow is consistent with Leon *et al.* [23] possibility that some of the hydrates observed in the mud volcanoes may be from biogenic origin, since the organic carbon and sedimentation rate appears to be sufficient as long as vertical fluid flow is considered.

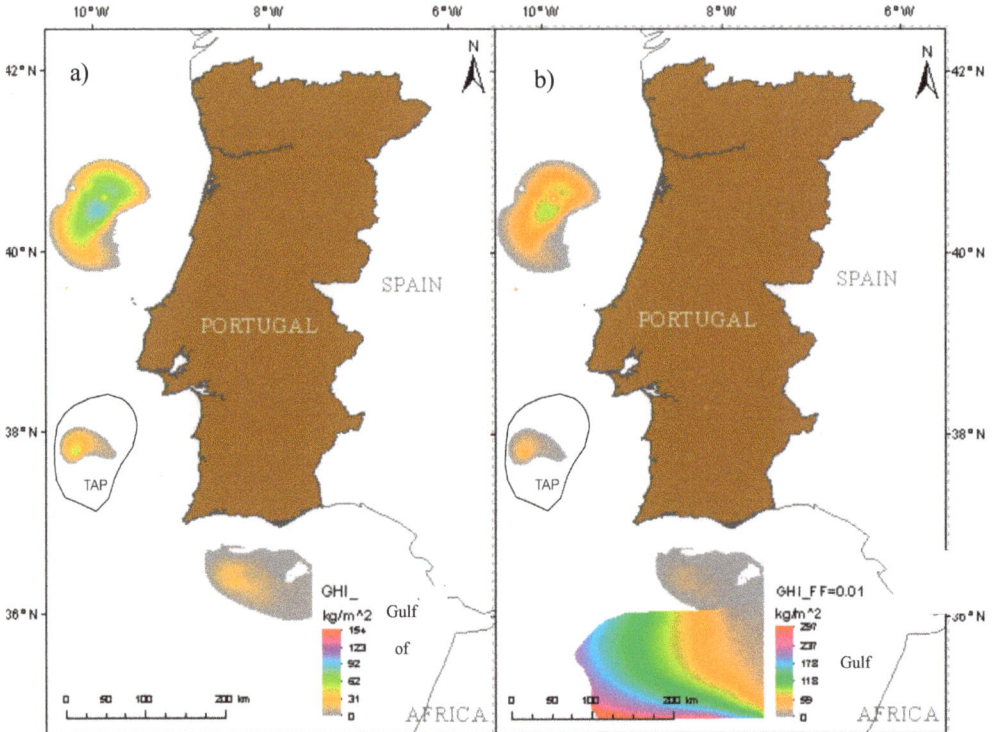

Figure 6. (a) GHI in the Full Compaction scenario; **(b)** GHI in the Full Compaction with fluid flow in the accretionary wedge ($q = 0.01$ cm/ky).

5. Discussion

Bernardes *et al.* [5] delimited the CO_2 hydrates stability zone in the Portuguese continental margin and applied four criteria to delineate the areas most suitable for CO_2 storage as hydrates, namely: (i) Distance from the main ports; (ii) Water column depth; (iii) thickness of the CO_2 hydrates stability zone; and (iv) Spatial variation of the hydrates stability zone thickness. Three preferential areas were defined as more suitable for CO_2 injection, possibly coupled with recovery of methane, if gas hydrates exist in those same areas.

CO_2 hydrates are stable only at more than 1100 m water depths, while gas hydrates are stable at around 600 m depths. Thus, it is the CO_2 stability that constrains the depths, and consequently the distance from the mainland, at which CO_2 storage coupled with methane recovery could be conducted.

Figure 7 overlaps the areas defined by Bernardes *et al.* [5], with the areas defined in the GHI conducted in this study. Two of the areas partly coincide. In a perspective of potential recovery of methane from gas hydrates and CO_2 storage, future research should focus in those overlapping two areas (Figure 7).

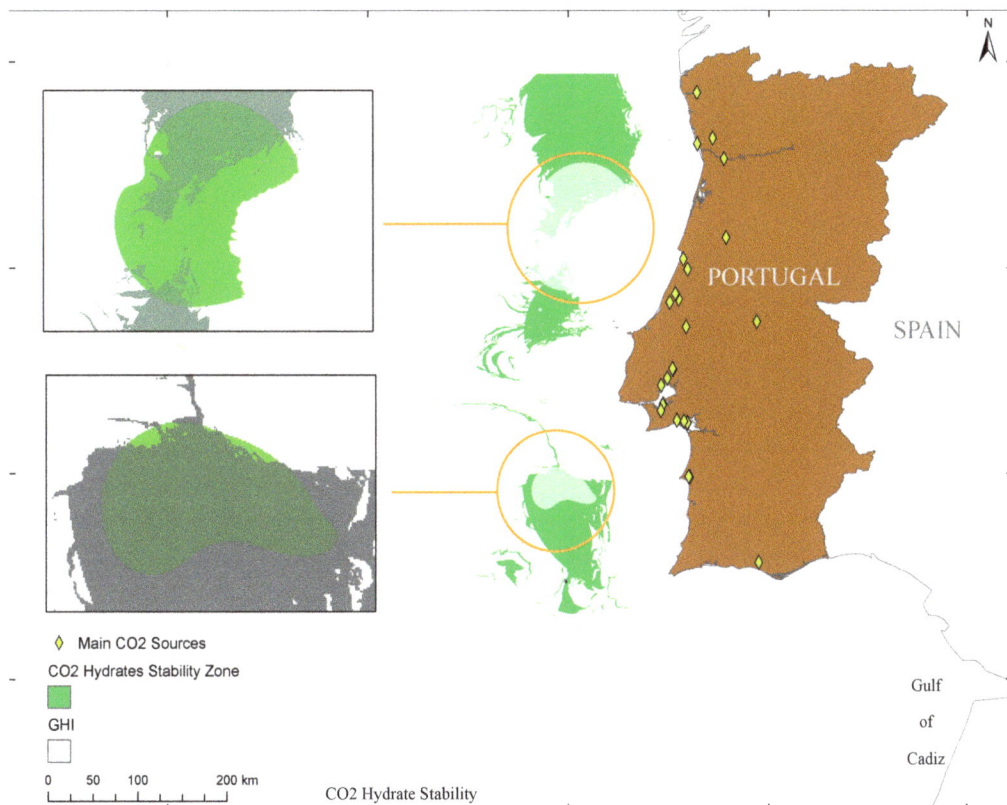

Figure 7. Overlap of preferential areas for CO_2 storage in hydrate form with GHI areas.

6. Conclusions

Countries with a deep continental margin not distant from the shoreline may envisage the reduction of greenhouse gas emissions from the storage of CO_2 as hydrates in sub-seabed sediments. Portugal is one of those countries, where the shallow offshore (bathymetry < 200 m) is, in some areas, less than 10 km wide. However, economic feasibility of that option can only come from added value of recovering CH_4 from existing gas hydrates. Storing CO_2 hydrates with simultaneous recovery of CH_4 from gas hydrates can contribute to mitigate climate change, while addressing the increasing energy demand. The goal of this article was to define areas, in the Portuguese geological continental margin, simultaneously with conditions favorable to the stability of the CO_2 hydrates and gas hydrates.

A GIS was implemented with data on bathymetry, pressure, seabed temperature, sediment thickness, organic carbon content, sedimentation rates and an updated geothermal gradient.

A non-linear approximation to the CSMHYD gas hydrate phase boundary was adjusted and implemented on a Fortran code, *CO2hydrate*, able to perform analytical calculations in the GIS data grid. The code delineates areas where hydrates are stable and computes the thickness of the gas hydrate stability zone (GHSZ).

Gas hydrates are stable at water depths over 600 m, contrasting to the 1100 m water depth required for the CO_2 hydrates stability (Bernardes *et al.*, [5]. The average thickness of the GHSZ is averaged in 528m, with a maximum value of 798 m.

An estimation of the Gas Hydrate Inventory (GHI), resorting to Piñero *et al.* [19] transfer functions, was conducted for a Full Compaction scenario and for the vertical fluid flow scenario in the accretionary wedge, South from Algarve, due to the known occurrence of gas hydrates in connection to mud volcanoes.

The Full Compaction scenario indicates that gas hydrates can form in the Portuguese continental margin in three main areas. Two of those areas coincide with the areas defined by Bernardes *et al.* [5] as suitable for storage of CO_2 in hydrate form. Future research focusing on those areas, should assess existing seismic sections in order to verify the existence of a BSR, and collect accurate data on the particulate organic carbon and the sedimentation rate. Furthermore, and in order to test the possibility of CO_2 injection, petrophysic and hydraulic characterization of the sub-seabed sediments should be conducted.

The simulations including vertical fluid flow indicate that it is possible for biogenic gas hydrates to form in the area of the accretionary wedge, even though the particulate organic carbon is not high. This corroborates Leon *et al.* [23] assertion that, in some of the mud volcanoes, the biogenic origin of part of the gas hydrates should not be discarded.

Acknowledgments

Special acknowledgement to Frederico Dias for the first major paper revision. Authors acknowledge the funding provided by the Task Group for the Extension of Continental Shelf (EMEPC) and the Institute of Earth Sciences (ICT) and the International Ocean Drilling Program (IODP) providing sub-seabed samples sent for us to analyze.

Author Contributions

Luís Bernardes had the idea of the work, managed the data, its GIS implementation and paper writing. Júlio Carneiro designed the *CO2Hydrate* computer model. Pedro Madureira was the contacts manager for data access and geological adviser. Filipe Brandão was the GIS adviser. Cristina Roque was the marine geology adviser.

Conflicts of Interest

The authors declare no conflict of interest.

References

1. IPCC. Climate Change 2007: Impacts, Adaptation and Vulnerability. In *Contribution of Working Group II to the Fourth Assessment Report of the Intergovernmental Panel on Climate Change*; Cambridge University Press: Cambridge, UK, 2007; p. 976.
2. IEA. *Key World Energy Statistics 2009*; IEA: Paris, France; 2009; p. 82.
3. Koide, H.; Takahashi, M.; Tsukamoto, H.; Shindo, Y. Self-Trapping Mechanisms of Carbon-Dioxide in the Aquifer Disposal. *Energy Convers. Manag.* **1995**, *36*, 505–508.
4. Li, X.; Ohsumi, T.; Koide, H.; Akimoto, K.; Kotsubo, H. Near-future perspective of CO_2 aquifer storage in Japan: Site selection and capacity. *Energy* **2005**, *30*, 2360–2369.
5. Bernardes, L.F.; Carneiro, J.; de Abreu, M.P. CO_2 hydrates as a climate change mitigation strategy: Definition of stability zones in the Portuguese deep offshore. *Int. J. Glob. Warm.* **2013**, *5*, 135–151.
6. Schoder, D.H.F.; Hester, K.; Howard, J.; Ratermant, S.; Lloyd Martin, K.; Smith B.; Klein, P. *ConocoPhilips (2013) Gas Hydrate Production Test Final Technical Report*; NETL: Houston, TX, USA, 2013.
7. Lorenson, T.D.; Kvenvolden, K.A. A Global Inventory of Natural Gas Hydrate Occurrence (Map). U.S. Geological Survey 2007. Available online: http://walrus.wr.usgs.gov/globalhydrate/ (accessed on 19 April 2015).
8. Milkov, A.V.; Sassen, R. Economic geology offshore gas hydrate accumulations and provinces. *Mar. Pet. Geol.* **2002**, *19*, 1–11.
9. Mazurenko, L.L.; Soloviev, V.A.; Gardner, J.M. Hydrochemical features of gas hydrate-bearing mud volcanoes, offshore Morocco. In Proceedings of the International Conference Geological Processes on European Continental Margins, IOC/UNESCO Workshop Report 168, Granada, Spain, 31 January–3 February 2000; pp. 18–19.
10. Hensen, C.; Nuzzo, M.; Hornibrook, E.; Pinheiro, L.M.; Bock, B.; Magalhães, V.H.; Brückmann, W. Sources of mud volcano fluids in the Gulf of Cadiz—Indications for hydrothermal imprint. *Geochim. Cosmochim. Acta* **2007**, *71*, 1232–1248.
11. Niemann, H.; Duarte, J.; Hensen, C.; Omoregie, E.; Magalhães, V.H.; Elvert, M.; Pinheiro, L.M.; Kopf, A.; Boetius, A. Microbial methane turnover at mud volcanoes of the Gulf of Cadiz. *Geochim. Cosmochim. Acta* **2006**, *70*, 5336–5355.
12. Thakur, N.K.; Sanjeev, R. *Exploration of Gas Hydrates*; Springer-Verlag: Berlin/Heidelberg, Germany, 2011; pp. 42–85.
13. Sloan, E.D. Fundamental Principles and Applications of Natural Gas Hydrates. *Nature* **2003**, *426*, 353–359.
14. Kvenvolvden, K.A.; Claypool, G.E. *Gas Hydrates in Oceanic Sediment*; U.S. Geological Survey: Reston, VA, USA, 1998; pp. 88–216.

15. Collet, T.S. *Gas Hydrates of the United States, in 1995 National Assessment of United States Oil and Gas Resources, USGS Digital Data Series, 30, on CD-ROM*; Gautier, D.L., Dolton, G.L., Takahasshi, K.I., Varnes, K.L., Eds.; USGS: Reston, VA, USA, 1995; p. 85.

16. Klauda, J.B.A.; Sender, S.I. Global distribution of methane hydrate in Ocean Sediment. *Energy Fuel* **2005**, *19*, 459–470.

17. Burwickz, E.; Rupke, L.H.; Wallmann, K. A new global gas hydrate budget based on global inventory of methane hydrates in marine sediments using transfer functions. *Biogeosciences* **2011**, *10*, 959–975.

18. Wallmann, K.; Pinero, E.; Burwicz, E.; Haeckel, M.; Hensen, C.; Dale, A.; Ruepke, L. The Global Inventory of Methane Hydrate in Marine Sediments: A Theoretical Approach. *Energies* **2012**, *5*, 2449–2498.

19. Piñero, E.; Marquardt, M.; Hensen, C.; Haeckel, M.; Wallmann, K. Estimation of the global inventory of methane hydrates in marine sediments using transfer functions. *Biogeosciences* **2013**, *10*, 959–975.

20. Pollack, H.N.; Hurter, S.J.; Johnson, J.R. Heat flow from the Earth's interior: Analysis of the global data set. *Rev. Geophys.* **1993**, *31*, 267–280.

21. Hamza, V.M.; Cardoso, R.R.; Ponte Neto, C.F. Spherical harmonic analysis of the earth's conductive heat flow. *Int. J. Earth Sci.* **2008**, *97*, 205–226.

22. Marquardt, M.; Hensen, C.; Piñero, E.; Haeckel, M.; Wallmann, K. A transfer function for the prediction of gas hydrate inventories in marine sediments. *Biogeosciences* **2010**, *7*, doi:10.5194/bg-7-2925-2010.

23. Leon, R.; Somoza, L.; Gimenez-Moreno, C.J.; Dabrio, C.J.; Ercilla, G.; Praeg, D.; Diaz-del-Rio, V.; Gomez-Delgado, M. A predictive numerical model for potential mapping of the gas hydrate stability zone in the Gulf of Cadiz. *Mar. Petrol Geol.* **2009**, *26*, 1564–1579.

24. Zitellini, N.; Gràcia, E.; Matias, L.; Terrinha, P.; Abreu, M.A.; DeAlteriis, G.; Henriet, J.P.; Dañobeitia J.J.; Masson, D.G.; Mulder, T.; *et al.* The quest for the Africa—Eurasia plate boundary west of the Strait of Gibraltar. *Earth Planet. Sci. Lett.* **2009**, *280*, 13–50.

25. Medialdea, T. *Estructura y Evolución Tectónica del Golfo de Cádiz Pub*; Serie Tesis Doctorales; Instituto Geológico y Minero de España: Madrid, Spain, 2007; Volume 8, p. 382. (In Spanish)

26. Gardner, J.M.; Vogt, P.R.; Somoza, L. The possible effect of the Mediterranean Outflow Water (MOW) on gas hydrate dissociation in the Gulf of Cadiz. *EOS Trans. AGU* **2001**, *82*, Abstracts OS12B-0418.

27. Somoza, L.; Díaz del Río, V.; Hernandez-Molina, F.J.; Leo. R.; Lobato, A.; Alveirinho, J.M.; Rodero, J.; TASYO Team. New discovery of a mud-volcano field related to gas venting in the Gulf of Cadiz: Imagery of multibeam data and ultra-high resolution data. In Proceedings of the Final Proceedings 3rd International Symposium Iberian Atlantic continental margin, Faro, Portugal, 25–27 September 2000; pp. 397–398.

28. Sloan, E.D.; Koh, C.A. *Clathrate Hydrates of Natural Gases*, 3rd ed.; CRC Press: Boca Raton, FL, USA, 2008; p. 720.

29. Benazzouz, O. Gas hydrates stability domains in the Portuguese Margin. The Gulf of Cadiz and the West Alboran SEA. Master's Thesis, Faculty of Science and Technology, Geosciences Department, University of Abdelmalek Essadi, Tangier, Marroco, October 2011.

30. Thiede, J.; Ehrmann, W.U. Late Mesozoic and Cenozoic sediment flux to the central North Atlantic Ocean. *Geol. Soc. Spec. Publ.* **1986**, *21*, 3–15.

31. Broecker, W.S.; Klas, M.; Clark, E.; Bonani, G.; Ivy, S.; Wolfli, W. The influence of $CaCO_3$ dissolution on core top radiocarbon ages for deep-sea sediments. *Paleoceanography* **1991**, *6*, 593–608.

32. Thomson, J.; Nixon, S.; Summerhayes, C.P.; Schönfeld, J.; Zahn, R.; Grootes, P.M. Implications for sedimentation changes on the Iberian margin over the last two glacial/interglacial transitions from (230Th-excess)0 systematics. *Earth Planet. Sc. Lett.* **1999**, *165*, 255–270.

33. Sarnthein, M.; Winn, K.; Jung, S.J.A.; Duplessy, J.; Labeyrie, L.D.; Erlenkeuser, H.; Ganssen, G.M. Changes in east Atlantic deepwater circulation over the last 30,000 years: Eight time slice reconstructions. *Paleoceanography* **1994**, *9*, 209–267.

34. Voelker, A.H.L.; Lebreiro, S.M.; Schönfeld, J.; Cacho, I.; Erlenkeuser, H.; Abrantes, F.F. Mediterranean outflow strengthening during northern hemisphere coolings: A salt source for the glacial Atlantic? *Earth Planet. Sc. Lett.* **2006**, *245*, 39–55.

35. Alt-Epping, U. Late Quaternary sedimentation processes and sediment accumulation changes off Portugal. Ph.D. Thesis, Elektronische Dissertationen an der Staats- und Universitätsbibliothek Bremen, Germany, 2008, pp. 1–169.

36. Suess, E. Particulate organic carbon flux in the oceans—Surface productivity and oxygen utilization. *Nature* **1980**, *288*, 260–263.

37. Martens, C.S.; Haddad, R.I.; Chanton, J.P. *Organic Matter Accumulation, Remineralization, and Burial in an Anoxic Coastal Sediment in Organic Matter: Productivity, Accumulation and Preservation in Recent and Ancient Sediments*; Whelan, J.K., Farrington, J.W., Eds.; Columbia University Press: New York, NY, USA, 1992; pp. 82–98.

38. Seiter, K.; Hensen, C.; Zabel, M. Benthic carbon mineralization on a global scale. *Glob. Biogeochem.* **2005**, *19*, doi:10.1029/2004GB002225.

39. Pangea Data Publisher & Environmental Science. Available online: http://www.pangea.de (accessed on 23 April 2014).

Theorizing for Maintenance Management Improvements: Using Case Studies from the Icelandic Geothermal Sector

Reynir Smari Atlason, Gudmundur Valur Oddsson and Runar Unnthorsson

Abstract: As renewable energy sectors evolve and grow within a country, the need for expertise to maintain its infrastructure grows. Such expertise is often provided by foreign industries. It is in the global interest to facilitate expertise to grow domestically, eventually leading to widespread clusters of industries around a renewable energy sector and a global growth of expertise. This ultimately fast tracks the development in the renewable energy sector since more players become active in developing solutions. In this article the factors influencing domestic development are identified from previous studies conducted within the Icelandic geothermal sector. The cause and effect relationships between the identified factors are then mapped. A system dynamics causal loop diagram based on Icelandic case studies is presented to visualise how the formation of industrial clusters in the renewable energy sector can be initiated. This visualisation, based on the Icelandic geothermal sector, can be of use for other industries in the renewable energy sector who are attempting to conduct their maintenance procedures domestically and increase the rate of innovation within a country.

Reprinted from *Energies*. Cite as: Atlason, R.S.; Oddsson, G.V.; Unnthorsson, R. Theorizing for Maintenance Management Improvements: Using Case Studies from the Icelandic Geothermal Sector. *Energies* **2015**, *8*, 4943–4962.

1. Introduction

It is estimated that easily reachable oil and gas will be depleted within the next fifty years. It is further anticipated that easily reachable coal will be depleted within the next century or so [1]. Fossil fuel energy sources have driven industrial processes and the ever increasing quality of life amongst the western nations over the past century. They have allowed for previously unknown levels of consumption and wealth relative to pre-industrial times. The depletion of these sources is of increasing concern because the effect of their depletion on human societies is virtually unknown. It is also known that on-going consumption of these sources is accompanied by severe environmental impacts [2]. One way to mitigate the effects of climate change and adapt to the ever increasing scarcity of non-renewable energy sources is to increase the use of renewable energy. The current development in the renewable energy sector is, however, not fast enough for renewable energy to compensate for a significant amount of fossil fuel use. British Petroleum (BP) expects global primary energy demand to increase by 41% before the year 2035, most of which will happen within the non-OECD (Organisation for Economic Co-operation and Development) countries [3]. By 2035, BP anticipates that renewable energy sources will grow from their current level of 2% to contribute about 8% of the world's energy supply [3]. An increase in the share of renewable energy sources can contribute to social and economic development, and can also accelerate energy access (especially in the developing world). It can increase energy supply security and some renewables can contribute to reduced greenhouse gas (GHG) emissions [4]. With this in mind, it should be of great importance to increase the rate of innovation and increase efficiency in the global renewable sector. These developments could potentially contribute to the benefits that were previously mentioned. Even though economic crisis is sometimes considered to have a negative impact on innovation, research, and development [5], crisis is also often seen to be the source of innovation. For example, the oil crisis in the 1970s pushed Icelandic society towards geothermal energy utilisation because domestic oil heating became very expensive.

During the 2008 global financial crisis, the cost of maintaining the Icelandic geothermal power plants grew immensely. This can be attributed directly to the devaluation of the Icelandic Krona, rising oil prices, and the relationship between power plant maintenance and energy prices. Even though some power plant maintenance had previously been conducted domestically, this crisis pushed many Icelandic energy companies to begin attempting to solve problems domestically that had previously been outsourced. This was mostly focused on expensive machinery such as turbines, that required costly, specialised, knowledge. Conducting this maintenance domestically required a build up of skills and knowledge within Iceland, since it had either not been present or was in shortage. This knowledge and skill transfer (KAST), originating to some extent in the 2008 global financial crisis, has resulted in a growth of expertise within the country. Icelandic industries are now almost fully capable of servicing the geothermal sector themselves. The knowledge and skills to conduct specialised, costly maintenance were previously sought internationally from a dispersed group of specialised industries. As the Icelandic geothermal companies began to seek solutions domestically, Icelandic industries, such as machine shops, began to address and solve problems that had previously been solved internationally. The KAST can also be regarded as an industrial cluster formation because the Icelandic industries who were beginning, or improving, their domestic services for the geothermal energy companies were all located within the same geographic region. It is simply a matter of time until these Icelandic industries enter global markets, providing domestically developed solutions.

During a time of crisis, problems may become to large to ignore. This was the case for the Icelandic geothermal industry, whose major maintenance activities that had previously been outsourced became simply to expensive. A method to facilitate this innovation without having an initial crisis would, however, be preferred in every case because a crisis may often lead to significant financial loss and cannot be controlled as desired. Even though various methods and tools are available to improve the rate of innovation, it is the intent of this article to show how the Icelandic geothermal sector managed to move major and expensive maintenance activities to Iceland. As a method to visualise the process of this cluster formation, a system dynamics causal loop diagram is presented, which is based on the Icelandic geothermal industry as it moved its major maintenance procedures to Iceland after the 2008 global financial crisis. The model provided was generated using a series of case studies within the geothermal industry, which were conducted by the authors and published in the scientific literature. The contents of this article are investigated in the context of cluster theory, where the theory is used to describe the concentration of industries around the renewable energy sector, in particular the Icelandic geothermal sector.

1.1. Cluster Theory

Industrial clusters can be defined as "geographic concentrations of interconnected companies, specialised suppliers, service providers, firms in related industries, and associated institutions in a particular field that compete but also cooperate [6]".

Players within a cluster include providers of specialised products and services, infrastructure providers, governmental institutions, think tanks, and trade associations who provide technical support that benefits or contribute to a specific sector. Clusters are an important competitive advantage because other factors that were previously important, such as access to non-scarce resources, are becoming less important as global logistics serve the need for resource transportation. For example, aluminium smelters are located in Iceland but the country lacks any bauxite resources. In addition, deploying sophisticated technology is not a factor because industries can freely use modern technology in their production. In pre-modern times, the technology that was available in one region was not so easily transferred or available to another region, today this is not the case. It then becomes clear that infrastructure, the legal environment, and the services that are located in geographic proximity to a particular industry have become a significant factor in how competitive the industry eventually becomes [7,8]. Being a part of a cluster increases productivity as access to inputs, information, technology and relevant institutions improves. As a cluster forms, the formation becomes self-reinforcing. This is further increased when the public sector is supportive and competition is present [7]. In cluster theory, the role of the government is to remove obstacles to industrial growth and achieve macroeconomic and political stability. It should, according to cluster theory, improve general microeconomic capacity "through improving the quality and efficiency of general-purpose inputs to business and the institutions that provide them" [6]. Regardless of the effectiveness of public policy, it has been shown that a cluster takes a decade or more to develop a competitive advantage [7].

A cluster's absorptive capacity is the "capacity of firms to establish intra- and extra- cluster knowledge linkages" [9]. This is the capacity of a cluster to gather knowledge from the outside and effectively distribute this knowledge on the inside. However, when digging deeper into cluster theory, it can be seen that the knowledge flow is not equally distributed between firms within a cluster. In fact, clustering may isolate some firms while others increase their collaboration. In addition, even though business flows are frequent between firms within a cluster, knowledge flow does not necessarily follow. This has been observed when wine clusters have been studied in Italy and Chile [10]. A sectoral system is in essence the same as a cluster. The players within such a system interact through cooperation, competition, exchange, and communication [8]. Clusters, or sectoral systems, are also a dynamic phenomena that is constantly changing [8]. This happens because firms who seek new markets tend to modify their business behavior or begin interacting with other components of a cluster in a different manner. It has furthermore been stated that a firm's value cannot only be seen from the patents issued, staff, and machines owned, but should also be seen in its participation and involvement within a cluster [11].

The literature is rich with information on the effects of cluster cooperation and how innovations are more likely to grow out of such environments [6,12,13]. Benefits have been shown to be partly due to the close geographic proximity of relevant industries, information, complementary relationships, and competitive pressure [14]. An example of this can be found in the technology industry in Silicon Valley, California. Industrial clusters are, however, not bound to be based on a single geographical location for each industry. A sector possibly benefits from having multiple clusters that are spread globally. If multiple clusters are operating around the same sector, an improved rate of innovation and development may possibly be experienced. The structure of a cluster is dependent on the "characteristics of technologies used, social norms and institutional factors that outline the rules followed" within the clusters [11]. Industrial clusters are a key factor in a nation's innovative capacity [13], it is therefore of great importance to facilitate clusters within a nation where the foundations for such a collaboration are sound. This importance is amplified when the need for faster development within a sector is desperately needed, such as within the renewable energy sector. Cluster formations, rather than the dispersion of industries, could potentially serve as the catalyst needed for the renewable sector to have the significant impact it needs to have within the global energy use portfolio. An impact that increases the access to renewable energy globally.

Even though the literature is rich with information about the benefits of cluster cooperation, it suffers from a lack of methods to initiate clusters where they do not currently operate. It has been attempted to fill this gap, with some success [14]. By conceiving a non-simulated system dynamics model, it has been shown how an industrial cluster effect may be achieved [14]. The previously proposed model is, however, based on a literature review rather than on case studies that are conducted by the authors. Others have attempted to simulate the agglomeration of industries within a region and show how knowledge and proximity effect the behaviour of a technology district [15]. These studies have, however, not shown how a cluster can be formed but rather the benefits of the proximity of players within an industrial cluster. The benefits of using a system dynamics causal loop diagram have been shown to be a convenient way of depicting a cluster behaviour [14,15]. Behaviours of industries within sectors, and the sectors themselves are not static. Industries evolve, compete, collaborate, perish and flourish. One way to describe a swift change in behaviour is to look at unforeseen, catastrophic, unwanted or any events forcing a change in behaviour for industries. Such events are often called disruptive events, leading to a so called quantum shift.

1.2. Quantum Shifts

Disruptive events often lead to a shift in behaviour, this is well known in natural evolution as well as in the corporate sector. Disruptive events are non-controllable events that force industries to modify their operational behaviour. Disruptive events can be felt in various forms for different industries, such as increased competition, modification in the legal or policy environment, or even ecological change, such as climate change. As mentioned, this process also happens in nature, but the focus of this article is on business. Disruptive events leading to adaptation are called quantum shifts or punctuated equilibria [16,17]. For industries to survive after a disruptive event, they must radically improve competitive innovation, new value creation, and create and distribute knowledge [16]. A graph depicting a business development before, during, and after a quantum shift can be seen in Figure 1. This graph, however, assumes that the industry has survived the disruptive event and continues to operate after the quantum shift. This is not necessarily correct because many businesses cease to exist after a disruptive event. A disruptive event can also be felt in the form of competition, such as an introduction of a service or a product on the marketplace by a competitor. However, a quantum shift does not necessarily lead to an elevated performance in the same business or operational direction prior to the disruptive event initiating the shift. Indeed, corporations may need, or see potentials, in shifting their direction of operation. This may include new market opportunities or introducing a new emphasis in their operations.

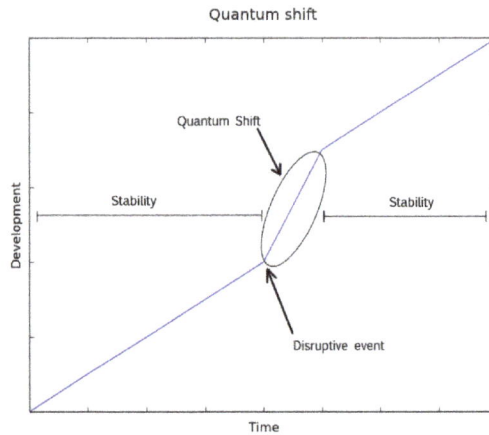

Figure 1. The effects of disruptive events on the operation and development of corporations.

Figure 2 depicts a quantum shift when the operational direction is subsequently altered. A disruptive event, such as a financial crisis, may facilitate a quantum shift for a whole industrial sector of a country rather than individual companies. In Iceland for example, the 2008 global financial crisis served as a disruptive event that forced the geothermal sector as a whole to face new challenges. As the Icelandic currency devalued dramatically, oil prices rose as did the price of spare parts for the power plants, since the price of oil and spare parts is linked. The value of Iceland's debts also rose because they were issued in foreign currency. This all lead to increased costs in the operation and maintenance of Icelandic power plants. The challenges faced were addressed effectively and this has led to further domestic collaboration that continues to thrive. To gain a good understanding of the Icelandic geothermal sector, one must understand who the main players are within the sector. The following section aims to provide such understanding.

2. Key Facts about the Icelandic Geothermal Sector

In this chapter the Icelandic geothermal power plants and the industrial cluster environment that they engage in are described.

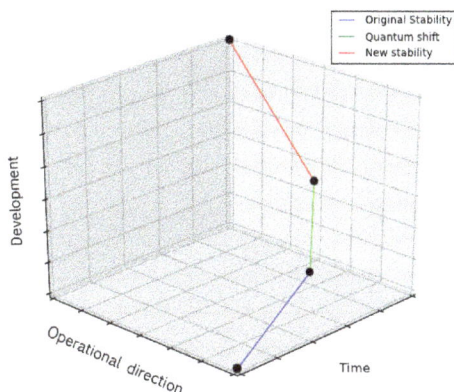

Figure 2. A quantum shift that includes a possible change in operational direction.

2.1. Icelandic Geothermal Power Plants

Six geothermal power plants operate within Iceland. In 2008, the accumulated production amounted to 4.038 GWh (Gigawatt hours) of electricity [18]. The plants are Bjarnarflag (3.2 MWe (Megawatt electrical)), Hellisheidarvirkjun (303 MWe), Krafla (60 MWe), Nesjavallavirkjun (120 MWe), Reykjanesvirkjun (100 MWe), and Svartsengi (76.4 MWe) [18]. Staff from all of the energy companies operating these power plants participated in this research. Icelandic power plants manage to keep a small amount of staff on site while keeping the plants operating without major problems. At Svartsengi and Reykjanesvirkjun (both operated by HS Energy), a staff of approximately 20 people are responsible for the plant maintenance and operations. This staff consists of mechanical and electrical engineers, and earth and environmental scientists. The employees attend to 12 turbines (five steam and seven Organic Rankine Cycle), 36 cooling fans, 17 geothermal wells, wellheads, 70 control valves, 100 pumps, 20 km of pipelines, and a vast amount of valves that all need regular maintenance [19]. A similar situation can be found in the other Icelandic plants, such as Nesjavellir and Hellisheidi, who operate with a total of only 26 employees in each plant [20]. The engineers operate the plant and conduct maintenance. The head of the plant engineering department answers directly under the plant manager. The company CEO is in all cases in direct contact with the plant managers. The plant engineers conduct a great amount of maintenance on site, in collaboration with domestic workshops. It has been stated publicly that the largest geothermal utilisers in Iceland are attempting to fully transfer all of their maintenance activities to Iceland. According to the plant manager at Nesjavellir, the efficient management of the power plants can be attributed to a great extent to the experience that their engineers acquired when working as marine engineers on fishing vessels [21]. Marine engineering education in Iceland is also more detailed than international standards require. At sea, marine engineers can only rely on themselves for repairs, regardless of the prevailing weather conditions, which requires them to adapt a certain mentality, this is especially true for the chief engineer. This mentality is much appreciated by the power plant chief engineers and it includes the notion of unity at site and resourcefulness in order to keep the vessel (in this case the plant) running at all costs. Consequently, the energy companies tend to prefer marine engineers that have worked as chief engineers.

2.2. Domestic Collaboration

Corporations in general tend to outsource more of their non-core activities than they have previously done [22]. This allows companies to focus on other important issues, such as differentiation in the market place or cost leadership [12]. However, industries may depend on non-core activities for successful operations. Non-core activities may become a financial burden that the corporation can solve with lower costs. Outsourcing can also lead to failures because the customer's expectations are not met. Outsourcing also creates dependency on the contractor. Not outsourcing allows the corporation to build knowledge and potential value; however, this is followed by an increase in other costs. Therefore, an industry must analyse the potentials fully before deciding to conduct previously outsourced procedures themselves. Furthermore, by not outsourcing. the corporation becomes independent and frees it from relying on a contractor for its operations. However, a balance is needed between outsourced services and those addressed by the organisation itself.

Various players within the Icelandic geothermal sector have engaged in a conversation on further collaboration in the operation and management of the Icelandic geothermal power plants [23]. These players include energy companies, utility companies, machine shops, banks, universities, research facilities, engineering consulting firms and others [24]. It has been shown that by domestically conducting the operational and maintenance tasks, sometimes overseen by foreign specialists, domestic know-how and experience could be increased and currency be kept within the country. Such procedures could potentially allow domestic service providers to enter foreign markets. Experiments have been conducted on geothermal steam turbine parts and the possibility for such parts to be produced domestically has been analysed. Such experiments have often proved successful. Some experiments were conducted fully by domestic machine shops but some were also conducted in collaboration with foreign specialists. Collaboration with foreign industries contributes to KAST to Iceland and can, therefore, be seen as beneficial to some extent.

Two mutual platforms have been established to smooth the communication platform between all parties: one platform is business driven, while the other is research driven. The business driven platform is called Iceland Geothermal and the research driven platform is called the Geothermal Research Group (GEORG). The business driven platform is organised by a private entity, which facilitates lectures and provides a communication platform.

Discussion has been ongoing in both the scientific and popular literature about the possibilities of a domestic cluster that builds on Iceland's knowledge of geothermal energy utilisation. This is expected to be a good way to improve the opportunities for domestic industries to serve the geothermal sector in the coming future [23]. The knowledge created within this cluster is expected to be of value to all participating companies. The Icelandic geothermal cluster has been defined as a business driven cluster. GEORG is also a geothermal cluster but is research driven and funds various geothermal based research [25]. GEORG consists of 22 multi-national research driven partners, ranging from universities and public companies to private partners. A connection, both internally and to private industries within the cluster, can also be considered to be a factor for the efficient operation of these power plants. Given the small scale of the Icelandic economy, relatively few service projects are conducted per year and domestic service providers are forced to use their employees for servicing other industries, such as fishing and aluminium. Together, these industries work together to help to support the buildup of domestic knowledge and expertise. By servicing various industries, Iceland's service providers can keep up the number of projects required to maintain their employees. In addition, solutions developed and implemented for one industry can be adapted to another.

The collaboration described above became increasingly visible after the global financial crash. The Porter report [23] can be viewed as a landmark for the visualisation of its benefits. The cluster and the collaboration were already forming before the 2008 financial crisis. However, the 2008 events pushed the Icelandic geothermal industry towards further developments domestically, especially with regards to the most expensive maintenance procedures, namely the turbines. The benefits after this quantum shift occurred are, however, attractive for industries involved in the renewable energy sector, although the initial seed is not. Theorizing for such developments is partially the focus of this article.

2.3. Article Intent and Content

The aim of this article is to gather vital information from a series of case studies that were conducted by the authors and then present a causal loop diagram depicting how cluster formation may partially be initiated. The introduction provides background information describing the importance of domestic industrial clusters to the global renewable energy sector. Subsequently, the methodologies used in each case study are briefly described. The results from the case studies are then used to form the results chapter. A causal loop diagram is shown to demonstrate the behaviour of the Icelandic geothermal industrial cluster. The limitations of this research, suggestions for future research, and general discussions about the developed diagram are provided in the discussions chapter.

3. Methodologies

This section will discuss theory building and describe how the research was planned, eventually providing a description of how and why each case study was conducted.

3.1. Theorizing and Case Studies

The construction of theory from case studies is a strategy that includes one or several case studies that are used to conceive theories or propositions [26]. Case research has been used as a powerful research tool in the development of new theory [27]. Calls have been made for more field-based research simply to keep up with technological changes in managerial methods within operations managements [28]. By conducting field case research, theory is not merely improved or enriched but, equally or more importantly, the researcher themselves are also enriched [27]. However, case research has various challenges, such as time consumption [29], the need for skilled interviewers, and the care needed to draw conclusions [27]. This is amplified when a series of case research studies are to be conducted in theory building. Case research assists with answering how and why questions and it is also good for theory building. Conducting case based research has several benefits [30]:

- The study is conducted within the natural environment of the phenomena, which allows for an observation of real practice.
- Why, what, and how questions can be answered with a great understanding of the phenomenon.
- Early investigations where variables are unknown and the phenomenon is unclear are suitable for case studies.

Various researchers have defined the methodology for case based theory building in a similar manner [27,31]. The steps are outlined in Table 1. Interestingly, even though case based research is praised by some researchers and calls have been made for such research, only a minority of operations management articles are actually case based. The research model has been defined as having five stages, as follows: (1) define the research question; (2) instrument development; (3) data gathering; (4) data analysis; and (5) dissemination [29,31]. In fact, the case based theory building process that is presented seems to be generally accepted by researchers in operations management [31]. Furthermore, it has been stated that service design is one of the potential areas for future operations management research, which is partially the focus of this article [29].

3.2. This Research in a Theoretical Context

In the context of this research, it was decided to follow the guidelines that are shown in Table 1 [31]. Case studies were conducted to answer the questions posed in each step. The theory building process used in this article is also found to be relevant to other theory building procedures demonstrated in the literature, hence its validity [32].

The results of each step were published in the scientific literature for approval, either in journals or peer reviewed conference proceedings.

Even though the methodology to fully construct a theory has been described [31,32], because the proposed model has not been tested, it is not our intent to fully form a theory in this article but rather to theorize. This includes the first three steps in Table 1. This article is a step on the way towards a theory.

3.3. Case Studies Conducted

The cluster formation will be visualised using a qualitative method. A system dynamics approach is used without simulation because it is not our intent to simulate the rate of formation of a cluster but rather to visualise which factors may initiate such formation.

To visualise the connection between different players within the Icelandic geothermal industry and to create a systems dynamics model, case studies were carried out that focused on various themes. The themes were selected after multiple meetings with chief maintenance engineers at the power plants and machine shops, and meeting with the staff at innovation and cluster centers. The themes that we studied required us to answer some questions that fitted within the theory building process as described in the literature [31].

Table 1. The theory building process as presented in the literature [31].

Purpose	Research Question	Research Structure	Examples of Data Collection Techniques	Examples of Data Analysis Procedures
1a. Discovery * Uncover areas for research and theory development	* What is going on here? * Is there something interesting enough to justify research	* In-depth case studies * Unfocused, longitudinal field study	* Observation * Interviews * Documents * Elite interviewing	* Insight * Categorization * Expert opinion * Descrptions
1b. Description * Explore territory	* What is there? * What are the key issues? * What is happening?	* In-depth case studies * Unfocused, longitudinal field study	* Observation interviews * Documents * Elite interviewing * Critical incident * Technique	* Insight * Categorization * Expert opinion * Descriptions * Content analysis
2. Mapping * Identify/describe key variables * Draw maps of the territory	* What are the key variables? * What are the salient/critical themes, patterns, categories?	* Few focused case studies * In-depth field studies * Multi-site case studies * Best-in-class case studies	* Observation * In-depth interviews * Diaries survey questionnaires * History * Unobtrusive measures	* Verbal protocol * Analysis * Cognitive mapping * Repertory grid technique * Effects matrix * Content analysis
3. Relationship building * Improve maps by identifying the linkages between variables * Identify the "why" underlying these relationships	* What are the patterns or links between variables? * Can an order in the relationship be identified? * Why should these relationship exist?	* Few focused case studies * In-depth field studies * Multi-site case studies * Best-in-class case studies	* Observation * In-depth interviews * Diaries survey questionnaires * History * Unobtrusive measures	* Verbal protocol * Analysis * Cognitive mapping * Repertory grid technique * Effects matrix * Content analysis * Factor analysis * Multidimensional * Scaling * Correlation analysis * Nonparametric analysis
4. Theory validation * Test the theories developed in the previous stages * Predict future outcomes	* Are the theories we have generated able to survive the test of empirical data? * Did we get the behavior that was predicted by the theory?	* Experiment * Quasi-experiment * Large scale sample of population	* Structured interviews * Documents * Open and closed-ended questionnaires * Lab experiments * Field experiments * Quasi experiments * Surveys	* Triangulation * Analysis of variance * Regression * Analysis * Path analysis * Survival analysis * Multiple comparison procedures * Nonparametric statistics
5. Theory extension/Refinement * Expand the map of the theory * Better structure the theories in light of the observed results	* How widely applicable/ genera-lizable are the theories developed? * Where do the theories apply? * Where do they not apply?	* Experiment * Quasi experiment * Large scale sample of population	* Structured interviews * Documents * Open and closed-ended questionnaires * Lab experiments * Field experiments * Quasi experiments * Surveys * Documentation * Archival research	* Triangulation * Analysis of variance * Regression * Analysis * Path analysis * Survival analysis * Multiple comparison procedures * Nonparametric statistics * Meta analysis

- Domestic service providers in the geothermal industry.

  This part describes how domestic industries are currently addressing major maintenance issues with regards to geothermal steam turbines in Iceland. It will examine how they collaborate and which products are being manufactured domestically [33,34]. When critical failures occurred, the repair process was visualised and the communication chain was identified. It was seen that domestic repairs were conducted in a more economical and faster way. This resulted in shorter down-time of the turbine and a subsequent lower loss of output. It was also shown that knowledge was building up within the machine shops and the energy company. During one study, where the current turbine operations and maintenance procedures were examined, it was seen that the operations and frequency of overhauls on geothermal turbines is changing as the staff becomes more experienced. Problems are analysed in collaboration with the Icelandic innovation center and also with the machine shops. If a solution is viable when looking at performance, then it was developed further. In this case, a faulty setting on a valve lead to the breakdown of the labyrinth packing. The problem was analysed and a repair, with an improved version of the labyrinth packing, was conducted on-site in collaboration with a domestic machine shop.

- Corporate culture with regard to innovation.

The development of a geothermal control valve was examined: first, from the corporate viewpoint of how the development process occurred within the company; and secondly, from the technical side of how the valve operates, is manufactured, and tested [35,36]. This allowed for a clear visualisation of a successful innovation process within the geothermal industry. A willingness to try the development of a solution posed by a staff member was observed because of a lack of solutions available to the problem observed and also because of the possible financial viability of the solution since it was to be produced domestically. It was shown that the CEO of the energy company allowed the staff member a certain degree of freedom to develop the proposed valve solution. This included funds for prototyping, and specialised consulting and testing. The valve became the standard for control valves within that particular company. A machine shop was included in the development process, allowing for the knowledge about the manufacturing side of the valve to be located in a geographic proximity to its final use. The valve is currently fully manufactured and used domestically.

- The effects of the operation engineer's previous experience.

This research also outlined the Icelandic geothermal cluster, and described who the main players are and how they are interconnected. It was seen that maintenance engineers operating within the energy companies possess a certain characteristic, perhaps because of their naval experience. The mentality brought to the geothermal power plants by naval engineers was statistically examined [24]. The engineers were found to be less considerate and less likely to seek supervisory opinion than regular workers on the market. It was noted that the naval engineers need to repair any failures on board while on the ocean. The same was seen with maintenance engineers in the geothermal power plants, who are very confident and willing to try developing domestic solutions.

- The effects of operational experience

The way that operational experience effects planning was visualised with regards to the wellheads at Hellisheidi geothermal power plants. Real data was gathered and statistically analysed [37]. This study was conducted to analyse how a maintenance pattern evolves with time as operational experience is gathered among staff. By using a Weibull survival distribution for the analysis, it was shown that the frequency of maintenance does in fact change with time, diverting from the original recommendations made by engineering consulting firms. This is also in line with the previous observations of the engineers' characteristics.

• Identification of future developments

A quantitative Kano model was used to identify the solutions sought after by Icelandic geothermal power plant maintenance engineers [38]. This study demonstrated how a model for customer satisfaction can be utilised within the renewable energy sector. When applied to the geothermal sector, it demonstrated which maintenance management tools are wanted by maintenance engineers. The problems to be addressed include the long documentation time and the uncertainty of postponing maintenance, among others. The Kano analysis tool can be used to visualise the needs of domestic industries while the other industries within the cluster can use the results to develop these solutions.

The results from the studies were then analysed and used to form a system dynamics causal loop diagram. The causal loop diagram can be used to visualise the connection between the factors examined. The causal loop diagram was not intended to be simulated but rather to demonstrate the behaviour of the industry before and after a quantum shift has occurred. A previously proposed theory building process was followed, which eventually led to this article. Table 2 outlines the case studies conducted and where they fit in with the theory construction process. The result from this article is essentially step 3, as defined in the theory building process [31]. This step includes relationship building, identifying an order in different relationships previously observed through field and case studies, eventually resulting in cognitive mapping. In this study, case studies have been conducted and the mapping was done using causal loop diagrams.

Table 2. A list of the case studies conducted, the intention of the studies and where they fit within the theory building process.

Case Study	Intention	Stage in Theory Building according to [31]	Research Structure	Method of Data Collection according to [31]
[35]	Explore how staff has influenced innovation within geothermal firms.	1a. Discovery	- Interviews - Observation	- Insight - Expert opinion - Descriptions
[24]	Get an overview of the Icelandic geoth. sector. Who are the players and what are the characteristics of plant maintenance engineers.	1a. Discovery	- In-depth case study	- Description - Insight
[38]	Visualise how operations management can be improved using staff knowledge.	1a. Discovery	- Interviews	- Expert opinion - Insight
[33]	Identify the current benefits of the geothermal cluster collaboration within Iceland. an acute repair of a geoth. turbine was examined	1b. Description	- In-depth case study	- Critical incident - Documents
[36]	Visualise how domestic industries can influence innovation process.	1b. Description	- In-depth field study	- Observation interviews - Documents - Technical specifications
[34]	Get an overview on how maintenance on geoth. turbines is conducted. Who are the main actors and what are the challenges ahead.	2. Mapping	- In-depth field study	- Verbal protocol - Analysis
[37]	Explore how experience influences the operations management in geothermal power plants.	2. Mapping	- In-depth field study	- Analysis

4. Results

In Figure 3, one can see the established causal loop diagram, which is coded with three colours. The initial condition before the quantum shift occurs is coloured in orange. As the quantum shift occurs, a process coded in blue is initiated. Third is the green graph, which has not been visualised throughout the case studies but has been stated to be a goal suitable for the Icelandic geothermal industry [23]. The relationship between the variables is explained in the following sections. Not all of the variables are equally weighed, and some can be regarded as key variables. The variables "willingness to try" and "KAST" are shown to be key variables in the system dynamics model.

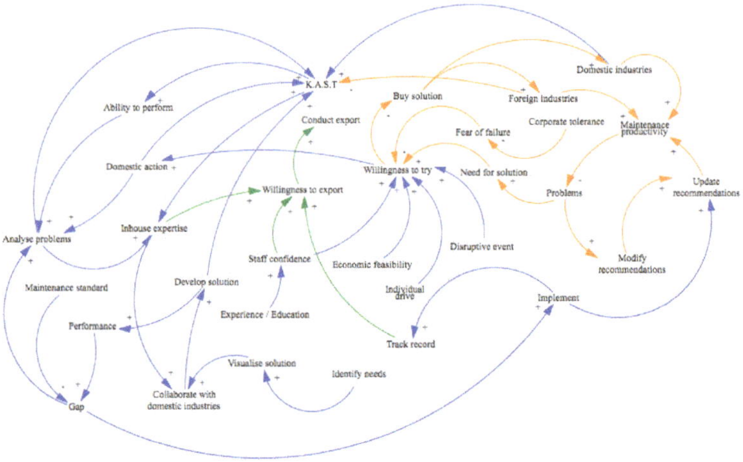

Figure 3. A causal loop diagram describing the behaviour leading to increased domestic development.

4.1. Initial Behaviour

The variable "maintenance productivity" can be seen as the efficiency (doing things the right way) or effectiveness (doing the right thing) of maintenance procedures within a given industry. The more frequent that the occurrence of an unexpected or unwanted maintenance is, the more problematic it will be. Therefore, maintenance recommendations should be reviewed and modified, and subsequently official recommendations should updated for the staff [34,37]. In many cases, a need to purchase an external service or solution (such as software) is needed. This leads to the "willingness to try" variable in Figure 3. Initially, there is not much willingness to try because the status quo has worked previously or a fear of trying a new method is prevalent. A solution is then purchased from an external company. Initially, this expertise comes from a foreign industry, leaving no build up of expertise or innovation within the country. However, in some cases domestic industries are able to provide the needed solution, especially if the risk / benefit ratio is favourable and tests can be made on site with minimal or acceptable risk. Domestic solutions are not necessarily custom made for the geothermal industry but may be a spinoff from other industries, such as the aluminium industry. When solutions are modified to serve the geothermal industry, the operations are shifted slightly for a certain industry that previously did not serve the geothermal sector. This leads to KAST domestically as previously non-related industries become familiar with problems faced in the geothermal sector and begin to focus on them. However, the fear of failure variable is very sensitive to corporate tolerance and especially top level tolerance. This initial behaviour, coloured in orange in Figure 3, can be seen as the first stability phase, as demonstrated in Figures 1 and 2. The industry has developed a certain behaviour that operates in a sufficient manner but does not initiate or support either cluster formation or knowledge buildup domestically.

4.2. Willingness to Try

When a problem occurs and the "willingness to try" variable in Figure 3 has reached a certain level within the organisation, domestic action will be initiated. The "willingness to try" variable is the willingness within an industry to seek the development or improvement of solutions domestically rather than using the solutions that were used in the past. This variable is mostly relevant among senior maintenance engineers or operations managers within the industry. There are certain ways for the willingness to try to reach high levels. This may either be a disruptive event (such as a financial crisis), individual drive within the organisation, or high confidence of the staff, in this case maintenance engineering staff. Staff confidence has been shown to be influenced by previous experience or education. In the case of the Icelandic geothermal industry, it was seen that marine engineers are preferred to engineers that have no marine experience [24] because of their high problem solving skills, ability to work under pressure, and previous nautical experience. Also, if the need for a solution is great but it is not available internationally, then the willingness to try variable increases greatly, which leads to domestic action [35,36]. There will always be a fear of failure linked to the "willingness to try" variable. "Corporate tolerance" is the only balancing variable influencing the fear of failure, underlining the importance of tolerance with regards to failure and experimentation within an organisation. As the tolerance increases within an organisation with regards to failures, the fear of failures decreases, giving in turn less impact to the willingness to try variable. The tolerance is, however, case by case dependent and is often the subject of a favourable cost / benefit ratio.

4.3. Domestic Action

As can be seen in Figure 3, only one variable ("willingness to try") leads to domestic action. When domestic action is initiated, a given problem is analysed. This increases not only in-house expertise but also KAST. In addition, KAST has been observed when foreign industries are hired to service the domestic geothermal sector but in-house staff are allowed to study the procedures. As more domestic expertise builds up, collaboration with domestic industries increases [24,33,36]. The variable "identify needs" influences the visualisation of solutions and collaboration with domestic industries to create such a solution. This variable was identified as a method to provide cascade industries with a method to visualise which solutions to develop to service the geothermal sector. Solutions are more rapidly developed when more domestic collaboration occurs. The performance of a solution that has been developed domestically is measured and a performance gap is seen between the current maintenance standard and the proposed solution that is developed domestically. If the gap is found to be favourable, then the new solution is implemented and the current recommendations are updated. This also increases the track record of domestic solutions and services. If the gap is found to be non-favourable, then this leads to the problem being analysed further and the cycle repeats itself. "Problem analysis", "domestic action" and "develop solution" are all variables leading to KAST. Furthermore, KAST leads to an increase in the in-house expertise as well as the domestic ability to perform. A need analysis serves as a good tool to take predetermined steps and service the observed sector [38]. Even though a "cluster formation" variable is not defined within the diagram, the environment for a cluster development is essentially facilitated for use when the blue portion of the model is initiated. Although the KAST is a key factor in the cluster formation, it is not the cluster formation itself. One can observe a negative effect on KAST when the cluster has been initiated but a solution is purchased from a foreign industry. This negative impact is observed because no build up of knowledge or skills occurs when foreign expertise is used to solve operational problems without the inclusion of any domestic observers, eventually domestic industries suffer.

4.4. Export

As in-house expertise increases, staff confidence increases and the track record of successful solution implementations grows. The variable "willingness to export" reaches a level where export of new solutions become viable. Although this connection has not been observed in the case studies, it is a logical continuation from previous behavior. As a final product, a cluster around the particular sector has grown domestically, serving global markets. A time delay between the start of the cluster initiative and the ability to start exporting solutions can be expected because domestic products need to be fully tested, their track record needs to be monitored, and expertise needs to built up domestically. A theoretical demonstration of such a delay is depicted in Figure 4.

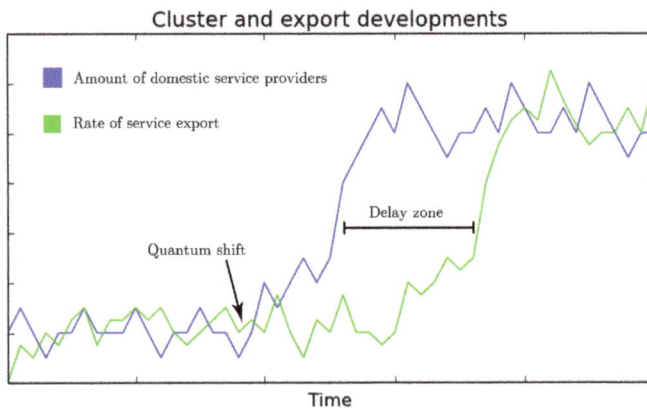

Figure 4. A theoretical relationship between a domestic cluster development and service exports. The delay zone shows a period between the cluster development and service exports.

4.5. Balancing Variables

Balancing (or negative) variables are likely to be important factors in the formation of industrial clusters. The "fear of failure" variable behaves as a balancing variable on the "willingness to try" because staff are less likely to try or develop new solutions if "corporate tolerance" is low. "Maintenance productivity" also serves as a balancing variable on the "problem" variable because fewer problems will be observed as the industry becomes more productive. The use of foreign industries influences the KAST variable in a negative way because foreign industries did not contribute towards KAST in Iceland. The variable "maintenance standard" balances the "gap" variable because a lower gap between previous solutions and developed solutions is going to be observed if the previous maintenance standard is high, thereby decreasing the possibility of new solutions being implemented.

5. Discussion

As demonstrated, three variables had a positive impact on the willingness to try variable. However, one of those variables, the disruptive event, is preferably avoided. Excluding the disruptive event puts an increased emphasis on the other two, staff confidence and individual drive. Individual drive can be impacted by an array of motivations, financial being only one of these. As has been shown, staff confidence is highly dependant on previous experience and education. It was found in Iceland that marine engineers are good candidates to oversee and conduct geothermal power plant maintenance. It is, however, not given that marine engineers possess the same qualities elsewhere as in Iceland. It is also not given that marine engineers are able to serve other renewable energy industries as efficiently. This may be contributed to by the use of steam turbines in geothermal power plants. Marine engineers might, therefore, have little knowledge about the operation and maintenance of solar or other types of power plants. Which characteristics are needed by the maintenance engineers to serve a given technology need to be identified. Although marine engineers are perhaps well suited to serve other industries, this needs to be researched in more depth. The "fear of failure" variable had a negative impact on the willingness to try. According to this model, fear of failure can only be minimised and does never increase the willingness to try. It is, therefore, in the role of corporate leaders to minimise this fear to minimise the impact on the willingness to try variable. This is especially true when the disruptive event is to be avoided. Methods to minimise the fear of failure can include improved prototyping and testing facilities, innovation seminars leading to organisational skill development, learning and growth [39].

A delay can be expected from when the cluster collaboration is initiated until exportation begins. This delay happens because products and services need to be developed and tested domestically until the track record, staff confidence and in-house expertise is sufficient and the willingness to export is enough for it to be conducted. This delay can be visualised in Figure 4. Given that the Icelandic geothermal sector has experienced a disruptive event, gone through, or is going through a quantum shift but has yet to enter global markets with products and services, it can be estimated that the Icelandic geothermal sector is currently located within the delay zone in Figure 4.

If the willingness to try variable reaches a significant level within different countries and domestic action is initiated, one can see that the resulting development increases the knowledge and skill within an industry in that particular country. However, it cannot be overlooked that by increasing the share of maintenance conducted domestically, the contractors that were previously used are no longer conducting those particular jobs. Under certain circumstances, foreign contractors keep consulting the domestic industry after the procedure has been transferred to a domestic industry. However, this consultation is merely marginal compared to previous procedures. This may have a negative effect in other regions as businesses are reduced. This could be observed as a disruptive event by international industries, pushing other industries to develop further because of the increased competition. The question also arises about how spread out clusters can actually become global before becoming a disoriented spread of small groups of companies with little connection to each other.

Building domestic clusters of adequate size, while not causing permanent harm to other clusters but rather increasing global competition, should be pursued by policymakers. These developments can potentially assist global society to develop renewable energy solutions at a faster pace, minimising the carbon footprint in energy generation and increasing the share of renewable energy technologies in the global energy portfolio.

6. Conclusions

This article uses the evidence from the case studies to demonstrate that the willingness to try domestic development within industries may be of importance when initiating industrial cluster development. If external factors will cause little willingness, then the particular industry will not participate in or initiate domestic developments but will rather continue using previous methods. The willingness to try variable needs to be triggered without a disruptive event. This may be initiated by a relevant corporate strategy (therefore minimising fear of failure within the industry), increasing the numbers of source staff with relevant experience (and therefore confidence), and facilitating individual drive. The economic benefits for the industry in question should also lead to an increased willingness to try domestic developments. Laying this groundwork should lead to domestic developments and assist with the formation of an industrial cluster around the sector in question. The model put forward, backed by case studies, shows the possible benefits of cluster collaboration for industries. Knowing the variables that influence the "willingness to try" variable may help leaders of industries to facilitate domestic development and innovation.

Future Work

It would benefit the literature if models such as the one presented in this article are quantified. This would make a practical tool available for policy makers who wish to facilitate the formation of domestic industrial clusters in the renewable energy sector. The model provided in this article demonstrates an idea, a framework, based on multiple observations, but it does not predict the rate of industrial cluster formations. Even though the role of government has been stated in the literature, governmental interventions are avoided in this article because governmental actions (such as incentives) have not been studied by the authors. It would benefit the model to include governmental actions because the importance of policy is likely to influence an industrial cluster formation.

Acknowledgments

Our gratitude goes to GEORG (Geothermal Research Group) for financial support. Saemundur Gudlaugsson, Gudmundur Hjaltalin at Reykjavik Energy and Thrandur Rognvaldsson and Steinn A. Stensson at Landsvirkjun for assisting with this research. Hreinn Halldorsson and Albert Albertsson at HS Energy for assistance. A special thanks goes to Harald Sverdrup for inspiration and assistanceion developing the causal loop diagram.

Author Contributions

Reynir S. Atlason conducted field observations, majority of analysis and the composition of the article. Gudmundur O. Valsson contributed to the design of the article, the theoretical background structure, analysis of data and model construction. Runar Unnthorsson contributed to the design of the article, oversaw the project and assisted with the data analysis and model building.

Conflicts of Interest

The authors declare no conflict of interests.

References

1. Shafiee, S.; Topal, E. When will fossil fuel reserves be diminished? *Energy Policy* **2009**, *37*, 181–189.
2. Field, C.; van Aalst, M. *Climate Change 2014: Impacts, Adaptation, and Vulnerability*; IPCC (Intergovernmental panel on climate change): Geneva, Switzerland, 2014; Volume 1.
3. *BP Energy Outlook 2035*; BP plc: London, UK, 2014.
4. Mitigation, C.C. *IPCC Special Report on Renewable Energy Sources and Climate Change Mitigation*; Technical Report; Cambridge University Press: Cambridge, UK, 2011.
5. Filippetti, A.; Archibugi, D. Innovation in times of crisis: National systems of innovation, structure, and demand. *Res. Policy* **2011**, *40*, 179–192.
6. Porter, M.E. Location, competition, and economic development: Local clusters in a global economy. *Econ. Dev. Q.* **2000**, *14*, 15–34.
7. Porter, M.E. Clusters and the New Economics of Competition. *Harvard Bus. Rev.* **1998**, *14*, 77–90.
8. Malerba, F. Sectoral systems of innovation and production. *Res. Policy* **2002**, *31*, 247–264.
9. Giuliani, E. Cluster absorptive capacity why do some clusters forge ahead and others lag behind? *Eur. Urban Regional Stud.* **2005**, *12*, 269–288.
10. Giuliani, E. The structure of cluster knowledge networks: Uneven and selective, not pervasive and collective; In Proceedings of DRUID Tenth Anniversary Summer Conference, Copenhagen, Denmark, 27–29 June 2005; pp. 27–29

11. Kogut, B. The network as knowledge: Generative rules and the emergence of structure. *Strateg. Manag. J.* **2000**, *21*, 405–425.

12. Porter, M.E. *Competitive Advantage: Creating and Sustaining Superior Performance*; Simon and Schuster: New York, NY, USA, 2008.

13. Furman, J.L.; Porter, M.E.; Stern, S. The determinants of national innovative capacity. *Res. Policy* **2002**, *31*, 899–933.

14. Lin, C.-H.; Tung, C.-M.; Huang, C.-T. Elucidating the industrial cluster effect from a system dynamics perspective. *Technovation* **2006**, *26*, 473–482.

15. Dangelico, R.M.; Garavelli, A.C.; Petruzzelli, A.M. A system dynamics model to analyze technology districts' evolution in a knowledge-based perspective. *Technovation* **2010**, *30*, 142–153.

16. Youngblood, M.D. Winning cultures for the new economy. *Strateg. Leadersh.* **2000**, *28*, 4–9.

17. Gould, S.J.; Eldredge, N. Punctuated equilibria: The tempo and mode of evolution reconsidered. *Paleobiology* **1977**, *3*, 115–151.

18. Orkustofnun. Jardvarmavirkjanir, Retrieved 09.01.2013. Available online: http://www. orkustofnun.is/jardhiti/jardhitanotkun/jardvarmavirkjanir/ (accessed on 9 January 2013).

19. Thorolfsson, G. Maintenance history of a geothermal plant: Svartsengi Iceland; In Proceedings of World Geothermal Congress 2005, Antalya, Turkey, 24–29 April 2005.

20. Geothermal Power Plant Technician. Keilir Technological Institute. Available online: http://en.keilir.net/kit/kit/education/geothermal-power-plant-technician (accessed on 10 February 2013).

21. Hagalin, G.; (Reykjavik Energy); Reynir S. Atlason (University of Iceland). Personal Communication, 2013.

22. Fuller, N. Beyond the core. *Supply Manag.* **2002**, *29*, 39.

23. Porter, M. The Icelandic geothermal cluster: Enhancing competitiveness and creating a new engine of Icelandic growth. In Proceedings of Icelandic Geothermal Conference, Reykjavik, Iceland, 1 November 2010.

24. Atlason, R.S.; Unnthorsson, R. Operation and maintenance in Icelandic geothermal power plants: Structure and Hierarchy. In Proceedings of the POWER Conference. American Society of Mechanical Engineers, San Diego, CA, USA, 15–21 November 2013.

25. Geothermal Research Group. Geothermal Research Group, Retrieved 01.08.2013. Available online: http://georg.hi.is/efni/georg_geothermal_research_group (accessed on 1 August 2013).

26. Eisenhardt, K.M.; Graebner, M.E. Theory building from cases: Opportunities and challenges. *Acad. Manag. J.* **2007**, *50*, 25–32.

27. Voss, C.; Tsikriktsis, N.; Frohlich, M. Case research in operations management. *Int. J. Oper. Prod. Manag.* **2002**, *22*, 195–219.

28. Lewis, M.W. Iterative triangulation: A theory development process using existing case studies. *J. Oper. Manag.* **1998**, *16*, 455–469.

29. Stuart, I.; McCutcheon, D.; Handfield, R.; McLachlin, R.; Samson, D. Effective case research in operations management: A process perspective. *J. Oper. Manag.* **2002**, *20*, 419–433.

30. Meredith, J.; Vineyard, M. A longitudinal study of the role of manufacturing technology in business strategy. *Int. J. Oper. Prod. Manag.* **1993**, *13*, 3–14.

31. Handfield, R.B.; Melnyk, S.A. The scientific theory-building process: A primer using the case of tqm. *J. Oper. Manag.* **1998**, *16*, 321–339.

32. Wallace, W.L.X. *The Logic of Science in Sociology*; Transaction Publishers: Piscataway, NJ, USA, 1971.

33. Atlason, R.S.; Gunnarsson, A.; Unnthorsson, R. Turbine repair at nesjavellir geothermal power plant: An icelandic case study. *Geothermics* **2015**, *53*, 166–170.

34. Atlason, R.S.; Unnthorsson, R.; Oddsson, G.V. Innovation and development in geothermal turbine maintenance based on Icelandic experience. *Geothermics* **2015**, accepted.

35. Atlason, R.S.; Unnthorsson, R. New design solves scaling problems on geothermal control valves. *Power* **2013**, *157*, 18–21.

36. Atlason, R.S.; Unnthorsson, R. *Wellhead Scaling Problems in Geothermal Power Plants Addressed Using a Needle Valve Derivative*; American Society of Mechanical Engineers: New York, NY, 2014.

37. Atlason, R.S.; Geirsson, O.; Elisson, A.; Unnthorsson, R. Geothermal wellhead maintenance: A statistical model based on documented icelandic experience. *Geothermics* **2015**, *53*, 147–153.

38. Atlason, R.S.; Oddsson, G.V.; Unnthorsson, R. Geothermal power plant maintenance: Evaluating maintenance system needs using quantitative kano analysis. *Energies* **2014**, *7*, 4169–4184.

39. Klein, K.J.; Knight, A.P. Innovation implementation overcoming the challenge. *Curr. Dir. Psychol. Sci.* **2005**, *14*, 243–246.

Industrial Energy Management Decision Making for Improved Energy Efficiency—Strategic System Perspectives and Situated Action in Combination

Patrik Thollander and Jenny Palm

Abstract: Improved industrial energy efficiency is a cornerstone in climate change mitigation. Research results suggest that there is still major untapped potential for improved industrial energy efficiency. The major model used to explain the discrepancy between optimal level of energy efficiency and the current level is the barrier model, e.g., different barriers to energy efficiency inhibit adoption of cost-effective measures. The measures outlined in research and policy action plans are almost exclusively technology-oriented, but great potential for energy efficiency improvements is also found in operational measures. Both technology and operational measures are combined in successful energy management practices. Most research in the field of energy management is grounded in engineering science, and theoretical models on how energy management in industry is carried out are scarce. One way to further develop and improve energy management, both theoretically as well as practically, is to explore how a socio-technical perspective can contribute to this understanding. In this article we will further elaborate this potential of cross-pollinating these fields. The aim of this paper is to relate energy management to two theoretical models, situated action and transaction analysis. We conclude that the current model for energy management systems, the input-output model, is insufficient for understanding in-house industrial energy management practices. By the incorporation of situated action and transaction analysis to the currently used input-output model, an enhanced understanding of the complexity of energy management is gained. It is not possible to find a single energy management solution suitable for any industrial company, but rather the idea is to find a reflexive model that can be adjusted from time to time. An idea for such a reflexive model would contain the structural elements from energy management models with consideration for decisions being situated and impossible to predict.

Reprinted from *Energies*. Cite as: Thollander, P.; Palm, J. Industrial Energy Management Decision Making for Improved Energy Efficiency—Strategic System Perspectives and Situated Action in Combination. *Energies* **2015**, *8*, 5694-5703.

1. Introduction

Improved industrial energy efficiency is a cornerstone in climate change mitigation. Research results suggest that there is major untapped potential for improved industrial energy efficiency. A vast amount of research has empirically and theoretically studied the fact that a large number of improvement measures are not implemented, even though the measures are seemingly cost-effective. The major model used to explain this discrepancy is the barrier model, which states that different barriers to energy efficiency inhibit the adoption of cost-effective measures. Moreover, the measures outlined in research and policy action plans are almost exclusively technology-oriented. Reference [1] questioned this technology paradigm, and stated that there is also a large untapped potential in the way technology is used in industry, *i.e.*, an energy management gap. In later empirical research [2] found that for energy-intensive industries, this potential was in parity with the technology potential, while for less energy-intensive companies, the potential for technology solutions was viewed as larger. An even larger absolute potential in percent was stated by non-energy-intensive industry (13%), compared with energy-intensive industry (6%). Brunke *et al.* [3], in their study of the Swedish iron- and steel industry, stated that the potential for management was about 2.4% of the total energy use, while that for technology was stated to be 7.3%. Following [1–4] empirically investigated more than 900 energy efficiency measures undertaken by 100 Swedish energy-intensive industrial companies showing that a large number of the adopted measures were not, in fact, technology implementations. This research challenges the existing view that (best-available) technology is the sole means by which improved energy efficiency is achieved in industry, and accentuates a knowledge gap in the way improved energy efficiency in general is viewed.

One way to overcome the extended energy efficiency gap is to work strategically with energy issues in companies and develop an energy management system. Energy management has become more and more important and more frequently discussed as a tool to achieve improved energy efficiency in companies. Most research in the field of energy management is however grounded in engineering science, and theoretical models on how energy management in industry is carried out are scarce. One way to further develop and improve energy management, both theoretically and in practice, is to explore how a socio-technical perspective can contribute to this understanding. In this article we will further elaborate this potential of cross-pollinating these fields. The aim of this paper is to relate energy management to two theoretical models: situated action and transaction analysis.

2. Models, Policies, and Energy Management

According to [5]: "Energy management can be defined as the procedures by which a company works strategically on energy, while an energy management system is a tool for implementing these procedures".

Oftentimes, these two terms, energy management and energy management systems, are used interchangeably. The major reason for this is that the research conducted and models used are often based on a model, the plan-do-check-act cycle (PDCA) [6], which views the operators based on the input-output model, *i.e.*, a signal goes into the operator in the form of information, and the signal is transformed by the operator into an action or activity which improves energy efficiency or reduces company energy use. This way of viewing energy management calls for a change.

One of the most cited market (failure) barriers to improved energy efficiency is information asymmetries and imperfections, and by reducing these asymmetries and imperfections by using energy information programs, a more perfect market is achieved, in terms of information. Oftentimes, the suggested and implemented actions for overcoming information asymmetries and imperfections are by the launch of energy information programs, the foremost of which are of energy assessment/energy audit programs. These programs are also based on the input-output model, by which the company is viewed as a utility-maximizing rational entity which, when information is provided, acts on this information and invests in new, more energy-efficiency technologies.

One of the most important policy activities to promote improved energy efficiency in energy-intensive industry is however by the use of Voluntary Agreements (VAs), Voluntary Agreement Programs (VAPs), or Long-Term Agreements (LTAs). The main idea, regardless of the name given to the policy, is a combination of energy assessment/energy auditing and energy management activities. To the authors' knowledge, the oldest VAP that exists today is the Japanese Keidanren [7]. Within the EU a large number of Member States (MS) have launched VAs aimed at their energy-intensive industrial sectors [8].

The energy management system standard and ISO 50001 were both designed according to the plan-do-check-act cycle. The standard is similar to quality and environmental management system standards [9], and is mostly implemented through energy policy programs, e.g., Voluntary Agreements among energy-intensive industries. Implementation of standardized energy management systems among industrial SMEs (Small- and Medium-Sized Enterprize) is limited [10]. In attempts to improve energy efficiency through energy management, simplified management systems have been developed, e.g., in Sweden [11] to promote energy management in industrial SMEs. Such initiatives have been inspired by the formal standard, but take a lighter approach than a standardized energy management system [11]. The input-output model is also the major model used within EU energy policy action plan formulation.

However, the underlying logic for this, *i.e.*, the models used to motivate both the spreading of information and the implementation of energy management practices, are only weakly linked with present theory building within the area of improved industrial energy efficiency. Almost exclusively, the scientific contributions in the field of energy management emanates from technical faculties. It is thus important to further explore the area of improved industrial energy efficiency in terms of the models used, and new models needed, not least from a socio-technical perspective.

3. Energy Management and the need to Delegate Leadership

One criticism of the input-output model and management systems is that they are based on a simplistic belief in rational actors choosing the best available technology. Improved industrial energy efficiency is multifaceted and an effect of that is the existence of an energy efficiency gap between the technical-economic potential for improved energy efficiency and what is actually implemented. If we acted as the theoretical rational woman or man, then this gap would not exist, but the gap obviously does exist according to numerous studies, and new tools and perspectives seem needed to approach these underlying problems.

First we need a multiple way to approach decision-making in industry, where management models need to be complemented by other decision models, apart from the "economic woman/man". The classic "garbage can" model [12] showed that decision-making in organizations does not necessarily need to be rational. How an industry understands a problem may be poor, due to the fact that people constantly enter and exit the organization, which makes learning processes complicated. The industry's "garbage can" consists of a collection of choices searching for problems, issues and appropriate decisions to attach themselves to. The idea is that people in the organization dump problems and solutions into an imaginary garbage can and the outcome is a result of when a solution randomly finds an appropriate problem. At times action is also taken without a plan, *i.e.*, without any stated intention or goal for those actions. A plan or an energy strategy does not necessary change a practice or infrastructure but according to [13] can likewise reproduce what already exists.

There are also different ways to manage organizations and lead change. In [14] the authors describe two different paths to follow, namely method or result governance. The different methods are described by an illustrative example. If the goal is to go from A to B, this location can be reached in different ways. One is to run along a sandy beach, which will be a rather quick way. The tracks in the sand will however also be washed away rather quickly, so it will not be possible for someone to follow the same path. The result is that each person will need to find their own way to B. The first path will then be based on all individuals solving the challenge and the burden to move from A to B will be dependent on individual capacity and external conditions. This is a description of result governance [5].

The second option is to construct a road. This will take a much longer time, and require more effort and organizational capacity. However, when the road is finished more people will travel easily from A to B and it will be possible to carry greater loads on the road [14]. Improved modes of transport can be developed and more cargo carried compared to the first way described. This last option is also less dependent on individual capacity and external conditions. This second option is that of standardization and improvements, which can support others who want to repeat a behavior. The second approach is method governance [5].

Result governance has the benefit that positive results can be achieved quite quickly, but then it is hard to uphold a continuum in behavior or measures. By this method members in the group need to achieve desired results on their own and solutions rely on individual approaches. Duplication of a solution is also hard to achieve, and structural capital does not accumulate in organizations. The organizational culture is not affected and if for example the leader changes job good results will not persist [14].

Method governance (the road example) influences behavior by using a method leading to standardization. It is possible by this method to influence how a group works. It establishes the conditions for continuous improvements. Behavior modified in such a way often provide more economically efficient results and creates conditions that make it possible to maintain or improve the work [14]. The drawback with this approach is of course if the method developed is not especially good or functional, as then the whole collective will follow the wrong path.

But in most cases it is also possible to improve methods, procedures and instructions. Unlike result governance, method governance creates structural capital and, with the right leadership, long-term improvement in group culture is possible [14]. This also reduces the risk of becoming too dependent on one individual and her or his capacity.

In accordance with this we can conclude that it is important to work with structures and to establish methods and procedures, which also is the idea with energy management systems. For an organization to achieve ambitious energy efficiency goals, empowering individuals in an organization to work on improved energy efficiency is as important, regardless of whether one takes the perspective for example of implementing standards or a more top-down management perspective. The managers also need to adopt a transparent strategy for who can take responsibility for what.

Another way to look at this is by applying transactional analysis (TA), which has its origin in psychiatry [15]. The adoption of an energy strategy is a way to structure energy work in a company. It is thus also a restriction of freedom to act for individuals and the organization. How this restriction is communicated is vital for how the energy management system will work in practice in an organization. Enforced measures are seldom looked upon with approval by employees, making how the measures are communicated vital. When taking a transactional analysis approach a communication strategy can be used that resembles peer-to-peer communication. The employees should be informed of changes or new procedures before adoption. They also should be given opportunities to provide their view on the changes. According to TA, this increases the chances that the organization will accept a new structure.

Delegating responsibility is at the same time connected with a risk, a risk that a task for example will be managed in a way that is not beneficial from a system perspective. On the other hand change management and the need to handle risks are important parts of leadership. Improvements in energy efficiency also include a certain degree of risk. A production manager that needs to close down machines or replace equipment in the production line risks facing a period during which production decreases. One strategy to handle this is to create stable systems, e.g., a flexible machine that is easy to turn on and off. The person in charge of the energy management program needs to encourage risk taking and support managers of production, quality, maintenance, *etc.* when taking investments that in the short run risk having adoption problems. An organization needs to accept a certain amount of risk. In industries with continuous production processes risk has proven to be a frequently mentioned barrier for energy efficiency. Batch production is less vulnerable, which is explained by the fact that a malfunction does not need to be as costly. In a continuous production process on the other hand equipment malfunction may cost several hundred thousand euros per hour [16].

The CEO of an industrial organization is most likely not the person responsible for an energy management program, but delegating that authority is praxis. However, delegation of authority also entails risk, even if of a slightly different kind. Delegation of authority may indicate that the top manager is not interested in an issue. If a more junior co-worker takes responsibility, that person may also lack power to advocate certain major steps necessary for improved energy efficiency. It is crucial that the person in charge for the energy management program have a formal or informal leadership position. At least a strong connection to the board of directors is needed [5].

There are good examples of successful leadership delegation. One example is when the CEO of a large multinational company decided to establish an in-house energy management program. An energy audit was conducted, which was followed by establishment of a group that met every month. The CEO delegated authority to the person in charge of the physical plant (e.g., HVAC, water and security system), but at the same time all managers were required to attend the meetings. The CEO also always attended the meetings. This work inspired another manager for the melting division to start working on energy efficiency. Quite soon several energy reduction measures were successfully implemented and in the end the savings equaled the sum of all undertaken measures suggested by the energy audit. This case is a successful example of an organizational change, where the CEO was able to empower individuals and mid-level managers to increase efficiency and even revenues [5].

4. Industrial Energy Management in the Perspective of Situated Action

Yet another way to understand decision-making in organizations is to have a situated action perspective [13]. When studying decision-making in this perspective, all activities are seen as situated and impossible to predict. In this perspective decisions on improved energy efficiency are made locally, in the practices where people meet, act and perform.

Imagine a meeting where the participants are supposed to decide on how to improve energy efficiency in the organization. From an energy management standard perspective the outcome will depend on existing policies, already decided goals and established procedures. From a situated action perspective the outcome of a meeting is a much more open issue. Rather than depending on a goal in a document or procedures in a standard it will be dependent on which actors participate in the meeting. The actors attending a meeting will most probably not have memorized all policies, standards and procedures that exist in the organization. They will base their input and contribution to the discussion on energy efficiency on their culturally embedded understanding of how to act, what choices are given in different contexts and what decisions seem to be suitable in different settings. In this way the outcome of an energy efficiency decision will be dependent on which actors participate on that occasion. If the meeting will be repeated but with totally new actors, then the discussions at the meeting will differ and thus also the outcome. Financial managers for example will discuss energy efficiency from their perspective which is different from say an HR manager or mechanical engineer.

The decisions made during a meeting are also a result of group dynamics and the participating actors' mutual relationships. The participants in meetings take different roles, and the roles actors have in one group will differ from their roles in another group. Actors take different roles, and in this sense too roles are situated. How a discussion goes will then depend both on which actors participate and also the mutual relations in the group. For that reason it is not unusual or even strange that one actor can have one opinion at one meeting and then change opinion at another meeting with another constellation of actors. The actor can simply have taken different roles at the meetings or the discussions have taken different turns which make holding what seem like opposing opinions by an individual very logical. The opinion must simply be understood in the perspective of situated action.

Going back to improved energy efficiency as an example, one energy efficiency measure can be interpreted as beneficial and valid in one situation while in another situation the same measure can be dismissed as inappropriate [17]. It depends on the situation, what problems and solutions become present in the discussion and what experience the involved actors have from energy efficiency in general and the measure in particular.

Actions are also something that we constantly do and are not necessarily intentional, reflected upon, or done to achieve a goal. According to Suchman, plans, for example, cannot be understood as instructions for targeted actions. Plans and strategies do not provide a solution for the problem, they simply relate it.

At the same time there is existing knowledge that an energy management model is a formal process, with its tools and procedures, which will have impact on the everyday work. Even if it is impossible for plans or standards to predict actions they are still influential. Some ideas from the plan or standard will interest many actors and be included in the discussions at meetings and integrated in ongoing processes. Other ideas and standard will simply not attract interest from anyone and these will not be made present at meetings or highlighted during a process, and these ideas will not have any impact in practice.

Ideas and goals attractive to many actors will have a better chance of survival, because it will be possible for these to be present on many occasions and at many meetings. Specific ways to handle issues in support of these ideas will be developed into procedures and will eventually be understood as the "right" way to do things. Ideas, working procedures, relations and roles that many support and maintain over time will in this way become formalized in specific contexts. But all this will not come automatically, but happen in situated actions. If an idea at the same time is supported with great emphasis by a (strong) leader the chances for its implementation are high. It seems vital to integrate this perspective into the discussion of how to improve industrial energy management both in practice and in theory.

Also, to be able to capture how these formal energy management tools are translated and changed, how phenomena are formalized and energy efficiency measures are shaped it is important to follow how this is done and constantly negotiated in local practices. Improved energy efficiency from this perspective is something that is done in interaction between actors and artefacts and it is in this interaction that the scope of energy efficiency is defined and decided and its ingoing parts are crystallized. It is also in this interaction that different actors have their roles set, as experts, advisors or decision-makers, and where some issues or artefacts become "energy efficient" to its characters. To understand energy efficiency in industry we need to understand not only energy management systems and the industrial technical energy systems, but also equally importantly how improved energy efficiency is achieved in situated actions.

5. Discussion

This paper attempted to provide further theoretical insights into the area of industrial energy management. Even though the scope of the study has been industry, the theoretical implications are generalizable to others sectors such as the transport sector as well. We conclude that the current model for energy management systems, the input-output model, is insufficient for understanding in-house industrial energy management practices as it does not fully take current scientific understanding of people into account. By incorporating situated action and transaction analysis as models to improve understanding of in-house energy management, we hope to have improved the understanding a step further. Energy management and energy management systems are not the same. Energy management demands leadership skills sufficient to be sensitive to employees' ideas and objections, but strong enough to dare to carry through ideas in the organization where people are hesitant.

By the incorporation of situated action and transaction analysis to the currently used input-output model, an enhanced understanding about the complexity of energy management is gained. In Figure 1, one example of such a revised model is given.

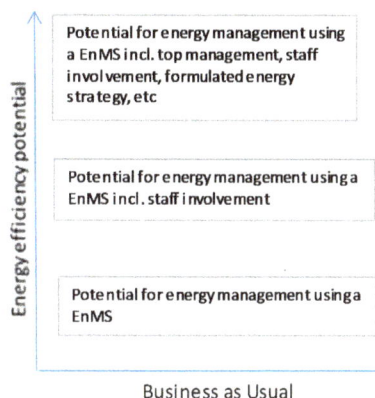

Figure 1. The energy efficiency potentials for various energy management approaches.

As illustrated in Figure 1, and which has also been elaborated upon by [1,3,5], it is clearly seen that solely implementing an energy management system, *i.e.*, a tool for improved energy efficiency, will not deploy the extended energy efficiency potential, *i.e.*, the energy efficiency potential outside of technology implementation. Rather, successful energy management means setting a clear strategy, having top management support, involving staff, *etc.*

We do not think it is possible to find a single energy management solution suitable for any industrial company, but rather the idea is to find a reflexive model that can be adjusted from time to time. An idea for such a reflexive model will contain structural elements from energy management models with consideration for decisions being situated and impossible to predict.

Our findings have serious implications for energy and climate change mitigation policy design. As stated in the introductory part of the paper, recent findings indicate a vast (often neglected) potential for energy management practices, which is one of the underlying major motivations for this type of policy. However, so far little attention has been paid to theoretical understanding of the growing number of industrial energy and climate change mitigation policies that are now being developed in different parts of the world. Our findings reveal that policies that have implementation of an energy management system as its main component, despite its vast potential, need to be sensitive to the persons implementing the management system.

For individual companies, our findings are key to understanding why some companies, despite having management systems in place, fail to deliver high energy efficiency improvement figures, while others do. In order to realize the vast potential for improved energy efficiency through energy management, the leaders of an organization implementing an energy management system need to be sensitive to the workers facing and carrying out the measures.

There is pressure from policy makers, the markets and internal processes behind the need for energy efficiency measures. This pressure is captured by the management level and if the demands

218

are in line with the management culture, the process can continue and measures to change behavior and activities will be initiated. This might in turn affect how energy-related practices are performed and whether this will lead to the expected results, for example, reduced energy use. If so, then a positive spiral has been created. It is also possible that the measures and ideas cannot come through existing cultures and get access to different situated actions. In such cases more interaction will be needed with both employees and managers at the company and an awareness of the importance of energy efficiency needs to be embedded in the culture.

This process can take more or less time, depending on whether the pressure for change is in line with the values of the existing organization and how easy it is to change activities and behavior, whether or not behavioral change leads to consistent change in practices, and, finally, whether or not the results are as expected or whether the process will have to start again.

Author Contributions

The responsibility of authorship of this research paper has been evenly distributed among the two authors, Patrik Thollander and Jenny Palm, who co-wrote the paper.

Conflicts of Interest

The authors declare no conflict of interest.

References

1. Backlund, S.; Thollander, P.; Palm, J.; Ottosson, M. Extending the energy efficiency gap. *Energy Policy* **2012**, *51*, 392–396.
2. Backlund, S.; Broberg, S.; Ottosson, M.; Thollander, P. Energy efficiency potentials and energy management practices in Swedish firms. In Proceedings of the ECEEE Industry Summer Study, Arnhem, The Netherlands, 11–14 September 2012.
3. Brunke, J.C.; Johansson, M.; Thollander, P. Empirical investigation of barriers and drivers to the adoption of energy conservation measures, energy management practices and energy services in the Swedish iron and steel industry. *J. Clean. Prod.* **2014**, *84*, 509–525.
4. Paramonova, S.; Thollander, P.; Ottosson, M. Quantifying the extended energy efficiency gap—Evidence from Swedish electricity-intensive industries. *Renew. Sustain. Energy Rev.* **2015**, submitted for publication.
5. Thollander, P.; Palm, J. *Improving Energy Efficiency in Industrial Energy Systems—An Interdisciplinary Perspective on Barriers, Energy Audits, Energy Management, Policies & Programs*; Springer-Verlag: London, UK, 2013.
6. Deming W.E. *Out of the Crisis*; MIT Press: Cambridge, UK, 1986.
7. Thollander, P.; Kimura, O.; Wakabayashi, M.; Rohdin, P. A review of industrial energy and climate policies in Japan and Sweden with emphasis towards SMEs. *Renew. Sustain. Energy Rev.* **2015**, *50*, 504–512.

8. Rezessy, S.; Bertoldi, P. Voluntary agreements in the field of energy efficiency and emission reduction: Review and analysis of experiences in the European Union. *Energy Policy* **2011**, *39*, 7121–7129.

9. International Organization for Standardization (ISO). *Environmental Management Systems—Requirements with Guidance for Use*; Swedish Standards Institute: Stockholm, Sweden, 2004; SS-EN ISO 14001:2004.

10. European Commission (EC). *Flash Eurobarometer 196—Observatory of European SMEs, Analytical Report*; The Gallup Organization Hungary: Brussels, Belgium, 2007.

11. Hrustic, A.; Sommarin, S.; Thollander, P.; Söderström, M. A simplified energy management system towards increased energy efficiency in SMEs. In Proceedings of the World Renewable Energy Congress, Linköping, Sweden, 8–13 May 2011.

12. Cohen, M.; March, J.; Olsen, J. A garbage can model of organizational choice. *Adm. Sci. Q.* **1972**, *17*, 1–25.

13. Suchman, L. *Human-Machine Reconfigurations. Plans and Situated Actions*; Cambridge University Press: Cambridge, UK, 2007.

14. Johansson, P.-E.; Thollander, P.; Moshfegh, B. Towards increased energy efficiency in industry—A manager's perspective. In Proceedings of the World Renewable Energy Congress, Linköping, Sweden, 8–13 May 2011.

15. Berne, E. *Games People Play: The Basic Handbook of Transactional Analysis*; Ballantine Books: New York, NY, USA, 1964.

16. Thollander, P.; Ottosson, M. An energy-efficient Swedish pulp and paper industry—Exploring barriers to and driving forces for cost-effective energy efficiency investments. *Energy Effic.* **2008**, *1*, 21–34.

17. Ryghaug, M.; Sørensen, K.H. How energy efficiency fails in the building industry. *Energy Policy* **2009**, *37*, 984–991.

Investors' Perspectives on Barriers to the Deployment of Renewable Energy Sources in Chile

Shahriyar Nasirov, Carlos Silva and Claudio A. Agostini

Abstract: In the last decade, the importance of exploiting Chile's Renewable Energy Sources (RESs) has increased significantly, as fossil fuel prices have risen and concerns regarding climate change issues grown, posing an important threat to its economy. However, to date, the advancement of Renewable Energy Technologies (RETs) in the country has been very limited due to various barriers. For this reason, identifying and mitigating the main barriers that hamper the advancement of RETs is necessary to allow the successful deployment of these technologies. Based on data collected from a questionnaire survey and interviews conducted among the major renewable project developers, the authors identify and rank the major barriers to the adoption of renewable energy technologies in Chile. Our findings show that the most significant barriers include "grid connection constraints and lack of grid capacity", "longer processing times for a large number of permits", "land and/or water lease securement" and "limited access to financing". Furthermore, we discuss the most critical barriers in detail together with policy recommendations to overcome them.

Reprinted from *Energies*. Cite as: Nasirov, S.; Silva, C.; Agostini, C.A. Investors' Perspectives on Barriers to the Deployment of Renewable Energy Sources in Chile. *Energies* **2015**, *8*, 3794-3814.

1. Introduction

Chile is one of the fastest-growing economies in South America. From 1990 until 2014, Chile's GDP has more than tripled, which has led to reductions in the poverty levels from 38.6% in 1990 to 7.8% in 2013, reaching an energy access rate of 99% [1]. However, this ongoing dynamism in the economy and the significant well-being improvement in the population has resulted in a steep increase in energy demand, dragging the power sector into a critical situation due to a lag in the energy supply. Currently, Chile urgently requires dealing with large energy needs for its growing economy in a continuous and secure manner. The projection of the National Energy Commission (CNE) of Chile shows that with energy consumption increasing at the current average annual growth rate of 6%, implying that the country needs to double its energy capacity by adding an approximately 15,000 MW additional installed capacity by 2030 [2], making it more dependent on external sources. In addition, even though Chile is a minor contributor to global CO_2 emissions (0.2%), the existing upward trend in the CO_2 growth rate—a 110% increases between years 1990 and 2011—has raised environmental concerns [1].

As a result, the government has increasingly recognized the importance of deploying the potential of Renewable Energy Sources (RESs) and now identifies them as an opportunity to address energy security and fulfill environmental goals. Due to several geographic characteristics, Chile has highly favorable conditions for the deployment of renewable energy generating plants, which was critical in the decision of the Chilean government to include renewables as part of the solution to meet its energy needs. At a glance, the country appears as the promised land for renewable energy developers:

large amounts of primary resources and a growing demand with high-energy prices. In addition, it is worth mentioning that Chile has very low tariffs for imported technology, reaching zero in most cases due to free trade agreements, and prides itself on having an open and transparent economy. However, despite the favorable conditions and the government goals, several obstacles still exist preventing the implementation of renewable projects on even more significant numbers. As a matter of fact, renewable energy projects with environmental approval reached 20,780 MW in 2014, but less than 10% of the capacity of these projects (2050 MW) has materialized to date [3]. It is then difficult to clearly understand why renewables do not seem to take off in the country as planned and there must be some barriers slowing and stopping the advancement of renewables. The focus of this paper is to contribute to the identification and description of the barriers to the development of renewable energy projects in Chile.

The discussion of barriers to RES deployment is certainly an important issue given the fact that most nations now strive to achieve higher shares of RES generation. While much has been said in this respect about existing barriers in the context of various countries, mostly developed ones, no related studies have been done about Chile. Even though barriers for the development and deployment of renewable energies might be quite situation specific in any given region or country, we believe that the identification of barriers from investors' perspectives in Chile and the discussion of their specific characteristics compared to other international experiences could provide valuable contributions to the literature and to other emerging economies. One reason for this is that there are some potential barriers that are specific to Chile and have not discussed in the literature. For example, the role of mining concessions on land where there is a significant potential for some renewable energies and the type of regulation for grid connection.

The methodology utilized in our study is based on a questionnaire survey (comprising quantitative and qualitative data collection) and a series of semi-structured interviews (qualitative data collection only). Personal interviews were conducted with various renewable energy developers to find out their perspective on the barriers to the diffusion of Renewable Energy Technologies (RETs). The rest of the paper is organized as follows: Section 2 provides a literature review on barriers to RES deployment. Section 3 describes RETs in Chile and their status; Section 4 outlines the research methodology; Section 5 presents results and discussions; and finally, Section 6 concludes the paper.

2. Literature Review on Barriers to RES Deployment

Despite remarkable progress of RETs over the past two decades, only a small fraction of their potential has been deployed, especially in developing countries. This is due to the existence of several types of barriers that influence penetration of RETs, preventing them from competing in the marketplace and achieving the necessary large-scale deployment. Significant research has already been carried out on the existence of barriers that hamper the adoption of these technologies [4–7]. The barriers can be generally classified into the following broad categories: institutional and regulatory barriers, economic and financial barriers, technical and infrastructure barriers, and public awareness and information barriers. For example, Lehmann et al. [8] examined major barriers to RES in EU member states from technological, economic, financial and institutional perspectives and proposed different measures to address the associated barriers.

2.1. Legal and Regulatory Framework Barriers

The absence of a comprehensive legal and regulatory framework poses a substantial barrier to the development of renewable energy. These mainly include the lack of a stable energy policy, lack of confidence in RETs, absence of policies to integrate RETs with the global market, and inadequately equipped governmental agencies. Byrnes *et al.* [9] explicitly highlighted these barriers in the context of Australia, adding administrative hurdles, delays in project approvals, and shortfalls in land securement. Another frequently encountered problem for the implementation of RETs is the lack of grid access regulation. Promotion of RETs requires objective, transparent and non- discriminatory criteria for grid access, however, it appears that the connection of these smaller scale technologies are still frequently blocked by many local electricity systems [10]. Open access to transmission is a significant challenge in the deployment of RETs. The existing electric regulations in many countries does not facilitate open access even under a public transmission system and this usually has the consequence of indefinite delays for new entrants. This is mainly explained by the existence of high concentration and vertical integration between generation and transmission, where companies may have incentives to block the entry of new market participants. Brown [11] also identified barriers in technology transfer of RETs, recognizing that barriers in various countries vary according to the political characteristics. Furthermore, a research by Butler and Neuhoff [12] show how existing renewable support mechanisms can be considered a regulatory barrier to renewable project developers. For this purpose, they analyzed the major renewable energy promotional policies including quota-based obligation and feed-in tariff systems in UK and Germany respectively. Based on comparative analysis, they suggest that in terms of "regulatory and market risk", the feed-in tariff is much favorable than the quota-based obligations. However, in other study, Ciarreta *et al.* [13] finds out that overall, feed-in tariffs can be considered as an effective instrument in the early stages of development, in the long term, even though such a system can become an obstacle to developing free market competition at the lowest production costs. For instance, due to a financially excessive burden originated by a feed-in-tariff program, the South Korean government implemented a Renewable Portfolio Standard in 2012, which replaced a feed-in-tariff program existing since 2002 [14].

2.2. Economic and Financial Related Barriers

Economic and finance related barriers act as a key stumbling block for the development of RET projects in many nations. This problem is especially critical for small scale RETs, since they lack the necessary guarantees to obtain funding and require high financial resources during the innovation and invention stages of development [15]. For this reason, compared to conventional energy production projects, RETs generally do not benefit from economies of scale. From a financial perspective, RETs have longer payback periods and higher investment costs [4,16]. These technologies not only have typically higher investment costs but are also, in some cases, considered to be riskier investments due to technology and resource uncertainties compared to the ones of conventional energy generation technologies. Mitchell *et al.* [17] and Banos *et al.* [18] have pointed out several diverse risks (e.g., performance and technical, contract risks, market risks) related to the suitability, reliability and solvency degree of the renewable project developers that add to the high risk profile of renewable investments. Arnold and Yildiz [19] introduced a Monte Carlo Simulation (MCS) approach to perform a risk analysis based on an entire life-cycle representation of RET-investment projects. Their results show that compared to standard Net Present Value estimation, MCS provides substantial value-added information with respect to the project's risk that is relevant for investors, lenders that would facilitate the feasibility of project financing. From another perspective, Owen [6] indicated that RETs would be able to be financially competitive, especially when compared to conventional plants, if the damage costs from combustion of fossil fuels were internalized. However, these externalities are not yet internalized through taxation and/or charges on energy generated from fossil fuels. This leads to additional costs which are not equal to the opportunity cost of electricity generated from these resources.

2.3. Technical and Infrastructure Barriers

Besides the abovementioned bottlenecks, there are a number of technical and infrastructure barriers that renewable project developers continuously face. The major technical barrier for the renewable industry regarding network integration is associated with the wide variety of requirements and standards, which vary from country to country [14]. Due to the intermittent characteristics of RETs, the optimal balancing of energy systems requires a high level of technical bases for their technological assessment. These include resource availability, magnitude, and system frequency. Hall and Bain [20] report the lack of energy-storage technologies as an important technical obstacle for electricity production from RES, which must be addressed through a significant technological change. Most of the electricity systems were primarily designed for conventional technologies and there exist an insufficient level of technical bases, in many cases even virtually non-existent, to evaluate RETs. This restriction makes the integration of renewables much slower.

2.4. Public Awareness and Information Barriers

Lack of public awareness has been recognized as a major barrier in the deployment of RETs in many countries [21–28]. The most common issues in this respect are insufficient knowledge regarding both environmental and economic benefits, inadequate knowledge of RETs [21,22], uncertainties about the economic viability of RES installation projects [4,23,24], and public opposition due to a number of reasons including seascape impacts, environmental damage, and lack of consultation concerns among local communities [25–28].

3. Renewables—Potential and Status in Chile

Historically, until the 1990s, electricity generation in Chile was 70%–80% based on hydro sources [29]. However, in the first decade of the century, recurring and severe droughts and growing demand obliged the Chilean government to search for external foreign fossil energy sources. The discovery of cheap natural gas resources in the neighboring country of Argentina during the late 90s, allowed the Chilean government to reach a long-term supply agreement with the Argentinean Government and direct almost all energy investments in building new gas infrastructure networks with Argentina [30]. The low cost of imported natural gas made combined-cycle plants more attractive compared to large hydro plants and coal. Consequently, traditional energy sources such as hydro and coal plants decreased their investments in the energy matrix and were replaced by combined-cycle gas plants. From 1995 to 2001, natural gas consumption from Argentina increased fourfold accounting for nearly 80% of Chile's total natural gas consumption [31]. However, since 2004, due to the introduction of price regulations that created significant domestic gas shortages in Argentina, gas exports to Chile were practically halted, forcing generators to replace gas-fired electricity plants with more expensive diesel and LNG sources.

As a result of an increasing energy demand, the rising costs of fossil fuel prices, the reductions in the available capacity of hydro-plants, and some environmental concerns, the government started slowly considering renewable energy sources. Chile is considered one of the most attractive countries for the development of RES, mostly because its geographic location and diversity provides abundant renewable energy resources. Particularly, a significant potential in biomass, hydropower, geothermal, solar, wave and wind have not been exploited yet [32]. There are over 4000 km of coast exposed to consistent and high Pacific swells which might boost wave energy, Southern Chile has significant areas of wind potential, and the Atacama Desert in northern Chile has excellent conditions for solar energy. The estimated potential of renewables in Chile is summarized in Figure 1.

The formulation of an explicit renewable energy policy in Chile only occurred a few years ago. The separation of RES from conventional sources and technologies in the Chilean energy matrix was introduced for the first time with the approval of Law No. 20257 in 2008 [33]. The law aimed to promote the generation of electricity from RES, considering for this purpose the following main renewable energy sources: biomass, small hydraulic energy (capacity is less than 20 MW), geothermal energy, solar energy, wind power and marine energy. The original design of the electricity system during the 80s together with the liberal economic tradition in the country, were the major factors considered by the Chilean government to establish a quota-based obligation (Renewable Portfolio Standard or RPS). According to the RPS, generators located in systems with more than 200 MW, need to incorporate a total of 10% of electricity from RES into their energy mix by 2024 [34]. As a transition period, between 2010 and 2014, generators need to start supplying at least 5% of their production from RES, then this percentage rises gradually by 0.5% each year, to reach 10% in 2024. This obligation is enforced as of January, 2010 for electricity generation by renewable installations. The law also establishes a fine to be paid by generators when their obligations are not met. The fine is approximately 28 US$/MWh and if the incompliance is repeated within the following three years, it raises to 42 US$/MWh [33]. However, the opinion of several experts is that, although the fine may seem significant compared to the marginal costs in the market, it is still cheaper for some generation companies not to comply with the quota and pay the fine instead of investing in RES [35]. In a recent modification to the Law No. 20257 [34], the government increased the promotion of electricity generation by RES in the energy matrix by doubling its renewable-energy target from the previous goal of 10% by 2024 to 20% by 2025. This new target obviously provides an even stronger incentive for the development of the renewable energy industry. To achieve the 20/25 target, a total of around 6500 MW of new renewable capacity should enter to the grid in the next 10 years, which means an average of around 650 MW every year.

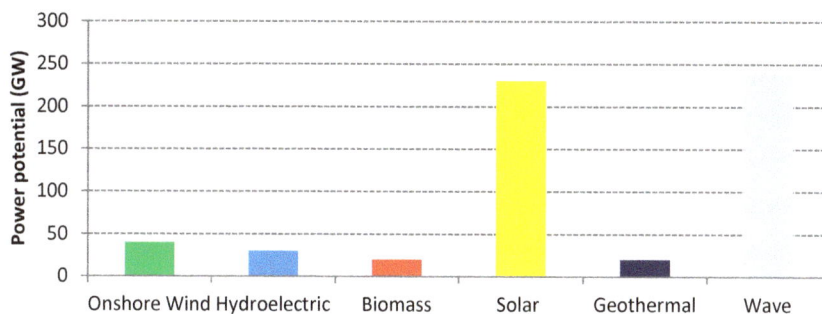

Figure 1. Estimated Potential of RES in Chile. Source: [34] and the authors' own elaboration.

Over the last few years Chile has witnessed significant interest from renewable energy developers looking to expand and diversify their operations in Chile. From the first time with the approval of the Renewable Law, Chile's renewable sector received an accumulated investment of around $6.7 bn between 2008 and 2013 (See Figure 2). Most of this investment has been made in wind, small hydro and solar resources.

The installed power capacity in renewable energy has also increased greatly from 470 MW in 2008 reaching to 2050 MW in 2014 (see Figure 3). As of 2014, installed energy capacity from RES has met and even surpassed the defined target. Moreover, renewable installed capacity added 940 MW in 2014 compared to 2013. RES is equivalent to 10% of the total capacity in the whole power system. Wind power leads the RES installed capacity with 840 MW, representing 40% of the share of renewables in the country. The rest of renewable technologies under operation are distributed in 25% biotechnologies (500 MW), 17% small hydro (350 MW) and 18% solar (60 MW).

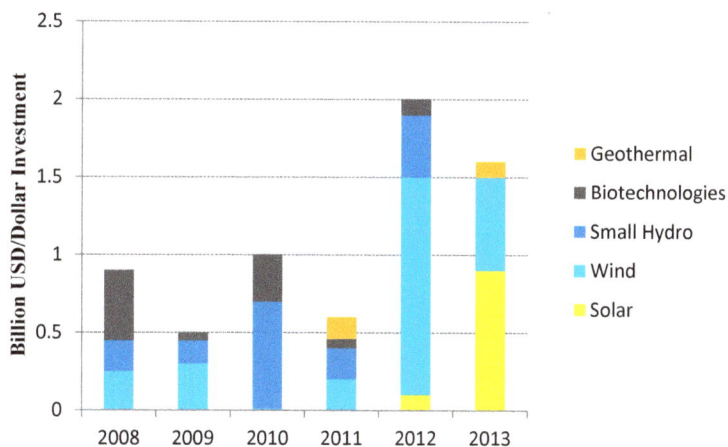

Figure 2. Annual renewable energy investments in Chile between 2008 and 2013. Source: [36].

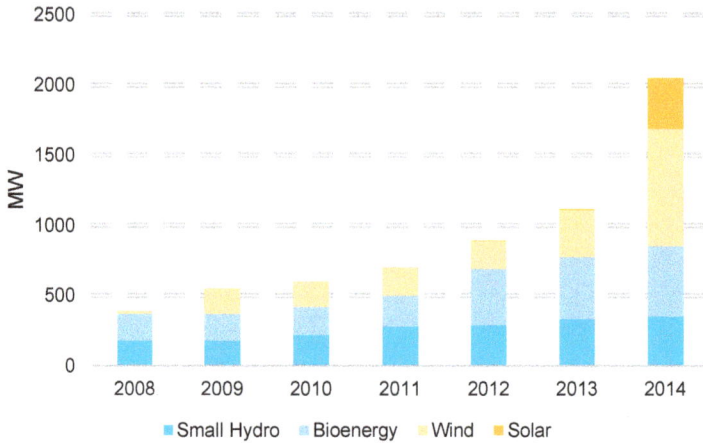

Figure 3. Evaluation of RES installed capacity in Chile between 2008 and 2014. Sources: [34] and the authors' own elaboration.

Even though the role and investments in renewable energy projects has increased in Chile over the last few years, its magnitude is small given the large potential the country has. Despite some improvements in the Chilean renewable industry, a serious number of projects under RES are still waiting to enter the market. In particular, renewable energy projects with environmental approval reached 14,500 MW in 2014. However, as can be seen in Table 1, only 1282 MW are under construction, which raises the question about what the barriers are that prevent the deployment of renewable energy projects in Chile and trying to answer that question is the goal of this paper.

Table 1. The status of Renewable energy projects in Chile in 2014.

Technology SIC + SING	Under Construction (MW)	Approved SEIA (MW)
Small Hydro	134	322
Solar PV	873	8,064
Solar CSP	110	760
Biomass	0	96
Biogas	0	1
Wind	165	5,195
Geothermal	0	120
TOTAL:	1,282	14,555

Sources: The data adopted from CIFES [34].

4. Research Methodology

The methodology utilized in the paper consists of two complementary methods: a questionnaire survey (comprising quantitative and qualitative data collection) and a series of semi-structured interviews (qualitative data collection only) with renewable project developers in Chile. The methodologies used by Zhang *et al.* [37] and Reddy and Painuly [38] provided the basis for our survey design. Eleftheriadis and Anagnostopoulou [39] identified the major barriers faced by solar and wind projects in Greece using a methodology similar to the one used in this paper.

The methodology for a questionnaire survey was split into two phases, which are described in the following subsection. The questionnaire was developed for conducting an online survey of investors' opinions concerning the barriers to the RETs in Chile. For this purpose, a selection of barriers and market actors were completed first. Then, a preliminary list of barriers was tested in a small pilot study to establish the extent to which the barriers found in the literature are applicable in Chile. This was followed by the implementation of questionnaire survey. Finally, the data collection from the online survey was complemented afterwards by five face-to-face interviews selected from survey. The purpose of these interviews is to provide important insights based on investors' extended opinions and experiences over the barriers they have faced in the marketplace.

4.1. Selection of List of Applicable Barriers in Chile

The findings from the previous section provide very useful insights about observed barriers in various countries. However, it is important to underline that the reported barriers in the literature may be very specific to a country or a region. In other to identify the barriers that are relevant for Chile, we initially considered the most common barriers identified in the international literature and a preliminary list of barriers was tested by a small pilot study to establish the extent to which the barriers found in other countries were applicable in Chile. Opinions and experiences were collected from experts with the goal of characterizing the most critical barriers, whom identified 18 barriers that were all included in the survey (see Table 2).

Table 2. Selected barriers to the advancement of renewable energy in Chile.

Category	Number	Barrier
Economic and Financial Barriers	1	Market design problems, that obstruct the integration of renewables
	2	High market concentration
	3	Difficulty in Power Purchase Agreement (PPA) negotiations
	4	Unstable prices in the spot market
	5	Longer economic recovery periods
	6	Lack of modeling of externalities
	7	Limited access to financing
	8	High initial investment costs
Technological and Infrastructure Barriers	9	Grid connection constraints and lack of grid capacity
	10	Inadequate infrastructure to accommodate renewables
	11	Longer processing times for large number of permits
	12	Lack of regulatory framework for land securement
	13	High risk of land speculation due to mining concessions
	14	Lack of coordination among relevant institutions
Institutional and Regulatory Barriers	15	Lack of political stability
Public Awareness and Information barriers	16	Local opposition to the development of projects
	17	Lack of dissemination and public awareness
	18	Lack of necessary scientific and technical skills in the workforce

228

In particular, the barrier dealing with "unstable prices in the spot market" is a common barrier as there is no dedicated mechanism to place the renewable energy production other than the local spot market. Chile also has significant barriers related to its infrastructure (especially, the transmission infrastructure) because the centralized planning has a limited reach, leaving on private hands the type, amount, and timing of the investment. The lack of maturity of the financial markets regarding renewable technologies creates economic and financial barriers that take a special relevance in the Chilean market. Something similar can be observed regarding the Institutional and Regulatory Barriers as the regulatory energy institutions and the local regulation have not been modernized to accommodate a variety of new technologies. In particular, Chile, as a mining country, has a direct conflict between mining concessions and energy infrastructure deployment. According to the literature review, this barrier is not present in other countries, probably because mining activity only thrives in a handful of countries and where the geographic areas with potential for renewable energies do not coincide with mining areas. Finally, the barriers regarding public awareness and information are rather common with new technologies in countries with an empowered population.

4.2. Identification and Selection of Market Actors

The targeted population for the survey was primarily the group of renewable energy project developers in Chile. A primary reason for this focus was the relatively small number of RET projects that have been deployed to date in Chile, which makes it relevant to perform a robust analysis of investors' perspectives on barriers to investing.

All respondents were directly involved in the development process of one or more renewable energy projects and were highly familiar with the barriers hindering the adoption in the country of these new technologies. A range of methods was utilized to identify the potential participants for the research. These included:

- Utilizing personal networks in industry
- Targeting attendees of relevant investment forums and conferences
- Approaching specific industry representatives directly

As a result, a total of 128 project representatives were invited to take part in the survey. The representatives were categorized based on the different technologies they were involved with (See Table 3). The possible technologies include: small hydro, wind, solar, biomass and geothermal. In order to obtain meaningful results, with a confidence interval of 90% and a margin of error of 0.1 [40], at least 46 responses were required. In the end, a total of 60 project representatives responded the survey. Table 3 provides a summary of the total invited respondents, required responses to be representative and achieved responses for each technology.

Table 3. Summary of respondents at the various renewable technologies.

Technology	Total Actors	Required Minimum Responses	Responses Achieved
Small Hydro	53	18.8	23
Wind	29	10.5	13
Solar	29	10.5	14
Biotechnologies	13	4.6	7
Geothermal	4	2.0	3
TOTAL:	128	46.4	60

4.3. Structure of the Questionnaire

The Quota Sampling Method (QSM) was utilized to carry out the descriptive analyses of the survey data. QSM is one of the more rigorous non-probability sampling methods that ensures representativeness by sampling individuals and guarantees to collect necessary information from targeted groups. Essentially it aims to obtain a representative sampling from a not necessarily random selection of individuals, where sampling continues until the quotas are achieved. In our study, individuals correspond to renewable energy investors and developers [21,38,41,42] who were targeted in the survey as the specific relevant groups [43]. One of the advantages of using a QSM is that it is considered to be one of the most effective methods to receive a significant rate of responses [4,44], given the particular difficulty that involves engaging the investment community for this purpose [45]. Another advantage of using quota sampling is that, it typically provides sufficient statistical power to detect group differences or to investigate trait or characteristics of a certain subgroup [46]. Although, QSM has been widely used in a range of research fields, there are several statistical problems inherent in the QSM [47]. Since QSM draws nonprobability samples from each group under investigation, it does not provide generalizable estimates of the target population or of subgroup differences within the target population and potentially creates complications for statistical inference [48]. Another problem with the quota method is that it uses proximity selection of subsequent respondent, presenting problems to estimate sampling errors because of the absence of randomness. QSM is the common practice of skipping eligible but absent respondents. This may potentially induce bias, which would be especially problematic if the bias is amplified rather than reduced by a larger sample size.

The questionnaire structure of our study is based on respondents' opinions on the significance of the different barriers to development of renewables in Chile. At the beginning of the survey, respondents were provided with background information about the aim of the research, the structure of the survey, and the list of the potential barriers. Next, the respondents were requested to rate the relative significance of each barrier on a five-point scale [4,37]. In addition, they also had an option to write down their comments and opinions regarding any rated barrier. The rates were from "5", meaning "extremely important" to them (indicating maximum impact on the development of the technology if the barrier is mitigated), to "1", meaning "least important" (least impact on the technology). A no-response received a "0" score indicating that the respondent does not consider the given barrier as an obstacle to the development of the renewable project.

Let, r_j^i denote the rate given by respondent i to barrier j (as described before, a score from 1 to 5 and n_j the total number of responders for barrier j. Scores were added across all respondents for a given barrier to calculate its average score:

$$R_j = \frac{\sum_i r_j^i}{n_j}$$

5. Results and Policy Discussion

The answers to the questionnaire provide useful information about the relative significance of each of the 18 barriers considered, which might guide the priorities in the policies considered by the government to promote the implementation of RES projects.

The mean scores, variance and standard deviation for each of the barriers are summarized in Table 4. The average score ranges from 4.35 to 2.45, with an overall mean of 3.47, implying that all of these barriers are somehow relevant. The most significant barriers include "Grid connection constraints and lack of grid capacity", with the highest mean value of 4.35, followed by "Longer processing times for large number of permits" (4.16), "Land and/or water lease securement (3.90), and "Limited access to financing" (3.71). These top-critical barriers will be discussed further in the following sub-sections. The detail discussion of the most critical barriers comes from face-to-face interviews with investors where we gathered data on their extended opinions and experiences over the barriers they have faced. The discussion is complemented by a literature review on identified barriers in other countries, which allows us to compare our results to these studies.

5.1. Grid Connection Constraints and Lack of Grid Capacity

In Chile, the current connection system presents significant limitations, resulting in a complex scenario and long delays for those who wish to invest in renewable generation. The connection procedure to the distribution systems does not make a distinction between renewable and conventional technologies.

Besides, the misalignment between planning and connection timescales is a critical issue experienced by many generators wishing to connect to the distribution systems in the country. In the opinion of interviewees, planning approvals go through long negotiations and can take an unpredictable period of time. The evaluation of the feasibility and profitability of the renewable projects then, depends critically on the distributors and there is little transparency and long delays in the process. All these uncertainties have an impact on the cost and complexity of the connection process and might prevent the implementation of a project, even if it is privately and socially profitable to do it.

Table 4. Mean scores variance and standard deviation of the barriers in the RES in Chile.

Barriers	Number of Responses					Average Score	Ranking	Variance	Standard deviation
	1	2	3	4	5				
1. Grid connection constraints and lack of grid capacity	1	2	5	13	30	4.35	1	0.91	0.96
2. Longer processing times for large number of permits	2	0	8	19	22	4.16	2	0.93	0.97
3. Land and/or Water Lease Securement	2	6	6	17	19	3.90	3	1.36	1.16
4. Limited access to financing	2	4	12	18	12	3.71	4	1.15	1.07
5. Difficulty in Power Purchase Agreements—PPAs negotiations	2	4	14	16	12	3.67	5	1.16	1.08
6. Market design problems in the integration of renewables	6	0	16	10	15	3.60	6	1.68	1.30
7. High concentration in the generation market	7	5	7	12	15	3.50	7	2.08	1.44
8. High initial investment costs	4	7	12	14	11	3.44	8	1.53	1.24
9. Lack of regulatory framework for land securement	4	7	13	11	12	3.43	9	1.60	1.26
10. Local opposition to the development of projects	2	11	15	12	12	3.40	10	1.38	1.18
11. Unstable prices in the spot market	4	6	16	11	10	3.36	11	1.45	1.21
12. Lack of modeling of externalities	5	10	11	15	11	3.33	12	1.64	1.28
13. Inadequate infrastructure to accommodate renewables	8	7	7	8	14	3.30	13	2.31	1.52
14. Lack of coordination among relevant institutions	5	10	14	11	12	3.29	14	1.66	1.29
15. Lack of dissemination and public awareness	8	4	9	8	9	3.16	15	2.14	1.46
16. Longer economic recovery periods	5	9	17	14	6	3.14	16	1.32	1.15
17. Lack of political stability	5	8	18	14	4	3.08	17	1.20	1.10
18. Lack of necessary scientific and technical skills in the workforce	12	9	11	5	3	2.45	18	1.59	1.26

Regulation in Chile mandates renewable generators to have a partial exemption for using the trunk transmission system. This exemption is calculated based on the generator size: plants producing less than 9 MW obtain full payment exemption and plants producing between 9 MW and 20 MW are subject to a partial exemption [31]. However, although the legal framework for the energy sector guaranties open access to any generator, in practice, access requests usually result in significant complications and delays, especially for new entrants. The complications include delays in the application process and excessive procedural requirements. This is mainly explained by a high concentration in the generation market, which creates market power than can be used to prevent entry, and also with vertical integration between some generators and transmission companies in Chile, which creates clear incentives to block the entry of new market participants. In addition, the absence of clarity on how the costs of connecting projects to the grid are shared between developers and grid owners generates additional uncertainty and further delays.

Depending on the wide variety of regulatory designs, electricity system requirements and norms, the grid connectivity concerns vary from country to country and in many cases even from utility to utility. For instance, the UK adopted a regulatory support mechanism (the renewable portfolio standard or RPS) that mandates grid operators to decide independently the optimal transmission capacity needed to economically and effectively distribute electricity generated from RETs [49]. Since developers do not obtain a special priority grid access, grid connectivity always remained a challenge and it was the cause of delays during several rounds of tenders. However, unlike in Germany, developers are not confronted with the rules and risks like observed in the UK and Chile. This is because under the feed-in tariff regime, the transmission operators have been mandated to provide grid connectivity to the nearest substation. The costs transmission are borne by the transmission system operators and charged to the federal grid agency of Germany [50].

5.2. Longer Processing Times for Large Number of Permits

Long and complicated bureaucratic procedures to obtain a large number of required permits have also been a major obstacle to the development of RETs in Chile. The process of obtaining permits and their requirements in Chile has a limited legal basis and it is not clearly reflected in any law or official administrative requirement. Although project developers are aware in advance of the required permits for a project, they lack the access to comprehensive information on how to obtain such permits. The existing high level of bureaucracy in the governmental bodies also makes the overall process excessively long and complicated, adding a significant risk to the project during the development phase. Several authorities and different administrative levels within each authority are usually involved in the process. As a result, in most cases, delays may exceed 700 days on average. Among required permits, obtaining the environmental approval by the *Environmental Impact Evaluation System (SEIA)* is the most critical one for renewable project developers in Chile. This is related to the structural problems of *SEIA* and the uncertainty about the required time needed to obtain them. According to opinions of interviewees, this timespan may vary between 90 and 210 days, depending on the nature of the projects. An extensive and time intensive set of application processes for renewable projects have been widely covered in international literature, such as Mizuno [51] presented similar experiences in Japan and relative different experiences in the state and municipal level from Australia were studied by Byrnes *et al.* [9]. In comparison to Chile, this procedure alone in various phases of the SEIA takes 570 days in Japan [51].

5.3. Land and/or Water Lease Securement

The major source of unsolved difficulties comes from the fact that many of the renewable projects (mainly hydro sources) have been submitted for approval to the SEIA in areas that are legally owned or claimed by indigenous communities in Chile, in particular by the Mapuche community. Historically, relations between indigenous communities and the Chilean government have been marked by conflict, primarily because of the expansion of industrial projects on lands that are part of their indigenous territory. Today, a common source of failure in the development of renewable projects, in particular hydroelectric projects in Latin America, is the lack of legal frameworks and adequate consultation with the directly impacted communities. Due to these reasons, various large hydro projects such as Garabí in 2011 in Argentina, Belo Monte project in 2012 in Brazil, and Hidro Aysén in 2014 in Chile, were all suspended [28]. The absence of compensation mechanisms to the communities, indigenous or not, for the impact of the projects and the lack of basis to ensure that surrounding communities can somehow explicitly benefit from the exploitation of these resources (a lower electricity bill for example), are among the critical reasons for the failure of some projects.

A multitude of barriers related to obtain land securement have also caused the market to move slowly. Major obstacles to obtain the use of land belonging to the state comes from the lack of a land inventory, including a geo referenced map, showing the current status and rights over all territories. Even for this purpose, submitting a request to the Ministry of National Assets can delay a project approximately six months. In the case of private land, the main complications arise if the land has numerous owners and the developer has to separately negotiate with each one of them. The risk of leasing land from third parties for the development of the renewable projects remains also a serious concern as an alternative. This is mainly because obtaining a mining concession for extensive territories is a very simple and fast process in a mining country such as Chile. Given this fact, speculators can request a mining concession with the sole purpose of trying to sell them to developers of energy projects later on. Mining concessions give property rights only to the underground land, which is not required for the installation of an energy project, but in the case of open-pit mines both projects are incompatible. This problem has been also identified in Mexico, which reports similar experiences as in Chile [52]. Many project developers have had the experience of purchasing land from the legal owner and later discovering that people are living illegally on the land but claim it as their own. Relocating these people has been difficult, expensive and time-consuming.

5.4. Limited Access to Project Financing

Due to the high levels of initial investment required by renewable technologies, access to financing is crucial to the development of projects using such technologies. Financing of renewable projects in Chile is new, developers have to spend a significant amount of time and effort in convincing the local financial institutions that are not familiar with the technologies and find them too risky. Therefore, considering a high risk premium associated to renewable energy in the domestic financial market, project developers face the problem of not being able to obtain funding from financiers or getting more expensive options that might make the project unprofitable. In practice, the financing options for renewable projects in Chile, particularly the role of microfinance sector, is very limited. Financial institutions in Chile are still very immature in the renewable industry and they are unwilling to finance large scale RE projects. To date, only a few projects have been able to obtain loans from local banks.

One of the major obstacles for obtaining local financing is associated with the particular characteristics of Chilean banks, including their conservative culture and a regulation that focus on bank solvency after the 1982 economic crisis, the lack of experience in the evaluation of renewable technologies, and the utilization of the "Project Finance" funding scheme [53]. High structuring costs of "Project Finance" create an additional obstacle to access to funding. The fixed cost is applied to all projects, independently of their investment requirements, which makes small-scale projects relatively more expensive. Finally, respondents also emphasized the lack of suitable longer-term financial incentives, as government subsidies or tax credits, as a major obstacle for the development of renewable energy sources. Although the Chilean government has introduced several low-interest loans, and also new instruments as guaranties through a government agency CORFO, which is in charge of promoting economic development and innovation, especially in small and medium firms, the respondents agreed that these measures are not sufficient to resolve the problem.

6. Conclusions and Policy Recommendations

Over the last years, a mix of high energy prices resulting from a severe multi-year drought, rising costs of fossil fuels and a steadily increasing energy demand, have put significant pressure on the Chilean economy to start looking into other sources of energy. On other hand, the country is endowed with resources that create an enormous potential for the use of renewable energy. As a result, the government has taken the first steps to significantly increase the role played by renewables in the energy matrix. For this purpose a new law provided an attractive incentive by establishing a renewable-energy target of 20% by 2025. However, the last trends in the development of renewable projects in the country have shown that renewables did not take off as well as planned. The evidence provided from a survey to developers shows that there are a series of barriers slowing and stopping the advancement of renewables.

The paper presents the most important barriers identified from the survey and follow up interviews conducted among the renewable energy developers and investors in Chile. The analysis of the results showed that the top five barriers ranked by the degree of importance given by the interviewees are "grid connection barriers", "administrative hurdles", "land and/or water securement problems" and "limited access to project financing". The analysis of each of these barriers provides valuable references to the Chilean government, contractors and other investors. Mitigating the identified barriers and creating further incentives remains a key challenge for the development of a major renewable energy sector. In this regard, it is clear that the Chilean government should play a key role in establishing additional incentive mechanisms and have a prioritized strategy to eliminate the main barriers that slow or even stop the development of renewable energies in the country.

As far as the need to reach the renewable-energy target of 20% by 2025, electricity grids will have to be upgraded and expanded. In the case of Chile, connecting the two major electric systems (SING in the north and SIC in the central region) that are currently separated would greatly enhance diffusion of RETs. The reason is that the north of Chile possesses excellent renewable power potential, especially for solar energy, and the largest part of the electricity demand—around 75%—is concentrated in the central region. Therefore, the connection between both systems would allow the solar energy sources to reach the demand centers. Australia faces a similar problem due to the remoteness of the solar energy sources and large initial investments are required in transmission to exploit solar energy [9], but in the case of Chile this can be solved at a much lower cost.

In theory, the construction of transmission lines has significant economies of scale and investments in constructing new transmission lines are considered to be very risky. To address this constraint, establishing coordinated common transmission lines for renewable projects may solve the problem and make the projects feasible. In addition, establishing a comprehensive national transmission planning process, creating standard interconnection procedures, regulating open access to transmission networks, strengthening pricing for transmission, are the most urgent measures to be taken.

With respect to the need for more financial resources and access to funding, the Chilean government must increase the accessible volume of public funding to the sector through different channels. Although the state development agency CORFO's role in offering financial incentives improves the situation, it is very unlikely that the different programs offered by CORFO become a sufficient mechanism to completely solve the problem in Chile as the resources provided are very limited and mostly target small firms. More capital injection through CORFO could play an active role in addressing these financing gaps through new operational mechanisms and adapted instruments. Besides, the government's engagements in offering loan guarantees or issuing "green bonds" for local and foreign commercial banks may help to mobilize long-term funding for promotion of renewable sources. Furthermore, as a part of a government policy, it is important to encourage energy-intensive industries in the country, especially the mining industry, to allocate funds for the promotion of technological innovation and renewable pilot projects. As an alternative option, in the initial phase of development, international financial institutions potentially may have also an important role to play in financing clean energy projects and particularly in accelerating market linkages. The investment flow from institutional investors may transmit a good sign of investor confidence and experience to the local banks, which should facilitate the awarding of loans for further projects. Additionally, monetizing positive and negative externalities and ensuring that they are included in energy prices would encourage renewable projects to have a fair competition with conventional sources in the market. This particularly important in the case of diesel, which is taxed at a very low rate given the negative externalities caused by its use [54], and coal, which is not taxed at all. Given that Chile does not have abundant coal reserves enact the corrective tax could be easier than in countries which face the same barrier but are coal producers like South Africa [55].

Regarding the streamlining of bureaucratic processes for permits and land and/or water lease securement issues, which is crucial to the deployment of RES systems, there is a compelling need to implement a sound legislation to contribute considerably to the simplification of the processes.

Furthermore, a clear and effective assessment methodology for SEIA studies should be established and standardization of all procedures should be implemented such that all licensing authorities would follow the same practices. This would avoid discretionary decisions for approving concessions and would reduce uncertainty for investors. The experience of many countries shows that a key factor in the successful development of a renewable project on territories where communities are against to the project development lies in the direct consultation and negotiations with these communities. In particularly, members of affected communities and the public in general should be given more of a chance to participate throughout the assessment process. It is important to mention however, that in the case of solar energy in Chile this might be a minor barrier compare to other countries, as the solar potential is in large areas of desert where almost nobody lives.

When looking at different options to mitigate the barriers, Chile can benefit from international experiences in implementing, policies that successfully reduce the barriers for renewable energies deployment in many aspects. There are relevant lessons related to technical advances and cost reductions resulting from large-scale market deployments of commercially mature RETs in first-mover countries. In solving some of the problems of grid connectivity, Germany is a good example of a country showing success with its strong and predictable policies and incentives that have spurred similar policy initiatives in many countries [50]. Moreover, regulatory incentives implemented in Australia have proved to be very effective and also an efficient way of eliminating administrative hurdles [9]. Finally, well-designed financial support to develop RETs has been fundamental in the success of market leaders, such as Spain, Germany and Japan [13,51]. Nevertheless, the application of successfully proven policies in other countries should be carefully adapted to the Chilean market conditions and complemented by rigorous actions to develop and improve the capacity of all local participants involved, including producers, regulators, the public, and the finance community.

Regarding the limitations of this research, it is important to note that the barriers to the implementation of renewables may vary across technologies and, therefore, some further analysis is needed to identify, understand and mitigate the barriers by considering the nature of each technology. It would also be relevant to further explore the role of the main barriers identified in this study, but from the perspective of the government and the financial institutions. Finally, in the case of financial institutions, it would be important to understand the role played in assessing the risk of each renewable project by the type technology, the existing regulation in the electric sector, and the administrative procedures required to implement a project.

Acknowledgments

This work was supported in Chile by the projects CONICYT/FONDAP/15110019 (SERC-CHILE), CONICYT/FONDECYT/1120490 and by the Center for the Innovation in Energy (UAI, Chile). We greatly appreciate help of our ex-student, Tania Rosales, in original data collection for our research. Finally, we are grateful to Hagani Karimov from George Washington University for his helpful comments.

Author Contributions

Shahriyar Nasirov and Carlos Silva designed and performed the research and wrote the paper with results checking. They were responsible for analyzing and interpreting the data. Claudio Agostini gave review suggestions and guidance of the manuscript on the whole writing process, reviewed the literature, and reviewed entire paper. All authors read and approved the final manuscript.

Conflicts of Interest

The authors declare no conflict of interest.

References

1. World Bank Group. *World Development Indicators, Chile*; World Bank Group: Washington, DC, USA, 2013.
2. Minister of Energy, Chile. National Energy Strategy 2012–2030: Energy for the Future. Available online: http://www.centralenergia.cl/uploads/2012/06/National-Energy-Strategy-Chile.pdf (accessed on 1 January 2012).
3. National Energy Commission (CNE). Statistics/Electricity. Available online: http://www.cne.cl/estadisticas/energia/electricidad (accessed on 29 April 2014).
4. Painuly, J. Barriers to renewable energy penetration; a framework for analysis. *Renew. Energy* **2001**, *24*, 73–89.
5. Mezher, T.; Dawelbait, G.; Abbas, Z. Renewable energy policy options for Abu Dhabi: Drivers and barriers. *Energy Policy* **2012**, *42*, 315–328.
6. Owen, A.D. Renewable energy: Externality costs as market barriers. *Energy Policy* **2006**, *34*, 632–642.
7. Foxon, T.J.; Gross, R.; Chase, A.; Howes, J.; Arnall, A.; Anderson, D. UK innovation systems for new and renewable energy technologies: Drivers, barriers and systems failures. *Energy Policy* **2005**, *33*, 2123–2137.
8. Paul, L.; Felix, C.; Melf-Hinrich, E.; Nele, F.; Clemens, H.; Lion, H.; Robert, P. Carbon lock-out: Advancing renewable energy policy in Europe. *Energies* **2012**, *5*, 323–354.
9. Byrnes, L.; Brown, C.; Foster, J.; Wagner, L. Australian renewable energy policy: Barriers and challenges. *Renew. Energy* **2013**, *60*, 711–721.
10. Oscar, M. Experience and new challenges in the Chilean generation and transmission sector. *Energy Policy* **2007**, *30*, 575–582.
11. Brown, M. Market failures and barriers as a basis for clean energy policies. *Energy Policy* **2001**, *29*, 1197–1207.
12. Butler, L.; Neuhoff, K. Comparison of feed-in tariff, quota and auction mechanisms to support wind power development. *Renew. Energy* **2008**, *33*, 1854–1867.
13. Ciarreta, A.; Gutierrez, C.; Nasirov, S. Renewable energy sources in the Spanish electricity market: Instruments and effects. *Renew. Sustain. Energy Rev.* **2011**, *15*, 2510–2519.

14. Chen, W.M.; Kim, H.; Yamaguchi, H. Renewable energy in eastern Asia: Renewable energy policy review and comparative SWOT analysis for promoting renewable energy in Japan, South Korea, and Taiwan. *Energy Policy* **2014**, *74*, 319–329.

15. Kellett, J. Renewable energy and the UK planning system. *Plann. Pract. Res.* **2003**, *18*, 307–315.

16. De Jager, D.; Rathmann, M.; Klessmann, C.; Coenraads, R.; Colamonico, C.; Buttazzoni, M. Financing renewable energy in the European energy market. Available online: http://www.buildup.eu/sites/default/files/content/2011_financing_renewable.pdf (accessed on 2 January 2011).

17. Mitchell, C.; Sawin, J.L.; Pokharel, G.R.; Kammen, D.; Wang, Z. Policy, financing and implementation. Available online: http://srren.ipccwg3.de/report/IPCC_SRREN_Ch11.pdf (accessed on 3 June 2012).

18. Banos, R.; Manzano-Agugliaro, F.; Montoya, F.G.; Gil, C.; Alcayde, A.; Gómez, J. Optimization methods applied to renewable and sustainable energy: A review. *Renew. Sustain. Energy Rev.* **2011**, *15*, 1753–1766.

19. Arnold, U.; Yildiz, Ö. Economic risk analysis of decentralized renewable energy infrastructures—A Monte Carlo simulation approach. *Renew Energy* **2015**, *77*, 227–239.

20. Hall, P.J.; Bain, E.J. Energy-storage technologies and electricity generation. *Energy Policy* **2008**, *36*, 4352–4355.

21. Zoellner, J.; Schweizer-Ries, P.; Wemheuer, C. Public acceptance of renewable energies: Results from case studies in Germany. *Energy Policy* **2008**, *36*, 4136–4141.

22. Sovacool, B.K. The cultural barriers to renewable energy and energy efficiency in the United States. *Technol. Soc.* **2009**, *31*, 365–373.

23. Zografakis, N.; Sifaki, E.; Pagalou, M.; Nikitaki, G.; Psarakis, V.; Tsagarakis, K. Assessment of public acceptance and willingness to pay for renewable energy sources in Crete. *Renew. Sustain. Energy Rev.* **2009**, *14*, 1088–1095.

24. Wolsink, M. Wind power implementation: The nature of public attitudes: Equity and fairness instead of "backyard motives". *Renew. Sustain. Energy Rev.* **2007**, *11*, 1188–1207.

25. West, J.; Bailey, I.; Winter, M. Renewable energy policy and public perceptions of renewable energy: A cultural theory approach. *Energy Policy* **2010**, *38*, 5739–5748.

26. Rogers, J.C.; Simmons, E.A.; Convery, I.; Weatherall, A. Public perceptions of opportunities for community-based renewable energy projects. *Energy Policy* **2008**, *36*, 4217–4226.

27. Yildiz, Ö. Financing renewable energy infrastructures via financial citizen participation—The case of Germany. *Renew. Energy.* **2014**, *68*, 677–685.

28. Varas, P.; Tironi, M.; Rudnick, H.; Rodriguez, N. Latin America goes electric: The growing social challenges of hydroelectric development. *IEEE Power Energy Mag.* **2013**, *11*, 66–75.

29. Bezerra, B.; Mocarquer, S.; Barroso, L.; Rudnick, H. Expansion pressure: Energy challenges in Brazil and Chile. *IEEE Power Energy Mag.* **2012**, *10*, 48–58.

30. Nasirov, S.; Silva, C. Diversification of Chilean energy matrix: Recent developments and challenges. Available online: http://www.iaee.org/en/publications/newsletterdl.aspx?id=256 (accessed on 28 April 2015).

31. International Energy Agency. *Chile Energy Policy Review 2009*; International Energy Agency: Paris, France, 2009.

32. Americas Society/Council of the Americas Energy Action Group (AS/COA). Toward energy security in Chile. Available online: http://www.as-coa.org/files/TowardEnergySecurity InChile.pdf (accessed on 15 February 2012).

33. National Energy Commission (CNE); Deutsche Gesellschaft für Technische Zusammenarbeit (GTZ). Non-conventional renewable energy (NCRE) in the Chilean electricity market. Available online: http://www.cne.cl/images/stories/public%20estudios/raiz/ERNCMercado Electrico_Bilingue_WEB.pdf (accessed on 1 October 2009).

34. The Centre for Innovation and Promotion of Sustainable Energy (CIFES). *NCRE Status in Chile Reports 2008–2014*; CIFES: Santiago, Chile, 2014.

35. Leyton, S. Chile Considers Bill to Boost Renewable Energy. Available online: http://www. renewableenergyworld.com/rea/news/article/2012/03/chile-considers-bill-to-boost-renewable-energy (accessed on 6 March 2012).

36. Bloomberg New Energy Finance (BNEF). Climatescope 2014: New frontiers for low-carbon energy investment in the Latin American and the Caribbean. Available online: http://www. climatefinanceoptions.org/cfo/node/3505 (accessed on 28 October 2014).

37. Zhang, X.; Shen, L.Y.; Chan, S.Y. The diffusion of solar energy use in HK: What are the barriers? *Energy Policy* **2011**, *41*, 241–249.

38. Reddy, S.; Painuly, J.P. Diffusion of renewable energy technologies—Barriers and stakeholders' perspectives. *Renew. Energy* **2003**, *29*, 1431–1447.

39. Iordanis, M.E.; Evgenia, G.A. Identifying barriers in the diffusion of renewable energy sources. *Energy Policy* **2015**, *80*, 153–164.

40. Chica, A.; Castejón, L. Elaboración, análisis e interpretación de encuestas, cuestionarios y escalas de opinión. Available online: http://rua.ua.es/dspace/bitstream/10045/20331/1/ Elaboraci%C3%B3n,%20an%C3%A1lisis%20e%20interpretaci%C3%B3n.pdf (accessed on 1 February 2006). (In Spanish).

41. Amigun, B.; Musango, J.; Brent, A. Community perspectives on the introduction of biodiesel production in the Eastern Cape Province of South Africa. *Energy* **2011**, *36*, 2502–2508.

42. Masini, A.; Menichetti, E. The impact of behavioural factors in the renewable energy investment decision making process: Conceptual framework and empirical findings. *Energy Policy* **2012**, *40*, 28–38.

43. Naresh, M. *Investigacion de Mercados*, 5th ed.; Pearson Educacion: Mexico City, Mexico, 2008. (In Spanish).

44. Dillman, D.A. *Mail and Internet Surveys: The Tailored Design Method Update with New Internet, Visual, and Mixed-Mode Guide*; John Wiley & Sons, Inc.: Hoboken, NJ, USA, 2007.

45. Wüstenhagen, R.; Menichetti, E. Strategic choices for renewable energy investment. Conceptual framework and opportunities for further research. *Energy Policy* **2012**, *4*, 1–10.

46. Ott, R.L.; Longnecker, M. *An Introduction to Statistical Methods and Data Analysis*, 6th ed.; Brooks/Cole, Cengage Learning: Belmont, CA, USA, 2009.

47. Schaeffer, R.L.; Mendenhall, W.; Ott, L. *Elementary Survey Sampling*, 4th ed.; PWS-Kent Publishing Company: Boston, MA, USA, 1990.

48. Brogan D.; Flagg, E.W.; Deming, M.; Waldman, R. Increasing the accuracy of the expanded programme on immunization's cluster survey design. *Ann. Epidemiol.* **1994**, *4*, 302–311.

49. Toke, D. The UK offshore wind power programme: A sea-change in UK energy policy? *Energy Policy* **2011**, *39*, 526–534.

50. Klessmann, C.; Nabe, C.; Burges, K. Pros and cons of exposing renewables to electricity market risks—A comparison of the market integration approaches in Germany, Spain and the UK. *Energy Policy* **2008**, *36*, 3646–3661

51. Mizuno, E. Overview of wind energy policy and development in Japan. *Renew. Sustain. Energy Rev.* **2014**, *40*, 999–1018.

52. Lokey, E. Barriers to clean development mechanism renewable energy projects in Mexico. *Renew. Energy* **2009**, *34*, 504–508.

53. Latin American Energy Organization (OLADE); United Nations Industrial Development Organization (UNIDO). Observatory of Renewable Energy in Latin America and Caribbean, Chile, Final Report, Component 3: Financial Mechanism. Available online: http://www.renenergyobservatory.org/uploads/media/Chile_Producto_3__Eng__07.pdf (accessed on 28 April 2015).

54. Agostini, C.A. Differential fuel taxes and their effects on automobile demand. *CEPAL Rev.* **2010**, *102*, 101–111.

55. Pegels, A. Renewable energy in South Africa: Potential barriers and options for support. *Energy Policy* **2010**, *38*, 4945–4954.

Carbon as Investment Risk—The Influence of Fossil Fuel Divestment on Decision Making at Germany's Main Power Providers

Dagmar Kiyar and Bettina B. F. Wittneben

Abstract: German electricity giants have recently taken high-level decisions to remove selected fossil fuel operations from their company portfolio. This new corporate strategy could be seen as a direct response to the growing global influence of the fossil fuel divestment campaign. In this paper we ask whether the divestment movement currently exerts significant influence on decision-making at the top four German energy giants—E.On, RWE, Vattenfall and EnBW. We find that this is not yet the case. After describing the trajectory of the global fossil fuel divestment campaign, we outline four alternative influences on corporate strategy that, currently, are having a greater impact than the divestment movement on Germany's power sector. In time, however, clear political decisions and strong civil support may increase the significance of climate change concerns in the strategic management of the German electricity giants.

Reprinted from *Energies*. Cite as: Kiyar, D.; Wittneben, B.B.F. Carbon as Investment Risk—The Influence of Fossil Fuel Divestment on Decision Making at Germany's Main Power Providers. *Energies* **2015**, *8*, 9620-9639.

1. Introduction

In late 2014, at the time of the Lima Climate Summit, Germany's largest electricity provider, E.ON, surprised the financial world with the spectacular announcement that the company would shed all of its nuclear and fossil fuel assets and focus exclusively on renewable energy generation. To do so, company activities will be split into two independent entities. From 2016, E.ON's fossil fuel and nuclear generation will be carried out by a new company called Uniper, while the original E.ON will concentrate on renewables, electricity distribution networks and customer service [1].

This announcement shows that the once so successful business model of Germany's electricity giants is becoming increasingly outdated. Pressure from the government's energy transition policy, the German *Energiewende*, has had a drastic effect on company strategy and finance. This policy renders nuclear power in Germany obsolete by 2022 and, more recently, has become a challenge for energy production from coal. This latest initiative has prompted a discussion on reducing emissions from the energy sector by 22 million tonnes of CO_2 by 2020, which is equivalent to shutting down about eight coal-fired power plants [2]. Furthermore, in order to reach its Energiewende and climate objectives, the German government has set several targets that should guide climate and energy policies up until 2050. By 2020, Germany is aiming to cut greenhouse gas emissions by 40% compared to 1990 levels, which means a reduction of about 500 Mt CO_2e. Between 1990 and 2014, Germany reduced emissions by about 24.7% [2]. However, given that over the past two years emissions have actually increased, the government plans to implement additional measures such as a climate levy.

As German energy giants have come under pressure from the government, the international financial sector has become increasingly concerned about a potential carbon bubble. This phenomenon is based on the notion that if the 2 °C global warming target is taken seriously by international politicians, "no more than one-third of proven reserves of fossil fuels can be consumed prior to 2050" [3]. This means leaving one third of proven global oil reserves, half of the gas reserves and some eighty per cent of coal reserves in the ground between 2010 and 2050 [4]. As a result, a large amount of fossil fuel companies' assets would be 'stranded' and these companies would be considered over-valued.

In this context, a recently published report from the Carbon Disclosure Project (CDP) analysed sixteen European electricity companies to assess their future low-carbon potential. The ranking focused on four key areas: carbon risk, renewable energy sources, coal exposure and water risk. The German companies noted above were rated poorly, with E.ON ranked nine, EnBW ranked 12 and RWE ranked 13 (Vattenfall was not included). The main reason for their poor performance was their high dependence on coal reserves and therefore high carbon cost exposure: a carbon price rising to 18 Euros would make these companies extremely vulnerable, the report stated. Based on their Earnings Before Interest and Taxes (EBIT) in 2013, 43% of RWE's earnings would be at risk, followed by EnBW (28%) and E.ON (18%). The CDP report argued that with climate policies becoming stricter, carbon costs would increase, causing these companies to become an investment risk [5].

The fossil fuel divestment campaign has emerged as an attempt to convince investors to rid themselves of fossil fuel-based company shares in their portfolios. This has led to company business models being questioned and branded unsustainable.

Given the most recent announcements by German energy giants to rid themselves of carbon assets, we ask the research question: are key strategic decisions at the four largest German energy companies being influenced by the global fossil fuel divestment campaign? Our answer is: "not quite yet".

Beginning with a methodology section showing that we base our insights on qualitative data collected through interviews and participatory observation as well as publicly available statistical data, we present a short overview of the fossil fuel divestment campaign in our findings section. We then illustrate that four factors other than the divestment campaign are having greater influence on the top level strategic management decisions at German energy giants. Our conclusions summarise our findings and suggest that while the influence of the campaign on German electricity companies may be less noticeable than in other countries, there is evidence that this could change in the future.

2. Methodology

The findings presented in this paper are based on the triangulation of data. We used primary data from expert interviews and participatory observations at two official company events to deepen our understanding of the general attitudes that prevail at the examined companies. We used secondary data provided by companies and government sources to underpin our arguments. In terms of the latter data, we extracted relevant figures and statistics directly from annual reports, official sustainability reports and government data. These are featured in the paper.

The qualitative data collection took place over two years and enabled us to gain an insight into decision making at the strategic level at the four German energy giants. The empirical work was carried out in three phases. Firstly, a desk review study of official company communication was conducted to establish the development of climate policy at the four companies and the companies' relationships with external stakeholders. Secondly, a pilot of three interviews was completed, to refine the questionnaire, gain a better understanding of the German electricity market and prepare for the analysis of the expert interviews. Finally, seven expert interviews with representatives from the four companies were conducted between August and October 2011. The interviews were on average 60 minutes long, were fully transcribed and analysed using a summarising qualitative content analysis. They covered the political situation in Germany, including the effects of the *Energiewende* and the government's promotion of renewable energy, and climate policy issues within the company. All interviews and the participatory observation were led by the first author of this paper. Much of the qualitative data forming the basis of this paper was published in an earlier, extensive study of corporate climate policy but is available in German only [6].

3. Findings

3.1. The Fossil Fuel Divestment Campaign

Over the past months and recent years, several organisations, such as fossilfreeindexes.com, the Carbon Tracker Initiative and the Stranded Assets Programme at the University of Oxford's Smith School of Enterprise and the Environment, have analysed the financial risk of fossil fuel companies. A social movement called the fossil fuel divestment campaign has been encouraging institutional and private investors to sell shares in companies that use fossil fuels in their energy or industrial production. Investors at universities and churches have been at the forefront of the campaign and now, this is becoming increasingly prominent among investment funds and banks. Table 1 shows examples of institutions reacting to this call for divestment.

The movement originated with a call by Bill McKibben of the 350.org project, published in Rolling Stone Magazine, to divest from fossil fuels. His 2012 article [10] argued that at current levels of fossil fuel extraction, the human race will face hardship as the climate becomes unpredictable and irregular. He asserted that carbon emissions from all known fossil fuel reserves about to be exploited would play havoc with the climate. McKibben subsequently founded the fossil fuel divestment campaign, which has become highly vocal in encouraging private and institutional investors to divest. Although it may not yet be influencing companies directly, the naming and shaming of oil, gas and coal-based businesses is stigmatising these firms. This could adversely affect their lending rates and political influence [11].

This fossil fuel divestment campaign has since received support from the United Nations climate secretariat [12]. The United Nations Secretary-General Ban Ki-moon declared at a press conference: "I have been urging companies like pension funds or insurance companies to reduce

244

their investments in coal and a fossil-fuel based economy to move to renewable sources of energy." [13].

Table 1. Examples of institutions implementing fossil fuel divestment [7–9].

Institution	Year	Amount Divested	Action
University of Stanford	2014	$21.4bn endowment	• After a petition by the student-led organisation 'Fossil Free Stanford' and the recommendation of Stanford's Advisory Panel on Investment Responsibility and Licensing (APIRL), the Stanford Board of Trustees decide against direct investments in coal mining companies
Rockefeller Brothers Fund	2014	$50bn	• In September 2014, the heirs to the Standard Oil fortune join the campaign and divest a total of $50bn from fossil fuels
The Bank of England	2015	-	• Warns insurance companies of "assets that could be left 'stranded' by policy changes which limit the use of fossil fuels" [7]
The Church of England	2015	£9bn fund (to divest £12m from tar sands oil and thermal coal)	• In May 2015, the Church of England announces, that it will divest its holdings of £12m in thermal coal and tar sands. The Church bans investments in any company that make more than 10% of their revenue from thermal coal and tar sands
Norway's Sovereign Wealth Fund	2015	Worth an estimated $850bn; Coal divestment: about $4.5bn is being discussed	• The $900-billion fund does not officially affiliate with the campaign • Nevertheless, the fund establishes higher climate change standards for companies in which it invests, and divests from more than 100 companies with unsustainable business models [8] • May 2015: announces that it will divest from companies which generate more than 30% of their revenue from coal
HSBC	2015	-	• Analyses the risks of stranded assets and concludes that due to climate change regulation, as well as technological advancements and innovations in efficiency, the energy mix is expected to alter in the coming years, "potentially resulting in further stranding of high carbon and high cost fossil fuels" [9] • Coal assets form the highest risk as they face greater regulatory risks, high emissions and substitution possibilities [9]

Critics, however, claim that divestment campaigns, like the one described above or the Guardian newspaper's 'Leave it in the Ground' initiative, may be doing more harm than good. Critical voices [14] note that shareholders, as the owners of companies, have the power to influence the management of these firms. In other words, investment equals control. Divestment campaigns therefore "turn an inside voice that can demand that a company listen into an outside voice that a company can easily ignore" [14]. Furthermore, Devinney [14] points out that divestment has no influence on the ability to invest, as the shares that are sold do not affect a company's cash flow.

3.2. The Campaign's Influence on Germany's Electricity Sector

For a long time, the majority of Germany's power sector, or more precisely, Germany's four biggest power companies, relied on gas and particularly on coal. Two of the companies (RWE and Vattenfall) relied heavily on lignite, the most CO_2-intensive fossil fuel, to help fulfil Germany's energy demand. To counterbalance their carbon account, these companies have recently based a significant part of their portfolio on nuclear energy and have slowly started investing in renewables. In the recent past, Carbon Capture and Storage (CCS) technology used to also be an important technology option for the four companies, to reduce their CO_2 emissions without giving up fossil fuels. However, the future of this technology in Germany is now doubtful (for more information on CCS in Germany, see e.g., [15,16]) and, in light of the German energy transition, nuclear energy is no longer an option: The remaining nuclear power plants will be phased out by 2022.

German electricity giants have recently taken high-level decisions to remove selected fossil fuel operations from their portfolio of company activity. This corporate strategy could be a response to the growing global influence of the fossil fuel divestment campaign. We find, though, that the following reasons are more plausible in explaining this situation and why, at the moment, the influence of the campaign might not be as strong as it is in other Western countries.

4. Discussion

The four companies examined here have recently taken high-level decisions to remove selected fossil fuel operations from their portfolio of company activity. We summarise implemented and scheduled plans briefly in Table 2.

These developments could be attributed to the divestment campaign, as they seem to be in line with the campaign's aims. However, we find that four alternative factors, which we call 'influencers', have had more impact on company decision making than the fossil fuel divestment campaign.

Influencer 1: Electricity Prices

Several considerations have prompted these shifts in the business strategies of the four companies, with the drop in electricity prices and subsequent fall in profit margins likely to be the primary driver. The rising share of renewable energies, illustrated in Figure 1, puts wholesale electricity prices under pressure, as the so-called merit order, which represents marginal supply costs of different technologies, is shifted to the right [19].

Table 2. The four companies' decisions to remove selected fossil fuel operations from their portfolio [17,18].

Company	Decisions
E.ON	• Plans to spin-off its fossil fuel and nuclear generation business in 2016 to a new company, Uniper • Focuses on renewables, electricity distribution networks and service for customers under the E.ON brand
RWE	• Decommissions several power stations (e.g., Goldenbergwerk, Westfalen C and Gersteinwerk K2) • Initiates the long-term mothballing of others (e.g. Gersteinwerk F+G and Weisweiler G+H) • Resolves several divestment decisions (selling off district heating sector of Essent, RWE's Dutch subsidiary, and RWE-DEA, the gas and crude oil production subsidiary)
Vattenfall	• Plans to sell the lignite generation plants and mines in eastern Germany • Has already divested in a number of operations, such as its heat and electricity network businesses in Poland and Finland
EnBW	• Expands its divestiture programme to a total of €2.6 billion for the period from 2013 to 2015 • Plans to shut down 686 MW in fossil power plants in Southern Germany (Marbach and Walheim) (As those power plants are deemed "system-relevant", the German Federal Network Agency (BNetzA) has prohibited their decommissioning until 2016)

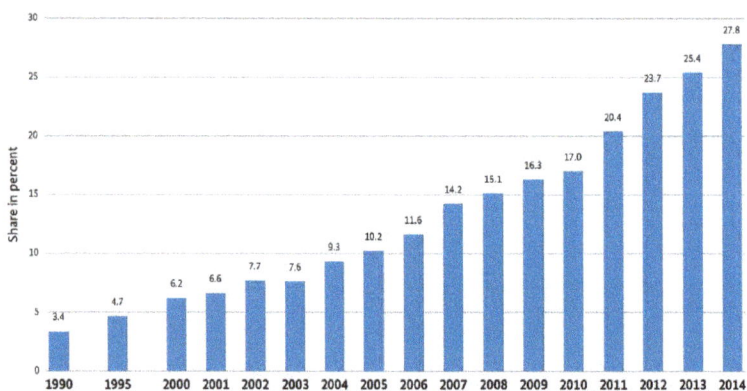

Figure 1. Development of renewable energy share of gross electricity consumption in Germany [20].

A study by Germany's Institute for Future Energy Systems (IZES), on behalf of the German Solar Industry Association (BSW-Solar), found that prices dropped despite power demand remaining unchanged. Furthermore, there was an observable alignment between the peak and base price, which used to be 20 to 25% apart. This was a direct result of the instalment of photovoltaic panels which led to a reduction of the electricity price on the power exchange by 520 to 840 million

Euro [21]. Figure 2 shows the overall development of the day-ahead price at the German power exchange between 2002 and 2014.

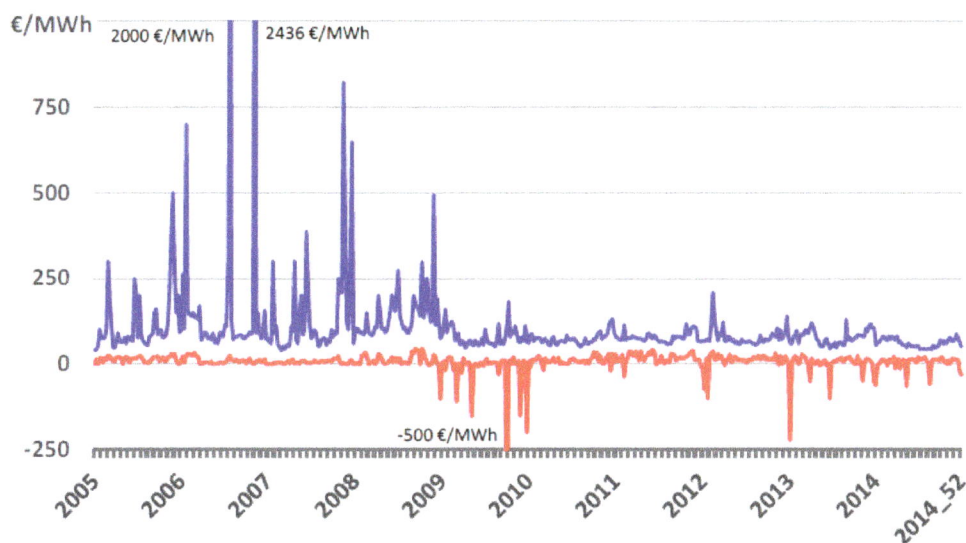

Figure 2. Weekly Day-Ahead maximum and minimum prices, Update: December 2014 [22].

Occasionally, the European Energy Exchange (EEX) even shows negative prices in times of low demand, reflecting high amounts of inflexible power generation, which cannot be shut down and restarted in a timely manner or at reasonable cost. "Frequent occurrences of negative prices in many European markets signal the need for better integration of renewables into the power grid. On a Sunday afternoon in mid-June wind and solar assured more than 60% of power generation in Germany, resulting in negative hourly prices in the whole CWE region" [23].

The Fraunhofer ISE Institute revealed that in Germany, hours with low prices (<10 €/MWh) quadrupled between the first half of 2013 compared to the first half of 2012, and that hours with negative prices have risen by nearly 50% [24]. They conclude that more flexibility is needed in the conventional power plant complex for a successful energy transition.

As a consequence of this trend, these companies have been, and are still, losing money from the operation of their conventional power plants. The new challenges, accompanied by the high percentage of fluctuating energy sources, demand new business models and strategies (key words are smart grids, load or demand-side management and electricity storage, as well as highly flexible fossil-fuel reserve power). Critics often blame high renewable subsidies for overcapacity in the market and decreasing electricity prices. A recent working paper from the IMF, however, puts this argument in perspective, showing that fossil fuels are actually subsidised by a stunning $5.3tn. This dwarfs global renewable energy subsidies, which are estimated at $120bn [25]. In fact, the IMF notes that the abolition of fossil fuel subsidies globally would reduce CO_2 emissions by more than 20% [26]. In Germany, subsidies for energy supply and consumption amounted to €21.6bn in 2010. The German *Umweltbundesamt*, the federal environmental agency, which provided these figures, points out that as these subsidies have a market-distorting effect, they simultaneously increase the need for renewable energy funding [27]. In Germany, renewable energies especially benefit through the German Renewable Energy Sources Act which provides remuneration not as a subsidy as such because it is not financed by taxes. Instead it is paid by every customer through a Renewable Energy Sources Act EEG surcharge, which is included in the electricity bill (in 2014: 6.24 ct/kWh, in 2015: 6.17 ct/kWh) [28].

Influencer 2: Company Ownership

The carbon divestment campaign mainly focuses on publicly listed companies but of the four companies examined here, only two, RWE and E.ON, are publicly listed and affected by the campaign. EnBW is de facto publicly listed, but the equity is actually divided between NECKARPRI Beteiligungsgesellschaft mbH (46.75% of the shares), which is 100% owned by the state of Baden-Württemberg (in south-west Germany) and OEW Energie-Beteiligungs GmbH (46.75% of the shares), a coalition of regional corporations and municipalities. EnBW holds about 2% of its own shares.

A closer look at the privately owned companies, EnBW and Vattenfall, shows that their business decisions seem to be in line with the divestment campaign's aim: Vattenfall is planning to sell its lignite generation plants and mines in eastern Germany, and EnBW has expanded its divestiture programme to a total of €2.6 billion for the period from 2013 to 2015. EnBW is also planning to shut down fossil fuel power plants. The reasons behind these companies' decisions are generally accepted to be government policy-related. The green-red coalition in the federal state of Baden-Württemberg is the driver for EnBW's change of course and the Swedish government, which owns Vattenfall AB, has prompted Vattenfall's restructuring plans.

Influencer 3: Government Regulation

Electricity generation is a highly regulated policy field in Germany. The federal government has mapped out its Energy Policy up to the year 2050 very precisely as can be seen in Figure 3:

Figure 3 shows that by 2020 renewables are to have a share of at least 35% in gross electricity consumption, a 50% share by 2030, 65% by 2040 and 80% by 2050. For gross final energy consumption, the targets are 18% by 2020, a 30% share by 2030, 45% by 2040 and 60% by 2050. Compared to 2008, heat demand in buildings is to be reduced by 20% by 2020, while primary energy demand is to fall by 80% by 2050. For greenhouse gases, a reduction of 80% until 2050 is the target.

Renewables have formed an important part of German energy policies for more than two decades. In 1991, Germany enacted the Grid Feed-In Law (German: Stromeinspeisungsgesetz, StromEinspG) obliging energy companies to purchase electricity from renewable energy sources at minimum prices. This law was replaced by the Renewable Energy Sources Act (German: Erneuerbare-Energien-Gesetz, EEG) in 2000, which aimed to increase the share of renewable energies in German electricity generation. The EEG was amended and revised in the years 2004, 2009, 2012 and 2014. The revised version of the EEG is a crucial part of the German energy transition, as it "aims to substantially slow any further rise in costs, to systematically steer the expansion of renewable energy, and to bring renewable energy more and more to the market" [30].

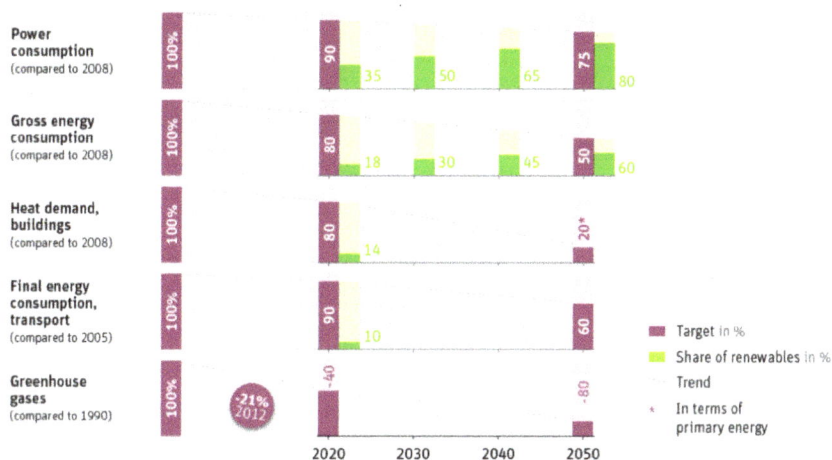

Figure 3. Long-term, comprehensive energy and climate targets set by the German government in 2010 [29].

Under this energy policy, the four large German electricity companies have had to reinvent their portfolio of activities as their nuclear assets and high shares of coal generation are incompatible with the political vision. This is unlike the UK, that introduced a capacity market system, where a payment is provided not only for electricity generation, but for the provision of reliable sources of capacity [31]. Such a model is unlikely to be introduced in Germany, as "overcapacities currently amount to roughly 60 gigawatts in the electricity market region relevant for Germany" [32]—largely due to the rising share of renewable energies. The share of nuclear energy in the gross electricity production has already decreased from 22.2% (before Fukushima) to 15.8%. Over the same period (2010 to 2014) the share of renewables has increased from 16.6% to 26.2% [33].

Although the capacity market is discussed in the Green Paper "An Electricity Market for Germany's Energy Transition" the option "Electricity Market 2.0", an optimised electricity market that provides enough incentive for the maintenance of sufficient capacity is favoured. "According to scientific studies, an optimised electricity market that allows undistorted price signals to reach market participants, and is safeguarded by a credible legal framework, is possible without an additional capacity market" [32]. In such a market the state is responsible for setting market rules: "through their specific demand, the electricity customers are independently responsible for determining the capacity level" [32].

With this amount of overcapacity, electricity companies lose their negotiation tactic of threatening governments with withholding investments in the energy infrastructure.

Policy is an important stakeholder of the company, but the owner or shareholder of a company still plays a significant role in business decisions. "Climate change has become one of the most financially significant environmental concerns facing investors" [34]. This is especially true for the electricity sector because the investment pressure on companies "is increasing as investors express their concerns about regulatory uncertainty and price volatility facing the industry" [34]. Roosevelt/Llewelyn point out that in recent years, investors have been the first to see the opportunities of a new environmental era and were among the first to invest in sustainable energy companies [35]. This trend is likely to continue: "As resource limitations, especially related to fossil fuels, increasingly impinge on corporate performance, more investors will include a company's environmental strategy as a variable in their analysis [36].

Concerning shareholder involvement in business strategies–the inside voice as Devinney frames it [14]–there are certainly institutional investors using their voice to demand sustainable business decisions [37,38]. Schmidt/Speich reveal that at RWE's 2009 general meeting, activist shareholders warned the company of the financial liability of emissions trading and of the high financial risk of involvement in Eastern European nuclear power plants like Belene.

However, it remains to be seen whether their influence is sufficient to bring about major changes, such as a complete fossil fuel phase-out. A gradually greener investment and alignment may be the result of shareholder activism, but a majority of shareholders might not support drastic changes in the business model of their company. This is especially true for a company like RWE, where municipalities are the largest single shareholder with more than 15% [39]. Municipalities depend heavily on annual dividends. Nevertheless, the questioning of the business model of electricity companies by an important shareholder like the Norwegian sovereign wealth fund might influence and impact other funds as well [40].

Influencer 4: Domestic Energy Sources

Through the *Energiewende*, the German government decided to abandon nuclear energy while maintaining high industrial output. The renunciation of coal, as proposed by the divestment campaign and parts of the German policy with view to national climate targets, faces a strong pro-coal lobby. Renewable energies amount to a quarter of Germany's gross electricity consumption yet lignite continues to be referred to as the only remaining domestic energy source in Germany. There are two large lignite deposits in the Rhineland (RWE) and in Lusatia (Vattenfall). Associated with this perspective is the national energy security agenda, as an energy self-sufficient Germany will generate jobs and wealth especially in regions that already face economic decline. Opponents of coal are facing powerful and diversified coal advocates, including not only the electricity companies, but also unions, local politicians and concerned citizens.

Lignite is the most important fossil fuel domestic energy source. The strong global coal lobby becomes even more vocal when political plans challenge lignite as a power source, just as it does when a potential climate levy for old coal-fired power stations is discussed. RWE's CEO Peter Terium predicted that this proposal "will introduce a total exit from lignite in the short run. Not only power plants, but also the associated open-cast mines and operations, would need to be closed down." As a result, "restructuring costs for the companies affected would run into the billions" [41]. A study commissioned by DEBRIV, the German association of all lignite producing companies and their affiliated organisations, found that that 40,000 jobs could be at risk if coal was phased out in Germany [42].

One reason behind the idea of a climate levy on coal is the high specific emission factor of lignite, which is listed in Table 3.

Table 3. Fossil fuel CO_2 emission factor in the German electricity mix [43].

FUEL/UNIT	CO_2 Emission Factor Related to Fuel Input (g/kWh)	Fuel Utilisation Rate in 2011 Related to Electricity Consumption (%)	CO_2 Emission Factor in 2011 Related to Electricity Consumption (g/kWh)	Comparison CO_2 Emission Factor Electricity Mix 2011 (g/kWh)
Gas	202	52%	388	
Hard Coal	339	38%	892	558
Lignite	404	35%	1,169	

Of Germany's ten power stations with the highest CO_2 emissions, nine use lignite [44]. The share of lignite in German electricity generation slightly increased after the first nuclear power plants were decommissioned, not least because of the low CO_2 certificate price under the European Emissions Trading Scheme (EU-ETS) [44]. A letter from 80 municipal electricity companies expressed their support for the plans of the Federal Ministry for Economic Affairs and Energy for the climate levy. The letter referred to the pending structural change in the energy sector, as the CO_2 reduction targets have been known for at least eight years. "The national and European climate protection targets can only be reached if, in addition to efficiency gains in the heating and transport sector, electricity generation in Germany is largely decarbonised by 2050" [45].

Despite this support, the lobby for lignite is strong. The original aim of the Ministry for Economic Affairs and Energy was to save 22 million tonnes of CO_2 emissions but this target was subsequently reduced to 16 million tonnes. The remaining 6 million tonnes of CO_2 emissions ought to be achieved using cogeneration power plants [46]. At the time of writing this paper, the responsible ministry is planning to mothball some lignite power plants (in total 2.7 GW) in order to keep them as a capacity reserve until 2020 [47], so instead of imposing an additional levy on these power plants–the government's original plan to help achieve international climate targets–the electricity companies now receive a cost-based compensation.

This example illustrates the broad support from different actors for Germany's most important fossil fuel domestic energy source. The fossil fuel divestment campaign is therefore facing resistance. NGOs, including *Urgewald* or *Germanwatch*, have faced similar obstacles in calling for a complete phase out of coal. That said, the transition of lignite power plants into a capacity reserve of limited duration can be viewed as a first step towards such a phase out.

5. Conclusions

In this paper we show that the current generation portfolios of the four largest German energy companies are heavily based on the fossil fuels gas, hard coal and lignite. E.ON's portfolio features 53% fossil fuels, RWE generates 76% of its energy from fossil fuels, Vattenfall 73% and EnBW 50% (see Table 4). In particular, the high share of energy generation from lignite, which is considered the dirtiest fossil fuel, justifiably labels those companies as ideal targets for the divestment campaign. Furthermore, the companies' current business models do not seem fit for the pending challenge of climate change. Neither are they in line with the political agenda of the German government.

Table 4 lists the structure of the power plant capacity in MW in Germany.

Table 4. Power Plant Capacity of the four German companies [48–51].

	E.ON	RWE	Vattenfall	EnBW
Gas	3887 (22%)	4411 (17%)	1707 (10%)	1191 (9%)
Hard Coal	4976 (28%)	5318 (20%)	2866 (17%)	4776 (35%)
Lignite	500 (3%)	10,291 (39%)	7767 (46%)	875 (6%)
Nuclear	5403 (30%)	3908 (15%)	771 (5%)	3333 (24%)
Renewables *	2104 (12%)	55 (3,107 **) (0.2%)	124 *** (1%)	2632 (19%)
Pumped Storage, Oil, Other	1136 (6%)	2537(10%)	3511 **** (21%)	941 (7%)
TOTAL	18,006 (100%)	26,520 (100%)	16746 (100%)	13,748 (100%)

* Comparatively high figures for E.ON and EnBW: E.ON only lists "Hydro", therefore these figures entail pumped-storage power plants as well; EnBW lists "Storage/pumped storage power plants using the natural flow of water" as Renewables; ** Renewables division, figure for whole RWE Group; *** Wind and Biomass (Waste); **** Hydro power amounts to 2,880 MW alone, which are mainly pumped-storage power plants.

Having said that, these companies have undertaken the first small steps to restructure: E.ON, for example, is now focussing on renewable energy while Vattenfall is selling lignite generation plants and mines in Eastern Germany. However, E.ON's restructure could mean it avoids liabilities should falling profits from fossil fuel and nuclear energy generation force its new company, Uniper, into insolvency. Under the E.ON brand, the company can focus on future-proofed renewable energy technologies but if the costs of decommissioning nuclear plants and nuclear waste management are excessive, Uniper could become insolvent leaving the German taxpayer to foot the bill.

In the most recent past, focussing on renewable energies has become an important business strategy. One indication is the establishment of new subsidiary companies by the four businesses examined here:

- E.ON Climate & Renewables (2007),
- RWE Innogy (2008),
- Vattenfall Europe New Energy GmbH (2007),
- EnBW Renewable Energies (2008).

Furthermore, these companies invest in renewable energy technology every year as part of their overall investments as Tables 5–7 show.

In the past, Vattenfall has formulated rolling investment plans for a five-year period. The company initially budgeted SEK 147 billion for the period 2012–2016 then in 2013 amended investment plans to SEK 105 billion for the period 2014–2018 [65]. More recently, as markets have become characterised by over-capacity in production and low prices, the company has decided to limit its most recent investment plan to two years (2015–2016) and to invest SEK 41 billion over that period, of which SEK 30.8 billion, or 75%, will be invested in electricity and heat production.

A large amount will be invested in wind energy (SEK 9.1 billion), whereas investments in fossil fuel production were determined several years ago under completely different market conditions [50]. The aim of the company is to reduce its CO_2 exposure to 65 million tonnes of absolute emissions by 2020 (in 2014 the CO_2 emissions amounted to 82.3 million tonnes, in 2010 they totalled 94 million tonnes). To achieve this target, the company has a two-part strategy: the divestment of fossil-based production and growth and investment in renewable production [50].

Table 5. E.ON–overall investment and share of investment in renewables [48,52–56].

Year	Investment (€ in millions)	Renewables
2008	18,406	1484 (8%)
2009	8655	1031(12%)
2010	8286	1163 (14%)
2011	6524	1114 (17%)
2012	6997	1791 (26%)
2013	8086	1028 (13%)
2014	4633	1222 (26%)

Table 6. RWE–overall investment and share of investment in renewables [49,57–61].

Year	Investment (€ in Millions)	RWE Innogy
2008	5693	1102 (19%)
2009	15,637	733 (5%)
2010	6643	709 (11%)
2011	7702	891 (12%)
2012	5544	1093 (20%)
2013	4624	1083 (23%)
2014	3440	738 (21%)

Table 7. EnBW–overall investment and share of investment in renewables [51,62–64].

Year	Investment (€ in millions)	Renewable Investment EnBW
2009	4374.1	153.7 (4%)
2010	2327.9	536.4 (23%)
2011	1319.0	216.6 (16%)
2012	877.4	121.6 (14%)
2013	1108.3	316.5 (29%)
2014	1956.7	610.8 (31%)

Even so, these figures somewhat pale in comparison to the total investment in renewable energies in Germany (see Figure 4). In fact, the lion's share of installed renewable capacity in Germany is not owned by the four energy giants. A study by Leuphana University of Luneburg, commissioned by the Agency for Renewable Energies [66], assessed the ownership structures of installed renewable energy production. Researchers concluded that the largest share (47%) of installations was owned by 'citizen energy' (community or local ownership, citizen-owned power stations and sole proprietors). An equally impressive 41% share was owned by what the study called institutional and strategic investors–companies involved in industrial production processes or

254

service providers. The study concluded that only 12% of installed renewable energy capacity was owned by energy suppliers, including the four companies examined in this paper. Figure 5 shows the distribution as put forward in the study.

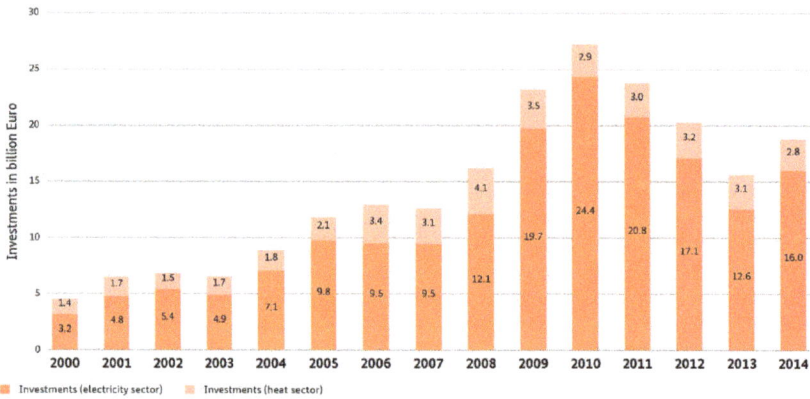

Figure 4. Development of investments in construction of renewable energy installations in Germany [20].

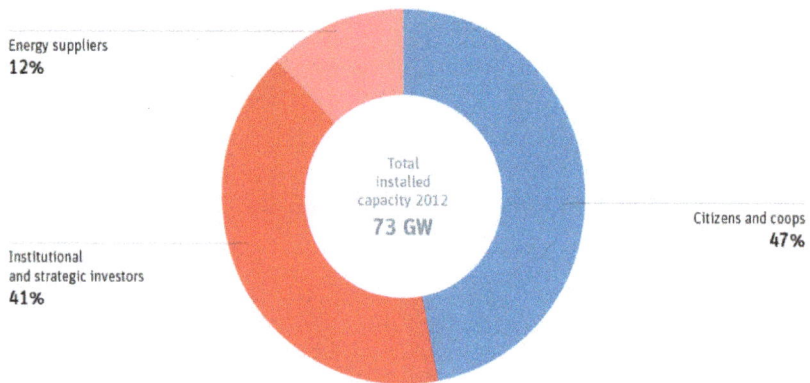

Figure 5. Installed capacity of renewable energies in Germany in 2012 as per property owner [29] * (* Coops stands for 'energy cooperatives': community-owned renewables projects).

The current business model featured by the four companies examined in this paper is often regarded as unsustainable and out of step with governmental climate targets. Furthermore, the carbon content of the companies' energy production is no longer only examined by governments and monitored by non-governmental organisations, but now also evaluated by the financial sector [6].

In 2003, the German WestLB, a former regional state bank, published a study with the title "Von Economics zu Carbonomics–Value at Risk durch Klimawandel" (From Economics to Carbonomics–Value at Risk caused by Climate Change) in which it analysed the response of the financial sector to expected climate regulation such as the Kyoto Protocol coming into force or the introduction of the European Emissions Trading Scheme (EU ETS). Organisations such as the Carbon Disclosure Project (CDP), which discloses the greenhouse gas emissions of major corporations, underline this interest. "When the financial services industry–which focuses like a laser on return on investment–starts worrying about the environment, something big is happening" [36].

In this paper, we have examined the influence of the fossil fuel divestment campaign on the four largest German electricity companies. We found that four entirely different influencers were more powerful in drawing attention to climate policies and their consequences at corporate board meetings than the aims of the divestment movement. These influencers were: electricity prices, company ownership, government regulation and domestic energy sources. We thus conclude that the campaign does not currently directly influence decision making at the German energy giants enough to alter the structure of the German electricity mix. Of course, our analysis provides only a snapshot and over time this campaign may well gain more influence in Germany. There is already some movement in the four big electricity companies towards renewable energy production. Up until a few years ago those companies were able to pursue their business model with fossil fuels and nuclear energy alone, despite Germany's strong civil movement which has been vociferously against nuclear energy. With the new social movement that is the fossil fuel divestment campaign another cause must be factored into company business decisions.

The possible influence the campaign actually has will depend on global climate policy. The consulting firm Mercer points out that: "It is clear from our survey that credible and consistent climate change legislation and regulation is required to drive greater integration of climate change into investment practices and to provide the major impetus for a shift from high carbon to low carbon investment. Without strong climate policy that provides transparency, longevity and clarity for investors, the revolution that is called for in transforming our energy systems will not be possible" [67]. A first step, a "golden opportunity", was described by the IMF's fiscal affairs head and former finance minister of Portugal, Vitor Gaspar [25]. He suggested taking advantage of low oil and coal prices and phasing out subsidies of global fossil fuels, which would reduce the global CO_2 emissions by more than 20% [26].

Acknowledgments

We would like to thank Stefan Thomas (Wuppertal Institute) for providing feedback on an earlier draft. We are indebted to Gail Whiteman (Erasmus University Rotterdam) and David Deephouse (University of Alberta) who helped shape our ideas at various stages of this work. Furthermore, we would like to thank two anonymous reviewers for encouraging us to improve our manuscript. Any remaining errors are ours.

Author Contributions

Both authors designed the work, analysed the data and interpreted the results. Dagmar Kiyar planned and conducted the qualitative research. She also selected the quantitative secondary sources. Both authors contributed to and edited the manuscript.

Conflicts of Interest

The authors declare no conflict of interest.

References

1. E.ON Press release 04/27/15: E.ON moves forward with transformation: key organizational and personnel decisions made. Available online: http://www.eon.com/en/media/news/press-releases/2015/4/27/eon-moves-forward-with-transformation-key-organizational-and-personnel-decisions-made.html (accessed on 15 July 2015).
2. BMUB: Aktionsprogramm Klimaschutz 2020. Kabinettsbeschluss vom 3. Dezember 2014. Available online: http://www.bmub.bund.de/fileadmin/Daten_BMU/Download_PDF/Aktionsprogramm_Klimaschutz/aktionsprogramm_klimaschutz_2020_broschuere_bf.pdf (accessed on 15 July 2015).
3. International Energy Agency (IEA). *World Energy Outlook 2012*; OECD: Paris, France, 2012.
4. McGlade, C.; Ekins, P. The geographical distribution of fossil fuels unused when limiting global warming to 2 °C. *Nature* **2015**, *517*, 187–190.
5. Carbon Disclosure Project (CDP). Flicking the switch. Are electric utilities prepared for a low carbon future. May 2015. Available online: https://www.cdp.net/Docs/investor/2015/electric-utilties-report-exec-summary-2015.pdf (accessed on 17 August 2015).
6. Kiyar, D. Um(welt)orientierung großer Energiekonzerne? Die großen Vier und ihr Beitrag zum Klimaschutz. Ph.D. Thesis, Westfälische Wilhelms-Universität Münster, Institut für Politikwissenschaft, Münster, 22 June 2012. http://repositorium.uni-muenster.de/document/miami/364834e7-9a8e-4ad8-ab4f-83feadd7e4ad/diss_kiyar_buchblock.pdf (accessed on 25 May 2015).
7. Bank of England. Confronting the challenges of tomorrow's world. Speech given by Paul Fisher, Deputy Head of the Prudential Regulation Authority and Executive Director, Insurance Supervision. Economist's Insurance Summit 2015, London. 3 March 2015. Available online: http://www.bankofengland.co.uk/publications/Documents/speeches/2015/speech804.pdf (accessed on 25 May 2015).
8. The Wall Street Journal. Norway Oil Fund Sheds More Coal Assets. 4 May 2015. Available online: http://www.wsj.com/articles/norway-oil-fund-sheds-more-coal-assets-1430765503 (accessed on 25 May 2015).
9. HSBC. Stranded assets: what next? How investors can manage increasing fossil fuel risks. 16 April 2015. Available online: http://www.businessgreen.com/digital_assets/8779/hsbc_Stranded_assets_what_next.pdf (accessed on 25 May 2015).
10. McKibben, B. Global Warming's Terrifying New Math. Rolling Stone. 19 July 2012. Available online: http://www.rollingstone.com/politics/news/global-warmings-terrifying-new-math-20120719 (accessed on 25 May 2015).

11. Ansar, A.; Caldecott, B.; Tilbury, J. Stranded assets and the fossil fuel divestment campaign: what does divestment mean for the valuation of fossil fuel assets? 2013. University of Oxford. Smith School Stranded Assets Programme. Available online: http://www.smithschool.ox. ac.uk/research-programmes/stranded-assets/SAP-divestment-report-final.pdf (accessed on 25 May 2015).

12. The Guardian. Climate change: UN backs fossil fuel divestment campaign. 15. March 2015. Available online: http://www.theguardian.com/environment/2015/mar/15/climate-change-un-backs-divestment-campaign-paris-summit-fossil-fuels (accessed on 25 May 2015).

13. UN. Ban Ki-moon urges more fossil fuel divestment. Statement 4. November 2014. Available online: http://newsroom.unfccc.int/financial-flows/ban-ki-moon-speaks-in-favour-of-divestment/ (accessed on 25 May 2015).

14. Devinney, T. The Guardian's fossil fuel divestment campaign could do more harm than good. 24 March 2015. Available online: http://theconversation.com/the-guardians-fossil-fuel-divestment-campaign-could-do-more-harm-than-good-39000 (accessed on 25 May 2015).

15. Fischer, W. No CCS in Germany Despite the CCS Act? In *Carbon Capture, Storage and Use. Technical, Economic, Environmental and Societal Perspectives*, 1st ed.; Kuckshinrichs, W., Hake, J.-F., Eds.; Springer: Heidelberg, Germany, 2015; pp. 255–286.

16. Pietzner, K.; Schwarz, A.; Duetschke, E.; Schumann, D. Media Coverage of Four Carbon Capture and Storage (CCS) Projects in Germany: Analysis of 1,115 Regional Newspaper Articles. *Energy Procedia* **2014**, *63*, 7141–7148.

17. RWE. Kraftwerksdaten 2015. 2. Quartal. Available online: http://www.rwe.com/web/cms/mediablob/de/59784/data/59766/103/transparenz-offensive/deutschland/kraftwerksausfaelle/Kraftwerksdaten-2015-Q2.pdf (accessed on 25 May 2015).

18. Bundesnetzagentur (BNetzA). Bescheid Transnet, Dezember 2013. Available online: http://www.bundesnetzagentur.de/SharedDocs/Downloads/DE/Sachgebiete/Energie/Unterneh men_Institutionen/Versorgungssicherheit/Erzeugungskapazitaeten/systemrelevante_KW/Besc heid_Transnet_13_12_2013.pdf?__blob=publicationFile&v=1 (accessed on 25 May 2015).

19. DG ENER. European Commission Quarterly Report on European Electricity Markets. Volume 7, Issue 4, Fourth quarter of 2014. Available online: https://ec.europa.eu/energy/sites/ener/files/documents/quarterly_report_on_european_electricity_markets_2014_q4.pdf (accessed on 25 May 2015).

20. Federal Ministry for Economic Affairs and Energy (BMWi). Development of renewable energy sources in Germany 2014. Charts and figures based on statistical data from the Working Group on Renewable Energy-Statistics (AGEE-Stat), as at February 2015. Available online: http://www.erneuerbare-energien.de/EE/Redaktion/DE/Downloads/development-of-renewable-energy-sources-in-germany-2014.pdf?__blob=publicationFile&v=6 (accessed on 25 May 2015).

21. Hauser, E.; Frantzen, J.; Leprich, U. Preissenkende Effekte der Solarstromerzeugung auf den Börsenstrompreis. Kurzstudie im Auftrag des Bundesverbandes Solarwirtschaft (BSW). Präsentation in Berlin, 31 January 2012. Available online: http://www.solarwirtschaft.de/fileadmin/media/pdf/120131_Präsentation_Preiseffekte_von_PV.pdf (accessed on 25 May 2015).

22. Fraunhofer Institute for Solar Energy Systems (ISE). Electricity Production and Spot-Prices in Germany 2014. 31 December 2014. Available online: http://www.ise.fraunhofer.de/de/downloads/pdf-files/data-nivc-/folien-electricity-spot-prices-and-production-data-in-germany-2014-engl.pdf (accessed on 16 July 2015).

23. DG ENER (2013). European Commission Quarterly Report on European Electricity Markets. Volume 6, Issue 2, Second quarter 2013. Available online: https://ec.europa.eu/energy/sites/ener/files/documents/20130814_q2_quarterly_report_on_european_electricity_markets.pdf (accessed on 25 May 2015).

24. Fraunhofer Institute for Solar Energy Systems (ISE). Kohleverstromung zu Zeiten niedriger Börsenstrompreise. August 2013. Available online: http://www.ise.fraunhofer.de/de/downloads/pdf-files/aktuelles/kohleverstromung-zu-zeiten-niedriger-boersenstrompreise.pdf (accessed on 25 May 2015).

25. The Guardian. Fossil fuels subsidised by $10m a minute, says IMF. 18 May 2015. Available online: http://www.theguardian.com/environment/2015/may/18/fossil-fuel-companies-getting-10m-a-minute-in-subsidies-says-imf (accessed on 25 May 2015).

26. International Monetary Fund (IMF). How Large Are Global Energy Subsidies? IMF Working Paper WP/15/105. May 2015. Available online: http://www.imf.org/external/pubs/ft/wp/2015/wp15105.pdf (accessed on 25 May 2015).

27. Umweltbundesamt. Umweltschädliche Subventionen in Deutschland. Aktualisierte Ausgabe 2014. Available online: http://www.umweltbundesamt.de/sites/default/files/medien/376/publikationen/umweltschaedliche_subventionen_in_deutschland_aktualisierte_ausgabe_2014_fachbroschuere.pdf (accessed on 15 July 2015).

28. Bundesnetzagentur. EEG-Umlage beträgt im kommenden Jahr 6,17 ct/kWh. Available online: http://www.bundesnetzagentur.de/SharedDocs/Pressemitteilungen/DE/2014/141014_PM_EEG_Umlage.html (accessed on 15 July 2015).

29. Energy Transition. The German Energiewende. Available online: http://energytransition.de (accessed on 25 May 2015).

30. Federal Ministry for Economic Affairs and Energy (BMWi). 2014 Renewable Energy Sources Act: Plannable. Affordable. Efficient. Available online: http://www.bmwi.de/EN/Topics/Energy/Renewable-Energy/2014-renewable-energy-sources-act,did=693154.html (accessed on 25 May 2015).

31. Gov.UK. Electricity Market Reform: Capacity Market. 24. July 2014. Last updated: 12. December 2014. Available online: https://www.gov.uk/government/collections/electricity-market-reform-capacity-market (accessed on 25 May 2015).

32. Federal Ministry for Economic Affairs and Energy (BMWi). An Electricity Market for Germany's Energy Transition. Discussion Paper of the Federal Ministry for Economic Affairs and Energy (Green Paper). Available online: http://www.bmwi.de/BMWi/Redaktion/PDF/G/gruenbuch-gesamt-englisch,property=pdf,bereich=bmwi2012,sprache=de,rwb=true.pdf (accessed on 25 May 2015).

33. AG Energiebilanzen. Bruttostromerzeugung in Deutschland ab 1990 nach Energieträgern. February 2015. Available online: http://www.ag-energiebilanzen.de/index.php?article_id= 29&fileName=20150227_brd_stromerzeugung1990-2014.pdf (accessed on 25 May 2015).

34. Labatt, S.; White, R.R. *Carbon Finance. The Financial Implications of Climate Change*; John Wiley & Sons, Inc.: Hoboken, NJ, USA, 2007.

35. Roosevelt, T.; Llewellyn, J. Investors hunger for clean energy. *Harvard Business Review* **2007**, *85*, 38–40.

36. Esty, D.C.; Winston, A.S. *Green to Gold. How Smart Companies Use Environmental Strategy to Innovate, Create Value, and Build Competitive Advantage*; John Wiley & Sons, Inc.: Hoboken, NJ, USA, 2009.

37. Schmidt, M.; Speich, I. Kritische Themen mit Unternehmen stärker diskutieren. Institutionelle Investoren hinterfragen immer mehr – Gut entwickelte Aufsichtsstrukturen stiften Anlegernutzen – Umwelt- und Sozialthemen integrieren. *Börsen-Zeitung. Zeitung für Finanzmärkte. Sonderbeilage: Nachhaltigkeit* 2011. Available online: https://www.boersen-zeitung.de/index.php?li=1&artid=2011045331&titel=Kritische-Themen-mit-Unternehmen-staerker-diskutieren (accessed on 25 May 2015).

38. Schmidt, M.; Speich, I. Risikomanagement durch Nachhaltigkeit. Michael Schmidt und Ingo Speich bewerten das Engagement aktiver institutioneller Investoren. *Deutsche Pensions- und Investmentnachrichten (dpn)* **2011**, 37–39. Available online: http://riskmanagement-conference.com/docme/presse/292ab40c68ad0193a10e47ce9ce17881.0.0/Risikomanagement_durch_Nachhaltigkeit_DPN_0411.pdf (accessed on 25 May 2015).

39. Verband der kommunalen Aktionäre VKA. RW Energie-Beteiligungsgesellschaft mbH & Co.KG. Available online: http://www.vka-rwe.de/index.php?id=4896 (accessed on 25 May 2015).

40. Theurer, M.; Bünder, H. Größter Staatsfonds der Welt plant Kohle-Boykott. *FAZ*. 28 May 2015. Available online: http://www.faz.net/aktuell/wirtschaft/klimaschutz-groesster-staatsfonds-der-welt-plant-kohle-boykott-13617340.html (accessed on 25 May 2015).

41. Energy Intelligence. EI New Energy. Vol. IV, No. 15. 16 April 2015. Available online: http://www.smithschool.ox.ac.uk/research-programmes/stranded-assets/EI%20New%20Energy%20Apr%2016.pdf (accessed on 25 May 2015).

42. DEBRIV. Aktuelle HWWI-Studie: Durch Sonderabgabe drohen massive Arbeitsplatzverluste in den Braunkohleregionen/ BMWi-Pläne gefährden alleine 40.000 Stellen in den Braunkohleregionen. April 2015. Available online: http://www.presseportal.de/pm/9341/2997191 (accessed on 25 May 2015).

43. Umweltbundesamt. Entwicklung der spezifischen Kohlendioxid-Emissionen des deutschen Strommix in den Jahren 1990 bis 2013. Climate Change 23/2014. Available online: http://www.umweltbundesamt.de/sites/default/files/medien/376/publikationen/climate_change_23_2014_komplett.pdf (accessed on 25 May 2015).

44. Öko-Institut. CO_2-Emissionen aus der Kohleverstromung in Deutschland. March 2014. Available online: http://www.oeko.de/oekodoc/1995/2014-015-de.pdf (accessed on 25 May 2015).

45. Energie-Chronik. 80 Stadtwerke untertsützen Gabriels Pläne für eine Klimaabgabe. Available online: http://www.udo-leuschner.de/energie-chronik/150404d1.htm (accessed on 25 May 2015).

46. Reuters. Klima-Abgabe – Gabriel knickt bei Kohle-Plänen ein. Handelsblatt. 18 May 2015. Available online: http://www.handelsblatt.com/politik/deutschland/klima-abgabe-gabriel-knickt-bei-kohle-plaenen-ein/11791272.html (accessed on 25 May 2015).

47. Federal Ministry for Economic Affairs and Energy (BMWi). July 2015. Eckpunkte für eine erfolgreiche Umsetzung der Energiewende. Politische Vereinbarungen der Parteivorsitzenden von CDU, CSU und SPD vom 1. Juli 2015. Available online: http://www.bmwi.de/BMWi/Redaktion/PDF/E/eckpunkte-energiewende,property=pdf,bereich=bmwi2012,sprache=de,rwb=true.pdf (accessed on 17 August 2015).

48. E.ON. Annual Report 2014. Available online: http://www.eon.com/content/dam/eon-com/ueber-uns/publications/150312_EON_Annual_Report_2014_EN.pdf (Accessed on 25 May 2015).

49. RWE. Annual Report 2014. March 2015. Available online: http://www.rwe.com/app/wartung/hv2014/bpk_docs/RWE-Annual-Report-2014.pdf (accessed on 25 May 2015).

50. Vattenfall. Toward a more sustainable energy portfolio. Annual and sustainability report 2014. Available online: http://corporate.vattenfall.com/globalassets/corporate/investors/annual_reports/2014/annual-and-sustainability-report-2014.pdf (accessed on 25 May 2015).

51. EnBW. Report 2014. Energiewende. Safe. Hands on. Available online: https://www.enbw.com/enbw_com/downloadcenter/annual-reports/enbw-report-2014.pdf (accessed on 25 May 2015).

52. E.ON. E.ON Company Report 2008. Annual Report I/II. Available online: http://www.eon.com/content/dam/eon-com/en/downloads/e/EON_Company_Report2008.pdf (accessed on 25 May 2015).

53. E.ON. E.ON AG Financial Statements pursuant to German GAAP and Combined Group Management Report for the 2010 Financial Year. Available online: http://www.eon.com/content/dam/eon-com/en/downloads/e/E.ON_2010_Jahresabschluss_en_.pdf (accessed on 25 May 2015).

54. E.ON. Annual Report 2011. Available online: http://www.eon.com/content/dam/eon-com/en/downloads/e/E.ON_2011_Annual_Report_.pdf (accessed on 25 May 2015).

55. E.ON. Annual Report 2012. Available online: http://www.eon.com/content/dam/eon-com/ueber-uns/GB_2012_US_eon.pdf (accessed on 25 May 2015).

56. E.ON. Annual Report 2013. Available online: http://www.eon.com/content/dam/eon-com/ueber-uns/publications/GB_2013_US_eon.pdf (accessed on 25 May 2015).

57. RWE. Annual Report 2009. Make big things happen. For our customers. Available online: http://www.rwe.com/web/cms/mediablob/en/575576/data/568444/3/rwe/investor-relations/agm/annual-general-meeting-2010/annual-report-2009.pdf (accessed on 25 May 2015).

58. RWE. Annual Report 2010. RWE straight talking. Available online: http://www.rwe.com/web/cms/mediablob/en/582360/data/609582/6/rwe/investor-relations/agm/annual-general-meeting-2011/Annual-report-2010.pdf (accessed on 25 May 2015).

59. RWE. Annual Report 2011. Starting new chapters. Available online: http://www.rwe.com/web/cms/mediablob/en/1338130/data/1310026/3/rwe/investor-relations/agm/annual-general-meeting-2012/Annual-report-2011.pdf (accessed on 25 May 2015).

60. RWE. Annual Report 2012. Available online: http://www.rwe.com/web/cms/mediablob/en/1802504/data/1798118/7/rwe/investor-relations/agm/annual-general-meeting-2013/Annual-Report-of-RWE-AG-2012.pdf (accessed on 25 May 2015).

61. RWE. Annual Report 2013. Shaping the future. Available online: http://www.rwe.com/web/cms/mediablob/en/1907134/data/1907462/5/rwe/investor-relations/agm/annual-general-meeting-2014/Group-Annual-Report-2013.pdf (accessed on 25 May 2015).

62. EnBW. Annual Report 2010. Powered by diversity. Available online: https://www.enbw.com/media/downloadcenter/annual-reports/annual-report-of-enbw-ag-2010.pdf (accessed on 25 May 2015).

63. EnBW. Annual Report 2011. Actively shaping the new energy concept. Available online: https://www.enbw.com/media/downloadcenter/annual-reports/annual-report-of-enbw-ag-2011.pdf (accessed on 25 May 2015).

64. EnBW. Report 2013. Energiewende. Safe. Hands on. Available online: https://www.enbw.com/enbw_com/investoren/investors_docs/news_und_publikationen/enbw-report-2013-condensed-version.pdf (accessed on 25 May 2015).

65. Vattenfall. Annual Report 2011. Available online: https://www.vattenfall.com/en/file/2011_Annual_Report.pdf_20332206.pdf (accessed on 25 May 2015).

66. Trend Research/ Leuphana Universität Lüneburg. Definition und Marktanalyse von Bürgerenergie in Deuschland. Im Auftrag der Initiative "Die Wende – Energie in Bürgerhand" und der Agentur für Erneuerbare Energien. Available online: http://100-prozent-erneuerbar.de/wp-content/uploads/2013/10/Definition-und-Marktanalyse-von-Bürgerenergie-in-Deutschland.pdf (accessed on 16 July 2015).

67. Mercer. Global Investor Survey on Climate Change – Annual Report on Actions and Progress 2010. June 2011. Available online: http://www.ceres.org/resources/reports/2010-global-investor-survey-on-climate-change-1/view (accessed on 25 May 2015).

Security of Supply in European Electricity Markets—Determinants of Investment Decisions and the European Energy Union

Saskia Ellenbeck, Andreas Beneking, Andrzej Ceglarz, Peter Schmidt and Antonella Battaglini

Abstract: The European Union and its Member States are seeking to decarbonize their energy systems, including the electricity sector and, at the same time, pursue market integration. However, renewable energy (RE) deployment and the liberalization of the energy-only market have raised concerns at the national level about the security of electricity supplies in the future. Some actors consider the lack of sufficient investments in generation capacities a threat to supply security. As a consequence, it was proposed that capacity markets solve these problems. The underlying assumption is that the market design is the only determining factor for investments in security of supply options. In this article, we question this narrow view and identify further determinants of the investment decisions of electricity market participants. Based on the insights of institutional sociology and economics, we understand the market to be a social institution that structures the behavioural expectations of market participants. Derived from the theoretical conceptualization and based on qualitative literature review and own work, we find four determinants for investment behaviour beyond the formal market design: Material opportunities, strategic actor behavior and identity, focusing events and discursive expectations about the future. With this perspective, we discuss the introduction of a European Energy Union as a possible tool that might have a great impact on the more informal determinants such as expectations about the future and the construction of a European energy narrative.

Reprinted from *Energies*. Cite as: Ellenbeck, S.; Beneking, A.; Ceglarz, A.; Schmidt, P.; Battaglini, A. Security of Supply in European Electricity Markets—Determinants of Investment Decisions and the European Energy Union. *Energies* **2015**, *8*, 5198-5216.

1. Introduction

The European Union and its Member States are seeking to decarbonize their energy systems including the electricity system. In order to achieve this goal, all European Member States have set renewable energy (RE) targets and introduced several support schemes for renewable energy deployment. However, no consensus has been reached regarding which specific energy transition pathway to follow, which policies and remunerations schemes are needed, what timing is appropriate and which responsibilities should rest on the EU and which on Member States.

Recent discussions focus on the introduction of capacity markets as a specific remuneration scheme for installed capacity in order to meet EU energy policy targets. Capacity payments would add to those for produced energy in order to make new generation capacities profitable or to keep old ones in the market. Policy-makers mainly justify this new compensation mechanism at the

Member State level, as there exists a perceived hesitation to invest in flexible capacities that could lead to an insufficient level of electricity supply in the future when aging assets go offline [1,2].

Thus, various countries have already introduced national capacity markets, remunerating generation capacity of different kinds of technologies including baseload capacities (see Figure 1). This is motivated by differing arguments, which can also be found in the available literature. Some authors [3,4] have questioned the ability of the liberalised energy-only market, in general, as a means to provide the right signals for appropriate long-term investments. Others refer to the changing needs of an electricity system driven by base-load, to a more residual load driven system that addresses the variability characteristic of renewable energy, especially wind and solar [5,6]. In this perspective, the increasing feed-in of variable renewable electricity with very low operating costs is changing the spot market price formation results under the merit order regime, which would call for a new market design to address these changes [7,8]. In contrast, some authors do not see any need to change the system at all and question the threat of a future shortage [9]. Policy-makers are thus confronted with conflicting ideas on how to stimulate the appropriate investment, and which changes in the market design are needed to deliver the wanted energy security capability of the system.

In this article, we broaden the perspective on how investment decisions in the electricity market are taken. We thus challenge the narrow view of the market design, and accordingly capacity mechanism, as being the only determinant for investments in security of supply options. Based on the insights of institutional sociology and economics, we understand the market to be a social institution that structures the behavioural expectations of market participants. In the following, we discuss the investment behavior of electricity market actors in the context of uncertain, diverging and sometimes conflicting strategic interests of different kinds of actors in European Member States. As a result, we find four determinants for investment behaviour beyond the formal market design.

The article is therefore structured in five parts. In the second section, we briefly describe the European energy sector, review the literature about investment decisions in the electricity sector and then present the theoretical foundation of understanding the electricity market as a social institution. Derived from that perspective, in the third section, we elaborate on four variables that have an impact on the investment decisions of market actors beyond the formal market design. In the fourth part of the article, we will discuss whether the recently proposed European Energy Union might be able to address these factors and if yes—How? Finally, we will discuss our results and hint to further research needs. Although national policies and strategies differ across the EU, we explicitly do not compare different remuneration schemes and market design options, but are rather interested in other—more informal—factors that influence the investment decisions of market actors. This should, on the one hand, point to the need for further research and on the other hand, help policy-makers explore new approaches to stipulate investments in security of supply capacities.

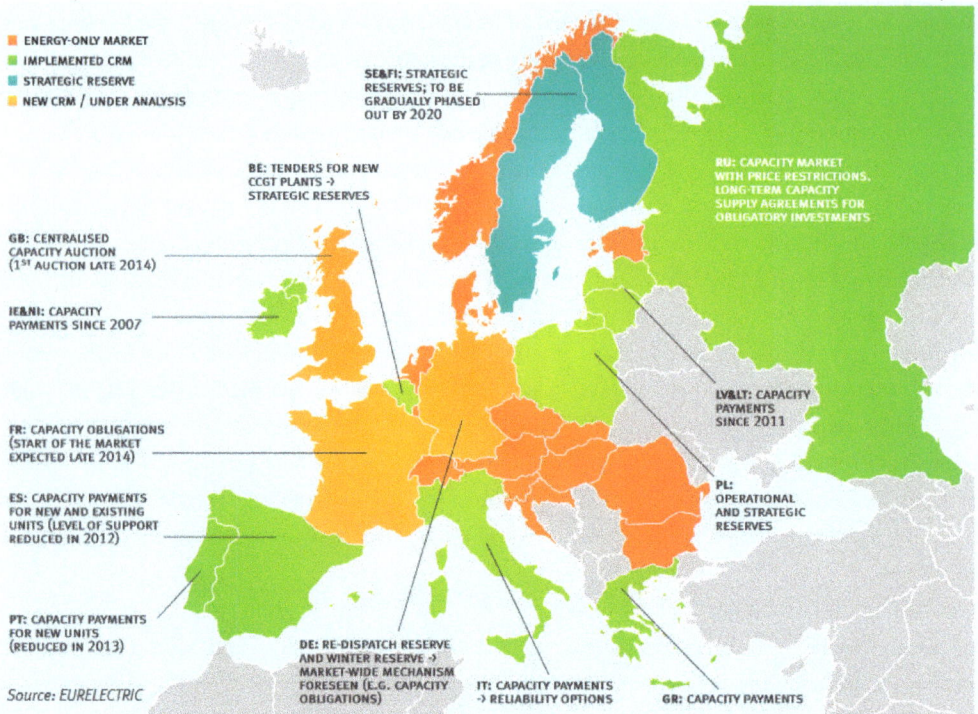

Figure 1. Implementation of capacity remuneration mechanisms (CRMs) across Europe as of June 2014 [10].

2. Investment Decisions in the Electricity Market

2.1. The European Electricity Market

Traditionally, electricity generation and distribution was a public domain executed by governmental agencies or publicly owned companies in most European countries. Furthermore, the electricity sectors were organized mainly nationally with very low regional interconnections. This has changed with the liberalization directives of the European Union, the first in 1996 [11] and the second in 2003 [11]. With the third energy package in 2009 [11] an ownership unbundling obligation was introduced that led to a legal separation of generation and sales activities from the transmission network. While investments in transmission networks remain highly regulated, investments in generation capacities are not. Although the internal market for electricity was aimed to be completed by 2014, today there are still several features of the market that lead to differing electricity prices and investment conditions within Europe. These features include differing energy policy targets and accompanying instruments across Member States, insufficient interconnector capacities, and related weak intra-European physical integration and the characteristics of the traded good electricity, such as the obligation to balance supply and demand at any time. Although most of the liberalization policies have been implemented in all Member States, some deficits mainly in the field of competition and actual unbundling remain. National electricity markets are still dominated by the successors of the former public utilities, distribution is still largely owned by utilities and new market players are not properly rewarded to actively participate in the market, thus, de facto, delaying market liberalization and the overall energy transition. In parallel to the liberalization process, European countries agreed on RE targets and are currently increasing the shares of renewable energy in their electricity mix. In view of this and despite of this, most countries face public and expert debates about underinvestment into security of supply options. Investments in renewable generation in 2013 account for 72% of total investments in electricity generation, while only ten years before 80% of investments were made in conventional fossil fuel capacities in the EU-27 (plus Norway and Switzerland). This is increasing concerns among energy actors about the ability of the system to deliver the same level of security as in the past dominated fossil fuel generation fleet.

2.2. The Market as an Efficiency Maximizer

While in regulated markets utilities are mostly paid on the basis of their costs and could therefore expect a stable return from investments in new generation capacities, in a market system, investment decisions have to be made under higher uncertainty [12]. For both systems, classical economic theory describes investment decisions as a purely rational economic action that is determined by economic factors and determinants such as for example price volatility [12,13], risks [14,15], costs [16–19], *etc.*, sometimes considering also regulatory and political aspects [20]. Some studies also take the behavior of market actors into account especially those analyzing the strategic interests of power companies as well as the changing objectives and interests following market liberalization [21]. A big part of the literature is purely quantitative and proposes a set of economic assumptions and optimality functions to calculate efficient behavior. However, some scholars also look at more qualitative factors. Masini and Menichetti [22] for example describe that "the investors' a priori beliefs, preferences over certain policy instruments and attitude towards technological risk affects the likelihood of investing in RE projects" ([22], page 1). Generally, the relatively young field of behavioral corporate finance research suggests that, due to various psychological factors, managers as well as investors act at least partly irrationally [23]. For energy efficiency investments of companies Cooremans [24] developed a model of the investment process, stating that it is applicable to all investment decisions. In this model the choice between different investments options, where the neo-classical perspective usually focuses on, is only one out of several steps of decision-making and in reality may or may not be based on rational tools for profit maximization like Internal Rate of Return (IRR) or Net Present Value (NPV). All steps are embedded in a bigger societal (environmental) and organizational context [24]. A study for the British Department of Energy & Climate Change [25] elaborates this model further defining a space of four dimensions in which a company makes its choices: "material", "regulatory-policy", "market" and "social-cultural". In the literature regarding investment decisions in the electricity market we find only limited research on issues that explicitly acknowledge the social context of investment decisions and thus do not understand efficiency and profit maximization as main driver of market decisions. In the next section we will use the theoretical concept of social institutionalism and apply it to the electricity sector.

2.3. The Market as a Social Institution

The market is understood here as a social institution that shapes the behaviour of actors in a way that expectations about the market result can be made [26,27]. Institutions are seen as social structures of meaning, as formal and informal systems of rules, which have an impact on the behaviour of actors and thus, the social interactions among themselves. Social interactions are enabled and facilitated by a common understanding of norms, language, identity and other patterns of social action to reduce the levels of complexity and uncertainty. As a consequence, social order is (re)produced by institutions [28–30]. Actors constantly evaluate their behaviour in the context of others: "Actors seek to fulfil the obligations encapsulated in a role, an identity, a membership in a political community or group, and the ethos, practices and expectations of its institutions" [31]. The electricity market as a social institution can thus be understood as a set of formal and informal rules, which shape the behaviour of actors. The course of action, identities and norms and as a result, the corresponding social order is thus established. Social order is defined here as all formal and informal rules that lead to the actual market result (not as the contingency of possible market outcomes). The market result includes all decisions regarding investments, pricing, contracts. These decisions are synonymic to what is called the behaviour of market actors later on. Market participants are defined as all actors that could potentially make a technically defined contribution to the security of supply. As the EU electricity markets are liberalized, these actors are mostly private companies or cooperative bodies. The market result of the electricity market can thus be understood as the total of all investment decisions in the market and their distributional consequences determined by the social order of the market. This view reflects the common interpretation of the market as a coordinating body between different participants in order to exchange services and goods. However, decisions on these exchange processes are embedded into a social, institutional-historical and cultural context [32]. The market as an institution therefore (re)produces a certain social order, without which the coordination of complex economic activities in the electricity sector would not be possible [26]. The market result (which is relevant here as a certain level of security of supply), is thus conceived as the dependent variable reliant on the social order and underlying rules. The set of rules that govern the market and coordinate the behaviour of actors in a certain way can be both formal and informal [33,34]. While prices may be considered as a central means of communication within the electricity market [34], market actors also inform themselves about the market context in other ways, such as via modelling and calculation of future developments. Formal rules in the current market are the regulatory framework of the remuneration scheme, such as in the assignment of property rights and definitions of products and responsibilities [33]. Informal rules include implicit factors that affect the market actor's behaviour such as the actor's assumptions on the level and volatility of prices and quantities, the behaviour of other market participants and the knowledge on the characteristics of the good and perception of market actors' own interests [32]. The electricity market is here limited to all factors that determine investments in security of supply options from a market participant's point of view. Other economic and policy areas that are affected or closely linked to the electricity market are either left out when they are not part of the technical or regulatory configuration of the electricity market itself, such as the financial crises or they are included into the broad categories of "expectations about the future" or "focusing events".

3. Factors that Influence the Market Result

In the following, we apply preliminary conceptual remarks on the market as a social institution to the case of security of supply in European electricity markets [35]. We define the behaviour of market actors as the dependent variable. If investments in flexible capacities are high enough, the threat of insufficient security of supply in the future is diminished. Capacities for security of supply in the electricity sector are defined here as the total of all investments in flexible options that are available to the respective electricity market at a certain point in time. Flexible options can be flexible generation capacities such as gas fired power plants, storage opportunities as well as demand side flexibility. We identify five determinants that may have an impact on the investment activity of a market actor: The current market design, the material opportunities in the national system and in coupled systems, the market participant's expectations about future developments, the strategic interest and identity of a market participant and focusing events that may pose a shock to the market, such as the Fukushima disaster. These variables are shown in Figure 2. We identified these five variables by literature review, by working together with practitioners in previous projects such as BETTER (http://www.better-project.net/), CLIM-RUN (http://www.climrun.eu/) and within RGI (http://www.renewables-grid.eu/) and by applying the theoretical concept of sociological institutionalism to the case of the electricity market. The market design variable is the most accepted one, which is discussed in most scientific and political analyses [1,3–6]. The material opportunities variable was identified by the work with practitioners within the above mentioned projects, as practitioners generally do not focus on theoretical cases, but on real existing ones and take real world settings into account. The market participant's expectations about future developments and the actor's strategic interest and identity variables are derived from the theoretical concept of sociological institutionalism that understands any type of actor as embedded into a social order where non-formalized factors can be equally important as formal ones. The focusing events factor is taken from the study of empirical cases in political science and according to the literature [36–40].

3.1. Electricity Market Design

The electricity market design is an explicitly created regulation consisting of legally defined rules for allocation, remuneration and the assignment of individual rights and obligations [41]. In theory, a change in the formal regulatory system can initiate a behavioural change in the market if specific cost-benefit ratios are restructured. This corresponds to calls for the introduction of capacity mechanisms as a means to secure future supply [1,5,42–44]. While the currently discussed proposals for new market designs within EU Member State countries differ from each other, they all intend to induce a change in the market actor's investment behaviour, either by introducing an administratively fixed capacity or an administered price [45]. Previous practical experience has shown that market design reforms did not always generate the intended effects or were created to achieve further political objectives such as saving jobs or keeping an industry alive. The consideration of context-specific factors as well as the design of the capacity mechanism does heavily influence the success or failure of the instrument [4]. Indeed, the global experience with capacity markets is very different; in some cases capacity mechanisms were even removed again because they didn't significantly improve security of supply compared to a situation without a capacity mechanism (e.g., New Zealand and Australia) [46]. Policy makers should therefore be aware of the complex design details as well as other factors that may have an impact on the behaviour of market actors before a capacity market is introduced.

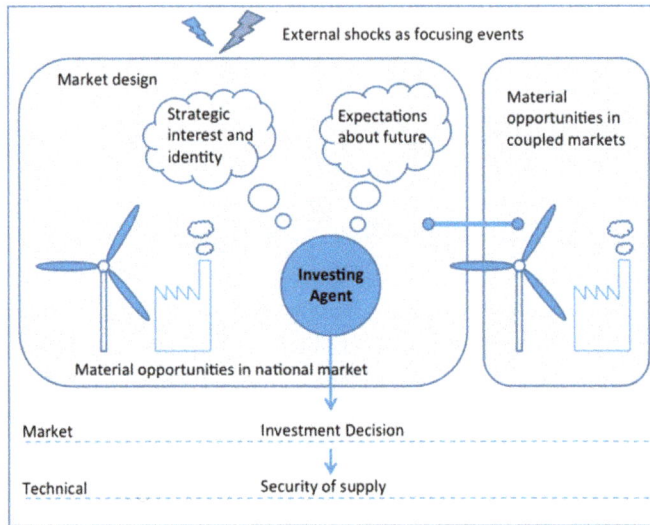

Figure 2. Factors that can influence investment decisions in the electricity market.

3.2. Characteristics of the Material Opportunities within the System and in Coupled Systems

In addition to the question of the formal market design, the material opportunities of the electricity system have a significant impact on the behaviour of market participants. In the electricity market, material opportunities are to be understood as the given technical equipment and capacities of the power system and the impact of these characteristics on the quantities supplied and the prices at any given time. Electricity markets are characterized by very long investment cycles, both in terms of cost calculations for new investments as well as the break-even point of energy infrastructure investments [47]. Conventional electricity generation capacities are also characterized by high fixed costs and high dismantling costs. Unused power plants cannot be easily used for other purposes or shipped elsewhere and are therefore typically characterized by high opportunity costs of a low utilization rate. In particular, inflexible base load power plants have the potential to hinder the market entry of flexible generation capacities after their depreciation, as they solely have to return the variable costs and are only able to lower their generation load to a certain level [41]. The existence of negative prices and rising export shares of the electricity produced, e.g., in Germany, are a result of these material opportunities in the system [48]. Power plants, which operate with higher-priced fuels such as gas power plants compared to lignite coal power plants, are therefore urged out of the market when over capacities of base load power plants exist. If market actors base their investment decision on the current market situation at this stage of the investment cycle, they will perceive possible investments in additional capacity as unprofitable.

The beginning of the investment cycle is characterized by the construction of new generating capacities. However, in the case of rising proportions of fluctuating renewable energy, there also exists a high demand for flexible options to meet the growing amount of variable residual load which could lead to high electricity prices in times of low wind and sun generation output [49]. This differs from those situations when numerous depreciated must-run units are still in place hindering electricity prices to send out these scarcity signals. The behaviour of market participants could therefore also change as a result of a modification in material opportunities. A large number of base load power plants going offline—for example by reaching the maximum lifetime or due to restrictive emission regulations—would massively increase the demand for flexibility options. This is especially the case in a system with a high proportion of fluctuating renewable energy as a pilot study commissioned by the Federal Ministry for the Environment 2012 calculated for Germany [49]. Thus, both the peak load electricity price and the volume of the residual load would significantly increase without the need to change the market design in case of discontinuation of base-load power plants [50]. Depending on the phase of the investment cycle and on how the material opportunities are set, market participant behaviour changes in relation to volume and price of peak (residual) load.

Mechanisms and characteristics of coupled systems also influence results in the electricity market [1,3,51]. For market participants involved in long-term investments, not only does the market situation in the respective market matter, but also the characteristics of physically coupled electricity systems that are thus able to provide electricity in shortage situations. The European Commission is making strong efforts to increase and facilitate cross-border electricity flows in Europe—And beyond—in its push for the implementation of the European internal electricity market [52]. For example, if sufficiently flexible (pumped storage) capacities in Norway, Switzerland and other electricity markets are available in coupled markets, or if this is at least expected by market actors, it could lead to decreased investments in flexible options in a regime of low electricity prices. Thus, the effects of coupled systems on security of electricity supply depend on a variety of factors such as the quantity and price for available flexible power in coupled systems, future interconnector capacities and market and grid regulations. Therefore, reluctance among national market participants to invest in flexibility options can also be explained by a perceived future possibility of coupled systems to satisfy demand in cases of domestic scarcity, thus lowering the economic incentive for those investments.

3.3. Strategic Actor Behaviour and Actor Identity

Conflicts of interest and information asymmetry between relevant actors can also impact the current market result [53]. In the context of the European electricity market, characterized by high technical complexity, economic uncertainty and sometimes a lack of transparency in decision-making, market actors may find a space to act strategically.

Despite liberalisation and unbundling strategies implemented in the past, the European electricity sector still exhibits oligopolistic features and is still largely dominated by operators of inflexible large power stations from the traditional energy industry. Their business activity is challenged by the politically desired transition to a flexible and mainly carbon-free generation park. Large market players in particular, who would be most able to invest in capital-intensive, more flexible capacities, could therefore have an interest in taking advantage of their influential position, which results from the possession of large capacities and further resources [54]. Market relevant information can be distortedly or selectively made public by major players in the market as to facilitate the pursuit of one's own interests in energy policy-making.

In view of the debate on the introduction of the so-called "strategic reserve" in Germany for example, the European Commission's Directorate-General for Energy communicated that there exists a "risk [in that] companies deliberately over exaggerate their intentions and taking capacities offline in order to make more profit" [55].

An undesired market performance can therefore also be interpreted as the result of a strategic reluctance in investing in flexible generation capacities in order to put pressure on policy-makers to shape the future market design in one's own interest.

These perceived "one's own" interests and responsibilities can differ widely depending on the actor's identity and network. Big oligopolistic energy producers with a long history in a specific field may have other priorities and targets as small independent power producers feeding into the grid and participating in the spot market. Publicly owned companies may follow other objectives than profitability such as those with industrial, environmental and social aspects. But even profitability targets can differ from each other depending on the shareholders' base and interests. For example, pension funds are seeking long-term stable returns on investments while others may focus more on high dividends in the short-term. Also, the expected and perceived appropriate rate of return can differ between actors depending on the discursive construction of a "successful" company strategy and thus, the actor's identity. Perceived responsibility for the functioning of the electricity system and the position within the economic system and society are important aspects for defining investment strategies.

3.4. Focusing Events

Investment decisions are also influenced by unexpected events, which make the concrete issue salient and concentrate the attention of the public, policy-makers and market actors. These often disastrous events called focusing events, are described as sudden and attention grabbing, lifting issues on the political agenda and therefore as triggers to induce policy change [36], or at least to open up "windows of opportunity" [37]. Moreover, the occurrence of a specific focusing event may also have an impact on the strategic actor's behavior and its success in co-shaping political decisions [38] (p. 40).

One focusing event for the energy sector was the Fukushima reactor meltdown in 2011 [39]. As empirical findings suggest, an affected policy is only likely to change when non-institutional interest groups or political parties are actively advocating this change [36,39]. This was certainly the case in Japan, where a big protest movement sprouted and all reactors were shut down.

Characteristically, focusing events generate ripple effects well beyond the immediate scope of the disaster itself [40]. Next to Italy, Belgium and Switzerland these effects were notably visible in Germany [39]. A grand coalition of all parties represented in the German national parliament decided to finally phase out all nuclear power plants by 2022, after it just had prolonged reactor run-time licenses one year before, arguing that nuclear energy is still needed for the energy transition as a "bridging technology". We do not want to address the reasons here (see for a discussion e.g., [39,56]); it is however apparent how focusing events can influence the formal market design on the one hand, but also societal debates and discursive aspects of the investment decisions of market actors on the other hand. The utility company E.ON even announced the intention to divest all its nuclear and fossil fuel assets [57]. It can be assumed that this strategy shift is partly due to the formal change in nuclear policy but the decision also hints to more informal aspects such as the company's expectations about the future. In the following section, we argue that discursive narratives are playing a role in decision-making. In this sense, the focusing event of Fukushima would have weakened the narrative of a nuclear future in Germany to a point where E.ON decided to (officially) not follow up on it any further.

It can be argued that another example of such a focusing event is the ongoing armed conflict in Eastern Ukraine. In the past, in 2006 and 2009, the interruptions of gas exports from Russia to Ukraine raised the question of the security of energy source supply to the EU [58,59]. These events were perceived by some actors as a sign of the EU's vulnerability. The political strife between Ukraine and Russia started in late 2013 and developed into military conflict in 2014. Once again, some market actors raised warning questions about the security of supply. Thus, these kind of events directly have an impact on investment decisions as it changes risk assessments and may effect profitability margins and even the feasibility of specific economic actions. Politically, the former Prime Minister of Poland, Donald Tusk, and now President of the European Council, introduced the idea of the European Energy Union as a reaction to these focusing events in 2014 (see Section 4 of this paper).

3.5. Uncertainties and Expectations

The expectations and actual investment behaviour of market participants are based on calculations and assumptions regarding future policies, demand, supply quantities, the characteristics of other market participants, resource availability and price developments. These assumptions pertain to timeframes spanning decades and are characterized by a high degree of uncertainty.

At the same time, market participants must constantly monitor the impact of current events, such as for example, relative price changes in global markets. Whatever actor-specific consequences and changes in expectations result from this, it is always dependent on the actor's interpretation and perspective. Due to limited information processing capacities of investment decision makers towards, for example policy change, future threats and consumer preferences, they have to rely on discursive narratives (discursive narratives constitute commonly shared patterns of perception and structures that can lead to common experience and reduce complexity when coming to a common understanding of and processing of new information. Discursive narratives give context to pure information to create an understanding of and to an issue and thus produce knowledge. Contextualized information can be interpreted and empowers actors to make decisions [60].) in order to reduce their individual uncertainty about future developments. Inter-subjective and commonly shared stable expectations regarding for example, the learning curves of technologies are crucial determinants of investment decisions. Authoritative actors (authoritative actors are those that are able to explicitly and implicitly influence the behaviour of other actors without coercion or pressure as they are attributed with particular expert knowledge. The knowledge of and decision recommendations from these authoritative actors are considered by their addressees to carry a high level of legitimacy, thus acknowledging the experts' authority (similar to Max Weber's concept of power) [61,62])) contribute to discursive narratives and reproduce their narratives within institutionalized stakeholder networks. This typically results in stakeholder group-specific expectations regarding developments in the future [63].

This can be seen in the varying perceptions of stakeholders regarding the consequences resulting from the nuclear energy phase-out in Germany. While predominantly representatives of local producers were concerned about the decision to extend the lifetime of nuclear power plants in 2010, major power supply companies (PSCs) welcomed this [64]. These differing perceptions also became apparent during the nuclear energy phase-out debates in 2011; municipal utility companies announced an investment offensive in efficient and flexible generating capacity while other big utilities started to sue the German government over the nuclear phase-out [65].

Based on the Social Construction of Technology (SCOT) approach [66,67], the technology assessment of a certain stakeholder group goes hand-in-hand with how the group is economically, ideologically and organizationally connected to that technology. Apart from economic interests, worldviews and networks, social and ideological factors also play a role such as discursive narratives about policies, prices, appropriate behavior and causal relationships between different events. The possible impact these differing expectations may have on model assumptions in energy and electricity market models (for example on the price of electricity or the electricity mix) was made clear in a study conducted by the German Institute for Economic Research (DIW). In particular, the overestimation of photovoltaic costs as well as the assumptions regarding carbon capture and nuclear power technology within the Energy Roadmap 2050 and the European Commission's Green Papers has led to distorted results in scenario developments and potential disincentives [68]. Conversely, a discursive change in contextual assumptions can also change profitability assessment results regarding investments in flexible capacities. This is especially valid for markets that are currently undergoing big systematic changes due to political targets and policies that make business-as-usual model assumptions unreasonable.

4. The European Energy Union as a Means to Stabilize Expectations

In early 2015, the new Commissioner for Energy Miguel Arias Cañete and the Vice-President of the Commission Maroš Šefčovič announced a framework strategy for the implementation of a European Energy Union [69]. It aims at merging the already existing 2030 Framework for Climate and Energy as well as the European Energy Security Strategy with a set of new measures into one coherent strategy. Central points include the further integration of the internal energy markets and more coordination between Member States. Already in the rationale behind the communication, two visions are expressed that are crucial to our context. The Commission illustrates "an integrated continent-wide energy system where energy flows freely across borders, based on competition and the best possible use of resources, and with effective regulation of energy markets at EU level where necessary" [69] (p. 2). Furthermore, investors should get confidence "through price signals that reflect long term needs and policy objectives" [69] (p. 2). Both points are fully in line with provisions foreseen in the third energy package and which have not been yet fully delivered, thus bringing possibly new energy into the implementation of already agreed policies.

Although the European Union only has limited competencies to shape the formal determinants of electricity market participant behaviour (as for example changing the market designs and introducing a European capacity market), it can play a crucial role in influencing the informal determinants of market behaviour. For all four factors identified above, the introduction of the Energy Union would have the potential to contribute to changes in investment behaviour and thus, might help trigger the desired investments.

Firstly, within a European Energy Union material opportunities, as the physical characteristics and capacities of the electricity system, could be better planned and publicly presented on a European level so that market participants are able to assess possible changes in the physical configuration of the market. This could not only deliver more transparency but also better planning mechanisms because national vested interests will need to be discussed in a broader context.

Secondly, strategic behaviour to advocate capacity market mechanisms for national (energy) industry objectives is less likely in a European approach due to a lower market share of actors in a European system. Furthermore, a European Energy Union could challenge the previously national oriented actor's strategies and might create a more European frame of acting and thus help to develop a more European energy identity of market actors in the long-term.

Thirdly, although it is not clear what effects focusing events would have, by having a European Energy Union in place, it could be expected that the effects of focusing events are no longer solely perceived and discussed on the national level, but on the European level as well, perhaps contributing to the creation of a more European energy identity through a more European debate. Furthermore, the focusing event of political instability and the armed conflict in the Ukraine is seen as a pivotal motivation for the proposal of the Energy Union in the first place and thus is perceived to address concerns which came up by the event itself.

Fourthly, contextual assumptions could be stabilized in a way that is politically agreed on due to long-term credible targets and accompanying measures. Depending on the specific design of the Energy Union, this undertaking has the potential to help overcome informal barriers in investment calculations. To actually reach this target, the Energy Union would have to fulfil at least two conditions:

a) The agreed targets and measures have to be concrete, transparent and credible. The governance on how to achieve them should be in place to decrease risks and thus, lower the risk premium. Due to renewable energy integration and economic and physical integration, investment decisions toward long-term participation will strongly benefit from clarity and transparency in the fast changing electricity market. Furthermore, targets and policies have to be agreed upon with a broad political consensus to also guarantee stability in different subsequent political constellations, as in the case of the German and Swiss nuclear phase-out after the Fukushima events.

b) A strong alignment between national policies and European targets needs to be developed to deliver confidence in the implementation and the long-term commitment of Member States. This alignment would avoid the multi-level governance trap [70] common to several areas in European policies. Further clarity could be added by setting not only fixed targets for RE generation but also for residual load.

In a nutshell, the European Energy Union could address the above mentioned informal barriers to security of supply investments in several ways, but most importantly it could trigger public debates on a European level and thus help to develop a grand narrative regarding the future common market, challenges and socio-technical solutions, which currently mainly exist at the national level.

5. Discussion and Conclusions

The perceived need to stimulate investment in flexible options and the surrounding public debates have translated into a call for capacity mechanisms that has even led to the introduction of capacity mechanisms in some countries. The underlying fear is that under the current market conditions, too few investments in storage, flexible power plants and transmission capacity expansions are being made. Some market actors consider the low prices and subsequent lack of incentives to invest in security of supply options as a failure of the energy-only market. In this paper, we have criticised the strong and exclusive link between the formal market design and the present market results and identified other factors of market actor behaviour. Thus, additional determinants apart from the market design can influence the behaviour of market actors.

Firstly, the material opportunities in the electricity system play a crucial role in the investment decisions of market actors. A low market price can thus be rather understood as the result of investment cycles when phases of over-capacity and insufficient capacity alternate. Incentives for investments in flexible capacity can therefore be present within the existing market design under modified material conditions. For example, less base load power plants in the case of high proportions of fluctuating renewable energy would result in higher prices and amounts of residual (peak) load and might made flexible options profitable on their own. Furthermore, market actors are taking the developments and determinants of coupled systems into account as well as the evolution of technologies when making their investment decisions. A lack in domestic investment activity can therefore be seen as an expected and sufficient foreign provision of security of supply options. Under these conditions, no domestic capacity mechanisms would be necessary.

Secondly, when it comes to capital-intensive investments in the energy sector, there is a possibility that dominant market actors, which benefit from the status quo, may have an interest in delaying investments in flexible options. A clear decision regarding the introduction—Or dismissal of a capacity mechanism could bring an end to the possibility of strategically delaying investments as a means of political pressure. Furthermore, the identity and perceived role in the market and society as a whole play an important part in defining the actor's "own interest" and thus, the investment strategy. Depending on the timeframe, the objectives of the company's business and the ownership structure, the definition of a "successful" investment strategy and subsequently, the perceived personal interest of a company differs widely between market actors.

Thirdly, focusing events have a tremendous potential to the discursive context around specific market decisions. Such events can change the political climate and development priorities by concentrating the attention of the broad public and many stakeholders, which as a result can influence the market actor's long-term decisions and investments.

Last but not least, it is possible that market actors revert back to narratives when they are highly uncertain about future developments. Following certain specific contextual assumptions, they might expect that investments in flexible options are not profitable. Conversely, this means that a change in market actor-specific contextual assumptions could stimulate investment. Policy-makers might achieve this change in perception by agreeing on a concrete energy roadmap along with credible targets regarding residual load and capacity development in base load power plants.

The Energy Union can play a crucial role in building up this credible narrative—Completely lacking until now (besides RE and climate targets)—On the energy transition and the (market) instruments that will support it. This narrative would have to be based on answers regarding the discontinuity of technologies and provide an unambiguous pathway that addresses all aspects of, for example, a possible fossil fuel phase-out. Accompanying a political and regulatory framework, it would have to address the conflict between a marginal cost-based electricity market and an electricity system with increasing proportions of RE electricity with almost no marginal costs. To be credible, this grand narrative would have to be in line with communications and decisions on other levels (national and regional) and other policy areas that are affected as, e.g., job market, technology and innovation policies or fiscal policies. This might be facilitated when a great proportion of stakeholders and policymakers from different levels and domains are informed, able to participate, and agree to the process and accept its result.

But also intra-European learning on the formal side could constitute an outcome of a European Energy Union. As the discussions on capacity mechanisms have evolved over time and moved from being dedicated to base-load energy generation options to more flexible options, new regulation proposals could be observed and evaluated by other Member States and a common learning process could be facilitated.

However, a systematic and comprehensive empirical study on the social and discursive determinants and other informal factors of market actor behaviour in the electricity market is still pending. Such an undertaking could provide detailed insight into expectation formation, knowledge production and thus, political actions regarding investments in security of supply. Having discussed the points above, it can be said however, that evidence of an urgent need of capacity mechanisms is equivocal. This of course in no way allows for the reversed conclusion: The formal market design with its set of legal remuneration and allocation schemes is of course an important determinant of investment behaviour; but it is not the only one. Thus, policy-makers can choose between different measures addressing formal or informal determinants to stipulate specific behaviour in the electricity market. We have substantiated that each of the five determinants discussed here may have an influence on investment decisions. Given the high levels of complexity and uncertainty, policy-makers must carefully consider the pros and cons in order to prevent the risk of generating unwanted disincentives. What specific instruments are available, at what cost, and the underlying political acceptance of these measures cannot be answered here. However, in light of the existence of several determinants of market actor investment behaviour, we do not see a compelling reason to focus solely on the implementation of capacity mechanisms to secure the desired investment behaviour.

Acknowledgments

The authors would like thank Ioana Bejan and Christian Reutter for providing valuable comments to the paper and supporting the research process in general.

Author Contributions

Saskia Ellenbeck as led the work and developed the concept and the theoretical framework. All other authors contributed to the entire paper, especially with data collection with practitioners and experts (Antonella Battaglini), with literature reviews on investment decisions in the electricity market (Andreas Beneking) and focusing events (Andrzej Ceglarz, Andreas Beneking) and market design and capacity markets (Peter Schmidt).

Conflicts of Interest

The authors declare no conflict of interest.

References

1. Energiewende, A. *Kapazitätsmarkt Oder Strategische Reserve: Was Ist der Nächste Schritt? Eine Übersicht Über Die in der Diskussion Befindlichen Modelle zur Gewährleistung der Versorgungssicherheit in Deutschland*; Oktoberdruck: Berlin, Germany, 2013. (In German)
2. European Commission. *State Aid SA.35980 (2014/N-2)—United Kingdom Electricity Market Reform—Capacity Market*; European Commission: Brussels, Belgium, 2014.
3. Tietjen, O. *Kapazitätsmärkte: Hintergründe und Varianten mit Fokus auf einen Emissionsarmen Deutschen Strommarkt*; Germanwatch: Bonn, Germany, 2012. (In German)
4. Cramton, P.; Ockenfels, A. Economics and design of capacity markets for the powersector. *Z. Energiewirtschaft* **2012**, *36*, 113–134.
5. Matthes, F.C.; Schlemmermeier, B.; Diermann, C.; Hermann, H.; von Hammerstein, C. *Fokussierte Kapazitätsmärkte. Ein Neues Marktdesign für den Übergang zu Einem Neuen Energiesystem*; Öko-Institut e.V.: Berlin, Germany, 2012. (In German)
6. Hach, D.; Spinler, S. Capacity payment impact on gas-fired generation investments under rising renewable feed-in—A real options analysis. *Energy Econ.* **2014**, doi:10.1016/j.eneco.2014.04.022.
7. Nicolosi, M.; Fürsch, M. The impact of an increasing share of RES-E on the conventional power market—The example of Gemany. *ZfE* **2009**, *3*, 246–254.

8. Sensfuß, F. *Analysen Zum Merit-Order Effekt Erneuerbarer Energien—Update für das Jahr 2010*; Fraunhofer Institut für Syste m und Innovationsforschung: Karlsruhe, Germany, 2011. (In German)

9. Haucap, J. *Braucht Deutschland Einen Kapazitätsmarkt für Eine Sichere Stromversorgung?* Düsseldorf University Press: Düsseldorf, Germany, 2013; Volume 51. Available online: http://www.uni-duesseldorf.de/home/fileadmin/redaktion/DUP/PDF-Dateien_/DICE/Ordnungspolitische_Perspektiven/051_OP_Haucap.pdf (accessed on 16 April 2015). (In German)

10. *Renewable Energy and Security of Supply: Finding Market Solutions*; Eurelectric Report; Eurelectric: Brussels, Belgium, 2014; p.19.

11. European Commission Energy Legislation. Available online: http://ec.europa.eu/competition/sectors/energy/overview_en.html (accessed on 21 May 2015).

12. Rothwell, G.; Gómez, T. *Electricity Economics: Regulation and Deregulation*, 1st ed.; John Wiley: Piscataway, NJ, USA, 2003.

13. Henriques, I.; Sadorsky, P. The effect of oil price volatility on strategic investment. *Energy Econ.* **2011**, *33*, 79–87.

14. Yang, M.; Blyth, W.; Bradley, R.; Bunn, D.; Clarke, C.; Wilson, T. Evaluating the power investment options with uncertainty in climate policy. *Energy Econ.* **2008**, *30*, 1933–1950.

15. Ben Ammar, S.; Eling, M. Common risk factors of infrastructure investments. *Energy Econ.* **2015**, *49*, 257–273.

16. Joskow, P.L. *Competitive Electricity Markets and Investment in New Generating Capacity*; AEI-Brookings Joint Center Working Paper No. 06–14; MIT Center for Energy and Environmental Policy Research (CEEPR): Cambridge, MA, USA, 2006.

17. Herrero, I.; Rodilla, P.; Batlle, C. Electricity market-clearing prices and investment incentives: The role of pricing rules. *Energy Econ.* **2015**, *47*, 42–51.

18. Hirshleifer, J. On the theory of optimal investment decision. *J. Polit. Econ.* **1958**, *66*, 329–352.

19. Boiteux, M. Peak-load pricing. *J. Bus.* **1960**, *33*, 157–179.

20. Gross, G.; Blyth, W.; Heptonstall, P. Risks, revenues and investment in electricity generation: Why policy needs to look beyond costs. *Energy Econ.* **2010**, *32*, 796–804.

21. Kuit, M.; Mayer, I.S.; de Jong, M. The INFRASTRATEGO game: An evaluation of strategic behavior and regulatory regimes in a liberalizing electricity market. *Simul. Gaming* **2005**, *36*, 58–74.

22. Masini, A.; Menichetti, E. The impact of behavioural factors in the renewable energy investment decision making process: Conceptual framework and empirical findings. *Energy Policy* **2009**, *40*, 28–38.

23. Nguyen, T.; Schüßler, A. How to make better decisions? Lessons learned from behavioral corporate finance. *Int. Bus. Res.* **2012**, *6*, 187–198.

24. Cooremans, C. Investment in energy efficiency: Do the characteristics of investments matter? *Energy Effic.* **2012**, *5*, 497–518.

25. What Are the Factors Influencing Energy Behaviours and Decision-Making in the Non-Domestic Sector? Available online: http://www.cse.org.uk/downloads/reports-and-publications/behaviour-change/factors_influencing_energy_behaviours_in_non-dom_sector.pdf (accessed on 21 May 2015).

26. Beckert, J. The social order of markets. *Theory Soc.* **2009**, *38*, 245–269.

27. Smelser, N.J.; Swedberg, R. *The Handbook of Economic Sociology*; Princeton University Press: Princeton, NJ, USA, 2010.

28. North, D.C. *Institutionen, Institutioneller Wandel und Wirtschaftsleistung*; Mohr Siebeck: Tübingen, Germany, 1992. (In German)

29. Risse, T. Konstruktivismus, rationalismus und theorien internationaler beziehungen—Warum empirisch nichts so heiß gegessen wird, wie es theoretisch gekocht wurde. In *Die neuen Internationalen Beziehungen*; Hellmann, G., Wolf, K.D., Zürn, M., Eds.; Nomos: Baden-Baden, Germany, 2003; Volume 10, pp. 99–132. (In German)

30. Berger, P.L.; Luckmann T. *Die gesellschaftliche Konstruktion der Wirklichkeit*; Fischer Verlag: Frankfurt am Main, Germany, 2003. (In German)

31. March, J.G.; Olsen, J.P. *The Logic of Appropriateness, Arena—Centre for European Studies*; University of Oslo: Oslo, Norway, 2004.

32. Granovetter, M. Economic action and social structure: The problem of embeddedness. *Am. J. Sociol.* **1985**, *3*, 481–510.

33. Fligstein, N. Markets as politics: A political-cultural approach to market institutions. *Am. Sociol. Rev.* **1996**, *61*, 656–673.

34. Storr, V. The social construction of the market. *Society* **2010**, *47*, 200–206.

35. Ellenbeck, S.; Schmidt, P.; Battaglini, A.; Lilliestam, J. Der Strommarkt als soziale Institution - Eine erweiterte Perspektive auf die deutsche Diskussion um Kapazitätsmechanismen. *DIW Vierteljahresheft* **2013**, *3*, 171–182. (In German)

36. Birkland, T.A. Focusing events, mobilization, and agenda setting. *J. Public Policy* **1998**, *18*, 53–74.

37. Kingdon, J. *Agenda, Alternatives, and Public Policies*; Little Brown: Boston, MA, USA, 1984.

38. Mahoney, C. Lobbying success in the United States and the European Union. *J. Public Policy* **2007**, *27*, 35–56.

39. Giger, N.; Kluever, H. Focusing events and policy change: The aftermath of Fukushima. In Proceedings of the European Political Science Association Conference, Berlin, Germany, 21–23 June 2012.

40. Leiserowitz, A. Editorial—Focusing events. *Environment* **2011**, *53*, 2–15.

41. Diermann, C. *Marktdesign für einen Kapazitätsmarkt Strom—Aktualisierte Zusammenfassung München*; Umweltministerium des Landes Baden-Württemberg, L. Beratungsgesellschaft: Berlin, Germany, 2012. (In German)

42. Elberg, C.; Growitsch, C.; Höffler, F.; Richter, J. *Untersuchungen zu Einem Zukunftsfähigen Strommarktdesign*; EWI: Cologne, Germany, 2012. (In German)

43. Leprich, U.; Hauser, E.; Grashof, K.; Grote, L.; Luxenburger, M.; Sabatier, M.; Zipp, A. *Kompassstudie Marktdesign—Leitideen für ein Design eines Stromsystems mit hohem Anteil fluktuierender Erneuerbarer Energien*; Ponte Press: Bochum, Germany, 2013. (In German)

44. Nicolosi, M. *Necessity of and Design Options for a Capacity Mechanism for Germany Ecofys*; Germany GmbH: Berlin, Germany, 2012.

45. The Agency for the Cooperation of Energy Regulators (ACER). *Capacity Remuneration Mechanims of the Internal Market for Electricity*; The Agency for the Cooperation of Energy Regulators: Ljubljana, Slovenia, 2013.

46. Winkler, J.; Sensfuß, F.; Keles, D.; Renz, L.; Fichtner, W. Perspektiven zur aktuellen Kapazitätsmarktdiskussion in Deutschland. In *Perspektiven für die langfristige Entwicklung der Strommärkte und der Förderung Erneuerbarer Energien bei ambitionierten Ausbauzielen*; Springer Fachmedien Wiesbaden: Wiesbaden, Germany, 2013. (In German)

47. Gaidosch, L. Zyklen bei Kraftwerksinvestitionen in Liberalisierten Märkten—Ein Modell des Deutschen Stromerzeugungsmarktes. Ph.D. Thesis, Technische Universität Berlin, Berlin, Germany, 2007. (In German)

48. Burger, B. *Stromerzeugung aus Solar- und Windenergie im Jahr 2013*; Frauenhofer Institut fürSolare Energiesysteme (ISE): Freiburg, Germany, 2013. (In German)

49. Nitsch, J.; Pregger, T.; Naegler, T.; Heide, D.; Scholz, Y.; de Tena, L.D.; Trieb, F.; Nienhaus, K.; Gerhardt, N.; Sterner, M.; *et al. Langfristszenarien und Strategien für den Ausbau der Erneuerbaren Energien in Deutschland bei Berücksichtigung der Entwicklung in Europa und global*; Final Report; Deutsches Zentrum für Luft- und Raumfahrt (DLR): Stuttgart, Germany; Institut für Technische Thermodynamik, K. Fraunhofer Institut für Windenergie und Energiesystemtechnik (IWES): Kassel, Germany; Ingenieurbüro für neue Energien (IFNE): Teltow, Germany, 2012. (In German)

50. Siegmeier, J.; von Hirschhausen, C. *Energiewende: Brauchen wir noch "Kapazitätsmärkte" für konventionelle Kraftwerke? Die Zukunft des Strommarktes—Anregungen für den Weg zu 100 Prozent Erneuerbare Energien*; Schütz, D., Klusmann, B., Eds.; Ponte Press Verlags GmbH: Bochum, Germany, 2011; pp. 108–131. (In German)

51. Baker, P.; Gottstein M. *Advancing Both European Market Integration and Power Sector Decarbonisation: Key Issues to Consider*; Briefing Paper by the Regulatory Assistance Project; Regulatory Assistance Project (RAP): Brussels, Belgium, 2011.

52. European Commission. *European Commission's Communication: Making the Internal Energy Market Work*; European Commission: Brussels, Belgium, 2012.

53. Williamson, O.E. *Transaction Cost Economics: Handbook of Industrial Organization*; Schmalensee, R., Willig, R., Eds.; North-Holland: Amsterdam, The Netherlands, 1989.

54. Dürr, A.; de Bièvre, D. The question of interest group influence. *J. Public Policy* **2012**, *27*, 1–12.

55. Schultz, S. Unrentable Kraftwerke EU Warnt vor Erpressung Durch Stromkonzerne, 2013. Available online: http://www.spiegel.de/wirtschaft/unternehmen/eu-papier-warnt-vor-panikmache-bei-stromversorgung-a-911798.html (accessed on 16 April 2015). (In German)

56. Bach, V.A. *Deutsche Atompolitik im Wandel: Welchen Unterschied machen die Parteien?* Diplomica Verlag GmbH: Hamburg, Germany, 2012. (In German)

57. E.ON. New Corporate Strategy: E.ON to Focus on Renewables, Distribution Networks, and Customer Solutions and to Spin off the Majority of a New, Publicly Listed Company Specializing in Power Generation, Global Energy Trading, and Exploration and Production. Available online: http://www.eon.com/content/eon-com/en/media/news/press-releases/2014/11/30/new-corporate-strategy-eon-to-focus-on-renewables-distribution-networks-and-customer-solutions-and-to-spin-off-the-majority-of-a-new-publicly-listed-company-specializing-in-power-generation-global-energy-trading-and-exploration-and-production.html/ (accessed on 16 April 2015).

58. Jordan, A.; Rayner, T. The evolution of climate policy in the European Union: An historical overview. In *Climate Change Policy in the European Union: Confronting the Dilemmas of Mitigation and Adaptation?*; Jordan, A., Huitema, D., van Asselt, H., Rayner, T., Berkhout, F., Eds.; Cambridge University Press: Cambridge, UK, 2010; pp. 52–80.

59. Kohl, W.L. Consumer country energy cooperation: The international energy agency and the global energy order. In *Global Energy Governance: The New Rules of the Game*; Goldthau, A., Witte, J.M., Eds.; Global Public Policy Institute: Berlin, Germany, 2010; pp. 195–220.

60. Aspers, P. Knowledge and valuation in markets. *Theory Soc.* **2009**, *38*, 111–131.

61. Ecker-Ehrhardt, M. "But the UN said so...": International organisations as discursive authorities. *Glob. Soc.* **2012**, *26*, 451–471.

62. Weber, M. *Wirtschaft und Gesellschaft: Grundriss der Verstehenden Soziologie*; Mohr Siebeck: Tübingen, Germany, 1980. (In German)

63. Mattes, A. *Potentiale für Ökostrom in Deutschland—Verbraucherpräferenzen und Investitionsverhalten der Energieversorger*; DIW econ GmbH: Berlin, Germany, 2012. (In German)

64. AKW-Laufzeitverlängerung—Stadtwerke Drohen mit Investitionsstopps, 2010. Available online: http://www.manager-magazin.de/unternehmen/artikel/a-683147.html (accessed on 15 April 2015). (In German)

65. Schlandt, J. Stadtwerke Nutzen Atomausstieg, 2011. Available online: http://www.fr-online.de/energie/stromproduktion-stadtwerke-nutzen-atomausstieg,1473634,8422348.html (accessed on 16 April 2015). (In German)

66. Bijker, W.E. *Of Bicycles, Bakelites and Bulbs: Towards a Theory of Sociotechnical Change*; The MIT Press: Cambridge, MA, USA; The MIT Press: London, UK, 1995.

67. Geels, F.W. Ontologies, socio-technical transitions (to sustainability), and the multi-level perspective. *Res. Policy* **2010**, *39*, 495–510.

68. Von Hirschhausen, C.; Kemfert, C.; Kunz, F.; Mendelevitch, R. Europäische Stromerzeugung nach 2020: Beitrag erneuerbarer Energien nicht unterschätzen. *DIW Wochenber.* **2013**, *29*, 3–13. (In German)

69. European Commission. *A Framework Strategy for a Resilient Energy Union with a Forward-Looking Climate Change Policy*; COM/2015/080 Final; European Commission: Brussels, Belgium, 2015.

70. Scharpf, F.W. The joint-decision trap: Lessons from German federalism and European integration. *Public Adm.* **1988**, *6*, 239–278.

Chapter 3:
Public Policy Issues

Temporal and Spatial Variations in Provincial CO_2 Emissions in China from 2005 to 2015 and Assessment of a Reduction Plan

Xuankai Deng, Yanhua Yu and Yanfang Liu

Abstract: This study calculated the provincial carbon dioxide (CO_2) emissions in China, analyzed the temporal and spatial variations in emissions, and determined the emission intensity from 2005 to 2015. The total emissions control was forecasted in 2015, and the reduction pressure of the 30 provinces in China was assessed based on historical emissions and the 12th five-year (2011–2015) reduction plan. Results indicate that CO_2 emissions eventually increased and gradually decreased from east to west, whereas the emission intensity ultimately decreased and gradually increased from south to north. By the end of 2015, the total control of provincial emissions will increase significantly compared to the 2010 level, whereas the emission intensity will decrease. The provinces in the North, East, and South Coast regions will maintain the highest emission levels. The provinces in the Southwest and Northwest regions will experience a rapid growth rate of emissions. However, the national emission reduction target will nearly be achieved if all provinces can implement reduction targets as planned. Pressure indices show that the South Coast and Northwest regions are confronted with a greater reduction pressure of emission intensity. Finally, policy implications are provided for CO_2 reductions in China.

Reprinted from *Energies*. Cite as: Deng, X.; Yu, Y.; Liu, Y. Temporal and Spatial Variations in Provincial CO_2 Emissions in China from 2005 to 2015 and Assessment of a Reduction Plan. *Energies* **2015**, *8*, 4549-4571.

1. Introduction

The international community has reached a consensus that global warming poses a serious threat. Consequently, greenhouse gas reduction plans have been successively promulgated in the major developed and developing countries. China, as the largest developing country, announced its program to reduce future carbon dioxide (CO_2) emissions during the 15th International Climate Conference held in Copenhagen in 2009. Moreover, achieving a national reduction target of CO_2 emissions depends on how intra-national units implement their own reduction targets. Therefore, China announced its 12th five-year (2011–2015) plan to reduce emissions in 2012, which explicitly stipulated the reduction target for each provincial unit in mainland China. Both of these announcements emphasized the decrease of CO_2 emission intensity. The first target, which was announced in Copenhagen, stated that the emission intensity in 2020 should be reduced by 40%–45% compared with the 2005 level. The second target, which was announced in Beijing, stated that the emission intensity in 2015 should be reduced by 17% compared with the 2010 level. Moreover, a binding reduction target was assigned for all provincial units in the second target. Nevertheless, no clear total control target is set for the nation and individual provinces. The total control of CO_2 emissions should be calculated to have a more intuitive emissions reduction target.

Understanding the amount of total control of CO_2 emissions is necessary for each province during the reduction period. This is the first issue discussed in this paper.

Calculating CO_2 emissions, one of the main greenhouse gasses, has been an increasing concern around the world in recent years. Research shows that the United States, China, the European Union, Japan, and India are considered as major emitters of CO_2 [1]. In 2006, China overtook the United States as the world's leading emitter of CO_2 [2]. Historical emissions provided the foundation for burden sharing of global CO_2 emissions responsibility in the future. Cumulative emissions [3] or per capita cumulative emissions [4] are adopted as the standard for burden sharing. The major emitters need to reduce emissions while maintaining economic development. As a result, the future emissions of these major emitters have become a research hotspot [5,6]. To calculate the CO_2 emissions of each province in the future, the emissions from previous years must be determined first. Many studies have reported on China's CO_2 emissions because China is the top emitting country in the world [2]. Energy consumption especially from fossil fuel combustion is considered to be the main source of CO_2 emissions. Carbon emissions from energy consumption usually account for more than 90% of the total carbon emissions [7]. Previous studies have used several calculation methods for the regional CO_2 emissions. Many of them examined CO_2 emissions from energy consumption instead of total regional emissions. For instance, Zhao [8] used a bottom-up inventory framework to calculate total annual CO_2 emissions based on detailed provincial economic and energy data. Liu [9] used energy consumption data to calculate China's regional and sectoral greenhouse gas emissions, including CO_2 emissions. Wang [10] used energy consumption data to calculate provincial CO_2 emissions from 1995 to 2011. Many examples used emissions from energy consumption in place of regional emissions, both at the national [11] and intra-national levels [12]. However, the manufacturing process of some industrial products also generates significant CO_2 emissions. The main raw material of the cement industry is calcium carbonate. The process of cement production emits large amounts of CO_2, along with calcium carbonate decomposition and coal combustion. Thus, some scholars began to consider CO_2 emissions from regional cement production. For instance, Xu [13] analyzed the changes in energy consumption and CO_2 emissions in China's cement industry based on the typical production process for clinker manufacturing. Kim and Worrell [14] analyzed carbon emissions from China's cement industry from 1980 to 1998. Ke [15] estimated CO_2 emissions from China's cement production from 2005 to 2009. Forests and green lands have the ability to absorb and store CO_2; This ability is called carbon sequestration [16]. From a regional perspective, carbon sequestration should be considered calculating total regional CO_2 emissions [17,18]. For instance, Fang [19] used a biomass method to estimate forest carbon stocks in five East Asian countries between the 1970s and 2000s. Guo [20] explored the spatial-temporal changes in forest biomass carbon stocks in China between 1977 and 2008. Most previous studies on carbon emissions in China considered only one aspect of the carbon emission accounting and focused on the national level. The regional total CO_2 emissions can be accurately calculated by considering all aspects of regional emission accounting.

Allocation results at the national level were found in many previous studies that investigated reduction targets or burden-sharing of CO_2 emissions. For instance, Wei [21] allocated permits of carbon emissions to 137 countries and regions on the basis of per-capita cumulative emissions.

Chakravarty [22] presented a framework for allocating a global carbon reduction target among nations. Some studies began to allocate the national reduction target to the industrial [23] or sectoral levels. For instance, Chen [24] disaggregated China's national CO_2 mitigation burden at the sectoral level. Some scholars recently studied the reduction burden-sharing of the inner regions of China, and most of this research is based on the targets of Copenhagen [25,26]. Yi [27] developed a comprehensive index and constructed an intensity allocation model for inner China provinces on the basis of the Copenhagen reduction target. Few relevant studies have been conducted on the second China reduction target in the 12th five-year (FY) period because most previous studies have focused on the target in the national level [28,29]. The results of previous studies regarding China's intra-nation or provincial emission reduction targets often provide the future controlling intensity but lack total emission control. This type of result is not a straightforward target. Central and local governments operate with difficulty. China is currently developing based on its 12th Five-Year Plan (FYP), which includes the second reduction target of CO_2 emissions. Thus, the present work calculates a reduction plan, conducted from 2011 to 2015. A target to reduce emission intensity yields pressure. Research on this emission reduction pressure is extremely scarce, and providing results at the provincial level is difficult. Moreover, the provinces formulate development plans in detail, setting the speed of economic growth. The second point of interest to guarantee the specific speed is to identify the amount of pressure every province receives to control emissions. The final point to consider is whether the national reduction target can be achieved if each province completes its own reduction target.

This paper reports the intra-national provincial-level CO_2 emissions of China. First, a more accurate framework is used to calculate the provincial CO_2 emissions from 2005 to 2010. Then, the spatial-temporal variations of emissions and emission intensity are analyzed. Second, the total control of provincial and national CO_2 emissions in 2015 is calculated. Third, two new indices are designed to assess the pressure of reduction for each province. Conclusions and policy implications are provided in the last two sections.

2. Methodology

2.1. Calculation of Provincial CO_2 Emissions

The CO_2 emissions of a province are mainly derived from energy consumption and the process of cement production, as well as the process of removing CO_2 via sequestration by forests and green lands. The provincial carbon emissions calculating framework was used in this paper (Figure 1).

Equation (1) was used to calculate the provincial CO_2 emissions:

$$C = C_e + C_c - S_f \tag{1}$$

where C is the total CO_2 emissions of a province (million tons, M_t), C_e represents the CO_2 emissions from energy consumption (Mt), C_c represents the CO_2 emissions from the process of cement production (M_t), and S_f is the CO_2 sequestration by forests and green lands (M_t).

The guidelines of Intergovernmental Panel on Climate Change (IPCC) were adopted to calculate CO_2 emissions from energy consumption using the following equation [30,31]:

$$C_e = \sum_j \frac{E_j \times NCV \times EF_j \times O_j \times 44}{12} \tag{2}$$

where E_j is the amount of energy consumption j (tons or m^3 for natural gas), NCV is the net calorific value (kJ/(kcal×kg)), EF_j is the carbon emission factor of energy j (kg/GJ), and O_j is the carbon oxidation rate of energy j (default value is 1).

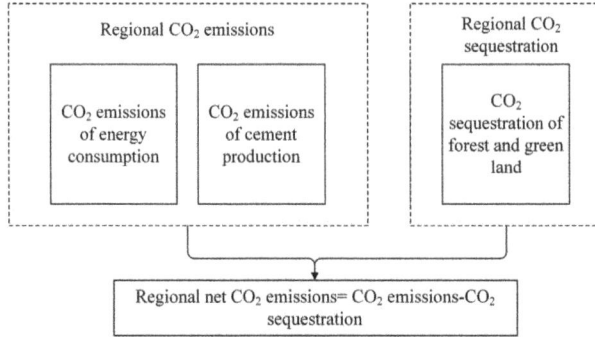

Figure 1. Regional net CO_2 emissions calculating framework.

IPCC guidelines and Zhai's article [32,33] to calculate CO_2 emissions from cement production using the following equation:

$$C_c = q \times r \times e \tag{3}$$

where q is the amount of cement (M_t), r is the proportion of clinker (%), and e is the CO_2 emission factor, which involves the three processes of decomposition of carbonate raw material, calcination of kiln dust, and decomposition of organic carbon in the calcined materials. The three processes have a total coefficient of 0.55 (tons per one ton).

Equation (4) was used to estimate the amount of CO_2 sequestration by forests and green lands [34]. Green lands refer to urban gardens and green spaces that are evergreen and have the absorption capacity of CO_2 such as forests:

$$S_f = (A_f + A_g) \times a_s \tag{4}$$

where A_f is the area of forests (hm^2), A_g is the area of green lands (hm^2), and a_s is the sequestration coefficient, for the sequestration amount per hectare per year. The carbon sequestration by Chinese forests has been estimated by many scientists. The sequestration amount per hectare per year is indirectly provided by these reports. Using the sequestration coefficient is inevitable in calculating the regional CO_2 absorption in a certain year by forests and green lands. A significant difference exists in the results of this sequestration coefficient (Table 1).

In the International Academic Symposium on Ecosystem Carbon Balance in 2005, Zhou precisely calculated, for the first time in the international academia, that each hectare of forests absorbed 0.5 tons of net carbon a year. That is, each hectare of forest a year can absorb 1.83 tons of CO_2. A total of 1.83 (tons per one hectare) was chosen as the calculation parameter of sequestration coefficient, which was also used in the article by Yu [40].

Table 1. Sequestration coefficient of forests in China.

Soures	Carbon Sequestration (t/hm²/year)	Carbon Dioxide Sequstration (t/hm²/year)
Ma and Wang [34]	0.61	2.24
Fang et al. [35]	0.53	1.93
Lai and Huang [36]	0.52	1.90
Zhou et al. [37]	0.50	1.83
Guo et al. [38]	0.39	1.42
Piao [39]	0.15	0.55

2.2. Forecasting Total Control of CO₂ Emissions

The emission intensity of CO_2 is usually defined as the unit of GDP emissions [29]. According to the definition of emission intensity, Equations (5–8) was used to forecast the future emissions of each province:

$$I_i = \frac{C_i}{GDP_i} \tag{5}$$

where I_i is the emission intensity (tons per 10,000 Chinese yuan, $t/10^4$ CNY) of province i, C_i is the CO_2 emissions (M_t) of province i, and GDP_i is the GDP of province i (CNY).

The total control of emissions in the target year is equal to the intensity in the target year multiplied by the GDP in the target year, which is expressed as follows:

$$C_i^t = I_i^t \times GDP_i^t \tag{6}$$

where C_i^t is the CO_2 emissions of province i in target year t, I_i^t is the emission intensity of province i in target year t, and GDP_i^t is the GDP of province i in target year t.

The emission intensity of the target year is calculated by the reduction plan, and the GDP of the target year is calculated by the economic growth setting as follows:

$$I_i^t = I_i^0 \times (1 - A_i) \tag{7}$$

$$GDP_i^t = GDP_i^0 \times (1 + G_i)^{t-0} \tag{8}$$

where 0 is the base year, t is the target year of forecasting, I_i^t is the emission intensity in the target year of province i, I_i^0 is the emission intensity in the base year of province i, G_i is the economic growth rate of province i (this value is generally fixed according to the annual economic development), and A_i is the intensity controlled target of province i in the 12th national reduction plan.

Given that the country is composed of provinces, the amount of national CO_2 emissions is equal to the sum of all provinces, as with the GDP. The national emission intensity of the target year can be calculated according to the definition of emission intensity. Unlike the base year national intensity, whether the national reduction target in this period can be achieved or not can be known. From the 12th FY from 2011 to 2015, according to the national reduction plan of China in this period, the base year is 2010, and the target year is 2015, which is expressed as follows:

$$I = \frac{\sum_i C_i}{\sum_i GDP_i} \tag{9}$$

$$GDP_i^t = GDP_i^0 \times (1 + G_i)^{t-0} \tag{10}$$

where I is the national intensity of CO_2 emissions, C_i is the CO_2 emissions of province i, GDP_i is the GDP of province i, F is the completion of the national reduction target (%), I^0 is the national base year intensity, I^t is the national target year intensity and A is the national reduction target of emission intensity in this period, which is 17%.

2.3. Assessment of Reduction Pressure

This study designed two indices to reflect the pressure of reduction for each province. One index is designed by forecasting future emissions based on historical emissions. The other one is based on the comparison of emissions over the same length of time; we choose 2010–2015 and 2005–2010.

2.3.1. Index of Forecasting Based on Historical Emissions

A logistic curve fit model of provincial CO_2 emissions is established based on the time sequence. The logistic equation is written as:

$$x_t = \frac{1}{c + ae^{bt}} \tag{11}$$

where x_t represents the CO_2 emissions of one province in year t, and a, b, and c are parameters of the logistic model. All of these parameters are estimated by statistical software Origin 9.0. Then, the index is designed as follows:

$$D_j = x_j^{2015} - C_j^{2015} \tag{12}$$

where D_j represents the pressure of province j (M_t), x_j^{2015} is the prediction value of CO_2 emissions of province j in 2015, and C_j^{2015} represents the emissions of total control by the reduction plan described in Section 2.2. A larger index corresponds to greater pressure.

2.3.2. Index of Same Length of Time

The ratio of the actual variation of CO_2 emissions in 2005–2010 and control variation in 2010–2015 is the index of reduction pressure. Expressed as a percentage to make the index more intuitive, the equation is written as follows:

$$P_j = \frac{\left(C_j^{2010} - C_j^{2005}\right)}{\left(C_j^{2015} - C_j^{2010}\right)} \times 100\% \tag{13}$$

where P_j represents the pressure of province j; C_j^{2005} and C_j^{2010} are the emissions of province j in years 2005 and 2010, respectively; and C_j^{2015} is the total control of emissions of province j by the reduction plan.

3. Study Area and Data

3.1. Study Area

Mainland China has 31 provincial-level administrative units, including 22 provinces, four municipalities (Beijing, Tianjin, Shanghai and Chongqing), and five autonomous regions (Guangxi, Ningxia, Tibet, Xinjiang, and Inner Mongolia). The study area comprises 30 provincial-level units of mainland China, except Tibet. In addition to missing related data, Tibet has China's lowest CO_2 emissions. The 30 units are referred to as provinces.

The provinces are divided into the following eight economic regions: (1) Northeast: Liaoning, Jilin, and Heilongjiang; (2) North Coast: Beijing, Tianjin, Hebei, and Shandong; (3) East Coast: Shanghai, Jiangsu, and Zhejiang; (4) South Coast: Fujian, Guangdong, and Hainan; (5) Middle Yellow River: Shaanxi; Shanxi, Henan, and Inner Mongolia; (6) Middle Yangtze River: Hubei, Hunan, Jiangxi, and Anhui; (7) Southwest China: Yunnan, Guizhou, Sichuan, Chongqing, and Guangxi; and (8) Northwest China: Gansu, Qinghai, Ningxia, and Xinjiang. This division is the latest regional classification officially provided by the Chinese government. The distribution of the eight regions is shown in Figure 2. For a long time, significant differences in the economic development levels have existed among the different regions of China because of the historic suffering and causal localization in districts. The North Coast, East Coast, and South Coast regions have higher levels of economic development, whereas the Northwest and Southwest regions are relatively backward.

3.2. Data sources and Description

Three aspects of data sources are involved in calculating the CO_2 emissions. The first aspect of energy consumption data and the net calorific values of each type of energy are obtained from the China Energy Statistic Yearbook [41–52]; the data include eight types of energy consumption in 30 provinces and each type of energy calorific value. The second aspect of cement production data is obtained from the China Statistical Yearbook [52–70]; The data included cement production in 30 provinces. The third aspect of forests and green lands data is obtained from the China Forestry Statistical Yearbook [71–88]; The data include an area of forests and green lands in 30 provinces. The parameters in the IPCC guidelines of greenhouse gas emissions are used to calculate CO_2 emissions. The proportion of cement clinker from industry report in recent years is used. All parameters for calculating CO_2 emissions are not listed in this paper. Moreover, the data on economic development setting are obtained. The data are found in the 12th FYP of each province. There are expectations and control of future economic development (Figure 3). The annual GDP figures of 30 provinces are obtained from the China Statistical Yearbook [52–70] and converted into comparable prices with the 2010 figure. In the section on assessment of reduction pressure, the data on historical emissions of provinces from 1995 to 2012, which were calculated by the method of provincial CO_2 emissions, were used.

Figure 2. Eight-region division of the study area.

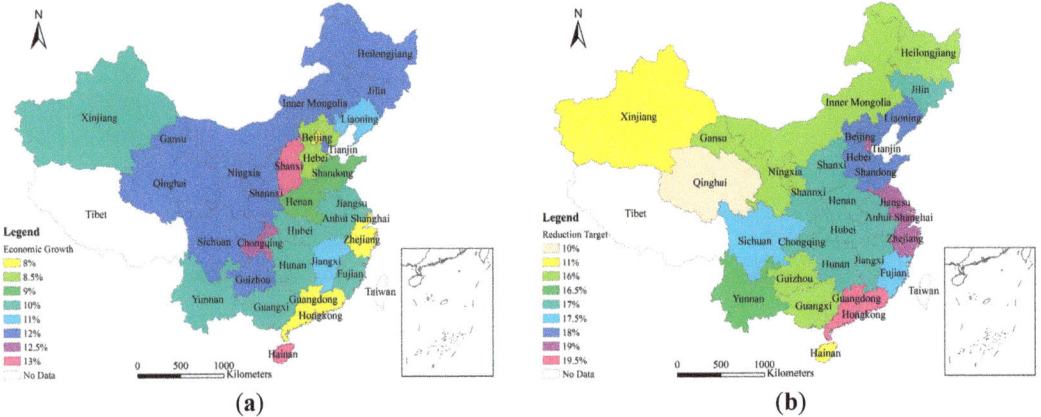

Figure 3. Target setting in 30 provinces (**a**) China's economic growth target in 12th FYP. (**b**) China's CO_2 emission reduction target in 12th FYP.

Provinces with a traditionally high economic level, such as Beijing, Shanghai, Zhejiang, and Guangdong, set a relatively low growth economic speed. The central and western provinces generally lag behind those of eastern China because of uneven development of the regional economies. In Figure 3a, the high setting growth provinces are nearly all in central and western China. The data from China's 12th FY work programs were used to control greenhouse gas emissions. This work plan is the only existing intra-national reduction plan for CO_2 emissions in China. The country and each province have their own CO_2 emission reduction targets (Figure 3b). According to this plan, the national CO_2 emission intensity should decrease by 17% in 2015 compared with the 2010 level. This target is spread into 30 provinces and incorporated into the development planning of each province. Unlike the economic growth setting, provinces in the high economy level (for instance, Tianjin,

Jiangsu, Shanghai, Zhejiang and Guangdong) are bonded with high reduction targets of emission intensity. The targets of the Middle Yellow River and Middle Yangtze River are consistent with the national target which has a rather low reduction target in the Northwest and Southwest regions. The reduction targets of Hainan, Xinjiang and Qinghai, as the lowest in all provinces, are significantly different.

4. Results and Discussion

4.1. Composition of Provincial CO_2 Emissions

The results of taking 2010 as an example to illustrate the composition of the three aspects in the calculation of provincial CO_2 emissions are shown in Figure 4. According to the results, energy consumption remains the main source of CO_2 emissions, accounting for more than 90% of CO_2 emissions in a province.

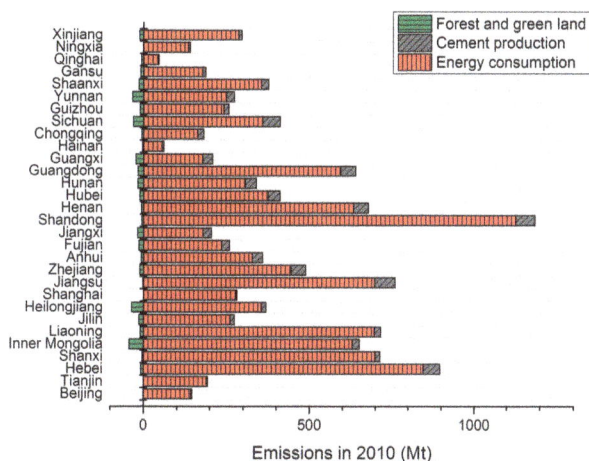

Figure 4. Three aspects of provincial CO_2 emissions.

The emissions from energy consumption are very close to the net provincial emissions. However, differences exist between the calculation results of regional emissions from pure energy consumption and the computing framework that is proposed by this paper. In some provinces, the emissions from cement production cannot be ignored. Figure 4 shows that, the amounts of emissions from cement production are evident in Hebei, Jiangsu, Shandong, Guangdong and Sichuan. In Guangxi, Sichuan, Shandong, Hunan, and Chongqing, the proportions of emissions from cement production are greater than 10%. According to the historical and statistical data, these provinces are the main cement producers in China. Although the amount of CO_2 sequestration is not obvious in most provinces, Inner Mongolia, Heilongjiang, Guangxi, Sichuan, and Yunnan are exceptions. The absorption by forests and green lands is also close to 10% of the emissions in 2010. Cement production and forests cannot be ignored to accurately calculate regional CO_2 emissions.

4.2. Temporal and Spatial Variations of Provincial CO_2 Emissions

The years 2005 and 2010 were chosen to analyze the variation in provincial CO_2 emissions. These two years represent the start and end years of China's 11th FYP. The data can better reflect the differences in CO_2 emissions over the past development stage. The temporal variation of CO_2 emissions is shown in Figure 5a. The temporal variation of emission intensity is shown in Figure 5b.

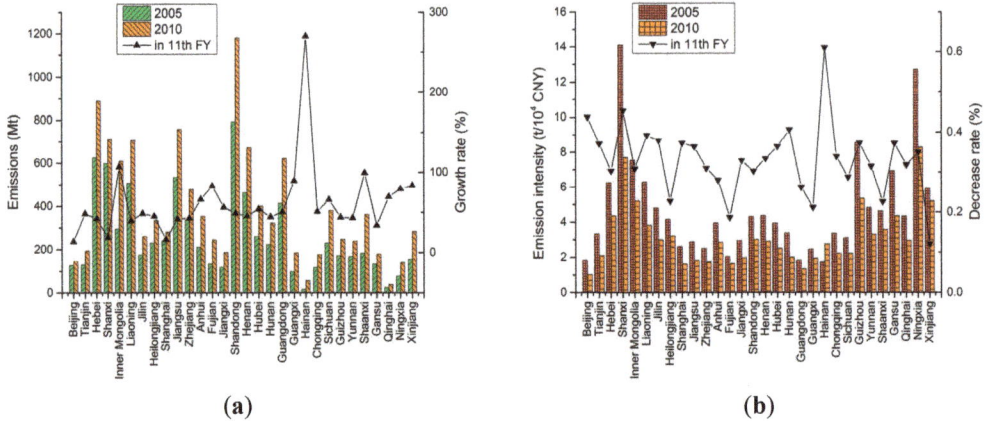

(a)   (b)

Figure 5. Temporal variation in provincial CO_2 emission (**a**) Variation of emissions in 2005–2010; and (**b**) Variation of emission intensity in 2005–2010.

The national emissions were calculated by summing the emissions from each province. The national emissions are 7778.35 million tons (M_t) in 2005 and 11,623.04 Mt in 2010. The national growth rate of CO_2 emissions in this period is nearly 50%. The gap in CO_2 emissions is clear among the provinces. The maximal provincial emission is dozens of times higher than the minimal one. The province with minimal emissions in 2005 is Hainan. Qinghai is the province with minimal emissions in 2010. Shandong always has the highest emissions in these two years. Hainan had the fastest growth rate of emissions because it had a low emission base in 2005 and a requirement of economic development during this period. Beijing and Shanghai had the lowest growth rate because these two metropolises paid more attention to controlling carbon emissions. Between these two years, the CO_2 emissions of most provinces indicated significant growth, with growth rates of more than 60% or even close to 100%. The national emission intensity is 3.91 tons per 10,000 CNY ($t/10^4$ CNY) in 2005 and 2.66 $t/10^4$ CNY in 2010. The decrease rate of the national emission intensity is nearly 32%. The emission intensities at the national and provincial levels exhibited dramatic decreases in this period. Only Hainan's intensity increased, and its decrease rate is indicated in Figure 5b as a positive value for convenience. Shanxi and Ningxia have the highest emission intensities at 14.10 and 12.74 in 2005 and 7.71 and 8.28 in 2010, respectively. They also had the most significant decrease in emission intensity but still maintained the highest intensity in 2010 of mainland China. Between these two years, the emission intensity of 29 provinces obviously decreased. The decreases were from 20% to 40%.

The variation in spatial distribution is also obvious. The unified grouping rules of emissions and emission intensity were adapted to different provinces into four groups, reflecting variations in time and space. Each group differs by 200 Mt in emissions and 2 t/10⁴ CNY in emission intensity. The distribution results are shown in Figures 6 and 7. The emissions of east China in 2005 and 2010 are the highest, decreasing gradually from east to west. The emissions of the provinces around Beijing, which belong to the North Coast region, are significantly higher than those of the other provinces. Moreover, Guangdong is one of the provinces where emissions are higher in 2005 and 2010. The emissions of the North Coast and Middle Yellow River regions are the highest according to the rule of the study area division in 2005 (Figure 6a). The previous two regions and East Coast are the highest in 2010 (Figure 6b). The CO_2 emissions of provinces in the North Coast region from 2005 to 2010 are becoming much higher than those of the other regions. This distribution of CO_2 emissions is more evident. Hebei and Shandong became the top two provinces with more than 800 Mt emissions. Other provinces, such as Xinjiang, Sichuan and Guizhou in the Southwest and Northwest regions, obviously increased their emissions during this period.

The emission intensity of northern China is the highest in both 2005 and 2010. Unlike emissions, the distribution of emission intensity rose from south to north in both 2005 and 2010. The provinces in the South Coast and East Coast regions have the lowest emission intensities. Although these provinces have high emissions, they also have high GDPs. A rule indicates that poor and backward provinces have higher emission intensities.

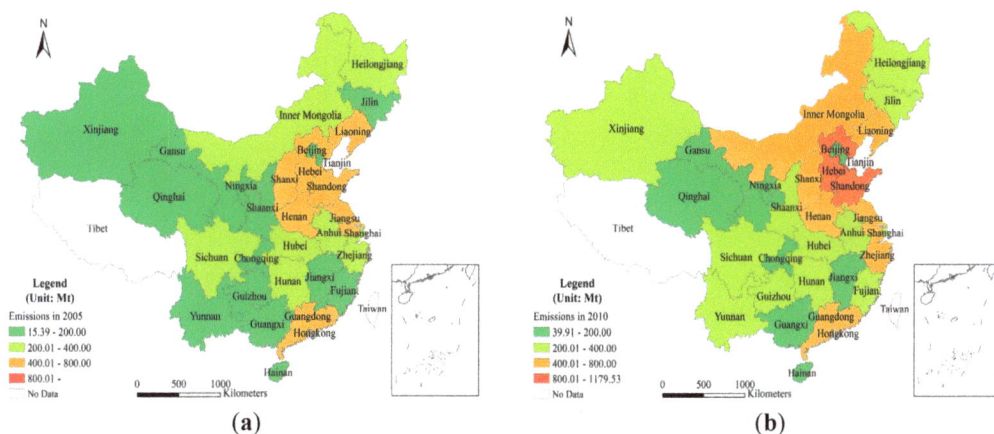

Figure 6. Provincial emission distribution (a) CO_2 emissions in 2005; and (b) CO_2 emissions in 2010.

The Northwest, Middle Yellow River and Southwest regions have the highest intensities of CO_2 emissions in 2005 (Figure 7a). The obviously high intensity in 2010 (Figure 7b) was distributed in the Northwest and Middle Yellow River regions. From 2005 to 2010, the intensity decreased for every province, except Hainan, mostly because of the rapid growth in GDP and the low level of energy utilization efficiency. The emissions in Hainan grew faster than its economic development. Shanxi and Ningxia in central mainland China are clearly the two provinces with the highest emission intensity. The overall spatial-temporal variation of CO_2 emissions in 2005 to 2010 can be summarized as follows: the emissions increased over time with a decreasing pattern from east to west, and the intensity decreased over time with a rising pattern from south to north.

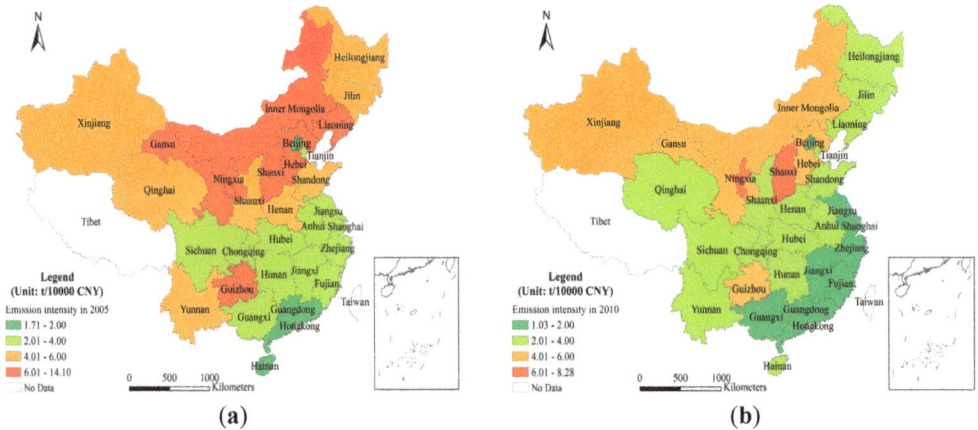

Figure 7. Provincial emission intensity distribution (**a**) Emission Intensity in 2005; and (**b**) Emission intensity in 2010.

4.3. Total Control of Emissions in 2015

Based on the reduction plan in the 12th FY of China, 2015 was set as the target year and 2010 as the base year. According to the model of forecasting CO_2 emissions in Section 2.2, the total control in 2015 and increments during the 12th FY of CO_2 emissions were calculated. The results are presented in Table 2 which includes the total control target of CO_2 emissions in 2015. This target was compared with the emissions in 2010 (Figure 8a).

The CO_2 emissions of each province in 2015 are still increasing compared with those in 2010. The provinces that have high emissions in the base year still have high emissions in the target year. Shandong, Hebei, and Shanxi are the top three emitting provinces. They will emit more than 1000 M_t emissions in 2015. The increment distribution of CO_2 emissions in the 12th FY is shown in Figure 8b. The increments are different among the provinces. Shanxi accounts for the maximum at 375.39 Mt, and Qinghai accounts for the minimum at 23.39 M_t. All provinces, except Shanghai, have 20% or even more than 50% growth in the 12th FY. Beijing, Shanghai, Zhejiang, and Guangdong are slower growth provinces in terms of CO_2 emissions. The high economic development levels in these regions force them to pay more attention to controlling CO_2 emissions. They have gradually limited their dependence on fossil energy contour carbon energy. By contrast, Hainan, Qinghai, Shanxi and other poor western and central provinces need to maintain a high growth of emissions at the present stage. On the right side of Figure 8a, a higher growth rate from Chongqing to Ningxia is shown. Provinces in this range belong to the west China region, which lags behind in economic development.

Table 2. Provincial total control and increments of CO_2 emissions.

Province	Emissions in 2015 (M_t)	Increments in 12th FY (M_t)	Province	Emissions in 2015 (M_t)	Increments in 12th FY (M_t)
Beijing	175.15	29.78	Henan	857.87	186.11
Tianjin	276.22	82.72	Hubei	534.22	134.57
Hebei	1095.54	207.02	Hunan	429.94	108.30
Shanxi	1084.70	375.39	Guangdong	736.11	113.77
Inner Mongolia	899.53	291.89	Guangxi	250.35	65.29
Liaoning	975.60	269.54	Hainan	93.42	36.45
Jilin	379.06	119.92	Chongqing	263.46	87.31
Heilongjiang	493.52	160.14	Sichuan	552.87	172.61
Shanghai	333.79	53.33	Guizhou	364.85	118.39
Jiangsu	987.63	230.54	Yunnan	321.42	82.41
Zhejiang	568.46	90.82	Shaanxi	529.89	167.63
Anhui	469.95	118.38	Gansu	265.12	86.03
Fujian	324.31	80.22	Qinghai	63.30	23.39
Jiangxi	260.57	74.26	Ningxia	207.15	67.22
Shandong	1488.18	308.65	Xinjiang	405.60	122.63

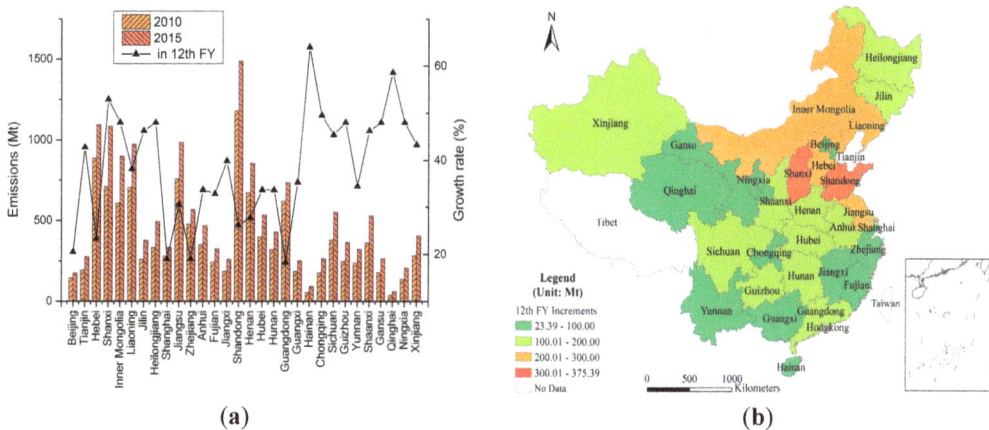

(a)　　　　　　　　　　　　　　　　　(b)

Figure 8. Provincial variation in CO_2 emission (a) Temporal variation of emissions in 2010–2015; and (b) Increments of emissions in 2010–2015.

CO_2 emissions and emission intensity were classified by the same group rules in Section 4.2 (Figure 9). The distribution of CO_2 emissions in 2015 is similar to 2010 and 2005 (Figure 9a). The overall distribution indicating that east China has higher emissions than west China has not changed. The North Coast and Middle Yellow River regions remain those with the highest emissions. The one obvious change is that the emissions of more provinces are over 800 Mt. The distribution of these high-emission provinces indicates that Beijing is a center of expansion. Moreover, Guangdong in the South Coast is now in the top emissions group. The Middle Yangtze River represents the region of higher growth in emissions for this period. Hubei, Hunan, and Anhui belong to this region. Many provinces in central and western China have significant emission growths, including Shanxi and

Henan in Middle Yellow River and Sichuan and Xinjiang in western China. The emission intensities of all provinces have decreased because the reduction plan in this period focuses on decreasing the emission intensity. The distribution is high in the north and low in the south (Figure 9b). Although the emission intensity is expected to decrease significantly after the provinces execute their own reduction targets, previous high-intensity provinces still have higher emissions than the others. Ningxia, Xinjiang, Inner Mongolia, Guizhou, and Shanxi are the best representatives of the provinces and still belong to the higher groups with intensities. Conversely, Gansu and Hebei no longer belong to the higher groups with intensities of 4–6 t/10^4 CNY. The Northeast, North Coast, East Coast, South Coast, and Middle Yangtze River regions belong to the lower-intensity groups. Emissions and there intensity are predicted to show large changes in their time and space after the reduction plans.

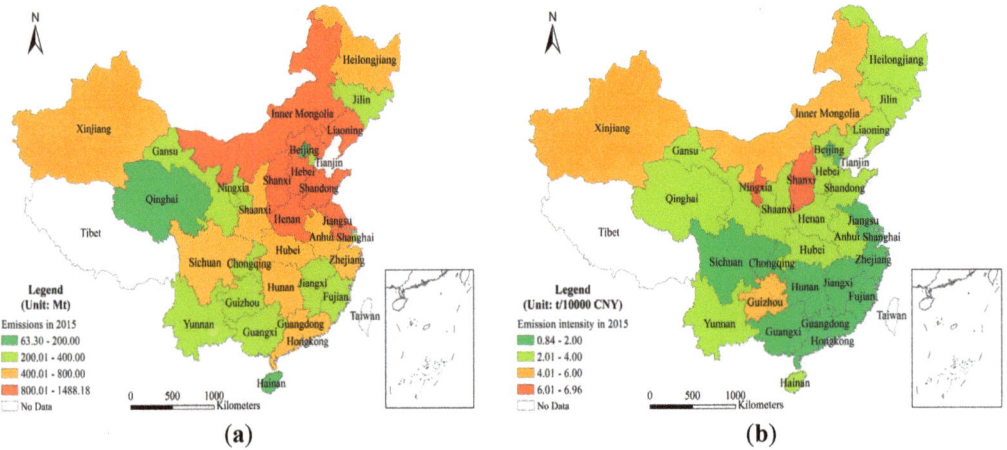

Figure 9. Provincial emissions and emission intensity distribution (**a**) CO_2 emissions in 2015; and (**b**) Emission intensity in 2015.

Based on the same assumptions described in Section 4.2, the national CO_2 emissions and emission intensity in 2015 can also be calculated. The results show that the national emissions are 15,687.76 M_t and intensity is 2.24 t/10^4 CNY. If fully in accordance with the national emission reduction target compared with the 2010 level, the national emission intensity needs to decrease to 2.21 t/10^4 CNY and the national emissions to 15,464.42 Mt. The national emissions increased by nearly 4000 Mt compared with the level in 2010. According to the definition of completion of the national reduction target, the end result is 92.95%. The national emission intensity decreased by 15.80%. This value is extremely close to achieving the national emission reduction plan.

4.4. Assessment of Reduction Pressure

The CO_2 emissions of every province were forecasted in 2015 using historical emission data from 1995 to 2012. These historical data on emissions were all calculated by the model described in Section 2.1. The forecasting values of 2015 and D index results are shown in Table 3.

Table 3. Forecasting emissions in 2015 and D index.

Province	Forecasting Value (M_t)	D Index (M_t)	Province	Forecasting Value (M_t)	D Index (M_t)
Beijing	140.29	−34.86	Henan	715.48	−142.39
Tianjin	272.91	−3.31	Hubei	533.75	−0.47
Hebei	1068.03	−27.52	Hunan	349.23	−80.71
Shanxi	762.57	−322.13	Guangdong	738.64	2.53
Inner Mongolia	917.59	18.06	Guangxi	394.00	143.65
Liaoning	830.40	−145.21	Hainan	79.65	−13.77
Jilin	323.14	−55.91	Chongqing	245.63	−17.83
Heilongjiang	404.04	−89.48	Sichuan	407.03	−145.84
Shanghai	295.52	−38.27	Guizhou	344.96	−19.89
Jiangsu	917.04	−70.59	Yunnan	258.07	−63.35
Zhejiang	507.38	−61.07	Shaanxi	557.72	27.83
Anhui	467.23	−2.72	Gansu	250.64	−14.49
Fujian	327.92	3.61	Qinghai	88.80	25.50
Jiangxi	227.08	−33.49	Ningxia	314.20	107.05
Shandong	1257.30	−230.88	Xinjiang	632.04	226.45

Logistic curves effectively describe the historical CO_2 emissions of each province. A difference exists between the forecasting value and total control value in 2015. According to the definition of index D, a positive value signifies that the total control emissions are less than the forecasted value, which indicates that total control is not adequate to normal development demand. A greater value D indicates greater pressure. Xinjiang, Guangxi, Ningxia, Qinghai, Shaanxi, Inner Mongolia, Fujian, and Guangdong are expected to have reduction pressure. These provinces are distributed in the Northwest and South Coast regions in mainland China. By comparing the provincial increments of CO_2 emissions in the 11th FY and 12th FY, the P index was calculated. A ratio of 50% was used as group rule for classification, and the P value was used to classify four groups. The distribution result is shown in Figure 10.

P value that is greater than 100% indicates that the control increments of emissions in the 12th FY are less than those at the 11th FY level. A greater P value corresponds to greater pressure to reduce emissions. The reduction plan in the 12th FY is difficult to produce. Zhejiang and Guangdong are in the first group of pressure, and they belong to the East Coast and South Coast regions, respectively. Hebei and Shandong in the North Coast region; Shaanxi, Henan and Inner Mongolia in the Middle Yellow River region; Hubei and Anhui in the Middle Yangtze River region; and Fujian and Hainan in the South Coast region are in the second group of pressure. The second group of pressure also includes Xinjiang and Guangxi, which belong to the Northwest and Southwest regions, respectively. The same groupings were found in the results of index D and P calculation. Xinjiang, Inner Mongolia, Shaanxi, Guangxi, Guangdong and Fujian have greater reduction pressures in the 12th FY. The South Coast and Northwest China regions have a greater reduction pressure of emission intensity.

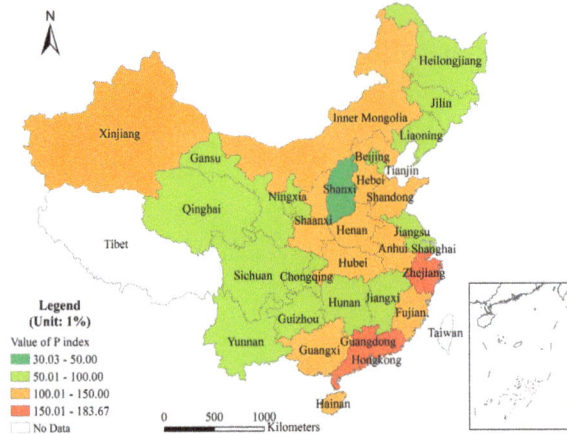

Figure 10. Distribution of P index.

The verification of the predictions is attempted based on the latest available data on provincial emissions in 2012. Provincial emission intensity is calculated in 2012 to estimate their completion of reduction plan compared with the level in 2010 (Table 4).

Table 4. Provincial completion of reduction plan in 2012 and incompletion before 2015.

Province	Completion in 2012	Incompletion before 2015	Province	Completion in 2012	Incompletion before 2015
Beijing	19.15%	−1.15%	Henan	14.92%	2.08%
Tianjin	16.63%	2.37%	Hubei	9.67%	7.33%
Hebei	6.47%	11.53%	Hunan	11.13%	5.87%
Shanxi	7.06%	9.94%	Guangdong	10.09%	9.41%
Inner Mongolia	−2.98%	18.98%	Guangxi	−11.01%	27.01%
Liaoning	9.78%	8.22%	Hainan	−2.34%	13.34%
Jilin	11.00%	6.00%	Chongqing	14.04%	2.96%
Heilongjiang	8.03%	7.97%	Sichuan	18.95%	−1.45%
Shanghai	12.70%	6.30%	Guizhou	4.11%	11.89%
Jiangsu	4.38%	14.62%	Yunnan	13.46%	3.04%
Zhejiang	12.69%	6.31%	Shaanxi	−0.22%	17.22%
Anhui	10.02%	6.98%	Gansu	3.99%	12.01%
Fujian	8.15%	9.35%	Qinghai	−17.48%	27.48%
Jiangxi	9.18%	7.82%	Ningxia	−13.72%	29.72%
Shandong	9.13%	8.87%	Xinjiang	−11.44%	22.44%

As of the end of 2012, Beijing and Sichuan had completed their reduction targets. Some provinces whose emission intensity reached a higher level than those in the 2010s are Qinghai, Ningxia, Xinjiang, Guangxi, Inner Mongolia, Hainan, and Shaanxi. Therefore, before the end of 2015, they faced great pressure to complete their reduction targets. Moreover, Fujian, Guangdong, Gansu, and Hebei had to decrease their emission intensity to nearly 10%. From 2013 to 2015, a greater reduction pressure is basically concentrated in the South Coast and Northwest China regions. These calculation results are consistent with those from the aforementioned two pressure indices.

5. Conclusions

A new computing framework was proposed to calculate provincial CO_2 emissions from 2005 to 2010 in China, and the composition of CO_2 emissions and spatial-temporal variations were analyzed. Then, China's 12th FY (2011–2015) reduction plan was then analyzed in depth, and the provincial total control of CO_2 emissions in 2015 was calculated. The pressure of reduction was finally assessed for every province. The main conclusions are presented as follows.

When calculating the provincial CO_2 emissions, energy consumption is the main source of China's provincial CO_2 emissions. However, emissions of cement production and absorption of forest and green land cannot be ignored. The overall spatial-temporal variations of CO_2 emissions in 2005–2010 reflected that emissions were increasing over time with a decreasing distribution from east to west, and intensity was decreasing over time with an increasing distribution from south to north. The North Coast and East Coast regions had high emissions, whereas the Middle Yellow River, Northwest, and Southwest regions had high emission intensities.

The total control of CO_2 emissions in 2015 was determined. After every province executed the 12th FY reduction plan to control CO_2 emissions, the CO_2 emissions of each province in target year 2015 is expected to exhibit a significant growth compared with the level in 2010. An increasing number of provinces will shift to the top emitting group. The Southwest and Northwest regions have higher emission growth rates although the amounts are still less than those in the North, East, and South Coast regions. In 2015, the overall distribution of provincial emissions and emission intensity will not change. The national emission intensity decreased by 15.80% when the provinces had reduced the emission intensity in accordance with the reduction plan. This value is extremely close to achieving the national emission reduction plan.

The reduction pressure on emission intensity was assessed when every province conducted its reduction target in the 12th FY. The provinces with greater pressure of reduction were chosen through two designed indices. The classifications by the indices provided the same results. They indicated that Xinjiang, Inner Mongolia, Shaanxi, Guangxi, Guangdong, and Fujian had greater reduction pressure in the 12th FY. The South Coast and Northwest regions had a greater reduction pressure on CO_2 emissions based on the division of provinces.

6. Policy Implications

According to the calculations and analysis conducted in the present study, the following superficial policy implications are presented to serve as references for decision makers to control CO_2 emissions in China. Several provinces are confronted with high pressures of CO_2 emission reduction during the 12th FY period. To solve this problem, the ratio of utilizing non-fossil or low-carbon energy should be improved. Some provinces with low economic levels have high pressure to reduce CO_2 emissions; for example, Xinjiang, Inner Mongolia, Shaanxi, and Guangxi have to propose more strategies of CO_2 emission reduction to guarantee economic development.

Some provinces that have high economic levels but low CO_2 emission intensity (e.g., Beijing, Shanghai, and Guangdong) should pay more attention to optimizing the energy consumption structure, increasing the ratio of technology-intensive industries, and promoting the development of low-carbon industry. Some provinces such as Shanxi, Inner Mongolia, Xinxiang, Ningxia, and Guizhou have abundant fossil energy resources, and their economy is dependent on energy consumption. To enforce them to have the same targets as other provinces is inappropriate. CO_2 emission reduction targets should be set according to local conditions. Moreover, central and western China, which are lagging behind economically, should actively promote low carbon development, and accelerate transformation of economic development patterns, ensuring sustainable economic growth.

Most provinces of China are still in the industrialization stage; therefore, a faster speed of economic growth expectations inevitably leads to greater energy demand and CO_2 emissions. Economic development that is overheated or too fast in a short period can induce significant growth of energy-intensive industries and products and, consequently, the increase of emission intensity. Thus, to achieve the national CO_2 emission reduction target, all intra-regions, especially Shaanxi, Hainan, and Chongqing, should consider the national reduction target of CO_2 emission intensity more than setting the speed of economic development. Eventually, maintaining economic growth at an appropriate and stable level will decrease the CO_2 emission intensity. Thus, the national CO_2 emission reduction target could likely be realized. The CO_2 emissions of the country and provinces will continue to increase while the intensity reduction plan is executed. If the future reduction plan chooses to reduce total emissions, the national and provincial targets will be extremely difficult to achieve. Moreover, economic development will be significantly affected. A future reduction plan should gradually implement more stringent emission reduction policies.

The amount and structure of energy consumption are the critical influencing factors of regional CO_2 emissions. The optimization of energy utilization mode, improvement of energy utilization efficiency, and the development and application of new energy and renewable energy favor CO_2 emission reductions. The most important point of realizing a low-carbon economy is a reasonable adjustment of the energy structure (e.g., exploiting hydroelectricity, wind power, solar energy, and biomass energy in the future based on the protection of the natural environment).

Acknowledgments

This research was supported in part by National "Twelfth Five-Year" Plan for Science & Technology Support (No. 2012BAH28B02), Ministry of Science and Technology of China. The authors would like to thank the reviewers for reviewing and correcting the paper.

Author Contributions

Xuankai Deng drafted the paper and contributed to data collection and calculation; Yanhua Yu contributed to data analysis; Yanfang Liu conceived and designed the research.

Conflicts of Interest

The authors declare no conflict of interest.

References

1. Raupach, M.R.; Marland, G.; Ciais, P.; Le Quere, C.; Canadell, J.G.; Klepper, G.; Field, C.B. Global and regional drivers of accelerating CO_2 emissions. *Proc. Natl. Acad. Sci. USA* **2007**, *104*, 10288–10293.
2. Gurney, K.R. Global change: China at the carbon crossroads. *Nature* **2009**, *458*, 977–979.
3. Raupach, M.R.; Davis, S.J.; Peters, G.P.; Andrew, R.M.; Canadell, J.G.; Ciais, P.; Friedlingstein, P.; Jotzo, F.; van Vuuren, D.P.; Le Quéré, C. Sharing a quota on cumulative carbon emissions. *Nat. Clim. Chang.* **2014**, *4*, 873–879.
4. Pan, X.; Teng, F.; Wang, G. Sharing emission space at an equitable basis: Allocation scheme based on the equal cumulative emission per capita principle. *Appl. Energy* **2014**, *113*, 1810–1818.
5. Pao, H.; Fu, H.; Tseng, C. Forecasting of CO_2 emissions, energy consumption and economic growth in China using an improved grey model. *Energy* **2012**, *40*, 400–409.
6. Auffhammer, M.; Carson, R.T. Forecasting the path of China's CO2 emissions using province-level information. *J. Environ. Econom. Manag.* **2008**, *55*, 229–247.
7. Zhang, M.; Mu, H.; Ning, Y.; Song, Y. Decomposition of energy-related CO_2 emission over 1991–2006 in China. *Ecol. Econ.* **2009**, *68*, 2122–2128.
8. Zhao, Y.; Nielsen, C.P.; McElroy, M.B. China's CO_2 emissions estimated from the bottom up: Recent trends, spatial distributions, and quantification of uncertainties. *Atmos. Environ.* **2012**, *59*, 214–223.
9. Liu, Z.; Geng, Y.; Lindner, S.; Guan, D. Uncovering China's greenhouse gas emission from regional and sectoral perspectives. *Energy* **2012**, *45*, 1059–1068.
10. Wang, Y.; Zhang, P.; Huang, D.; Cai, C. Convergence behavior of carbon dioxide emissions in China. *Econom. Model.* **2014**, *43*, 75–80.
11. Wang, C.; Chen, J.; Zou, J. Decomposition of energy-related CO_2 emission in China: 1957–2000. *Energy* **2005**, *30*, 73–83.
12. Zhang, Y.; Wang, H.; Liang, S.; Xu, M.; Liu, W.; Li, S.; Zhang, R.; Nielsen, C.P.; Bi, J. Temporal and spatial variations in consumption-based carbon dioxide emissions in China. *Renew. Sustain. Energy Rev.* **2014**, *40*, 60–68.
13. Xu, J.; Fleiter, T.; Eichhammer, W.; Fan, Y. Energy consumption and CO_2 emissions in China's cement industry: A perspective from LMDI decomposition analysis. *Energy Policy* **2012**, *50*, 821–832.
14. Kim, Y.; Worrell, E. CO_2 Emission trends in the cement industry: An international comparison. *Mitig. Adapt. Strategies Glob. Chang.* **2002**, *7*, 115–133.
15. Ke, J.; Zheng, N.; Fridley, D.; Price, L.; Zhou, N. Potential energy savings and CO_2 emissions reduction of China's cement industry. *Energy Policy* **2012**, *45*, 739–751.
16. Münnich Vass, M.; Elofsson, K.; Gren, I. An equity assessment of introducing uncertain forest carbon sequestration in EU climate policy. *Energy Policy* **2013**, *61*, 1432–1442.
17. Houghton, R.A.; Hackler, J.L.; Lawrence, K.T. The US carbon budget: Contributions from land-use change. *Science* **1999**, *285*, 574–578.

18. Houghton, R.A. Carbon emissions and the drivers of deforestation and forest degradation in the tropics. *Curr. Opin. Environ. Sustain.* **2012**, *4*, 597–603.

19. Fang, J.; Guo, Z.; Hu, H.; Kato, T.; Muraoka, H.; Son, Y. Forest biomass carbon sinks in East Asia, with special reference to the relative contributions of forest expansion and forest growth. *Glob. Chang. Biol.* **2014**, *20*, 2019–2030.

20. Guo, Z.; Hu, H.; Li, P.; Li, N.; Fang, J. Spatio-temporal changes in biomass carbon sinks in China's forests from 1977 to 2008. *Sci. China Life Sci.* **2013**, *56*, 661–671.

21. Wei, Y.; Wang, L.; Liao, H.; Wang, K.; Murty, T.; Yan, J. Responsibility accounting in carbon allocation: A global perspective. *Appl. Energ.* **2014**, *130*, 122–133.

22. Chakravarty, S.; Chikkatur, A.; de Coninck, H.; Pacala, S.; Socolow, R.; Tavoni, M. Sharing global CO_2 emission reductions among one billion high emitters. *Proc. Natl. Acad. Sci. USA* **2009**, *106*, 11884–11888.

23. Chih Chang, C.; Chia Lai, T. Carbon allowance allocation in the transportation industry. *Energy Policy* **2013**, *63*, 1091–1097.

24. Chen, W.; He, Q. Intersectoral burden sharing of CO_2 mitigation in China in 2020. *Mitig. Adapt. Strategies Glob. Chang.* **2014**, doi:10.1007/s11027-014-9566-3.

25. Yuan, J.; Hou, Y.; Xu, M. China's 2020 carbon intensity target: Consistency, implementations, and policy implications. *Renew. Sustain. Energy Rev.* **2012**, *16*, 4970–4981.

26. Wang, K.; Zhang, X.; Wei, Y.; Yu, S. Regional allocation of CO_2 emissions allowance over provinces in China by 2020. *Energy Policy* **2013**, *54*, 214–229.

27. Yi, W.; Zou, L.; Guo, J.; Wang, K.; Wei, Y. How can China reach its CO_2 intensity reduction targets by 2020? A regional allocation based on equity and development. *Energy Policy* **2011**, *39*, 2407–2415.

28. Meng, M.; Niu, D.; Shang, W. CO_2 emissions and economic development: China's 12th Five-Year plan. *Energy Policy* **2012**, *42*, 468–475.

29. Lu, Y.; Stegman, A.; Cai, Y. Emissions intensity targeting: From China's 12th Five Year Plan to its Copenhagen commitment. *Energy Policy* **2013**, *61*, 1164–1177.

30. Geng, Y.; Tian, M.; Zhu, Q.; Zhang, J.; Peng, C. Quantification of provincial-level carbon emissions from energy consumption in China. *Renew. Sustain. Energy Rev.* **2011**, *15*, 3658–3668.

31. Li, L.; Chen, C.; Xie, S.; Huang, C.; Cheng, Z.; Wang, H.; Wang, Y.; Huang, H.; Lu, J.; Dhakal, S. Energy demand and carbon emissions under different development scenarios for Shanghai, China. *Energy Policy* **2010**, *38*, 4797–4807.

32. Zhai, S.; Wang, Z.; Ma, X.; Huang, R.; Liu, C.; Zhu, Y. Calculation of regional carbon emission: A case of Guangdong Province. *Chin. J. Appl. Ecol.* **2011**, *22*, 1543–1551. (In Chinese)

33. Eggleston, S.; Buendia, L.; Miwa, K.; Ngara, T.; Tanabe, K. *IPCC Guidelines for National Greenhouse Gas Inventories*; IGES: Kanagawa, Japan, 2006.

34. Ma, X.; Wang, Z. Estimation of provincial forest carbon sink capacities in Chinese mainland. *Chin. Sci. Bull.* **2011**, *56*, 433–441.

35. Fang, J.; Guo, Z.; Piao, S.; Chen, A. Estimation of carbon sinks by terrestrial vegetation in China from 1981 to 2000. *Sci. China* **2007**, *37*, 804–812.

36. Lai, L. The effect of land-use on carbon emission in China. Ph.D. Thesis, Nanjing University, Nanjing, China, December 2010. (In Chinese)

37. Academy of Science, China. Experts estimate the accurate data of carbon absorption by each hectare of forest. Available online: http://www.cas.cn/ky/kyjz/200507/t20050710_1028594.shtml (accessed on 15 November 2014). (In Chinese)

38. Guo, Z.; Hu, H.; Li, P.; Li, N.; Fang, J. Spatio-temporal changes in biomass carbon sinks in China's forests during 1977–2008. *Sci. China Life Sci.* **2013**, *43*, 421–431.

39. Piao, S. Forest biomass carbon stocks in China over the past 2 decades: Estimation based on integrated inventory and satellite data. *J. Geophys. Res.* **2005**, *110*, G01006.

40. Yu, Y.; Chen, C. Spatial and temporal characteristics of net emission of carbon dioxide in China. *Sci. Geogr. Sin.* **2013**, *33*, 1173–1179. (In Chinese)

41. National Bureau of Statistics of People's Republic of China. *China Energy Statistical Yearbook 1991–1996*; China Statistics Press: Beijing, China, 1998.

42. National Bureau of Statistics of People's Republic of China. *China Energy Statistical Yearbook 1997–1999*; China Statistics Press: Beijing, China, 2001.

43. National Bureau of Statistics of People's Republic of China. *China Energy Statistical Yearbook 2000–2002*; China Statistics Press: Beijing, China, 2004.

44. National Bureau of Statistics of People's Republic of China. *China Energy Statistical Yearbook 2004*; China Statistics Press: Beijing, China, 2005.

45. National Bureau of Statistics of People's Republic of China. *China Energy Statistical Yearbook 2005*; China Statistics Press: Beijing, China, 2006.

46. National Bureau of Statistics of People's Republic of China. *China Energy Statistical Yearbook 2006*; China Statistics Press: Beijing, China, 2007.

47. National Bureau of Statistics of People's Republic of China. *China Energy Statistical Yearbook 2007*; China Statistics Press: Beijing, China, 2008.

48. National Bureau of Statistics of People's Republic of China. *China Energy Statistical Yearbook 2008*; China Statistics Press: Beijing, China, 2009.

49. National Bureau of Statistics of People's Republic of China. *China Energy Statistical Yearbook 2009*; China Statistics Press: Beijing, China, 2010.

50. National Bureau of Statistics of People's Republic of China. *China Energy Statistical Yearbook 2010*; China Statistics Press: Beijing, China, 2011.

51. National Bureau of Statistics of People's Republic of China. *China Energy Statistical Yearbook 2011*; China Statistics Press: Beijing, China, 2012.

52. National Bureau of Statistics of People's Republic of China. *China Energy Statistical Yearbook 2012*; China Statistics Press: Beijing, China, 2013.

53. National Bureau of Statistics of People's Republic of China. *China Statistical Yearbook 1996*; China Statistics Press: Beijing, China, 1996.

54. National Bureau of Statistics of People's Republic of China. *China Statistical Yearbook 1997*; China Statistics Press: Beijing, China, 1997.

55. National Bureau of Statistics of People's Republic of China. *China Statistical Yearbook 1998*; China Statistics Press: Beijing, China, 1998.

56. National Bureau of Statistics of People's Republic of China. *China Statistical Yearbook 1999*; China Statistics Press: Beijing, China, 1999.

57. National Bureau of Statistics of People's Republic of China. *China Statistical Yearbook 2000*; China Statistics Press: Beijing, China, 2000.

58. National Bureau of Statistics of People's Republic of China. *China Statistical Yearbook 2001*; China Statistics Press: Beijing, China, 2001.

59. National Bureau of Statistics of People's Republic of China. *China Statistical Yearbook 2002*; China Statistics Press: Beijing, China, 2002.

60. National Bureau of Statistics of People's Republic of China. *China Statistical Yearbook 2003*; China Statistics Press: Beijing, China, 2003.

61. National Bureau of Statistics of People's Republic of China. *China Statistical Yearbook 2004*; China Statistics Press: Beijing, China, 2004.

62. National Bureau of Statistics of People's Republic of China. *China Statistical Yearbook 2005*; China Statistics Press: Beijing, China, 2005.

63. National Bureau of Statistics of People's Republic of China. *China Statistical Yearbook 2006*; China Statistics Press: Beijing, China, 2006.

64. National Bureau of Statistics of People's Republic of China. *China Statistical Yearbook 2007*; China Statistics Press: Beijing, China, 2007.

65. National Bureau of Statistics of People's Republic of China. *China Statistical Yearbook 2008*; China Statistics Press: Beijing, China, 2008.

66. National Bureau of Statistics of People's Republic of China. *China Statistical Yearbook 2009*; China Statistics Press: Beijing, China, 2009.

67. National Bureau of Statistics of People's Republic of China. *China Statistical Yearbook 2010*; China Statistics Press: Beijing, China, 2010.

68. National Bureau of Statistics of People's Republic of China. *China Statistical Yearbook 2011*; China Statistics Press: Beijing, China, 2011.

69. National Bureau of Statistics of People's Republic of China. *China Statistical Yearbook 2012*; China Statistics Press: Beijing, China, 2012.

70. National Bureau of Statistics of People's Republic of China. *China Statistical Yearbook 2013*; China Statistics Press: Beijing, China, 2013.

71. National Bureau of Statistics of People's Republic of China. *China Forestry Statistical Yearbook 1996*; China Statistics Press: Beijing, China, 1996.

72. National Bureau of Statistics of People's Republic of China. *China Forestry Statistical Yearbook 1997*; China Statistics Press: Beijing, China, 1997.

73. National Bureau of Statistics of People's Republic of China. *China Forestry Statistical Yearbook 1998*; China Statistics Press: Beijing, China, 1998.

74. National Bureau of Statistics of People's Republic of China. *China Forestry Statistical Yearbook 1999*; China Statistics Press: Beijing, China, 1999.

75. National Bureau of Statistics of People's Republic of China. *China Forestry Statistical Yearbook 2000*; China Statistics Press: Beijing, China, 2000.

76. National Bureau of Statistics of People's Republic of China. *China Forestry Statistical Yearbook 2001*; China Statistics Press: Beijing, China, 2001.
77. National Bureau of Statistics of People's Republic of China. *China Forestry Statistical Yearbook 2002*; China Statistics Press: Beijing, China, 2002.
78. National Bureau of Statistics of People's Republic of China. *China Forestry Statistical Yearbook 2003*; China Statistics Press: Beijing, China, 2003.
79. National Bureau of Statistics of People's Republic of China. *China Forestry Statistical Yearbook 2004*; China Statistics Press: Beijing, China, 2004.
80. National Bureau of Statistics of People's Republic of China. *China Forestry Statistical Yearbook 2005*; China Statistics Press: Beijing, China, 2005.
81. National Bureau of Statistics of People's Republic of China. *China Forestry Statistical Yearbook 2006*; China Statistics Press: Beijing, China, 2006.
82. National Bureau of Statistics of People's Republic of China. *China Forestry Statistical Yearbook 2007*; China Statistics Press: Beijing, China, 2007.
83. National Bureau of Statistics of People's Republic of China. *China Forestry Statistical Yearbook 2008*; China Statistics Press: Beijing, China, 2008.
84. National Bureau of Statistics of People's Republic of China. *China Forestry Statistical Yearbook 2009*; China Statistics Press: Beijing, China, 2009.
85. National Bureau of Statistics of People's Republic of China. *China Forestry Statistical Yearbook 2010*; China Statistics Press: Beijing, China, 2010.
86. National Bureau of Statistics of People's Republic of China. *China Forestry Statistical Yearbook 2011*; China Statistics Press: Beijing, China, 2011.
87. National Bureau of Statistics of People's Republic of China. *China Forestry Statistical Yearbook 2012*; China Statistics Press: Beijing, China, 2012.
88. National Bureau of Statistics of People's Republic of China. *China Forestry Statistical Yearbook 2013*; China Statistics Press: Beijing, China, 2013.

China's Low-Carbon Scenario Analysis of CO₂ Mitigation Measures towards 2050 Using a Hybrid AIM/CGE Model

Wei Li, Hao Li and Shuang Sun

Abstract: China's emissions continue to rise rapidly in line with its mounting energy consumption, which puts considerable pressure on China to meet its emission reduction commitments. This paper assesses the impacts of CO_2 mitigation measures in China during the period from 2010 to 2050 by using a computable general equilibrium method, called AIM/CGE. Results show that renewable energy makes a critical difference in abating emissions during the period from 2010 to 2020. The scenarios with emission trading would drive more emission reductions, whereby the emission-cutting commitment for 2020 would be achieved and emission reductions in 2050 would be more than 57.90%. Meanwhile, the share of non-fossil energy increases significantly and would be more than doubled in 2050 compared with the BAU scenario. A carbon tax would result in a significant decline in emissions in the short term, but would have an adverse effect on economic growth and energy structure improvements. It is also observed that the integrated measures would not only substantially decrease the total emissions, but also improve the energy structure.

Reprinted from *Energies*. Cite as: Li, W.; Li, H.; Sun, S. China's Low-Carbon Scenario Analysis of CO_2 Mitigation Measures towards 2050 Using a Hybrid AIM/CGE Model. *Energies* **2015**, *8*, 3529-3555.

1. Introduction

There is a growing awareness that global climate change caused by the increase in greenhouse gas emissions into the biosphere is one of the main challenges in the 21st century. In order to meet the challenge of global climate change and to achieve energy-environment-economy development, the building of a low carbon society is a priority in the national development strategies of many countries. The concept of a low carbon society concerns all aspects of the society, including economy, culture and life [1]. The European Union (EU) has officially declared that it would cut emissions by at least 80% by 2050 [2]. Meanwhile, the U.S. and Japan have also made it clear that substantial investment in clean energy development is expected to prompt the reduction of CO_2 emissions by 80% in 2050 compared with those in 2005 [3]. Developing countries represented by the BRICS nations (Brazil, Russia, India, China and South Africa) have also expressed a strong willingness to control emissions, while putting forward their own specific plans or programs for emission mitigation. According to China's decarbonisation roadmap, the emissions per unit of gross domestic product (GDP) are required to be reduced by 40% to 45% compared with those in 2005, and renewable energy consumption should account for 15% of the total by 2020 [4]. However, China's emissions are more likely to continue rising rapidly in line with its ongoing industrialization and urbanization, which will result in more energy consumption [5], hence, China will face tremendous challenges in achieving its mitigation targets provided effective measures are not taken.

In response, the Chinese central government has made unremitting efforts towards exploring mitigation policies and other measures.

In the context of pursuing climate policy targets for 2020, a number of studies are underway to deeply probe different emission abatement policies and assess the corresponding compliance costs. Economists exceptionally value market-based climate management tools, to strive to minimize the economic costs in realizing the given emission reduction goals, and to provide dynamic incentives for long-term development of cheaper abatement technologies. The legislative package is designed to curb carbon emissions and to maintain low-cost abatement incentives in which carbon renewable, taxes and emission trading may all play a prominent role.

There is evidence that a broader set of energy and climate policies must be implemented to lower the abatement costs as much as possible. Research on imposing duties on carbon emissions has been done in the past few decades, mainly to identify the right tax rate that would help reduce carbon emissions with the lowest adverse impact on economic operations, and consequently alleviate climate change [6–8]. Emission trading is the most important regulatory instrument that makes emission allowances obligatory for CO_2 emissions from electricity generation as well as energy-intensive industries [9–11]. There is extensive literature on emission reduction targets, possible mitigation policies and costs in developed countries [12,13], but few studies are focused on the impacts of energy and climate policies in the context of emission mitigation in China. Undoubtedly, it is very necessary to quantitatively assess the potential results of the deployment of mitigation policies to reach the objective of impacting both the national economy and the energy mix to a large extent. Impact assessments of emission reduction commitments and non-fossil energy target plans for 2020, have been performed in [5,14].

Currently, renewable energy, considered as the main way to achieve energy savings and emission cuts, will see a substantial expansion over the next several decades. Subsequently, under a reasonable policy guidance for key economic sectors, technological changes will have great economic implications through improving energy efficiency whilst reducing the reliance on fossil fuels and cutting emission intensity [15]. Different from the carbon market in developed countries, an absolute cap on emissions needs to be set in China, which is eventually to be implemented by the different sectors and enterprises. Both carbon taxes and ET encourage enterprises to mitigate emissions in the production process by means of market forces. It is expected that timely introduction of a carbon tax from 2015 to 2020 which will achieve its impact through price measures will curb mounting emissions, while emission trading, a cap mechanism based on market competition, provides cost-effective and flexible environmental compliance for energy systems.

Increases of carbon emissions and their possible negative effects are major concerns in the imposition of any policy strategy [16]. Considering that climate policies including renewable energy plans, carbon taxes and emission trading are particularly helpful to combat climate change, it is very necessary to investigate the implications of alternative climate policies on economic and social development, so as to consequently provide policy recommendations for the state.

In recent years, as a popular policy simulation tool, computable general equilibrium (CGE) models have been widely employed in the analysis of climate policy, in particular, the assessment of long-term effects on economic consequences, sector productivity and energy systems [14,17,18].

This study aimed to provide an in-depth analysis of these mitigation measures, especially the impacts of renewable energy, ET and carbon taxes, on economic development, sectoral output, energy systems, *etc.* by establishing a new and revised hybrid static Asia-Pacific Integrated Model/computable general equilibrium (AIM/CGE) model.

The analysis is organized as follows: Section 2 briefly describes the AIM/CGE model and design scenarios. Section 3 presents the acquired simulation results, including the impacts of mitigation commitment on GDP, sectoral output, energy supply, and energy consumption by sectors and by types, energy intensity, carbon intensity and carbon prices and the impacts of different time levels to cancel the electricity price subsidies, increasing carbon taxes and limited emission trading. Finally, the results are summarized in brief, which also shows the useful policy implications of the study.

2. Methodology

2.1. AIM/CGE Model

The AIM/CGE model was developed by the National Institute for Environmental Studies (NIES), Japan. The AIM/CGE model is a recursive dynamic model, used to investigate the influence of climate policy at both the global [19,20] and country levels [21,22].

In this study, based on the "Standard CGE model", an improved AIM/CGE model is established to evaluate the impact of China's climate commitments, incorporating alternative energies and climate policies. The structure of China's AIM/CGE is approximately the same as that of the standard version [23]. It consists of four parts: production block, income block, expenditure block and market block. The present version of AIM/CGE is shown in Figure 1.

2.1.1. Production Block

In the standard AIM/CGE model, the default classification has 19 non-energy sectors and 19 energy sectors. In this model, in consideration of the study purpose and convenience, the sector classifications are disaggregated to 24 sectors, including 17 non-energy sectors and seven energy sectors. There are three basic types of inputs, including intermediate commodities, energy commodities and primary factors of capital and labor, regardless of land. It is assumed that the labor and capital are fully employed, perfectly competitive and freely mobile. The aggregation of intermediate commodities is in Leontief form while the composite of energy is a constant elasticity of substitution (CES) aggregation of fossil fuel (the bundle of six kinds of primary energy) and electricity. For the specific value of elasticity of substitution for inputs and outputs, see [5]. The following mathematical terms describes the production module at the lowest level:

$$F_i = A_i^f \left[\sum_{k=1}^{8} \alpha_{k,i} EC_{k,i}^{\frac{\sigma_{f,i}-1}{\sigma_{f,i}}} \right]^{\frac{\sigma_{f,i}}{\sigma_{f,i}-1}} \tag{1}$$

where eight kinds of energies, including coal, coke, crude oil, gasoline, kerosene, diesel, fuel oil, and natural gas, with the sum of share coefficients being 1. Moreover, produced emissions owing to the consumption of fossil energy would be calculated as follows:

$$EM_i = \sum_{k=1}^{8} c_k EC_{i,k} \tag{2}$$

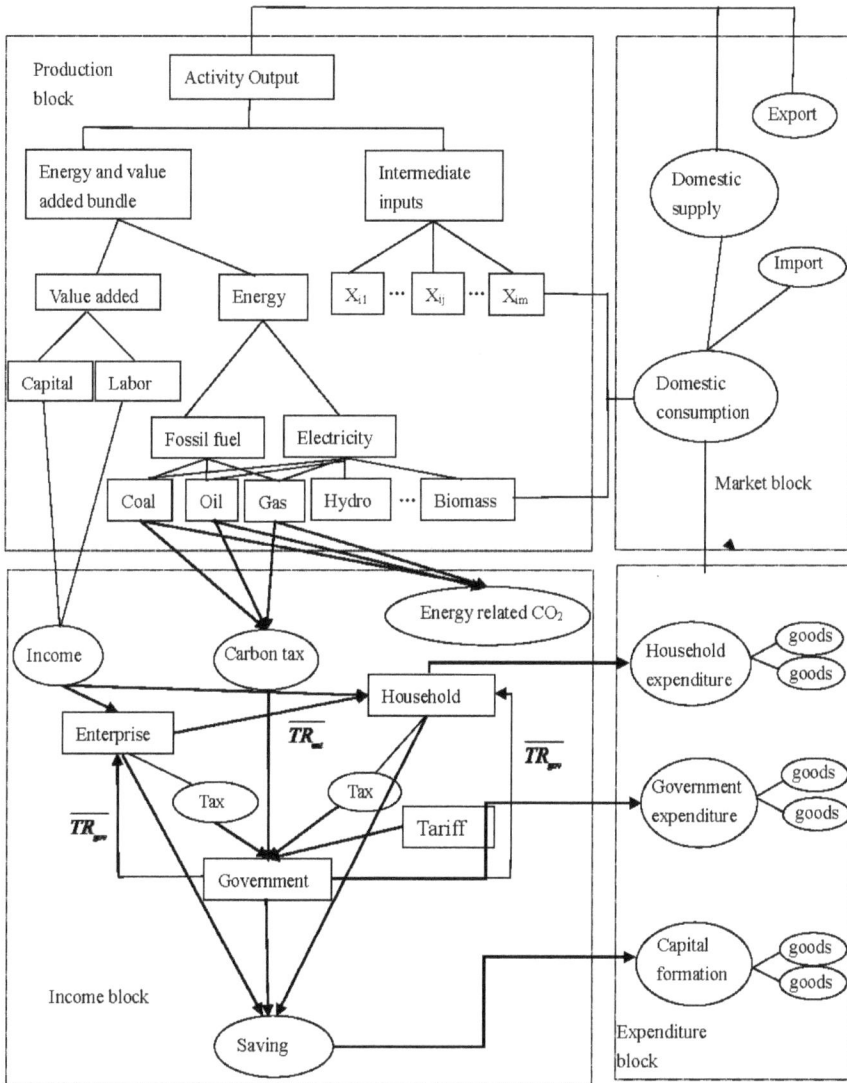

Figure 1. Overview of China's AIM/CGE model structure (for one region).

At the second level, the CES function represents the substitution possibilities between the demand for aggregation of fossil fuels and electricity as well as the value added composite of capital and labor:

$$E_i = A_i^e [\alpha_{F,i} F_i^{\frac{\sigma_{E,i}-1}{\sigma_{E,i}}} + (1-\alpha_{F,i})EL_i^{\frac{\sigma_{E,i}-1}{\sigma_{E,i}}}]^{\frac{\sigma_{E,i}}{\sigma_{E,i}-1}} \tag{3}$$

$$KL_i = A_i^{kl}[\alpha_{L,i}K_i^{\frac{\sigma_{KL,i}-1}{\sigma_{KL,i}}} + (1-\alpha_{L,i})L_i^{\frac{\sigma_{KL,i}-1}{\sigma_{KL,i}}}]^{\frac{\sigma_{KL,i}}{\sigma_{KL,i}-1}} \tag{4}$$

Subsequently, the value-add and energy aggregates a value added energy bundle with the CES composite, subject to a constant elasticity of substitution σ_{KLE}:

$$KLE_i = A_i^{kle}[\alpha_{KLE,i}KL_i^{\frac{\sigma_{KLE,i}-1}{\sigma_{KLE,i}}} + (1-\alpha_{KLE,i})E_i^{\frac{\sigma_{KLE,i}-1}{\sigma_{KLE,i}}}]^{\frac{\sigma_{KLE,i}}{\sigma_{KLE,i}-1}} \tag{5}$$

However, the composite of intermediate inputs trades off with the production function of Leontief forms with a fixed proportion of input factors, which is different from others:

$$M_i = \sum_{j=1}^{m}\alpha_{i,j}X_{i,j} \tag{6}$$

At the last level, the output comes from a CES nesting indicating the demand of sectors for value-added energy bundle and intermediate products:

$$Q_i = A_i^q[\alpha_{KLEM,i}KLE_i^{\frac{\sigma_{KLEM,i}-1}{\sigma_{KLEM,i}}} + (1-\alpha_{KLEM,i})M_i^{\frac{\sigma_{KLEM,i}-1}{\sigma_{KLEM,i}}}]^{\frac{\sigma_{KLEM,i}}{\sigma_{KLEM,i}-1}} \tag{7}$$

2.1.2. Income Block

The income module consists of three sectors, namely, households, government and enterprises. The representative household acquires income from the rental of basic factors and the lump-sum transfer from the government and enterprises. The enterprises or corporations mainly derive their income from capital income and lump-sum transfers from the government. The government collects its earning mainly from various taxes, including carbon taxes and uses the income to purchase commodities for its consumption and for transfers to other institutions. This block mainly describes the income allocation and distribution. The equations below give an excellent description of the earnings that institutions acquire:

$$Y_h = \sum_{i=1}^{24}W_i \cdot L_i + r_i \cdot HK + \beta \cdot \overline{TR_{gov}} + \overline{TR_{ent}} \tag{8}$$

$$Y_{ent} = \sum_{i=1}^{24}r_i \cdot K_i + \gamma \cdot \overline{TR_{gov}} \tag{9}$$

$$Y_{gov} = \sum_{i}^{24}(\tau_i \cdot P_i \cdot Q_i) + \tau_r \cdot Y_h + \tau_{ent} \cdot Y_{ent} + TARIFF + TAX_{carbon} \tag{10}$$

It must be pointed out that institutions, including households and enterprises, should pay the carbon tax based on their CO_2 emissions, seen in Figure 1. The CO_2 emissions by each sector and household are estimated according to their fossil energy consumption. In the basic model version, the revenues are assumed to be received by the households. However, in this model, the revenues collected through carbon taxes are distributed to households and enterprises in the form of temporary tax cuts and transfer payments. Meanwhile, the returns are consistent with the calculation of confirmed taxable income rate, respectively. This study ignores the impact of carbon tax on households and focuses on the impact on sectors.

In this study, we follow the assumption that direct taxes and transfers to other domestic institutions are defined as fixed shares in the basic model version. In addition, the assumption of the balance of government financial revenues and expenditure is applied in this model. The taxation tariffs are also an important tax revenue and a control means for the central government. Products brought into one country are often subject to import taxes, or customs duties. Accordingly, imports are controlled by the adjustment of customs duty rates.

2.1.3. Expenditure Block

The expenditure block describes household expenditures, government spending and capital formation, whose final consumption is goods. The income of the representative household is spent on commodities or investments. It is natural that households intend to maximize commodity utility, which is subject to the constraints of income and prices. In terms of investment, household expenditures are assumed to grow linearly at the same pace as the GDP growth rate. Government spending and capital formation is assumed as a constant coefficient function. Both purchases and transfer of the government are introduced as an exogenous variable, so that the consuming quantity is fixed.

2.1.4. Market Block

The market block represents the process of commodity trading. Domestic commodities are disaggregated into domestic and export markets. Based on the Armington assumption, the aggregation of domestically produced and imported commodities with an imperfect substitute supplies the domestic market, derived on the assumption that domestic demanders minimize cost. The derived demands for imported commodities are met by international supplies that are infinitely elastic at given world prices. The import prices paid by domestic demanders also include import tariffs (at fixed *ad valorem* rates) and the cost of a fixed quantity of transactions services per import unit, covering the cost of moving the commodity from the border to the demander. Similarly, the derived demand for domestic output is met by domestic suppliers. The key difference with the standard version is that certified emission reductions (CERs) are regarded as a commodity in this model. We are working on the assumption that CERs are a commodity produced by enterprises. The amount of CERs owned by enterprises is determined by the emissions gap between the limits set by the government and the real value. The CERs can be traded in the carbon market. The carbon market demands are for a composite commodity made up of imports and domestic output. Developers of emission reduction projects act as a buyer transacting the entity of emissions reduction at a lower cost.

The AIM/CGE model treats the volume of all commodities as monetary units. Domestic energy sectors produce energy commodities and ensure supplies if not imported. The energy sectors in this study are classified into seven kinds. In the basic version, the share of imported and domestic consumption is determined by the ratio of the current price to the previous year's price. In this study, we solve the CES aggregation function to find the optimal value. The Equations (11)–(14) show how the optimal allocation of all commodities including energy between the imports and domestic supply:

$$QC_l = A_l^{am}[\alpha_{m,l}QM_l^{\sigma_{m,l}} + (1-\alpha_{m,l})QAM_l^{\sigma_{m,l}}]^{\frac{1}{\sigma_{m,l}}} \tag{11}$$

$$\frac{PM_l}{PAM_l} = \frac{\alpha_{m,l}}{(1-\alpha_{m,l})}\left(\frac{QAM_l}{QM_l}\right)^{1-\sigma_{m,l}} \tag{12}$$

$$PC_l \cdot QC_l = PM_l \cdot QM_l + PAM_l \cdot QAM_l \tag{13}$$

$$PM_l = (1+\tau_{m,l}) \cdot \overline{pwm_l} \cdot \overline{EXR} \tag{14}$$

2.2. Dataset

The dataset used in this model includes the 2005 input-output table [24], energy balance table [25], and CO_2 emission factors of fossil fuels [26]. The year of 2005 is thus used as a base year. In this study, we use a single country social accounting matrix (SAM) which covers the typical sectors. The SAM in 2005 in this model is from the prolonged input-output table. In China's input output table, petroleum and natural gas are treated as one sector, the disaggregation of which in this study is done by using the consumption information of petroleum and natural gas among sectors in the energy balance table. The SAM describes physical volume transaction data as physical unit volumes. The volume of energy consumption is calculated based on the monetary volume transaction data and the energy price in the base year.

In the CGE model, energy consumption is commonly treated as a monetary term. However, in this paper, energy consumption is discussed in physical terms. Data from the energy balance table in 2010 is used to turn monetary terms into physical counterparts. SAM is formulated based on the energy balance table, whose balance is adjusted by a cross-entropy method [27]. Relevant coefficients include carbon emission factors and elasticity of substitution. It is assumed that the annual GDP growth rate is 7.5% from 2011 to 2015 in accordance with the policy guidance [28]. For the total GDP and growth rate after 2015, we refer to [29]. The current population growth rate in China is slow. If the total fertility rate is held at 1.8, the population in China will reach a peak of 1500 million persons around 2033, but then drop to 1383 million in 2050 [30]. Consequently, the annual growth rate of population can be roughly calculated (see Table 1).

The labor grows at exactly the same rate as population change. The capital stock is dynamically determined by the previous year's stock, capital formation and depreciation. The relevant data are based on [31]. The efficiency coefficients in the CES energy and value-added function, known as the total factor productivity (TFP), stand for the technology status. The efficiency coefficients change autonomously at 1.5% per year in agriculture, construction and the service sector. However, the efficiency coefficients in the industrial sector are determined by labor and capital stocks and will be calculated every year [14]. The direct subsidies to the production of renewable electricity are used to assess the impact of the renewable plan. The feed-in tariff and subsidy rate for the renewable during the study period are displayed in Table 2 [15].

Table 1. GDP and population growth.

Year	GDP		Population	
2005 (base year)	Billion USD	2903.75	Million persons	1307.56
2006–2010		9.57%		0.80%
2011–2015		7.50%		0.61%
2016–2020		6.15%		0.61%
2021–2025		5.82%		0.14%
2026–2030	Growth rate	5.50%	Growth rate	0.12%
2031–2035		5.20%		−0.08%
2036–2040		4.90%		−0.13%
2041–2045		4.61%		−0.33%
2046–2050		3.05%		−0.52%

Table 2. The feed-in tariff and subsidy rate for the renewable during the study period.

Renewable electricity	Renewable electricity feed-in tariff (Yuan/kwh)	Fossil fuel electricity sale price (Yuan/kwh)	Subsidy rate (%)
Wind	0.55	0.40	38
Solar	0.95	-	138
Biomass	0.75	-	88

Looking ahead, the non-fossil energy share is predicted to be at 11.4% in 2015 and 15% in 2020, the specific target of which is shown in Figure 2 [32]. The total CO_2 trading volume is likely to exceed 24 million tons this year, which will surge to about 227 million tons in 2015, nearly nine times higher than the current level. Furthermore, the implementation of subsidies for renewable electricity will occur between 2010 and 2020, and possibly be phased out in 2020, referring to the corresponding data [15].

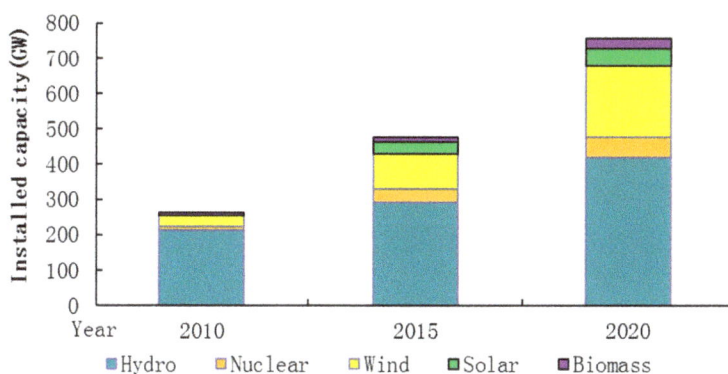

Figure 2. Published situation and targets for installed capacity of renewable energy towards 2015 and 2020 in China.

2.3. CO2 Mitigation Scenarios

In this paper, a total of 11 scenarios are analyzed to investigate the impacts of three key policy measures. In the business as usual (BAU) scenario, it is assumed that the three CO_2 mitigation countermeasures or policies have not been introduced during the study phase. The GDP growth and population change follow the assumptions of Table 1. In addition, some variables including labor productivity changes, energy efficiency and capital productivity improvement are introduced to indicate technical progress and the corresponding values refer to [5]. The fossil fuel consumption from 2010 to 2050 is adopted from [33]. The other ten scenarios are designed to assess the impacts of policy combinations in fulfilling CO_2 mitigation commitments in contrast to the BAU scenario by using the AIM/CGE model. China has made an ambitious commitment to cut down its carbon dioxide emissions of GDP per unit from 40% to 45% by 2020 compared with that in 2005. Accordingly, all scenarios are divided into CM40 and CM50, which are roughly consistent with the national mitigation targets, *i.e.*, 40% and 50% reduction of the total CO_2 emissions. In this paper, all sectors should cut their emissions by 40% or 50%, respectively, compared with those in 2005. The CO_2 emissions come from the fossil energy consumption when enterprises produce their products. All sectors should take emission reduction into account and then decide the appropriate activity level. The CO_2 emissions will restrict the activity level as an exogenous variable. Since laws and regulations promoting the development of renewable energy were formulated in 2006, all scenarios have included the renewable energy, excluding the BAU scenario. The National Development and Reform Commission (2014) announced that a carbon market would probably be launched during 2016–2020 [34]. Considering the stage and uncertainty of this policy implementation, it is estimated that emission trading will be established throughout the country in 2016. A carbon tax is to be collected by 2020. When the national emission trading market is started in 2016, CERs transactions will be conducted freely on a global scale. In this study, carbon tax rate is designed as a constant, and always remains at 5 \$/t. The "E", "T", and "C" in the end of scenario names mean renewable energy, emission trading and carbon tax, respectively. For example, the emission trading policy will be taken into account in the CM40ET, CM40ECT, CM50ET and CM50ECT scenarios. Detailed information about all scenarios is shown in Table 3.

Table 3. CO_2 emission reduction measures in all scenarios.

Scenario	Reduction level (%)			Renewable energy	Emission trading	Carbon tax
	0%	40%	50%			
BAU	+	−	−	−	−	−
CM40	−	+	−	−	−	−
CM50	−	−	+	−	−	−
CM40E	−	+	−	+	−	−
CM50E	−	−	+	+	−	−
CM40ET	−	+	−	+	+	−
CM40EC	−	+	−	+	−	+
CM50ET	−	−	+	+	+	−
CM50EC	−	−	+	+	−	+
CM40ETC	−	+	−	+	+	+
CM50ETC	−	−	+	+	+	+

Note: "+" indicates the given scenario containing the appropriate policy and "−" indicates that corresponding measure is not considered.

3. Simulation Results

Our model allows us to assess the impacts of emission mitigation measures on macroeconomics, sector output, energy system and carbon prices. The results of each scenario are presented under the background of three alternative mitigation policies to support the exploration of China's future mitigation strategy.

3.1. Carbon Intensity

3.1.1. Carbon Intensity in This Model

China's sky-rocketing energy consumption causes high emissions. The current situation shows that emission mitigation is relatively difficult. Nevertheless, China's government has made an ambiguous commitment to lower emissions per unit of GDP from 40% to 45% compared with those in 2005. Over the past five years, the total CO_2 emissions have increased from 5170 million tons in 2005 to 8332 million tons in 2010, a 61.16% rise. In terms of emissions for GDP per unit from 2005 to 2010, they only dropped from 1.36 to 1.29 tons per 1000 US dollars (t/1000 USD, a 5.14% drop) meaning that there are enormous pressures in achieving reduction targets.

Results of CO_2 emission intensity during 2020–2050 are illustrated in Figure 3. It is shown that CO_2 emission intensity changes significantly after the deployment of renewable energy, a drop of 33.33% in CM40E and 36.43% in CM50E by 2020, respectively, compared with 1.36 t/1000 USD in 2005. On the basis of renewable energy, emission trading will reduce emission intensity by 43.31% in CM40ET and 46.53% in CM50ET by 2020, which obviously shows a high probability of achieving China's CO_2 emission mitigation target commitments made in Copenhagen.

The CO_2 emissions in the BAU scenario remain on a downward trend, but the value in 2050 is apparently higher than that in other scenarios. The carbon intensity is less than 0.70 t/1000 USD in the scenarios with renewable energy and emission trading, which means that the targets proposed in Copenhagen are achievable. It is apparent that the carbon tax overwhelmingly contributes to lower carbon emissions when implemented. However, the effect of emission trading measures on emission intensity reduction obviously surpasses that of the carbon tax. Besides, the impacts of carbon tax on reducing emission intensity work in the short term, for example, the annual average rate of decrease of emission intensity is 1.1%–1.4% in CM40EC, CM40ETC, CM50EC, and CM50ETC during 2020–2030, while it is only 0.2%–0.5% during 2030–2050. In the short run, energy intensive enterprises have to reduce their production to fulfill the emission reductions in the face of the absolute increase in production costs because fixed costs remain constant and the introduction of new technology or switching to other production may be unrealistic. Therefore, the emission reduction and carbon intensity drop are obvious. However, in the long run, introduced low carbon technology and turning over to other sector makes the carbon tax less efficient. Moreover, energy efficient enterprises lack any motivation to reduce emissions because they pay lower carbon taxes. Hence, the carbon intensity would not sink rapidly in the long term.

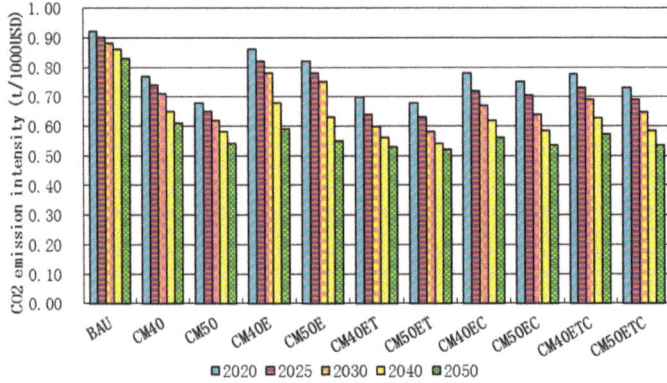

Figure 3. CO_2 emission intensity from 2020 to 2050.

In contrast, the emission intensity in scenarios without a carbon tax falls steadily during the research period. The estimated value of emission intensity falls to 0.54 t/1000 USD in CM40ET and 0.52 t/1000 USD in CM50ET, respectively, by 2050. The CM50ECT scenario sees roughly the same effect as CM50ET.

3.1.2. Renewable Energy Subsidies and Carbon Intensity

In the early stages, the development of renewable energy is faced with immense difficulties due to the high costs. Given that renewable energy plays a key role in the development of any national energy strategy, most advanced countries and regions have established appropriate capital subsidy mechanisms. As the largest developing country, China has adopted effective measures to support the development of renewable energy, mainly including renewable energy feed-in tariffs. The continuing financial support for promoting the development of renewable energy will unquestionably increase the government's financial burden. The tariff subsidies are estimated to be 16.7 billion dollars in 2015 and 25.3 billion dollars in 2020. Tariff subsidies are predictably eliminated based on the renewable energy development trend. Nevertheless, a sober assessment is needed to establish the best specific time to cancel subsidies.

In this study, diverse temporal patterns are used to assess the impacts when the government removes its tariff subsidies at different times. It is assumed that subsidies are removed in 2015, 2020 and 2025, respectively. It shows a marked comparison of carbon intensity in 2050 after removing the tariff subsidies at different times. Results reveal that the early cancellation of subsidies in 2015 curbs carbon intensity reduction in the future, while the abolition of subsidies in 2025 makes little difference. The following reasons may provide a general understanding of this phenomenon. Although the cost of renewable energy is higher than that of fossil energy, renewable energy is the major means to reduce the emissions before 2015. Over the ten years, the development cost of renewable energy has dropped by more than 50%, in particular, the cost of photovoltaic power generation has decreased by 15% every year. The renewable electricity will be cheaper after 2015, which could be acceptable for some enterprises. However, the thermal power generation cost in China is likely to rise because of the increasing coal price. Moreover, in order to achieve the mitigation goals, enterprises are bound to use renewable energy even if it is more expensive because it is still cheaper than the introduction of new technology.

By comparing the scenario of removing subsidies in 2020 with that in 2025, it is found that the margin of carbon intensity reduction in all reduction measures is less than 0.2 t/1000 USD. In other words, the result coincides with experts' predictions that after 2020, the removal of subsidies is relatively appropriate at least from the perspective of reducing carbon emissions.

3.1.3. Limited Emission Trading and Carbon Intensity

Free emission trading implies that no trade credit and scope will be captive. However, emission trading may be restricted to a certain level. China's enterprises will be in trouble in free emission trading with other countries, such as the EU, stemming from the gap of production technology. China's enterprises are always at a disadvantage in market competition due to their technical level. In order to protect domestic firms, limited emission trading may play an important role in accelerating the development of native industries. Therefore, it is significant to explore the impacts of trade constraints on activity level and mitigation.

After a comprehensive analysis and calculation, results convincingly indicate the inappropriate expectations. For protectionist reasons, 50% of the trade is merely restricted to domestic sources but this is unable to produce the desired effect, and worse still, it prevents sectors from raising the level of output. Protectionism prevents enthusiasm for innovation, but is conducive to production technology development. The decline in GDP reaches 1.3% compared with that in the free trade level, unwilling to address this aspect, which means sector activity will be undoubtedly affected, partly due to the technical gap. Surprisingly, carbon emission intensity is not affected much by the carbon taxes deployed in the CM40ETC and CM50ETC scenarios, compared with the average increase of 0.15 t/1000 USD in other scenarios. Obviously, the emission trading for limiting the amount of the transaction does not lower the carbon emission intensity as expected, and instead it causes some adverse effects on economic development and technological progress.

3.1.4. Increasing Carbon Tax and Carbon Intensity

In this study, the carbon tax rate is held at 5 $/t for simplicity sake, however, it is changeable in the practical economic and social context based on the government policies. Therefore, in this section, the carbon tax rate is assumed to increase from 2 $/t to 10 $/t within a decade, after which it remains unchanged and consistent with the current policy direction. The output in the industrial sector accounts for a large proportion of the GDP. As expected, a carbon tax inevitably has an adverse impact on the activity level in the industrial sector. However, the effects of carbon tax can be significantly reduced if a lower tax rate is followed initially in accordance with the deployment of government norms. In addition, the increase of marginal abatement costs is less pronounced. A lower carbon tax serves as a guide in improving the production technology and energy efficiency to some extent. With the increase of tax rate, enterprises will shoulder the pressure of reducing the energy consumption per unit of output to meet the cost increases due to energy consumption.

Simulation results indicate CO_2 emission intensity changes with the carbon tax rate gradually increasing from 2 \$/t to 10 \$/t according to the hypothesis. Compared with fixed rates, incrementing the tax rate contributes to a more substantial reduction in carbon emissions and the consequential drop in carbon intensity. From a practical point of view, higher carbon tax indeed can greatly reduce carbon emissions instead of carbon intensity. Based on the simulation results, emission intensity will see a big decline in CM50EC and CM50ETC, reducing to 0.51 t/1000 USD in CM50ETC, a drop of 0.2 t/1000 USD, however, a higher carbon tax rate has a huge impact on China's GDP. At the carbon tax rate of 10 \$/t, the GDP loss in CM40EC and CM50EC reaches 3.47% and 5.06%, respectively.

3.2. CO_2 Mitigation and CO_2 Prices

CO_2 prices are changeable with the increase of imposed CO_2 emission constraints, as presented in Table 4. In the CM40 and CM50 scenarios, it is assumed that carbon emissions per unit of GDP are reduced by 40% and 50%. The CO_2 prices in 2050 are consistent with emissions reduction.

Table 4. Carbon prices and the corresponding CO_2 mitigation.

Scenarios	CO_2 mitigation (%)	Carbon prices USD/t-CO_2
CM40	40	51.6
CM50	50	67.9
CM40E	56.6	98.1
CM50E	59.6	118.2
CM40ET	61.0	143.4
CM40EC	61.8	92.8
CM50ET	58.8	138.4
CM50EC	60.7	100.8
CM40ETC	57.9	84.75
CM50ETC	60.5	110.2

This occurs in the scenarios with renewable energy policies, *i.e.*, the CM40E and CM50E, in which CO_2 prices are much higher than those in the CM40 and CM50 scenarios. The reason is that higher emission reduction constraints result in higher cost of CO_2 mitigation procedures and high carbon prices. The CO_2 price is the most significant driving force of emission reduction. Therefore, higher prices help to drive more CO_2 emission reduction.

Emission trading results in high CO_2 prices and encourages more emission reductions. Emission trading brings about the highest CO_2 prices in the CM40ET and CM50ET scenarios. In the CM40ET scenario, the carbon emission reduction level is higher than that in the CM50ET, but with nearly the same CO_2 price. The key reason is that the relationship between the CERs supply and its price roughly shows an inverted U-shaped curve. When the CERs price is lower, the price increase will result in more CERs supply. The price of CERs is unlikely to go up unboundedly. Too high a CERs price is uneconomic or unacceptable for enterprises. In this case, the enterprises prefer to pay the penalty for uncompleted emissions reduction rather than take the initiative in mitigating the emissions. When all enterprises achieve their optimal reductions based on profit function, they lose the desire to cut the emissions unless the CERs prices fall. If not, mitigation target could not be achieved. Therefore, only a depressed CERs price would drive more emissions reduction. Therefore, it is possible that a certain price corresponds to two different emission reduction scenarios.

It can be seen that the carbon price in CM40E and CM50E is higher than that in CM40 and CM50. Generally, more renewable energy should reduce the demand for carbon permits and reduce their price. However, renewable energy availability promotes the output growth in sectors compared with mandatory emission reductions. The emissions increase due to production expansion is more than the decrease due to the use of renewable energy. In fact, the CO_2 emissions in the scenarios with renewable energy are relatively higher than in the mandatory reduction scenario, which would drive the rise of carbon price.

3.3. Macroeconomic Impacts

Deployment of emission mitigation policies will inevitably exert a far-reaching influence on Chinese macroeconomic operation. Mandatory carbon emission reduction requirements are equivalent to imposing carbon emission constraints on the macroeconomics. As a result, economic operation costs will increase, and economic agents will make necessary adjustments to concepts, technologies, business models and even consumption patterns. Affected by this mechanism, the government as organizers and managers have assigned emission limits to the enterprises. This would limit how much CO_2 can be emitted in production activities. The enterprises will be penalized for excessive emissions, which may severely disrupt production. Therefore, enterprises are bound to adjust their production plans or introduce low-carbon technology to meet the emission targets, which would raise the product price as likely as not. Besides, the decrease of fossil energy consumption is probably inevitable due to the law of conservation of carbon. The energy consumers must spend time to understand and find alternative energy sources. In this study, we focus on the direct influence on the yield of all sectors. By definition, the GDP would be calculated according to the output level and fixed base year price. Reasonable mechanism of emission reduction is helpful for promoting technological innovation and industrial upgrading, and attracting more investments, which contributes to the improvement of the trading environment, particularly for developing countries.

GDP is deemed to be changeable in line with the implementation of mitigation policies, which is usually used as the indicator of evaluating the economic impacts. Table 5 shows the effect of mitigation policies on GDP growth under the corresponding targets. Compared with the BAU scenario, cumulative GDP loss in 2020 is relatively mild in the CM40 and CM50 scenarios, *i.e.*, 2.54% in CM40 and 3.52% in CM50, respectively. Higher GDP loss in CM50 shows that when a higher mitigation target is set, the loss of GDP will increase greatly due to the rapid expansion of marginal abatement costs. However, economic costs are relatively lower in the short run under the emission reduction scenario, which is ascribed to constant progress in technology and the improvement of labor productivity. However, GDP in CM40 and CM50 shows a considerably different growth rate in 2050. The loss of cumulative GDP is 3.61% in CM40, while a larger loss of 6.91% occurs in CM50, indicating that mandatory emission reductions will have substantial influence in the long run. This is probably due to an unreasonable adjustment of industry structure caused by exceeding emission reductions that makes economic growth flabby.

Table 5. GDP losses due to CO_2 mitigation policies (%).

Year	Emission reduction scenarios									
	CM40	CM50	CM40E	CM50E	CM40ET	CM50ET	CM40EC	CM50EC	CM40ETC	CM50ETC
2010	−0.83	−1.53	−1.22	−1.47	−1.22	−1.47	−1.22	−1.47	−1.22	−1.47
2015	−1.53	−2.01	−1.65	−1.83	−1.65	−1.83	−1.65	−1.83	−1.65	−1.83
2020	−2.54	−3.52	−2.11	−2.12	−1.82	−1.92	−2.11	−2.12	−1.82	−1.92
2025	−2.73	−3.85	−2.23	−2.36	−1.87	−2.16	−2.48	−3.11	−2.19	−2.54
2030	−3.25	−4.50	−2.36	−2.58	−1.95	−2.35	−2.81	−3.54	−2.37	−2.82
2040	−3.42	−5.42	−2.49	−2.71	−2.11	−2.51	−2.96	−4.07	−2.52	−3.10
2050	−3.61	−6.91	−2.57	−2.79	−2.23	−2.60	−3.09	−4.85	−2.63	−3.41

Note: The value of GDP is estimated based on commodity prices in 2005.

Expansion of renewable energy use plays an absolutely critical role in cutting emissions during the period from 2010 to 2020. However, it would have an adverse effect on GDP growth. As shown in Table 5, the measurable cumulative reduction of GDP is about 1.53% in CM40E, and 2.79% in CM50E by 2050. This is mainly because developing renewable energy sources is more costly than fossil energy in the short run. Moreover, the government will invest much in developing it. All these efforts increase the economic operation costs. However, in the long run, the costs will fall because of technology maturity and renewable energy will become as cheap as fossil energy. The relatively abundant energy decreases the cost of activity level, and consequently, boosts the economy.

Emission trading, expected to be fully implemented in 2016, significantly promotes economic growth. Table 5 shows that from 2015, the GDP cumulative loss compared with the BAU scenario has been continuously declining as time goes on. However, from 2018 in CM40ET and 2033 in CM50ET, GDP starts to exceed the BAU scenario, ultimately up to 1.85% in CM40ET and 1.04% in CM50ET compared to the BAU scenario by 2050. This gives rise to accumulative total GDP growth of 1.44% in CM40ET and 0.98% in CM50ET. Compared with CM40E and CM50E, the GDP loss in CM40ET and CM50ET is also lower. Under emission trading policy, CERs would be regarded as a kind of good to trade. According to the assumption and the definition of GDP, the GDP would increase compared with the scenarios without emission trading and this situation will continue.

A carbon tax means increasing the revenue for the government, which is stored before serving as production subsidies in various sectors. A carbon tax adversely affects the GDP in the short-term, leading to a large gap of 0.9% in CM40EC compared with CM40E in 2025. Energy costs in all sectors would increase if a carbon tax is introduced, which lowers the activity level in sectors and suppresses GDP growth. The analysis shows that the trend of GDP growth rates with carbon tax in the scenarios of 50% CO_2 mitigation is similar to that in CM40EC, the difference is that the GDP loss in CM50EC is much greater. In addition, the negative impacts of carbon tax on GDP decline in the long run. Table 5 shows that in CM40EC, CM50EC, CM40ETC and CM50ETC, the GDP loss continues to decline from 2020 to 2050, which shows that the impact of a carbon tax weakens gradually. Carbon taxes have an adverse impact on industrial structure optimization, which impedes economic development. Especially in the short term, the impact of carbon tax on economic growth is heavy. In the long run, the enterprises have enough time to decide where the money is going and then manufacture and expand the activity level. As a result, the impact of carbon tax is weakened. Comparatively speaking, emission trading have a positive impact which constantly drives GDP growth. On the contrary, the impact of carbon tax is negative and is significant in the short term.

We prefer to lower the GDP loss as much as possible when carrying out mitigation measures. In 2050, a GDP loss above 3% is seen in CM40, CM50, CM40EC, CM50EC and CM50ETC. However, in the scenario of BAU, GDP growth is assumed to be 3.05% in 2050. The above scenarios probably result in economy stagnation or negative growth and are unreasonable. A lower GDP loss is enjoyed in CM40E, CM40ET and CM40ETC. For this reason the above three policy combinations are acceptable.

Table 6. Change of activity level in various scenarios in 2050 compared with BAU (%).

Sectors	Scenarios									
	CM40	CM50	CM40E	CM50E	CM40ET	CM50ET	CM40EC	CM50EC	CM40ETC	CM50ETC
Agriculture	0.85	3.89	1.02	4.21	1.25	4.31	1.06	4.36	1.05	4.19
Mineral mining	−0.42	−2.63	−0.38	−2.44	−0.34	−2.01	−0.42	−2.65	−0.39	−2.43
Foods and Tobacco	0.81	5.26	0.96	5.32	1.00	6.47	0.99	5.38	0.94	5.61
Textiles and clothing	0.4	2.23	0.51	2.65	0.67	3.21	0.52	2.45	0.53	2.64
Wood and furniture	0.35	1.98	0.43	2.33	0.65	2.86	0.47	2.45	0.48	2.41
Paper products	−0.08	−0.64	−0.06	−0.54	−0.05	−0.45	−0.08	−0.62	−0.07	−0.56
Chemical	−0.32	−1.85	−0.25	−1.52	−0.18	−1.34	−0.32	−1.79	−0.27	−1.63
Cement	−0.93	−3.05	−0.76	−2.54	−0.72	−2.43	−0.88	−3.11	−0.82	−2.78
Non-metallic Mineral Products	−0.30	−1.67	−0.22	−1.54	−0.16	−1.41	−0.28	−1.98	−0.25	−1.65
Iron and steel	−1.06	−4.66	−0.94	−3.59	−0.88	−3.31	−1.26	−4.58	−1.04	−4.04
Nonferrous products	−0.56	−2.06	−0.43	−1.68	−0.39	−1.56	−0.55	−2.31	−0.48	−1.90
Water production	0.22	0.86	0.32	1.02	0.36	1.25	0.39	1.08	0.32	1.05
Machinery	−0.35	−2.34	−0.32	−2.14	−0.26	−2.04	−0.35	−2.38	−0.32	−2.23
Other manufacturing	0.12	0.96	0.15	0.98	0.18	1.25	0.89	0.15	0.34	0.84
Construction	0.02	−0.05	0.05	−0.03	0.08	0.01	0.03	−0.05	0.05	−0.03
Transportation	−0.21	−0.65	−0.16	−0.47	−0.16	−0.36	−0.12	−0.42	−0.16	−0.48
Service sector	0.2	0.78	0.35	1.14	0.46	1.45	0.38	1.32	0.35	1.17

3.4. Non-Energy Sector Output Impacts

The integrated policy of promoting emission reductions has different impacts at the sectoral level. With the mitigation measures being introduced, energy intensive enterprises will be faced with increased costs and shrinking profits [35]. Energy costs increases have an important impact on activity levels. The activity level in energy-intensive sectors, such as steelmaking, and cement production shall suffer the most (Table 6). For instance, the output in the steel sector drops by 1.06% and 4.66% in the CM40 and CM50 scenarios, respectively, suffering severely from mandatory emission reductions, followed by the cement sector with a decline in output by 0.93% in CM40 and 3.05% in CM50, respectively. The decrease of activity level means that CO_2 emissions from production will fall consequently.

The carbon tax increases enterprises' production costs due to the raising prices of fossil energy, which suppresses the activity level. The feed-in subsidies make renewable energy competitive in the energy market. Renewable energy becomes an significant force to drive the fossil energy price drop. Consequently, the production costs in all sectors will decline due to the deployment of renewable energy sources. The decline of production costs helps increase production, in particular for the energy-intensive industries such as the power, steel, chemical and transport sectors because the expenditures on energy use account for a large part of their total costs [15,36]. Under an emission trading mechanism, enterprises don't need to follow the specified emissions limits strictly and pay heavy penalties for excessive emissions. The emissions gap would be offset by purchasing CERs at lower cost. Compared with the mandatory emission reduction scenario, the production costs will fall when emissions exceed the limits. The enterprises can rearrange their production schedules and expand the production to obtain greater profits, therefore, emission trading would improve the activity level.

Nevertheless, Table 6 shows that mitigation measures cannot significantly alleviate the decline in output level. First, the implementation of a carbon tax increases the production costs both in the short term and long term, in particular to energy intensive sectors. The increased costs with hamper the production expansion. Second, renewable energy is more favored than fossil energy. Due to government feed-in tariff subsidies, the price of renewable electricity will be reduced which will make renewable energy highly competitive. However, given the assumption of subsidy removal in 2020, this high competitiveness would decrease. In addition renewable energy is in short supply and cannot fully meet the demand. Consequently, the motivation for energy intensive sectors is limited. Third, under a carbon trading mechanism, enterprises need to purchase CERs in the carbon market which increases their total production costs. Although the enterprises are able to meet emission reduction quotas, they cannot expand their production unconditionally due to the emission limits and only do so if the government forecasts their output and allocates reasonable emissions. Unfortunately, the carbon trading would then become meaningless.

Energy efficient sectors, including the service, agriculture and food industry shall enjoy a drastic increase. The energy costs accounts for a small share of costs in energy efficient sectors and the impact of energy cost changes on output is limited. On the contrary, the introduction of new technology improves production efficiency. The most striking discovery is that the output levels in energy efficient sectors go up in pace with the rise in emission reductions For example, the output of service production increases the most under any scenario, in particular, in CM40ETC (3.11%) and CM50ETC (4.59%). Even so, only a small amount of the increase in emissions comes from service, foods and agriculture sectors.

3.5. Fossil Energy Supply and Final Consumption

The impacts of carbon abatement on fossil energy supply in all the scenarios in 2050 are illustrated in Figure 4. Despite the mitigation measures, coal supply by 2030 is expected to suffer a blow, which is due to the fact that coal combustion releases more CO_2 than other fossil energies.

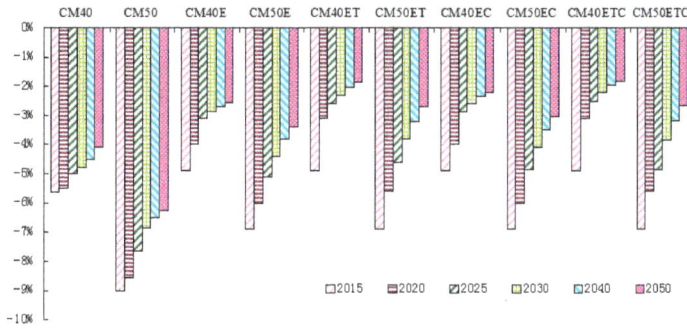

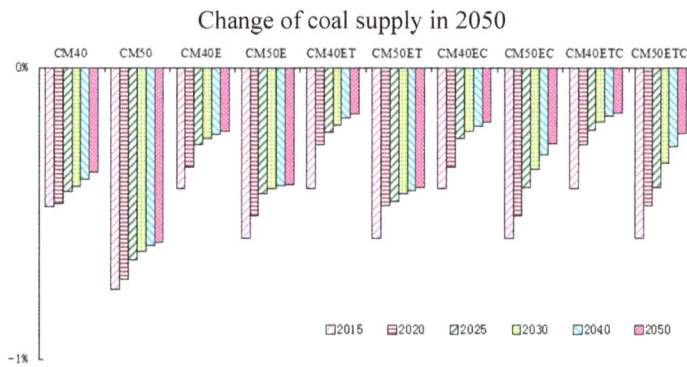

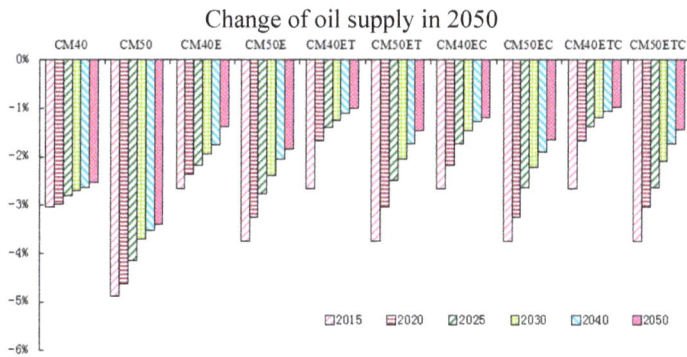

Figure 4. Change of fossil energy supply in the varied scenarios compared with BAU in 2050.

Coal supply will be 3158 million tonnes coal equivalent (Mtce) in CM40ETC and 2960 Mtce in CM50ETC, accounting for 45.2% and 43.2% respectively in the total energy supply in 2050, while oil production will be almost unaffected, *i.e.*, a slight drop of 0.78% in CM50, whose share will roughly remain unchanged in all scenarios. Limited resources and production capacities result in China's heavy dependence on imported counterparts. The share of oil consumption remains steady, but oil consumption will severely depend on imports in 2050, with a degree of dependence of up to 80%, which is bound to increase energy security risks. China's renewable energy development of wind, solar, and non-traditional biomass will replace fossil energy use. Meanwhile, renewable electricity generation increases significantly with the decreasing cost of renewable electricity.

Renewable electricity in both the CM40 and CM50 scenarios sees relatively moderate increases, which indicates binding emission reduction commitments promote renewable electricity development to a certain extent (Figure 5). Current emission reduction policies result in significant growth in renewable electricity generation. Furthermore, the total and renewable electricity in general are higher under the CO_2 emission reduction target of 50% than that at 40%. It is strongly noticeable that renewable electricity increases by approximately six times in CM50ET and the total electricity generation reaches 2584.52 Mtce, including 555.80 Mtce in thermal power, 730.02 Mtce in hydro power, 848.70 Mtce in nuclear power and 450.00 Mtce in renewable power, respectively. In contrast to CM40ET, the renewable energy contribution grows faster than that in CM40EC and the same situation occurs in CM50EC and CM50ET. In addition, the contributions of nuclear and hydro power in the scenarios with emission trading increase significantly that with a carbon tax. In comparison, emission trading is more efficient than a carbon tax in term of promote renewable energy development.

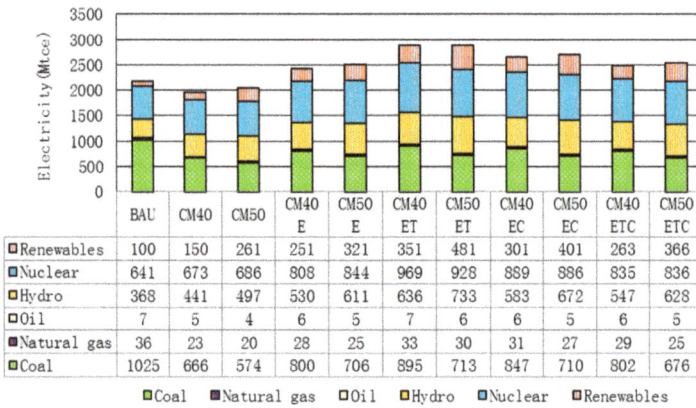

	BAU	CM40	CM50	CM40 E	CM50 E	CM40 ET	CM50 ET	CM40 EC	CM50 EC	CM40 ETC	CM50 ETC
□Renewables	100	150	261	251	321	351	481	301	401	263	366
■Nuclear	641	673	686	808	844	969	928	889	886	835	836
□Hydro	368	441	497	530	611	636	733	583	672	547	628
□Oil	7	5	4	6	5	7	6	6	5	6	5
■Natural gas	36	23	20	28	25	33	30	31	27	29	25
□Coal	1025	666	574	800	706	895	713	847	710	802	676

□Coal　■Natural gas　□Oil　□Hydro　■Nuclear　□Renewables

Figure 5. The total electricity generation mix through 2050.

Figures 6 and 7 show the total primary energy consumption by sector and by type in 2050. Primary energy consumption decreases dramatically in all countermeasure scenarios. Compared with the BAU scenario, primary energy consumption sees a decrease of 3.01% and 5.50% in CM40 and CM50 respectively. A substantial fall in primary energy consumption is observed in the scenario with emission reduction measures, whose trend is more obvious when a higher emission reduction proportion is adopted. For instance, in the scenario, the renewable and emission trading and the primary energy consumption drop significantly by 17.35% in CM40ET, 37.12% in CM40ETC, 27.13% in CM50ET, and 41.61% in CM50ETC, respectively. Meanwhile, the total primary energy consumption reduces to 4835.20 Mtce in CM40ETC and 4490 Mtce in CM50ETC, in which the share of non-fossil energy is 29.19% and 32.33%, respectively. The primary energy structure in the CM40 and CM50 scenarios is not significantly different from that in the BAU scenario, implying that compulsory emission reduction measures do not improve the energy mix. The oil and natural gas consumption remain rather stable in CM40 and CM50, while the consumption in all countermeasure scenarios declines by 10%–40%. Coal consumption experiences dramatic decreases with the amount being 44.45% and 46.67% of BAU's level in CM40ETC and CM50ETC. Cumulative energy savings is estimated to be 27,625 Mtce in CM50ETC and 15,301 Mtce in CM50ETC from 2020 to 2050.

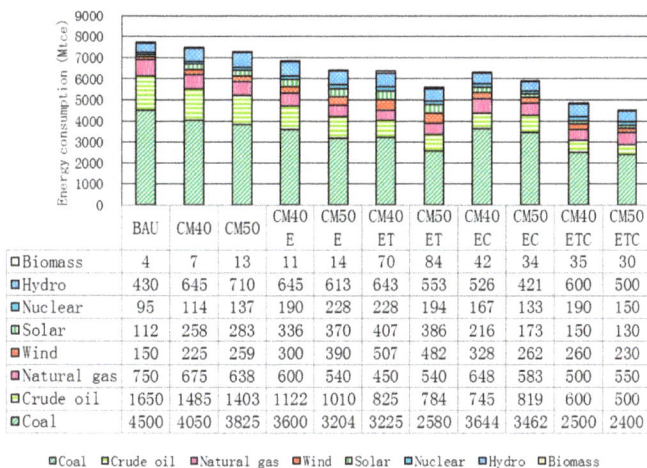

	BAU	CM40	CM50	CM40 E	CM50 E	CM40 ET	CM50 ET	CM40 EC	CM50 EC	CM40 ETC	CM50 ETC
Biomass	4	7	13	11	14	70	84	42	34	35	30
Hydro	430	645	710	645	613	643	553	526	421	600	500
Nuclear	95	114	137	190	228	228	194	167	133	190	150
Solar	112	258	283	336	370	407	386	216	173	150	130
Wind	150	225	259	300	390	507	482	328	262	260	230
Natural gas	750	675	638	600	540	450	540	648	583	500	550
Crude oil	1650	1485	1403	1122	1010	825	784	745	819	600	500
Coal	4500	4050	3825	3600	3204	3225	2580	3644	3462	2500	2400

Coal   Crude oil   Natural gas   Wind   Solar   Nuclear   Hydro   Biomass

Figure 6. The total energy consumption by fuel types through 2050.

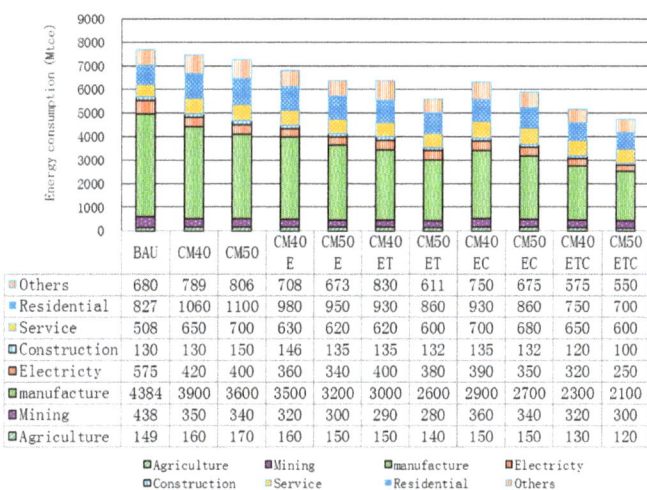

	BAU	CM40	CM50	CM40 E	CM50 E	CM40 ET	CM50 ET	CM40 EC	CM50 EC	CM40 ETC	CM50 ETC
Others	680	789	806	708	673	830	611	750	675	575	550
Residential	827	1060	1100	980	950	930	860	930	860	750	700
Service	508	650	700	630	620	620	600	700	680	650	600
Construction	130	130	150	146	135	135	132	135	132	120	100
Electricty	575	420	400	360	340	400	380	390	350	320	250
manufacture	4384	3900	3600	3500	3200	3000	2600	2900	2700	2300	2100
Mining	438	350	340	320	300	290	280	360	340	320	300
Agriculture	149	160	170	160	150	150	140	150	150	130	120

Agriculture   Mining   manufacture   Electricty
Construction   Service   Residential   Others

Figure 7. The total energy consumption by sector through 2050.

In the countermeasure scenarios, the energy use in energy-intensive sectors declines significantly. It is visible that manufacturing, whose energy consumption decreases by 31.5% in CM40ETC and 35.2% in CM50ETC, respectively, is the highest energy consuming sector which suffers the greatest decline. The energy consumption in the service sector increases from 18%–34%.

3.6. Sensitivity Analysis

The sensitivity analysis of this AIM/CGE model should be tested to identify whether the result will be heavily influenced by parameter assumption. In this study, a sensitivity analysis is conducted with respect to the elasticities between the production factors labor and capital (σ_{KL}), fossil energy and electricity (σ_E). We focus on the impacts of a 5% increase or drop from the base value on GDP and carbon intensity in CM40E and CM40ET.

Higher elasticity of σ_{KL} means higher flexibility for producers to choose the factors of labor and capital when fulfilling their production quotas. Consequently, production cost drops away relatively, which results in higher activity levels and GDP growth. Results show that GDP loss is reduced by 4.36% from the base value. However, the carbon intensity do not change dramatically.

The σ_E have a significant impact on carbon intensity and GDP. A 5% increase of σ_E leads to a 11.49% drop of carbon intensity and a 5% drop of σ_E leads to a 6.38% increase. Higher σ_E put pressure on energy and carbon prices. Energy consumers have greater choice and flexibility to choose which kind of energy to consume. Hence, lower cost energy will be chosen, *i.e.*, natural gas and electricity. The lower energy price stimulates production and drives GDP growth. Meanwhile, the carbon intensity will drop due to habit changes.

4. Conclusions

This study introduces the AIM/CGE model to investigate the effects of alternative mitigation policies, including renewable energy, emission trading and a carbon tax during 2010–2050. The simulation results show the effect on carbon intensity, activity level, GDP, energy system, and carbon prices in detail.

The emission reduction rate in the BAU scenario is approximately 28.68% in 2020 and 46.8% in 2050, based on the level of 2005. Absolute emission limits show negative impacts on future economic growth and activity level in energy intensive sectors. Activity level in non-energy intensive sectors increases slightly. In addition, the impact is larger when a higher emission reduction target level is set. In contrast, mitigation measures will curb CO_2 emissions, promote economic growth and optimize the energy mix. More importantly, energy-intensive sectors will improve energy utilization technology as far as possible to lower their costs. The cumulative energy consumption reduction is estimated to be 27,625 Mtce in CM50ETC from 2020 to 2050. Moreover, climate policies lead to lower emissions, reduced energy intensity, and improved energy and industrial mixes. It is necessary to take all the measures into account to reduce emissions.

Among the mitigation measures, renewable energy makes a critical difference in abating emissions during the period from 2010 to 2020, reducing emissions per unit of GDP by 33.33% in CM40E and 36.43% in CM50E. After 2020, according to the hypotheses, abolishing subsidies on renewable energy will slow down the pace of development of renewable energy, which means that energy efficiency improved by renewable energy will affect the accompanying emission mitigation. Nevertheless, the renewable introduced reduces about 36% of the emissions per GDP, but still fails to meet the goals. Therefore, it is necessary to pay more attention to the development of renewable energy and the determination of the appropriate subsidy rates for renewable energy.

The emission trading introduced in 2016 will bring about high carbon prices, *i.e.*, 143.4 USD/t in CM40ET, which encourages emissions reduction. Limited emission trading does not lower the carbon emission intensity as expected. Free emission trading must be integrated with mitigation measures to achieve the emission reduction targets. The government should try its best to plan the implementation of emission trading to avert unfair and unreasonable factors.

Carbon taxes would reduce carbon emissions, however, they does not work very well in the long term. Most noticeably, carbon taxes have a considerable negative impact on the sustainability, optimization of energy and economy system in the short term. It is necessary to give perfect rein to carbon tax by setting different tax rates.

Acknowledgments

This research was financially supported by the National Natural Science Foundation of China (NSFC) (71,471,061), Hebei Province Science and Technology Plan Project (14,457,694D) and the Fundamental Research Funds for the Central Universities (2014ZP11).

Author Contributions

Wei Li designed this research and gave the whole guidance; Hao Li largely wrote the research, including data simulation, result display and analysis; Shuang Sun. collected all the data. All authors read and approved the final manuscript.

Appendix: Definition of Symbols Used in this Model

The variable symbols and meanings used in this study are as follows:

F_i	the fossil energy aggregate in i th sector
A_i^f	the efficiency coefficient of fossil energy input in i th sector
$\alpha_{k,i}$	the share coefficient of k th energy in i th sector
$EC_{k,i}$	the input of k th energy in i th sector
$\sigma_{f,i}$	the elasticity of substitution of various fossil energy in i th sector
c_k	the CO_2 emissions factor of k th energy
EM_i	the CO_2 emissions in i th sector
EL_i	the renewable electricity input in i th sector
A_i^e	the efficiency coefficient of energy input in i th sector
$\alpha_{F,i}$	the share coefficient of fossil energy in i th sector
$\sigma_{E,i}$	the elasticity of substitution between fossil energy and electricity in i th sector
E_i	the energy aggregate in i th sector
K_i	the capital input in i th sector
L_i	the labor input in i th sector
A_i^{kl}	the efficiency coefficient of capital and labor input in i th sector
$\sigma_{KL,i}$	the elasticity of substitution between capital and labor in i th sector
$\alpha_{L,i}$	the share coefficient of labor input in i th sector
KL_i	the capital-labor aggregate in i th sector
$\sigma_{KLE,i}$	the elasticity of substitution between energy and capital-labor in i th sector

$\alpha_{KLE,i}$	the share coefficient of capital-labor input in ith sector
A_i^{kle}	the efficiency coefficient of capital-labor and energy input in ith sector
KLE_i	the capital-labor and energy aggregate in ith sector
$\alpha_{i,j}$	the direct consumption coefficients of jth intermediate inputs in ith sector
$X_{i,j}$	the jth intermediate inputs per unit of outputs in ith sector
M_i	the intermediate inputs aggregate in ith sector
A_i^q	the efficiency coefficient of output in ith sector
$\sigma_{KLEM,i}$	the elasticity of substitution between capital-labor-energy and intermediate inputs in ith sector
$\alpha_{KLEM,i}$	the share coefficient of capital-labor-energy input in ith sector
Q_i	the output in ith sector
Y_h	the gross income of household sector
W_i	the wage rate in ith sector
r_i	the rate of return on capital in ith sector
HK	the capital stock in household sector
K_i	the capital input in ith sector
$\overline{TR}_{gov}$	the transfer payment from government to household and enterprise
$\overline{TR}_{eni}$	the transfer payment from enterprise
β	the share of transfer payment from government to household
γ	the share of transfer payment from government to enterprise
Y_{gov}	the gross income of government
Y_{ent}	the gross income of enterprise
τ_i	the ad valorem rate of duty in ith sector
P_i	the product prices in ith sector
τ_h	the rate of individual income tax
τ_{ent}	the rate of corporate income tax
$TARIFF$	tariff revenue
TAX_{carbon}	the carbon tax revenue
QC_i	the aggregate supply in domestic market in ith sector
A_l^m	the scale parameter of the aggregate supply in domestic market of lth commodity
$\alpha_{m,l}$	the share parameter of imported commodities of lth commodity
$\sigma_{m,l}$	the elasticity of substitution between imported commodities and domestic production of lth commodity
QM_l	the imports of lth commodity
QAM_l	the domestic production of lth commodity
PC_l	the consumer price of lth commodity
PM_l	the imports price of lth commodity
PAM_l	domestic production price of lth commodity

$\tau_{m,l}$	the import tariffs of lth commodity
$\overline{pwm_l}$	the world price of lth commodity
$\overline{EXR}$	the exchange rate

Table A1. The description of sector classifications used in this model.

Non-Energy Sectors		Energy Sectors	
1	Agriculture	1	Coal mining
2	Mineral mining	2	Extraction of petroleum
3	Foods and Tobacco	3	Natural gas
4	textiles and clothing	4	Coking
5	Wood and furniture	5	Petroleum and nuclear fuel processing
6	Paper products	6	Gas production and supply
7	Chemical	7	Electricity
8	Cement		
9	Non-metallic mineral products		
10	Iron and steel		
11	Nonferrous products		
12	Water production		
13	Machinery		
14	Other manufacturing		
15	Construction		
16	Transportation		
17	Service sectors		

Table A2. The estimated feed-in subsidies for the renewable during the study period (Billion USD).

Year	CM40E			CM50E			CM40ET			CM50ET		
	Wind	Solar	Biomass	Wind	Solar	Biomass	Wind	Solar	Biomass	Wind	Solar	Biomass
2010	0.75	0.02	0.13	0.83	0.17	0.22	0.75	0.02	0.13	0.83	0.17	0.22
2015	2.42	0.84	0.31	2.78	1.25	0.56	2.42	0.84	0.31	2.78	1.25	0.56
2020	4.89	1.21	0.74	5.71	1.88	1.17	5.70	2.47	1.60	6.30	2.98	2.16

Table A3. The estimated revenues of carbon tax during 2020–2050 (Billion USD).

Year	CM40EC	CM50EC	CM40ETC	CM50ETC
2020	8.39	8.06	6.84	6.59
2025	9.32	8.98	7.27	7.14
2030	10.07	9.73	7.61	7.52
2035	10.79	10.34	7.89	7.85
2040	11.45	11.04	8.36	8.14
2045	11.93	11.42	8.57	8.43
2050	12.37	11.88	8.70	8.32

Table A4. The government receipts in all scenarios during 2010–2050 (Billion USD).

Year	CM40	CM50	CM40E	CM50E	CM40ET	CM50ET	CM40EC	CM50EC	CM40ETC	CM50ETC
2010	1336.03	1315.59	1299.54	1280.43	1264.81	1246.22	1231.02	1212.92	1198.12	1180.51
2015	2229.65	2184.83	2148.78	2109.46	2074.65	2036.69	2003.08	1966.43	1933.98	1898.59
2020	3091.58	2982.76	2919.82	2857.92	2805.91	2752.03	2693.96	2636.85	2588.86	2539.16
2025	3766.12	3621.13	3540.38	3456.82	3392.18	3318.91	3236.60	3135.94	3067.26	2989.36
2030	4922.17	4700.67	4589.74	4471.32	4384.13	4281.11	4160.81	4013.51	3918.39	3807.89
2040	8055.88	7619.25	7429.53	7228.19	7075.68	6898.08	6693.89	6421.45	6259.63	6065.58
2050	11,091.77	10,325.33	10,059.97	9779.29	9561.21	9312.62	9024.86	8587.16	8361.31	8076.19

Table A5. Change of coal supply in 2050.

Year	CM40	CM50	CM40E	CM50E	CM40ET	CM50ET	CM40EC	CM50EC	CM40ETC	CM50ETC
2015	−5.62%	−9.00%	−4.90%	−6.90%	−4.90%	−6.90%	−4.90%	−6.90%	−4.90%	−3.74%
2020	−5.50%	−8.56%	−4.00%	−6.00%	−3.10%	−5.60%	−4.00%	−6.00%	−3.10%	−3.04%
2025	−5.00%	−7.66%	−3.10%	−5.10%	−2.60%	−4.60%	−2.85%	−3.82%	−2.53%	−2.63%
2030	−4.80%	−6.85%	−2.85%	−4.40%	−2.30%	−3.80%	−2.58%	−3.30%	−2.20%	−2.09%
2040	−4.50%	−6.50%	−2.68%	−3.80%	−2.03%	−3.20%	−2.36%	−2.88%	−1.96%	−1.73%
2050	−4.10%	−6.25%	−2.55%	−3.40%	−1.85%	−2.70%	−2.20%	−2.61%	−1.82%	−1.44%

Table A6. Change of oill supply in 2050.

Year	CM40	CM50	CM40E	CM50E	CM40ET	CM50ET	CM40EC	CM50EC	CM40ETC	CM50ETC
2015	−0.47%	−0.76%	−0.41%	−0.58%	−0.41%	−0.58%	−0.41%	−0.58%	−0.41%	−0.58%
2020	−0.46%	−0.72%	−0.34%	−0.51%	−0.26%	−0.47%	−0.34%	−0.51%	−0.26%	−0.47%
2025	−0.42%	−0.65%	−0.26%	−0.43%	−0.22%	−0.39%	−0.24%	−0.41%	−0.21%	−0.41%
2030	−0.40%	−0.58%	−0.24%	−0.37%	−0.19%	−0.32%	−0.22%	−0.35%	−0.19%	−0.33%
2040	−0.38%	−0.55%	−0.23%	−0.32%	−0.17%	−0.27%	−0.20%	−0.29%	−0.17%	−0.27%
2050	−0.35%	−0.53%	−0.21%	−0.29%	−0.16%	−0.23%	−0.19%	−0.26%	−0.15%	−0.22%

Table A7. Change of gas supply in 2050.

Year	CM40	CM50	CM40E	CM50E	CM40ET	CM50ET	CM40EC	CM50EC	CM40ETC	CM50ETC
2015	−3.05%	−4.88%	−2.66%	−3.74%	−2.66%	−3.74%	−2.66%	−3.74%	−2.66%	−3.74%
2020	−2.98%	−4.64%	−2.17%	−3.25%	−1.68%	−3.04%	−2.17%	−3.25%	−1.68%	−3.04%
2025	−2.71%	−4.15%	−1.68%	−2.76%	−1.41%	−2.49%	−1.55%	−2.63%	−1.37%	−2.63%
2030	−2.60%	−3.71%	−1.55%	−2.39%	−1.25%	−2.06%	−1.40%	−2.22%	−1.19%	−2.09%
2040	−2.44%	−3.52%	−1.45%	−2.06%	−1.10%	−1.73%	−1.28%	−1.90%	−1.06%	−1.73%
2050	−2.22%	−3.39%	−1.38%	−1.84%	−1.00%	−1.46%	−1.19%	−1.65%	−0.99%	−1.44%

Table A8. CO_2 emission intensity from 2020 to 2050.

Year	BAU	CM40	CM50	CM40E	CM50E	CM40ET	CM50ET	CM40EC	CM50EC	CM40ETC	CM50ETC
2020	0.92	0.77	0.68	0.86	0.82	0.70	0.68	0.78	0.75	0.78	0.73
2025	0.90	0.74	0.65	0.82	0.78	0.64	0.63	0.72	0.71	0.73	0.69
2030	0.88	0.71	0.62	0.78	0.75	0.60	0.58	0.67	0.64	0.69	0.65
2040	0.86	0.65	0.58	0.68	0.63	0.56	0.54	0.62	0.59	0.63	0.58
2050	0.83	0.61	0.54	0.59	0.55	0.53	0.52	0.56	0.54	0.57	0.54

Conflicts of Interest

The authors declare no conflict of interest.

References

1. Yuan, H.; Zhou, P.; Zhou, D. What is low-carbon development? A conceptual analysis. *Energy Procedia* **2011**, *5*, 1706–1712.
2. European Union (EU). Roadmap for a Low-Carbon Economy in 2050. Available online: http://ec.europa.eu/clima/consultations/articles/0005_en.htm (accessed on 8 December 2010).
3. Japan-U.S. Summit Meeting. Japan-U.S. Joint Message on Climate Change Negotiations. Available online: http://www.mofa.go.jp/region/n-america/us/pv0911/summit.html (accessed on 13 October 2010).
4. National Development and Reform Commission (NDRC). National Climate Change Plan (2014–2020). Available online: http://www.sdpc.gov.cn/zcfb/zcfbtz/201411/t20141104_642612.html (accessed on 19 September 2014).
5. Dai, H.; Masui, T.; Matsuoka, Y. Assessment of China's climate commitment and non-fossil energy plan towards 2020 using hybrid AIM/CGE model. *Energy Policy* **2011**, *39*, 2875–2887.
6. Bandyopadhyay, G.; Bagheri, F.; Mann, M. Reduction of fossil fuel emissions in the USA: A holistic approach towards policy formulation. *Energy Policy* **2007**, *35*, 950–965.
7. Hammar, H.; Sjöström, M. Accounting for behavioral effects of increases in the carbon dioxide (CO_2) tax in revenue estimation in Sweden. *Energy Policy* **2011**, *39*, 6672–6676.
8. Lin, B.; Li, X. The effect of carbon tax on per capita CO_2 emissions. *Energy Policy* **2011**, *39*, 5137–5146.
9. Szabó, L.; Hidalgo, I.; Ciscar, J.C.; Soria, A. CO_2 emission trading within the European Union and Annex B countries: The cement industry case. *Energy Policy* **2006**, *34*, 72–87.
10. Lee, C.F.; Lin, S.J.; Lewis, C. Analysis of the impacts of combining carbon taxation and emission trading on different industry sectors. *Energy Policy* **2008**, *36*, 722–729.
11. Sugino, M.; Arimura, T.H.; Morgenstern, R.D. The effects of alternative carbon mitigation policies on Japanese industries. *Energy Policy* **2013**, *62*, 1254–1267.
12. Deetman, S.; Hof, A.F.; Pfluger, B.; van Vuuren, D.P.; Girod, B.; van Ruijven, B.J. Deep greenhouse gas emission reductions in Europe: Exploring different options. *Energy Policy* **2013**, *55*, 152–164.
13. Hübler, M.; Löschel, A. The EU Decarbonisation Roadmap 2050—What way to walk? *Energy Policy* **2013**, *55*, 190–207.
14. Thepkhun, P.; Limmeechokchai, B.; Fujimori, S.; Masui, T.; Shrestha, R.M. Thailand's Low-Carbon Scenario 2050: The AIM/CGE analyses of CO_2 mitigation measures. *Energy Policy* **2013**, *62*, 561–572.
15. Qi, T.; Zhang, X.; Karplus, V.J. The energy and CO_2 emissions impact of renewable energy development in China. *Energy Policy* **2014**, *68*, 60–69.
16. Wang, Y.; Liang, S. Carbon dioxide mitigation target of China in 2020 and key economic sectors. *Energy Policy* **2013**, *58*, 90–96.

17. Li, Y.P.; Huang, G.H.; Li, M.W. An integrated optimization modeling approach for planning emission trading and clean-energy development under uncertainty. *Renew. Energy* **2014**, *62*, 31–46.

18. Hermeling, C.; Löschel, A.; Mennel, T. A new robustness analysis for climate policy evaluations: A CGE application for the EU 2020 targets. *Energy Policy* **2013**, *55*, 27–35.

19. Orlov, A.; Grethe, H. Carbon taxation and market structure: A CGE analysis for Russia. *Energy Policy* **2012**, *51*, 696–707.

20. Matsumoto, K.; Masui, T. Economic impacts to avoid dangerous climate change using the AIM/CGE model. *Procedia Environ. Sci.* **2010**, *6*, 162–168.

21. Okagawa, A.; Masui, T.; Akashi, O.; Hijioka, Y.; Matsumoto, K.; Kainuma, M. Assessment of GHG emission reduction pathways in a society without carbon capture and nuclear technologies. *Energy Econ.* **2012**, *34*, S391–S398.

22. Xu, Y.; Masui, T. Local air pollutant emission reduction and ancillary carbon benets of SO_2 control policies: Application of AIM/CGE model to China. *Eur. J. Oper. Res.* **2009**, *198*, 315–325.

23. Fujimori, S.; Tu, T.T.; Masui, T.; Matsuoka, Y. *AIM/CGE [Basic] Manual*; National Institute for Environmental Studies: Tsukuba, Japan, 2011.

24. China Input-Output Association (CIOA): 2005 Input-Output Table. Available online: http://www.iochina.org.cn/Download/xgxz.html (accessed on 1 January 2006).

25. National Bureau of Statistics (NBS). *China Statistical Yearbook 2010*; China Statistics Press: Beijing, China, 2010.

26. IPCC. IPCC Guide Lines for National Green House Gas Inventories. 2006. Available online: http://www.ipcc-nggip.iges.or.jp/public/2006gl (accessed on 1 April 2007).

27. Peng, Z. *In-Output Tables of China*; China Statistics Press: Beijing, China, 2007.

28. Wei, S.; Huang, Y. Keqiang economics. *Western Development.* **2013**, *9*, 22–25.

29. Wang, K.; Wang, C.; Chen, J. Analysis of the economic impact of different Chinese climate policy options based on a CGE model incorporating endogenous technological change. *Energy Policy* **2013**, *37*, 2930–2940.

30. The Central People's Government of the People's Republic of China. National Population Development Strategy Research Report. Available online: http://www.gov.cn/gzdt/2007-01/11/content_493677.htm (accessed on 11 January 2007).

31. National Bureau of Statistics of China. *China Statistical Yearbook*; China Statistics Press: Beijing, China, 2008–2010. (In Chinese)

32. National Energy Administration (NEA). Energy Development "Twelfth Five Year Plan". Available online: http://www.nea.gov.cn/2013–01/28/c_132132808.htm (accessed on 1 January 2013).

33. Jiang, K.; Hu, X. China's low carbon development scenario in 2050. In *2050 China Energy and CO_2 Emissions Report. Study Team of 2005 China Energy and CO_2 Emissions Report*; Science Press: Beijing, China, 2009.

34. National Development and Reform Commission (NDRC). Interim Measures on Management Carbon Trading. Available online: http://www.in-en.com/finance/html/energy07360736802226927.html (accessed on 22 December 2014).

35. Yang, Y.; Qiu, W.; He, D. The game analysis between the government and the enterprises under mandatory emission reduction mechanism. *Syst. Eng.* **2012**, *2*, 110–114.

36. Wang, X.; Pang, J. The challenges, opportunities and recommendations for Chinese energy intensive enterprises to cope with climate change. *Clim. Chang. Res. Prog.* **2009**, *2*, 110–116.

Energy's Shadow Price and Energy Efficiency in China: A Non-Parametric Input Distance Function Analysis

Pengfei Sheng, Jun Yang and Joshua D. Shackman

Abstract: This paper extends prior research on energy inefficiency in China by utilizing a unique shadow price framework allocation in 30 Chinese provinces. We estimate the shadow price for energy input using the framework of production, and use the ratio of the shadow price to the market price to describe energy utilization. Using Chinese provincial-level data from 1998 to 2011, the results of the analysis reveal that shadow prices in China have grown rapidly during the sample period, which signifies that China has improved its performance in energy utilization since 1998. However, there are eighteen provinces whose shadow prices are lower than market prices. This result suggests that energy utilization is at a low level in these provinces and can be improved by a reallocation of inputs.

Reprinted from *Energies*. Cite as: Sheng, P.; Yang, J.; Shackman, J.D. Energy's Shadow Price and Energy Efficiency in China: A Non-Parametric Input Distance Function Analysis. *Energies* **2015**, *8*, 1975-1989.

1. Introduction

According to data from the BP Statistical Review of World Energy 2011, the energy intensity of China in 2011 was 2.31 tons of oil equivalents per $100 million GDP (current U.S. dollars). This figure is not only higher than the United States, OPEC, Japan and other developed countries, but also much higher than the world average. Meanwhile, prior research has demonstrated that China is highly inefficient in energy utilization from multiple political and economic perspectives [1,2]. As China's economy and population continue to grow, energy inefficiency is likely to be a burden on both economic growth and environmental protection.

Prior research has tried to explain the reasons for China's low energy efficiency from the perspective of economic growth, government behavior, market segmentation and resource allocation [1–3]. However, little research has been conducted from the perspective of energy scarcity. Energy scarcity refers to the gap between the shadow price of energy inputs and the market price, serving as an important basis for private enterprise to make energy decisions. When the shadow price of energy is not equal to its market price, private enterprise will reallocate energy and other inputs in the production process to maximize profits [4]. Thus energy scarcity is not just a basis for energy pricing, but also an important mechanism to reduce energy consumption and stimulate enterprises to improve energy efficiency.

The current literatures utilize both parametric and non-parametric methods to estimate the shadow price of inputs, outputs, and undesirable output (such as the emission of sulfur dioxide and carbon dioxide). A wide variety of parametric and non-parametric shadow price estimation methods have been used in the energy economics literature, primarily to estimate shadow prices of undesirable outputs, such as pollutants. For the parametric analysis, the shadow price of undesirable output has been estimated by a traditional distance function [5,6]. Pittman [7] estimated the shadow price of undesirable outputs based on a maximum revenue approach. Inspired by the input distance function proposed by Shephard [8], Hailu and Veeman [9] and Lee [10] both calculated the shadow price of industrial pollutants by using the dual function. Färe *et al.* [11] adopt Shephard's output distance function to deduct the shadow prices of environmental pollutants, combining the translog production function and the directional distance function. Färe *et al.* [12] established a quadratic direction distance function, and estimated the shadow price of pollutants using parametric linear programming methods. Using similar methods, Murty *et al.* [13] estimated the shadow price of pollutants of an Indian thermal power plant.

The non-parametric analysis adopts Data Environment Analysis (DEA) to estimate the shadow price of inputs, outputs, and undesirable outputs. According to Shephard's output distance function, Boyd *et al.* [14] used the gradient vector of the output distance function to estimate the shadow price of pollutants. Based on the directional distance function [15], Lee *et al.* [16] added non-efficiency factors into the linear programming model to evaluate the shadow price of pollutants.

While much research has been done on the estimation of shadow price of pollutants, relatively little research has been done on the shadow price of energy. Khademvatani and Gordon [4] present a theoretical model of shadow price of energy as a relevant measure of the marginal efficiency of energy, and also stress the importance of shadow prices for policy makers for such matters as determining energy taxes. By using the dataset of Chinese industrial sector from 2001 to 2009, Ouyang and Sun [17] find that China has been relatively undervalued energy's price due to the requirements of industrial development and social stability. We contribute to this small but growing literature through a non-parametric framework of energy's shadow price and a detailed analysis of both economy-wide and provincial-level shadow prices in China using data from 30 provinces 1998 to 2011.

2. Method and Materials

2.1. Framework of Energy's Shadow Price and Energy Efficiency

For the parametric approach, the different formation of the production function would significantly affect the estimated result. Meanwhile, continuous data processing can only generate an average shadow price rather than reflecting the identical contribution of energy on outputs for single decision-making unit (DMU). Therefore, this paper adopts a non-parametric analysis method based on Shepard's input distance function to estimate the shadow price of energy.

As a first step we assume that there are N production DMUs and each DMU uses energy input (E) and other inputs M ($X = (x_1, x_2, x_3... x_M)$) to produce the outputs P ($Y = (y_1, y_2, y_3 ..., y_M)$). Production technology can be defined as follows:

$$T = \{(Y),(E,X)\} \in R^N \tag{1}$$

T represents the production technology, with E and X representing factors used to produce Y. According to production theory, T should meet the following conditions: (1) Bound, which means that increasing expected output is limited due to current technologies and input constraints;

(2) Strong Disposability, if $\{(Y), (E, x)\} \in T$, then $\{(Y), ((1 + \theta_1)E, (1 + \theta_2)X)\} \in T$ or $\{((1 - \theta_3)Y), (E, X)\} \in T$, where $\theta_1 > 0$, $\theta_2 > 0$, $\theta_3 > 0$; and (3) Convexity, which means that the production function should obey the law of the diminishing returns.

Under the constraints of production technology T, the input-oriented production distance function could be defined as follows:

$$\text{Max } w_1\beta_1 + w_2\beta_2$$
$$\{(Y), (E(1-\beta_1), X(1-\beta_2))\} \in R^N \tag{2}$$

Equation (2) indicates that the DMU would reach the production frontier by reducing energy inputs and other inputs, where w_1 and w_2 are the weights of E and X, respectively.

Based on Equation (2), the DMU's profit maximization function is as follows:

$$\text{Max } P_Y Y - P_X X - P_E E$$
$$st. \quad D((1-\beta_1)E, (1-\beta_2)X, Y) = 1 \tag{3}$$

P_Y refers to the outputs' price vector, P_E is energy's price, and P_X is the vector of other inputs' prices.

Model (3) can be optimized using the following Lagrange equation:

$$\text{Max } P_Y Y - P_X X - P_E E + \lambda(D((1-\beta_1)E, (1-\beta_2)X, Y) - 1) \tag{4}$$

In Equation (4), λ is the Lagrange multiplier.

The first-order conditions of Equation (4) are listed as follows:

$$P_Y + \lambda \times \frac{\partial D(E(1-\beta_2), X(1-\beta_1), Y)}{\partial(Y)} = 0 \tag{5}$$

$$-P_X + \lambda \times \frac{\partial D(E(1-\beta_2), X(1-\beta_1), Y)}{\partial(X)}(1-\beta_1) = 0 \tag{6}$$

$$-P_E + \lambda \times \frac{\partial D(E(1-\beta_2), X(1-\beta_1), Y)}{\partial(E)}(1-\beta_2) = 0 \tag{7}$$

$$D(E(1-\beta_2), X(1-\beta_1), Y) - 1 = 0 \tag{8}$$

Equations (5)–(7) are the first-order conditions of the Lagrange multiplier corresponding to the output vector Y and the two input vectors X and E. Equation (8) indicates that the DMU is on the production frontier.

The above equations can be used to obtain the relative shadow price of energy inputs:

$$\frac{P_E}{P_Y} = -\frac{\partial D(E(1-\beta_2), X(1-\beta_1), Y) \Big/ \partial(E)}{\partial D(E(1-\beta_2), X(1-\beta_1), Y) \Big/ \partial(Y)}(1-\beta_2) \tag{9}$$

Setting $P_Y = 1$, the absolute shadow price of energy is presented as follow:

$$P_E = -\frac{\partial D(E(1-\beta_2), X(1-\beta_1), Y)\Big/\partial(E)}{\partial D(E(1-\beta_2), X(1-\beta_1), Y)\Big/\partial(Y)}(1-\beta_2)$$

$$= F(E,X,Y)(1-\beta_2)$$

$$= F(E,X,Y) - F(E,X,Y)\beta_2$$

(10)

From Equation (10), it can be found that P_E includes two parts, where $F(E, X, Y)$ is the shadow price of energy on the optimal path, and the $F(E, X, Y)$ β_2 represents the reduction of energy shadow price affected by the change of marginal outputs due to inefficiency. In accordance with Hu and Wang [1], $(1 - \beta_2)$ is DMU's total-factor energy efficiency. Thus, Model (2) can be represented by the following linear programming model:

$$\text{Max } W_1\beta_1 + W_2\beta_2$$

$$\sum_{i=1}^{N}\lambda_i Y_i \geq Y_j; \quad \sum_{i=1}^{N}\lambda_i X_i \leq (1-\beta_1)X_j;$$

$$\sum_{i=1}^{N}\lambda_i E_i \leq (1-\beta_2)E_j; \quad \sum_{i=1}^{N}\lambda_i = 1, \quad \lambda_i \geq 0$$

(11)

2.2. Data Description

Supported by the classical production theory, every DMU produces gross domestic product (Y) by using capital stock (K) and labor (L), apart from energy (E). In this paper, data from 30 provinces covering 1998 to 2011, which was gathered from "China Statistical Yearbook" [18], "Statistical Yearbook of the Chinese Investment in Fixed Assets" [19] and "China Energy Statistical Yearbook" [20]. The stock of physical capital and GDP is adjusted based on 1998 prices.

If we only use primary end-use energy consumption as the measurement of regional energy consumption, the allocation of secondary energy resources is ignored. This may result in an underestimation or overestimation of regional energy consumption. Therefore, this study uses the sum of the net addition of secondary energy in the region and primary end-use energy as the actual amount of energy consumption. For this calculation, the net addition of secondary energy consumption is converted to coal equivalent. For physical capital stock, the perpetual inventory method and the Zhang *et al.* [21] method is used to calculate Chinese provincial physical capital stock, and all data are deflated according to the 1998 price levels. Labor is measured by the mean of the number of workers employed at the end of the year and the end of the previous year.

The descriptive statistics for these variables are presented in Table 1.

Table 1. Descriptive statistics.

Variable	Descriptive statistics					Pearson correlation			
	Mean	Median	Max	Min	Standard deviation	L	K	E	Y
L	2393.15	2046.50	6547.75	254.84	1601.767	1.00	--	--	--
K	6880.71	4469.63	46,916.31	230.03	7340.121	0.62	1.00	--	--
E	8495.40	6619.40	35,978.00	409.30	6467.278	0.69	0.89	1.00	--
Y	3338.61	2259.18	22,118.75	117.60	3422.192	0.69	0.96	0.87	1.00

2.3. Energy Consumption in China

In addition to rapid economic growth in China since 1998, energy consumption also has also increasing dramatically during this time from 1.36 billion tons of standard coal in 1998 to 3.48 billion tons in 2011. This annual growth rate in energy consumption is 7.48 percent, which is only a little smaller than the GDP growth rate during this time. On a regional basis, Eastern China has the largest growth rate by energy consumption and is responsible for 46 percent of total energy consumption in China. Figure 1 below illustrates the regional growth in energy consumption across regions in China.

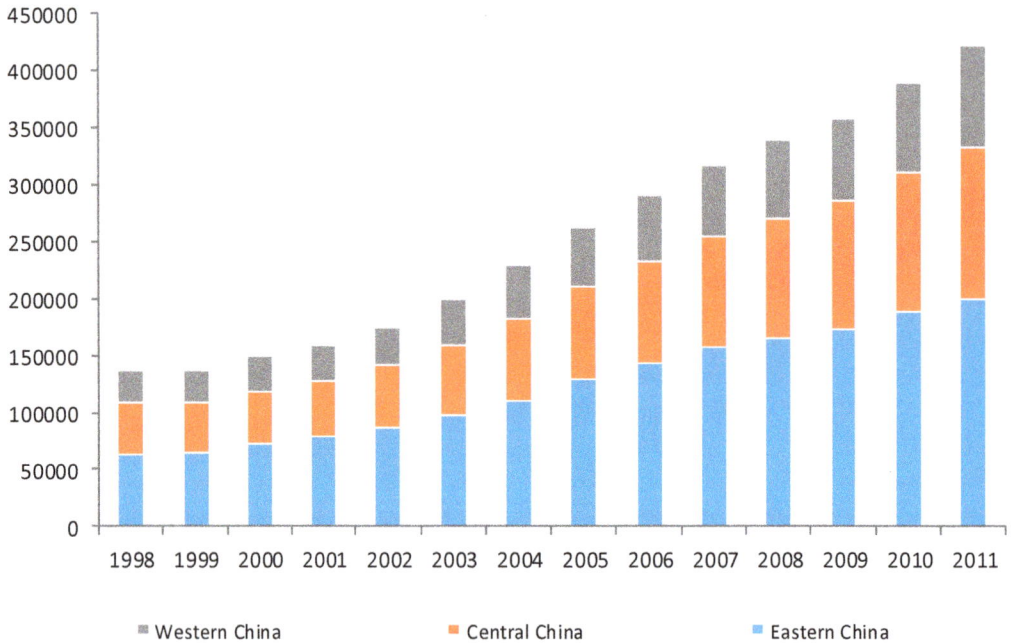

Figure 1. Energy consumption in China. Source: China Energy Statistical Yearbook [20].

While growth in energy consumption has closely followed GDP growth in China, energy intensity has shown a downward trend during this same time period. Energy intensity is the ratio of energy consumption to GDP, which is a common index to describe energy utilization. Figure 2 illustrates the national and regional level trends in energy intensity in China. Energy intensity has decreased from 1.61 of 1998 to 1.19 of 2011, which means that the energy consumption to produce 10,000 RMB GDP has been reduced by 0.42 ton of standard coal in China. However, this downward trend was briefly interrupted from 2002 to 2005, a period that saw no significant reduction in energy intensity. This is because Chinese industrial development entered a new round of heavy industrial development during this period and the share of heavy industry's gross output to total industry increased from 60.9% in 2002 to 69.5% in 2006. Although Eastern China is the biggest consumer of energy, it has also experienced the largest drop in energy intensity. This indicates that Eastern China consumes the least amount of energy to produce the same GDP as

Western China and Central China. As showed in Figure 2, energy intensity has decreased in all regions in China but the gaps between the three regions have remained constant.

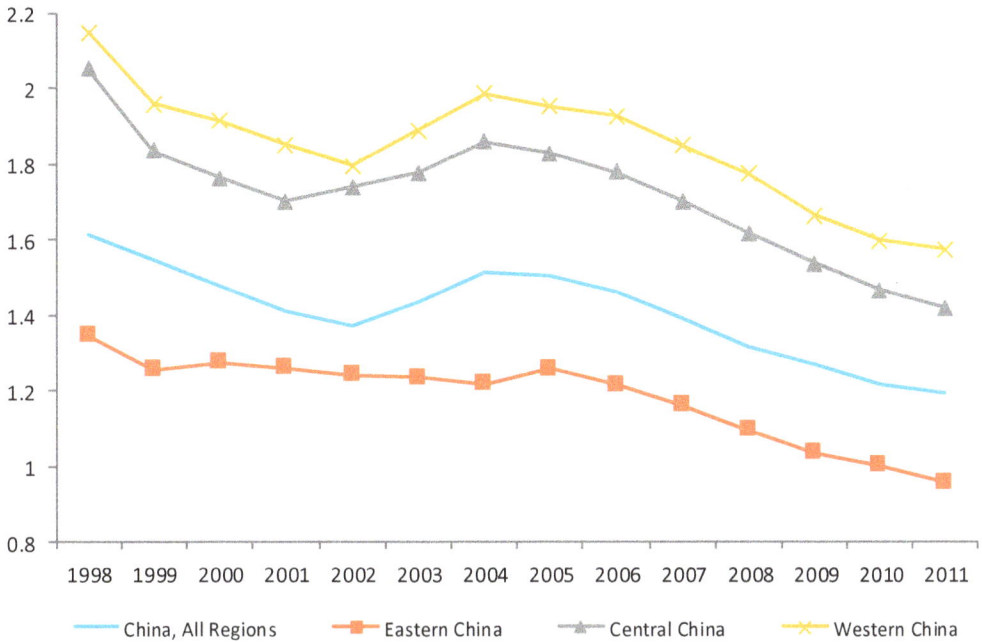

Figure 2. Energy intensity in China. Source: China Energy Yearbook and China Statistical Yearbook [20].

3. Results and Discussion

3.1. Energy Efficiency

For our analysis of energy efficiency in China, we use the definition of Hu and Wang [1], who define energy efficiency as the ratio of the optimal energy input to actual energy input. Figure 3 illustrates energy efficiency for China's three regions (see Appendix C for a breakdown of provinces in each of the three regions). In 2011, overall energy efficiency in China is 0.64, but regionally Eastern China has the highest energy efficiency of 0.84 compared to 0.63 and 0.40 for Central and Western China, respectively. This indicates that actual energy inputs could be decreased by 36 percent for the whole of China and 16, 37 and 60 percent for Eastern, Central, and Western China, respectively, without a decrease in output. During the period 1998 to 2003, there is no significant improvement in energy efficiency at both the national and regional level. This is likely due to weak economic development following the Asia financial crisis in 1998.

After 2003, energy efficiency shows an obvious upward trend in China. This corresponds to several other trends. First, China's economy began to grow rapidly again at this time. Second, stricter environmental regulations were enacted around this time. Finally, the service sector began to grow relative to other sectors at this time. Given the lower energy use and higher energy efficiency associated with the service sector, this could account for some of the growth in energy efficiency (see Appendix B for data on service sector growth in China). Regarding regional differences, Central and Western China show a slight increase in energy efficiency between 1998 and 2011, which closely follows the national trend. Eastern China shows the largest improvement in energy efficiency, well above the national trend. Interestingly, Eastern China also shows the largest growth in the service sector during this time, whereas Eastern and Central China's service sector did now grow.

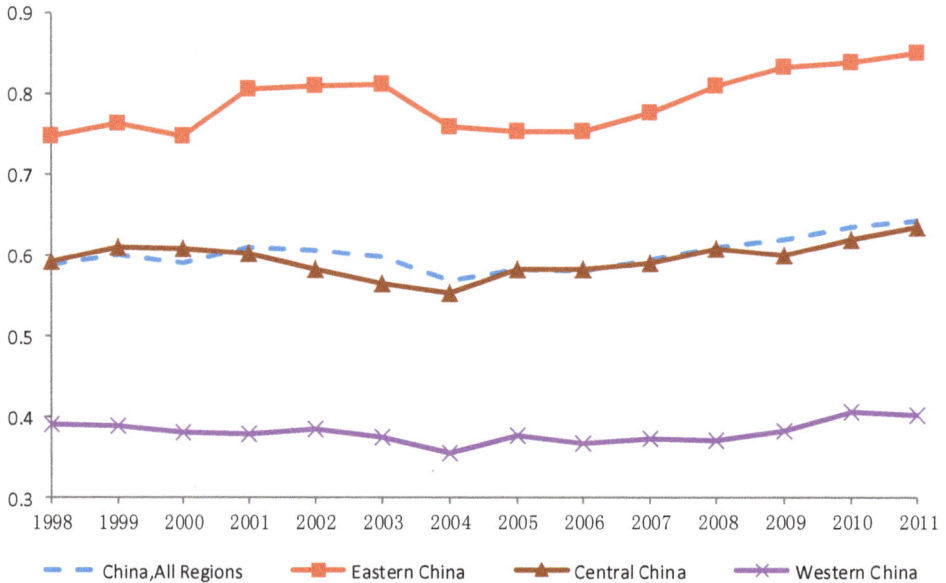

Figure 3. China's regional energy efficiency (1998–2011).

3.2. Shadow Price of Energy

Figure 4 below illustrates the trends in shadow prices across China as measured by our non-parametic methods. In the beginning of the sample period, the shadow price for energy is 0.1595, 0.1931, 0.1631 and 0.1167 for China overall, Eastern China, Central China, and Western China, respectively. These figures suggest that energy input can produce 1595 RMB, 1931 RMB, 1631 RMB and 1167 RMB with each additional energy input as measured by metric tons of standard coal when using 1998 shadow energy prices. However, using 2011 shadow energy prices the additional metric ton of standard coal can produce 2019 RMB, 2305 RMB, 1934 RMB and 1765 RMB. From 1998 to 2001, the shadow price in Western China shows rapid growth, but Eastern China and Central China show no obvious increase. Unlike energy efficiency in Figure 3, the shadow prices for the whole of China and the three main regions show no obvious upward trend from 2001 to 2005. This is contrary to the rapid GDP growth experienced during this same time period. However, following the implementation of stricter environmental regulations shadow prices start to rise in 2006 in all three regions in China. Another explanation for the impact of environmental regulation on shadow prices of energy may be the role of technology and new manufacturing equipment. Fujii *et al.* [22] find that investments in new manufacturing technology may reduce total factor productivity at least in the short-run, but if the goal is to increase productivity while minimizing undesirable outputs, such as CO_2 emissions from energy usage, then new manufacturing technology is highly effective. It may be the case that environmental regulations had the effect of encouraging investment in new manufacturing technologies.

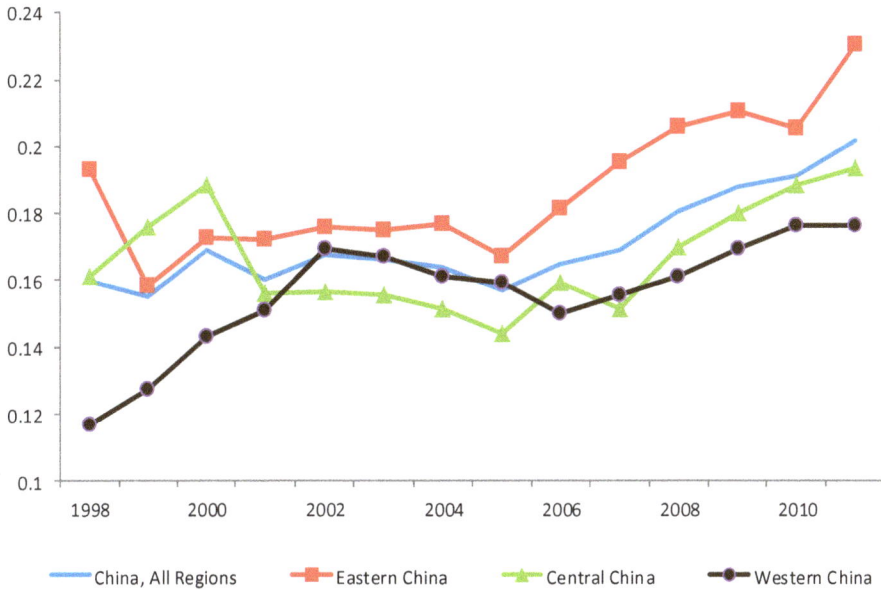

Figure 4. China's regional shadow price of energy.

In patterns similar to those shown in Figure 3, Eastern China also has a larger shadow price of energy input than Central China and Western China, and this indicates that Western China and Central China have worse energy utilization performance than Eastern China. Interesting, though, is the large increase in shadow prices in Western China from 1998 to 2002. This corresponds to a similar increase in the proportion of the service sector in Western China during the same time period (see Appendix B). Likewise, the largest growth in shadow prices after 2006 occurs in Eastern China, which also experienced growth in the service sector proportion of their economy after 2006.

Figure 5 below illustrates the ranking results of Chinese provincial shadow prices of energy input in 2011. Jiangsu, Hainan, Jiangxi, Fujian and Henan provinces have the best performance by the shadow price of energy, while the provinces with the lowest shadow price are Sichuan, Guangxi, Zhejiang, Ningxia and Anhui. Among the top performing provinces, Jiangsu, Hainan and Fujian are the provinces that belong to the Eastern regions with a higher level of economic development, but Jiangxi and Henan have lower levels of economic development and are located in Central China. Among the lowest performing provinces, Zhejiang is located in Eastern China, and the other low performing provinces are located in Western China and Central China. Thus, the shadow price of energy is not necessarily determined solely by higher economic development, but likely by other factors, such as industry structure, technological change, and resource endowments.

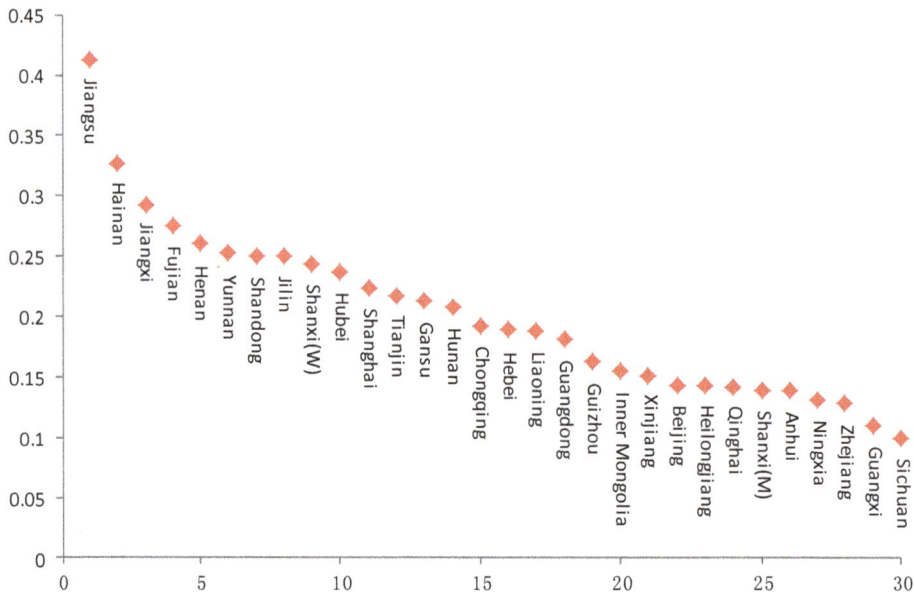

Figure 5. Chinese provincial shadow price of energy in 2011.

3.3. Shadow Price versus Market Price of Energy

Comparing the shadow price of energy input with energy efficiency, the shadow price of energy represents the marginal productivity of energy input under the given technology and energy efficiency as defined by Hu and Wang [1] presents the relative distance from the actual energy input to the optimal energy input with the assumption that all the provinces have the same technology and can reach the production frontier by resource allocation. However, every province has a diverse industrial structure, economic development level, and factor endowments. Thus, the assumption of identical technology across provinces would not be met. But the shadow price is the marginal productivity, which is computed without the assumption of identical technologies across provinces, and thus reflects provincial characteristics, such as industry structure and technology change. Thus we follow Khademvatani and Gordon [4] using the gap between the shadow price and the market price as the preferred method to measure the performance of energy.

As the market price of energy cannot be collected from official statistics in China, we use the retail price indices of fuel (REF) as the proxy variable for the change in the market price of energy. In order to capture the nominal change in the shadow price for energy, the GDP deflator is used to adjust the constant shadow price to be the current shadow price since the shadow price is measured by the additional GDP produced by the additional input of energy, and we use the ratio of the annual growth of shadow price to REF to describe the relationship between the shadow price and market price of energy. We then use a t-test for mean-comparison to analyze whether the shadow price is larger or smaller than the market price of energy.

Table 2 presents the ratio of the shadow price to the market price for energy for China's thirty provinces. There are eighteen provinces whose shadow price is lower than the market price confirmed by the t-test, which suggests that the provinces could increase the marginal productivity and improve energy efficiency by a reallocation of energy input and other inputs. For the remaining twelve provinces, we cannot reject the null hypothesis that the shadow price is equal to the market price of energy, and thus these provinces may be at their Pareto-optimal levels of energy input. Regarding Table 2, some of the provinces whose shadow prices are equal to the market price for energy have higher levels of energy efficiency, such as Jiangsu, Tianjin, and Zhejiang. But there are also some provinces with shadow prices equal to the market price that have low energy efficiency, such as Chongqing and Sichuan. Even though the shadow prices of energy are equal to the market price in these provinces, their fast economic development relies heavily on the rapid growth of the second industry, which is a larger consumer of energy than other industries.

Table 2. Ratio of shadow price to market price for energy in China.

Province	1999	2001	2003	2005	2007	2009	2011	$T(R<1)$	$T(R \neq 1)$	$T(R>1)$
Beijing	0.93	0.92	0.92	0.83	1.00	0.77	1.25	0.08	0.15	0.92
Tianjin	1.09	0.92	0.95	0.84	1.50	0.95	0.97	**0.45**	0.90	0.55
Hebei	1.06	0.93	0.93	0.81	1.04	1.02	0.97	0.01	0.01	0.99
Liaoning	0.28	0.06	1.10	0.95	0.99	0.94	1.53	**0.88**	0.24	0.12
Shanghai	1.03	0.89	1.00	0.88	1.03	1.02	1.00	0.02	0.04	0.98
Jiangsu	1.05	1.03	0.92	0.82	1.11	1.26	1.05	**0.49**	0.98	0.51
Zhejiang	0.95	1.03	0.93	0.86	1.02	1.10	1.42	**0.21**	0.42	0.79
Fujian	0.85	1.00	0.91	0.73	1.05	1.17	0.94	0.03	0.07	0.97
Shandong	1.06	0.89	0.95	0.81	0.99	1.02	0.99	0.01	0.02	0.99
Guangdong	1.01	1.03	0.90	0.84	1.00	1.08	0.81	0.01	0.01	0.99
Hainan	1.00	1.06	0.89	0.85	1.00	1.08	0.98	0.10	0.22	0.89
Shanxi	1.09	0.39	0.89	0.81	0.34	0.99	0.94	0.02	0.05	0.98
Inner Mongolia	1.28	1.00	1.01	0.81	1.05	1.00	1.00	**0.19**	0.39	0.81
Jilin	1.00	0.87	0.88	0.85	1.07	1.10	0.93	0.02	0.04	0.98
Heilongjiang	1.01	0.97	1.02	0.96	1.04	0.95	0.94	**0.15**	0.29	0.85
Anhui	1.09	1.03	0.99	0.31	1.09	0.94	0.80	**0.65**	0.70	0.35
Jiangxi	1.22	1.07	0.98	0.84	0.98	1.10	1.03	**0.56**	0.88	0.44
Henan	1.03	0.95	0.95	0.83	1.06	0.95	0.89	0.00	0.00	1.00
Hubei	1.06	1.04	0.92	0.87	1.08	1.12	0.96	**0.15**	0.30	0.85
Hunan	1.06	1.03	0.92	0.78	1.04	1.11	0.93	0.05	0.11	0.95
Guangxi	1.24	1.10	0.86	0.76	0.97	1.03	0.95	**0.73**	0.54	0.27
Chongqing	1.15	1.15	1.00	0.91	0.97	1.01	0.87	**0.76**	0.47	0.24
Sichuan	1.26	1.08	0.91	0.83	0.96	1.01	0.95	**0.64**	0.72	0.36

344

Table 2. *Cont.*

Province	1999	2001	2003	2005	2007	2009	2011	$T(R < 1)$	$T(R \neq 1)$	$T(R > 1)$
Guizhou	1.15	1.06	0.90	0.96	0.99	0.97	0.92	0.10	0.21	0.90
Yunan	1.04	1.04	0.96	0.82	1.05	0.92	0.97	0.05	0.09	0.95
Shannxi	1.18	0.98	0.93	0.84	0.99	0.96	0.95	0.07	0.15	0.93
Gansu	1.01	0.98	0.96	0.82	1.02	0.95	0.92	0.05	0.11	0.95
Qinghai	0.90	1.05	0.97	0.79	1.00	0.93	0.86	0.03	0.05	0.97
Ningxia	1.09	1.00	0.80	0.82	1.09	0.96	0.85	0.02	0.04	0.98
Xinjiang	1.07	0.89	1.00	0.87	0.96	0.93	0.83	0.08	0.16	0.92

Notes: $T(R < 1)$ is the one tailed t-test for the alternative hypothesis of the shadow price being less than the market price, $T(R \neq 1)$ the two tailed test for the alternative hypothesis of the shadow price not equaling the market price, and $T(R > 1)$ the one tailed t-test for the alternative hypothesis of the shadow price being greater than the market price. The value below each t-test is the p-value. For 18 provinces the shadow price is significantly lower than the market price. Highlighted in bold are the p-values for the 12 provinces where no significant difference between the shadow price and market price was found.

4. Conclusions and Policy Recommendations

Overall, our analysis of energy shadow prices and energy efficiency in China leads us to three main findings. First, based on the measure of energy efficiency proposed by Hu and Wang [9] we conclude that energy efficiency is at a low level in China and can be improved by reallocation of the energy input and other inputs. Energy efficiency does not show any substantial changes from 1998 to 2003 both in the whole China and the three main regions, but does show a significant increase after the stricter environmental policies were implemented after 2006. With regards to the gaps among the three main regions, Eastern China performs better than Central and Western China by energy efficiency, and the gaps between provinces show no obvious reduction during the sample period. These results are similar to prior studies, such as Hu and Li [1], Chang and Hu [23] and Wang and Feng [24], but show some differences with Zhang *et al.* [25] who takes undesirable outputs, such as carbon dioxide, sulfur dioxide emission and the chemical oxygen demand, into account.

Second, our results also indicate that shadow prices of energy experienced rapid growth after 2006 in both the whole of China and in the three main regions. This corresponds to a period of growth in energy efficiency during the same period. As for the provincial ranking of energy's shadow prices, Jiangsu, Hainan, Jiangxi and Fujian have shadow prices close to market price, which corresponds closely with energy efficiency in their regions. However, other provinces, such as Henan and Shandong, have shadow prices close to market prices but low energy efficiency as measured by the ratio of the annual growth of energy's shadow price to REF. Eighteen provinces have shadow prices lower than the market price for energy, which indicates that they can increase the marginal productivity of energy and improve energy efficiency through reallocation of inputs.

Finally, analysis of energy scarcity serves as an important basis for formulating national energy policy. More than half of the provinces have a lower shadow price than the market price, which suggests a degree of market failure in Chinese energy markets. Khademvatani and Gordon [4] suggest that changes in tax policy can help raise the shadow price of energy to move shadow prices closer with market prices. In addition to tax policy, environmental regulations may also improve energy efficiency. Our results show only a slight increase in energy efficiency after 2006 when stricter environmental regulations were enacted, but sharply rising shadow prices during this time. This indicates that environmental regulations may be more effective than previously believed, and overtime may lead to shadow prices more closely in line with market prices. In addition, we find that Eastern China experienced strong growth in both shadow prices and energy efficiency. This may be due to the relative growth of the service sector in Eastern China. This suggests that future efforts by policy makers to balance China's economy away from heavy industry towards consumption and services may also lead to improved energy efficiency and a closer alignment between shadow prices and market prices of energy.

A promising area for future research would be to examine shadow prices at the industry level to further examine the relationship between service industries and shadow prices compared to other sectors. In addition, micro-level analysis of shadow prices and new manufacturing technologies would help further examine how environmental regulations after 2006 lead to increases in shadow prices across China. Finally, additional research should be done on shadow prices of energy both in rapidly developing economies, such as India, as well as highly developed economies with strong environmental protection laws, such as Western Europe or Japan.

Acknowledgments

Support from National Science Foundation of China under Grant No. 71373297 and No. 71133007 are greatly acknowledged.

Author Contributions

Pengfei Sheng took the lead in drafting "the way forward", Jun Yang provided the methodological design and statistical analysis, and Joshua D. Shackman assisted with the literature revision and conclusions.

Appendix A. Energy Efficiency across China's Thirty Provinces

Table A. Energy Efficiency across China's Thirty Provinces.

Province	1998	1999	2000	2001	2002	2003	2004	2005	2006	2007	2008	2009	2010	2011
Beijing	0.68	0.78	0.77	0.81	0.82	0.86	0.86	0.89	0.89	0.89	0.89	1.00	1.00	1.00
Tianjin	0.59	0.60	0.62	0.74	0.78	0.83	0.81	0.83	0.84	0.87	0.90	0.89	0.88	0.85
Hebei	0.43	0.44	0.39	0.38	0.37	0.37	0.37	0.38	0.38	0.39	0.49	0.47	0.48	0.47
Liaoning	0.38	0.40	0.41	1.00	1.00	1.00	0.46	0.47	0.47	0.50	0.56	0.58	0.61	0.60
Shanghai	1.00	1.00	1.00	1.00	1.00	1.00	1.00	1.00	1.00	1.00	1.00	1.00	1.00	1.00
Jiangsu	0.84	0.87	0.89	0.92	0.95	0.94	0.88	0.83	0.84	1.00	1.00	1.00	1.00	1.00
Zhejiang	0.83	0.83	0.80	0.78	0.77	0.77	0.78	0.82	0.83	0.84	0.85	0.88	0.88	1.00
Fujian	1.00	0.99	0.99	0.97	0.93	0.91	0.94	0.80	0.80	0.81	0.81	0.95	0.97	0.93
Shandong	0.74	0.77	0.65	0.58	0.55	0.55	0.60	0.58	0.58	0.59	0.75	0.73	0.75	0.75

Table A. *Cont.*

Province	1998	1999	2000	2001	2002	2003	2004	2005	2006	2007	2008	2009	2010	2011
Guangdong	1.00	1.00	1.00	1.00	1.00	1.00	1.00	1.00	1.00	1.00	1.00	1.00	1.00	1.00
Hainan	0.71	0.72	0.70	0.68	0.72	0.70	0.63	0.65	0.65	0.65	0.65	0.65	0.65	0.75
Shanxi	1.00	1.00	1.00	0.99	0.93	0.91	0.93	0.96	0.94	0.96	0.92	0.92	0.96	0.98
Inner Mongolia	0.23	0.28	0.24	0.25	0.21	0.22	0.24	0.25	0.25	0.25	0.30	0.30	0.31	0.30
Jilin	0.32	0.32	0.31	0.30	0.29	0.27	0.25	0.32	0.31	0.31	0.31	0.32	0.33	0.31
Heilongjiang	0.40	0.40	0.45	0.45	0.43	0.45	0.41	0.51	0.51	0.52	0.63	0.63	0.65	0.63
Anhui	0.32	0.34	0.40	0.41	0.42	0.46	0.44	0.50	0.54	0.57	0.58	0.49	0.55	0.59
Jiangxi	1.00	1.00	1.00	1.00	1.00	1.00	1.00	1.00	1.00	1.00	1.00	1.00	1.00	1.00
Henan	0.77	0.75	0.67	0.67	0.66	0.64	0.66	0.69	0.69	0.70	0.71	0.71	0.73	0.72
Hubei	0.53	0.53	0.53	0.53	0.53	0.50	0.47	0.50	0.50	0.50	0.51	0.52	0.53	0.61
Hunan	0.55	0.55	0.55	0.59	0.60	0.57	0.54	0.58	0.57	0.58	0.60	0.61	0.62	0.71
Guangxi	0.81	0.94	0.93	0.82	0.75	0.61	0.58	0.51	0.50	0.50	0.50	0.51	0.51	0.51
Chongqing	0.67	0.57	0.56	0.58	0.61	0.64	0.54	0.59	0.51	0.51	0.48	0.53	0.55	0.56
Sichuan	0.61	0.62	0.61	0.56	0.54	0.50	0.49	0.51	0.53	0.56	0.55	0.60	0.63	0.68
Guizhou	0.21	0.23	0.21	0.22	0.24	0.21	0.22	0.27	0.27	0.27	0.27	0.27	0.28	0.27
Yunan	0.43	0.45	0.45	0.43	0.42	0.43	0.41	0.40	0.40	0.40	0.40	0.41	0.42	0.41
Shannxi	0.47	0.55	0.56	0.54	0.52	0.52	0.48	0.48	0.48	0.48	0.49	0.49	0.59	0.57
Gansu	0.32	0.30	0.31	0.33	0.34	0.35	0.35	0.36	0.36	0.37	0.37	0.38	0.39	0.44
Qinghai	0.28	0.23	0.25	0.26	0.27	0.27	0.26	0.24	0.23	0.23	0.23	0.24	0.29	0.25
Ningxia	0.24	0.25	0.19	0.20	0.19	0.15	0.14	0.15	0.15	0.18	0.19	0.18	0.19	0.15
Xinjiang	0.28	0.29	0.30	0.30	0.31	0.31	0.30	0.37	0.36	0.36	0.35	0.33	0.33	0.29

Appendix B. Proportion of the Service Sector's Added Value to GDP

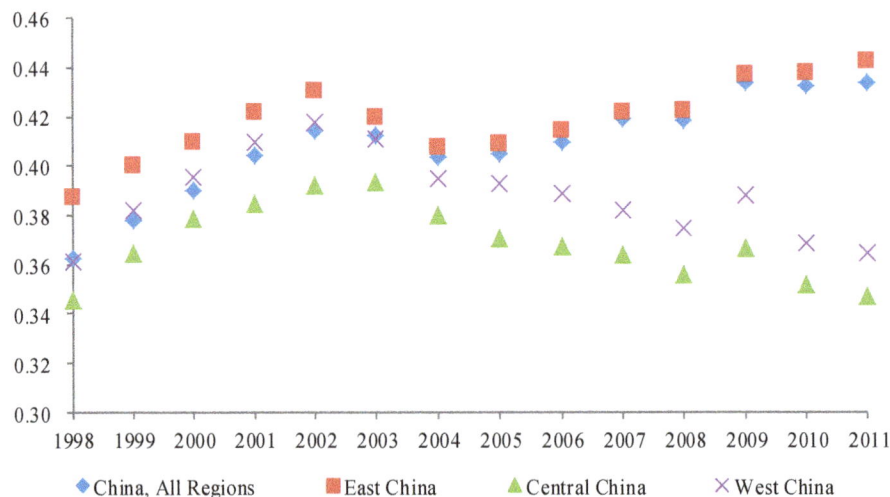

◆ China, All Regions   ■ East China   ▲ Central China   × West China

Appendix C. Domain Division in China

East China consists of 11 provinces, which are Beijing, Tianjin, Hebei, Liaoning, Shanghai, Jiangsu, Zhejiang, Fujian, Shandong, Guangdong and Hainan. Central China contains Shanxi, Inner

Mongolia, Jilin, Heilongjiang, Anhui, Jiangxi, Henan and Hunan. West China covers Guangxi, Chongqing, SiChuan, Guizhou, Yunnan, Gansu, Ningxia and Xinjiang.

Conflicts of Interest

The authors declare no conflict of interest.

References

1. Hu, J.L.; Wang, S.C. Total-factor energy efficiency of regions in China. *Energy Policy* **2006**, *34*, 3206–3217.
2. Zhang, X.P.; Cheng, X.M.; Yuan, J.H.; Gao, X.J. Total-factor energy efficiency in developing countries. *Energy Policy* **2011**, *39*, 644–650.
3. Chen, Z.C.; Lin, Z.S. Multiple timescale analysis and factor analysis of energy ecological footprint growth in China 1953–2006. *Energy Policy* **2008**, *36*, 1666–1678.
4. Khademvatani, A.; Gordon, D.V. A marginal measure of energy efficiency: The shadow value. *Energy Econ.* **2013**, *38*, 153–159.
5. Aigner, D.; Chu, S.F. On the estimating the industry production function. *Am. Econ. Rev.* **1968**, *58*, 226–239.
6. Schmidt, P. On the statistical estimation of parametric frontier production functions. *Rev. Econ. Stat.* **1976**, *58*, 238–239.
7. Pitman, R.W. Multilateral productivity comparisons with undesirable outputs. *Econ. J.* **1983**, *93*, 883–891.
8. Shephard, R.W. *Theory of Cost and Production Functions*; Princeton University Press: Princeton, NJ, USA, 1970.
9. Hailu, A.; Veeman, T.S. Environmentally sensitive productivity analysis of the Canadian pulp and paper industry, 1959–1994: An input distance function approach. *J. Environ. Econ. Manag.* **2000**, *40*, 251–274.
10. Lee, M. The shadow price of substitutable sulfur in the US electric power plant: A distance function approach. *J. Environ. Manag.* **2005**, *77*, 104–110.
11. Färe, R.; Grosskopf, S.; Lovell, C.A.K.; Yaisawarng, S. Derivation of shadow prices for undesirable outputs: A distance function approach. *Rev. Econ. Stat.* **1993**, *75*, 374–380.
12. Färe, R.; Grosskopf, S.; Noh, D.; Weber, W. Characteristics of a polluting technology: Theory and practice. *J. Economet.* **2005**, *126*, 469–492.
13. Murty, M.N.; Kumar, S.; Dhavala, K.K. Measuring environmental efficiency of industry: A case study of thermal power generation in India. *Environ. Resour. Econ.* **2007**, *38*, 31–50.
14. Boyd, G.; Molburg, J.; Prince, R. Alternative methods of marginal abatement cost estimation: Nonparametric distance functions. In Proceedings of the 17th North American Conference of the International Association for Energy Economics, Boston, MA, USA, 26–30 October 1996; pp. 86–95.
15. Chung, Y.H.; Fare, R.; Grosskopf, S. Productivity and undesirable output: A directional distance function approach. *J. Environ. Manag.* **1997**, *51*, 229–240.

348

16. Lee, J.D.; Park, J.B.; Kim, T.Y. Estimation of the shadow prices of pollutants with production/environment inefficiency taken into account: A nonparametric directional distance function approach. *J. Environ. Manag.* **2002**, *64*, 365–375.
17. Ouyang, X.; Sun, C. Energy savings potential in China's industrial sector: From the perspectives of factor price distortion and allocative inefficiency. *Energy Econ.* **2015**, *48*, 117–126.
18. *China Statistical Yearbook*; Chinese Statistical Bureau: Beijing, China, 1998–2011.
19. *Statistical Yearbook of the Chinese Investment in Fixed Assets*; Chinese Investment Press: Beijing, China, 1998–2011.
20. *China Energy Statistical Yearbook*; China Statistics Press: Beijing, China, 1998–2011.
21. Zhang, J.; Wu, G.; Zhang, J. The estimation of China's provincial capital stock: 1952–2004. *Econ. Res. J.* **2004**, *10*, 35–44. (In Chinese)
22. Hidemichi, F.; Shinji, K.; Shunsuke, M. Changes in environmentally sensitive productivity and technological modernization in China's iron and steel industry in the 1990s. *Environ. Dev. Econ.* **2010**, *15*, 485–504.
23. Chang, T.; Hu, J. Total-factor energy productivity growth, technical progress, and efficiency change: An empirical study of China. *Appl. Energy* **2010**, *87*, 3262–3270.
24. Wang, Z.; Feng, C.; Zhang, B. An empirical analysis of China's energy efficiency from both static and dynamic perspectives. *Energy* **2014**, *74*, 322–330.
25. Zhang, N.; Kong, F.; Yu, Y. Measuring ecological total-factor energy efficiency incorporating regional heterogeneities in China. *Ecol. Indic.* **2015**, *51*, 165–172.

A Supply-Chain Analysis Framework for Assessing Densified Biomass Solid Fuel Utilization Policies in China

Wenyan Wang, Wei Ouyang and Fanghua Hao

Abstract: Densified Biomass Solid Fuel (DBSF) is a typical solid form of biomass, using agricultural and forestry residues as raw materials. DBSF utilization is considered to be an alternative to fossil energy, like coal in China, associated with a reduction of environmental pollution. China has abundant biomass resources and is suitable to develop DBSF. Until now, a number of policies aimed at fostering DBSF industry have been proliferated by policy makers in China. However, considering the seasonality and instability of biomass resources, these inefficiencies could trigger future scarcities of biomass feedstocks, baffling the resilience of biomass supply chains. Therefore, this review paper focuses on DBSF policies and strategies in China, based on the supply chain framework. We analyzed the current developing situation of DBSF industry in China and developed a framework for policy instruments based on the supply chain steps, which can be used to identify and assess the deficiencies of current DBSF industry policies, and we proposed some suggestions. These findings may inform policy development and identify synergies at different steps in the supply chain to enhance the development of DBSF industry.

Reprinted from *Energies*. Cite as: Wang, W.; Ouyang, W.; Hao, F. A Supply-Chain Analysis Framework for Assessing Densified Biomass Solid Fuel Utilization Policies in China. *Energies* **2015**, *8*, 7122-7139.

1. Introduction

Due to the rapid development of economies and society in recent years, the demand for energy increases significantly in the global world, while the limited fossil energy and high emission of pollutants makes government begin to emphasize and encourage the use of biomass energy [1]. The World Energy Council (WEC) predicted that the proportion of biomass in the total global renewable energy will be increased to 60% by 2020 [1,2].

Generally speaking, the final products of biomass include three forms, *i.e.*, gaseous, liquid and solid ones [2], among which Densified Biomass Solid Fuel (DBSF) is a typical representative of the solid form of biomass. After being processed and treated under a certain temperature and pressure, many loose raw materials such as straw, twigs, sawdust, *etc.* can be squeezed into special shapes and changed into DBSF with high density and low ash contents, which makes it convenient for transportation and combustion, and cost-effective to be utilized [3]. DBSF can be widely used for steam, in hot water boilers as fuel, and for power generation by direct combustion or CHP (combined heat and power). Table 1 shows the main differences between DBSF and traditional fossil fuels. Among them, Datong Coal as one kind of high quality coal in China, was selected as a typical fossil fuel representative with low ash content.

Table 1. The property difference between the Densified Biomass Solid Fuel (DBSF) and Datong Coal.

Fuel Type	ρ (g/cm³)	S_{ad} (%)	A_{ad} (%)	FC_{ad} (%)	FT (°C)
DBSF	1.0–1.4	0.05–0.2	1.0–13	13–20	1000–1200
Coal (Datong)	1.25–1.5	1.78	12.04	47.82	1500

ρ means density, and DBSF's density has less difference with that of coal. S_{ad} is Sulphur content (Air dried basis). S_{ad} in per unit of DBSF is much less than that of Datong Coal. Generally, S_{ad} in DBSF from straw is not more than 0.2%, and that from woody material is only about 0.05%. A_{ad} refers to ash content (Air dried basis). The ash content of corn straw DBSF accounts for about 8%, and that of woody DBSF is roughly 5%. Only the ash content of DBSF made from Rice straw and rice husk reach 13%. FC_{ad} means fixed carbon (Air dried basis). FC_{ad} in DBSF amounts to roughly one third of that of Datong Coal. FT means ash fusion temperature.

Hong Hao et al. [2] utilized 2 × 1400 kW hot-water boilers and 2 × 1400 kW steam boilers as pilot targets, and the results showed that the actual measured data of hot conversion efficiency accounted for about 84% using DBSF as fuel, and the emission indicators of actual data was much lower than that of National Boiler Ambient air quality standards. Particulate pollutant emission concentration of DBSF boilers is about 30 mg/m³, much less than that of coal boiler emission standard as 80 mg/m³, and equal to that of natural gas combustion. The SO_2 emission concentration was about 41.3 mg/m³, much less than that of coal boiler emission, which has a standard of 400 mg/m³.

Meanwhile, Zong Yi et al. pointed out that the combustion of DBSF in the stove had the high thermal conversion efficiency of 60%–80% due to more full combustion, compared to that of 5%–8% using traditional firewood combustion [4]. DBSF combustion produces less black smoke with less concentration of C particle caused by incomplete combustion compared to the traditional direct combustion of biomass material [4]. In general, the emission concentration of SO_2, NO_x all decrease using DBSF as fuel, and DBSF is considered as a kind of clean fuel in China [2]. Therefore, DBSF promotion and application in China can play a major role in environmental improvement.

Recently, some regions of China have experienced severe threats by fog and haze, among these, in situ concentrated combustion of agricultural residues was considered to partly contribute to haze formation in harvest season [4]. Studies have shown that agricultural residue seasonal burning can contribute to around 30% of particle generation as concentrated combustion in one or two days in large cropland area [4]. DBSF utilization is considered to be an ideal solution for the disposal of agricultural and forestry residues, associated with reduction of land occupy and air pollutant generation. Meanwhile, DBSF as alternative to fossil energy, the emission of SO_2 and ash can reduce by 80% compared to coal under the same heat value.

In the past decade, DBSF utilization has been broadly expanded in many countries across the world including China. According to "the national renewable energy medium and long-term development plan in China", the annual consumption of DBSF in China would have reached 50 million tons by 2020 [5]. In order to achieve the goal of DBSF utilization, industrial development should inextricably link with the incentives and preferential policies with relevant departments of the central and local governments. Therefore, it is meaningful for China to analyze the industrial policies to support the development of DBSF utilization. However, there are a number of issues associated with DBSF exploitation and utilization, such as its supply chain, availability and distribution.

Focusing on the particular stages of the supply chain may enable policy advocates or policy makers to identify particular interventions to target bottlenecks to utilization, interaction effects of policies, or to assess the degree to which current policy practices are conducive to objectives. Such a framework may contribute to increase understanding of the factors critical to DBSF development in China. Although the main focus of the paper is on China, it would be relevant to other developing countries, which would be looking to further develop their DBSF industry [6]. Therefore, this review paper focuses on DBSF policies and strategies in China based on the supply chain classification, in particular by addressing the following issues with a holistic approach: (1) what is the potential of DBSF in China; (2) what are the DBSF industrial strategies and policies in China, and how were the strategies articulated and executed for attaining policy objectives in each supply chain stage, and (3) what is the deficiency of current DBSF industry policies in China, and how can it be improved?

2. Introduction of Densified Biomass Solid Fuel (DBSF) in China

2.1. The Resources of DBSF

The majority of raw materials of DBSF come from crop straw, agricultural processing and forestry residues [2,3]. Among them, agricultural and forestry residues are the primary sources for DBSF. The distribution of agricultural and forestry residues in China are illustrated in Figures 1 and 2.

As China's basic farmland protection system does not change and the basic agricultural planting structure is relatively consistent, annual output of agricultural residue is correspondingly stable [3,5]. The total amount of agricultural residue was 630 million tones recently [5,7]. Among the generated agricultural residues, crop and wheat straws are occupying the largest proportion, reaching 270 million tons. Rice husk, corn cob, peanut shells and other processing residues can also be used as the raw materials of DBSF. In addition to the loss of collection and transportation, the use of returning to field, feed or other stuffs, the best estimate suggests that the consumption of agricultural residues as fuel is about 168 million TCE (ton of standard coal equivalent) [8].

Meanwhile, forestry biomass resources are mainly from forestry logging, tending and processing residues. The available amount of China's annual tree branches and forestry residues can reach about 900 million tons, while about 300 million tons can be utilized as energy equivalent to 200 million TCE [9]. Although agricultural and forestry residues have some different features, both of them are important DBSF sources in different regions of China. Abundant agricultural and forestry residues make it suitable for DBSF development in China.

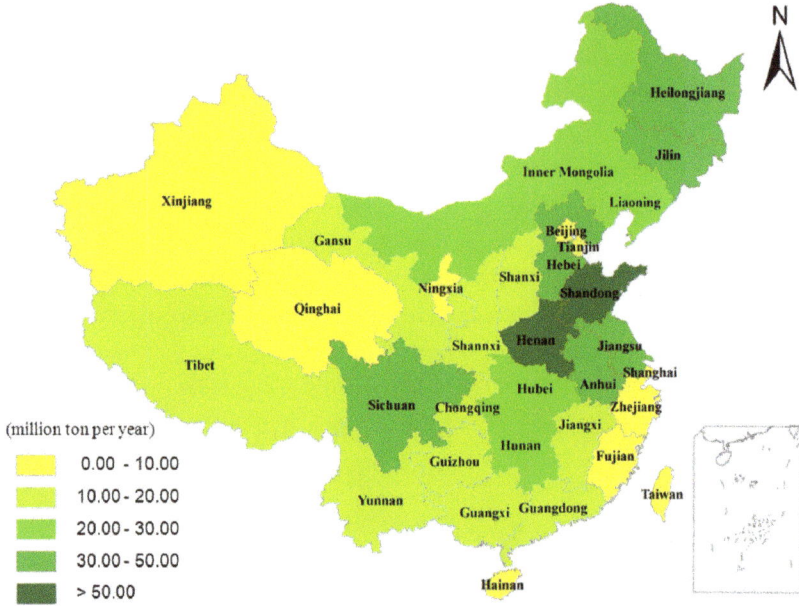

Figure 1. The distribution of the average output of agricultural residues in China.

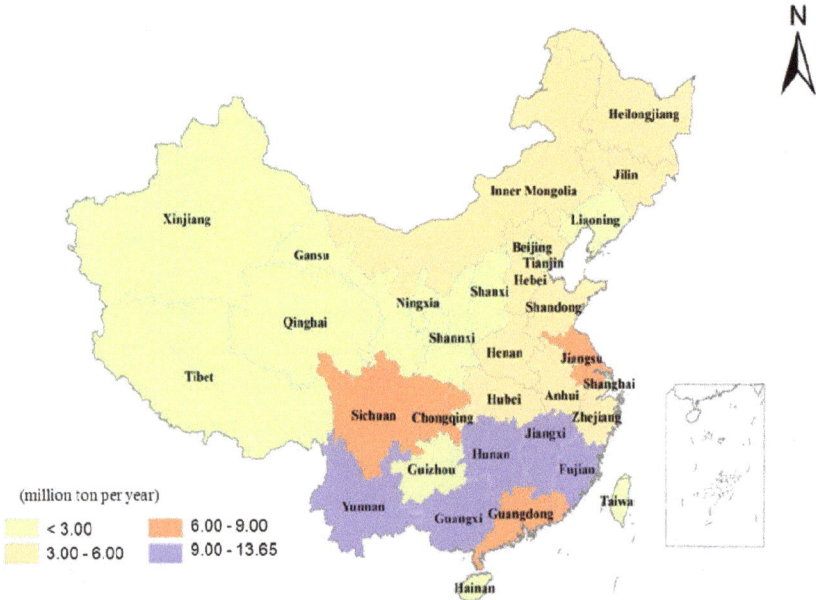

Figure 2. The distribution of the average output of forestry residues in China.

2.2. The Classification of DBSF Supply Chain

DBSF supply chain has been defined as the integrated management of DBSF production from harvesting biomass resources to energy conversion facilities. Generally speaking, in order to facilitate the subsequent policy analysis, the long DBSF supply chain can be divided into three stages. The complete DBSF supply chain is shown in Figure 3, while the products of DBSF are illustrated in Figure 4. The entities or enterprises involved in the DBSF processing stages may include:

Stage I: The companies or entities related to raw materials collection, *i.e.*, harvest, storage and transportation;

Stage II: The enterprises or companies related to DBSF production process, including the production of DBSF, the design and manufacture of DBSF production equipment, storage and transportation of DBSF products, *etc*;

Stage III: The enterprises or entities related to market application of DBSF, including the market-oriented fuel use, the design and manufacture of DBSF boilers and other terminal utilization equipment, the design and production of energy conversion equipment.

Major customers of stage III include household use, as well as commercial heating, steam supply, *etc.* As the policy support is hard to cover all aspects of the DBSF industry chain, and the support for transport and storage is relatively weak compared with other links of DBSF industry chain. In order to simplify the analyses of policy on the complicated supply chain, the policy on transport and storage is not considered as a key link to analyze.

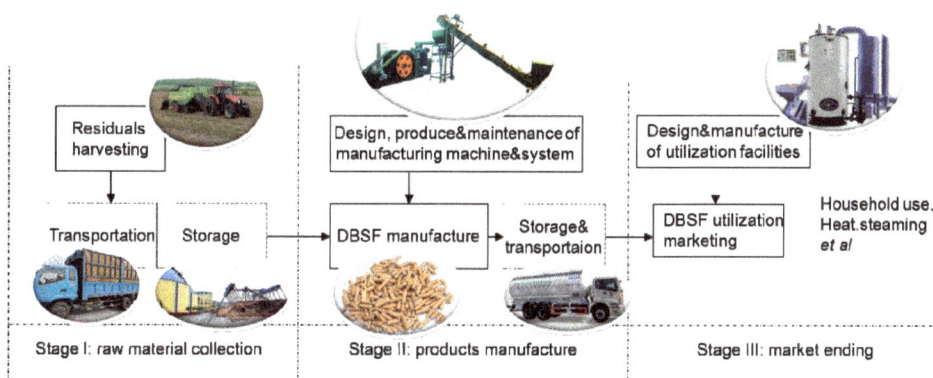

Figure 3. The DBSF supply chain classification.

Figure 4. The main products of DBSF.

2.3. Issues of DBSF Supply Chain

Since 2004, the DBSF industry has grown more than 18% per year at the global scale. As yet, DBSF industry in China is in the initial stage of industrial development, with small market scale [3–5]. The main issues of DBSF utilization in China have been concluded in full view of supply chain as below.

2.3.1. The Issue of Resource Collection

There always exists a contradiction between the scattered bio-resource distribution and the concentrated industrial production of DBSF in medium and large scale, together with seasonal harvest and successive production demand. At the same time, the district expansion of supply and demand inevitably leads to the raw material price soaring. As the dispersion and seasonal instability of feedstock, collecting agricultural and forestry residues in traditional ways cannot meet the large-scale commercial production. Actually, feedstock process should have a balanced mechanism to achieve a good dynamic equilibrium, so that the enterprise can receive sufficient raw materials with reasonable and stable price [6].

2.3.2. The Issue of Production Technology

In terms of DBSF production, the biomass solidification technology is considered to be the core issue. Generally, these technologies can be divided into thermoforming and compressed-forming according to the temperature required for the processing. Roller die extrusion, piston stamping and screw extrusion are the three major types of solidified fuel processing machines. Nowadays, the technologies used in some developed countries are highly specialized and automated and are characterized by high thermal efficiency and less pollution. However, because of high equipment price, high power consumption, improper combustion parameters that may induce slag in stoves and limited variety of feedstock suitable for processing, these technologies are not suitable for China. China is a latecomer in the research of biomass solidification technologies, and there exists a large gap between the technologies used in China and in developed countries. In order to encourage the development of biomass solidification technology, the government has moved to improve its incentives for DBSF research and production. The detailed R&D projects are described in Section 4.2.

2.3.3. The Issue of Market Application

Market demand for DBSF as the last part of supply chain is considered to be insufficient in China. The fundamental way to promote DBSF in large-scale applications is in place of fossil energy, e.g., coal. However, it is restricted by the obstacle of DBSF price fluctuation with unstable raw material supply. Actually, the issues in different links of the supply chain interact and transfer from the beginning to the end. On the other hand, a number of heating enterprises cannot accept the concept of DBSF as an alternative energy due to a lack of correct understanding of DBSF as a clean and renewable energy. In comparison, DBSF has already been successfully applied in extensive fields in foreign countries, particularly in Europe. In addition to the central heating boiler, DBSF application has entered into ordinary home use. In China, the domestic and commercial applications have not been fully developed, and the market field is extremely limited.

Many studies suggest that the main issues representing the DBSF utilization and promotion in China focus on the manufacturing technologies in stage II. Yet in this paper, the main obstacles are not just concentrated on production technology, but also exist in the collection phase and application market. Therefore, we adopt the analysis on the overall policy contents combined with the supply chain.

3. China's DBSF Development Strategies

3.1. Policy Conceptual Framework

The framework of DBSF related policies and regulations can be divided into three levels, shown in Figure 5. The first level is the basis of the legal documents, such as the "Renewable Energy Law" which is a comprehensive legal framework of renewable energy including DBSF. The second level is determined by the national governments in accordance with the general legal framework and basis for planning the development of industry specific milestones, such as the renewable energy planning in the period of the "12th Five-Year Plan". The third level is identified as a set of policy instruments, e.g., positive incentive and reverse punishment to ensure the realization of milestones [10,11]. Detailed policy instruments built upon the DBSF supply chain are analyzed thoroughly in Section 4.

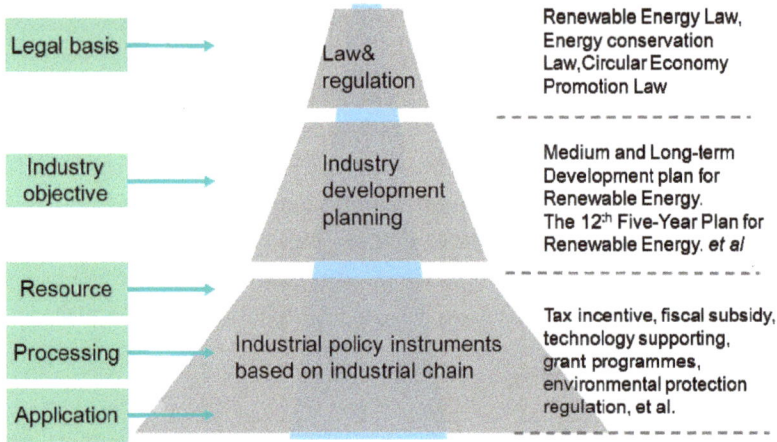

Figure 5. The conceptual framework of DBSF related policy.

3.2. Laws Related to DBSF Development

On the first level, law and regulation at the state level is shown in Table 2, e.g., the Renewable Energy Law, and the Energy Conservation Law. The legal basis laid the basis for the development of the DBSF industry aiming to limit the high-polluting industries, to encourage the development of clean energy, like DBSF utilization.

Table 2. The main laws related to DBSF in China.

No.	Name of Law	Date of Issue	Relative Items
1	"Renewable Energy Law of the People's Republic of China"	28 February 2005	Article 16: The state shall encourage clean and efficient development and use of biomass fuels.
2	"Energy Conservation Law of People's Republic of China"	28 October 2007	Article 7: The State encourages and supports development and utilization of new energy resources and renewable energy resources. Article 59: The State encourages and supports vigorous development of marsh gas, and popularizes biomass, and other renewable energy in rural areas...
3	"The Circular Economy Promotion Law of the People's Republic of China"	29 August 2008	Article 34: The state encourages and supports agricultural producers and relevant enterprises to employ advanced or applicable technologies to make a comprehensive utilization of straws of crops by-products from the processing of agricultural products. Article 35: The people's governments above county level and their departments of forestry shall ... encourage and support forestry producers and relevant enterprises to employ timber-saving and alternative technologies to make a comprehensive utilization of forestry wastes, inferior woods, short ends, fuel woods and sand shrubbery *etc.*, and improve the comprehensive utilization rate of timbers.

3.3. Plans Related to DBSF Development

On the basis of laws and regulations, the country has established a set of development planning tools, such as the Medium and Long-term Development Plan for Renewable Energy, and the Five-Year Plan for Renewable Energy in the "Eleventh Five-year" and "Twelve Five-year" periods. The core items of each plan are described in Table 3.

Table 3. The main plans to DBSF development in China.

No.	Name of Plan	Date of Issue	Relative Items
1	Medium and Long-term Development plan for Renewable Energy in China	4 September 2007	By 2020, the use of DBSF in China nationwide will reach 50 million tons. By that time, DBSF will have become a commonly used form of high quality fuel.
2	The 11th Five-Year Plan for Renewable Energy	18 March 2008	Accelerate the development of biomass energy, enlarge production capacity of DBSF...
3	The 12th Five-Year Plan for Renewable Energy	6 August 2012	By 2015, the use of DBSF in China nationwide reaches 10 million tons.

In the 11th five-year plan, there was no specific objective or indicator for DBSF industry development, except some descriptive requirements. However, the 12th five-year plan has explicitly put forward the development target of DBSF industry which illustrates DBSF can speed up development during the 12th five-year period.

4. Policy Instrument Analysis Based on DBSF Supply Chain

Central and local governments have developed and implemented a series of policy instruments to support and motivate the development of DBSF industry, e.g., financial subsidies of straw energy utilization, preferential income tax and Value Added Tax (VAT) policies, science and technology projects supporting policies, the coal-fired limits policies, *etc.* These policies have played a decisive role in promoting the production and use of DBSF. The policy instruments based on the supply chain are illustrated in Figure 6.

4.1. Policy Related to Resource Collection

The first phase of DBSF supply chain is defined as the collection and storage of agricultural and forestry residues. Among them, the existing policies mainly focus on the harvesting of agricultural residues. Forestry residue, as another available DBSF resource, is so far absent in the consideration of policy makers.

4.1.1. Subsidy on the Purchase of Agricultural Machines

For enterprises engaged in the harvest of agricultural residues, especially small businesses, the cost of agricultural machinery and equipment is a huge burden. In order to accelerate the pace of agricultural mechanization in China, an agricultural machinery purchase subsidy policy has been implemented in recent years. The list of equipment types in agricultural purchase subsidy policy, includes corn harvester, cotton harvester, tall crops swather, peanut harvester, rapeseed harvester, binder machine, baling machine, pick-up baler, straw chopper, peanut shelling machine, feed crusher, dryer, *etc.* [12]. A great many of these mechanical functions are coincident with those of DBSF's collection and pretreatment features. Financial subsidies on the purchase of these machines can help reduce the DBSF resource collection cost.

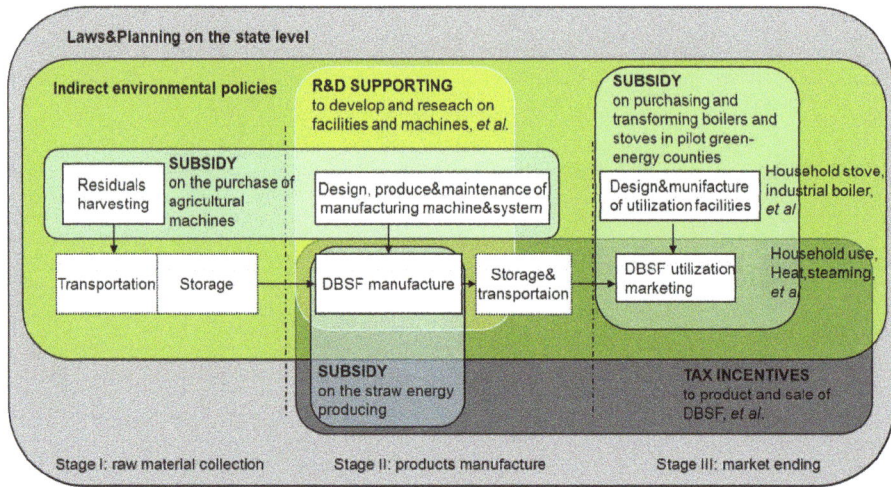

Figure 6. The policy instruments based on DBSF industry chain.

4.1.2. Prohibition of Burning Straw

Recent studies have shown that centralized straw burning in farmland can cause serious air pollution event. National Satellite Meteorological Center (NSMC) of China Meteorological Administration would regularly publish the information on Satellite-Remote-Sensing-Based Monitoring of Straw Burning in harvest season. It is shown that the air pollution problem is very serious caused by burning straw in Anhui, Henan, Jiangsu and other provinces in harvest season [13].

Therefore, some local governments have developed a set of corresponding local prohibiting policies of burning straw to reduce the air pollution especially in harvest season. The amount of straw returning to the field and other losses only accounts for about 20% of the total straw generation [2]. DBSF is accepted as an efficient solution for agricultural residues. In fact, the prohibition of burning straw in rural areas contributes to the local straw harvest and pretreatment, and is considered to be the driving force for the comprehensive utilization of straw and the production of DBSF.

4.2. Policy Related to DBSF Producing

The subsequent phase of the DBSF supply chain is mainly about DBSF production and processing, as well as the manufacture of DBSF production equipment. The subsidy and incentive policies are mainly inclined to this section.

4.2.1. Subsidy Funds for Straw Energy Utilization

The Ministry of Finance established funds to support the straw industrial development, and formulated "Interim Measures on financial subsidies for straw energy utilization" [14] (referred to as the "Interim Measures") in October 2008. "Interim Measures" explained that this fiscal subsidy only supports enterprises engaged in straw energy production, such as DBSF producers; enterprises applying for grant funds shall have more than 10,000 tons of annual straw usage capacity. In 2011, a total of 50 enterprises in straw energy production received the principal subsidy funds, in which 46 enterprises engaged in the production of DBSF.

4.2.2. Subsidy Policy on the Purchase of Agricultural Machines

As mentioned in Section 4.1, agricultural machinery can receive subsidy. The equipment of DBSF briquette and pellet production in the second stage of DBSF supply chain is also included in the agricultural machinery to enjoy farm machinery purchase subsidy.

4.2.3. The policy on VAT of Resource Comprehensive Utilization Products

On 21 November 2011, the Ministry of Finance and the State Administration of Taxation issued "A notice on the adjustment and improvement VAT policy on the resources comprehensive utilization and services" [15]. Enterprises such as DBSF producers can enjoy the support of immediately returning VAT, if the production uses rice husks, peanut shells, corncobs, camellia shells, cottonseed hulls, forestry residues, and small fuel wood as raw materials. In order to encourage the enterprises comprehensive utilization of agricultural and forestry residues, this policy has clarified VAT would be refundable for those DBSF production enterprises as soon as it is imposed.

4.2.4. Science and Research Project Support

Currently, the support of science and technology projects, mainly concentrated for DBSF production equipment and molding technology research and development in stage II of supply chain [16,17]. The main R&D programme on DBSF production is listed in Table 4.

4.3. Policy Related to DBSF Application

The equipment for DBSF utilization is quite different from normal combustion equipment using coal or other fossil energy. The companies involved in stage III are classified as those who provide heat or steam using DBSF as energy and the manufacturers of DBSF boilers and stoves.

4.3.1. The Preferential Policies on Enterprise Income Tax and VAT

On 20 August 2008, "Preferential Catalogue for Enterprise Income Tax for Comprehensive Utilization of Resources" was promulgated collectively by the Ministry of Finance, the State Administration of Taxation, and the National Development and Reform Commission [18]. In this document, the types of heating companies generated by DBSF from straws and other agriculture residues were listed, which means preferential policies on Enterprise Income Tax can apply to these companies.

Table 4. R&D programme on DBSF production in China.

Starting Year	Supporting Organization or Project	Name of Programme	Main Contents or Outputs
2006	National Key Technology R&D Program in the 11th Five year Plan	"DBSF products and equipment development" programme	R&D of the movable DBSF equipment with integrated function of the raw material pretreatment, grinding, molding process on a large scale with low energy consumption.
2011	Chinese government, the World Bank (WB) and the Global Environment Facility (GEF)	China Renewable Energy Scale-up Programme CRESP)	(1) Develop the PM485-II biomass pellet molding machine with high-efficiency and low-cost; (2) Optimization of biomass pellet fuel molding technology and equipment with low-cost; (3) R&D of biomass briquetting technology and equipment at room temperature with low power consumption, which can solve the problem of high moisture of straw materials; (4) R&D of Biomass compact molding equipment, which should complete the split ring molded briquetting machine design;
2013	National Key Technology R&D Program in the 12th Five year Plan	"Low-cost DBSF equipment research and application" programme	Within this programme, "Large-scale DBSF technology integration and industrialization demonstration" projects should form 100,000 tons of annual DBSF production capacity and 300 sets of DBSF molding equipment.

4.3.2. Subsidy Fund for Green Energy Counties

On 10 December 2009, the National Energy Administration published a "Notice of Recommending Green Energy Counties". The National Energy Administration, Ministry of Finance, and Ministry of Agriculture collectively organized the construction of Green Energy Demonstration Counties. Demonstration subsidy funds mainly focused on DBSF utilization project and rural energy service system.

Theoretically speaking, each project should provide heat for the daily cooking of 1000 rural households and public institutions as hospitals, schools, government institutions, kindergartens, and nursing homes. The demonstration subsidy funds should only be used for purchasing and transforming boilers and stoves using DBSF as fuel [19,20].

4.3.3. The Promotion of BDSF Boilers

Local governments established compulsive policies against burning coal or low-quality coal in urban districts, including Zhengzhou, Kunming, Shijiazhuang, Changsha, Suzhou, Ningbo, Luoyang, Weifang, *etc.* Therefore, regional environmental policies promoted the application of DBSF indirectly. For example, on 20 June 2007, Zhengzhou government published "Notice of proceeding the transformation of clean fuel boilers". The document declares the following aspects: coal-burning boilers below 10 T/h should be removed or transformed to use clean energy or clean fuel. With the implementation of the policies above, Zhengzhou greatly increased the market demand for DBSF. It has become a common scene that coal-burning boilers below 10T/h are replaced by DBSF boilers in recent years, which promoted the producing and selling of DBSF.

5. Effects of Policies on DBSD Development

5.1. The Effect on Technical and Technological Development

Researchers considered that the foremost technical bottleneck in DBSF supply chain is DBSF production [7,9], namely biomass solidification technology. In order to promote the development of biomass solidification technology, Chinese government has extended the supporting on production technology through policy incentives and R&D projects [21]. Currently, the gap between the technologies used in China and developed countries has been gradually reduced, and the direction of DBSF production technologies can be highly automatic with high thermal efficiency, less pollution, and more reasonable process.

Many studies suggest that the main technical barriers of DBSF development in China focus on the manufacturing technologies in stage II. However, the main technical issues also exist in the collection phase and application market. One of the major challenges in DBSF supply chains is to ensure that raw material is exploited in a resource efficient manner [22]. There exist evidences that resource consumption for energy production is often implemented without setting a proper plan of replacement planting [23,24]. Considering the seasonality and instability of biomass resources, these inefficiencies could trigger future scarcities of biomass feedstocks, baffling the resilience of biomass supply chains. The research proved that the current feedstock management systems find it extremely difficult to meet the requirements of large scale bioenergy developments, because they are only designed for small-to-medium scale handling and logistics requirements. Therefore, resource and supply chain efficiencies and process productivity consist of the main technical or technological factors of biomass supply chains, and current studies should consider the whole DBSF supply chain.

5.2. The Effect on DBSF Production Enterprises

Through data collection of the companies who gained grant fund of straw energy utilization, and trade information statistics from the Industry Bureau registration website, this paper has compiled a DBSF enterprise distribution map in China, as shown in Figure 7. Enterprises with an annual DBSF production capacity of less than 10 thousand tons are not involved in the statistical data.

It is illustrated that the production enterprises are mainly situated in Hebei, Shandong, Henan, Heilongjiang, Jiangsu, and Guangdong provinces. The district setting of production Enterprises in China mainly follows the proximity-based principle, to reduce transportation and storage costs of raw materials. Henan, Hebei, and Shandong are China's major crop production regions, and Heilongjiang Province is among the main national grain bases, that all could provide abundant agricultural biomass resources. Meanwhile, the environmental protection policy and the incentive subsidy could be the main driving forces behind DBSF production in these three provinces. The local prohibition of straw burning guides the peasants to find a reasonable disposal method of agricultural residues. Subsidy on the manufacturing machines and tax incentives like VAT actually has positive effect on enterprises establishment and development.

Figure 7. The distribution of DBSF production enterprise in China.

Otherwise, the impetus of the regions in Jiangsu, Guangdong and other provinces for the most part relies on the vigorous demand for energy, as they are all located in coastal areas and classified as developed areas with higher industrial production output values. These regions need the multiple schemes of fossil energy alternatives, and DBSF is selected as energy supplement. In the same way, medium-density agricultural crops and forest resources in Jiangsu and Guangdong province reflect sufficient raw materials could meet the DBSF production demands. The subsidy and tax incentive cannot be considered as the core elements of positive effects but the energy demands, that a little different from above regions.

5.3. The Effect on DBSF Market Utilization

The market is the major benchmark of industrial development; therefore, only the products can meet the market demand and gain market recognition and acceptance, which will ultimately drive the rational industry development. The annual DBSF consumption amount in China has been concluded and illustrated in Figure 8. DBSF industry is among the latecomers to the Chinese market, and the initial stage could be identified from 2004 to 2010. Until 2010, the annual consumption amount of DBSF was only 2.5 million tons [9]. Resent policy instruments seem not to play a powerful role in pushing the DBSF utilization in stage III of supply chain.

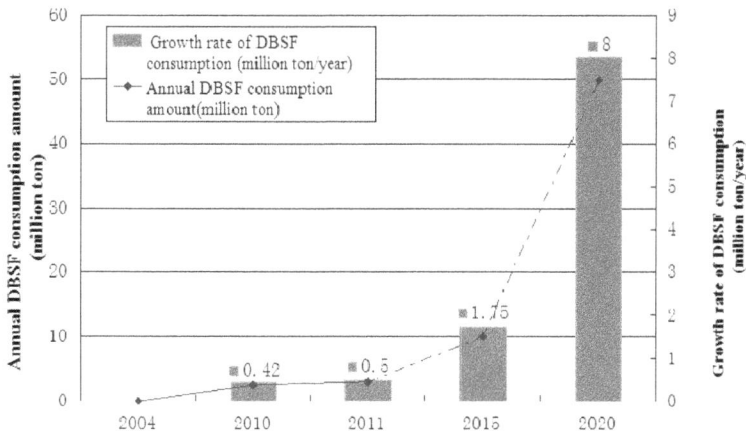

Figure 8. The annual DBSF consumption amount and growth rate in China.

According to "Medium and Long-term Development plan for Renewable Energy in China", the annual consumption of DBSF in China should reach 10 million tons until 2015 and 50 million tons in 2020. That indicates an annual growth rate of DBSF consumption rises up to 8 million tons per year from 2016 to 2020. In order to achieve this speed of development, China needs to enter the stage of rapid development of DBSF industry during the year of 2016 to 2020.

5.4. The Effect on Environment and Economy Improvement

DBSF production and application has solved the growing problem that tons of agricultural and forestry residues stack *in situ*, and tend to eliminate chaotic and direct burning of biomass resources. Promotion policies have made DBSF a substitute for coal as energy resource with less ash and lower SO_2/NO_x [20,21]. DBSF utilization would decrease the atmosphere pollution caused by the combustion of coal or other fossil energy. According to the plan data in 2015, the annual consumption amount of DBSF is up to 1.75 million tons, equivalent to reducing 875 thousand tons of coal usage. To some extent, DBSF utilization in most rural families can also decrease the use of firewood to improve indoor air-quality and peasants' living conditions.

Moreover, local peasants would increase their incomes through the harvesting and collecting of agricultural and forestry residues. Coming to the fee of 150 Yuan per ton of straw, it would increase local farmers' income by about 150 million Yuan when the amount of straw harvesting reaches 1 million tons. During the process of DBSF harvesting and production, plenty of job opportunities would be provided that is beneficial to absorb social surplus labors. Obviously, through the promotion policies of DBSF utilization, it is proven to bring positive economic, social and ecological benefits in China.

6. Suggestions

By sorting and analyzing the policies on DBSF development, we submit some suggestions as follows.

(1) The resources of DBSF are classified as wood and agricultural residues. Among these, wood material only occupies a small portion of DBSF utilization. Unlike energy utilization of straws, which has subsidies from central government, the production cost of wood (pellet) molding fuel is relatively high. Unless the users have strict demands on the quality of molding fuel, general users cannot afford the high cost of wood particles molding fuel. As a result, developing a specialized subsidy policy for DBSF that use forestry residues as raw material is necessary. There are subsidies, awards, and tax preference and reduction policies on the production and utilization of DBSF that come from agricultural residues. Now these preferential policies should also apply to DBSF that use forestry residues as raw materials.

(2) By analyzing each preferential policy of the supply chain, we conclude that these policies tend towards intermediate link—DBSF production. Moreover, expanding production scale and improving productivity of DBSF is considered to be the vital themes for China's DBSF development by government and research institutes. In contrast, there exists few influential policies on resource harvesting, collection and market utilization. In the field of scientific research projects, however, it is particularly imbalanced and extremely centralized. Input for scientific research is restricted in the upstream and downstream of the supply chain. From this view, state-supporting science and technology projects in DBSF should concern the technical bottlenecks from the whole supply chain, combining the strength of enterprises, universities and research institutions.

(3) In fact, the promotion driving dynamics differ from province to province and the incentives play a different role of DBSF market in different provinces. The central government has established industrial development goals, so it is necessary to distribute these goals to each province and autonomous region based on resources, markets, funds, development foundations and market demands. Provincial governments should have specific implementing regulations for each plan on the development of DBSF, such as a comprehensive utilization plan for straw, a forestry energy plan, a circular economy plan, *etc*. One of the conditions in the implementation of a specific plan is to enhance the management of the plan. Concretely speaking, the management should include the distribution of the plan goals and the establishment of implementing regulations.

(4) Once the supporting projects or incentives suspend, how is the sustainable development of biomass utilized, especially DBSF? This is a realistic question for most developing countries. As for the construction project of green energy demonstration counties in China, there are a few unsolved problems after the construction of the DBSF project. The first is how to make the sales of DBSF continuous in rural market and another is to ensure companies that invest on these projects have sustainable proper profits. Therefore, that makes the policy makers contemplate the idea that the present policy should be sustainable.

7. Conclusions

China has endowed with abundant biomass resources, such as forestry and agricultural residues, to promote DBSF utilization. However, due to the seasonality and instability of agricultural and forestry resources, imbalance in resources such as feedstocks could trigger future scarcities of DBSF productivity, baffling the resilience of biomass supply chains. In this research, China's DBSF policies were reviewed and analyzed based on the views of supply chain.

The policy framework has been categorized as three levels. Policies on laws and plans have an important directing impact on the strategy and schedule of DBSF industrial development. The third level is defined as a set of policy instruments, e.g., financial subsidy, tax incentive, and R&D project support, which would ensure the realization of DBSF development milestones. Policy has a straightforward boosting or restraining influences on company operations.

Through the analysis of current policies based on the supply chain, weak links in contemporary policy systems have been described. The present DBSF market is on a small scale, lack of balance in supply chain, and short of market drive, which severely restrains the healthy development of DBSF industry. The effects of policies on R&D, production enterprises distribution, market utilization and environment and economy have been analyzed, and then proposed some suggestions. The policy makers also should be concerned with the bottlenecks from the view of whole supply chain, combining the strength of local governments, enterprises, universities and research institutions. Overall, the policy framework is important for authorities to manage and achieve the DBSF development objective in the future.

Acknowledgments

This paper was financially supported by the Supporting Program of the "Twelfth Five-year Plan" for Sci & Tech Research of China (2012BAD15B05).

Conflicts of Interest

The authors declare no conflict of interest.

References

1. Fang, H. The status and development prospect of renewable energy in China. *Renew. Energy* **2010**, *28*, 137–142.
2. Hong, H.; Ye, W.H.; Song, B.; Zhang, X.X. An empirical study on industrialization of biomass briquette in China. *Resour. Sci.* **2010**, *32*, 2172–2178.
3. Chen, Y.S.; Sun, Y.F. Countermeasures and Suggestions on the Development of Biomass Briquetting Densification Fuel Industry. In Proceeding of the International Conference on Renewable Energy Scale-up Development and the Third Energy Technical Forum in Far-Yangtze River Triangle Area, Nanjing, China, 16–18 November 2006; pp. 545–549.
4. Zong, Y.; Zhu, M.D.; Ma, L.; Wang, Y.C.; Sun, D.T.; Yan, R.X.; Hu, X.W.; Bai, A.F.; Xu, C.; Zhu, R.Z. The effect and suggestion of straw *in situ* burning on atmospheric environment. *Public Commun. Sci. Technol.* **2014**, *3*, 113–116.
5. Medium and Long-term Development Plan for Renewable Energy in China. Available online: http://www.sdpc.gov.cn/zcfb/zcfbtz/2007tongzhi/t20070904_157352.htm (accessed on 18 June 2013).
6. Dennis, R.B.; Cassandra, M.; Christine, L. A supply chain analysis framework for assessing state-level forest biomass utilization policies in the United States. *Biomass Bioenergy* **2011**, *35*, 1429–1439.

7. Chen, L.J.; Li, X.; Han, L.J. Renewable energy from agro-residues in China: Solid biofuels and biomass briquetting technology. *Renew. Sustain. Energy Rev.* **2009**, *13*, 2689–2695.

8. Wang, X.W.; Bai, J.M. *Assessment of Biomass Resource Availability in China*; Li, J.J., Bai, J.M., Eds.; China Environmental Science Press: Beijing, China, 1998.

9. Wei, W.; Zhang, X.K. Development status and prospect of solid biofuel-preparation in China. *Guangdong Agric. Sci.* **2012**, *5*, 135–138.

10. Zhang, D.L. Prospect on research and development of biomass briquette technologies. *Mod. Agric.* **2007**, *12*, 98–103.

11. Liu, J.W.; Lei, T.Z; Han, G.; Bai, W. The analysis on policies and regulations of biomass power generation in China. *Solar Energy* **2007**, *11*, 8–10.

12. Agriculture Machinery Catalogue. Available online: http://www.agri.gov.cn/V20/ZX/hxgg/201112/t20111213_2434849.htm (accessed on 15 May 2013).

13. Satellite-Remote-Sensing-Based Monitoring of Straw Burning. Available online: http://www.secmep.cn/secPortal/portal/indexLogin.faces (accessed on 1 May 2014).

14. Interim Measures on Financial Subsidies for Straw Energy Utilization. Available online: http://www.mof.gov.cn/zhengwuxinxi/caizhengwengao/caizhengbuwengao2008/wengao200811qi/200903/t20090304_118500.html (accessed on 3 May 2014).

15. A Notice on the Adjustment and Improvement VAT Policy on the Resources Comprehensive Utilization and Services. Available online: http://szs.mof.gov.cn/zhengwuxinxi/zhengcefabu/201111/t20111123_609990.html (accessed on 1 May 2013).

16. Li, J.M.; Xue, M. Current situation and development prospects of the utilization of biomass energy in China. *Manag. Agric. Sci. Technol.* **2010**, *4*, 1–4.

17. Zhao, X.; Wang, J.; Liu, X.; Feng, T.; Liu, P. Focus on situation and policies for biomass power generation in China. *Renew. Sustain. Energy Rev.* **2012**, *16*, 3722–3729.

18. Preferential Catalogue for Enterprise Income Tax for Comprehensive Utilization of Resources. Available online: http://www.sdpc.gov.cn/tztg/t20081014_240694.htm (accessed on 12 July 2013).

19. Green Energy Counties. Available online: http://www.gov.cn/gongbao/content/2011/content_1967422.htm (accessed on 14 March 2014).

20. Zhang, P.D.; Yang, Y.L.; Tian, Y.S.; Yang, X.T.; Zhang, Y.K.; Zheng, Y.H.; Wang L.S. Bioenergy industries development in China: Dilemma and solution. *Renew. Sustain. Energy Rev.* **2009**, *13*, 2571–2579.

21. Chen, L.P.; Li, D.K. Development Situation of the Industry of Densified Biomass Solid Fuel in China. In Proceedings of the International Conference on Materials for Renewable Energy and Environment (2012 ICMREE), Beijing, China, 18–21 May 2012; pp. 476–480.

22. Mafakheri, F.; Nasiri, F. Modeling of biomass-to-energy supply chain operations-applications, challenges and research directions. *Energy Policy* **2014**, *67*, 116–126.

23. McKendry, P. Energy production from biomass (Part1): Overview of biomass. *Bioresour. Technol.* **2002**, *83*, 37–46.

24. Adams, P.W.; Hammond, G.P.; McManus, M.C.; Mezzullo, W.G. Barriers to and drivers for UK bio-energy development. *Renew. Sustain. Energy Rev.* **2011**, *15*, 1217–1227.

The Impact of a Carbon Tax on the Chilean Electricity Generation Sector

Carlos Benavides, Luis Gonzales, Manuel Diaz, Rodrigo Fuentes, Gonzalo García, Rodrigo Palma-Behnke and Catalina Ravizza

Abstract: This paper aims to analyse the economy-wide implications of a carbon tax applied on the Chilean electricity generation sector. In order to analyse the macroeconomic impacts, both an energy sectorial model and a Dynamic Stochastic General Equilibrium model have been used. During the year 2014 a carbon tax of 5 US$/tCO₂e was approved in Chile. This tax and its increases (10, 20, 30, 40 and 50 US$/tCO₂e) are evaluated in this article. The results show that the effectiveness of this policy depends on some variables which are not controlled by policy makers, for example, non-conventional renewable energy investment cost projections, natural gas prices, and the feasibility of exploiting hydroelectric resources. For a carbon tax of 20 US$/tCO₂e, the average annual emission reduction would be between 1.1 and 9.1 million tCO₂e. However, the price of the electricity would increase between 8.3 and 9.6 US$/MWh. This price shock would decrease the annual GDP growth rate by a maximum amount of 0.13%. This article compares this energy policy with others such as the introduction of non-conventional renewable energy sources and a sectorial cap. The results show that the same global greenhouse gas (GHG) emission reduction can be obtained with these policies, but the impact on the electricity price and GDP are lower than that of the carbon tax.

Reprinted from *Energies*. Cite as: Benavides, C.; Gonzales, L.; Diaz, M.; Fuentes, R.; García, G.; Palma-Behnke, R.; Ravizza, C. The Impact of a Carbon Tax on the Chilean Electricity Generation Sector. *Energies* **2015**, *8*, 2674-2700.

1. Introduction

Chile is a minor contributor to global greenhouse gas (GHG) emissions (0.2%). However, according to the last official GHG national inventory, the emissions from the energy sector have grown 101% between 1990 and 2010. Figure 1 shows the national inventory by sectors [1]. These statistics also show that the electricity generation sector is responsible for the highest amount of emissions. A recent study led by the government shows this situation has not changed. The electricity generation sector will continue to be the main GHG emitter, followed by the transport and industry sectors. These three sectors represent 77.2% of total emissions in 2013 [2].

Chile's motivation to contribute to worldwide emissions reductions stems from the United Nations Framework Convention on Climate Change (UNFCCC) and its principle of common but differentiated responsibilities. The country intends to contribute to achieving the ultimate objective of the Convention by undertaking mitigation actions, as well as taking advantage of the potential environmental and social benefits and improvements in the quality of growth that can be directly derived from mitigation actions. Following this line of action, Chile will take nationally appropriate mitigation actions to achieve a 20% deviation below the "Business as Usual" emissions growth trajectory by 2020, as projected from the year 2007 according to Chile's commitment at Copenhagen in 2010. During the last Climate Summit in New York in September 2014, the current president reaffirmed this voluntary commitment, subject to international support.

A more comprehensive analysis will help to establish exactly how Chile will fulfil its commitment to achieve its desired reduction in emissions. This paper aims to analyse the impact of a specific economic instrument: a carbon tax applied to the electricity generation sector. In 2014 Chile had a change of government following the presidential elections. The new government proposed a tributary reform which includes a carbon tax of 5 US$/tCO2 applied on fixed sources with installed capacity up to 50 MW (therefore, the carbon tax is not applied to the industry, transport, commercial, and residential sectors). A carbon tax would reduce the GHG emissions through two broad influences—a demand effect, reducing energy demand due to higher prices in the electricity sector as well as in the whole economy, and a substitution effect, that is switching from more to less carbon intensive fuels. The second effect is analysed in this paper. In addition, this instrument is compared with other energy policies such as an increase in non-conventional renewable energy sources and a cap on the total emissions of the electricity generation sector.

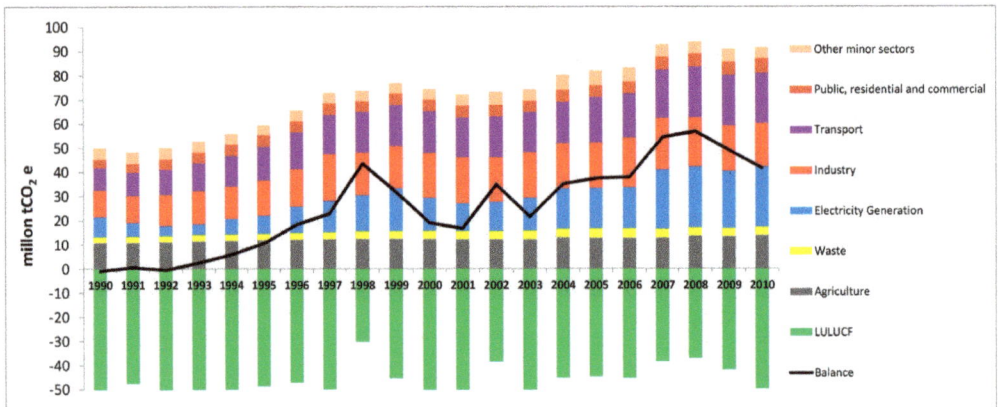

Figure 1. Chilean national GHG inventory by sector [1]. The blue bar is the electricity generation sector which is analysed in this paper.

Carbon taxes have mostly been implemented in Scandinavian countries, Australia (it was abolished in 2014), and a few other European countries. Finland (1990), Sweden (1991), Norway (1991) and Denmark (1992) led the way in implementing a carbon tax [3].

Economic and energy models have been used in previous studies. Large-scale energy-economy models have been extensively used in the European Union (EU) climate and energy policies [4–6], and combining energy and economic models has been a standard in the EU. The link is necessary when sectorial models cannot provide all the answers that policy makers are looking for, for example, the impact on GDP pathway of an energy policy [7]. Most previous studies have used computable general equilibrium (CGE) models to analyse the economic implication of this kind of environmental policy. The GEM-E3 (a CGE model called General Equilibrium Model for Energy–Economy–Environment interactions) and the Energy–Environment–Economy macro-econometric (E3ME) model are being used to evaluate policy issues for the European Commission. These models are linked to the Price-Induced Market Equilibrium System (PRIMES) energy model [6]. In the Long Term Mitigation Scenario process [8,9] and in [10] a CGE model is used to evaluate the impact of a carbon tax in South Africa. In [11] the macroeconomic impact in Thailand of introducing an emission trading system and carbon capture storage technologies was evaluated. In [12,13] a CGE model is used to evaluate how a carbon tax policy impacts on energy consumption and GHG emissions in China. In [14] a multi-sector and multi-region CGE model is used to quantify the economic impacts of EU climate and energy package in Poland. In [15] two scenarios are evaluated to quantify the macroeconomic impacts in Pakistan, the first scenario only included different levels of carbon tax, and in the second scenario a carbon tax and energy efficiency improvements that have been jointly simulated are evaluated. Another approach is possible to find in [16], in this reference a Dynamic Stochastic General Equilibrium (DSGE) is developed to evaluate the macroeconomic impact of different mitigation action in Poland. In [17] an Input-Output matrix is used to estimate the short-term effects of a carbon tax in Italy, which includes the percentage increase in prices and the increase in the imports of commodities to substitute domestically produced ones as intermediate input.

Energy sectorial models have also been used in previous studies. TIMES (The Integrated MARKAL-EFOM System) Integrated Assessment Model (TIAM) is a bottom-up energy system model with a detailed description of different energy forms, resources, processing technologies and end-uses [18]. Model for Energy Supply Systems and their General Environmental impact (MESSAGE) is a supply energy model. Both models have been used by the Intergovernmental Panel on Climate Change (IPCC) to project global energy and emission scenarios. Also these models have been used to analyse local options with more details. For example, in [19] renewable energy options are analysed to mitigate climate change in India, and in [20] the MESSAGE model was used to evaluate 12 strategies to reduce the Malaysia's carbon footprint of the energy sector. In [21] the Asian Pacific Integrated Model (AIM) is used for scenario analysis of GHG emission and the impacts of global warming in the Asian Pacific region. This model comprises 4 discrete models which are linked: The emission model and the global warning impact model are linked to two global physical models.

In [22] a fuzzy mixed-integer energy planning model under carbon tax policy is developed. In [23] an end-use energy model is presented for assessing policy options to reduce greenhouse gas emissions. This model evaluates the effects of imposing a carbon tax on various carbon emitting technologies in order to reduce CO_2 emissions. In [24] the impact of renewable energy source incentives and mitigation policies (feed-in tariffs, quota obligation, emission trade, and carbon tax) are considered in the framework of the generation planning problem to be solved by a generation company. Renewable energy quota and emission limits result in a set of new constraints to be included in a traditional generation expansion planning model. In [25] an integrated power generation expansion planning model towards low-carbon economy is proposed. In [26] a short term optimization method is proposed to determine the optimal tax rate among generating units.

The literature review shows that macroeconomic models (CGE, DSGE, I/O matrix, *etc.*) do not represent endogenously all the details of the power sector to simulate the generation expansion planning under carbon tax scenarios. Due to this fact, in this paper both energy sectorial and macroeconomic models have been implemented, and an approach to link these two is presented. In [6] the GEM-E3 model cannot produce energy system simulations as accurately as the PRIMES model, therefore the GEM-E3 model is calibrated according to projections obtained by the PRIMES model [27]. The PRIMES model is more aggregated than engineering models and far more disaggregated than econometric models. Also in [6], the power generation mix was treated as exogenous in E3ME model and adapted to the results of the PRIMES model. In [14] a detailed bottom-up representation is used to model the electricity generation sector, however, this representation is less detailed in comparison to sectorial models. In [10] a long-term electricity investment plan of a previous study is used to calibrate a CGE multi-sectorial model. Also in [14] the power development plan is an input for the CGE model. In [28] a technological and the time period detail is introduced in a CGE framework to represent electricity demand and electricity generation by power plants. In a typical CGE model the energy sector is represented using one load block or stage per year. However, in this work [28] it was represented until 180 load blocks.

In this paper a DSGE model is selected instead of a CGE model. What DSGE and CGE modelling have in common is that they belong to the micro founded macroeconomic models of general equilibrium, but they differ in two important issues regarding modelling results: dynamics and uncertainty. DSGE models are strictly dynamic models, while CGE models are comparatively static ones. The dynamic characterization of the models allows for optimal decision rules that are not policy invariant, and where time is directly considered. This characterization has gained in terms of allowing analyses of the paths of the variables behavior instead of comparing different equilibriums, and by analysing the reachability of them, relative to the static characterization. The CGE type of model allows comparing steady state equilibriums but not the trajectories toward the equilibriums.

The treatment of uncertainty in the DSGE models makes them superior to CGE models, which are deterministic, because they allow a better fit of the theoretical models with the data. These features are causing a shift from CGE to DSGE modeling, which is becoming a valuable tool for assessing policy analysis, mechanisms analysis and projections [29,30]. However, the CGE models include a large variety of sectors, so they make possible to analyze sectoral composition of output, employment, capital, *etc.*, while the DSGE models focus in more aggregated analysis.

In Chile there are two main independent power systems, the Central Interconnected System (in Spanish, Sistema Interconectado Central, SIC), and the Norte Grande Interconnected System (in Spanish, Sistema Interconectado del Norte Grande, SING). Figures 2 and 3 show the historical generation by source for the SIC and SING, respectively. The hydroelectricity generation is one of the main sources in the SIC, whereas in the SING coal is one of the main energy sources. The maximum demand in SIC was 7,535 MW and SING was 2,300 MW in 2014.

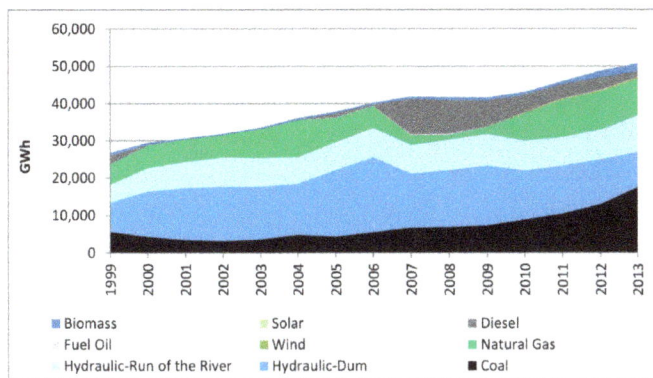

Figure 2. Electricity generation by sources, SIC 1999–2013.

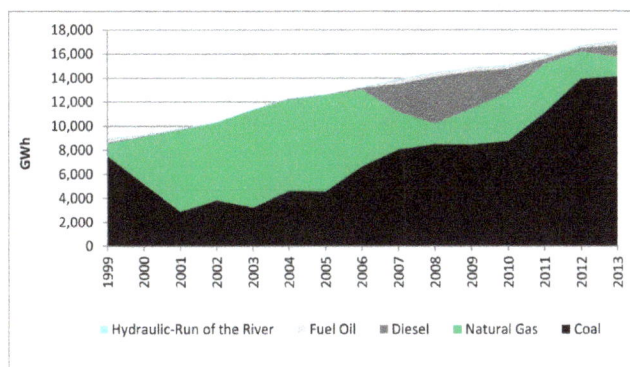

Figure 3. Electricity generation by sources, SING 1999–2013.

Chile has great potential to produce electricity with renewable energy sources, such as hydroelectric (12,000 MW), solar (1,000,000 MW), wind (40,000 MW) and geothermal (16,000 MW) sources [31]. However, some issues have affected the development of this technology in Chile. In the case of solar and wind energy, there is uncertainty related to the evolution of the future investment cost and the lack of access to long term contracts constitutes a barrier for project developments. In the case of hydroelectric sources, environmental problems have faced some projects. For example, the hydroelectric generation potential of the Aysén region has been estimated to be more than 7,000 MW. Two specific projects have been evaluated in this zone: HidroAysen (2,750 MW) and Cuervo (640 MW). However, these projects have faced the opposition of several groups due to the fact that these would be installed in a pristine region of Patagonia, known for glaciers and lakes. In addition, these projects require a transmission line of more than 2,000 km to inject its energy to SIC power system. The first project presented its environmental evaluation in year 2008, and was approved in May of 2011. However an action complaint against the environmental resolution was presented which had to be resolved by a Minister Committee. The final resolution of this was extended for more than three years. Finally, the environmental evaluation was rejected by the current Minister Committee, and it is not clear if the company will present the environmental evaluation again. The environmental evaluation of the transmission line has not been presented yet.

Figures 2 and 3 show that natural gas was one of the main energy sources between 2000 and 2006. Most of the natural gas was imported from Argentina, however this supply experienced many shortfalls. To overcome this energy problem in Chile, two Liquid Natural Gas (LNG) terminals were built: Mejillones and Quintero. The first began to operate in 2009. However, it is currently operating below its maximum capacity due to the high price of LNG in comparison to electricity generation from coal (see Figure 3). In the case of the Quintero LNG terminal, it is operating at full capacity. Four companies share ownership of this terminal: British Gas; the National Petroleum Company of Chile (ENAP), which is a state refiner; Chilean Distributor for Natural Gas for Metropolitan Region (METROGAS), and National Electricity Company (ENDESA), which is one of the biggest private electricity generation companies in Chile. Apart from ENDESA, there are other companies which have natural gas power plants (for example Nehuenco (785 MW) and Nueva Renca (305 MW)), however, these do not have open access to the terminal. At times METROGAS or ENAP have sold gas surpluses to these companies, but at a high price. There are uncertainties about access to the terminal to get gas at competitive prices or the access using their own terminal.

To evaluate the above uncertainties, a sensitivity analysis is proposed to manage the following aspects: solar photovoltaic technology investment cost, projection for LNG prices, and potential use of hydroelectric resources in the extreme south of Chile. Different baseline scenarios are built considering these variables. Other uncertain sources could be included in this analysis but the focus of this work has been limited to these in order to show the impact on the projection of GHG emission reduction and macroeconomic results.

The main contributions of this paper are: to analyse the impact of the carbon tax at the Chilean electricity sector using both a sectorial model and a DSGE model. A novel approach is proposed to integrate results from the sectorial model and a DSGE model. This work shows that the effectiveness of reducing GHG emission depends on some variables which could not be controlled by policy makers. Finally, the carbon tax is compared to other energy policies and interesting results are found.

The paper is organized in four sections. In Section 2 the methodological approach and implementation are presented. Section 3 presents the results and analysis of the Chilean case. Finally, in Section 4 the main conclusions are summarized.

2. Methodology

2.1. Overview Description

Figure 4 shows an overview of the approach to evaluate the carbon tax. A generation expansion planning model is developed to project the installed capacity and electricity generation for the period 2013–2030. The models for industry, transport and commercial, public and residential (CPR) sectors project the electricity demand which is an input for the generation expansion planning model. In this iteration, the sectorial models are run considering a base value for the GDP projection and considering a carbon tax applied to the electricity generation sector. The electricity price obtained from the electricity generation sector, expressed at constant prices, is an input for the DSGE model. The new electricity price projection is introduced in the DSGE model as a price shock. Then, the DSGE model will estimate the effects on GDP path due to the electricity price shock.

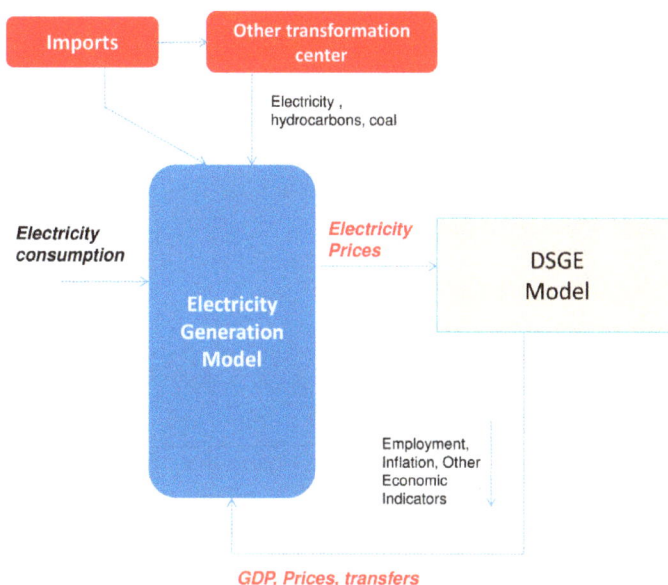

Figure 4. General description of the Chilean approach. The price of the electricity is output of the sectorial model and is an input for the DSGE model.

The DSGE model, as its name indicates, is a general equilibrium model, which means that it takes into account all interactions occurring among all the economic agents, who adjust their decisions according to changes in relative prices. Under this framework, an increase in the relative price of electricity encourages firms to substitute from electricity toward capital and labour, given the technological possibilities, making the electricity demand in the DSGE endogenous. However, this substitution is not perfect, but rather limited, given the complementarity of energy with capital and labour which negatively affects production possibilities. After the energy prices rise, the economy must move to a new long-run equilibrium, with lower growth rates during the period of convergence. The size and number of periods of lower GDP growth rates during the transition period will depend on how big the increase in the electricity price due to the carbon tax is and how fast the tax is implemented. If it is slightly and gradual, the GDP growth rate will be affected for a short time and not significantly. Conversely, if the rise in the tax is stronger the GDP growth rate will be strongly affected and for longer periods of time. When the economy gets back to the equilibrium, it will grow at rates similar to those before the imposition of the tax rates. This is thanks to the exogenous improvements in productivity and population growth. However, this growth will be based on a lower level, due to lower growth rates occurred in the transition period.

This will result in a permanent deviation of the GDP level regard the baseline level (without tax). The approach proposed considers an iteration procedure, where the new path of GDP is introduced into the sectoral models in order to estimate again a new lower electricity demand. It is expected that with a lower demand, the emission projection would reduce, as it is analysed in [32].

A limitation of this approach should be recognized. The literature review shows that both CGE and DSGE models represent in a simplified manner the productive sectors of the economy. In particular our DSGE model does not explicitly include an electricity generation sector; the electricity sector works as an imported commodity which is sold to the economy at the prices given by the linkage with the electricity generation sectorial model. In this context, electricity is the only energy factor included in the DSGE model, which implies that the firms are not able to substitute electricity with another type of fuel when they face electricity price increases. In this context, it is possible that the results might slightly overestimate the negative impact on GDP growth rate due to the implementation of a carbon tax.

2.2. Generation Expansion Planning Model

The objective of this model is to determine the optimal combination of power generation to meet the projected demand for the time horizon 2013–2030. We formulate the problem as if planning was matched by a central or state government. The objective function is to minimize the capital costs in new plants, the variable cost related to fuel consumption, variable cost associated to non-fuel consumption, and the cost of unserved energy. In addition, the objective function includes a penalty for the GHG emitted emissions:

$$\sum_{t}^{T}\sum_{i}^{N}\frac{I_{it}P_{it}}{(1+r)^{t}} + \sum_{t,s,b}(\sum_{i}^{N}\frac{\Delta_{tsb}C_{it}G_{itsb}}{(1+r)^{t}} + \sum_{i}^{N}\frac{\Delta_{tsb}f_{t}E_{i}G_{itsb}}{(1+r)^{t}}) \qquad (1)$$

where the optimization problem variables P_{it}(MW) is the additional installed capacity by year, and G_{it}(MWh) is the electricity generation in the stage s and load block b of the year t. Every year is divide into monthly (12) stages, and every stage is represented by five load blocks (the load duration curve is divided into five blocks). Δ_{tsb} is the duration of the load block b. I_{it} is the capital cost annuity (US$/MW), C_{it} is the variable cost (fuel and non-fuel cost) (US$/MWh), E_{i} (ton CO_2/MWh) is the emission factor for each kind of technology, f_{t}(US$/ton CO_2) is the carbon tax and r is the private discount rate (10%). The factor f_{t} will take the amount of 5, 10, 20, 30, 40, and 50 US$/tCO$_2$e between 2017 and 2030. The problem is subject to the following constraints:

- Energy balance between the electricity generations and demand. A multi-nodal formulation is used to represent the interconnection between SIC and SING. The electricity demand is exogenous in this model.
- Upper and lower bounds on the electricity generation of power plants.
- Maximum feasible amount of investment for each kind of technology that could happen in a year.
- Maximum feasible amount of investment for each kind of technology for the total period. The potential installed capacity for hydroelectric run-of-river power plants is set according

potential projects which are currently under evaluation. In the case of geothermal energy, a conservative expectative is assumed in comparison with previous works in Chile [33].

- Quota obligation to renewable energy generation according to new renewable energy law (20% by 2025).

Other assumptions:

- Technical parameters of the electricity generation plants were obtained from the ISO webpages of SIC [34] and SING [35]. The model represents 290 power plants.
- Information regarding power plants which are being built was collected from reports published by the National Regulatory Institution [36,37].
- Hydroelectric-dam plants can regulate their generation during every stage. On the other hand, hydraulic-run of river and small hydroelectric plants are not able to regulate their energy generation.
- Minimum technical power is considered for coal and LNG power plants. This constraint avoids starts up and shut down between blocks in the same stage. This constraint is included to avoid infeasible solutions when high penetration of non-conventional renewable energy sources with an intermitted generation (e.g., solar and wind generation) is introduced to the power system.
- Minimum sizes for new power plants are considered. The minimum size for a coal, natural gas, and hydroelectric power plant were 250 MW, 200 MW and 100 MW, respectively.
- Investment costs for the first year were obtained from [36,37]. The cost projection is based on growth rate used in [33]. The investment costs projection used in this work are in the Appendix.
- Currently there is transmission project that will connect the two main power systems of Chile (SINC and SING). It is supposed that the interconnection between SIC and SING will happen by the year 2019.
- The baseline scenario does not consider nuclear energy as an energy option.

2.3. Macroeconomic Model

A DSGE model calibrated for the Chilean economy is used for this exercise. This model is based on a previous work by Medina and Soto [38,39]. Figure 5 represents the structure of the main agents involved in the model.

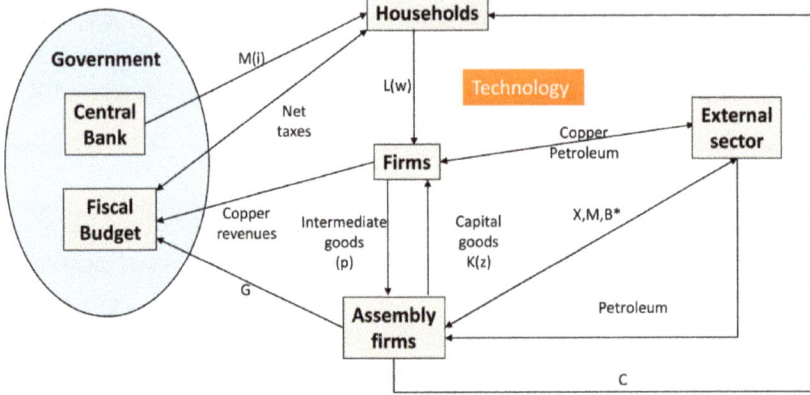

Figure 5. General description of DSGE model.

There is a continuum of households and different types of firms in the economy. Households live infinitely, take decisions on consumption and savings, and set wages in a staggered way. There is a set of firms that produce differentiated varieties of tradable intermediate goods using labour and capital. They have monopoly power over the varieties they produce and set prices in a staggered way. Another set of firms are importers that distribute domestically different varieties of foreign intermediate varieties. These firms have monopoly power over the varieties they distribute, and also set prices in a staggered fashion. Setting wages and prices in a staggered way to the equilibrium price is an assumption used by previous macroeconomic literature [40–42] which implies introducing of prices rigidities in order to model the gradual adjustment of the economy to shocks. These firms have monopoly power over the varieties they distribute, and also set prices in a staggered fashion. There is a third single firm that produces a commodity good which is completely exported abroad. This firm has no market power in the international market price. Production by this firm is exogenously determined and requires no inputs. Its revenues are owned by the government and by foreign investors. Domestic and foreign intermediate varieties are used to assemble two final goods: home and foreign goods. Electricity is included in the foreign goods, which is sold in the domestic economy at a given price determined by the exogenous electricity generation model. These two final goods are combined into a bundle consumed by households, another bundle consumed by the government and a third bundle that corresponds to new capital goods that are accumulated to increase the capital stock.

A previous DSGE model [38] for the Chilean economy was adapted in this paper. The optimization problem of the firms consists in maximizing the total income less the cost:

$$Max\ Y \times P_y - L \times P_l - K \times P_k - E \times P_E \tag{2}$$

where Y is the production of the firm, K is the capital, L is number of employees, E is the electricity consumption of the firm, P_y is the price of the production, P_l is the salary, P_k is the price of the capital, and P_E is the price of the electricity. The production function can be expressed as:

$$Y = A((1-\gamma)(K^\alpha L^{1-\alpha})^{\frac{\varepsilon-1}{\varepsilon}} + \gamma(E)^{\frac{\varepsilon-1}{\varepsilon}})^{\frac{\varepsilon-1}{\varepsilon}} \tag{3}$$

The elasticity ε was calibrated by using the historical data of the National Energy Balance from 1990 to 2012. An energy price index is estimated using historical data of fossil fuel primary sources (coal, diesel, and natural gas):

$$E_t = \sum \alpha_j E_j \tag{4}$$

In the last equation E_t is the total energy from fossil fuels sources resulting from the sum of each "j" energy source E_j weighted by of their relative prices (α_j). The elasticity of substitution ε is calibrated using historical data of E_t. This is a strong assumption, which means that the only energy input used in the economy is electricity. This assumption allows estimating with the DSGE model the effect of the carbon tax imposed using the electricity generation model.

Moreover, the share of the energy by the total output of the economy would be modified with the productivity according the indirect productivity of the expenditure of energy by output E_t/GDP. The electricity price obtained from the electricity generation sector for these different scenarios is an input for the DSGE model. The following expression is used to calculate the increase in the price of electricity (ΔPE):

$$\Delta PE \left(\frac{US\$}{MWh}\right) = \frac{\Delta CAPEX + \Delta OPEX + TAX}{\sum_{t=tini}^{T} G_t/(1+r)^{tini-t}} \tag{5}$$

where $\Delta CAPEX$ corresponds to the present value of the variation of the capital expenditure (first term of objective function in the Equation (1)) in power plants in comparison with the baseline scenario, *i.e.*, the case without carbon tax; $\Delta OPEX$ represents the variation of the operation expenditure (second term of objective function in the Equation (1)), TAX is the carbon tax that the generation companies should pay due to their GHG emissions (third term of objective function in the Equation (1)), G_t is the total electricity generation in the year t, $tini$ is the starting year for the carbon tax application. $\Delta CAPE$, $\Delta OPEX$, and TAX are outputs of the electricity generation model. Finally, r is the private discount rate used in the electricity generation sector. It is assumed that the cost that electricity generation companies will pay due to the carbon tax will be transferred to their customers. The discount rate used in that equation the private discount rate and it is equal to 10%. This is the typical discount rate used in the Chilean electricity sector which it is a near 100% a private sector. For example, the regulation of the transmission system states that the annuity of the transmission asset is calculated with a discount rate of 10%. Also, the electricity distribution regulation states a discount rate of 10% to calculate the annuity of the investments. Consequently, the selected discount rate intends reflecting the decision dynamic of the electricity. This approach implies a constant and permanent increase in the electricity price after the carbon tax is imposed.

The price of electricity in the DSGE model presents the following autoregressive process of order 1:

$$PE_t = PE_{t-1} + \varepsilon_t \; ; \; \varepsilon_t \sim N(0, \sigma^2) \tag{6}$$

Equation (6) works as the linkage between the electricity generation sectoral model and the DSGE. This structure allows introducing the increase of electricity price resulting from the sectoral model into the DSGE model as an electricity price shock.

The variance of the normal distribution (σ^2) is calibrated according to the historical data of electricity prices. The simulation procedure seeks to replicate the new exogenous path of electricity prices, with a permanent increase due to carbon tax, using the shock ε_t of Equation (6) to mimic this effect. More details of the model can be found in [38,39]. After changing energy prices the economy must move to a new long-run equilibrium. The resulting steady state will show a shift in the growth rates during the convergence period. Upon reaching the new steady state the economy will grow at similar rates before imposing the tax because of the improvements in productivity and population growth, but this growth will result in a lower GDP level due to lower growth rates in the transition period.

Regarding the implementation of the proposed analysis framework, the generation expansion planning model was programmed in MATHPROG using a GLPK distribution [43] and solved by CPLEX MIP solver [44]. The DSGE model was programmed using DYNARE and MATLAB [45]. The MIP problems were solved with a gap below 0.03%.

3. Case Study

The impacts of a carbon tax on the Chilean electricity generation sector are evaluated considering six levels: 5, 10, 20, 30, 40, and 50 US$/tCO$_2$e. The carbon tax is applied between 2017 and 2030 to both the SING and SIC power systems. For this exercise a simplification of the framework explained in Section 2 will be used. In particular, there will be no iteration process from the new GDP path estimated by the DSGE to the electricity generation model. Due to the lack of specific information, scenarios with gradual carbon tax rate increases are not explored. Nevertheless, the effect of any gradual implementation scheme should be within the observed overall simulation results.

For this simulation exercise several assumptions are made. An annual exogenous growth rate of 1.5% and 1.0% for productivity and employment is assumed, respectively. For every scenario we assume that the convergence time of electricity prices to the new equilibrium is 12 quarters and gradual increases will be linear. The current price of the original steady state energy is 105 US$/MWh. This price does not include the distribution cost.

A sensitivity analysis is performed for three main parameters which are inputs for the optimization problem (the reason for selecting these parameters was discussed in the Introduction Section): projection of investment cost in solar photovoltaic (PV) technology, projection for LNG prices, and potential use of hydroelectric resource in the extreme south of Chile, see Table 1. Considering these uncertainties four scenarios are projected (see Table 2). We define these scenarios as baseline scenarios.

3.1. Electricity Generation Sector Results

Figure 6 shows the GHG trajectory for the scenarios defined in Table 2. These scenarios are the four baseline scenarios defined above (without carbon tax). The carbon tax is applied in every one of these four scenarios. Figure 6 shows that the GHG trajectories are different due to the different assumptions defined in Table 2. For example, the GHG emissions of Baseline #3 begin to fall from the year 2022 in comparison to the other baseline scenarios. In this case it is supposed that the hydroelectric generation potential of the Aysén region will be available from 2022. The planning expansion solution obtained from the optimization model shows that 2,750 MW would be installed between 2022 and 2026. This hydroelectric potential is not available in the other baselines scenarios (see Tables 1 and 2). The electricity generation of hydroelectric power plants depends on the hydrological conditions. In the model used in this paper a hydrological trend based on historical series, is supposed. It means that, in the future, some years will be wet, others will be dry, and others will be normal. It explains some ups and downs in the GHG emissions. For example, in the year 2024, the GHG emissions grow due to the fact that this year is wetter than previous years. All the baselines scenarios were evaluated with the same hydrological trend.

Table 1. Sensitivity analyses considering three sources of uncertainties.

Solar Investment Cost	LNG Prices	Hydroelectric Resources in the Extreme South of Chile
Case 1: Base situation	Case 1: Base situation	Case 1: No exploitation of additional hydroelectric resources.
Case 2: More optimistic projection of solar investment cost, see Annex	Case 2: More optimistic LNG prices projection for power plants which have not open access to LNG terminal, see Annex	Case 2: An additional potential of 2,750 MW of hydroelectric source is considered.

Table 2. Evaluated scenarios built from the sources of uncertainties. These scenarios are called baseline scenarios.

Scenario "S" or Baseline Scenarios	Solar Investment cost	LNG Prices	Exploit the Hydroelectric Resource in the Extreme South of Chile
S1 (Baseline #1): Moderate Scenario in terms of the development of solar investment cost projection and LNG price. This scenario does not consider the development of big hydroelectric projects in the south of Chile.	Case 1	Case 1	Case 1
S2 (Baseline #2): Equal to scenario 1 but more optimistic regarding the solar investment cost projection.	Case 2	Case 1	Case 1
S3 (Baseline #3): Equal to scenario 1 but it considers the development of big hydroelectric projects in the south of Chile.	Case 1	Case 1	Case 2
S4 (Baseline #4): Equal to scenario 1 but more optimistic regarding the projection of LNG prices.	Case 1	Case 2	Case 1

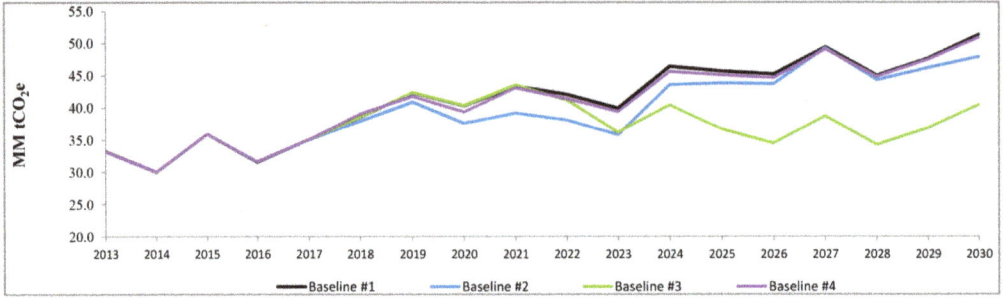

Figure 6. GHG trajectory for different baseline scenarios (MM = million).

Figures 7–10 show the total GHG emissions (SING and SIC) for the baseline scenarios, in comparison to different carbon tax levels (5–50 US$/tCO$_2$e). The results show that the emission trajectory depends on the evaluated baseline scenario. In Scenario #1, from the year 2023 the results show a big difference between a tax of 30 and 40 US$/tCO$_2$. This is due to the fact that when the carbon tax of 40 US$/tCO$_2$ is applied the coal technology is less competitive in comparison to the next cheapest technology, in this case, solar energy. In 2023, the Levelized Cost of Energy (LCE) of coal technology without carbon tax is 84 US$/MWh, and with a carbon tax of 10, 20, 30, and 40 US$/tCO$_2$ are 92, 101, 109, and 118 US$/MWh, respectively, while the LCE of solar technology is 110 US$/MWh. Therefore, with a carbon tax of 40 US$/tCO$_2$e the solar technology is cheaper than coal technology, and then there are less GHG emissions. However, note that with a carbon tax of 30 US$/tCO$_2$e the difference is very small. Probably a value higher than 30 and lower 40 would have the same result. This is the reason why in Scenario #2 (in this case the solar investment cost is more optimistic) the carbon tax of 30 US$/tCO$_2$e has a major impact. It is important to emphasize that the methodology used an optimization problem with technical constraints of the system operation to project the investment in new power plant, but, by simplification, we have calculated the LCE to give an explanation of the big difference between 30 and 40 US$/tCO$_2$e.

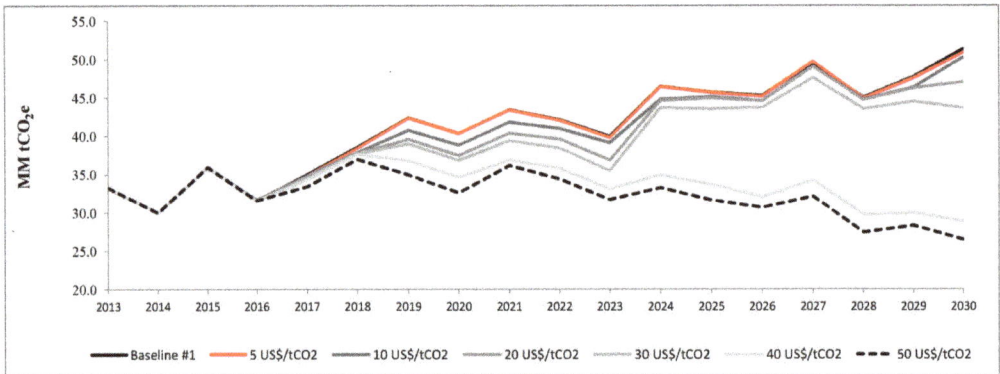

Figure 7. Emissions for baseline scenario #1 in comparison to different scenarios of carbon tax (MM = million).

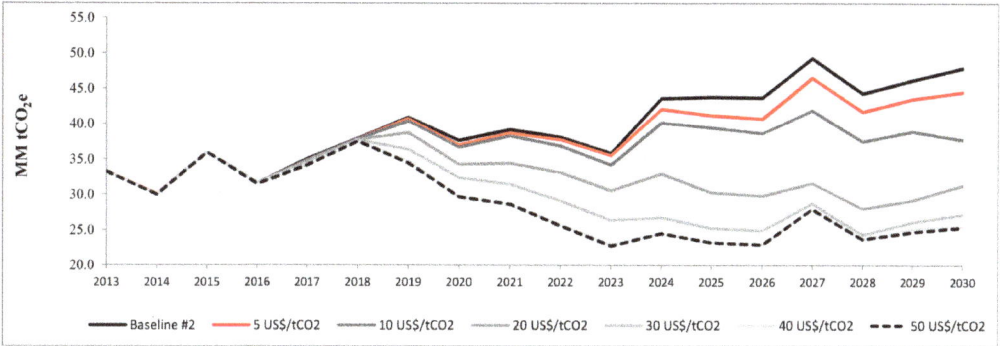

Figure 8. Emissions for baseline scenario #2 in comparison to different scenarios of carbon tax (MM = million).

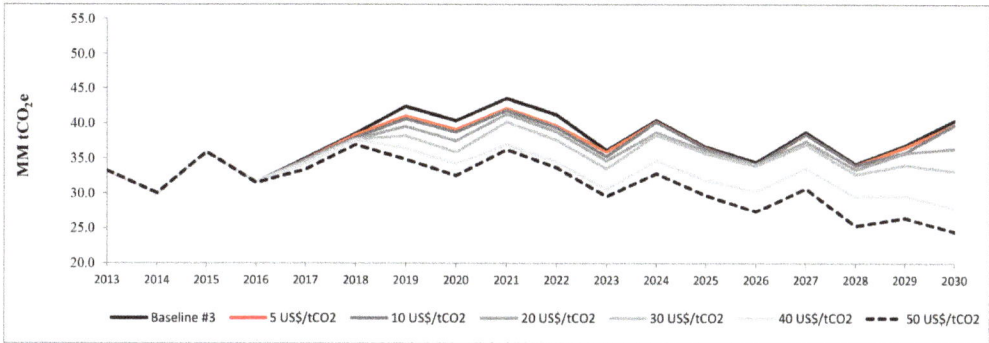

Figure 9. Emissions for baseline scenario #3 in comparison to different scenarios of carbon tax (MM = million).

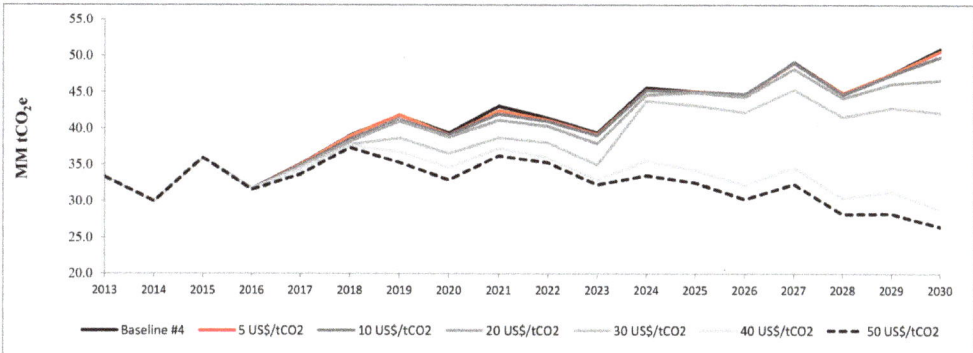

Figure 10. Emissions for baseline scenario #4 in comparison to different scenarios of carbon tax (MM = million).

In addition, different indicators are proposed to compare the impact of the carbon tax on the electricity generation sector with the baseline scenario: OPEX, CAPEX, and carbon tax revenue. These indicators are expressed in present value and as deviation with respect to the baseline scenario. The carbon tax revenue is the sum of the whole period. All the monetary values are expressed in constant prices. Also cumulative emission reduction, annual average emission reduction, and average increase of the electricity price at the national level are reported (see Table 3). The average increase of the electricity price is calculated using Equation (5) and it is an output of the sectorial model. It is supposed that the average increase of the price will be constant between 2017 and 2030.

Table 3. Resulting indicators to assess the impact of the carbon tax on the electricity generation sector. These indicators are expressed in present value and as deviation with respect baseline scenario. The negative values mean a reduction with respect to baseline scenario (MM = million). OPEX, operational expenditure; CAPEX, capital expenditure.

Unit Scenario	Carbon tax	Δ OPEX	Carbon tax revenue	Δ CAPEX	Cumulative Δ emission	Average Δ Annual Emission	Δ Electricity Price
Unit	US$/tCO$_2$	MM US$	MM US$	MM US$	MM tCO$_2$e	MM tCO$_2$e	US$/MWh
S1	5	−58.3	1168.8	57.8	−1.1	−0.1	2.1
	10	−136.9	2288.3	178.9	−13.4	−1.0	4.2
	20	−644.1	4488.0	765.5	−25.0	−1.8	8.2
	30	−970.5	6581.5	1226.8	−40.9	−2.9	12.2
	40	−2039.0	7644.0	3314.3	−139.8	−10.0	15.9
	50	−2906.1	9147.4	4564.2	−162.4	−11.6	19.3
S2	5	−407.1	1085.5	430.9	−19.5	−1.4	2.0
	10	−762.8	2109.8	837.5	−43.6	−3.1	3.9
	20	−1807.3	3748.9	2228.8	−118.9	−8.5	7.5
	30	−2681.2	5328.4	3349.2	−150.3	−10.7	10.7
	40	−126.9	6815.3	4051.9	−170.5	−12.2	13.8
	50	−3692.1	8043.6	5044.7	−195.3	−13.9	16.8
S3	5	−81.2	1054.0	93.6	−7.9	−0.6	1.9
	10	−241.1	2095.5	265.6	−10.9	−0.8	3.8
	20	−533.8	4112.1	624.6	−22.3	−1.6	7.5
	30	−966.0	6023.8	1200.7	−36.0	−2.6	11.2
	40	−1490.8	7500.5	2189.9	−76.5	−5.5	14.7
	50	−2359.5	8919.6	3493.2	−104.4	−7.5	18.0
S4	5	−106.3	1156.5	110.3	−2.2	−0.2	2.1
	10	−231.1	2300.6	248.6	−5.0	−0.4	4.1
	20	−448.3	4534.2	523.3	−15.7	−1.1	8.2
	30	−973.0	6485.4	1345.4	−46.4	−3.3	12.3
	40	−2161.0	7686.1	3387.5	−130.5	−9.3	15.9
	50	−2753.9	9233.7	4338.8	−152.6	−10.9	19.3

The results of Table 3 show that as the carbon taxes go up, the electricity sector's GHG emissions come down. The operational costs also come down, while the total investment in new power plants increases and the cost of electricity rises, relative to a business-as-usual scenario. The total investment increases due the fact that more renewable energy sources are introduced which have higher investment cost (US$/kW) than coal plants. The results show that the effectiveness of the carbon tax depends on certain variables which are not controlled by policy makers. The highest emission reduction happens in Scenario 2. This is because the solar investment cost projection is more optimistic, and the carbon tax promotes the introduction of new non-conventional renewable energy sources to happen earlier in comparison to Scenario 1. On the contrary, the impact on the carbon tax is low in Scenario 3. This is because this baseline scenario includes more hydroelectric sources than Scenario 2, and then there is less thermoelectric generation with coal sources. In Scenario 4, there is more electricity generation from LNG in comparison to Scenario 1 (and less with coal generation). The electricity price increase is explained by two factors: the switch from new power plants emitting GHG to sources that do not emit GHG, and by the tax the companies (operating and new) will pay for emitting GHG gases. However, the main factor is the second, and it explains more than 85% of the electricity price increase in all the cases (see carbon tax revenue of Table 3). The carbon tax of 5 US$/tCO$_2$e approved during 2014 in Chile would have a low impact on emission reduction. All the scenarios have average emission reductions below 0.6 million tCO$_2$e, except for the case with optimistic solar investment cost projection (1.4 million tCO$_2$e). In [2] a national Baseline Scenario for all the sectors (transport, energy, forestry, waste, *etc.*) was projected for the period 2013–2030. This reference shows that the total emission would vary between 110 and 125 million tCO$_2$e by 2020. Then, the average emission reduction for a carbon tax of 5 US$/tCO$_2$e would represent less than 1.4% of emission reduction. Also for a tax of 10 and 20 US$/tCO$_2$e it would not have a big impact in terms of reducing GHG emissions. For these levels of tax, the minimum emission reductions are 0.4 and 1.1 million tCO$_2$e, respectively. Table 4 summarizes the emission reduction range and the increase of the price of electricity for different levels of carbon tax.

Table 4. A summary of the emission reduction and electricity price rise range of Table 3. These values are compared with the baselines scenarios (without carbon tax).

Carbon Tax Level (US$/tCO$_2$e)	Average Annual Emission Reduction Range (Million tCO$_2$e)	Increase of Price of Electricity Range (US$/MWh)
5	[0.1, 1.4]	[1.9, 2.1]
10	[0.4, 3.1]	[3.8, 4.2]
20	[1.1, 8.5]	[7.5, 8.2]
30	[2.6, 10.7]	[10.7, 12.3]
40	[5.5, 12.2]	[13.8, 15.9]
50	[7.5, 13.9]	[16.8, 19.3]

3.2. Macroeconomic Results

The effect of taxing CO$_2$ emissions in the electricity sector on the gross domestic product (GDP) path is evaluated using the DSGE model described above. The electricity price rise observed in the sectoral model is evaluated as a price shock in the DSGE model. Figure 11 shows the GDP changes for the six CO$_2$ tax scenarios in comparison to the baseline scenarios.

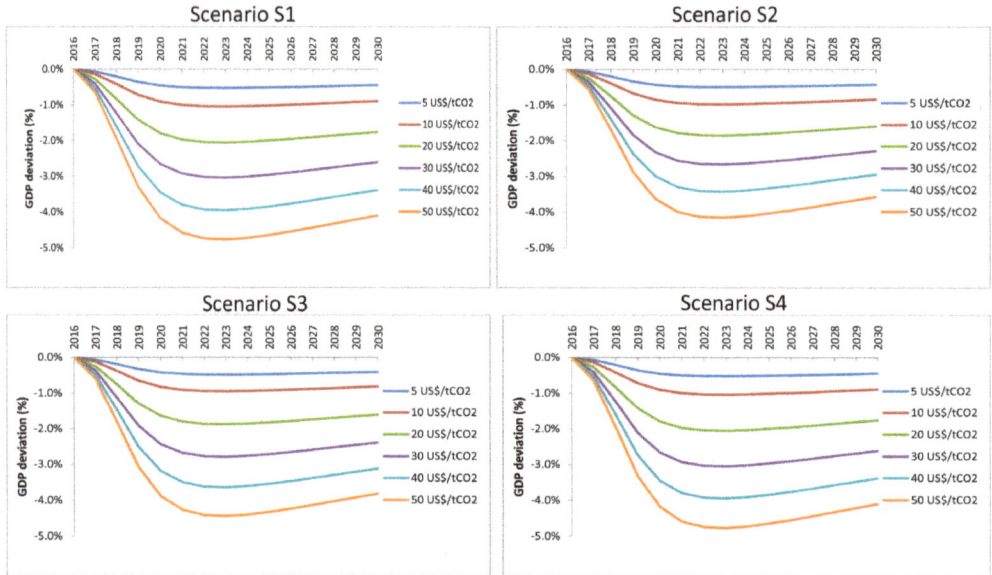

Figure 11. Effect on the level of GDP (% over baseline deviations).

In all the cases a negative impact on GDP pathway is observed and it depends on the level of the tax. For example, for a carbon tax of 5 US$/tCO2e the maximum GDP reduction would be 0.4% by 2030. While for a carbon tax of 20 US$/tCO2e the maximum GDP reduction would be between 1.6% and 1.8% by 2030. Table 5 shows the equivalent average annual GDP growth rate observed after the carbon tax is applied (for the period 2017–2030). For example, in the case of a tax of 5 US$/tCO2e, the GDP growth rate would be between 3.47% *versus* 3.5% in the baseline scenario, and in the case of a tax of 20 US$/tCO2e, the GDP growth rate would be between 3.37% and 3.38% *versus* 3.5% in the baseline scenario.

Table 5. Impact of carbon tax on the annual GDP growth rate.

Carbon Tax (US$/tCO2e)	Scenario 1		Scenario 2		Scenario 3		Scenario 4	
	GDP Growth Rate (%)	GDP Growth Rate Reduction (%)	GDP Growth Rate (%)	GDP Growth Rate Reduction (%)	GDP Growth Rate (%)	GDP Growth Rate Reduction (%)	GDP Growth Rate (%)	GDP Growth Rate Reduction (%)
5	3.47%	0.03%	3.47%	0.03%	3.47%	0.03%	3.47%	0.03%
10	3.43%	0.07%	3.44%	0.06%	3.44%	0.06%	3.43%	0.07%
20	3.37%	0.13%	3.38%	0.12%	3.38%	0.12%	3.37%	0.13%
30	3.30%	0.20%	3.33%	0.17%	3.32%	0.18%	3.30%	0.20%
40	3.25%	0.25%	3.28%	0.22%	3.27%	0.23%	3.25%	0.25%
50	3.19%	0.31%	3.23%	0.27%	3.21%	0.29%	3.19%	0.31%

It is important to note this analysis is not considering recycling of the results (tax collection is not reinvested in the economy), so the cases presented below can be considered as the worst case scenario (maximum negative effect) of this instrument over the GDP path. In addition, the emission reduction of these cases could be higher if the fiscal revenue of the carbon tax supports the implementation of mitigation actions or energy programs. However, in Chile these revenues are not necessary reallocated for these purposes. In fact, the revenues of the carbon tax will be used to funded part of the new tributary reform.

3.3. Comparison to Other Policies

The carbon tax is compared with other policies that stakeholders can apply in order to reduce GHG emissions: introduction of non-conventional renewable energy sources and a sectorial cap. Currently Chile has a non-conventional renewable energy (NCRE) law, based on a quota system, which states that the 20% of energy sales has to be provided by NCRE sources by 2025. This law was approved in 2013 (a previous version of this law, approved in 2008, stated that 10% of energy sales had to be provided by NCRE sources by 2024). We evaluated this potential increase up to 25% by 2030 (25/30 case) and 30% by 2030 (30/30 case). Table 6 shows the results of this evaluation.

The results show that a similar emission reduction can be obtained by increasing the NCRE quota, but the impact on the electricity price would be lower in comparison to the carbon tax scenarios. For example, for Scenario 1, the average annual emission reduction of the 25/30 case (1.0 million tCO₂e) is similar to the emission reduction of 10 US$/tCO₂e carbon tax case, and the emission reduction of the 30/30 case (1.9 million tCO₂e) is similar to the emission reduction of 20 US$/tCO₂e carbon tax case (1.8 million tCO₂e).

Table 6. Evaluation of a change to the current non-conventional renewable energy law.

Unit Scenario	NCRE Case	Δ OPEX MM US$	Carbon Tax Revenue MM US$	Δ CAPEX MM US$	Cumulative Δ Emission MM tCO₂e	Average Δ Annual Emission MM tCO₂e	Δ Electricity Price US$/MWh
S1	25/30	−124.9	0.0	202.1	−13.9	−1.0	0.1
	30/30	−334.4	0.0	516.6	−27.2	−1.9	0.3
S2	25/30	−91.9	0.0	101.8	−5.4	−0.4	0.0
	30/30	−275.0	0.0	307.3	−20.6	−1.5	0.1
S3	25/30	−289.8	0.0	381.8	−13.2	−0.9	0.2
	30/30	−349.3	0.0	570.7	−34.8	−2.5	0.4
S4	25/30	−177.5	0.0	249.1	−13.0	−0.9	0.1
	30/30	−292.3	0.0	469.2	−28.0	−2.0	0.3

However, the electricity price increase is lower in both cases. In the first case the price of electricity would increase 0.1 US$/MWh *versus* 4.2 US$/MWh in the carbon tax case, and the second it would increase 0.3 US$/MWh *versus* 8.2 US$/MWh in the carbon tax case. Therefore, the impact on GDP will be lower. In this work the economic instruments to implement an increase of the NCRE quota are not analyzed. Currently, in Chile the introduction of renewable energy sources is financed by the private sector. The effect of NCRE law modifications over the path of gross domestic product (GDP) with respect to the baseline scenario is shown in the Figure 12.

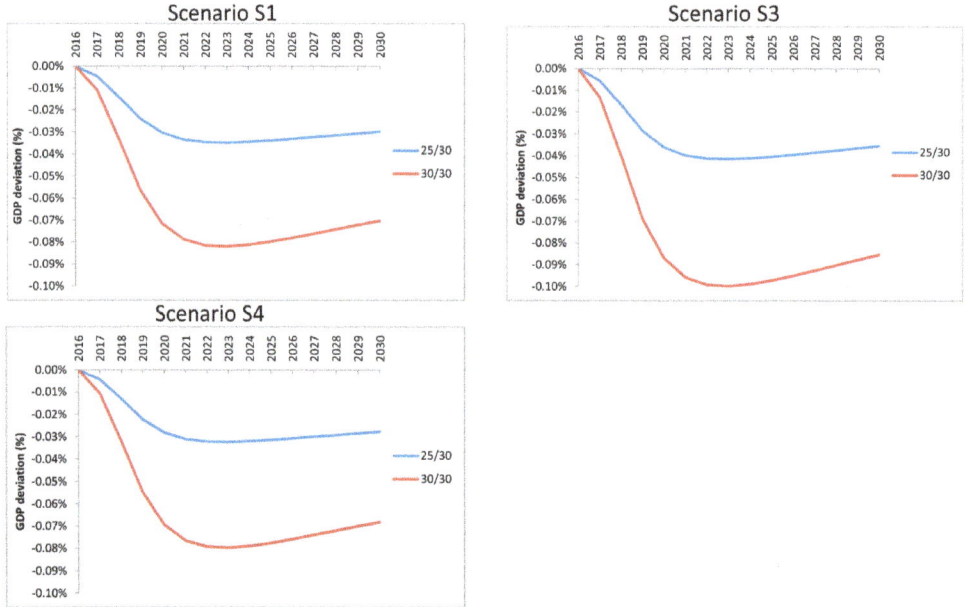

Figure 12. Effect of NCRE law modifications on the level of GDP (% over baseline deviations).

In all the evaluated scenarios the GDP reduction is lower than 0.1% (Scenario S2 is not reported due to the low impact on the GDP). Additionally, a sectorial cap in the electricity generation sector is evaluated. The simulations were done considering an emission cap (maximum level of GHG emission) equals to the emission trajectory when the carbon tax is applied. The cap is introduced in the optimization problem as a constraint. The results of applying this cap are shown in Table 7. where:

- Cap5: in this case the emission cap is equal to the emission trajectory that we got when the carbon tax of 5 US$/tCO₂e is applied;
- Cap10: in this case the emission cap is equal to the emission trajectory that we got when the carbon tax of 10 US$/tCO₂e is applied;
- Cap20: in this case the emission cap is equal to the emission trajectory that we got when the carbon tax of 20 US$/tCO₂e is applied;
- Cap30: in this case the emission cap is equal to the emission trajectory that we got when the carbon tax of 30 US$/tCO₂e is applied;
- Cap40: in this case the emission cap is equal to the emission trajectory that we got when the carbon tax of 40 US$/tCO₂e is applied;

▪ Cap50: in this case the emission cap is equal to the emission trajectory that we got when the carbon tax of 50 US$/tCO₂e is applied.

Table 7. Indicators associated to the application of sectorial cap on the Chilean electricity generation sector in the four scenarios (MM = million).

Unit Scenario	Cap Scenario	Δ OPEX	Carbon Tax Revenue	Δ CAPEX	Cumulative Δ Emission	Average Δ Annual Emission	Δ Electricity Price
		MM US$	MM US$	MM US$	MM tCO₂e	MM tCO₂e	US$/MWh
S1	Cap5	−58.7	0.0	62.4	−1.2	−0.1	0.0
	Cap10	−149.3	0.0	189.2	−13.6	−1.0	0.1
	Cap20	−622.0	0.0	745.7	−25.0	−1.8	0.2
	Cap30	−932.2	0.0	1195.4	−40.9	−2.9	0.5
	Cap40	−1999.6	0.0	3279.2	−139.8	−10.0	2.3
	Cap50	−2926.9	0.0	4590.0	−162.4	−11.6	3.0
S2	Cap5	−459.2	0.0	488.0	−21.2	−1.5	0.1
	Cap10	−857.4	0.0	955.9	−48.6	−3.5	0.2
	Cap20	−1916.3	0.0	2371.4	−125.2	−8.9	0.8
	Cap30	−3121.5	0.0	4057.0	−170.9	−12.2	1.7
	Cap40	−3708.9	0.0	5084.8	−195.6	−14.0	2.5
	Cap50	−3741.6	0.0	5140.5	−197.0	−14.1	2.5
S3	Cap5	−60.0	0.0	75.4	−8.3	−0.6	0.0
	Cap10	−236.1	0.0	261.9	−11.0	−0.8	0.0
	Cap20	−499.4	0.0	594.8	−22.5	−1.6	0.2
	Cap30	−968.8	0.0	1201.2	−36.0	−2.6	0.4
	Cap40	−1516.3	0.0	2209.5	−76.5	−5.5	1.2
	Cap50	−2335.5	0.0	3470.8	−104.4	−7.5	2.0
S4	Cap5	−104.3	0.0	107.8	−2.4	−0.2	0.0
	Cap10	−227.9	0.0	242.5	−5.3	−0.4	0.0
	Cap20	−457.0	0.0	533.9	−15.7	−1.1	0.1
	Cap30	−913.4	0.0	1286.7	−46.4	−3.3	0.7
	Cap40	−2160.3	0.0	3388.9	−130.5	−9.3	2.2
	Cap50	−2738.0	0.0	4325.9	−152.6	−10.9	2.8

Similar to the previous analysis, the results show that the same GHG emission reduction can be obtained by applying a cap in comparison to the carbon tax scenarios, however, the price of electricity would increase less than in the carbon tax cases. For example, in the Scenario 1, the Cap10 and Cap20 cases would increase the price of the electricity by 0.1 and 0.2 US$/MWh, respectively, *versus* 4.2 and 8.2 US$/MWh in the cases where the carbon tax is applied. The effect of the emission cap on the gross domestic product (GDP) path with respect to the baseline scenario is shown in the Figure 13. In all the scenarios the GDP reduction is lower than 1%, which is lower than the observed reductions in the carbon tax scenarios.

388

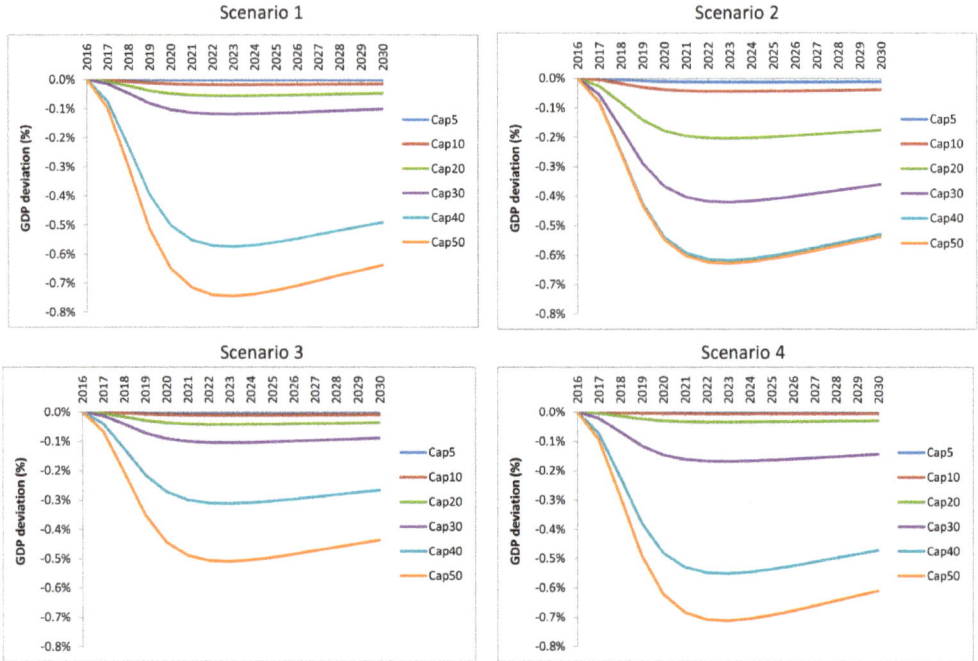

Figure 13. Effect of sectoral cap on the level of GDP (% over baseline deviations).

4. Conclusions

The economy-wide implications of a carbon tax applied on the Chilean electricity market were successfully evaluated. A novel approach is proposed to integrate results from the electricity generation model and a macroeconomic DSGE model. The results show that the carbon tax has a negative impact on the GDP pathway, and the effectiveness of this policy, in the case of Chile, depends on some variables which are not controlled by policy makers such as non-conventional renewable energies investment cost projections, prices of LNG, and the exploitation of hydroelectric resources. Therefore, it is recommended that different development scenarios be evaluated in order to estimate the impact on GHG emission reduction of this policy. In a scenario with a carbon tax of 20 US$/tCO$_2$e, the annual average emission reduction would be between 1.1 and 9.1 million tCO$_2$e. However, the price of the electricity (electricity generation level) would increase between 8.3 and 9.6 US$/MWh, which is equivalent to a 7.4% and 8.5% with respect to the current price of electricity. This price shock decreases the annual GDP growth rate by a maximum value of 0.13%. It means that the average yearly GDP growth rate will be 3.37% *versus* 3.5% in the baseline scenario. On the other hand, alternative policies such as an increase to the quota of non-conventional renewable energy sources and a sectorial cap were evaluated. The results show that the same emission reductions can be achieved with these policies, but with a lower impact on the electricity price and the GDP pathway. Future work considers extending the proposed approach including the theoretical behavior of the iterative linkage between the two models (sectoral and DSGE) and the simulation of gradual tax rate penetration.

Acknowledgments

The paper was supported by Climate and Development Knowledge Network (CDKN), Mitigation Action Plans and Scenarios (MAPS) Programme and CONICYT/FONDAP/15110019.

Author Contributions

Carlos Benavides proposed the general methodology, developed the sectoral model to simulate the impact of the carbon tax on the electricity generation sector, and managed the paper project. Luis Gonzales contributed to develop the economic approach, designed the link between the sectoral and macroeconomic models, calibrated and performed the simulations on the DSGE model. Gonzalo Garcia contributed to develop the economic approach and gave technical support to do the simulation on the DSGE model. Rodrigo Fuentes supervised the results and methodology to evaluate the macroeconomic impacts of the carbon tax. Manuel Diaz helped with the overall coordination and result analysis. Rodrigo Palma supervised and contributed to the general methodology and analyzed the results. Catalina Ravizza helped to design the methodology. All the authors contributed to the state of the art, preparation, and approval of the manuscript.

Annex

Figure 14 shows the investment costs in the power generation sector that were considered for the preliminary calibration for the SIC and the SING.

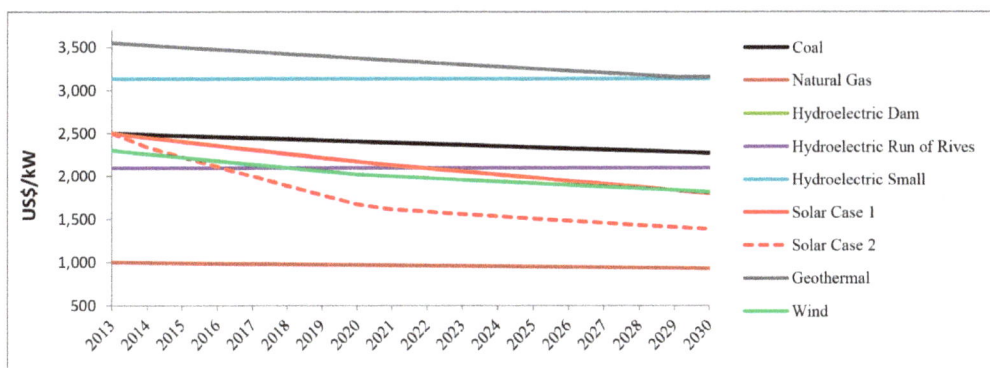

Figure 14. Investment cost (US$/kW).

Figure 15 shows the LNG prices projection considered in this paper.

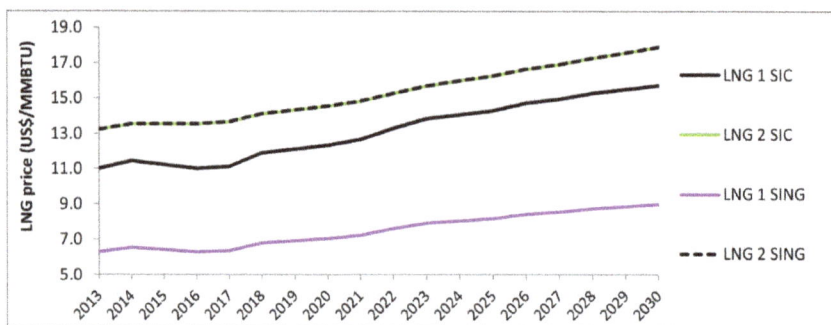

Figure 15. LNG prices projection. LNG 1 SIC and LNG 1 SING are the prices of LNG for plants which have open access to the LNG terminal in SIC and SING, respectively. LNG 2 SIC and LNG 2 SING is the price for plant which have not open access to LNG terminal.

Conflicts of Interest

The authors declare no conflict of interest.

References

1. *Chile's First Biennial Update Report to the United Nations Framework Convention on Climate Change*; Ministry of Environment: Santiago, Chile, 2011.
2. MAPS-Chile Project. Opciones de mitigación para enfrentar el cambio climático: resultados de Fase 2. Ministry of Environment: Santiago, Chile, 2014. Available online: http://mapschile.cl/files/Resultados_de_Fase_2_mapschile_2910.pdf (accessed on 4 January 2015). (In Spanish)
3. SBS. Factbox: Carbon Taxes around the World, 2013. Available online: http://www.sbs.com.au/news/article/1492651/Factbox-Carbon-taxes-around-the-world (accessed on 4 January 2015).
4. Capros, P.; Paroussos, L.; Fragkos, P.; Tsani, S.; Boitier, B.; Wagner, F.; Busch, S.; Resch, G.; Blesl, M.; Bollen, J. European decarbonisation pathways under alternative technological and policy choices: A multi-model analysis. *Energy Strategy Rev.* **2014**, *2*, 231–245.
5. Capros, P.; Paroussos, L.; Fragkos, P.; Tsani, S.; Boitier, B.; Wagner, F.; Busch, S.; Resch, G.; Blesl, M.; Bollen, J. Description of models and scenarios used to assess European decarbonisation pathways. *Energy Strategy Rev.* **2014**, *2*, 220–230.
6. European Commision (EC). *Impact Assessment on Energy and Climate Policy up to 2030*; European Commission (EC): Brussels, Belgium, 2014.
7. Moyo, A.; Winker, H.; Wills, W. The Challenges of Linking Sectoral and Economy-Wide Models. MAPS-Programme. Available online: http://www.mapsprogramme.org/wp-content/uploads/Challenges-of-Linking_Brief.pdf (accessed on 5 January 2015).
8. Vorster, S.; Winkler, H.; Jooste, M. Mitigating climate change through carbon pricing: An emerging policy debate in South Africa. *Clim. Dev.* **2011**, *3*, 242–258.

9. Winkler, H. *Taking Action on Climate Change: Long-term Mitigation Scenarios for South Africa*; UCT Press: Cape Town, South Africa, 2010.

10. Alton, T.; Arndt, C.; Davies, R.; Hartley, F.; Makrelov, K.; Thurlow, J.; Ubogu, D. Introducing carbon taxes in South Africa. *Appl. Energy* **2014**, *116*, 344–354.

11. Thepkhun, P.; Limmeechokchai, B.; Fujimori, S.; Masui, T.; Shrestha, R. Thailand's Low-Carbon Scenario 2050: The AIM/CGE analyses of CO_2 mitigation measures. *Energy Policy* **2013**, *62*, 561–572.

12. Chi, Y.; Guo, Z.; Zheng, Y.; Zhang, X. Scenarios analysis of the energies' consumption and carbon emissions in China based on a dynamic CGE model. *Sustainability* **2014**, *6*, 487–512.

13. Guo, Z.; Zhang, X.; Zheng, Y.; Rao, R. Exploring the impacts of a carbon tax on the Chinese economy using a CGE model with a detailed disaggregation of energy sectors. *Energy Econ.* **2014**, *45*, 455–462.

14. Böhringer, C.; Rutherford, T. Transition towards a low carbon economy: A computable general equilibrium analysis for Poland. *Energy Policy* **2013**, *55*, 16–26.

15. Mahmood, A.; Marpaung, C. Carbon pricing and energy efficiency improvement—Why to miss the interaction for developing economies? An illustrative CGE based application to the Pakistan case. *Energy Policy* **2014**, *67*, 87–103.

16. Bukowski, M.; Kowal, P. *Large Scale, Multi-sector DSGE Model as a Climate Policy Assessment Tool—Macroeconomic Mitigation Options (MEMO) Model for Poland*; IBS Working Paper; Institute for Structural Research (IBS): Warsaw, Poland, 2010.

17. Mongelli, I.; Tassielli, G.; Notarnicola, B. Carbon Tax and its Short-Term Effects in Italy: An evaluation through the Input-Output model. In *Handbook of Input-Output Economics in Industrial Ecology*; Springer: Dordrecht, The Netherlands, 2009.

18. Syri, S.; Lehtilä, A.; Ekholm, T.; Savolainen, I.; Holttinen, H.; Peltola, E. Global energy and emissions scenarios for effective climate change mitigation—Deterministic and stochastic scenarios with the TIAM model. *Int. J. Greenh. Gas Control* **2008**, *2*, 274–285.

19. Anandarajah, G.; Gambhir, A. India's CO_2 emission pathways to 2050: What role can renewables play? *Appl. Energy* **2014**, *131*, 79–86.

20. Fairuz, S.M.C.; Sulaiman, M.Y.; Lim, C.H.; Mat, S.; Ali, B.; Saadatian, O.; Ruslan, M.H.; Salleh, E.; Sopian, K. Long term strategy for electricity generation in Peninsular Malaysia—Analysis of cost and carbon footprint using MESSAGE. *Energy Policy* **2013**, *62*, 493–502.

21. Matsuoka, Y.; Kainuma, M.; Morita, T. Scenario analysis of global warming using the Asian Pacific Integrated Model (AIM). *Energy Policy* **1995**, *23*, 357–371.

22. Zang, H.; Xu, Y.; Li, W. An uncertain energy planning model under carbon taxes. *Front. Environ. Sci. Eng.* **2012**, *6*, 549–558.

23. Kainuma, M.; Matsuoka, Y.; Morita, T.; Hibino, G. Development of an end-use model for analyzing policy options to reduce greenhouse gas emissions. *IEEE Trans. Syst. Man Cybern. C Appl. Rev.* **1999**, *29*, 317–324.

24. Careri, F.; Genesi, C.; Marannino, P.; Montagna, M. Generation expansion planning in the age of green economy. *IEEE Trans. Power Syst.* **2011**, *26*, 2214–2223.

25. Chen, Q.; Kang, C.; Xia, Q.; Zhong, J. Power generation expansion planning model towards low-carbon economy and its application in China. *IEEE Trans. Power Syst.* **2010**, *25*, 1117–1125.

26. Wei, W.; Liang, Y.; Liu, F.; Mei, S.; Tian, F. Taxing strategies for carbon emissions: A bilevel optimization approach. *Energies* **2014**, *7*, 2228–2245.

27. National Technical University of Athens. The GEM-E3 Model Reference Manual. Available online: http://www.e3mlab.ntua.gr/manuals/GEMref.PDF (accessed on 5 March 2015).

28. Rodrigues, R.; Linares, P. Electricity load level detail in computational general equilibrium—Part I—Data and calibration. *Energy Econ.* **2014**, *46*, 258–266.

29. Del Negro, M.; Schorfheide, F. DSGE Model-Based Forecasting; Elliott, G., Timmermann, A., Eds.; In *Handbook of Economic Forecasting SET 2A-2B*; Elsevier: Amsterdam, The Netherlands, 2013; Volume 2, pp. 57–140.

30. Edge, R.; Gurkaynak, R. How Useful are Estimated DSGE Model Forecasts? Available online: http://www.federalreserve.gov/pubs/feds/2011/201111/ (accessed on 7 January 2015).

31. Energy Ministry, GIZ. Renewable Energy in Chile, wind, solar and hydroelectric potential in Chile. 2014. Available online: http://www.minenergia.cl/archivos_bajar/Estudios/Potencial_ ER_en_Chile_AC.pdf (accessed on 4 January 2015). (In Spanish)

32. Mercure, J.F.; Pollitt, H.; Chepreecha, U.; Salas, P.; Foley, A.M.; Holden, P.B.; Edwards, N.R. The dynamics of technology diffusion and the impacts of climate policy instruments in the decarbonisation of the global electricity sector. *Energy Policy* **2014**, *73*, 686–700.

33. Chile 2030 Energy Scenario Platform. Santiago, Chile. 2014. Available online: http://escenariosenergeticos.cl/wp-content/uploads/Escenarios_Energeticos_2013.pdf (accessed on 7 January 2015). (In Spanish)

34. CDEC-SIC (Central Interconnected System Load Economic Dispatch Center). Available online: http://www.cdecsic.cl/ (accessed on 6 January 2015). (In Spanish)

35. CDEC-SING (Center for Economic Load Dispatch of Northern Interconnected System). Available online: http://cdec2.cdec-sing.cl/ (accessed on 6 January 2015). (In Spanish)

36. National Energy Comission. April 2014 Node Price Report of the Central Interconnected System. Santiago, Chile. 2014. Available online: http://www.cne.cl/tarificacion/electricidad/ precios-de-nudo-de-corto-plazo/abril-2014 (accessed on 6 January 2014). (In Spanish)

37. National Energy Comission. April 2014 Node Price Report of Norte Grande Interconnected System. Santiago, Chile. 2014. Available online: http://www.cne.cl/tarificacion/electricidad/ precios-de-nudo-de-corto-plazo/abril-2014 (accessed on 6 January 2014). (In Spanish)

38. Medina, J.; Soto, C. Model for Analysis and Simulations: A Small Open Economy DSGE for Chile. Central Bank of Chile. 2006. Available online: http://www.bcentral.cl/conferencias-seminarios/otras-conferencias/pdf/modelling2006/soto_medina.pdf (accessed on 7 January 2015).

39. Medina, J.P.; Soto, C. The Chilean Business Cycle through the lens of Stochastic General Equilibrium Model. Working Papers 457, Central Bank of Chile, 2007. Available online: http://www.bcentral.cl/conferencias-seminarios/otras-conferencias/pdf/variables/Medina_Soto.pdf (accessed on 7 January 2015).

40. Calvo, G. Staggered prices in utility-maximizing framework. *J. Monet. Econ.* **1983**, *12*, 383–398.

41. Christiano, L.; Eichenbaum, M.; Evans, C. Nominal rigidities and the dynamic effects of a shock to monetary policy. *J. Political Econ.* **2005**, *113*, 1–45.
42. Erceg, C.; Henderson, D.W.; Levin, A.T. Optimal monetary policy with staggered wage and price contracts. *J. Monet. Econ.* **2000**, *46*, 281–313.
43. GLPK. Available online: https://www.gnu.org/software/glpk/ (accessed on 7 January 2015).
44. CPLEX Optimizer. Available online: http://www-01.ibm.com/software/commerce/optimization/cplex-optimizer/ (accessed on 7 January 2015).
45. Cerda, R. *Implementation of DSGE Model in MATHLAB*; Institute of Economics, Pontifical Catholic University of Chile: Santiago, Chile, 2010.

A Study to Improve the Quality of Street Lighting in Spain

Alberto Gutierrez-Escolar, Ana Castillo-Martinez, Jose M. Gomez-Pulido, Jose-Maria Gutierrez-Martinez, Zlatko Stapic and Jose-Amelio Medina-Merodio

Abstract: Street lighting has a big impact on the energy consumption of Spanish municipalities. To decrease this consumption, the Spanish government has developed two different regulations to improve energy savings and efficiency, and consequently, reduce greenhouse-effect gas emissions. However, after these efforts, they have not obtained the expected results. To improve the effectiveness of these regulations and therefore to optimize energy consumption, a study has been done to analyze the different devices which influence energy consumption with the intention of better understanding their behavior and performance. The devices analyzed were lamps, ballasts, street lamp globes, control systems and dimmable lighting systems. To improve their performance, they have been analyzed from three points of view: changes in technology, use patterns and standards. Thanks to this study, some aspects have been found that could be taken into account if we really wanted to use energy efficiently.

Reprinted from *Energies*. Cite as: Gutierrez-Escolar, A.; Castillo-Martinez, A.; Gomez-Pulido, J.M.; Gutierrez-Martinez, J.-M.; Stapic, Z.; Medina-Merodio, J.-A. A Study to Improve the Quality of Street Lighting in Spain. *Energies* **2015**, *8*, 976-994.

1. Introduction

Street lighting is an integral part of the municipal environment, promoting comfort, as well as enhancing the safety and security of its users [1]. This kind of lighting has the greatest impact on energy consumption in most Spanish municipalities, and may account for up to 80% of the electricity consumed by municipalities [2]. Furthermore, the average lamp power used in Spain, with an average of 157 W per lamp, is one of the highest in the European Union, well above the United Kingdom (76 W) or The Netherlands (61 W) [3]. This is perhaps due to the fact that 20% of street lighting lamps are based on outdated and inefficient technologies [4]. Table 1 shows the percentage of each kind of lamp in use in 2007.

Table 1. Percentage of each kind of lamp in some European countries in 2007.

Country	HPM (%)	HPS (%)	LPS (%)	MH (%)	FL (%)	LED (%)
Netherlands	4.76	44.76	28.57	2.85	19.04	0
UK	0	41	44	0	15	0
Spain	20	70	0	0	10	0

HPM = high pressure mercury; HPS = high pressure sodium; LPS = low pressure sodium; MH = metal halide; FL = fluorescent; LED= light-emitting diode.

To improve this situation, the Spanish Government put forth the Royal Decree 1890/2008 [5] and its corresponding Complementary Technical Instructions. Its objectives are: (1) to improve energy savings and efficiency, and consequently, reduce greenhouse-effect gas emissions; (2) to limit glare and light pollution; and (3) to reduce intrusive or annoying light levels. Moreover a strategy, known as the Energy Saving and Efficiency Strategy (E4) was also defined [6], which established a series of standard actions aimed at improving the energy system. The target set in this Plan was to achieve a consumption of 75 kWh per inhabitant per year. Its main measures were: (1) to establish a program of replacement of existing external public lighting equipment, based on obsolete technologies, with other more up-to-date and efficient equipment; (2) implementation of energy audits; and (3) to set up and run energy training courses for municipal technicians and the maintenance managers of municipal installations.

Sanchez de Miguel [7], who defined a procedure to estimate the energy consumption in public electric lighting in Spain from 1992 to 2012, came to the conclusion that the most populated provinces appeared to have begun to stabilize the growth of their expenditure on public lighting, but that this had no occurred in the less populated provinces where this expense continued to rise at a similar rate despite the economic crisis. The general trend of Spain during the last eighteen years had been one of nearly constant growth. One of the purposes defined by the strategy was to promote the use of more efficient equipment. Analysing the lighting level control devices installed in the Community of Andalusia, for example, it is possible to observe that 64% of the installations do not have voltage regulators [8]. This sort of devices allows the amount of energy consumed to be reduced when the conditions are appropriate, for example when the number of vehicles does not exceed some predetermined quantity.

Neither of the regulations mentioned before have obtained the expected results. To help with this issue, the aim of this manuscript is to detect any aspects that the previous regulations might have overlooked. These aspects have been analysed form the point of view of energy efficiency and energy consumption. If these new considerations were to be included in future updated versions of the regulations, we are sure that the quality of street lighting would be similar to the standards of other European countries which are doing quite well in this area. The remainder of this paper is organized as follows: Section 2 presents the related work, Section 3 describes the main elements of our study and finally Section 4 contains the conclusions.

2. Related Work

The related work is divided into two parts: the first part analyses the different proposals to measure energy efficiency and the second part shows the strengths of some foreign street lighting standards. There are only a few European countries that have provisions addressing the energy efficiency of the whole street lighting system, among them Spain and The Netherlands [9]. Hence, the first part shows the different methodologies used to measure the energy efficiency. The way proposed by the Slovenian government in 2007 [10] consisted in measuring the annual energy consumption per citizen per year. The main disadvantage of this proposal is that for areas with high population density it is easier to achieve lower level values than for areas with low population density. Another way was presented by Silva [11], who developed a tool which can assess street lighting performance in the context of energy efficiency. This tool uses three indicators: one to evaluate lighting performance and two others to evaluate energy performance, one being luminaire coverage and efficiency and the other lighting control devices. The only shortcoming of this tool is that the score used for lighting control devices has only two values—zero or one—depending on whether the installation has (one) or does not have (zero) this kind of devices. In the research carried out by Pracki [12], a new classification system based on the installed and normalised power densities was proposed. Although he claimed that energy consumption depends on the burning hours, he did not take that into account in his proposal. Different criteria were used in a German road lighting competition [13], where the energy efficiency of the street lighting was defined as the amount of energy consumption per kilometre per year in kWh/(km × Y), and the energy used (kWh) to produce a certain luminous flux over time. Besides, in the research carried out by Kyba [14] the same definition of efficiency in urban street lighting (kilowatt hours per kilometre per year) was also proposed because this measure allows assessment of all the elements that influence energy consumption. For the case of Spain [5], energy efficiency is based on the lit-up surface, average illuminance and the total active power installed.

There are a lot of options to define energy efficiency, but it seems to be impossible to use just one measure to describe the energy efficiency of street lighting systems [15], although all of them have the same goal of reducing the energy consumption without sacrificing the visibility conditions and comfort. This article is focused on the main devices which influence energy consumption with the purpose of improving the energy efficiency considering the current Spanish regulations.

The second part shows the strengths of different street lighting standards compared with the Spanish Regulation already implemented. The Technical Regulation of Lighting and Street Lighting (RETILAP) [16] from Colombia incorporates a section to establish the coexistence between luminaires and trees. The Public Lighting Design Manual from Hong Kong [17] defines the design layout (single-sided, staggered, opposite and twin-central) regarding the mounting height of the luminaire. Another strength of this Manual is that it defines the distance between luminaires and fire hydrants in order to not to block their operation. Minnesota's Energy Law [18] establishes that a lamp with initial efficiency less than 70 lumens per watt must be replaced when worn out by light sources using lamps with initial efficiency of at least 70 lumens per watt. The Spanish regulations established that the new lamps shall have an initial efficiency of at least 65 lumens per watt.

3. Main Elements

An analysis of the main elements is necessary to understand how each component affects the final energy consumption. These elements are divided as follows: lamps, ballast, street lamp globes, hours of operation, lighting level control devices and renewable energies. To be sure about their involvement in the final energy consumption, each component was studied individually. Then, after studying how each component influences the final energy, different criteria to save energy were established. According to Boyce [19], there are four options to save energy: changes in technology, in patterns of use, standards and basis of design, but from our point of view, changes to the basis of design require a careful reconsideration of what such lighting is for and how it might best be achieved, so each element was studied excluding the fourth option.

3.1. Lamps

There is no doubt that lamps are the most representative component of street lighting. There are several types of lamps on the market which can be used on this kind of installation, including, among others, high pressure mercury (HPM), high pressure sodium (HPS), low pressure sodium (LPS), metal halide (MH) and light-emitting diode (LED) lamps. At present, in street lighting applications, HPS and MH lamps are the most widely used light sources. LEDs are fast developing light sources and are considered a promising light source for general lighting, although this kind of source on the market is not that cheap yet. Currently, HPS lamps are the dominant light source used in road lighting because of the long lamp lifetime and high luminous efficacy. MH lamps offer high luminous efficacy and good color rendering properties [20].

There are two different options to save energy in the case of lamps: changing the standards and changing the technology; for example, the British Standard BS 5489 [21] allows reducing the required lighting class when the color rendering index (CRI) of the lamp is higher than 60 (white light) [22]. On the other hand, the Hong Kong regulations only allow reducing it if the lamp has a CRI equal to or greater than 80 [17]. This reduction is only permitted on subsidiary roads. If the current Spanish standard took into account this reduction, the illuminance level would be reduced by at least 25%. Table 2 shows the reduction of illuminance level.

Analysing Table 2, we notice that there are two more lighting classes in the British standard than in the Spanish standard. These lighting classes are S5 and S6. In this respect, we must agree with the Royal Decree because 40% of night-time street crime occurs when lighting levels are at 5 lux or below [23]. Before recommending the incorporation of this advantage into the Spanish standard, it is necessary to be sure that this change does not decrease the quality of the installations. There are several researches that confirm the benefits of white light. One of them is the study conducted by Godfrey [24], who concluded that driver reactions with cool white light are more efficient than with "warm" coloured light.

Table 2. Illuminance savings by reducing lighting class.

Lighting class (Spain)	Lighting class (BS)	Illuminance level (lux)	New class	New illuminance level (lux)	Illuminance savings (%)
S1	S1	15	S2	10	33%
S2	S2	10	S3	7.5	25%
S3	S3	7.5	S4	5	33%
S4	S4	5	S5	3	40%
-	S5	3	S6	2	33%
-	S6	2	-	-	-

Lewis [25] also reported the results of reaction time tests where detection of a pedestrian was conducted using MH, HPM, HPS and LPS. He found an approximately 50% increase in reaction time for sodium sources *versus* MH, at a luminance level of 0.1 cd/sq.m. At a relatively high lighting level of 1 cd/sq.m, he reported an increase in reaction time of approximately 15% of HPS *versus* MH, and 25% for LPS *versus* MH. In our opinion, there are several evidences that prove the benefits of the white light yet the current Spanish Standard does not include it. This should be incorporated in order to improve the energy efficiency.

Related to changes in the technology, it is necessary to guarantee that these changes do not decrease the amount of light output. The best parameter to compare two kinds of lamps without decreasing the luminous flux is the luminous efficacy of the lamp. This parameter is the quotient luminous flux emitted by the power consumed by the source, unit lumen per Watt [26]. Table 3 shows the main features of the different kinds of lamps.

Table 3. Main features of the different kind of lamps.

Lamp type	Luminous efficacy (lum/W)	Colour rendering index (CRI)
HPM	50	15
HPS	130	25
MH	80–108	75–90
LED	90–130	>80

Luminous efficacy is also used by the Spanish regulations [5], where the minimum values specified is 65 lum/W. As it can be seen in Table 3, HPM lamps do not comply with the requirements, so it does not make sense for this kind of lamp to appear in the Spanish standard. In the case that lighting designers wanted to change the kind of lamp, they may follow this criterion because it is possible to find lamps with the same or higher luminous flux and less power consumption. For example, by simply replacing common bulbs with energy-saving LED lamps one can reduce energy consumption by up to 80% [27].

3.2. Ballast

All kinds of lamps require a ballast to operate correctly. For this reason, the presence of this device in street lighting systems is indispensable to ignite the discharge and control the lamp. Ballast devices can be divided mainly into two types: electromagnetic and electronic. Electronic ballasts are considered more energy efficient than electromagnetic ballasts, and for this reason they have been promoted as replacements the latter, to the point that some countries have changed their regulations to encourage their use. Other advantages are that electronic ballasts produce no flicker effects and provide an instantaneous startup [28]. Due to the fact that electromagnetic ballasts have high power loss from the iron and copper losses in the magnetic choke, they are 10%–15% less efficient than electronic ballasts [29,30]. To verify that the power of electronic ballasts is lower than that of electromagnetic ballasts, different ballasts were studied of Philips [31]. Table 4 shows the power savings for two different LPS lamp powers.

Table 4. Power savings using electronic ballasts with LPS lamps.

Lamp type & power (W)	Electromagnetic ballast power (W)	Electronic ballast power (W)	Power saving
1 × SOX 35 W	11.7	3.7	8
1 × SOX 55 W	19.5	5.5	14

These power savings are under nominal conditions and although they might be considered insignificant, they should be taken into account because the power saving percentage in the case of a 55 W SOX is 18%. To analyze the benefit of this replacement under normal conditions, the research carried out by Omar [32] was studied. They examined the energy consumption of 277 units of 250 W HPS for a month. The energy consumption with electromagnetic ballasts was 30,913.2 kWh and the energy used with electronic ballasts was 20,172.7 kWh. Therefore in this case the energy saving was 34.74%. Besides, there are other researchers that have studied the benefits regarding the supply voltage. A good example is the research done by Dolora [33], who studied the savings for HPS 150 W lamps. This research concluded that the supply voltage bears on in the final energy consumption. Table 5 shows the power variation regarding the supply voltage.

Table 5. Power variation between electronic and electromagnetic ballasts [33].

Supply voltage	Electromagnetic ballast		Electronic ballast		Power variation
(Vac)	Power (W)	Illuminance (lx)	Power (W)	Illuminance (lx)	(%)
200	122.7	161	154.2	256	25.6%
210	136.7	190	156.8	256	14.7%
220	152.3	220	157.1	256	3.1%
230	168.3	255	156.0	256	−7.3%
240	187.1	291	154.8	256	−17.2%
250	204.2	327	154.3	256	−24.4%
260	224.6	367	154.2	256	−31.3%

As it can be seen, when the supply voltage is 250 V, the percentage of power variation is 24.4%, this means that the luminaire power can vary by up to 49.9 W. The problem with the Royal Decree [5] is that it only specifies the maximum power per luminaire, when in our opinion the maximum ballast power should be specified because designers sometimes are not aware if the kind of ballast that satisfies the requirements. Table 6 shows a good example.

Table 6. Luminaire power for different kind of ballasts.

Nominal lamp (W)	Maximum lamp power allowed (R.D 1890) [5]	Lamp power plus electromagnetic ballast (W)	Lamp power plus electronic ballast power (W)
1 × SOX 35 W	42 W	46.7 W	38.7 W
1 × SOX 55 W	65 W	74.5 W	60.5 W

As it can be appreciated, lighting designers must pay attention when choosing the ballast because although the maximum power is defined, the luminaire power must be checked because in the analyzed case the installation of electromagnetic ballasts would not satisfy the minimum requirements.

3.3. Street Lamp Globes

Although people believe that street lamp globes do not influence energy consumption, the choice of this part is very important because it influences the upward reflected light and thereby light pollution. Light pollution is not simply any astronomical or ecological light pollution, because enormous amounts of energy are wasted with light pollution. For example, at the end of the 1990s the amount of sky glow over Sapporo, Japan was equivalent to 15 million kWh of energy, 29 million kWh over London, UK and 38 million kWh over Paris, France [34]. The total amount used for public outdoor lighting in Helsinki, Finland is roughly 170 million kWh, meaning that all Helsinki could be illuminated with just five days of the "waste light" of Paris. The light sent upward is thus estimated to produce economic losses worth billions of euros every year [35]. The best option to save energy regarding the light pollution is by changing standards. The current Croatian regulation establishes lower levels than the Spanish regulations. Table 7 compares the maximum upward light ratio of the installation (ULR) for Croatia [36] and Spain [5].

Table 7. Maximum percentage of ULR for Croatia and Spain.

Croatia standard		Spanish standard	
Classification zone	**Maximum ULR (%)**	**Classification zone**	**Maximum ULR (%)**
EO	0%	Not exist	0%
E1	0%	E1	1%
E2	2.5%	E2	5%
E3	5%	E3	15%
E4	15%	E4	25%

As it can be appreciated, the maximum percentage of ULR in Spain is higher than in Croatia. Although Croatia is not the country with the strictest regulations, in our opinion the Spanish regulation should incorporate at least the minimum level established in the Croatian rules.

The Chilean D.S.N° 686/98 regulation [37] defines that a lamp with a luminous flux equal to or less than 15,000 Lm cannot emit more than 0.8% of its nominal flux above horizontal level when installed in a luminaire. Lamps with a luminous flux of more than 15,000 Lm should not emit more than 1.8% of their nominal flux above horizontal level when installed in a luminaire.

In 2007 Slovenia adopted a law (Official Gazette of the Republic of Slovenia, No. 81/2007) aimed at tackling light pollution. The law requires that 0% of the output of a luminaire should shine above the horizon (90°) [38].

To analyze how ULR influences this kind of installation, several simulations were done with the DIALux software. The analysis consisted in studying what happens if the luminaire has the same kind of lamp and the street lighting globes are different. The model of the studied luminaire was the CitySpirit Modern (Philips, Amsterdam, The Netherlands), the street lighting globes were four and the kind of lamp was LED. Figure 1 shows the average illuminance regarding the ULR for 22 X XR-E-PE/WW, 22 X XR-E-Q3/NW and 22 X XR-E-Q5/CW lamps.

As it can be seen for this luminaire model, if the ULR increases, the average illuminance decreases, but the lamp power and the lit-up surface were the same for the three simulations, therefore the energy efficiency bears upon the ULR. Another model of lamp analyzed was the Urbana (Philips) and again ULR was studied and the same performance can be appreciated, the street lighting globes were two in this case and the lamp was an HPL-N80W. Table 8 shows the results.

Figure 1. Illuminance regarding the ULR.

Table 8. ULR regarding the street lamp globes.

Globes	Total lamp flux	System flux	Luminaire power	Lit-Up surface	E average (Lux)	Overall uniformity	ULR (%)
	3600	684	90	80	5.02	0.2	4.5
	3600	792	90	80	4.74	0.273	16

It is possible to think that as the system flux is higher in the second option than in the first option, the average illuminance would be higher than the first one, but the reality is that as this sort of street lighting globes does not have any device to avoid the light pollution, and thus the average illuminance is lower than in the first case. From our point of view, ULR magnitude should be taken into account for the energy label, because with the current systems only assess the illuminance on the lit-up surface.

3.4. Hours of Operation

The current Spanish standard [5] includes three possible devices for that purpose: astronomic time switches, twilight switches and remote management systems for electrical boards. Astronomic time switches turns lights on and off with a fixed time offset from sunrise and sunset. To estimate the daily hours of sunrise and sunset the latitude and longitude are needed because of the movement of the Sun, as it can be seen on the sunrise sunset calculator program tool [39].

Twilight switches measure the amount of natural light available to turn on and off the lamps regarding this level. As happens with astronomic time switches, it is possible to establish an approximation of the number of burning hours using the latitude and the level of natural light required to turn the system on or off [40]. An option to decrease the hours of operation and therefore to save energy with this kind of device is by changing the use pattern. Angus Council (U.K.) [41] studied the trimming of photocells; the factory setting of the switch on/switch off levels are 70 lux on and 35 lux off (70/35). By reducing the switch ratio to 35/18 they could typically save 92 burning hours per year per luminaire. The Institution of Lighting Professionals (ILP) [42] estimated that if the switching levels were reduced 35/16, a saving of 1%–2% per luminaire could be achieved. This regulation is not recommended for older lamp types such as LPS and HPM operating on conventional ballasts. Such installations should be operated at 70 lux on and 35 lux off as a minimum to allow the lamps to fully run up by the time the lighting is required. The only drawback of the previous studies is that they did not specify the latitude. This lack of information was solved in the study carried out by American Electric Lighting (AEL) [43] because the latitude was taken into account in the results. Table 9 shows the hours of operation at latitude 35° (Los Angeles, California) for various photocell settings.

Table 9. Hours of operation regarding the twilight settings [39].

On (lux)	Off (lux)	Hours of operation
8.6	10.7	4113
10.7	12.9	4130
32.2	19.3	4187
16.1	24.7	4167
27.9	33.3	4204
10.7	32.2	4265
21.5	107.6	4340

Remote management systems are composed by a server-client architecture system for monitoring, detecting, controlling and communicating problems instantly to a central control room or directly to maintenance technicians [44]. Telemanagement integration in street lighting networks of small cities has hardly been developed both in a conceptual and applicative way, especially due to limited economical resources of local communities which have become responsible for too many new tasks, public illumination being one of them [45].

The hours of operation depend on these devices which also consume energy. Analyzing the data of the manufacturer ORBIS [46], it can be appreciated that the power consumptions are very similar independently of the kind of device. Table 10 shows the power consumption.

Table 10. Power consumption of street lighting control systems.

Type	Model	Self-Consumption (VA)
	DATA ASTRO	5
Astronomic	ASTRO NOVA CITY	6
	ASTRO UNO	6
	ORBIFOT	8
Twilight	VEGA	8
	ORBILUX	3.4
Remote Management	XEO LUM	4.8

As each device uses different technology and criteria to turn on and off, the hours of operation established for each device will be different. We have measured the natural light level during different days with the purpose of understanding the operation of each device. Figure 2a shows the natural light level several days at sunrise and Figure 2b shows the natural light level of several days at sunset, where the data of both figures were measured in Madrid (Spain) in September 2014. A PCE-174 (ORBIS) digital illuminance meter was used to obtain the data.

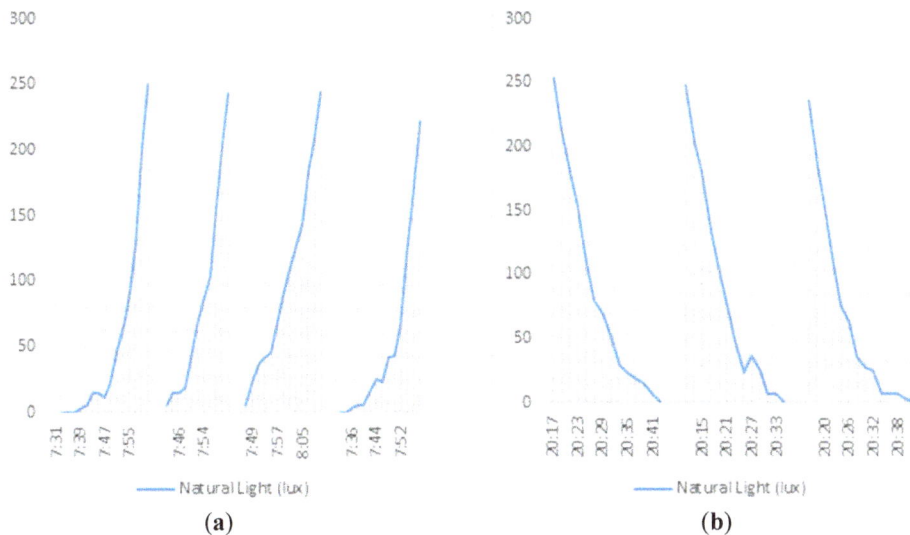

(a)   (b)

Figure 2. Natural light level during the sunrise (**a**); and during the sunset (**b**) in Madrid.

As it can be seen, the tendency is different for each day because of the weather conditions are different and therefore climate bears upon the natural light levels. In that aspect we agree with Howell [40] that the weather conditions are even more significant than latitude in determining days. Hence the main drawback of astronomic time switches is that they do not take into account the real level of natural light. Besides, analyzing in detail the data of the previous trimming, Table 11 shows the time when the natural light reached a certain value. It can be seen, trimming the photocells allows decreased the hours of operation, while on the other hand natural light level reached 35 lux twice on 21 September. This issue is the main problem of photocells because undulations in light level can cause erratic operation, but this can be solved with the controller.

Table 11. Time when the natural light reached a certain value.

Day	Sunrise			Sunset		
	18 (lux)	35 (lux)	Savings Minutes	70 (lux)	35 (lux)	Savings Minutes
19 September 2014	7:46	7:48	2 min	20:29	20:32	3 min
20 September 2014	7:46	7:47	1 min	20:28	20:31	3 min
21 September 2014	7:46	7:48	2 min	20:21	20:24 and 20:27	-------
22 September 2014	7:47	7:48	1 min	20:25	20:28	3 min
23 September 2014	7:41	7:45	4 min	20:24	20:27	3 min

As it can be seen, photocell trimming could save approximately 4 min per day. This means that the amount of burning hours may reduce by 24 per year.

3.5. Lighting Level Control Devices

There are three different types of level control devices contemplated in the Spanish standard [5]: series inductive type ballasts for dual power level, power controlled electronic ballasts and regulators and stabilizers in the head of the line.

The main problem of using ballasts for dual power levels is that these systems act locally, requiring an adjustment device attached to each of the individual charges and also a general control system to control all of them [47]. Regulators and stabilizers are able to control the voltage according to different parameters such as number of vehicles per hour [48], weather conditions or the presence of pedestrians [49]. Their operation consists of hanging the input mains voltage to a variable voltage within the range from 220 to 170 V [50]. Those changes are accompanied by variations of illuminance and lamp power. Figure 3 shows the working of these sort of systems, where it can be seen their potential on energy savings.

The main advantage of stabilizer lighting systems is that they are able to avoid overvoltage situations. The research carried out in China [51] showed how, despite the fact the nominal voltage is established at 230 V like in Spain, it reached values as high as 246 V. This overvoltage situation is the main reason for the shortened lifetime of lamps.

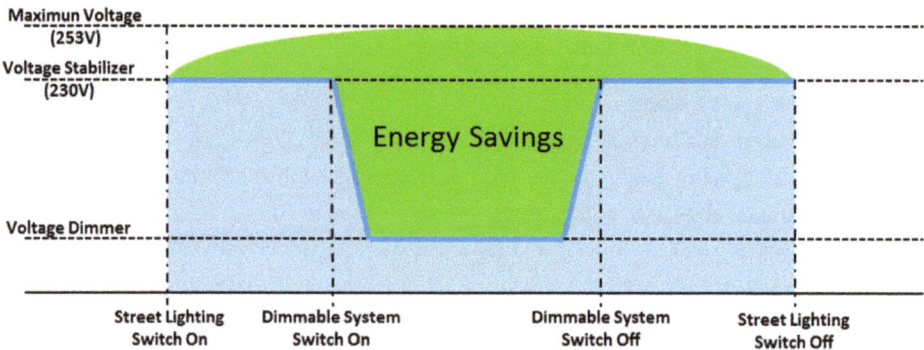

Figure 3. Regulator and stabilizer devices.

Taking into account that the energy savings depend on input voltage, it is necessary to define the input voltage in order to satisfy the minimum luminous flux level allowed. According to Bacelar [52], the minimum luminous flux level should be established at 50%, because it was shown that this dimming does not seem to have a great influence to the visibility of observers nor drivers. Furthermore this minimum level coincides with the current standard [5]. Following the recommendations of General Electric [53], the minimum voltage regarding the kind of lamp is shown in Table 12.

Analyzing in detail the research conducted by Yan [50], who studied the characteristics of HPS lamps of 50, 70, 100, 150, 250 and 400 W dimming the voltage. It can be observed that the percentage of light output decreases more than 50% for 180 Vac in the case of HPS and MH. Figure 4 shows the percentage of light output for the case of 50 and 70 W HPS lamps.

Table 12. Minimum voltage regarding the kind of lamp according to GE [53].

Kind of lamp	Minimum voltage (Vac)
HPM	200
HPS	180
LPS	190
MH	180

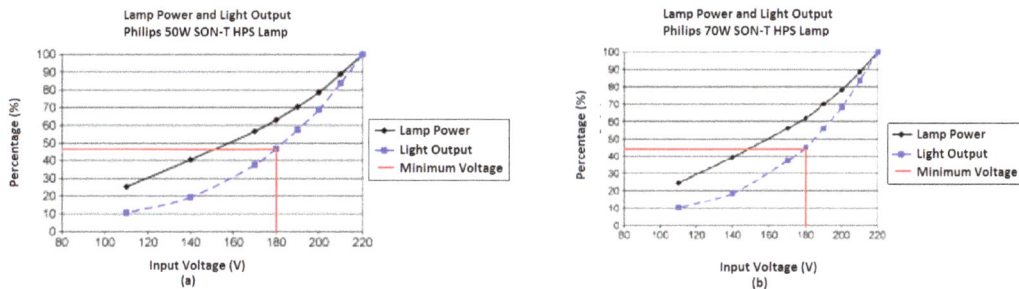

Figure 4. Lamp power, light output and minimum voltage for HPS lamps (50 and 70 W).

Therefore the minimum voltages showed in Table 12 are not completely right because they do not satisfy the minimum requests impose by the current standard. In our opinion the minimum voltage for each kind of lamp should be the values shown in Table 13.

Table 13. Minimum voltage to decrease the light output 50%.

Kind of lamp	Minimum voltage (Vac)	Decrease luminosity flux (%)
HPM	200	30%
HPS	190	50%
LPS	190	10%
MH	190	50%

From our point of view, the unique shortcoming of Spanish standard [5] regarding lighting level control devices is that it does not specify when it can be used. If we followed the recommendations of the Dutch ministry, dimmable road lighting systems could operate at 20% when the density of traffic at night is low, at 100% when the traffic density is high and 200% when there is a combination of high traffic density and exceptional conditions such as fog. The conclusions were that 20% light level has no negative safety effects and is sufficient for low traffic density but 200% light level is not justified because the cost is high and the safety improvements are marginal at best [54]. Another project [55] also investigated the effect of dimming, the lighting level setting were determined as follows; 100% when there are more than 3000 vehicles per hour, 75% when the range of vehicles is 3000–1500 and 50% when the number of vehicles per hour is lower than 1500. Following both projects and observing the behavior of Spanish roads, Figure 5 shows the number of vehicles per hour of a road in the Community of Murcia.

406

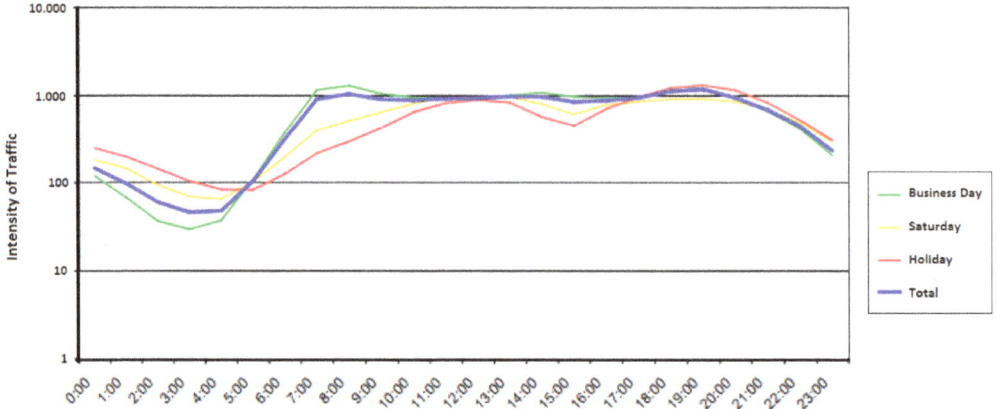

Figure 5. Number of vehicles per hour in a road from the Community of Murcia [56].

As it can be seen, lighting level control devices can operate perfectly from 1:00 am to 5:00 am, because the number of vehicles decreases considerably. In this aspect, the Croatian normative [57] specifies that if the local government does not prescribe a schedule, the street lighting must be turned off or reduced by 50% at least at 1:00 am. In our opinion, it should be mandatory within the Spanish normative that lighting level control devices reduce the light levels at least from 1:00 am, because most of the time the conditions allow it.

3.6. Renewable Energies

The global necessity for energy savings requires the usage of renewable sources in many applications and outdoor lighting installations are no exception. Spain, owing to its location and climate, is one of the countries in Europe with the most abundant solar resources [58]. Global solar irradiation on a horizontal plane is estimated as being between 1.48 and 3.56 kW/m^2 day in Spain.

The solar energy option may be the best solution in the case of an autonomous street lighting system because of the long life time, easy installation and modularity [59]. This sort of renewable energy allows reducing the CO_2 emissions considerably and thus the energy consumption. A good example of the benefits of solar energy in street lighting is the research carried out by Nunoo [60], who achieved energy savings per day of 603 kWh. Analysing in detail the research carried out by Constantinos [61], who optimized a photovoltaic system for street lighting, the total autonomous days of operation may reach up to 315 per year. In other words, in this case the energy savings were about 86%.

On the other hand, maintenance of the photovoltaic panels is very important, because dust effects reduce the performance of solar panels. The research carried out by Al-Almmri [62] shows that the losses of the output power of the fixed solar panel can reach 26% for one month. As well, their orientation can cause a considerable loss of efficiency. Likewise, the slope of the panel should be changed two to four times a year to maximize the solar absorption, since the optimum slope in the summer is not the same as the optimum one in the winter [63]. These drawbacks can be solved with regular maintenance.

Outdoor lighting can be supplied with other kinds of renewable sources or even a combination of several types of renewable sources like the research performed by Al-Fatlawi [64], who combined solar and wind energy. Power systems which include photovoltaic systems and wind turbines typically include energy storage devices so that loads can be operated when solar energy is not available or when wind velocities are too low to generate power [65].

Nowadays, renewable energies are indispensable to satisfy the normative for buildings, however the Royal Decree [5] overlooks this subject in the field of street lighting. Previous research shows that the incorporation of solar energy for street lighting is an incredible opportunity to reduce energy consumption and improve the quality.

4. Conclusions

Following the completion of this paper, this study has shown some aspects that they should be incorporated into the Spanish Standards to improve the quality of street lighting. The related work allows us to know that there are other important aspects that the Spanish normative does not contemplate, such as the minimum distance between luminaires and trees, or the minimum distance between luminaires and fire hydrants. These recommendations could be considered irrelevant but any step forward makes headway.

Regarding lamps, white light is a new concept that benefits when lamps have a color rendering index higher than 60. The incorporation of this subject could reduce the illuminance level at least 25% for subsidiary roads. This advantage has been incorporated within the British Standard and in Hong Kong, now it is the time for Spain. Furthermore, we have noted that the British Standard considers two more lighting classes than the Spanish Standard.

In relation to ballasts has been corroborated that electronic ballasts consume less energy than electromagnetic ones. Although this power saving may be considered insignificant, the example analyzed obtained a power savings of 18% regarding the luminaire power with electromagnetic ballast. The weakness of the Spanish standard is that it only specifies the maximum luminaire power. Hence designers must take into account the choice of the kind of ballast because although the maximum power is defined, it is very easy exceed the maximum luminaire power value.

Concerning light pollution, Spain is not very strict and should be more rigorous. The simulations done with DIALux verify that ULR bears upon the illuminance and therefore if the Spanish regulation were stricter regarding light pollution, street lighting systems would improve in quality.

Related to hours of operation each device works using a different technology and therefore the hours of operation are different in each device. Moreover, it has been corroborated thanks to the measures of natural light using a digital illuminance meter that weather conditions are even more significant than latitude in determining days. The recommended trimmings have been corroborated and photocell trimming may save 24 h per year. This action allows one to decrease the energy consumption while maintaining good service.

Finally, the benefits of lighting level control devices is shown, while on the other hand it is required to be careful with the input voltage value because if the trimming is too low, the illuminance would not satisfy the minimum requirements and could affect the visibility of drivers and observers.

408

Moreover 1:00 am to 5:00 am was defined from as the best period to use them. We wish to highlight that the Spanish normative should encourage the use of renewable energies for street lighting.

Acknowledgments

The authors want to thank the effort and the support that the Ferrovial Company deposited in the Department of Computer Sciences for the Ciudad 2020 project. Besides, we are grateful to PCE Iberica S.L. for the donation of the digital illuminance meter (PCE-174) and acknowledge the help of the people who revised the English language of this manuscript.

Author Contributions

Alberto Gutierrez-Escolar has contributed to the sections on lamps and renewable energies, Ana Castillo-Martinez has developed the ballast section, Jose Maria Gutierrez-Martinez has developed the hours of operations section and Jose M. Gomez-Pulido has obtained the natural light data and finally Zlatko Stapic has been the person in charge of finding the differences between the Spanish Standard and the rest of the regulations. Jose-Amelio Medina-Merodio has contributed to the section on street lamps globes. All the authors were involved in preparing the manuscript.

Nomenclature

CRI	Color Rendering Index
HPM	High Pressure Mercury
HPS	High Pressure Sodium
LPS	Low Pressure Sodium
LED	Light-Emitting Diode
MH	Metal Halide
R.D.	Royal Decree
ULR	Upward Light Ratio

Conflicts of Interest

The authors declare no conflict of interest.

References

1. New York State Energy Research and Development Authority (NYSERDA). How to Guide to Effective Energy-Efficient Street Lighting. Available online: http://www.rpi.edu/dept/lrc/nystreet/how-to-officials.pdf (accessed on 5 January 2015).
2. Elejoste, P.; Angulo Perallos, A.; Chertudi, A.; Zuazola, I.J.G.; Moreno, A.; Azpilicueta, L.; Astrain, J.J.; Falcone, F.; Villadangos, J. An easy to deploy street light control system based on wireless communication and LED technology. *Sensors* **2013**, *13*, 6492–6523.

3. Sanchez de Miguel, A. Differential photometry study of the European Light Emission to the space. In Proceedings of the World Conference in Defence of the Night Sky and the Right to Observe the Stars, La Palma, Spain, 20–23 April 2007; pp. 379–383.

4. Van Tichelen, P.; Geerken, T.; Jansen, M.; vanden Bosch, M.; van Hoof, V.; vanhooydonck, L.; Vercalsteren, A. Final Report Lot 9: Public Street Lighting. Available online: http://amper.ped. muni.cz/jhollan/light/EuP/VITOEuPStreetLightingFinal.pdf (accessed on 5 January 2015).

5. Royal Decree 1890/2008. Regulation in outdoor lighting installations and their complementary instructions EA-01 and EA-07. Available online: https://www.boe.es/boe/dias/2008/11/19/pdfs/ A45988-46057.pdf (accessed on 5 January 2015).

6. *Estrategia de ahorro y eficiencia energética en España*; Ministerio de Industria, Turismo y Comercio: Madrid, Spain, 2011. (In Spanish)

7. Sanchez de Miguel, A.; Zamorano, J.; Gomez Castaño, J.; Pascual, S. Evolution of the energy consumed by street lighting in Spain estimated with DMSP-OLS data. *J. Quant. Spectrosc. Radiat. Transf.* **2014**, *139*, 109–117.

8. Agencia Andaluza de la Energía. Guía de Ahorro y Eficiencia Energética en Municipios. Available online: http://www.agenciaandaluzadelaenergia.es/sites/default/files/guia_de_ ahorro_y_eficiencia_energxtica_web_def1.pdf (accessed on 5 January 2015). (In Spanish)

9. European Commission. Green Public Procurement Street Lighting and Traffic Lights Technical Background Report. European Commission, DG Environment-C1, BU 9, 1160 Brussels, 2011. Available online: http://ec.europa.eu/environment/gpp/pdf/tbr/street_lighting_tbr.pdf (accessed on 5 January 2015).

10. Bizjak, G.; Kobav, M.B. Consumption of electrical energy for public lighting in Slovenia. In Proceedings of the 5th ILUMINAT, Cluj-Napoca, Romania, 20 February 2009.

11. Silva, J.; Mendes, J.F.; Silva, L.T. Assessment of energy efficiency in street lighting design. *WIT Trans. Ecol. Environ.* **2010**, *129*, 705–715.

12. Pracki, P. A proposal to classify road lighting energy efficiency. *Light. Res. Technol.* **2011**, *43*, 271–280.

13. Bundeswetbewerb. Energieeffiziente Stadtbeleuchtung. Available online: http://www. bundeswettbewerb-stadtbeleuchtung.de/ (accessed on 5 January 2015).

14. Kyba, C.C.M.; Hänel, A.; Hölker, F. Redefining efficiency for outdoor lighting. *Energy Environ. Sci.* **2014**, *7*, 1806–1809.

15. Stockmar, A. Energy efficiency measures for outdoor lighting. *Light Eng.* **2011**, *19*, 15.

16. Ministerio de Minas y Energía. Anexo General. *Reglamento Técnico de Iluminación y Alumbrado Público*; RETILAP: Colombia, 2010. Available online: http://www.minminas. gov.co/minminas/downloads/archivosSoporteRevistas/7853.pdf (accessed on 5 January 2015). (In Spanish)

17. The Government of the Hong Kong Special Administrative Region. Public Lighting Design Manual. Available online: http://www.oshc.org.hk/others/bookshelf/WB112003E.pdf (accessed on 5 January 2015).

18. *Roadway Lighting Design Manual*; Minnesota Department of Transportation: Saint Paul, MN, USA, May 2006.

19. Boyce, P.R.; Fotios, S.; Richards, M. Road lighting and energy saving. *Light. Res. Technol.* **2009**, *41*, 245–260.

20. Simpson, R.S. *Lighting Control: Technology and Applications*; Focal Press: Oxford, UK, 2003.

21. British Standard Institution (BSI). *BS 5489-1:2003, Code of Practice for Design of Road Lighting—Part 1: Lighting of Roads and Public Amenity Areas*; BSI: London, UK, 2003.

22. Fotios, S.; Goodman, T. Proposed UK guidance for lighting in residential roads. *Light. Res. Technol.* **2012**, *44*, 69–83.

23. Australian Capital Territory Government (ACT) Crime Prevention & Urban Design. Resource Manual. Available online: http://apps.actpla.act.gov.au/tplan/planning_register/register_docs/resmanual.pdf (accessed on 5 January 2015).

24. Bridger, G.; King, B. Lighting the way to road safety: A policy blindspot? In Proceedings of the Australian Road Safety Research Policing Education Conference, Wellington, New Zealand, 4–6 October 2012.

25. Lewin, I.; Box, P.C.; Stark, R.E. Roadway Lighting: An Investigation and Evaluation of Three Different Light Sources. Available online: http://ntl.bts.gov/lib/24000/24600/24606/AZ522.pdf (accessed on 5 January 2015).

26. Murphy, J.T.W. Maximum spectral luminous efficacy of white light. *J. Appl. Phys.* **2012**, *111*, 104909.

27. Mullner, R.; Riener, A. An energy efficient pedestrian aware Smart Street Lighting system. *Int. J. Pervasive Comput. Communi.* **2011**, *7*, 147–161.

28. Chung, H.H.; Ho, N.M.; Yan, W.; Tam, P.W.; Hui, S.Y. Comparison of dimmable electromagnetic and electronic ballast systems—an assessment on energy efficiency and lifetime. *IEEE Trans. Ind. Electron.* **2007**, *54*, 3145–3154.

29. Gil-de-Castro, A.; Moreno-Munoz, A.; de la Rosa, J.J.G.; Arias, J.F.; Pallares-Lopez, V. Study of harmonic generated by electromagnetic and electronic ballast used in Street Lighting. In Proceedings of the International Symposium on Industrial Electronics (ISIE), Gdansk, Poland, 27–30 June 2011; pp. 425–430.

30. Gil-de-Castro, A.; Moreno-Munoz, A.; Larsson, A.; de la Rosa, J.J.G.; Bollen, M.H.J. LED street lighting: A power quality comparison among street light technologies. *Light. Res. Technol.* **2013**, *45*, 710–728.

31. Philips. Available online: http://www.ecat.lighting.philips.es/l/ (accessed on 5 January 2015).

32. Omar, M.H.; Rahman, H.A.; Majid, M.S.; Rosmin, N.; Hassan, M.Y.; Omar, W.W. Design and simulation of electronic ballast performance for high pressure sodium street lighting. *Light. Res. Technol.* **2013**, *45*, 729–739.

33. Dolara, A.; Faranda, R.; Guzzetti, S.; Leva, S. Power quality in public lighting systems. In Proceedings of the 14th International Conference on Harmonics and Quality of Power (ICHQP), Bergamo, Italy, 26–29 September 2010; pp. 1–7.

34. Isobe, S.L.; Hamamura, S. Light pollution and its energy loss. *Astrophys. Space Sci.* **2000**, *273*, 289–294.

35. City of Helsinki. Kaupungin Valot: Helsingin Valaistuksen Kaupunkikuvalliset Periaatteet (In Finnish) City Lights: Urban Principles of the Lighting of Helsinki. City of Helsinki: Helsinki, Finland, 2003. Available online: http://www.hel.fi/static/rakvv/kaupungin_valot.pdf (accessed on 5 January 2015).

36. *Nacrt prijedloga uredbe o standardima upravljanja rasvijetljenošću s konačnim prijedlogom uredbe* (In Croatian). In *The Proposal of Act on Standards in Lightning Management with Final Proposal of the Act* (In English); Ministry of Environmental and Nature Protection: Zagreb, Croatia, May 2013.

37. D.S. N° 686/98 Norma de emisión para la regulación de la contaminación lumínica. Available online: http://www.vialidad.cl/areasdevialidad/medioambiente/Documents/Normativa/Normas %20de%20Calidad/DS86CONTAMINACIONLUMINICA.pdf (accessed on 5 January 2015).

38. Official Gazette of the Republic of Slovenia no. 81/2007. Available online: http://www.uradni-list.si/1/objava.jsp?urlid=200781&stevilka=4162%20 (accessed on 5 January 2015).

39. SunriseSunset. Available from: http://www.sunrisesunset.com (accessed on 5 January 2015).

40. Howell, E.K. Photoelectric controls for street lights. *Electr. Eng.* **1961**, *80*, 780–785.

41. Angus Council. Infrastructure service committee 19 April 2011, CHRISTMAS LIGHTING—PREPARATION FOR 2011 DISPLAYS; REPORT NO 292/11. Available online: http://archive.angus.gov.uk/ccmeetings/reports-committee2011/Infrastructure/292.pdf (accessed on 27 January 2014).

42. Institution of Lighting Professionals (ILP). Street Lighting—Invest to Save, reduction or Removal of Street Lighting—Interim Advice Note LB1. Available online: https://www.theilp. org.uk/documents/street-lighting-invest-to-save/ (accessed on 5 January 2015).

43. American Electric Lighting (AEL). Lighting system cost impacted by photocontrol choice. Available online: http://www.americanelectriclighting.com/products/dtlphotocontrol/ framework2_2.asp (accessed on 5 January 2015).

44. Baenziger, T.D. Effective lighting control system for public spaces. *Light Eng.* **2007**, *15*, 45–52.

45. Popa, M.; Cepisca, C. Energy consumption saving solutions based on intelligent street lighting control system, U.P.B. *Sci. Bull. Ser. C* **2011**, *73*, 297–308.

46. ORBIS. Available online: http://www.orbis.es/inicio.aspx?inPkyIdi=2 (accessed on 5 January 2015).

47. Blanquez, F.R.; Rebollo, E.; Blanquez, F.; Platero, C.A.; Frias, P. High efficiency voltage regulator and stabilizer for outdoor lighting installations. In Proceedings of the 13th International Conference on Optimization of Electrical and Electronic Equipment (OPTIM), Brasov, Romania, 24–26 May 2012; pp. 136–142.

48. Moghadam, M.H.; Mozayani, N. a street lighting control system based on holonic structures and traffic system. In Proceedings of the 3rd International Conference on Computer Research and Development (ICCRD), Shanghai, China, 11–13 March 2011; pp. 92–96.

49. Zotos, N.; Stergiopoulos, C.; Anastasopoulos, K.; Bogdos, G.; Pallis, E.; Skianis, C. Case study of a dimmable outdoor lighting system with intelligent management and remote control. In Proceedings of the International Conference on Telecommunications and Multimedia (TEMU), Chania, Crete, Greece, 30 July 2012; pp. 43–48.

50. Yan, W.; Hui, S.Y.R.; Chung, H.H. Energy saving of large-scale high-intensity-discharge lamp lighting networks using a central reactive power control system. *IEEE Trans. Ind. Electron.* **2009**, *56*, 3069–3078.

51. Chung, H.S.H.; Ho, N.M.; Hui, S.Y.R.; Mai, W.Z. Case study of a highly-reliable dimmable road lighting system with intelligent remote control. In Proceedings of the European Conference on Power Electronics and Application, Dresden, Germany, 11–14 September 2005.

52. Bacelar, A. The influence of dimming in road lighting on the visibility of drivers. *J. Light. Vis. Environ.* **2005**, *29*, 44–49.

53. GE Energy Industrial Solutions. Available online: http://www.gepowercontrols.com/es/resources/literature_library/catalogs/downloads/GRADILUX_cat_Spain_ed01-12_4659.pdf (accessed on 5 January 2015).

54. Van Hoek, K.T. Dutch approach to energy efficient street lighting. In Proceedings of the 8th European Lighting Conference Lux Europa, Amsterdam, The Netherland, 11–14 May 1997.

55. Collins, A.; Thurrell, T.; Pink, R.; Feather, J. Dynamic dimming the future of motorway lighting? *Light. J.* **2002**, *67*, 25.

56. Dirección General de Carreteras. Región de Murcia. Consejería de Obras Públicas y Ordenación del Territorio. Available online: http://www.carm.es/web/pagina?IDCONTENIDO=31421&IDTIPO=100&RASTRO=c399$m578,31401 (accessed on 5 January 2015).

57. NN 114/11 (Oficial Gazette) Zakon o zaštiti od svjetlosnog onečišćenja. Available online: http://www.zakon.hr/z/496/Zakon-o-za%C5%A1titi-od-svjetlosnog-one%C4%8Di%C5%A1%C4%87enja (accessed on 27 January 2014). (In Croatian)

58. Diez-Mediavilla, M.; Alonso-Tristan, C.; Rodríguez-Amigo, M.C.; García-Calderón, T. Implementation of PV plants in Spain: A case study. *Renew. Sustain. Energy Rev.* **2010**, *14*, 1342–1346.

59. Costa, M.A.D.; Costa, G.H.; dos Santos, A.S.; Schuch, L.; Pinheiro, J.R. A high efficiency street lighting system based on solar energy and LEDs. In Proceedings of the Power Electronics Conference (COBEP), Bonito, Brazil, 27 September–1 October 2009.

60. Nunoo, S.; Attachie, J.C.; Abraham, C.K. Using solar power as an alternative source of electrical energy for street lighting in Ghana. In Proceedings of the Innovative Technologies for an Efficient and Reliable Electricity Supply (CITRES), Waltham, MA, USA, 27–29 September 2010.

61. Bouroussis, C.A.; Topalis, F.V. Optimization of potential and autonomy of a photovoltaic system for Street lighting. *WSEAS Trans. Circuits .Syst.* **2004**, *3*, 1392–1397.

62. Al-Ammri, A.S.; Ghazi, A.; Mustafa, F. Dust effects on the performance of PV street light in Baghdad city. In Proceedings in Renewable and Sustainable Energy Conference (IRSEC), Ouarzazate, Morocco, 7–9 March 2013.

63. Fernández-Infantes, A.; Contreras, J.; Bernal-Agustín, J.L. Design of grid connected PV systems considering electrical, economical and environmental aspects: A practical case. *Renew. Energy* **2006**, *31*, 2042–2062.

64. Wadi Abbas Al-Fatlawi, A.; Abdul-Hakim, S.R.; Ward T.A.; Rahim, N.A. Technical and economic analysis of renewable energy powered standalone pole street lights for remote area. *Environ. Prog. Sustain. Energy* **2014**, *33*, 283–289.

65. Sperber, A.N.; Elmore, A.C.; Crow, M.L.; Cawlfield, J.D. Performance evaluation of energy efficient lighting associated with renewable energy applications. *Renew. Energy* **2012**, *44*, 423–430.

The Future of Solar Power in the United Kingdom

Gerard Reid and Gerard Wynn

Abstract: We used detailed industry data to analyse the impacts of expected further cost reductions on the competitiveness of solar power in Britain, and assess whether the solar market can survive without support in the near future. We investigated three solar power markets: large-scale, ground-mounted "solar farms" (defined in our analysis as larger than a 5000 kilowatt system); commercial roof-top (250 kW); and residential rooftop (3 kW). We found that all three would be economic without support in the next decade. Such an outcome assumes progressively falling support under a stable policy regime. We found that unsubsidised residential solar power may be cheaper with battery storage within the next five to 10 years. Unsupported domestic solar battery packs achieve payback periods of less than 10 years by 2025. That could create an inflexion point driving adoption of domestic solar systems. The variability of solar power will involve some grid integration costs at higher penetration levels, such as more frequent power market scheduling; more interconnector capacity; storage; and backup power. These costs and responses could be weighed against non-market benefits including the potential for grid balancing; lower carbon and particulate emissions; and energy security.

Reprinted from *Energies*. Cite as: Reid, G.; Wynn, G. The Future of Solar Power in the United Kingdom. *Energies* **2015**, *8*, 7818-7832.

1. Introduction

To date, European countries have supported the growth of solar photovoltaic (PV), with the goals of cutting carbon emissions, boosting energy security and nurturing a clean technology sector. As these countries cut support, the industry may appear at a cross-roads. Evidence from rapid cost reductions and capacity growth suggests that solar power will prosper without support. The last few decades shows solar module costs have fallen by about 20% for every doubling in installed capacity [1]. Recent cost reductions have reduced the share of solar modules in full system costs. Further reductions will increasingly depend on other, so-called balance of system costs, or "soft costs". The rate of cost reductions may therefore fall.

Recent market growth shows the emergence of solar power as a serious global energy player. In the last 10 years, cumulative installed capacity has grown at an average rate of 49% annually [2]. In 2013, about 37 gigawatts (GW_p) of new PV capacity were added globally, bringing cumulative capacity to more than 135 GW_p. On the human scale, electricity is no longer generated exclusively by huge, centralised utilities, instead by hundreds of thousands or millions of households, with 1.5 million solar installations in Germany and more than 600,000 in Britain [3,4].

1.1. What is Grid Parity?

The three main configurations of solar PV are small-scale, residential rooftop; commercial rooftop; and large-scale, ground-mounted solar farms. Large-scale solar delivers electricity into the medium-voltage, transmission network. Once large-scale solar is competitive with wholesale power prices, called grid parity, it will be economic without government support. In this report, we use British government projections for wholesale power prices [5]. It is noted, however, that wholesale power prices may fall faster than these projections, as a result of more wind and solar power, or rise, depending on fossil fuel prices and energy technologies going forward. The notion of "support-free" large-scale solar may be less relevant in an increasingly regulated power market where all technologies are supported, as we are seeing in Britain. In this event, parity with gas may be the target.

Roof-top solar delivers electricity into the home or business, at the low-voltage, distribution end of the electric grid, called distributed generation. It is sometimes assumed that once roof-top solar is cheaper than residential power prices, it is cost-competitive without support. In fact, competitiveness depends on the proportion of solar power that households use ("self-consumption"); retail power prices; and the proportion that they feed into the grid instead. In Europe, households with roof-top solar presently consume about 30% of the solar power they generate, feeding the remainder back into the grid.

Self-consumption of solar power is already competitive in many countries without subsidy, compared with the alternative of using mains electricity. However, roof-top solar power is still more expensive than wholesale electricity prices. As a result, exporting surplus power to the grid is still not competitive without a supported, "export tariff", which is well above the wholesale power price. If solar users had to export power at wholesale power prices, roof-top installations would only be starting to break-even now in central and southern Europe [6].

Maximising self-consumption is therefore critical for subsidy-free, rooftop solar. Going forward, we see this issue being resolved by continuing cost reductions, and trends which drive self-consumption rates to well above 50% (see Section 3). These trends include smart energy devices in our homes, which coordinate home appliances with solar power generation, plus cost reductions in battery storage.

1.2. The UK Market

Cumulative solar PV capacity is already above 5,000 megawatts (MW_p) [7], compared with total generating capacity in Britain of about 71,200 MW [8]. Britain's Department of Energy and Climate Change (DECC) has estimated the cost of large-scale electricity generation for different technologies, commissioned from 2015–2030 [9]. The study uses both assumed, average costs of capital, and technology-specific costs of capital, the latter taking into account factors such as construction time and planning permitting risk. Where technology-specific costs of capital are used, the study found that large-scale solar in Britain is already cheaper than offshore wind power; is in the same ballpark as nuclear; and will be able to compete with gas and onshore wind by 2025 [9]. Meanwhile, the rooftop market in Britain is nearing cost-competitiveness with domestic mains electricity, as solar costs fall and residential power prices rise.

Germany now provides a possible glimpse of Britain's electric power system in 2020. Solar photovoltaic (PV) power accounts for about 7% of the country's final electricity demand. Solar accounts for most peak demand in summer, and as much as half of all electricity demand on summer weekends. It has up-ended power markets, pushing wholesale power prices lower. Having zero fuel costs and a guaranteed right to sell power into the grid, it can displace gas and coal-fired power, leading even to negative wholesale power prices, and threatening utility profits.

1.3. Comparisons between Britain and Germany

Germany is a good benchmark for Britain, given its similar energy mix (fossil fuels, nuclear and renewables); standard of living; level of power demand; and solar irradiance [10]. The big difference at present is that Germany is the world's biggest market for solar, with an installed capacity of some 37.2 gigawatts across about 1.5 million installations [3]. Britain, in contrast, has about 5 GW_p installed across 0.6 million installations [4,7]. In 2014, solar power accounted for nearly 7% of total final electricity consumption in Germany [11], compared with 1.3% in Britain [12].

As the British solar market develops we expect it to go through many of the changes seen in Germany, including growing competitiveness across the solar value chain. Solar installation costs are lower in Germany than the UK because of greater efficiency, particularly in financing but also in development and installation. We believe that German and UK full installed solar prices will converge over the next years. Lower German costs are reflected in differences in feed-in tariffs and installation costs for rooftop solar. In Britain, the support for solar power generation by 0–4 kW systems is 14.38 pence per kWh for 20 years, plus inflation, plus an export tariff of 4.77 pence [13]. The German feed-in tariff for small systems is 10.1 pence (12.69 euro cents) per kWh, over the same period. The differences can also be seen in the installation costs, which were £1,580 per kW in Q1 2014 in Britain, compared with £1,310 (€1,640) per kW in Germany [14,15]. Germany's Fraunhofer Institute calculated the most cost-efficient solar farms were now competitive with onshore wind and well ahead of offshore wind [16].

1.4. Cost Calculation: Levelised Cost of Energy (LCOE) and Payback Periods

One common measure of the cost of generating solar power is the levelised cost of energy (LCOE), which divides the lifetime cost of a solar installation by lifetime power generation, measured in pence per kilowatt hour (kWh). For the sake of simplicity, LCOE excludes important costs, such as grid integration; waste disposal; and pollution. LCOE is a useful way to account for important factors such as capacity factor and the weighted average cost of capital (WACC), two critical variables.

Capacity factor is the actual output of a power plant as a percentage of its theoretical maximum. In the case of solar, it will take into account local solar irradiance and day length. In very sunny countries, such as Australia, solar panel load factors can reach 30% or more. Britain's Department of Energy and Climate Change (DECC) calculated an average capacity factor for solar PV in Britain of

10.3% in 2013 [17]. Weighted cost of capital (WACC) reflects the average cost of financing for a project. WACC will be higher for less mature technologies, because investors require a higher return on equity to compensate for the higher risk. The WACC is used to discount future cash flows, and so critically affects the cost calculation. British government estimates for large-scale solar LCOE illustrate the point. Using a 10% WACC across all energy technologies, DECC ranks solar costs higher than wind, nuclear and gas. However, using a lower, technology-specific WACC for large-scale solar power of 6.2%, reflecting the maturity of the technology and speed of construction, DECC ranked large-scale solar as the cheapest form of UK power generation before 2025 [9].

Most rooftop solar consumers assess solar investments in terms of payback periods, rather than LCOE. As a result, we use LCOE as a measure for the economics of large-scale, ground-mounted solar, and payback periods for the economics of commercial and residential rooftop solar power. The payback period is defined as the length of time it takes to recoup the upfront investment, based on annual savings as a result of reduced utility bills. We assumed steadily rising domestic power prices, using the latest DECC projections; rising self-consumption rates; and we discounted the revenues and expenses according to a discount rate or WACC (see Section 5. Methodology). We expect that most customers would require payback periods around 10 years or below before considering an investment. Payback periods were calculated using the same approach as for LCOE, including estimates for cost reductions in solar hardware and balance of systems over the next decade.

Increasing the self-consumption rate is critical for the economics of unsubsidised residential systems, as described above. With higher self-consumption, households avoid selling surpluses at a very low wholesale power price (presently about 5 pence per kWh in Britain), and buying mains electricity at much higher retail power prices (about 16 pence). While self-consumption rates in Europe are presently about 30%, home management systems are emerging which can boost these to 45%. As support is withdrawn, the incentive for self-consumption will rise. We assume steadily rising self-consumption rates (see Section 5. Methodology). Critically, battery storage increases self-consumption above 50%, and may therefore be the cornerstone of unsubsidised residential systems.

2. UK Solar Economics

2.1. Solar Module Selling Costs and Prices

Solar module prices have fallen sharply over the past four decades. Solar module cost reductions are driven by a combination of innovation in the efficiency of material use; light conversion; and production. Regarding light conversion efficiencies, for example, U.S.-based First Solar expects to reach efficiencies of 19.5% in 2017, referring to its Cadmium Telluride (CdTe) thin film cells, from 13% in 2013 [18]. Such numbers refer to the proportion of light energy striking a solar module that is converted to electricity. Recent cost reductions additionally reflect global commoditization of solar cells and modules, and in particular a ramp-up of manufacturing capacity in China, leading to global surpluses. A decade ago, solar panel (module) prices were as high as £4.00/Watt and the global market for solar was 500 megawatts (MW_p) installed per year. Today modules prices are well below £0.40/Watt and the global market in 2014 is expected to be over 40 GW_p.

Our own predictions, based on in-depth conversations with manufacturers, suggest best-in-class module costs falling from £0.32/Watt in 2014 to £0.20/Watt in 2020. Figure 1 shows our expectations for changes in module production costs from 2014 to 2020, taking into account small cost increases expected in operation and materials, more than offset by savings as a result of innovation in manufacturing and economies of scale.

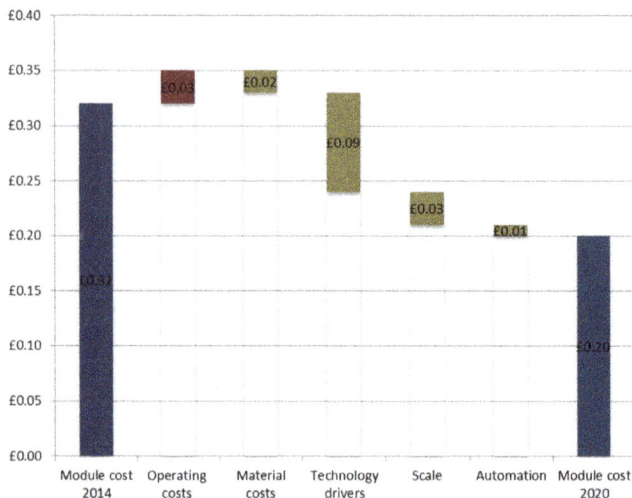

Figure 1. Module production cost reductions, 2014–2020. Sources: First Solar, and unpublished industry cost maps.

Full solar system costs may not maintain the same pace of reductions as seen in the past five years. That is because the swiftest reductions have come from solar modules, which now account for a smaller share of the total. The remaining, so-called balance of system costs, include inverters, installation and financing. Inverters convert direct current electricity generated by solar modules into alternating current required by many machines and household appliances. Inverter costs are continuing to fall, and Britain will in addition benefit from continuing reductions in installation and financing, as the supply chain matures. However, these cost reductions may be more gradual. Meanwhile, we expect soft costs such as installation and financing to fall, as discussed above, as Britain converges with other more mature markets such as Germany.

We see the pace of full system cost reductions in Britain moderating for the rest of this decade, compared with the previous five years. Nevertheless, we still expect full installed costs to fall by about one third between now and 2020 (see Figure 2). This is slightly more ambitious than some estimates. For example, the International Energy Agency recently estimated that global average full installed solar costs (including equipment, labour and financing) would halve by 2040 or sooner [2].

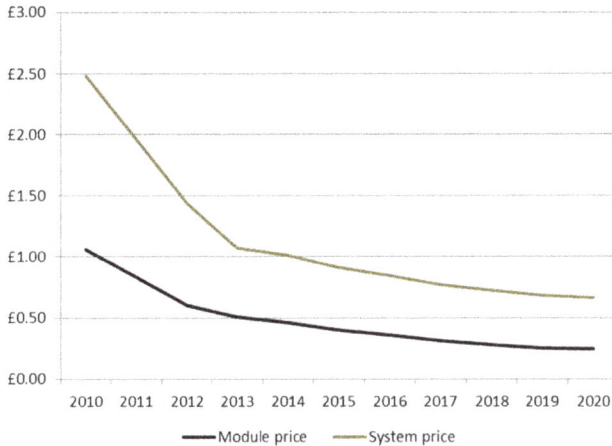

Figure 2. Full installed costs, UK ground-mounted systems, 2010–2020. Source: Our research, industry experts.

2.2. Solar Battery Pack Economics

Various potential remedies exist for the variability of solar power. Battery storage is one of these, where solar battery pack products are now emerging. Storage solutions available today are expensive. Electricity must be converted into another form of energy and then converted back into electrical energy. Lithium-ion is one promising battery storage technology currently under development. Lithium ion battery packs are still costly, at around £320/kWh [19]. Battery costs are falling, however, partly as a result of production and innovation in the automotive sector. With its planned "gigafactory", Tesla Motors believe that their battery packs could reach £100–130/kWh in 2020 [20,21]. See Figure 3 for our projection of battery pack costs, taking into account published Tesla projections and our unpublished interviews with the German battery developer, Younicos.

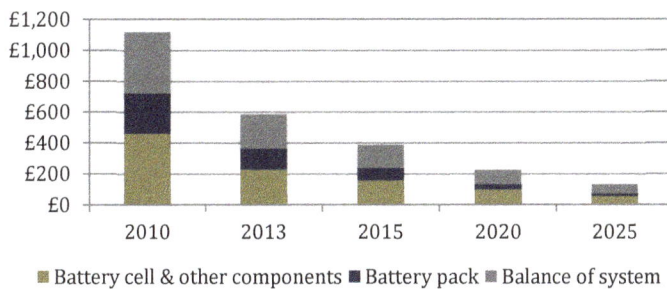

Figure 3. Battery pack production cost reductions, 2010–2020 (£/kWh). Sources: Published Tesla cost data; conversations with industry experts, including Younicos.

At present in Germany, the problem for unsubsidised solar is the very low wholesale power price at which solar surpluses must be sold into the grid. The Swiss investment bank, UBS, last year calculated that unsupported rooftop solar in southern Germany already breaks even (defined as total annual electricity costs with and without solar panels), assuming a grid export price of 3 cents, and 30% self-consumption [6]. The regulated, "subsidised" export price at present is up to 12.69 euro cents per kilowatt hour, in Germany, compared with domestic power prices of about 29 cents, and spot wholesale power prices of about 3 cents.

Without supported grid export prices, it becomes critical to maximise self-consumption. Households can first change their behaviour by using more self-generated electricity in the daytime, called load shifting. Leading global inverter manufacturer SMA Solar has developed software which matches the operation of household appliances and heating systems with forecast home solar output, through radio-controlled switches. This can increase self-consumption to 45%, the company estimates [22]. Batteries can make a bigger difference. Households can use deliberately small battery packs, minimising extra costs, and extend home-generated solar power past sunset, and increase self-consumption beyond 50% (see Section 5). Unsupported residential battery-pack solar PV systems are already becoming a cost-effective option in Germany, Italy and Spain, according to UBS. UBS sees a particular benefit from combining solar PV with a static battery, plus an electric vehicle (EV). That is because of a natural fit, where the static (non-EV) battery would mop up surplus daytime solar supply, and use this to charge the EV battery at night. In Germany, unsubsidised solar battery/ EV packages would deliver a return on investment of more than 7% by 2020, compared with a conventional car and no solar panels, according to UBS [21].

2.3. Cost Trajectories: Fossil Fuels

Gas is what is called the "marginal provider" in Britain, meaning that power prices are determined most of the time by gas plants, as opposed to much of the continent where it is determined by coal and power prices in neighbouring countries. Depending on the cost trajectory for advanced turbine design, gas-fired power may have more limited scope for reductions, given that the largest cost element in terms of its LCOE is fuel cost (the natural gas price), which is both difficult to predict and hedge. There are huge differences in global gas prices, with Japan (in 2013) paying on average $17 per Mbtu as opposed to $10 in Europe and $3 in the US. The major reason for this is the difficulty and high cost of transporting gas (see Figure 4). In Britain, domestic resources are dwindling, with little hope of UK shale gas coming online for another decade, meaning that other, more expensive sources need to be found.

Partly, as a result of expected rises in gas prices, as well as the growing cost of support for environmental policies and grid network upgrades, the Department of Energy and Climate Change (DECC) projects rising British residential power prices for the rest of this decade and beyond [5]. If British gas and power prices rise, it will become easier for solar PV to compete, and market penetration will continue to grow.

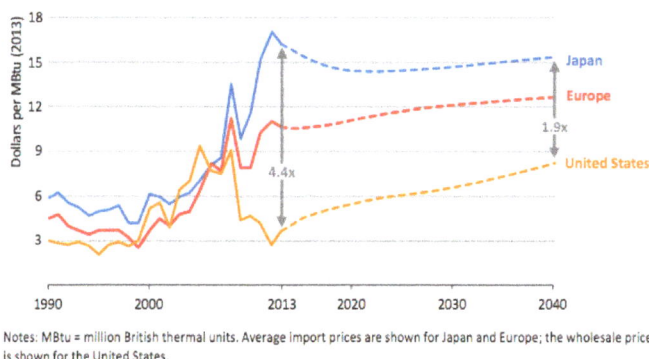

Notes: MBtu = million British thermal units. Average import prices are shown for Japan and Europe; the wholesale price is shown for the United States.

Figure 4. Natural gas price by region in the New Policies Scenario, World Energy Outlook 2014, IEA/OECD. Source: International Energy Agency (2014) [23]; Based on IEA data from IEA WEO 2014 © OECD/IEA, IEA Publishing; modified by GWG Energy and Alexa Capital. Licence: http://www.iea.org/t&c/termsandconditions/.

3. Projected Solar Costs: Our Findings

We conducted an analysis of projected costs of solar hardware and balance of systems. Underlying cost data were based both on published commercial projections, and interviews with multiple installers, developers and manufacturers in the industry. The analysis enabled us to make estimates for cost, assuming no government support, across different parts of the UK over the next 10 years (see Section 5. Methodology). Below we present the findings for the LCOEs of large-scale ground-mounted solar, and payback periods for commercial and residential rooftop solar.

3.1. Large-Scale Ground-Mounted Solar LCOEs

In the southern half of England, we estimate that large-scale solar farms will reach parity with onshore wind power in 2015, and full parity with fossil fuels and wholesale prices by 2025 at the latest (see Figure 5). These comparisons are with forecasts for rising fossil fuel and wholesale power prices, as projected by the UK government [5].

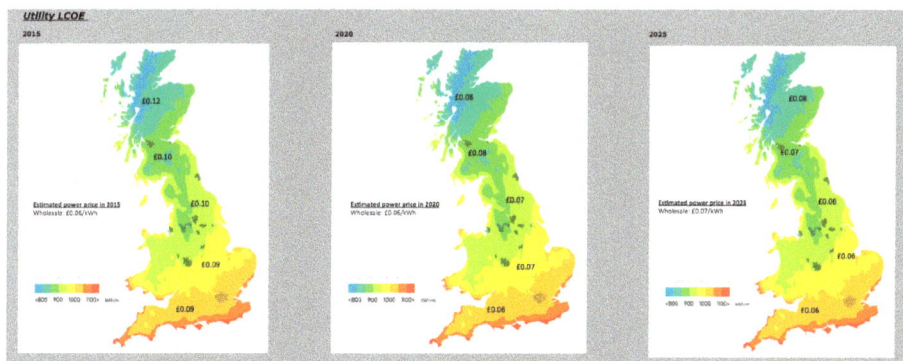

Figure 5. UK large-scale solar farm LCOEs, £/kWh, 2015–2025.

422

3.2. Commercial Rooftop Payback Periods

Our analysis finds that commercial rooftop solar systems reach payback periods of well below 15 years in southern England by 2020. These findings assume that commercial business can use 70% of the power produced (see Figure 6). Payback periods fall below 10 years across Britain more generally by 2025, using government assumptions for irradiation and power prices. These estimated payback periods could be a substantial driver for this market, depending on the minimum payback periods required to trigger investments by individual businesses.

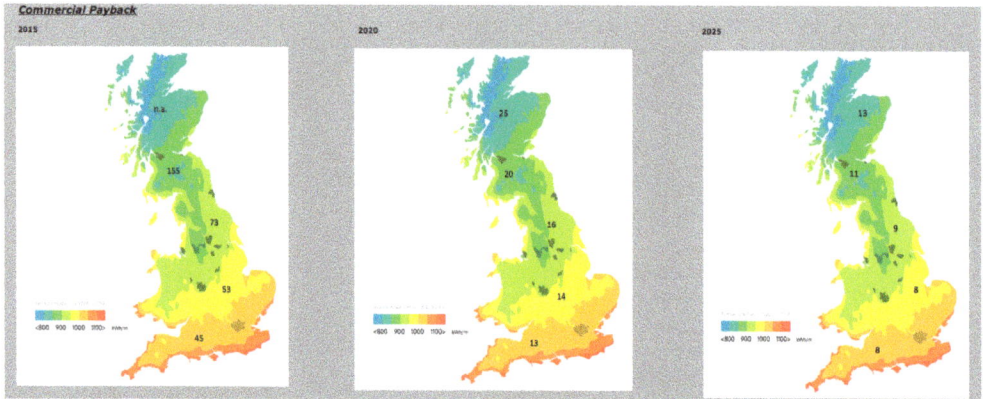

Figure 6. UK commercial rooftop payback periods, number of years, 2015–2025.

3.3. Residential Rooftop Payback Periods

By 2020, paybacks of 16 years are reached in southern England, under our various assumptions (see Figure 7). By 2025, payback periods are as low as eight years in southern England, and even in northern Scotland only 14 years. At these levels, residential solar may be viable without government help. These findings assume steadily rising consumption rates of home-generated solar power, and therefore bigger savings on avoided utility bills. If self-consumption remained at present rates of about 25%–30%, unsubsidised residential solar may struggle even in 2025 (see Table 1). On the other hand, a lower cost of capital would bring forward parity without support. We assumed much higher residential financing costs, for example, than the bottom end of the 1%–12% range used by Britain's National Audit Office [24].

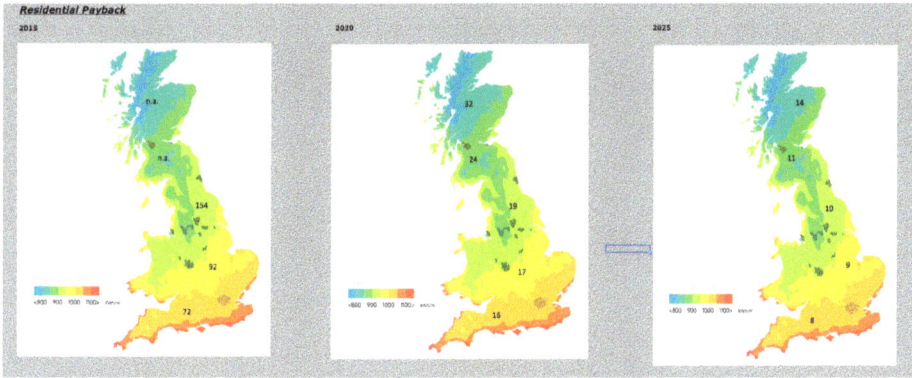

Figure 7. Residential rooftop payback periods, number of years, 2015–2025.

Table 1. UK average payback period for unsupported residential solar, according to self-consumption ratio, 2015–2025.

2015					
Self consumption ratio	100%	75%	50%	25%	0%
Payback period, years	19	29	69	(190)	(40)
2020					
Self consumption ratio	100%	75%	50%	25%	0%
Payback period, years	8	11	17	40	(106)
2025					
Self consumption ratio	100%	75%	50%	25%	0%
Payback period, years	6	7	10	18	70

3.4. Solar Battery Pack Payback Periods

Our analysis of the economics of solar battery pack systems suggests that these could achieve payback periods of below 15 years in 2020 (see Figure 8).

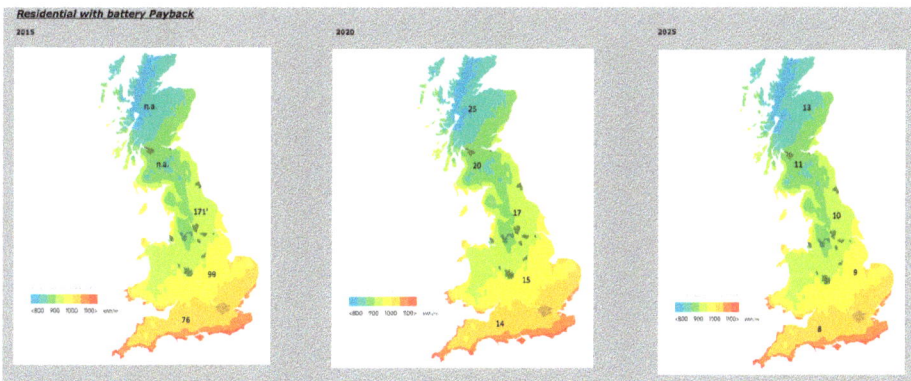

Figure 8. Projected solar battery pack payback periods, number of years, 2015–2025.

That is a little more competitive than our analysis of residential solar without batteries, reflecting higher self-consumption rates. Our modelling suggests payback periods of 10 years or less for solar battery pack systems across England and Wales by 2025, at which point financial support may no longer be needed.

4. Discussion

Solar PV is likely to be a critical technology in the 21st century. It is already a technology which is nearing maturity. At this stage, it makes sense that governments continue to support the industry until it is fully economic without subsidies. Progressive and predictable reductions in support over the next decade will help build a more mature, low-cost supply chain, while maintaining value for money and preventing developers from inflating prices. Getting the right support level is critical to driving sustained cost reductions.

The analysis presented in this study suggests that support for solar power in Britain can be cut progressively to zero over the next five to 10 years. The trick is finding the right balance, between driving efficiencies which create a new, low-margin business model, and killing a British solar industry which has export potential, particularly in finance and project development. Support for solar power to date has led to capacity increases which have cut costs as supply chains matured [25]. The International Energy Agency showed that Britain now has one of the most cost-effective markets for solar systems, with lower costs than major markets including the United States, Japan and France [2]. Public policy advisers such as the Global Commission on Economy and Climate have stressed that governments should reduce support for renewable energy progressively, but in a predictable way [26].

Policymakers could reduce the impact of solar support on domestic power prices, by shifting some price-based support towards alternatives such as low interest rate credit, and subsidies for batteries. Britain's Green Investment Bank, for example, has so far excluded solar power from loans of £1.6 billion for renewable energy. The government can support measures to optimise the grid integration of renewables, including the rollout of smart meters, and a more computerised grid, which uses digital technologies for faster, deeper, more responsive network communication and control. Near-term support for domestic solar battery packs, for example through grants or low-cost credit, could also aid a shift away from price-based support while preserving the economics of solar power.

This report shows a cost trajectory where both large-scale and rooftop solar will be able to survive increasingly without direct support over the next decade. Such a definition of support excludes wider policy measures which would indirectly benefit solar power, such as carbon pricing, or capacity payments for gas-fired power plants, which would support the grid integration of variable renewables. Such less visible support may be justified in the context of clear, un-priced and under-priced non-market benefits of solar power.

5. Methodology

Cost data were gleaned from a mix of interviews with leading companies and experts in the solar and battery space. Regarding projections, there is uncertainty about each of these factors and their values which can vary regionally and across time as weather patterns change and technologies evolve. Regarding technology change, and its impact on capital costs, the authors undertook unattributed interviews with five leading global manufacturers, and five solar developers operating in the UK. The results of these interviews were validated using the cost roadmaps of three listed companies: First Solar, Sun Edison and Trina Solar. Soft costs were based on interviews with leading installers. A similar process was followed to evaluate the prospects for commercially competitive batteries. The literature reveals that costs are falling, but with large uncertainties on past, current and future costs of the dominating Li-ion technology. In this study, interviews were conducted with the leading Asian manufacturers, on an unattributed basis. These interviews were validated with the publicly available cost roadmap of Tesla; third party research from investment banks such as UBS; and interviews with buyers of the battery technology, such as Germany's Younicos.

These data were used to derive cost estimates using two approaches. Levelised cost of electricity (LCOE) analysis was used to determine the cost of power generation by large-scale, ground-mounted solar systems. Payback periods were calculated for commercial and residential systems. LCOE is a standard measure used globally to compare the costs of differing generation technologies, also used by institutional investors to determine the valuation of generation assets. However, it is not used by the average household when making investment decisions. Householders often think in terms of how long it takes to recover their upfront investment. Payback periods are similarly appropriate for commercial businesses which also think in these terms.

Under LCOE analysis, the costs for constructing and operating a plant over its lifetime are summed and divided by the amount of power that the plant produces. The resulting LCOE is expressed in pounds per KWh. The key inputs to calculating the LCOE are: capital costs, operations and maintenance costs (O&M); financing costs; fuel costs; and an assumed utilization rate for each area. The latter was determined based on so-called isolation rates. This study used insolation data from the UK Met Office. Such data show the amount of electricity that can be generated by an optimally positioned 1kW rated PV solar panel. For example, in Leeds a 1kW solar module should produce 825 kWh of electricity in a typical year whereas a module in Aberdeen only 700 kWh. To calculate the net present value of costs, we used a weighted average cost of capital (WACC) to discount revenues and expenses, as calculated below. The discount rate (WACC) is very important, because the rate reflects the riskiness of the investment, and small changes can lead to relatively large changes in investment returns. For simplicity, we assumed a WACC of 6.2%, which is in line with the UK government's assumptions for large-scale solar power:

$$WACC = \left(\frac{Debt}{Total\ Capital}\right)k_d(1-t) + \left(\frac{Equity}{Total\ Capital}\right)k_s = (w_d)(k_d)(1-t) + (w_s)(k_s)$$

where w_d = Weight of debt proportion to total capital, w_s = Weight of equity proportion to total capital,
k_d = Cost of debt, k_s = Cost of equity and t = Corporate tax rate.

The study used payback periods for the economics of commercial and residential rooftop systems. The payback period is defined as the length of time it takes to recoup the upfront investment, based on annual savings as a result of reduced utility bills. To this it was assumed that consumer power prices would rise in line with the latest DECC projections (see assumed self-consumption rates below). Payback periods were calculated using the same analysis as for LCOE, including estimates for cost reductions in solar hardware and balance of systems over the next decade. In our LCOE and Payback Period calculations, we have assumed the following:

- Module prices for large scale systems of £0.46 pence today falling to £0.24 in 2020 and then £0.22 in 2025;
- Zero support for solar power generation; export tariff at the level of the wholesale power price;
- Load factors of 9.3% to 12.5%, depending on latitude;
- Increased efficiency across the whole solar value chain;
- Cost of debt: 4% and cost of equity 6% (German levels);
- Debt to equity ratios of 60:40 for large-scale and commercial solar, and 80:20 for residential;
- The latest (October 2014) UK projections for residential, commercial and wholesale power prices, from the Department of Energy and Climate Change (DECC) [27];
- For roof-top systems, self-consumption rates for commercial users of 70%; for residential systems, 32% in 2015 rising to 45% in 2025; and 55% for residential systems with batteries;
- Depreciation: 10 years for batteries, 25 years for solar panels.

Acknowledgments

Thanks to Ellis Acklin, financial consultant, for his analysis of the solar pricing and system data; to Stefano Ambrogi, for his guidance in originating this report; and Chloe Battle for her assistance in graphics production.

Author Contributions

Gerard Reid led data collection and modelling, and Gerard Wynn was responsible for literature review and drafting of the paper.

Conflicts of Interest

The authors declare no conflict of interest.

References

1. *Global Overview on Grid-Parity Event Dynamics.* Available online: https://www.q-cells.com/uploads/tx_abdownloads/files/11_GLOBAL_OVERVIEW_ON_GRID-PARITY_Paper_02.pdf (accessed on 30 November 2014).
2. International Energy Agency (IEA). *Technology Roadmap: Solar PV*; IEA: Paris, France, 2014. Available online: http://www.iea.org/publications/freepublications/publication/TechnologyRoadmapSolarPhotovoltaicEnergy_2014edition.pdf (accessed on 30 November 2014).
3. Fraunhofer Institute for Solar Energy Systems. *Photovoltaics Report* 24 October 2014. Available online: http://www.ise.fraunhofer.de/de/downloads/pdf-files/aktuelles/photovoltaics-report-in-englischer-sprache.pdf (these data are as of end-August 2014) (accessed on 30 November 2014).

4. Department for Energy and Climate Change (DECC). Solar Photovoltaics Deployment. DECC: London, UK. Available online: https://www.gov.uk/government/statistics/solar-photovoltaics-deployment (these data are as of end-October 2014) (accessed on 30 November 2014).

5. Department of Energy and Climate Change (DECC). Updated energy and emissions projections 2014: Growth assumptions and prices. 2014. Available online: https://www.gov.uk/government/publications/updated-energy-and-emissions-projections-2014 (accessed on 30 November 2014).

6. UBS Investment Research. The unsubsidized solar revolution. 2013. Available online: http://www.qualenergia.it/sites/default/files/articolo-doc/UBS.pdf (accessed on 30 November 2014).

7. Solarbuzz. UK Solar PV Deployment Reaches 5 GW. 2014. Available online: http://www.solarbuzz.com/resources/articles-and-presentations/uk-solar-pv-deployment-reaches-5-gw (accessed on 30 November 2014).

8. National Grid. Winter outlook 2014/15. 2014. Available online: http://www2.nationalgrid.com/UK/Industry-information/Future-of-Energy/FES/Winter-Outlook/ (accessed on 30 November 2014).

9. Department of Energy and Climate Change (DECC). Electricity Generation Costs: December 2013. 2013. Available online: https://www.gov.uk/government/uploads/system/uploads/attachment_data/file/269888/131217_Electricity_Generation_costs_report_December_2013_F inal.pdf (accessed on 30 November 2014).

10. Solar and Wind Energy Resource Assessment (SWERA). Renewable Energy Data Exploration. Available online: http://maps.nrel.gov/swera?visible=swera_dni_nasa_lo_res&opacity=50&extent=-74.01,-33.74,-29.84,5.27 (accessed on 5 November 2014).

11. Fraunhofer Institute for Solar Energy Systems. Recent Facts about Photovoltaics in Germany. Last update: 19 May 2015. Available online: http://www.ise.fraunhofer.de/en/publications/veroeffentlichungen-pdf-dateien-en/studien-und-konzeptpapiere/recent-facts-about-photovoltaics-in-germany.pdf (accessed on 20 July 2015).

12. Department of Energy and Climate Change. Energy Trends June 2015. 2015. Available online: https://www.gov.uk/government/uploads/system/uploads/attachment_data/file/437455/Energy_Tren ds_June_2015.pdf (accessed on 20 July 2015).

13. Energy Saving Trust. Feed-in tariff scheme. Available online: http://www.energysavingtrust.org.uk/domestic/content/feed-tariff-scheme (accessed on 30 November 2014).

14. DECC. Solar PV cost data. 2014. Available online: https://www.gov.uk/government/statistics/solar-pv-cost-data (accessed on 30 November 2014).

15. BSW. Statistikpapier "Photovoltaik". 2014. Available online: http://www.solarwirtschaft.de/unsere-themen-photovoltaik/zahlen-und-fakten.html (accessed on 30 November 2014).

16. Fraunhofer Institute for Solar Energy Systems ISE. Levelized cost of electricity: Renewable energy technologies. 2013. Available online: http://www.ise.fraunhofer.de/en/publications/veroeffentlichungen-pdf-dateien-en/studien-und-konzeptpapiere/study-levelized-cost-of-electricity-renewable-energies.pdf (accessed on 30 November 2014).

17. Department of Energy and Climate Change (DECC). Renewable Energy in 2013. DECC: London, UK, 2014. Available online: https://www.gov.uk/government/uploads/system/uploads/attachment_data/file/323429/Renewable_energy_in_2013.pdf (accessed on 20 July 2015).

18. Goldman Sachs. *Global Clean Energy Monthly: Spotlight on Continued Upside in the Solar Technology Roadmap, Plus Lower Cost per Watt*; Goldman Sachs: New York, NY, USA, 31 March 2014. Unpublished report.

19. Joint Research Centre. PV Status Report 2013. 2013. Available online: http://iet.jrc.ec.europa.eu/remea/pv-status-report-2013 (accessed on 30 November 2014).

20. Morgan Stanley. Solar power and energy storage: policy factors versus improving economics. UBS, 2014. Global Utilities, Autos and Chemicals: Will solar, batteries and electric cars re-shape the electricity system. UBS, Zurich. 2014. Available online: https://neo.ubs.com/shared/d1fXWW5AKk6 (accessed on 30 November 2014).

21. UBS Investment Research. Global Utilities, Autos and Chemicals: Will solar, batteries and electric cars re-shape the electricity system. 2014. Available online: https://neo.ubs.com/shared/d1fXWW5AKk6 (accessed on 30 November 2014).

22. SMA Solar. The Simplified Storage Solution. Available online: http://www.sma.de/en/products/solarinverters/sunny-boy-3600-5000-smart-energy.html (accessed on 20 July 2015).

23. International Energy Agency (IEA). World Energy Outlook 2014. IEA: Paris, France, 2014. Available online: https://www.iea.org/publications/freepublications/publication/WEO_2014_ES_English_WEB.pdf (accessed on 20 July 2015).

24. National Audit Office. The modelling used to set feed-in tariffs for solar photovoltaics. 2011. Available online: http://www.nao.org.uk/wp-content/uploads/2011/11/NAO_briefing_FiTs_Nov11.pdf (accessed on 20 July 2015).

25. UK Energy Research Centre. UKERC Technology and Policy Assessment Cost Methodologies Project: PV Case study. 2012. Available online: http://webcache.googleusercontent.com/search?q=cache:oJb6yQIqPhAJ:www.ukerc.ac.uk/asset/F7903594-3C61-4778-8216B4104E0A5765/+&cd=1&hl=en&ct=clnk&gl=uk (accessed on 20 July 2015).

26. Global Commission on Economy and Climate. Global Action Plan. In: Better Growth, Better Climate. 2014. Available online: http://static.newclimateeconomy.report/wp-content/uploads/2014/09/NCE_Chapter9_GlobalActionPlan.pdf (accessed on 30 November 2014).

27. European Commission. Guidelines on State aid for environmental protection and energy 2014-2020. 2014. Available online: http://eur-lex.europa.eu/legal-content/EN/TXT/PDF/?uri=CELEX:52014XC0628%2801%29&from=EN (accessed on 30 November 2014).

Forecasting Fossil Fuel Energy Consumption for Power Generation Using QHSA-Based LSSVM Model

Wei Sun, Yujun He and Hong Chang

Abstract: Accurate forecasting of fossil fuel energy consumption for power generation is important and fundamental for rational power energy planning in the electricity industry. The least squares support vector machine (LSSVM) is a powerful methodology for solving nonlinear forecasting issues with small samples. The key point is how to determine the appropriate parameters which have great effect on the performance of LSSVM model. In this paper, a novel hybrid quantum harmony search algorithm-based LSSVM (QHSA-LSSVM) energy forecasting model is proposed. The QHSA which combines the quantum computation theory and harmony search algorithm is applied to searching the optimal values of σ and C in LSSVM model to enhance the learning and generalization ability. The case study on annual fossil fuel energy consumption for power generation in China shows that the proposed model outperforms other four comparative models, namely regression, grey model (1, 1) (GM (1, 1)), back propagation (BP) and LSSVM, in terms of prediction accuracy and forecasting risk.

Reprinted from *Energies*. Cite as: Sun, W.; He, Y.; Chang, H. Forecasting Fossil Fuel Energy Consumption for Power Generation Using QHSA-Based LSSVM Model. *Energies* **2015**, *8*, 939-959.

1. Introduction

Since the implementation of the policy of reform and opening to the outside world in 1979, China has experienced outstanding economic development, with an average annual growth rate of 10%. Due to the continuous sustainable positive economic growth rate and large scale industrialization, electricity consumption is rising quickly [1]. Electricity plays an important role in socio-economic development, which is considered as the backbone for national economy's prosperity and progress. It is estimated that the electricity demand in China will continue to grow, since the Chinese government aims to raise its GDP with an incredible speed in the next 30 years. Affected by energy resource endowments, China's power generation sector relies heavily on fossil fuel energy and its products. In 2011, national thermal power plants generated 3.8975 trillion kWh, accounting for 82.54% of the total electricity generation [2]. The fossil fuel-based generation pattern will not change in a short term. In the future, the fossil fuel consumption for power generation will inevitably increase. Accurate forecasting of fossil fuel energy consumption for power generation is therefore important and fundamental for rational energy planning formulation in electric power industry.

In energy prediction field, some traditional forecasting approaches have been adopted in the last decades, such as regression model, time series model, Grey forecasting technique and bottom up long-range energy alternatives planning system (LEAP) model. Zhang [3] and Limanond [4] applied partial least square regression model and log-linear regression model to forecast the transport energy demand in China and Thailand, respectively. Kumar [5] applied three time series models, namely Grey–Markov model, Grey–Model with rolling mechanism and singular spectrum analysis (SSA) to forecast the consumption of conventional energy in India. Amarawickrama [6] estimated the electricity demand functions for Sri Lanka using six econometric techniques. Hsu [7] proposed an improved GM (1, 1) model to predict the power demand in Taiwan. Huang [8] used a LEAP model to forecast long-term energy supply and demand and compared future energy demand and supply patterns, as well as greenhouse gas emissions in Taiwan. The conventional statistical methods usually require the assumption of the normal distribution of energy consumption data [9]. Moreover, they are inherently limited in the presence of nonlinearities in data, which partially results from the use of linear model structures or the static nonlinear function relationships. An alternative way to deal with nonlinearities is to use Neural Networks (NN), which a powerful data modeling tool that is able to capture and represent complex input/output relationships [10–13]. It is considered that the NN could perform better than the traditional models from a statistical aspect [14,15]. However, estimating the network weights requires large amounts of data, which may be very computer-intensive. Another, it always yields limited generalization capability and unpredictably large errors since the NN usually implements the empirical risk minimization (ERM) principle; and it may show slow convergence speed, arriving at local minimum and overfitting issues [16].

Support Vector Machine (SVM), pioneered by Vapnik in 1995 [17], can effectively solve the learning problems of small sample size, higher dimension and nonlinearities [18–20]. The main advantage is that the solution of SVM is global and unique since SVM implements the structural risk minimization (SRM) principle [21–23]. Moreover, the SVM often outperforms aritificial neural network (ANN) in practice and it is less prone to overfitting [24]. The Least Squares Support Vector Machine (LSSVM), proposed by Suykens and Vandewalle in 1999 [25], is a variant of SVM. LSSVM adopts a least squares linear system as a loss function instead of the quadratic program in original SVM which is time consuming in training process [26–30]. The LSSVM shows manifest advantages, such as good nonlinear fitting ability, strong generalization capability, fast computing speed, dealing with small samples, not relying on the distribution characteristics of the samples and so on [31–34]. The performance of the LSSVM model is largely dependent on the selection of the parameters. Therefore, to obtain better forecasting accuracy, how to set the parameters of the LSSVM is very crucial. So far, there is no effective way to guide the parameters selection process. The contribution of this paper is to develop a hybrid quantum harmony search algorithm-based LSSVM (QHSA-LSSVM) energy forecasting model. In our work, the QHSA is applied to determining the optimal parameters of LSSVM model in order to enhance the learning and generalization ability.

The harmony search algorithm (HSA), a novel evolutionary algorithm, is a metaheuristic technique mimicking the improvisation behavior of musicians [35–38]. The QHSA combines the quantum computation theory [39,40] and harmony search algorithm, which adopts concepts and principles of the quantum mechanism, such as quantum bit (qubit) and superposition of states [41,42] to the HSA. It can effectively improve the performance of HSA. The practical annual energy consumption data of power generation is employed to test the performance of the proposed QHSA-LSSVM forecasting model.

The remainder of this paper is structured as follows: Section 2 introduces the LSSVM model, QHSA theory, and then a hybrid QHSA-LSSVM model is proposed in detail. Section 3 shows the data source for simulation. The empirical simulation and results analysis are presented in Section 4. Finally Section 5 gives our main conclusions.

2. Methodologies

2.1. Least Squares Support Vector Machine (LSSVM)

SVM represents a relatively new computational learning method based on statistical learning theory [17]. The basic idea of SVM is that the original input data space is mapped into a higher dimensional dot product space called a feature space. The actual problem is transformed into a quadratic programming problem with inequality constraints. LSSVM is an alternate formulation of SVM regression. In LSSVM, the inequality constraints are replaced with equality constraints; and a least squares linear system is adopted as a loss function instead of the time-consuming quadratic program in original SVM [25,26,29,33,43,44]. In the following, LSSVM algorithm is described briefly.

Let $\{x_i, y_i\}_{i=1}^m$ be a given training set of m data points, where $x_i \in R^n$ is the input vector and $y_i \in R$ is the corresponding output vector. LSSVM maps the input vector x_i to a higher dimensional feature space, using a nonlinear kernel function $\varphi(\cdot)$, shown as Equation (1). Through Equation (1), the nonlinear estimation in original space can be transformed into linear function estimation in feature space:

$$y(x) = w^T \varphi(x) + b \tag{1}$$

where w and b denote the weights vector and the bias respectively, which should be estimated from the training data. For LSSVM function estimation, the formulation of optimization issue can be written as follows:

$$\min J(w,b,\xi_i) = \frac{1}{2} w^T w + \frac{1}{2} C \sum_{i=1}^m \xi_i^2 \tag{2}$$

$$s.t. \quad y_i = w^T \varphi(x_i) + b + \xi_i, i = 1, 2, \cdots, m$$

where C is the regularization factor, and ξ_i is the slack variable which expresses the difference between the desired output and the actual output. Next, the corresponding Lagrangian function $L(w,b,\xi,a)$ is constructed to solve the above optimization problem, shown in Equation (3):

$$L(w,b,\xi,a) = \frac{1}{2} w^T w + \frac{1}{2} C \sum_{i=1}^m \xi_i^2 - \sum_{i=1}^m a_i \{w^T \varphi(x) + b + \xi_i - y_i\} \tag{3}$$

where $a_i \in R$ is a Lagrangian multiplier. From the Karush–Kuhn–Tucker (KKT) conditions, the following equations must be satisfied:

$$\frac{\partial L}{\partial w} = 0 \rightarrow w = \sum_{i=1}^{m} a_i \varphi(x_i)$$

$$\frac{\partial L}{\partial b} = 0 \rightarrow \sum_{i=1}^{m} a_i = 0$$

$$\frac{\partial L}{\partial \xi_i} = 0 \rightarrow a_i = C\xi_i$$

(4)

$$\frac{\partial L}{\partial a_i} = 0 \rightarrow w^T \varphi(x_i) + b + \xi_i - y_i = 0$$

After eliminating the variables w and ξ_i, the solution is given by the following set of linear equations, shown in Equation (5):

$$\begin{bmatrix} 0 & \vec{1}^T \\ \vec{1} & Q+C^{-1}I \end{bmatrix} \begin{bmatrix} b \\ A \end{bmatrix} = \begin{bmatrix} 0 \\ Y \end{bmatrix}$$

(5)

where $\vec{1} = [1,1,\cdots,1]^T$, $Y = [y_1, y_2, \cdots, y_m]^T$, $A = [a_1, a_2, \cdots, a_m]^T$, and I is the identity matrix. Define $Q_{ij} = K(x_i, x_j) = \varphi(x_i)^T \varphi(x_j)$, which is satisfied with Mercer's condition. In our work, the Gaussian radial basis function (RBF) is selected as the kernel function, as is expressed in Equation (6):

$$K(x_i, x_j) = \exp\left\{ -\|x_i - x_j\|^2 / 2\sigma^2 \right\}$$

(6)

The LSSVM regression model becomes:

$$y(x) = \sum_{i=1}^{m} a_i K(x, x_i) + b$$

(7)

The kernel function width coefficient σ and the regularization factor C have manifest influence on the performance of the LSSVM model. The width coefficient σ affects the width of RBF, and the regularization factor C influences the complexity of the model and the penalty degree. In order to avoid selecting the parameters arbitrarily, the optimal values of these two parameters are determined by QHSA, which can effectively enhance the prediction accuracy.

2.2. Quantum Harmony Search Algorithm (QHSA)

The harmony search algorithm (HSA), proposed by Geem *et al.* [35], is a novel phenomenon-mimicking algorithm inspired by the improvisation process of musicians [38]. The HSA has been successfully applied to many computational optimization problems, such as the traveling salesman problem [35], continuous engineering optimization [45], the layout of pipe networks [46] and course timetabling issues [47], due to its advantages of fewer parameters, excellent effectiveness and robustness compared with other heuristic optimization algorithms [48]. However, the performance of the HSA seems to be affected by the initial harmony memory (HM) parameters and solution vectors, and it usually falls into local searching for complicated numerical optimization issues. Inspired by quantum theorem, the quantum harmony search algorithm (QHSA) is proposed in our work to solve the aforementioned problems and effectively enhance the convergence rate, generalization ability and optimization capability.

Quantum theory is considered as one of the greatest scientific achievements in the twentieth century. The concept of quantum computing is presented by Benioff [49] and Feynman [50] in the 1980s through combing quantum theory and information science. In quantum computing, the qubit is adopted to express the information; and the theory of superposition, entanglement and collapse of states is employed for information computation [39,40,42,51]. Quantum computing greatly improves the computational efficiency and has attracted widespread interest and research. QHSA merges the quantum computing theory with HSA. In QHSA, the qubit is employed to express the harmony vector in HM; and the new harmony vector is introduced to adjust the bandwidth (BW) dynamically. This algorithm can improve the global search ability and optimization speed through combining superposition of qubits. The concrete optimization procedures of QHSA are as illustrated in Figure 1.

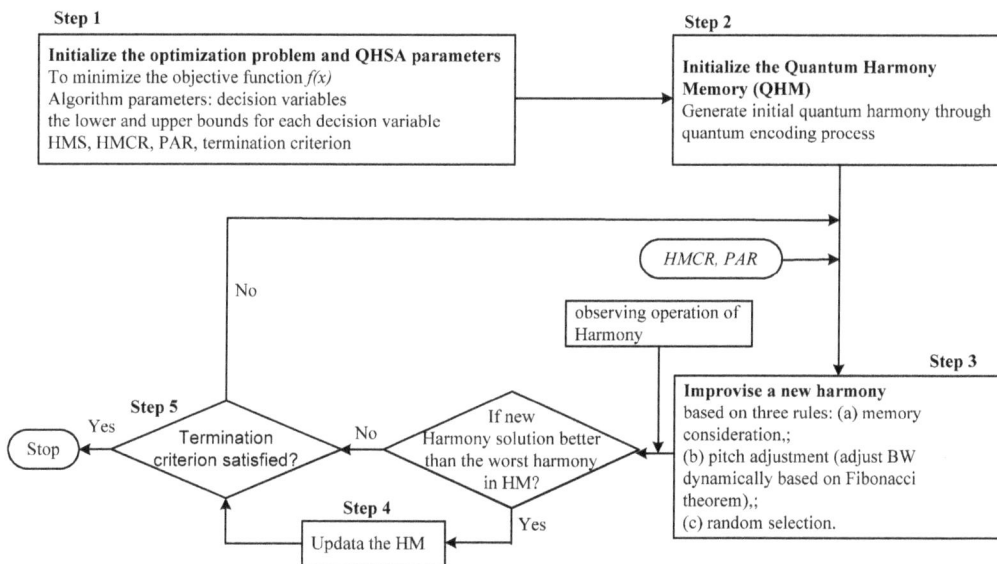

Figure 1. Quantum Harmony Search Algorithm (QHSA) optimization procedures.

Step 1. Initialize the optimization problem and algorithm parameters.

Minimize $f(x)$, s.t. $x_i \in X_i$ $i = 1,2,\cdots,N$

where $f(x)$ is the objective function; x is the set of each design variable (x_i); X_i is the set of the possible range of values for each design variable; N is the number of design variables. In addition, the HS algorithm parameters including harmony memory size (HMS), harmony memory considering rate (HMCR), pitch adjusting rate (PAR) should also be specified in this step.

Step 2. Initialize the Quantum Harmony Memory (QHM).

The HM is a location storing all the solution vectors. The HM matrix is filled with randomly generated solution vectors and sorted by the values of the objective function $f(x)$. Inspired by the concept of states superposition in quantum computing, qubit is adopted to express the harmonies in HM.

In quantum computing, the qubit is the smallest quantum piece of information, which may be in the $|1\rangle$ state, $|0\rangle$ state, or in any superposition of the two [39,40,42]. Each qubit state can be represented as the linear superposition of the two basic states, shown in Equation (8):

$$|\varphi\rangle = \alpha|0\rangle + \beta|1\rangle \tag{8}$$

where α and β is a pair of complex numbers that specify the probability amplitudes of the corresponding state; $|\alpha|^2$ and $|\beta|^2$ give the probabilities that the qubit will be in the state "0" and "1" respectively, with $|\alpha|^2 + |\beta|^2 = 1$. Compared with the single bit in classical computing, the qubits with superposition of states can carry more information and improve the computational efficiency.

In QHSA, the individual harmony can be written as Equation (9) or Equation (10):

$$q^t_i = \begin{bmatrix} q^t_{i1} & q^t_{i2} \cdots & q^t_{in} \end{bmatrix} = \begin{bmatrix} \alpha^t_{i1} & \alpha^t_{i2} & \cdots & \alpha^t_{in} \\ \beta^t_{i1} & \beta^t_{i2} & \cdots & \beta^t_{in} \end{bmatrix} \qquad i = 1, 2, \cdots, m \tag{9}$$

$$q^t_i = \begin{bmatrix} \theta^t_{i1} | \theta^t_{i2} | \cdots \theta^t_{in} \end{bmatrix} \tag{10}$$

where q^t_i is the ith quantum harmony individual at generation t in HM denoting a potential solution vector; α^t_{ij} and β^t_{ij}, initialized with $1/\sqrt{2}$, are the probability amplitude for q^t_i with $|\alpha^t_{ij}|^2 + |\beta^t_{ij}|^2 = 1$ ($i = 1, \cdots, m$, $j = 1, \cdots, n$); θ^t_{ij} is the quantum rotation angle of q^t_i which satisfies $\alpha^t_{ij} = \cos\theta^t_{ij}$, $\beta^t_{ij} = \sin\theta^t_{ij}$; m is the size of HM; n is the dimension of the problem concerned.

Step 3. Improvise a new harmony from the HM based on HMCR and pitch adjustment.

The new harmony vector is randomly selected from the historical values stored in the HM with searching probability HMCR and the possible range of values with searching probability (1-HMCR). The harmony vector is pitch-adjusted with PAR, which is shown as follows:

$$x_i = x_i \pm rand() \cdot BW \tag{11}$$

Since BW and PAR have great influence on the precision of solutions and the ability of fine-tuning [52], how to determine the suitable values of these two parameters is an important issue. A small BW value can enhance the local optimization capacity around the new harmony vector, while, a large BW can enlarge the search area of new vectors in the process of pitch adjusting and is good to escape the local optima. In classical HSA, BW is taken a fixed value and usually selected by practical experience, not considering the effect of BW on global or local optimization. In our work, to enhance the performance of the QHSA, BW is adjusted dynamically based on Fibonacci theorem.

To use the new harmony information, we adjust BW dynamically and decrease the number of parameters chosen in the initialization process. The new harmony is adopted to calculate BW, as shown in Equation (12):

$$q'_{ijnew} = q_{ijnew} + Rand(0,1) \times \left(\frac{\pi}{2} - q_{ijnew}\right), Rand(0,1) > 0.618$$

$$q'_{ijnew} = q_{ijnew} - Rand(0,1) \times q_{ijnew}, Rand(0,1) \leq 0.618 \tag{12}$$

where q'_{ijnew} is the new vector after pitch adjusting; q_{ijnew} is the new vector before pitch adjusting; $Rand(0,1)$ is a uniform random number. In classical HSA, the algorithm chooses an upper neighboring value with 50% probability and lower with 50% around new vector. Inspired by Fibonacci theorem, 0.618 is chosen as the border of vector adjusting direction, shown in Equation (12) [51,53].

Step 4. Update the HM.

First, transform the new harmony vector to a solution with real values and use the objective function to get the fitness value. The quantum harmony collapses to a single state through Equation (13). The harmony observing operation is repeated during the process of updating procedure:

$$q'_{ij} \rightarrow |1\rangle \quad , \text{when} \quad rand(0,1) > |\alpha_{ij}|^2 \quad i = 1,2,\cdots,m$$

$$q'_{ij} \rightarrow |0\rangle \quad , \text{others} \quad\quad\quad\quad j = 1,2,\cdots,n \tag{13}$$

Second, on condition that the new harmony vector shows better fitness function than the worst harmony in the HM, the new harmony is included in the HM and the existing worst harmony is excluded from the HM.

Step 5. The Quantum Harmony Search Algorithm is terminated when there is no significant improvement in the best found solution after some predetermined number of iterations or the maximum number of iterations is reached. If not, repeat Steps 3 and 4.

2.3. QHSA Based LSSVM Model

The QHSA-based LSSVM (QHSA-LSSVM) forecasting model is described in this section. In QHSA-LSSVM, the optimization objective function is specified as the mean absolute percentage error (MAPE). The MAPE index is the most widely used accuracy measurement in forecasting. It expresses the forecasting errors from different measurement units into percentage errors on actual observations [54], shown in Equation (14):

$$\min f(C,\sigma) = \min\left\{\frac{1}{T}\sum_{t=1}^{T}\left|\frac{y_t - \hat{y}_t}{y_t}\right|\right\}$$

$$s.t. \quad C \in [C_{\min}, C_{\max}] \tag{14}$$

$$\sigma \in [\sigma_{\min}, \sigma_{\max}]$$

where y_t is the actual value for the tth period; $\hat{y}_t$ represents its forecasting result for the same period; $C_{\min}$ and $C_{\max}$ are the lower and upper bound for the regularization factor C; $\sigma_{\min}$ and $\sigma_{\max}$

are the lower and upper bound for the kernel function width coefficient σ; T is the number of data used for the MAPE calculation. The basic idea of parameters optimization in LSSVM is to search the optimal values of C and σ through iterative algorithm. The concrete procedures of QHSA-LSSVM are described as follows:

Step 1. Initialize the optimization problem and algorithm parameters.

In our study, the MAPE serves as the fitness function to identify the suitable parameters in the LSSVM model. Algorithm parameter selection is an unavoidable issue in the intelligent optimization field process. The parameters are usually selected according to trial-and-error methods, not avoiding blindness and randomness. The uniform design technique is adopted to realize the selection of initial parameters of QHSA, which can make the combination of parameters uniformly distribute in value range space during part experiment process [55]. It can greatly reduce the number of the trials and ensure the representativeness of the test results. There are two design parameters, σ and C; and the upper bound and lower bound should be set in this step. Moreover, HMS, HMCR and PAR will also be initialized.

Step 2. Quantum Encoding of Harmony Memory.

According to QHSA theory, each harmony vector is firstly transformed into a quantum state, namely the quantum harmony vector (QHV). For QHSA-LSSVM, C and σ are encoded into quantum harmony vectors which are formulated by a quantum bits encoding process.

Step 3. Improvise a new harmony from the HM.

A new harmony vector is generated based on three rules: memory consideration, pitch adjustment and random selection.

Step 4. Observation of Harmony and Update the HM.

Collapse the new quantum harmony vector to a single state with real values by Equation (13); and then calculate the objective function. Update the HM on condition that the new harmony vector shows better fitness than the worst harmony in the HM.

Step 5. Stopping criterion.

If there is no significant improvement in the best found solution or the maximum number of iterations is reached, the QHSA stopping criterion satisfies. The optimal values of C and σ in the LSSVM model can be obtained. Otherwise, go back to Step 2.

3. Data Sources

This section describes how to apply the QHSA to searching for the optimal values of C and σ in LSSVM and then establish the QHSA-LSSVM forecasting model for energy consumption to validate the performance of the aforementioned method. The annual data of the fossil fuel energy consumption for power generation from 1992 to 2012 were collected from China Energy Statistical Yearbooks [56]. The primary energy consumption data were converted to standard coal consumption (10^8 tce) through using conversion coefficients from the General Principles for

Calculation of the Comprehensive Energy Consumption (Chinese National Principle, GB/T 2589-2008) [57]. The converted standard coal consumption data are shown in Table 1.

Table 1. Annual standard coal consumption of China between 1992 and 2012 (unit: 10^8 tce).

Year	Energy Consumption	Year	Energy Consumption	Year	Energy Consumption
1992	2.2397	1999	3.7304	2006	7.1388
1993	2.3788	2000	3.9590	2007	8.2251
1994	2.6089	2001	4.2523	2008	9.2669
1995	2.8411	2002	4.4687	2009	9.1993
1996	3.1704	2003	4.9264	2010	9.6327
1997	3.5003	2004	5.6848	2011	10.3205
1998	3.7539	2005	6.2886	2012	11.7500

4. Empirical Simulation and Results Analysis

4.1. The Selection of Comparison Models

Several comparative forecasting models for energy consumption prediction were selected so as to compare the results with the proposed QHSA-LSSVM model. It can be seen from Table 1 that the annual energy consumption series exhibits an obvious increasing trend, so the regression model and GM (1, 1) model were employed as the classical methods to capture the rising trend. The BP and LSSVM were adopted as the artificial intelligent techniques to simulate the relationship between the current data and a number of its previous values. The forecasting performance of the proposed QHSA-LSSVM model will be compared with linear regression model, GM (1, 1) model, BP model and LSSVM model in the next section.

4.2. The Network Structure and Parameters Setting

The selected data were the annual standard coal consumption of China during the period of 1992–2012. A total of 21 data points are divided into the training set (1992–2007) and the testing set (2008–2012).

In our work, according to the roll-based forecasting technique, three previous values are selected as the input variables of the BP, LSSVM and QHSA-LSSVM for each current output value. That means the inputs are $X_{n-3}, X_{n-2}, X_{n-1}$ and the corresponding output is X_n. Firstly, the top three energy data (from 1992 to 1994) were fed into BP, LSSVM and QHSA-LSSVM model, and then the first forecasting value of 1995 could be obtained. Secondly, the next roll-top three energy data (from 1993 to 1995) were employed for forecasting value of 1996. Similarly, the processes are cycling until all the forecasting values are obtained.

In LSSVM and QHSA-LSSVM model, the MAPE indicator was adopted as the optimization objective function $f(x)$ to measure the accuracy in a fitted time series value in statistics, which is expressed as Equation (14). The two parameters of σ and C for the proposed QHSA-LSSVM model could be optimized using QHSA. A flowchart of the QHSA algorithm for parameter initialization is shown in Figure 1. The details of the initial parameters are set as: HMS = 35,

HMCR = 0.99, PAR = 0.6, BW = 1, lb = 0.001, ub = 250; that is, the lower bound (lb) and upper bound (ub) for σ and C are set 0.0001 and 250, respectively. In the typical LSSVM model, the default values of σ and C were set 20 and 35, respectively. And the structure of LSSVM is the same with QHSA-LSSVM. For BP network, the neuron number of hidden layer is chosen with experience. In our work, the BP model contains three layers with three input neurons, eight hidden neurons and only one output neuron. The maximum number of training epochs (iterations) is 5000.

The BP, LSSVM and QHSA-LSSVM model are all realized in MATLAB 7.6.0 (R2008a) on Windows 7 with a 32-bit operating system. After training, the BP, LSSVM and QHSA-LSSVM can be used to forecast the future energy consumption value.

4.3. Experimental Results and Analysis

With the given data (1992–2012), about two-thirds (1992–2007) of it were used as training data to calculate the corresponding parameters of these five models, and the remaining one-third were used as testing data to validate the forecasting performance of the models. The solution parameters in regression model and GM (1, 1) model were determined using the training data. In our work, 30 independent runs were implemented in order to test the stability of model for BP, LSSVM and QHSA-LSSVM. In the proposed QHSA-LSSVM model, Gaussian radial basis function (RBF) is selected as the kernel function. According to the roll-based technique, the two parameters σ and C can be optimized step by step, and until the QHSA reaches the stopping criterion. The finally obtained optimal values for σ and C are 23.8564 and 150.00, respectively.

Due to the use of the roll-based technique, only 13 (1995–2007) data are suggested, *i.e.*, the simulation results for year 1992, 1993 and 1994 cannot be obtained. Therefore, the error evaluation is conducted from 1995 to 2012 to ensure the same comparison condition. The forecasting results and errors of regression, GM (1, 1), BP, LSSVM and QHSA-LSSVM are listed in Table 2.

Figure 2 shows the corresponding forecasting curves of these five models in order to make a clear comparison between the proposed QHSA-LSSVM model and other four comparative models. We tested the performance of the proposed QHSA-LSSVM model with other four comparative models (regression model, GM (1, 1), BP, and LSSVM) for training data, testing data and all the sample data, respectively. The details are shown in the next sections.

4.3.1. Simulation Performance Comparison for Training Data

In our work, the simulation data points and the simulation errors for five models from 1992 to 2007 (training set) have been conducted to assess the models' fit performance. Table 3 lists the comparison of simulation performance among five models for training set, and Figure 3 shows the corresponding error histograms for direct observation.

Table 2. Forecasting results and errors of Regression, GM (1, 1), BP, LSSVM and QHSA-LSSVM (unit: 10^8 tce).

Year	Actual	Regression		GM (1, 1)		BP		LSSVM		QHSA-LSSVM	
		Forecast	Error (%)	Forecast	Error (%)	Forecast	Error (%)	Forecast	Error (%)	Forecast	Error (%)
1995	2.8411	2.7573	−2.9496	2.6636	−6.2490	3.6039	26.8492	3.0052	5.7759	2.9767	4.7728
1996	3.1704	3.1052	−2.0565	2.9054	−8.3600	3.6864	16.2750	3.1217	−1.5361	3.1894	0.5993
1997	3.5003	3.4531	−1.3485	3.1691	−9.4621	3.9730	13.5047	3.3180	−5.2081	3.4760	−0.6942
1998	3.7539	3.801	1.2547	3.4568	−7.9148	4.1609	10.8425	3.5915	−4.3262	3.8063	1.3959
1999	3.7304	4.149	11.2213	3.7706	1.0773	4.3371	16.2637	3.9208	5.104	3.5591	−4.5920
2000	3.9590	4.4969	13.5868	4.1129	3.8868	4.1735	5.4182	4.1392	4.5517	3.8160	−3.6120
2001	4.2523	4.8448	13.9336	4.4862	5.5015	4.9150	15.5851	4.3516	2.3352	4.4611	4.9103
2002	4.4687	5.1927	16.2016	4.8935	9.5061	5.0721	13.5038	4.5897	2.7077	4.7478	6.2457
2003	4.9264	5.5406	12.4675	5.3377	8.3494	5.2183	5.9253	4.9523	0.5257	5.0392	2.2897
2004	5.6848	5.8886	3.5850	5.8223	2.4183	6.0260	6.0027	5.4627	−3.9069	5.4183	−4.6879
2005	6.2886	6.2365	−0.8285	6.3508	0.9893	7.0366	11.8949	6.2335	−0.8762	5.9874	−4.7896
2006	7.1388	6.5844	−7.7660	6.9273	−2.9622	7.2799	1.9770	7.1550	0.2269	6.9851	−2.1530
2007	8.2251	6.9323	−15.7177	7.5562	−8.1325	8.2276	0.0302	8.0985	−1.5392	7.5299	−8.4522
2008	9.2669	7.2803	−21.4375	8.2421	−9.6503	8.9703	−3.2003	8.7695	−5.3675	8.9352	−3.5794
2009	9.1993	7.6282	−17.0783	8.9903	−0.3547	9.1812	−0.1969	8.9244	−2.9883	9.3967	2.1458
2010	9.6327	7.9762	−17.1974	9.8065	4.1871	8.8555	−8.0690	8.5952	−10.7706	9.9820	3.6262
2011	10.3205	8.3241	−19.3441	10.6967	6.4669	9.3514	−9.3901	8.2691	−19.8769	10.3205	0.0000
2012	11.7500	8.6720	−26.1958	11.6677	2.3828	9.3913	−20.0746	8.1174	−30.9157	11.5587	−1.6281

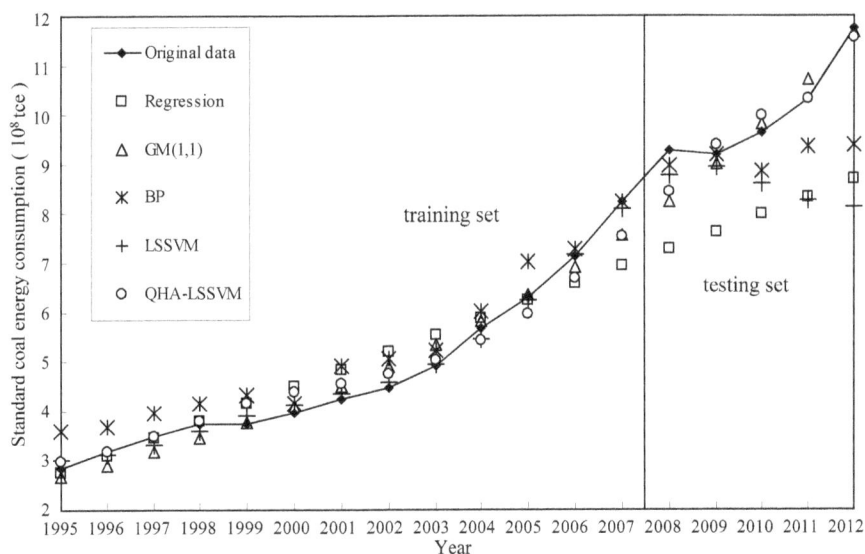

Figure 2. The curves of original data and forecasting results of five models.

Table 3. The comparison of simulation performance among five models for training set.

Model	Regression	GM (1, 1)	BP	LSSVM	QHSA-LSSVM
MAPE	0.0792	0.0575	0.1108	0.0297	0.0378
RMSE	0.5375	0.3094	0.4989	0.1391	0.3314
MAE	0.4026	0.2627	0.4439	0.1226	0.2701
AAE	0.0845	0.0551	0.0932	0.0257	0.0567
MaxAPE	0.1620	0.0951	0.2685	0.0578	0.0845

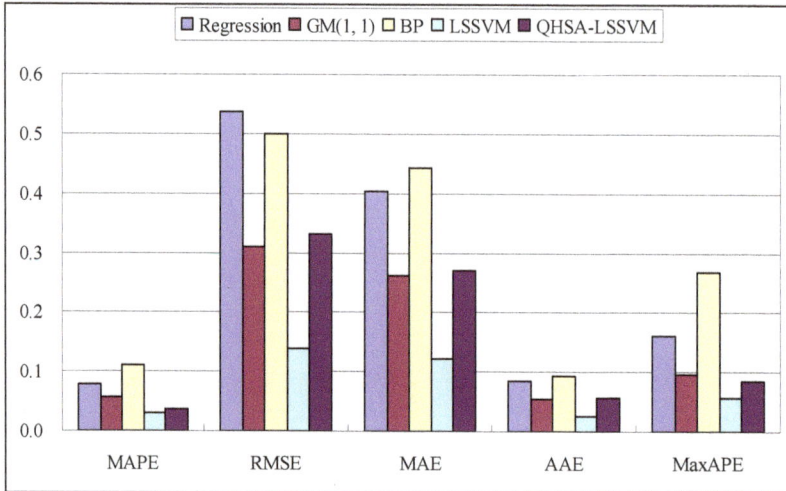

Figure 3. The error histogram for QHSA-LSSVM and other comparative models for training data.

(1) Training Data Point Analysis

In this part, the simulation data points are validated and compared among the five models. The deviations between the simulation results and the actual values are calculated, shown in Table 2. For annual data forecasting, the error range [−5%, +5%] is considered as a satisfactory and practical error bound. For 13 data points, there are five simulation data points larger than 5%, and two points smaller than −5% for the regression model. The maximum error is 16.2016% in 2002 and the minimum relative error is −15.7177% in 2007. In the GM (1, 1) model, there are three points larger than 5% and five points lower than −5%. The corresponding maximum and minimum errors are 9.5061% and −9.4621% in 2002 and 1997, respectively. For BP, there are only two data points within the error bound, and the maximum and minimum errors are 26.8492% in 1995 and 0.0302% in 2007. In LSSVM, there is only one simulation data point larger than 5%, and only two points smaller than −5%. The maximum relative error is 5.7759% in 1995, and the minimum relative error is −5.2081% in 1997. In QHSA-LSSVM model, there are only two data points that exceeds the relative error range [−5%, +5%], one is larger than 5% and the other is smaller than

−5%. The maximum relative error is 6.2457% in 2002, and the minimum relative error is −8.4522% in 2007.

Next, the maximum absolute percentage error (MaxAPE) is compared to measure the simulation risk. The MaxAPE indicator is calculated according to Equation (15), and the results are shown in Table 3.

$$\text{MaxAPE} = \max_{t}\left(\left|\frac{y_t - \hat{y}_t}{y_t}\right|\right) \times 100, t = 1, 2, \cdots, N \tag{15}$$

where y_t is the energy consumption value in the tth year; $\hat{y}_t$ represents its simulating (or forecasting) result for the same period; and N is the number of data.

MaxAPE concentrate on the maximal absolute percentage error which reflects the forecasting risk of choosing one certain model. For the training set, the MaxAPE values are 0.1620, 0.0951, 0.2685, 0.0578 and 0.0845 for regression model, GM (1, 1), BP, LSSVM and QHSA-LSSVM, respectively. Compared with other four models, the MaxAPE of LSSVM is the smaller than that of other comparative models, especially the proposed QHSA-LSSVM model. Will it really be less risky to choose LSSVM than to choose QHSA-LSSVM not only to simulate the historical training data but also to future unknown data? It is not necessary the case. We should assess the risk both in the simulation process and in prediction process to check the possible overfitting problem in models.

(2) Simulation Performance Comparison and Analysis

In this section, several common-used accuracy measures, including MAPE, root mean square error (RMSE), mean absolute error (MAE) and average absolute error (AAE), are employed to o assess the simulation performance in a comprehensive aspect. These error criterion indicators are expressed as follows:

$$\text{MAPE} = \frac{1}{N}\sum_{t=1}^{N}\left|\frac{y_t - \hat{y}_t}{y_t}\right| \times 100, t = 1, 2, \cdots, N \tag{16}$$

$$\text{RMSE} = \sqrt{\frac{1}{N}\sum_{t=1}^{N}(y_t - \hat{y}_t)^2}, t = 1, 2, \cdots, N \tag{17}$$

$$\text{MAE} = \frac{1}{T}\sum_{t=1}^{T}\left|y_t - \hat{y}_t\right| \tag{18}$$

$$\text{AAE} = \frac{1}{N}\sum_{t=1}^{N}\frac{\left|y_t - \hat{y}_t\right|}{\frac{1}{N}\sum_{t=1}^{N}y_t}, t = 1, 2, \cdots, N \tag{19}$$

Table 3 provides the MAPE, RMSE, MAE and AAE values of all models, which can reflect how well the models, with estimated parameters, fit the data. The non-scaled error metric MAPE is the mean of the absolute percentage errors of forecasts, providing the errors in terms of percentage. It can avoid the problem of positive and negative errors canceling each other out. It can be seen from Table 3 that the MAPE values are 0.0792, 0.0575, 0.1108, 0.0297 and 0.0378 for the regression model, GM (1, 1), BP, LSSVM and QHSA-LSSVM. The RMSE values of regression,

GM (1, 1), BP, LSSVM and QHSA-LSSVM are 0.5375, 0.3094, 0.4989, 0.1391 and 0.3314. The MAE and AAE for these five models are 0.4026, 0.2627, 0.4439, 0.1226, 0.2701 and 0.0845, 0.0551, 0.0932, 0.0257, 0.0567, respectively. For the training data, the LSSVM model shows smaller error values of MAPE, RMSE, MAE and AAE than those of QHSA-LSSVM and GM (1, 1) model. The worst two models in simulation performance are the regression model and BP neural network model. Although, the RMSE, MAE and AAE error values of GM (1, 1) model are smaller than those of QHSA-LSSVM, the values of MAPE and MaxAPE are larger than those of QHSA-LSSVM, so it is a little difficult to determine the overall simulation performance difference between GM (1, 1) and QHSA-LSSVM. The forecasting performance of these two models will be compared in the next section. It seems that the QHSA-LSSVM model for the training data shows no obvious advantages over the LSSVM and GM (1, 1) models, but the most important aim for model construction is to forecast future data accurately. Whether a model is good or not lies in the predictive ability or the extrapolation ability, not merely the simulation ability for training data. Next, the forecasting performances of these models are compared in details for testing data.

4.3.2. Forecasting Performance Comparison for Testing Data

The purpose for building a model is to apply it to extend future forecasts, so it is necessary to identify which model may prove accurate and be useful for forecasting. Table 4 lists the different error criterion indicator values of these five models for the testing set, and the corresponding error histogram is drawn in Figure 4.

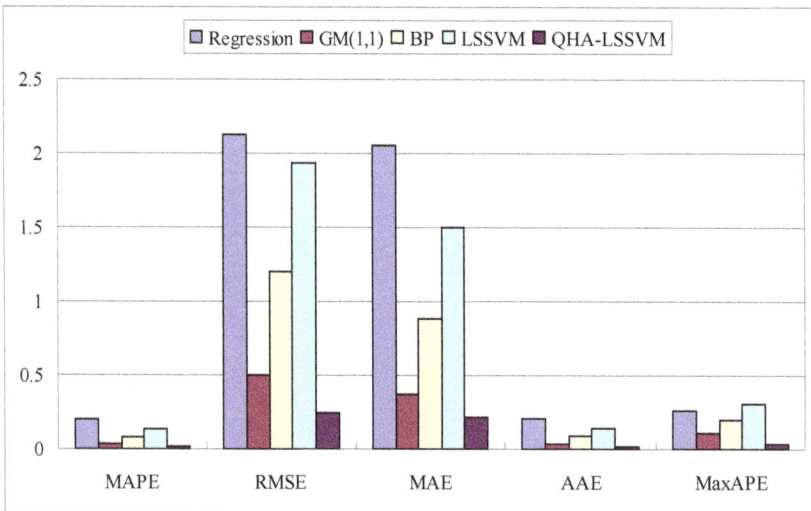

Figure 4. The error histogram for QHSA-LSSVM and other comparative models for testing data.

Table 4. The comparison of forecasting accuracy indicators among five models for testing set.

Model	Regression	GM (1, 1)	BP	LSSVM	QHSA-LSSVM
MAPE	0.2025	0.0390	0.0819	0.1398	0.0220
RMSE	2.1270	0.5045	1.1996	1.9392	0.2480
MAE	2.0578	0.3732	0.8840	1.4988	0.2139
AAE	0.2051	0.0372	0.0881	0.1494	0.0213
MaxAPE	0.2620	0.1106	0.2007	0.3092	0.0363

(1) Testing Data Point Analysis

It can be seen from Table 2 that all five forecasting data points are all within the error bound [−5%, +5%] in the proposed QHSA-LSSVM model. There are three data points exceeding the −5% lower bound in BP. In the GM (1, 1) model, there is one point larger than +5% and one point smaller than −5%. There is only one satisfactory forecasting data point within the error bound for LSSVM and none for the regression model. The maximum error is 3.6262% in 2010 and the minimum error is 0 in 2011 for QHSA-LSSVM. The maximum errors and minimum errors for other four comparative models are larger than the proposed model. For testing set, the MaxAPE value of QHSA-LSSVM model is 0.0363, which is much smaller than that of other models. Therefore, the proposed QHSA-LSSVM model shows better forecasting ability on data point analysis.

(2) Forecasting Performance Comparison and Analysis for Testing Data

Table 4 lists the MAPE, RMSE, MAE and AAE error values of the five models. The MAPE values are 0.2025, 0.0390, 0.0819, 0.1398 and 0.0220 for the regression model, GM (1, 1), BP, LSSVM and QHSA-LSSVM, respectively. The RMSE error values are 2.1270, 0.5045, 1.1996, 1.9392 and 0.2480; the MAE values are 2.0578, 0.3732, 0.8840, 1.4988 and 0.2139; the AAE values are 0.2051, 0.0372, 0.0881, 0.1494 and 0.0213. It can be seen that the MAPE, RMSE, MAE and AAE values obtained by QHSA-LSSVM are much smaller than those obtained by the other four comparative models. It indicates that the proposed QHSA-LSSVM model shows better prediction accuracy and satisfactory forecasting performance for testing data. At the meantime, it is found that the MAPE, RMSE, MAE and AAE error values of LSSVM model are worse for the testing data than those for the training set. We can infer that the overfitting phenomenon occurs in the LSSVM model with small training samples, which may decrease the model's forecasting performance. The LSSVM model looks better because it's making absolutely small errors on the training (or learning) data, shown in Table 3, but when the LSSVM model is applied to a new dataset (the testing set), it does not perform as well on the testing data, with larger errors, shown in Table 4. The generalization ability of LSSVM model is poor. In this case, the LSSVM model overfits the training data; thus, it is not suitable for future prediction. The proposed QHSA-LSSVM model just makes good balance on training set and testing set. It has a relatively good accuracy on the learning data and the best accuracy on the testing data. Overall, the proposed QHSA-LSSVM model is suitable for both data simulation and future trend prediction. The prediction performance of GM (1, 1) for testing data is worse than the simulation performance for the training data. This model has ideal predictive effect for approximate homogenous exponential sequence; otherwise, it may result in larger forecasting error. Moreover, GM (1, 1) is not suitable for long time prediction periods. The forecasting performances of the regression model and BP for the testing data are the worst ones among these five models.

4.3.3. Overall Accuracy Comparison for All Sample Data

Besides, a comprehensive accuracy comparison of these five models for all sample data has been conducted. The values of five error criterion indicators, *i.e.*, MAPE, RMSE, MAE, AAE and MaxAPE for all sample data are listed in Table 5; and the corresponding error histogram is shown in Figure 5.

Table 5. The comparison of forecasting accuracy indicators among five models for all sample data.

Model	Regression	GM (1, 1)	BP	LSSVM	QHSA-LSSVM
MAPE	0.1134	0.0524	0.1028	0.0603	0.0334
RMSE	1.2105	0.3739	0.7613	1.0289	0.2563
MAE	0.8624	0.2934	0.5661	0.5049	0.2018
AAE	0.1385	0.0471	0.0909	0.0811	0.0324
MaxAPE	0.2620	0.1106	0.2685	0.3092	0.0845

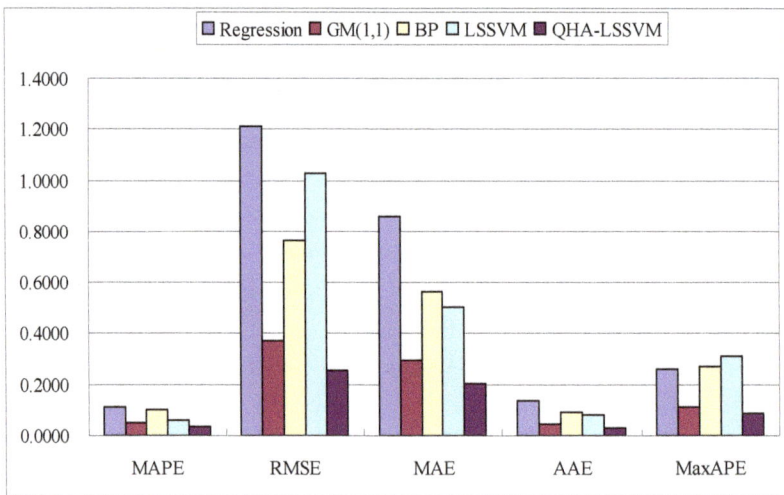

Figure 5. The error histogram for QHSA-LSSVM and other comparative models for all sample data.

As shown in Table 5, the MAPE for regression, GM (1, 1), BP, LSSVM and QHSA-LSSVM are 0.1134, 0.0524, 0.1028, 0.0603, and 0.0334. Compared with MAPE which aim is to assess average values of absolute percentage errors, RMSE is the square root of MSE; and MSE is the average value of the "total square error" which is the sum of the individual squared errors. The stated rationale for squaring each error is to remove the sign so that the "magnitude" of the errors influences the average error measure. For this indicator, large errors normally have a relatively greater influence on the "total square error" than do the smaller errors. In our work, the RMSE values of regression model, GM (1, 1), BP and LSSVM are 1.2105, 0.3739, 0.7613, 1.0289; while the RMSE of QHSA-LSSVM is only 0.2563. It means that other four comparative models may concentrate within a decreasing number of increasingly larger individual errors, thus resulting in larger MSE and RMSE values.

Similarly, MAE involves summing the magnitudes (absolute values) of the errors to obtain the "total error" and then calculating the average value of the "total error". It is an average of the absolute errors, which summarizes performance in the way that disregards the direction of over- or under-prediction and does place emphasis on the mean signed difference. After calculating according to Equation (18), the MAE values are 0.8624, 0.2934, 0.5661, 0.5049 and 0.2018 for the regression model, GM (1, 1), BP, LSSVM and QHSA-LSSVM, respectively. This also demonstrates there are no obvious larger individual absolute errors between the actual data series and the forecasting data series in the QHSA-LSSVM model.

AAE is a more comprehensive indicator since it can assess the deviation of individual absolute errors from the average value of actual data. The total "deviation" is divided by T, thus obtaining an AAE value. For the four comparative models, the AAE values are 0.1385, 0.0471, 0.0909 and 0.0811, respectively, while the AAE value of QHSA-LSSVM is only 0.0324, which is smaller than that of the four preceding comparative models.

The MaxAPE values of regression model, GM (1, 1), BP and LSSVM model are 0.2620, 0.1106, 0.2685 and 0.3092, respectively; and that of the proposed QHSA-LSSVM is only 0.0845, which means that selecting QHSA-LSSVM has less forecasting risk.

Through errors comparison, the regression errors are almost higher than other three models (*i.e.*, GM (1, 1), BP and LSSVM). It means that the regression model is not suitable to capture the fluctuation increasing trend without constant growth rate in our work. The forecasting performance of GM (1, 1) for all sample data is better than the regression model since the errors of the GM (1, 1) model are lower; and it shows better performance for the training data than the testing data. The errors of BP are almost the highest, which indicates this model is the worst one, except for the regression model. Why does BP show such poor simulation performance not only in training data but also in testing data? The possible reasons are analyzed as follows: the network structure, especially the number of neurons in the hidden layer, has a tremendous influence on the final output. Both the number of hidden layers and the number of neurons in the input and output layers must be carefully considered. In our work, the neurons in the input layers are three, according to the rolling-based method, and the output neuron is the corresponding certain year data. The number in the hidden layer is only adjustable; and there is no theory yet to determine how many hidden neurons are reasonable for better model fitting. Therefore, the fixed number of hidden neurons is the main reason that leads to larger simulation error. Another, training neural network is a complex task using a large number of sample data, with probable long training time and local minimum. A second problem may occur even when the training data is insufficient. It indicates that BP needs a large number of samples to train the network, and is not suitable for small data samples.

For LSSVM, the errors of LSSVM are satisfactory for the training set (1992–2007), and are even smaller than those of our proposed QHSA-LSSVM model. However, the errors for the testing data become much larger, which means it has poor performance in the testing set and is not suitable for future prediction. Moreover, the errors of LSSVM for all sample data are larger than that of QHSA-LSSVM. The most possible reason is that LSSVM model can easily fall into an over-fitting problem, so the prediction ability for future trends is not satisfactory. Another is that the use of the parameters σ and C with fixed values may result in large errors.

For QHSA-LSSVM, it can be observed from Table 5 and Figure 5 that MAPE, RMSE, MAE, AAE and MaxAPE of the proposed QHSA-LSSVM model is the lowest compared with the regression model, GM (1, 1), BP and LSSVM for all the sample data. The comparison findings prove that QHSA-LSSVM performs better than the other four comparative models in terms of forecasting accuracy for the whole data. The proposed QHSA-LSSVM is also suitable for foreseeable forecasting.

In conclusion, we used MAPE, RMSE, MAE, AAE and MaxAPE to validate the forecasting performance of the five forecasting models. Through calculating the values of error criterion indicators for training data, testing data and all the sample data, the overall performance of QHSA-LSSVM is the best with the lowest error values. The results prove that the optimal parameters σ and C determined by QHSA can effectively improve the model's performance and the forecasting accuracy.

5. Conclusions

The difficulty in the LSSVM forecasting technique is how to choose suitable values for parameters σ and C since this can affect the model's learning performance and generalization ability. In order to avoid selecting the parameters randomly and reduce the adjustment process, a novel hybrid Quantum Harmony Search Algorithm-based LSSVM (QHSA-LSSVM) forecasting model is proposed for the annual power generation standard coal consumption forecasting. The QHSA effectively combines the quantum computation and harmony search algorithm. Based on the concepts and principles of the quantum mechanism, QHSA can improve the global search ability and optimization speed through combining superposition of qubits. In LSSVM, the optimal values of the two parameters are determined through QHSA, which can solve the problem of parameters selection. To validate the performance of the proposed QHSA-LSSVM model, four other comparative models (regression model, GM (1, 1), BP and LSSVM) were employed in our work. Through calculating the error values of MAPE, RMSE, MAE, AAE and MaxAPE of the five models, the errors for QHSA-LSSVM are the smallest, not only for the testing data, but also for all the sample data. This indicates that the QHSA-LSSVM model is suitable for small sample forecasting and can effectively enhance the prediction accuracy. Moreover, the QHSA can determine the optimal parameters values of the LSSVM model, which could solve the random parameter selection problem and decrease the forecasting errors. In all, the proposed QHSA-LSSVM model is satisfactory for future data trend prediction.

Acknowledgments

This work was supported by "Philosophy and Social Science Research of Hebei Province", "Soft Science Research Base of Hebei Province" and "the Fundamental Research Funds for the Central Universities (12MS137)".

Conflicts of Interest

The authors declare no conflict of interest.

References

1. Lindner, S.; Liu, Z.; Guan, D.; Geng, Y.; Li, X. CO_2 emissions from China's power sector at the provincial level: Consumption *versus* production perspective. *Renew. Sustain. Energy Rev.* **2013**, *19*, 164–172.
2. National Bureau of Statistics of China. *China Statistical Yearbook*; China Statistics Press: Beijing, China, 2012. (In Chinese)
3. Zhang, M.; Mu, H.; Li, G.; Ning, Y. Forecasting the transport energy demand based on PLSR method in China. *Energy* **2009**, *34*, 1396–1400.
4. Limanond, T.; Jomnonkwao, S.; Srikaew, A. Projection of future transport energy demand of Thailand. *Energy Policy* **2011**, *33*, 2754–2763.
5. Kumar, U.; Jain, V.K. Time series models (grey-Markov, grey model with rolling mechanism and singular spectrum analysis) to forecast energy consumption in India. *Energy* **2010**, *35*, 1709–1716.
6. Amarawickrama, H.A.; Hunt, L.C. Electricity demand for Sri Lanka: A time series analysis. *Energy* **2008**, *33*, 724–739.
7. Hsu, C.; Chen, C. Applications of improved grey prediction model power demand forecasting. *Energy Convers. Manag.* **2003**, *44*, 2241–2249.
8. Huang, Y.; Bor, Y.J.; Peng, C. The long-term forecast of Taiwan's energy supply and demand: LEAP model application. *Energy Policy* **2011**, *39*, 6790–6803.
9. Yu, S.; Wei, Y.-M.; Wang, K. A PSO-GA optimal model to estimate primary energy demand of China. *Energy Policy* **2012**, *42*, 329–340.
10. Gurney, K. *An Introduction to Neural Networks*; UCL Press: London, UK, 1997.
11. Anderson, J.A. *An Introduction to Neural Networks*; MIT Press: Cambridge, MA, USA, 1995.
12. Hassoun, M.H. *Fundamentals of Artificial Neural Networks*; MIT Press: Cambridge, MA, USA, 1995.
13. Geem, Z.W.; Roper, W.E. Energy demand estimation of South Korea using artificial neural network. *Energy Policy* **2009**, *37*, 4049–4054.
14. Patterson, D.W. *Artificial Neural Networks*; Prentice Hall: Singapore, Singapore, 1996.
15. Zhang, G.; Patuwo, B.E.; Hu, M.Y. Forecasting with artificial neural networks: The state of the art. *Int. J. Forecast.* **1998**, *14*, 35–62.

16. Kecman, V. *Learning and Soft Computing: Support Vector Machines, Neural Networks, and Fuzzy Logic Models*; MIT Press: Cambridge, MA, USA, 2001.

17. Vapnik, V.N. *The Nature of Statistical Learning Theory*; Springer-Verlag: New York, NY, USA, 1999.

18. Burges, C. A tutorial on support vector machines for pattern recognition. *Data Min. Knowl. Discov.* **1998**, *2*, 121–167.

19. Cristianini, N.; Shawe-Taylor, J. *An Introduction to Support Vector Machines and other Kernel-Based Learning Methods*; Cambridge University Press: Cambridge, UK, 2000.

20. Cortes, C.; Vapnik, V. Support vector networks. *Mach. Learn.* **1995**, *3*, 273–297.

21. Vapnik, V.N. An overview of statistical learning theory. *IEEE Trans. Neural Netw.* **1999**, *10*, 988–999.

22. Vapnik, V.; Golowich, S.E.; Smola, A. Support vector method for function approximation, regression estimation, and signal processing. In *Advances in Neural Information Processing Systems 9*; MIT Press: Cambridge, MA, USA, 1997; Volume 9, pp. 281–287.

23. Suykens, J.A.K. Nonlinear modeling and support vector machines. In Proceedings of the 18th IEEE Instrumentation and Measurement Technology Conference, Budapest, Hungary, 21–23 May 2001.

24. Li, Y.B.; Li, Y. Survey on uncertainty support vector machine and its application in fault diagnosis. In Proceedings of the 3rd IEEE International Conference on Computer Science and Information Technology (ICCSIT), Chengdu, China, 9–11 July 2010.

25. Suykens, J.A.K.; Vandewalle, J. Least squares support vector machine classifiers. *Neural Process. Lett.* **1999**, *9*, 293–300.

26. Suykens, J.A.K.; van Gestel, T.; de Brabanter, J.; de Moor, B.; Vandewalle, J. *Least Squares Support Vector Machines*; World Scientific: Singapore, Singapore, 2002.

27. Suykens, J.A.K.; Lukas, L.; Vandewalle, J. Sparse approximation using least squares support vector machines. In Proceedings of the IEEE International symposium on circuits and systems, Geneva, Switzerland, 28–31 May 2000.

28. Suykens, J.A.K.; de Brabanter, J.; Lukas, L.; Vandewalle, J. Weighted least squares support vector machines-robustness and sparse approximation. *Neurocomputing* **2002**, *48*, 85–105.

29. Suykens, J.A.K.; Vandewalle, J. Recurrent least squares support vector machines. *IEEE Trans. Circuits Syst. I IEEE Trans. Fundam. Theory Appl.* **2000**, *47*, 1109–1114.

30. Suykens, J.A.K.; Lukas, L.; van Dooren, P.; de Moor, B.; Vandewalle, J. Least squares support vector machine classifiers: A large scale algorithm. In Proceedings of the European Conference on Circuit Theory and Design (ECCTD'99), Stresa, Italy, 29 August–2 September 1999.

31. Zhao, X.H.; Wang, G.; Zhao, K.K.; Tan, D.J. On-line least squares support vector machine algorithm in gas prediction. *Min. Sci. Technol.* **2009**, *19*, 194–198.

32. Espinoza, M.; Suykens, J.A.K.; de Moor, B. Fixed-size least squares support vector machines: A large Scale application in electrical load forecasting. *Comput. Manag. Sci.* **2006**, *3*, 113–129.

33. Lin, K.P.; Pai, P.F.; Lu, Y.M.; Chang, P.T. Revenue forecasting using a least-squares support vector regression model in a fuzzy environment. *Inf. Sci.* **2011**, *14*, 196–209.

34. Zhou, D. Estimation of GM (1, 1) model parameter based on LS-SVM algorithm and application in load forecasting. *Mod. Manag.* **2012**, *2*, 45–49.

35. Geem, Z.W.; Kim, J.H.; Loganathan, G.V. A new heuristic optimization algorithm: Harmony search. *Simulation* **2001**, *76*, 60–68.

36. Omran, M.G.H.; Mahdavi, M. Global-best harmony search. *Appl. Math. Comput.* **2008**, *198*, 643–656.

37. Geem, Z.W.; Tseng, C.L. New methodology, harmony search and its robustness. In Proceedings of the Late-Breaking Papers of Genetic and Evolutionary Computation Conference, New York, NY, USA, 9–13 July 2002.

38. Mahdavi, M.; Fesanghary, M.; Damangir, E. An improved harmony search algorithm for solving optimization problems. *Appl. Math. Comput.* **2007**, *188*, 1567–1579.

39. Hey, T. Quantum computing: An introduction. *Comput. Control Eng. J.* **1999**, *10*, 105–112.

40. Nielsen, M.A.; Chuang, I.L. *Quantum Computation and Quantum Information*, 1st ed.; Cambridge University Press: Cambridge, UK, 2000.

41. Han, K.H.; Kim, J.H. Quantum-inspired evolutionary algorithms with a new termination criterion, H gate, and two-phase scheme. *IEEE Trans. Evol. Comput.* **2004**, *8*, 156–169.

42. Han, K.H.; Kim, J.H. Quantum-inspired evolutionary algorithm for a class of combinational optimization. *IEEE Trans. Evolut. Comput.* **2002**, *6*, 580–593.

43. Esen, H.; Ozgen, F.; Esen, M. Modeling of a new solar air heater through least square support vector machine. *Expert Syst. Appl.* **2009**, *36*, 10673–10682.

44. Li, H.; Guo, S.; Zhao, H.; Su, C.; Wang, B. Annual electric load forecasting by a least squares support vector machine with a fruit fly optimization algorithm. *Energies* **2012**, *5*, 4430–4445.

45. Lee, K.S.; Geem, Z.W. A new meta-heuristic algorithm for continues engineering optimization: Harmony search theory and practice. *Comput. Method Appl. Mech. Eng.* **2005**, *194*, 3902–3933.

46. Geem, Z.W.; Kim, J.-H.; Loganathan, G.V. Harmony search optimization: Application to pipe network design. *Int. J. Model. Simul.* **2002**, *22*, 125–133.

47. Al-Betar, M.A.; Khader, A.T.; Gani, T.A. A harmony search algorithm for university course timetabling. *Ann. Oper. Res.* **2012**, *194*, 3–31.

48. Lee, K.S.; Geem, Z.W. A new structural optimization method based on the harmony search algorithm. *Comput. Struct.* **2004**, *82*, 781–798.

49. Benioff, P. The computer as a physical system: A microscopic quantum mechanical Hamiltonian model of computers as represented by Turing machines. *J. Stat. Phys.* **1980**, *22*, 563–591.

50. Feynman, R.P. Simulating physics with computers. *Int. J. Theor. Phys.* **1983**, *21*, 467–488.

51. Chang, H.; Sun, W.; Gu, X. Forecasting energy CO_2 emissions using a Quantum Harmony Search Algorithm-Based DMSFE combination model. *Energies* **2012**, *6*, 1456–1477.

52. Wang, C.M.; Huang, Y.F. Self-adaptive harmony search algorithm for optimization. *Expert Syst. Appl.* **2010**, *37*, 2826–2837.

53. Chang, H.; Gu, X.S. Multi-HM adaptive harmony search algorithm and its application to continuous function optimization. *Res. J. Appl. Sci. Eng. Technol.* **2012**, *4*, 100–103.

54. Goodwin, P.; Lawton, R. On the asymmetry of the symmetric MAPE. *Int. J. Forecast.* **1999**, *15*, 405–408.
55. Fang, K.T. *Uniform Design and Uniform Design Table*; Scientific Press: Beijing, China, 1994. (In Chinese)
56. *China Energy Statistical Yearbook*; National Bureau of Statistics of China, National Development and Reform Commission: Beijing, China, 1993–2013. (In Chinese)
57. *General Principles for Calculation of the Comprehensive Energy Consumption*; GB/T 2589-2008; China Standard Press: Beijing, China, 2008. (In Chinese)

Chapter 4:
Global Phenomena and Global Governance

Global Energy Development and Climate-Induced Water Scarcity—Physical Limits, Sectoral Constraints, and Policy Imperatives

Christopher A. Scott and Zachary P. Sugg

Abstract: The current accelerated growth in demand for energy globally is confronted by water-resource limitations and hydrologic variability linked to climate change. The global spatial and temporal trends in water requirements for energy development and policy alternatives to address these constraints are poorly understood. This article analyzes national-level energy demand trends from U.S. Energy Information Administration data in relation to newly available assessments of water consumption and life-cycle impacts of thermoelectric generation and biofuel production, and freshwater availability and sectoral allocations from the U.N. Food and Agriculture Organization and the World Bank. Emerging, energy-related water scarcity flashpoints include the world's largest, most diversified economies (Brazil, India, China, and USA among others), while physical water scarcity continues to pose limits to energy development in the Middle East and small-island states. Findings include the following: (a) technological obstacles to alleviate water scarcity driven by energy demand are surmountable; (b) resource conservation is inevitable, driven by financial limitations and efficiency gains; and (c) institutional arrangements play a pivotal role in the virtuous water-energy-climate cycle. We conclude by making reference to coupled energy-water policy alternatives including water-conserving energy portfolios, intersectoral water transfers, virtual water for energy, hydropower tradeoffs, and use of impaired waters for energy development.

Reprinted from *Energies*. Cite as: Scott, C.A.; Sugg, Z.P. Global Energy Development and Climate-Induced Water Scarcity—Physical Limits, Sectoral Constraints, and Policy Imperatives. *Energies* **2015**, *8*, 8211-8225.

1. Introduction

Globally, increasing demand for energy continues to outpace rates of population and economic growth [1]. The quest for sustainable energy futures will depend significantly on water-resource availability and quality impacts associated with energy development [2,3]. Both energy and water are inextricably linked to climate change, which tends to heighten the use of both resources [4] while increasing the variability of water availability for energy development, other human uses, and ecosystem processes. Drought and water scarcity in particular have direct effects for energy development [5,6], principally electrical power generation [7] but also the rapidly expanding production of biofuels [8]. The nexus between energy and water—both the water needed for energy development as described in this paper and energy for water pumping, conveyance, treatment, and other operations [9]—has important implications for climate change. For example, energy development and use generate greenhouse gases that significantly contribute to global warming. Additionally, adaptation to the effects of climate change [10] and mitigation of its anthropogenic causes are fundamentally centered on the use and management of energy and water—separately as resources and increasingly in tandem as the water-energy nexus [11].

Climate change and variability are now firmly linked to anthropogenic drivers via greenhouse gas emissions, in particular, carbon dioxide from a range of human activities including electricity generation and land use, the two processes we are concerned with in this paper. Policy-makers are increasingly called on to adopt and incentivize programs that mitigate CO_2 while at the same time adapting to the effects of climate change [12]. In this context, the water-energy nexus plays a critical role in resource-use policy [13]. The availability and quality of water resources greatly influence energy options, and conversely, water management has an appreciable impact on CO_2 emissions [14].

Water is required for a range of energy development processes. The environmental quality impacts of fossil-fuel development, e.g., petroleum, coal, and natural gas, are increasingly being factored into water-energy nexus assessments [15]. Here we focus on water use for: (a) hydroelectric and thermoelectric power generation, and (b) biofuel production (chiefly feedstock irrigation but also other life-cycle processes). Even with the technological shift from once-through cooling to evaporative cooling of thermoelectric generation, water consumption (depletion through evaporation) per unit of power generated represents an increasing demand on water resources. Additionally, irrigation is required for biofuel feedstocks (e.g., sugarcane or corn for ethanol and soy or rapeseed for biodiesel), and consumes significant amounts of water although some feedstocks are raised under rainfed conditions. In river basins with physically stressed water resources or in locations where water is allocated for other human uses (often with secure water rights) or environmental flows, energy demands for water are of growing concern [16,17].

Several regional and national assessments of water requirements for energy development have been published [9,18–21]. Yet, limited work has addressed key components of the water-for-energy challenge at the global scale [8,22,23]. Most recently, Spang et al. [8] developed a metric for water consumption for energy production portfolios including various fuel types, then used it to calculate the water-for-energy footprint for 158 countries. These data were normalized based on several other indicators in order to rank countries according to different metrics [24]. While useful for developing a comprehensive assessment of the consumptive use of water for a given country's energy sector, such analyses are thus far temporally limited, providing a snapshot of water consumption for a given year. They should be augmented with current energy production trends including biofuels, analyses of technological innovation, and policy alternatives to address water-resource constraints. In order to develop a more complete picture, this paper quantitatively evaluates physical and sectoral (allocative) water scarcity resulting from thermoelectric generation and biofuels production trends at the global scale using current data, and identifies and assesses policy options to address these challenges. The goal of these analyses is to highlight current and future challenges with meeting water demands for energy generation.

This paper is organized as follows. Above, we have briefly framed the need to consider water and energy interlinkages in the context of climate change. While an increasing number of studies are available, most are constrained by regional or local focus or they are temporally limited. Next, we present our approach and methods for a global assessment of time-series trends of the water-resource use implications of electrical energy and biofuel feedstock irrigation. In the discussion section that follows, we consider the implications of climatic trends plus adaptive management and technological options to address these challenges. We identify and discuss "flashpoint" countries that are expected to face increasing constraints of water availability for energy development. Finally, we conclude with an assessment of policy alternatives for expanding energy requirements while also accounting for climate change and variability.

2. Methods and Data

The mapping and coupled energy-water resource analyses presented here are based on robust global datasets, specifically, 2010 electrical power and biofuel production and trends to 2020 from the U.S. Energy Information Administration [25], and freshwater availability and sectoral allocations from the U.N. Food and Agriculture Organization [26] and World Bank [27]. Newly available data on water consumption and life-cycle impacts of electrical generation [28] and biofuel production were also incorporated [29–31], including projections to 2020 for ethanol and biodiesel for several countries [32].

The cooling tower process for thermal electricity generation requires approximately 45 times lower withdrawals than once-through cooling [22]. Because cooling type for individual power plants globally is not widely reported, we estimated annual freshwater withdrawals for combined thermal and nuclear electricity based on both cooling tower and once-through technologies. We recognize that once-through cooling continues to be used in electricity generation, e.g., in some European countries, Canada, and U.S. (where once-through cooling has become less common since 1970). However, the results reported here for all countries are based on assumed cooling tower technology, which we derived by multiplying EIA generation values by a median withdrawal intensity of 3.8 m^3 per MWh. This assumption results in estimates of freshwater withdrawals that are lower than actual, *i.e.*, if data existed to accurately account for once-through cooling.

Projected future energy generation was based on compound annual growth rates (CAGR) (Equation 1) where $V(t0)$ is the earliest available electricity generation value and $V(t_n)$ is the most recent value within the period 2000–2010. No CAGR was calculated for lack of adequate data if fewer than five years of generation data were available for a given country.

$$CAGR\ (t_0, t_n) = (V(t_n)/V(t_0))^\wedge(1/(t_n\text{-}t_0)) - 1 \qquad (1)$$

Irrigation water applied for ethanol feedstock production was calculated by multiplying EIA ethanol production data by country-specific irrigation quantity coefficients for ethanol-producing nations obtained from de Fraiture *et al.* [33]. We also sought to include estimates of life-cycle water use associated with specific ethanol and biodiesel feedstocks. However, there is no robust global database on water use for the cultivation and processing of the many different biofuel feedstocks, although some estimates of "water consumption for energy production" [8] and life-cycle water use [29] have been derived for a few feedstocks. Use of water coefficients for biofuel production is further complicated by the fact that water consumption of biofuel feedstocks varies widely depending on the particular feedstock and climatic conditions [34]. For example, while soybeans under rainfed conditions consume no "blue water" as irrigation [34], in locations where they are irrigated, water consumption estimates can reach up to 844 m^3/GJ [8]. Additionally, while some biodiesel feedstocks are grown under rainfed conditions, water is still used in the total life cycle during the processing stage. Because robust data on life cycle water usage for the various biodiesel feedstocks are not available, we assumed a minimum 0.031 l/MJ of water for biodiesel processing of all feedstocks under all climates of using the estimate reported in Spang et al. [8]. This represented biodiesel production using feedstocks grown under rainfed conditions, imported from other countries (as in the United Kingdom and South Korea), or from sources such as recycled cooking oils that are not primarily produced via large-scale agricultural production. Because the USA's soybean feedstocks for biofuels are increasingly produced under irrigated conditions, a higher coefficient of 21.71 L/MJ was used for USA biodiesel [29]. Utilizing water use coefficients obtained from Mulder et al. [29] and Spang et al. [8], we assumed ethanol feedstocks (corn and sugarcane) were irrigated, with the exception of Canada. For Brazil and the USA, we present a range of values (discussed below).

The derived estimates of sectoral withdrawals for combined thermal and nuclear electricity generation and for biofuels production (from EIA data) were then taken as a percentage of total available freshwater for the industrial and agricultural sectors, respectively, found in FAO and World Bank databases, just a year apart from EIA energy data. Electricity generation, water withdrawal and availability, and biofuel production data were compiled into a GIS geodatabase for mapping.

3. Results

3.1. Spatial and Temporal Trends in Energy Generation

Presenting multiple dimensions of energy-generation analyses in a single graphic each for thermoelectric generation and biofuels production results in figures that require some explanation. Recent (2000–2010) growth rates in total electricity production comprising conventional thermoelectric, non-hydropower renewables, hydropower, and nuclear generation for 199 countries are presented in Figure 1, which also shows projected increases for the period 2010–2020 in total thermal electricity generation. Additionally, the percentage mix of fuel ethanol to total biofuel production for those countries producing more than 5000 barrels (795,000 liters) of biofuels per day in 2010 is shown in Figure 2, which also shows recent (2000–2010) growth rates in total biofuel production.

The water requirements for current and future thermoelectric generation and biofuels production, as detailed in the Methods section above, are presented here. Hydropower was not assessed due to inherent methodological difficulties in attributing evaporative losses resulting solely from power generation by multi-purpose reservoirs [35]. Additionally, non-hydropower renewables such as solar and wind energy, use minimal quantities of water (with the exception of concentrated solar and geothermal using steam cycles) and were not explicitly assessed here. Thus, our analysis focuses on water requirements for conventional thermal and nuclear generation.

Finally, water demand for biofuels was attributed to feedstock production (with irrigation volumes for sugarcane and ethanol reported separately by major producing countries), in addition to water for processing associated with life cycle and consumptive water use analyses. Based on figures reported by de Fraiture et al. [33], fuel ethanol feedstocks were assumed to be irrigated except in the case of Canada, where corn and wheat are not typically irrigated; instead we applied an estimate of the non-irrigation water use for corn ethanol from Mulder et al. [29]. However, because the amount of water used for corn cultivation for ethanol varies widely from state to state in the USA [36], and between the major sugarcane growing regions in Brazil [37], we present a range of estimates for water consumption for ethanol feedstocks for those two countries that included lower bounds of zero irrigation (Table 1). Although not all sugarcane cultivation is irrigated in Brazil, the percentage is increasing, especially in the northeast region [37] and particularly as a result of drought and related climate effects. Because robust country-level data are not available for biodiesel feedstocks, we assumed a minimum amount of water use for all biodiesel processing for all countries except for irrigated soybeans in the USA [38] as mentioned.

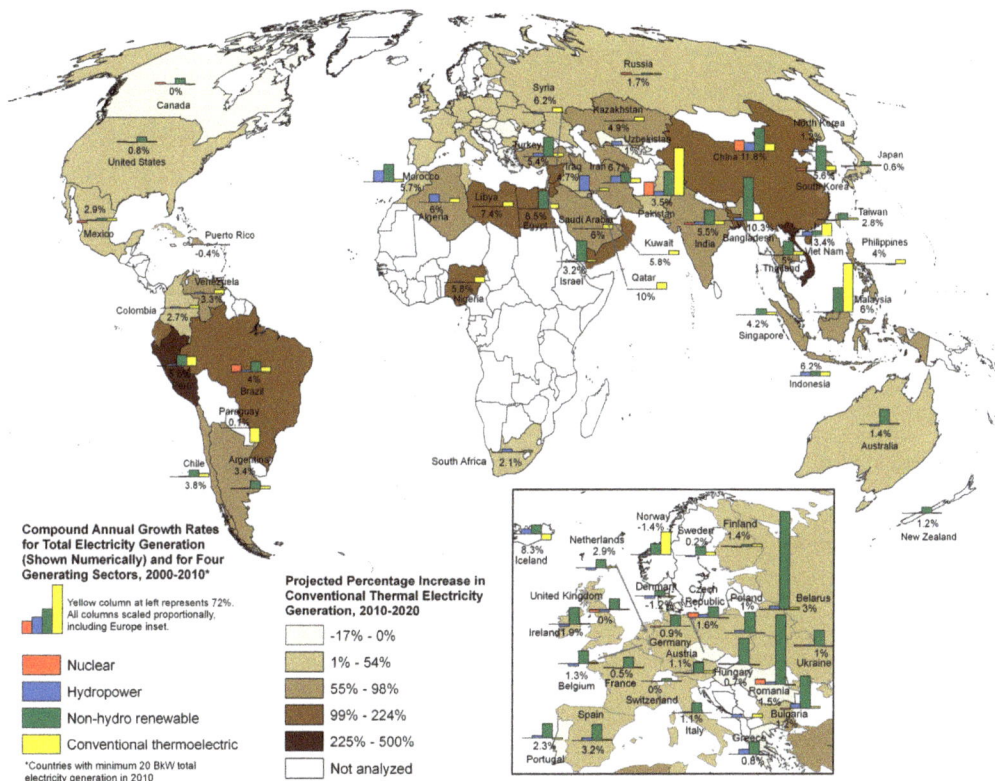

Figure 1. Growth in electricity generation by country (total, nuclear, hydropower, non-hydropower renewable, and conventional thermal), 2000–2010; and percentage increase in conventional thermal electricity generation, 2010–2020.

Canada 26.4
United States 889.9
China 43.0
South Korea 6.5
India 7.0
Thailand
Indonesia
Colombia 12.0
Brazil 527.1
Australia 7.9
Argentina 38.1

Percentage Fuel Ethanol of Total Biofuel Production, 2010*

Numeric values in the map are total biofuel production (thousand BPD) for year 2010.

1% - 26%
27% - 51%
52% - 75%
76% - 100%

*For countries with total biofuel production at least 5,000 BPD.

Compound Annual Growth Rates in Biofuel Production Based on 2000-2010 Annual Totals**

All bars including inset drawn to same scale. Black bar below denotes 200% growth rate.

200%

Fuel Ethanol

Biodiesel

**If production data were not reported for the full period, growth rates were calculated if at least five years of data were available.

Sweden 5.9
Finland 7.5
United Kingdom 9.5
Netherlands
Poland 11.0
Belgium
Germany 62.0
Czech Republic
France 55.0
Austria 6.0
Italy
Hungary
Spain 15.5
Portugal 6.0
24.0

Source: U.S. Energy Information Administration

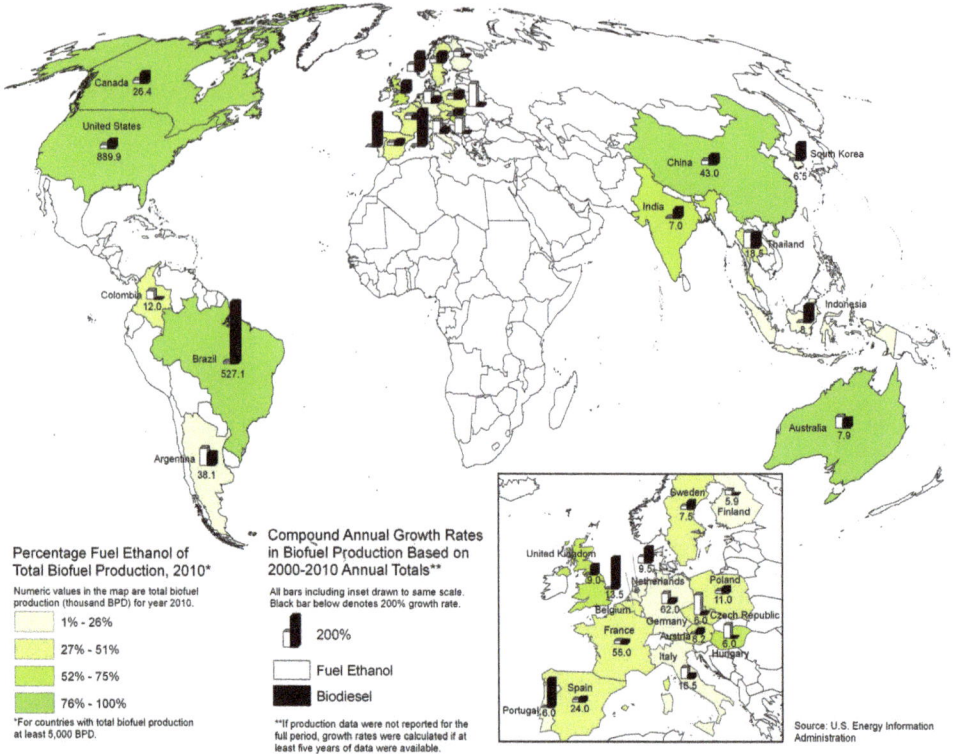

Figure 2. Share of ethanol in total biofuel production, 2010; and growth in total biofuel production.

3.2. Growth in Thermal Electrical Production

With increasing energy development comes increasing water demand. Results shown in Figure 1 indicate that most industrialized Western nations have exhibited low or flat growth rates in conventional thermoelectric power generation along with concurrent rapid growth in the development of non-hydropower renewable energy. This is especially true, for example, in Western European countries, consistent with renewable energy targets adopted in EU directives and individually by EU member states [39]. Most of the recent growth in thermoelectric power generation at the global level has come from countries in the Middle East, East and Southeast Asia, and South America. The BRIC countries show diverse energy portfolios; Brazil, Russia, India, and especially China show positive growth rates in all four categories of electricity generation. At 3595.5 billion kWh in 2011, China is orders of magnitude greater than most other countries (data not shown). While China and USA currently have comparable total net electricity generation, USA has a CAGR of only 1% based on the period 2000–2010. By contrast, the CAGR for Chinese thermoelectric generation is 11.4%.

Table 1. Water withdrawals for thermoelectric and nuclear power generation as fractions of industrial water withdrawals; and total water withdrawals for energy (thermoelectric and nuclear power generation, and biofuels feedstock irrigation), 2010 and 2020. Water withdrawals are defined as diversions from freshwater bodies (FAO AQUASTAT) [26], not depletion through evaporation. Color coding shown at bottom of table.

Country	Current Thermo & Nuclear Water Withdrawal/ Industrial Water Withdrawal [%, Fraction], 2010	Future Thermo & Nuclear Water Withdrawal/ Industrial Water Withdrawal [%, Fraction], 2020	Current Irrigation Withdrawals for Ethanol/ Agricultural Water Withdrawals [%, Fraction], 2010 *	Current Thermo & Nuclear Water Consumption + Lifecycle Water (Ethanol & Biodiesel)/Total Internal Renewable Water [%, Fraction], 2010 **	Future Thermo & Nuclear Water Consumption + Lifecycle Water (Ethanol & Biodiesel)/Total Internal Renewable Water [%, Fraction], 2020 **
Australia	32.6%	37.3%		0.1%	0.1%
Brazil	2.8%	6.2%	0%–7.7%	0.01%–0.3%	0.02%–0.4%
Canada	2.7%	2.8%	0%–8.9%	0.0%	0.0%
China	10.0%	28.4%	0%–1.6%	0.4%	2.2%
Egypt	11.8%	24.2%		15.7%	32.3%
India	17.1%	29.1%	0%–0.2%	0.1%	0.3%
Mexico	10.7%	14.8%		0.1%	0.2%
Pakistan	15.7%	22.4%		0.2%	0.4%
S. Korea	57.4%	101.4%		1.7%	3.8%
Saudi Arabia	113.2%	202.5%		20.1%	35.9%
South Africa	120.4%	148.9%	0%–0.1%	1.2%	n/a
Thailand	18.6%	30.6%	0%–4.7%	0.2%	0.6%
Turkey	12.9%	21.3%		0.1%	0.2%
UK	28.9%	28.2%		0.5%	0.5%
USA	6.3%	6.8%	0%–11%	0.42%–0.8%	0.6%–1.1%
Venezuela	21.6%	40.1%		0.0%	0.0%
[Value:]	>10%	>10%	>10%	>10%	>10%
[Value:]	>30%	>30%	>30%	>30%	>30%

* Lower values assume no irrigation of ethanol feedstock. Upper values assume some irrigation based on estimates reported by de Fraiture *et al.* (2008) [33]. ** Lower values for Brazil and USA assume water consumption for ethanol processing but no irrigation of feedstock.

3.3. Increasing Water Demands for Conventional Electricity Generation

This anticipated global increase in electricity generation from conventional, *i.e.*, non-renewable, sources will be accompanied by greater water demands. But while these nations are all expected to expand conventional thermal electricity generation capacity, they differ in the amount of overall industrial water usage that can and will be devoted to such development. The results shown in Table 1 indicate that all major countries with the exception of the UK are projected over 2010–2020 to increase the fraction of industrial water withdrawn for use in nuclear and conventional thermoelectric power generation. For example, in China, in 2010 water withdrawn for nuclear and conventional thermoelectric power generation accounted for an estimated 10.0% of all industrial water withdrawals and is projected to increase to 28.4% by 2020. India may also increase from 17% to almost 30% of its industrial water supply for conventional thermal electricity by that same future date. In contrast, water withdrawals for these same uses only account for 2.8% of Brazil's total industrial water withdrawals in 2010 and are expected to increase to about 6% by 2020. This is related to the significant contribution of hydropower to Brazil's overall portfolio.

Of the 16 major energy-producing nations included in Table 1, all but Brazil, Canada, and USA were devoting at least 10% of total industrial water available to nuclear and conventional thermoelectric power generation, with four using more than 30%, and South Africa and Saudi Arabia each over 100%. In this context, it should be noted that seawater used for cooling is not included in the current definition of industrial withdrawals of (fresh) water. Nevertheless, use of seawater and other waters not suitable for irrigation or other human purposes, e.g., inland brackish water, "produced" water from oil and gas development, effluent, *etc.*, will increasingly need to be used in energy generation and other industrial processes.

3.4. Growth Trends in Water for Biofuel Production

The major fuel ethanol producing countries as shown in the map—USA and Brazil at 889.9 and 527.1 thousand BPD, respectively—were by far the largest producers of total biofuel (fuel ethanol and biodiesel) in 2010. At least 76% is fuel ethanol, not biodiesel. The next highest is China at 43.0 thousand BPD. For the purposes of this analysis, our results quantify the relative proportions of agricultural water withdrawals used to irrigate ethanol feedstocks. In Table 1 we report the current (2010) withdrawals for irrigation for ethanol as a percentage of total agricultural water withdrawals for each country. None of the countries is above 10% (one of our thresholds) except for the USA, the world's largest ethanol producer, where large volumes of irrigation water are used for corn production as an ethanol feedstock [40], especially in more arid western states [36]. Irrigation requirements for corn per liter of ethanol produced vary widely geographically, from 5 L L^{-1} in Ohio to 2138 L L^{-1} in California [36]; thus, the relative share of agricultural water devoted to corn may be much lower at a state or regional scale. The second largest producer of biofuels, Brazil, applies less water than the USA, 7.7%, for irrigating energy feedstocks because sugarcane is largely rain-fed. While growth in biodiesel production has been rapid in recent years in Brazil, ethanol still comprises by far the larger share of total biofuel production.

We also estimated the future amount of total agricultural water devoted to ethanol crop production based on recent trends. These estimates assume that total agricultural water withdrawals do not begin increasing. This is based on data showing that total freshwater withdrawals for agriculture have remained steady or have not increased appreciably during the period 2002–2011 for the countries shown in Table 2, with the exceptions of India and Saudi Arabia. We also assume that the recent rates of expansion of energy crop acreage requiring irrigation continue. While in 2010 the irrigation requirements for ethanol feedstock production were relatively low, fuel ethanol production in these major producing countries has increased although less rapidly than in earlier decades. For example, in USA the contribution of ethanol to the renewable fuels standard is near its maximum while other biofuel feedstocks do not yet have appreciable market share. Additionally, the European Union has cut back its demand for biofuels in order to minimize impacts on developing countries. With uncertainty in energy security coupled with climate change impacts on energy demand and water availability, however, these feedstocks may demand a markedly greater proportion of the total water available for agriculture. Applying a 10% reduction in total water available for agriculture due to effects of climate change while assuming the percentage of all available agricultural water applied for ethanol feedstock cultivation remains the same, Canada and the USA would devote 10% and 12% of all agricultural water to ethanol feedstock cultivation, respectively. If we apply a further reduction due to allocative scarcity (*i.e.*, other sectors adapting to water scarcity by reallocating water currently used in agriculture), totaling 25% reduction, the percentage of all agricultural water applied for ethanol feedstock production in the USA increases to 15%, Canada to 12%, and Brazil to 10%. It should be noted that drought conditions in California and Australia, for example, exemplify how reductions in water allocated to agriculture frequently result in such drastic cuts.

Table 2. Increases in carbon dioxide emissions, agricultural freshwater withdrawals, and irrigation freshwater withdrawals based on reported data (FAO AQUASTAT) [26]. Color coding shown at bottom of table.

Country	CO_2 Emissions Increase [%/yr], 1999–2009	Total Freshwater Withdrawals Increase [%/yr], 2002–2011	Agricultural Freshwater Withdrawals Increase [%/yr], 2002–2011	Industrial Freshwater Withdrawals Increase [%/yr], 2002–2011
Australia	2.1%	0.0%	0.0%	0.0%
Brazil	1.4%	−0.2%	−1.6%	−0.5%
Canada	0.0%	0.0%	0.0%	0.0%
China	8.8%	0.6%	−1.4%	3.7%
Egypt	5.6%	0.0%	0.0%	0.0%
India	5.6%	2.5%	2.3%	6.1%
Mexico	1.6%	1.1%	1.0%	0.8%
Pakistan	4.9%	0.7%	0.6%	−9.6%
S. Korea	2.5%	0.0%	0.0%	0.0%
Saudi Arabia	6.7%	3.7%	3.5%	15.6%
South Africa	3.0%	0.0%	0.0%	0.0%
Thailand	3.3%	0.0%	0.0%	0.0%
Turkey	3.5%	−0.5%	−0.7%	0.5%
UK	−1.2%	−2.0%	−0.2%	−5.6%
USA	−0.4%	0.1%	−0.2%	0.4%
Venezuela	0.7%	0.0%	0.0%	0.0%
[Value:]	>1% /yr	>1% /yr	>1% /yr	>1% /yr
[Value:]	>3% /yr	>3% /yr	>3% /yr	>3% /yr

We also combined 2010 water consumption for nuclear and thermoelectric electricity generation with lifecycle water use for all biofuels for each country and report as a percentage of total internal renewable water (Table 1). Egypt and Saudi Arabia are highlighted as already using a relatively high percentage (>10%) of freshwater resources. As shown in the far right column of Table 1, assuming growth rates continue, these two countries, with Thailand and USA added, project to withdraw an increasing percentage of freshwater for these combined purposes.

4. Discussion

It is evident that water withdrawals for energy production are increasing, a challenge that poses difficult policy questions for climate adaptation and carbon mitigation, as well as for the water-energy nexus as a management tool to meet future demands for these resources. While our analysis presents conservative estimates of water withdrawals, it is evident is that: (a) water demands for energy are increasing, (b) few robust estimates exist, (c) climate impacts are expected to exacerbate current and future trends, and (d) water and energy planners have taken little notice of these trends at least until very recently. We compare the results reported here to previous related work and then briefly demonstrate the implications of these results for several "flashpoint" countries.

4.1. Climate Adaptation in the Water and Energy Sectors

The principal climate-change processes that are projected to intensify globally—warming temperature and increasing variability of precipitation resulting in drought and flood extremes [12]—drive increased demand for energy and water separately as resources, and via nexus effects that each exerts on the other. Urban adaptation to climate change, for example, tends to raise electricity requirements for (a) air-conditioning resulting from warming, (b) pumping and infrastructure management under conditions of both drought and flooding, and (c) redundant power supplies in transportation, emergency response, and medical systems planned for under conditions of power-grid tripping or more catastrophic failure. Cities are also implementing a range of green-infrastructure interventions to address urban heat island effects of warming, e.g., urban water bodies and landscaping vegetation, which tend to raise water diversions and consumption. In agricultural systems, climate change has a multiplier effect for water and electricity demand as well as adaptive response—that is, warming temperatures significantly increase water requirements for crop growth that can be met through increasing irrigation applications, which in turn can increase power demand for pumping and reduce hydropower generation from storage reservoirs as infrastructure operators are forced to decide on tradeoffs among multiple uses of water. Alternately, as considered above, agriculture may experience allocative water scarcity, resulting in lower yields, reduced area planted, and in general, loss of output, financial returns, and farm labor.

Perhaps more significant, however, are the carbon implications of conventional fossil fuel-based generation of electricity. The first column in Table 2 shows rapidly escalating CO_2 emissions at the country and global levels, for which a leading cause is the rising demand for electricity. The IPCC [12] indicates that economic growth is a more potent driver than population growth alone. Heightened emissions in turn translate into warming and a speeding up of the hydrological cycle with greater variability in drought and flood cycles. Carbon-mitigation efforts aiming to decarbonize economic activity and future growth consider alternative fuels, including hydropower and biofuels among other sources—all of which portend future increases in water consumption.

4.2. Relevance to Other Estimates of Intensity of Water Demand for Energy

As Spang *et al.* [8] point out, there is a global shortage of detailed estimates of the water consumption of energy generation. Still, to interpret the results reported here on geographic water availability on a per-country basis, it is helpful to consider how they relate to recent work in a similar vein. In particular, Spang *et al.* [8] developed the first country-level comparison of water consumption for fuels and electricity production using a derived metric of 'water consumption for energy production'. They calculated water consumption for production of various sub-types of fossil, nuclear, and biomass fuels and then applied them to the global scale, generating national energy portfolios for 158 countries. In their companion paper [24], they normalized these earlier per-country water consumption results by various other indicators (GDP, population, total energy production, and regional water availability). The results reported here expand on this approach, using more recent data as inputs (2010) and by examining temporal trends—in the form of compound annual growth rates—rather than a single snapshot in time, as was done in other previous studies [20,21,41,42]. Based on the results described above, we have identified several flashpoint countries that warrant further discussion.

4.3. Comparative Analysis of Flashpoint Countries

We observe increasing water demands for conventional and nuclear electricity generation at alarming levels for several countries. As shown, at least 13 countries are already using a relatively high percentage—10% or more—of the total industrial water withdrawals for these purposes. Many of the very countries projected to increase thermoelectric generation are arid and already using relatively high amounts of freshwater resources for these power sources. Surprisingly, we find that a few countries, e.g., Saudi Arabia and South Africa as shown above, already appear to be diverting more water for thermoelectric and nuclear power generation than the total reported industrial water withdrawals. Our analysis does not account, however, for dry cooling systems, e.g., for coal-based generation as increasingly implemented in South Africa. The results for Saudi Arabia may seem counterintuitive, but Spang *et al.* [24] made a similar observation that both the United Arab Emirates and Qatar were using over three times the total amount of water naturally internally available in those countries.

We find a rapid expansion in recent years of irrigation of ethanol crops in the U.S. and associated water use. Upper bound estimates place Brazil, Canada, and the USA at close to 10% of total agricultural water applied to ethanol crop production. If recent growth rates in ethanol crop production under irrigation were to continue, the associated water withdrawals would escalate to unrealistic levels. Therefore, planted acreage will not increase indefinitely. However, even a more modest gradual increase would be accompanied by an increase in the use of irrigation to intensify production in certain regions, depending on climatic conditions. This appears to be the case for Brazilian sugarcane production. Additionally, while we assumed soybean production for Brazilian biodiesel was cultivated under rainfed conditions, FAO (AQUASTAT) [26] reports that 624,000 ha were irrigated in 2006; 11.7% of all irrigated cropland. Increases in the production of biofuels based on irrigated feedstocks are highly concerning because, as others have pointed out, biofuel feedstock cultivation is the most water-intensive compared to other fuel sources [8]. Chiu *et al.* [36] observed that the continued expansion of corn cultivation for ethanol in the Great Plains and Western USA is likely to exacerbate the expected water challenges in those regions. Mulder *et al.* [29] analysis of water use efficiency led them to conclude that "the development of biomass energy technologies in scale sufficient to be a significant source of energy may produce or exacerbate water shortages around the globe and be limited by the availability of fresh water."

5. Conclusions

We have assessed current and future trends in energy production, specifically electricity generation and biofuel feedstocks and processing, in relation to the consumption of water under changing climatic conditions. Despite ongoing energy diversification, fossil fuels remain the principal energy source and will for some time to come. Policy options to address these challenges can be difficult and complex [43] and are often overlooked in sectorally focused planning [44]. Technology enhancement and the means to spur innovation are crucial choices [45]. Technological change tends to be most dynamic in countries with low installed capacity, which can allow for leap-frogging in the adoption of technologies. However, access and cost to new technologies can be formidable challenges that the global community must address, through funding of adaptation tied to verifiable benchmarks. Particularly for gains in efficiency, technology substitution has already resulted in progress. While this allows for better input-output conversions, e.g., reducing the coefficient values used and cited above, rebound and take-back effects [46] that tend to increase, instead of limiting resource use, must be explicitly addressed through programmatic interventions, incentives for conservation tied to efficiency, and low-carbon adaptive strategies.

Adaptation of water use under climate change and the implications this holds for energy demand are often not explicitly considered in climate or energy policy. Various coupled energy-water policy measures have been identified. These include water-conserving energy portfolios as described, e.g., in the United States by Scott *et al.* [13]. Such options will be increasingly adopted, given the financing and public-resistance pressures against large, new energy and water infrastructure. We have referred above to allocative water scarcity, yet intersectoral water transfers can be used to enhance energy production while intensifying agriculture (invariably the source of water transferred) and assuring food security. The long-distance conveyance of energy through electricity grids allows for generation that can be distant from the location of acute water scarcity—an example of virtual water for energy. Policy-makers must be cognizant the reverse does not occur, *i.e.*, locations with adequate water for power generation must not convey electricity from generation sources in water-scarce locations, even though financial advantages for such virtual exchange may exist. The use of impaired waters (effluent, saline and brackish waters) for energy production will become increasingly common, just as seawater is used for thermoelectric cooling. Finally, hydropower is a unique water-energy nexus technology and policy domain in which tradeoffs must be explored [47] and rights and regulations must be explicitly accounted for [48]. As with the other options discussed above, integrating technology and policy options to address water, energy, and climate challenges in an integrated manner is above all a question of institutional arrangements.

Acknowledgments

The authors gratefully acknowledge the support of the Inter-American Institute for Global Change Research, via project CRN3056, which is supported by the US National Science Foundation (NSF) Grant No. GEO-1128040; NSF Grant DEB-1010495; and the Morris K. and Stewart L. Udall Foundation. The comments received from three anonymous reviewers were invaluable in improving the analyses.

Author Contributions

Christopher Scott conceived of the research and undertook initial data exploration. Zachary Sugg contributed significantly to the data analysis and interpretation of results. Both authors worked together on the discussion and conclusions.

Conflicts of Interest

The authors declare no conflict of interest.

References

1. Chu, S.; Majumdar, A. Opportunities and challenges for a sustainable energy future. *Nature* **2012**, *488*, 294–303.
2. Ackerman, F.; Fisher, J. Is there a water–energy nexus in electricity generation? Long-term scenarios for the western United States. *Energy Policy* **2013**, *59*, 235–241.
3. Rio Carrillo, A.M.; Frei, C. Water: A key resource in energy production. *Energy Policy* **2009**, *37*, 4303–4312.
4. Shah, T. Climate change and groundwater: India's opportunities for mitigation and adaptation. *Environ. Res. Lett.* **2009**, *4*, 035005.
5. Hightower, M.; Pierce, S.A. The energy challenge. *Nature* **2008**, *452*, 285–286.
6. King, C.W.; Holman, A.S.; Webber, M.E. Thirst for energy. *Nat. Geosci.* **2008**, *1*, 283–286.
7. Van Vliet, M.T.H.; Yearsley, J.R.; Ludwig, F.; Vögele, S.; Lettenmaier, D.P.; Kabat, P. Vulnerability of US and European electricity supply to climate change. *Nat. Clim. Chang.* **2012**, *2*, 676–681.
8. Spang, E.S.; Moomaw, W.R.; Gallagher, K.S.; Kirshen, P.H.; Marks, D.H. The water consumption of energy production: an international comparison. *Environ. Res. Lett.* **2014**, *9*, 105002.
9. Siddiqi, A.; Díaz Anadón, L. The water energy nexus in Middle East and North Africa. *Energy Policy* **2011**, *39*, 4529–4540.
10. Chandel, M.K.; Pratson, L.F.; Jackson, R.B. The potential impacts of climate-change policy on freshwater use in thermoelectric power generation. *Energy Policy* **2011**, *39*, 6234–6242.
11. Scott, C.A. The water-energy-climate nexus: Resources and policy outlook for aquifers in Mexico. *Water Resour. Res.* **2011**, *47*, W00L04.

466

12. IPCC. *Climate Change 2014: Synthesis Report. Contribution of Working Groups I, II and III to the Fifth Assessment Report of the Intergovernmental Panel on Climate Change. Intergovernmental Panel on Climate Change*; Pachauri, R.K., Allen, M.R., Barros, V.R., Broome, J., Cramer, W., Christ, R., Church, J.A., Clarke, L., Dahe, Q., Dasgupta, P., *et al.*, Eds.; IPCC: Geneva, Switzerland, 2014; p. 151.

13. Scott, C.A.; Pierce, S.A.; Pasqualetti, M.J.; Jones, A.L.; Montz, B.E.; Hoover, J.H. Policy and institutional dimensions of the water–energy nexus. *Energy Policy* **2011**, *39*, 6622–6630.

14. Shrestha, E.; Ahmad, S.; Johnson, W.; Batista, J.R. The carbon footprint of water management policy options. *Energy Policy* **2012**, *42*, 201–212.

15. Vengosh, A.; Jackson, R.B.; Warner, N.; Darrah, T.H.; Kondash, A. A critical review of the risks to water resources from unconventional shale gas development and hydraulic fracturing in the United States. *Environ. Sci. Technol.* **2014**, *48*, 8334–8348.

16. Averyt, K.; Fisher, J.; Huber-Lee, A.; Lewis, A.; Macknick, J.; Madden, N.; Rogers, J.; Tellinghuisen, S. *Freshwater Use by US Power Plants: Electricity's Thirst for a Precious Resource*; Union of Concerned Scientists: Cambridge, MA, USA, 2011.

17. Scanlon, B.R.; Duncan, I.; Reedy, R.C. Drought and the water-energy nexus in Texas. *Environ. Res. Lett.* **2013**, *8*, 045033.

18. Sovacool, B.K.; Sovacool, K.E. Identifying future electricity-water tradeoffs in the United States. *Energy Policy* **2009**, *37*, 2763–2773.

19. Flörke, M.; Teichert, E.; Bärlund, I. Future changes of freshwater needs in European power plants. *Manage. Environ. Quality* **2011**, *22*, 89–104.

20. DOE. Energy demands on water resources: report to Congress on the interdependency of energy and water. US Department of Energy (DOE): Washington, DC, USA, 2006. Available online: http://energy.sandia.gov/wp/wp-content/gallery/uploads/dlm_uploads/121-RptToCongress-EWwEIAcomments-FINAL.pdf (accessed on 25 June 2013).

21. Elcock, D.; Future, U.S. water consumption: The role of energy production. *J. Am. Water Resour. Assoc.* **2010**, *46*, 447–460.

22. Vassolo, S.; Döll, P. Global-scale gridded estimates of thermoelectric power and manufacturing water use. *Water Resour. Res.* **2005**, *41*, W04010.

23. Gerbens-Leenes, P.W.; van Lienden, A.R.; Hoekstra, A.Y.; van der Meer, T.H. Biofuel scenarios in a water perspective: the global blue and green water footprint of road transport in 2030. *Global Environ. Change* **2012**, *22*, 764–775.

24. Spang, E.S.; Moomaw, W.R.; Gallagher, K.S.; Kirshen, P.H.; Marks, D.H. Multiple metrics for quantifying the intensity of water consumption of energy production. *Environ. Res. Lett.* **2014**, *9*, 105003.

25. International energy statistics. U.S. Energy Information Administration [EIA]. Available online: http://www.eia.gov/cfapps/ipdbproject/IEDIndex3.cfm (accessed on 23 January 2013).

26. AQUASTAT database. Food and Agriculture Organization of the United Nations [FAO]. Available online: http://www.fao.org/nr/water/aquastat/main/index.stm (accessed on 14 March 2013).

27. Indicators. World Bank. Available online: http://data.worldbank.org/indicator (accessed on 12 March 2013).

28. Macknick, J.; Newmark, R.; Heath, G.; Hallett, K.C. Operational water consumption and withdrawal factors for electricity generating technologies: a review of existing literature. *Environ. Res. Lett.* **2012**, *7*, 045802.

29. Mulder, K.; Hagens, N.; Fisher, B. Burning water: A comparative analysis of the energy return on water invested. *Ambio* **2010**, *39*, 30–39.

30. Bailey, R. Trouble with Biofuels: Costs and Consequences of Expanding Biofuel Use in the United Kingdom. In *Energy, Environment and Resources EER PP*; 2013/01; Chatham House: London, UK, 2013; p. 24.

31. Mittal, A. Energy-water nexus: Many uncertainties remain about national and regional effects of increased biofuel production on water resources. Government Accountability Office: Washington, DC, USA, 2010. Available online: http://www.gao.gov/assets/300/299103.pdf (accessed on 12 February 2015).

32. OECD-FAO Agricultural Outlook Database. Organisation for Economic Co-operation and Development; Food and Agricultural Organization of the United Nations. Available online: http://www.agri-outlook.org/database.html (accessed on 25 May 2015).

33. De Fraiture, C.; Giordano, M.; Liao, Y. Biofuels and implications for agricultural water use: Blue impacts of green energy. *Water Policy* **2008**, *10*, 67–81.

34. Kanellakis, M.; Martinopoulos, G.; Zachariadis, T. European energy policy—A review. *Energy Policy* **2013**, *62*, 1020–1030.

35. Wurbs, R.A.; Ayala, R.A. Reservoir evaporation in Texas, USA. *J. Hydrol.* **2014**, *510*, 1–9.

36. Chiu, Y.-W.; Walseth, B.; Suh, S. Water embodied in bioethanol in the United States. *Environ. Sci. Technol.* **2009**, *43*, 2688–2692.

37. Herreras Martinez, S.; van Eijck, J.; Pereira da Cunha, M.; Guilhoto, J.J.M.; Walter, A.; Faaij, A. Analysis of socio-economic impacts of sustainable sugarcane–ethanol production by means of inter-regional Input–Output analysis: Demonstrated for Northeast Brazil. *Renew. Sust. Energ. Rev.* **2013**, *28*, 290–316.

38. Chiu, Y.W.; Wu, M. Assessing county-level water footprints of different cellulosic-biofuel feedstock pathways. *Environ. Sci. Technol.* **2012**, *46*, 9155–9162.

39. Mekonnen, M.M.; Hoekstra, A.Y. The green, blue and grey water footprint of crops and derived crop products. Value of water research report series no. 47. UNESCO: Delft, The Netherlands, 2010. Available online: http://waterfootprint.org/media/downloads/Report47-WaterFootprintCrops-Vol1.pdf (accessed on 20 February 2015).

40. King, C.W.; Webber, M.E.; Duncan, I.J. The water needs for LDV transportation in the United States. *Energy Policy* **2010**, *38*, 1157–1167.

41. Kenny, J.F.; Barber, N.L.; Hutson, S.S.; Linsey, K.S.; Lovelace, J.K.; Maupin, M.A. Estimated use of water in the United States in 2005. United States Geological Survey: Reston, VA, USA, 2009. Available online: http://pubs.usgs.gov/circ/1344/ (accessed on 18 January 2015).

42. Orcutt, M. Water power. *MIT Tech. Rev.* 19 April 2011. Available online: http://www.technologyreview.com/graphiti/423762/water-power/ (accessed on 22 February 2015).

468

43. Siddiqi, A.; Kajenthira, A.; Díaz Anadón, L. Bridging decision networks for integrated water and energy planning. *Energy Strateg. Rev.* **2013**, *2*, 46–58.
44. Schoonbaert, B. The water-energy nexus in the UK: Assessing the impact of UK energy policy on future water use in thermoelectric power generation. Master's thesis. Kings College, University of London, UK, 2012. Available online: http://www.kcl.ac.uk/sspp/departments/geography/study/masters/dissertations/Dissertation-2012-Schoonbaert.pdf (accessed on 20 February 2015).
45. Rhodes, A.; Skea, J.; Hannon, M. The global surge in energy innovation. *Energies* **2014**, *7*, 5601–5623.
46. Sorrell, S. Energy substitution, technical change and rebound effects. *Energies* **2014**, *7*, 2850–2873.
47. Koch, F.H. Hydropower—The politics of water and energy: Introduction and overview. *Energy Policy* **2002**, *30*, 1207–1213.
48. Bauer, C.J. Slippery property rights: multiple water uses and the neoliberal model in Chile, 1981–1995. *Nat. Resour. J.* **1998**, *38*, 109.

Horizontal and Vertical Reinforcement in Global Climate Governance

Martin Jänicke

Abstract: This paper is dealing with mechanisms that can accelerate the global diffusion of climate-friendly technologies. The accelerated diffusion of low-carbon technology innovation can possibly be achieved by interactive processes such as: (1) mutually reinforcing cycles of policy-induced domestic market growth, innovation, and policy feedback; (2) lead markets and political lesson-drawing, the reinforced international adoption of innovations from pioneer countries; and (3) interaction between the vertical and horizontal dynamics in multi-level systems of governance. The three mechanisms are not exclusive. They can overlap and reinforce each other. After a theoretical introduction they will be described. The empirical focus is on the European system of multi-level climate governance. The paper draws some final conclusions for policy makers.

Reprinted from *Energies*. Cite as: Jänicke, M. Horizontal and Vertical Reinforcement in Global Climate Governance. *Energies* **2015**, *8*, 5782-5799.

1. Introduction

The increase of greenhouse gases and the scientific consensus on the consequences of manmade changes to the atmosphere of the Earth is a dramatic challenge to the governance of necessary climate mitigation. What is needed is a high speed of technological change towards a low-carbon economy, comparable to the industrial revolutions of past centuries, and it can be asked what strategic options exist that can accelerate mitigation efforts. Evidence shows that indeed, there have been cases of accelerated change in the last decade. The international diffusion of renewable energy technologies is a prominent example. This paper deals with mechanisms that can accelerate the diffusion of climate-friendly technologies.

Three types of interactive processes seem to be interesting in this regard:

(1) Mutually reinforcing cycles: the interactive reinforcement of policy-induced domestic market growth, induced innovation, and policy feedback.

(2) Reinforced international diffusion of innovations from pioneer countries, which can be both (a) the diffusion of low-carbon technologies from lead-markets and (b) the diffusion of the supporting policy resulting from "lesson-drawing" by other countries.

(3) Reinforced diffusion by multi-level governance: Multi-level governance can stimulate vertical and horizontal learning at all levels of the global system. This has become particularly relevant regarding horizontal dynamics on the sub-national level being induced by higher levels.

These mechanisms are characterized by a multi-factorial interactive reinforcement of innovation and diffusion processes. A reinforced diffusion of climate-friendly technology can be observed at different levels of the multi-level system of global governance. The following analysis will refer to selected cases of best practice (a pragmatic methodological decision, which excludes the discussion of failures).

2. Economic and Political Mechanisms of Accelerated Diffusion

Mechanisms of acceleration and self-reinforcement are not unknown in economics and in political science. Brian Arthur presented a theoretical discussion on "dynamical systems of the self-reinforcing or autocatalytic type" both in the natural sciences and in economics. According to Arthur, self-reinforcing mechanisms in economics are related to four "generic sources":

- *Large set-up or fixed costs*, giving advantage to increasing economies of scale.
- *Learning effects*, which act to improve products or lower their costs.
- *Coordination effects*, which confer advantages to "going along" with other economic agents.
- *Adaptive expectations*, where increased prevalence in the market enhances beliefs of further prevalence [1].

Arthur mentions "virtuous cycles" and the option of "strategic action" as well as the possible role of policy "to 'tilt' the market" toward certain dynamics [1]. He also mentions an important condition for a new equilibrium: "self-reinforcement (that) is not offset by countervailing forces" but supported by "local positive feedbacks" [1]. Although this is not extended and lacks discussion or empirical analysis, Arthur gives a remarkable early theoretical view of a phenomenon that has become highly important, particularly in environment and climate policy research. We will present empirical cases, which are compatible with the typology of his "generic sources", but the picture is different if the mechanism of policy-feedback is included.

Modern *innovation research*, particularly on eco-innovation, has brought new theoretical and empirical insights into the phenomenon of accelerated technical change [2–5]. Political science has added the dimension of policy feedbacks to the interpretation of interactive dynamics in modern policy-making [6,7]: Policies generate resources, incentives, and information for political actors, which can reinforce the policy.

The present author has contributed to this research by adding the policy cycle to the reinforcing cycles of market growth and innovation in an analytical model for the diffusion of clean energy technology [8]. The policy cycle (agenda setting-policy formulation-decision-implementation-policy outcome-evaluation-new agenda setting, *etc.*) is a mechanism of policy learning and change. It is particularly open to policy feedback, for instance if there are unexpected co-benefits of the policy.

"Lesson-drawing" [9] is another potential mechanism of political reinforcement. It can support the diffusion of policy innovations, for instance if there is a certain "group dynamics" of countries: a collective learning leading to the broad adoption of a certain "trendy solution" [10].

There may be more types of acceleration. Economic but also regulatory competition [11] can reinforce the diffusion of goods or policies. Both economists and political scientists are familiar with the purposeful use of a window of opportunity [12]. Here we find an incidental convergence of "multiple streams" providing a situational opportunity for decision makers [13]. However, this does not necessarily produce a stable result. Windows of opportunity (such as the situation after the Chernobyl catastrophe) often close after a while. Therefore this type of acceleration is excluded from consideration here. This article deals with an accelerated transformation, *i.e.*, change with stable long-term effects [7].

The diffusion of innovative low-carbon technologies and innovative supporting policies are typically interlinked. There is, however, no clear causal relationship, but a pattern of multiple interactions between technology and policy [14]. Policy can support the innovators of a low-carbon technology, and the innovators may provide new technology-based policy options for climate policy. Policy may act as a first mover, and its diffusion by lesson-drawing may support the diffusion of the technology. Often, the technological innovation comes first (as in the case of wind power) and governmental support can reinforce its success in national and global markets. In any case, the interaction between policy and technology can contribute to a reinforced diffusion of both the low-carbon technology and the supporting policy. This is a "coordination effect" in terms of Arthur's classification [1].

In recent times there has been a rejuvenation of industrial policy [15,16]. It seems that green growth strategies and the designing of environmental and climate protection in terms of industrial policy are prominent examples of this tendency [17–20]. The translation of environmental and climate policy goals into the language of a technology-based economic strategy has become a success story in Germany and other countries such as China. Many governments regard themselves as actors in a highly competitive global market for clean technologies, in which innovation is considered the core of competitiveness [21,22]. From the perspective of climate policy this means that this policy has been able to mobilize economic interests. The following analysis will show that this ability can be observed at all levels of the global multi-level governance system.

3. Interactive Cycles of Climate-Friendly Innovation

It is a basic economic truth that growing markets induce demand for further innovation, which reduces production costs, improves the quality of the end-product and often reinforces market growth again. This is the learning effect in Arthur's classification [1]. Markets for climate-friendly technologies, however, are characterized by the fact that they are typically policy-driven [23]. Therefore, a third dynamic mechanism is relevant: not only the market and the technical innovation cycle, but also the policy cycle (see also [24]). It is essentially a political learning process from agenda-setting and policy formulation to the final outcomes and their evaluation.

Reinforcement by interactive cycles of low-carbon innovation can therefore be described as follows:

- Ambitious targets based on a clean-energy innovation plus effective policy implementation.
- Market growth of the supported clean-energy technology.

- Induced technological learning (secondary innovation).
- More ambitious targets: policy feedback from the new economic interests.

It has been shown that cases of accelerated diffusion of low-carbon technologies can be explained by the interaction of the three cycles (Figure 1). The author has studied 15 empirical cases in which these kinds of dynamic interactions can be observed [8,21]. The example of green power in Germany alongside the successive increase of targets (2020) is shown in Figure 2. As in certain other cases, the policy starts with an ambitious target inducing an unexpected market growth, which again induces innovation and finally a positive policy feedback in the form of an increase in the policy's targets. In 2000, the ambitious (and originally contested) German target was 20% green power by 2020. It was increased after nine years and again only one year later. The target for 2025 is now 40%–45% (compared with 1990).

Even more remarkable is the example of wind and solar energy in China. The target for wind power to be installed by 2020 was several times increased, from 20 GW to 200 GW, due to an unexpected rapid diffusion. The example of installed PV capacity is shown in Figure 3. Here the target increased from 1.8 to 100 GW.

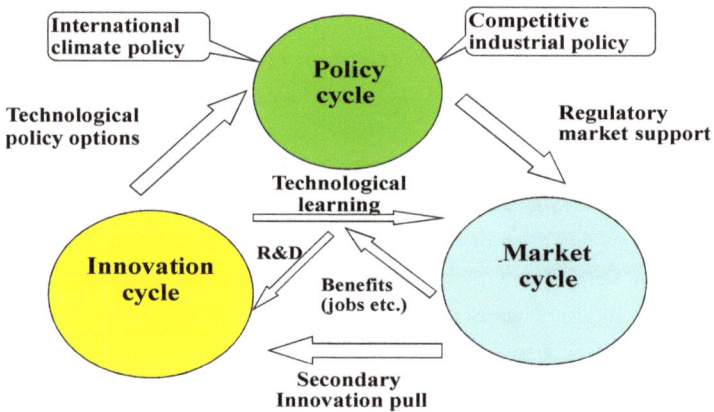

Figure 1. Mutually reinforcing cycles of clean-energy innovation [8].

The Intergovernmental Panel on Climate Change in its Special Report on Renewable Energy Resources and Climate Change Mitigation has drawn the policy conclusion regarding the "virtuous cycles" of clean-energy innovation: "that long-term objectives for renewable energies and flexibility to learn from experience would be critical to achieve cost-effective and high penetrations of renewable energies" [5].

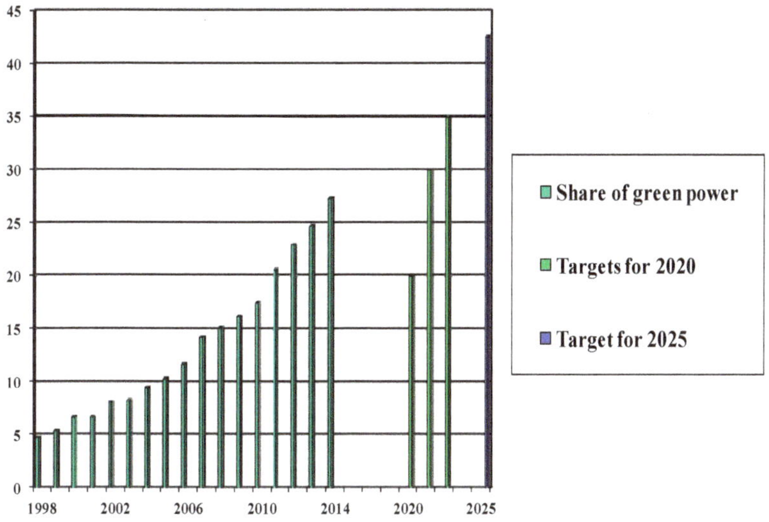

Figure 2. Share of green power 1998–2014 and targets for 2020/2025 in Germany (BMUB 2015).

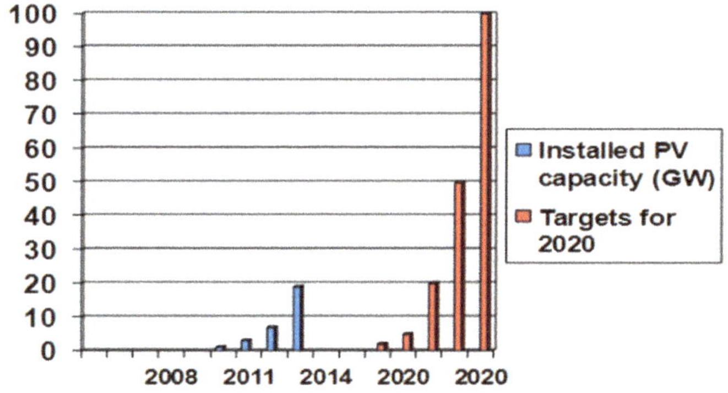

Figure 3. Installed PV capacity in China and targets for 2020 (Data basis REN21 2014).

4. Enforced Diffusion from Pioneer Countries: Lead Markets and Political Lesson-Drawing

A second mechanism of enforced diffusion is provided by national pioneers and trend setters [25]. The creation of a lead market for low-carbon technologies in a pioneer country together with political lesson-drawing [9] by other countries has been a prominent mechanism for the international diffusion of such technologies. Both mechanisms are independent, but they can reinforce each other.

Enforced diffusion of clean-energy innovation by lead markets can be described as follows:

- Lead markets are the national "runway" to start into the global market, where an innovative technology finds supportive conditions such as price, demand, or market structure.
- National lead markets for clean-energy innovations are specific because they are "policy-driven" and provide a regulatory advantage by political support.

- The international diffusion of the supporting policy ("lesson-drawing") can create an additional transfer advantage.

The economic mechanism is the enforced diffusion of climate-friendly technologies via lead markets. A national lead market is, according to Beise *et al.*, "the core of the world market where local users are early adopters of an innovation on an international scale" [26]. Well-known general cases are lead markets for mobile phones (Scandinavia), fax (Japan) or the Internet (USA). They originated in markets with special market advantages, such as price, market structure, demand, or export advantages.

Lead markets in pioneer countries have played a special role in the diffusion of low-carbon technologies. They re-financed the costs for technological learning until the product was sufficiently cheap and effective to diffuse into international markets. In addition, they had a demonstration effect showing how a certain climate-related problem could be solved, often including an economic advantage. This mechanism has become an important pathway for translating climate policy objectives into global markets. Examples encompass the development of wind power in Denmark and Germany, photovoltaic installations in Japan and Germany, heat pumps in Sweden, hybrid motors in Japan, and fuel-efficient diesel cars in Germany [26]. Examples for lead markets in emerging economies include solar water heating in China and bio-fuel technology in Brazil.

Lead markets for climate-friendly technologies arise in countries with a "regulatory advantage" and a "transfer advantage" [27]. This means that the technology and their international diffusion are supported by policy [28]. "Lesson-drawing" by other countries supports policy diffusion. This political "lesson-drawing" is the second mechanism of reinforced international diffusion. In the context of lead markets it refers to the process of learning how to support markets for a specific climate-friendly technology and results in the diffusion of a specific supporting instrument or policy mix. Lesson-drawing is similar to Arthur's mechanism of "adaptive expectations"—although it is *policy* learning. Similar to enforced technology diffusion, reinforced policy diffusion depends to a high degree on expectations, where increased prevalence in the global policy arena "enhances beliefs of further prevalence" [1].

Reinforced international diffusion by "lesson-drawing" can therefore be described as follows:

- "Trendy solutions" of pioneer countries are adopted by other countries, e.g., as a strategy to avoid domestic trial-and-error.
- "Adaptive expectations": increased diffusion enhances beliefs in further diffusion.
- Effect of "critical mass", *i.e.*, the stage in the process at which diffusion becomes self-perpetuating.

The anticipated probability that a certain regulation will become an international standard (also supported by international harmonization) has become a strong driver of policy diffusion [29]. A critical mass of countries adopting a certain trendy solution [10] reinforces the diffusion (see also [30]). At this stage, the process achieves sufficient momentum to become self-perpetuating.

The speed of diffusion and lesson-drawing in technology-based climate policy has been in many cases remarkable. The diffusion of the instrument of feed-in tariffs may be used as an illustration

(Figure 4 [31]). The diffusion of targets for green electricity occurred even faster. By early 2014, 144 countries had introduced targets for green power, a number that doubled since 2007 [32]. Even policies to support energy efficiency, which is often regarded as the more difficult part of climate policy, can have a high speed of international diffusion: out of 85 countries analyzed by the French institute ADEME, the share of countries with national targets for energy efficiency doubled within only five years to 80% [33]. This speed of diffusion is in clear contrast to the slow progress in international climate negotiations. Lesson-drawing has been characterized as "governance by diffusion" [34]. It is remarkable that it is a completely voluntary process, significantly different from global climate governance through legally binding international obligations [22].

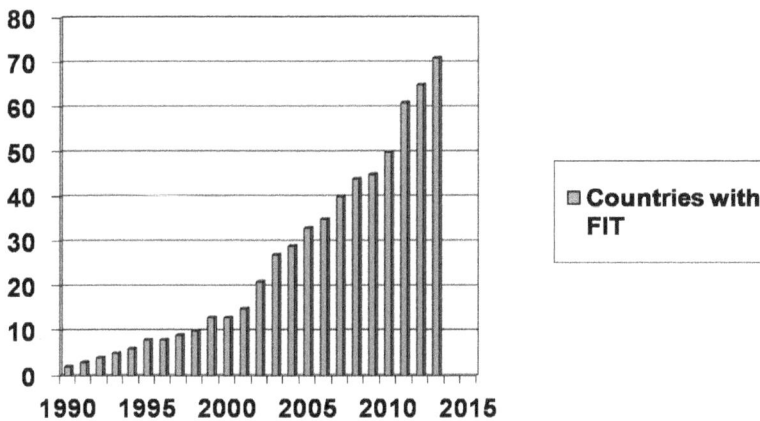

Figure 4. International diffusion of feed-in tariffs 1990–early 2013 (data basis [31]).

A special reinforcement of a lead-market process takes places when feedback from the international markets occurs, which is driven by second-mover countries now entering the original lead market by successfully producing similar products at lower prices. The Chinese solar industry and its booming exports to Europe may be taken as an example [35]. The case marks a situation where a former lead market has to find a new role in the competition for innovation. This may create difficulties for the former pioneer. However, in terms of climate protection, this reinforcement of diffusion based on lower prices is a clear advantage.

So far, lead markets in rich countries have provided the basis for clean technologies to diffuse from industrialized and emerging economies into international markets. A more recent development is the role of lead markets in emerging countries like India, where the lag markets are developing countries. Most interestingly for a sustainable energy future are lead markets for frugal innovations [36]. Frugal innovations are cheap, simple and robust. Beyond that they also try to save resources at all stages of the supply chain [37]. They are worth mentioning here, because, due to a generally low profit share, they depend on large-scale markets. The existence of such large markets in emerging economies can lead to the advantage of falling unit costs to increased output as a mechanism of reinforcement [1].

476

5. Multi-Level Governance: the Vertical Reinforcement of Horizontal Diffusion

5.1. The Multi-Level System of Global Climate Governance

Multi-level governance "characterizes the mutually dependent relationships—be they vertical, horizontal, or networked—among public actors situated at different levels of government" ([38], see also [39,40]). The multi-level perspective (MLP) is "a middle-range theory that conceptualizes overall dynamic patterns in socio-technical transitions" [41]. Multi-level reinforcement is a particularly interesting aspect (Figure 5). Schreurs and Tiberghien have used this terminology to explain the dynamics of climate policy in the European Union and its member states ([42], see also [43]). However, it is also relevant in the global context. Here it is used to explain the dynamic interaction between the national and the sub-national levels.

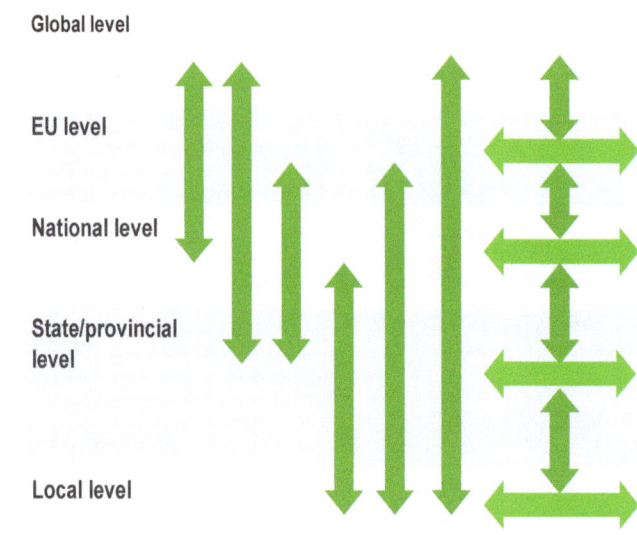

Figure 5. Multi-level governance: possible horizontal and vertical interactions.

Multi-level governance—from the global to the local level—can be regarded as a general mechanism of reinforcement. The broad variety of possible vertical and horizontal interactions makes it possible that innovation take place at different parts of the governance system (Figure 5). Furthermore, they can be adopted in other parts of the system. Lesson-drawing (peer-to-peer learning) from pioneers is possible at all levels. Horizontal up-scaling of de-central innovations and best practice is another mechanism. The resulting policy of higher levels (nation state, the European Union) can stimulate horizontal dynamics at lower levels. This aspect of multi-level reinforcement will be exemplified by the case of the European Union.

The multi-level system of climate governance is a global system because it has a global base of climate-related knowledge, motivation, and legitimacy. Moreover, it is a global system because a global market for climate-friendly technology has been established together with the global arena of climate policy. An important condition of its innovation dynamics is the leadership role of the higher levels.

At each level of the multi-level system of global climate governance, a broad variety of motives and opportunities can be observed. At the level of provinces/regions or federal states, the following motives to support or to adopt climate-friendly technologies exist: rich regions can be motivated to transfer their successful economic policy to the new field of climate policy. Poor regions, on the other hand, can try to support renewable energies or energy-saving investments in the housing sector to overcome unemployment. Another driver may be opposition of the region against the national government (as in the case of Scotland, Quebec or California). Geographical advantage might provide another condition to support renewals (as with wind energy in coastal zones). Political scientists often point to the party constellation of a certain regional/state government [44]. In the EU, there are several responsibilities for climate and energy—beyond emission trading—at the regional level [45]. The EU has a regional commissioner and a regional committee, which has recently issued a multi-level governance convention (2014). There exist international horizontal networks such as the R20 Regions, or the Network of Regional Governments for Sustainable Development, which has been established at the World Summit in Johannesburg (2002).

Cities and local communities have important responsibilities in policy areas that are relevant to climate policy. Housing and the energy consumption of households, transport regulations and infrastructures, land-use and urban planning, or waste policy are important policy fields in this regard. Most important is the responsibility for local energy supply, where cities in Europe or the US can have strong influence [38]. The fact that 80% of EU greenhouse gases emissions are related to urban activities illustrates the importance of the local level. Thus, cities are also important places for climate policy experiments and innovations [46]. Horizontally active international networks such as ICLEI or the Covenants of Mayors play an important role [47]. In addition, national networks such as the City Energy Project (USA) or the Chinese Low Carbon Eco-Cities Association can play a role [48]. The German "100%-Renewable Energy" network (Figure 6) is a remarkable example of horizontal dynamics at the lowest level being supported by climate policy activities at higher levels.

Local climate mitigation and horizontal lesson-drawing between cities is being generally supported by the EU Commission and also by the central government in India and China.

5.2. Horizontal Dynamics Induced by Higher Levels

The reinforcement at lower levels by multi-level governance can be described as follows:

- Experimentation, innovation and best practice at different levels.
- Local and regional best practice being scaled up and supported by the higher level.
- Support from the higher level inducing horizontal dynamics at the lower level: pioneers become relevant as benchmark, partners or competitors.

Political leadership on the higher levels can impose the up-scaling and generalization of experiments, innovations and best practices from lower levels. If the higher levels take the lead, providing regulatory financial or informational support to the lower levels, they will strengthen the

478

role of pioneers at the lower levels and induce horizontal lesson-drawing, cooperation, or competition at the same level (Figure 7). Pioneer cities or provinces/states at the lower levels become benchmarks for others. Support from above therefore provides new means and opportunities for the diffusion of climate-friendly innovation. From an economic perspective this includes the potential broadening of markets to a supra-regional scale. The mobilization of economic interests and the translation of climate policy goals into the language of market dynamics is an integrating common factor at all levels.

- **2010: 72 Renewable-Energy Regions, 7,8 Mio. inhabitants,**

- **2014: 146 Renewable-Energy Regions (60 starter regions), 25 Mio inhabitants**

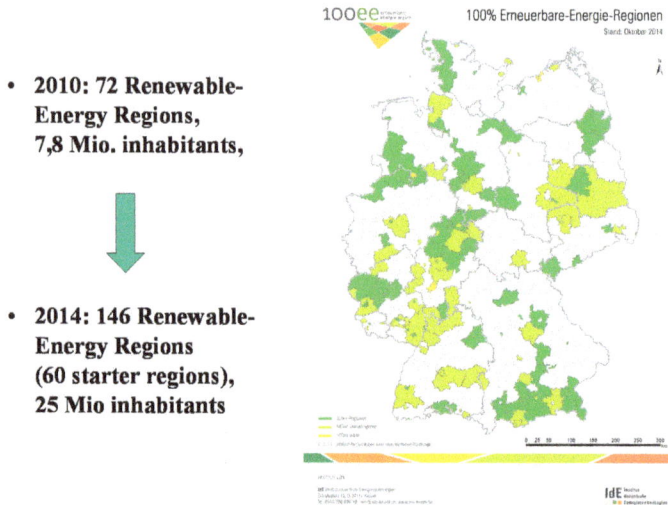

Figure 6. "100%-Renewable-Energy" Regions in Germany 2014 (2010) (Umwelt 12/2012, IdE 2015).

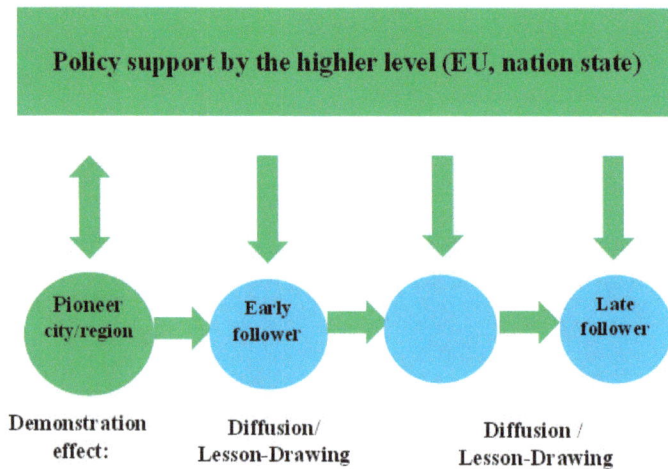

Figure 7. Horizontal dynamics induced by the higher policy level.

6. The Case of the European Union

The EU has provided the best practice in climate mitigation and multi-level climate governance. As a regional system of climate governance it is unique compared with other world regions (NAFTA, African Union, ASEAN *etc.*). Regarding to the global 2° K target the achievements may not be sufficient. Nevertheless, they are remarkable because they have not been expected before. Greenhouse gases have been reduced by nearly 20% from 1990 to 2012. Moreover, the target for 2020 has already been nearly achieved (Figure 8). The accelerated speed of greenhouse gas reduction may be partly explained by the economic downturn. However, the diffusion of renewable energies had a similar tendency. By 2013 renewable energies accounted for 70% of new electric power capacity (Figure 9). One year later, the share was 79%—a significant increase from 57% only five years before [31,32].

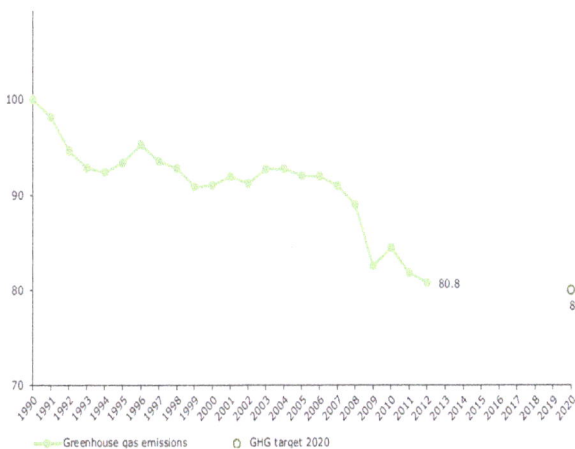

Figure 8. EU-28 greenhouse gas emissions, 1990–2012 (EEA 2014).

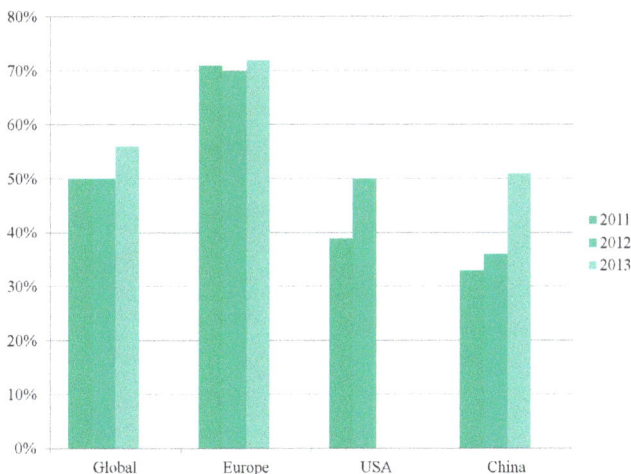

Figure 9. Green power as share of new power capacity, 2011–2013 [31].

The effect of the EU climate package (2008) cannot be disputed. It seems remarkable that this package was partly motivated by ideas of a policy-driven dynamic market for low-carbon technologies. Already in 2007, the EU Commission proposed a comprehensive lead market and innovation strategy "to create a virtuous cycle of growing demand, reducing costs by economies of scale, rapid product and production improvements and a new cycle of innovation that will fuel further demand and a spinout into the global market" [49]. After 2007 the European market was indeed the lead market for wind and solar power. Innovation and lesson-drawing took place at a high level throughout Europe. In some sense an interaction of the described mechanisms of reinforcement—a syndrome of reinforcement—could be observed at that time. Although the EU top level lost some of its dynamics in recent time, the European system of multi-level climate governance remained a strong driver for climate-friendly modernization [43]. This needs to be explained more broadly.

Multi-level climate governance was a purposeful strategy of the European institutions. There is a special institutional framework for regions/provinces and also a climate governance strategy for cities. Other characteristics of EU countries that provide a green opportunity structure include green political parties and public media. The EU has turned a "free market" into a market with strong environmental framework conditions. The World Bank recently confirmed that the EU has a specific "environmentally sustainable growth model" [50].

Beyond that there is a specific European mechanism of multi-level reinforcement: the interaction of national environmental policy innovation with the European harmonization mechanism in the context of the common market. The EU commission can under certain conditions authorize member states to maintain or introduce stricter measures of environmental policy. When a member state is authorized to do this, "the Commission shall immediately examine whether to propose an adaptation to that measure" (Treaty on the Functioning of the European Union, Art. 114.7; 193). Environmental policy innovation in member states can with a certain probability become a European regulation, which follows the principle of a "high level of environmental protection". Climate policy is part of the EU environmental policy (Art 191). This mechanism can stimulate regulatory competition between member states to become the frontrunner of a European regulation [11,42]. It can be regarded as lesson-drawing by the EU Commission: learning from empirical best practice, avoiding time consuming experimentation and being supported by certain national governments, which can also provide competent advice. The UK emissions trading scheme (2002) can be taken as an example. It was also intended to deliver "first mover advantages" to UK companies before the introduction (2005) of the EU emissions trading scheme [51]. Other examples are the UK Energy Efficiency Commitment (2002) and the German Renewable Energy Law (2000), both being followed by EU Regulations (2001 and 2012).

Climate policy as a process started in the EU at the national and sub-national level. Pioneer countries like Germany, Denmark, and the United Kingdom (UK) generalized and integrated many political and economic experiments and best practices that had already taken place at lower levels, paving the way for their adoption at a higher level. Thus, the process of climate policy has moved bottom-up to the European and global levels. Extending the national policy innovations to the European Union has often been a governmental strategy of member states to stabilize the national pioneer role, but also to create a European market for domestic innovations in climate-friendly technologies.

The Europeanization of climate policies was accompanied by the establishment of lobby organizations, which articulated an economic interest for clean energy at the EU level. Examples include the European Renewable Energy Council, the European Alliance to Save Energy, the European Insulation Manufacturers Association, Lighting Europe and the European Heat Pump Association.

Meanwhile feedback can be observed at the local level, reinforcing earlier initiatives: cities and local communities, often organized as networks [47], use national and European policies and incentives—whether regulations, subsidies, or public procurement—to mobilize economic interests for climate-friendly technologies. These can be investments in the form of renewable energies or low-energy buildings.

Most remarkable is the role of the Covenant of Mayors with about 6400 (2015) participating local communities. It was launched by the European Commission together with the EU climate and energy package in 2008. Within this network, the participating local authorities have to present action plans and a GHG reduction target of at least 20%. The economic dimension is underlined by the fact that the European Investment Bank is involved in the financing of implementation measures. The Smart Cities Partnership Initiative of the EU Commission is a similar economic mechanism. The horizontal dynamics—particularly the competition between cities—are stimulated by an official Benchmark of Excellence, which is also a database of best practice [52].

Private ownership of green power seems to be a strong driver of change at the local level in several countries. In Germany, more than half of the green power installations are owned by private person. Europe, when compared with other global regions, has not only the advantage of a strong supra-national level of climate governance, but it also started early with a high proportion of decentralized and local ownership of green power installations (Bloomberg New Energy Finance 4/2014).

It seems that the local level now has become the most dynamic driver of technical change towards a low-carbon energy system. An evaluation of the Covenant of Mayors shows that 63% of the local communities being assessed by the EU are planning to reduce GHG emissions by more than 20%. A reduction of about 370 million tons is expected by 2020 (ENDS Europe 24 June 2013). The database of the Covenant provides empirical evidence that in recent years, the climate policy process has mobilized strong economic interests at the local level, mainly in the building sector (44% of the activities) and in local energy production.

The former policy initiative at the higher levels has created the necessary preconditions for this booming development at the sub-national level. The EU Directive on Energy Performance of Buildings, for instance, has stimulated a strong activity among local communities with pioneer cities such as Freiburg, Manchester, Copenhagen, and Malmö playing an important role.

It seems that pioneer countries like Germany, Denmark, and the United Kingdom are also leading countries when local dynamics are concerned. These three countries have achieved the highest GHG reduction rates. They also have the most ambitious GHG reduction targets for the period 1990–2025 (Germany 40%–45%, UK 50%; Denmark 40% by 2020). In all three countries this is the result of policy-induced reinforcing cycles of innovation and market growth with policy feedback. They are also cases of best practice regarding the mobilization of economic interests for climate governance at the sub-national level.

7. The Advantage of a Polycentric Approach

As we have seen, it is no disadvantage that the implementation of climate policy takes place under the condition of a broad variety of actors, dimensions, and levels. On the contrary: a "polycentric approach" [53] can be a real opportunity [54]. It should be mentioned that this polycentric approach includes not only governments and businesses, but also societal actors. Civil society—with networks of all kinds and at all levels of the multi-level game—seems to be the indispensable context of the energy transformation, although its highly complex causality is not easy to assess in terms of empirical research.

The extremely high complexity of multi-level climate governance may cause the problem of final responsibility: if everybody is responsible, in the end there might be a situation in which nobody actually takes responsibility. So far, reaching a solution is still primarily the final responsibility of national governments acting within broad networks, often as collective players (e.g., G20). National governments, if compared with the small administration of global regimes, such as under the United Nations Framework Convention on Climate Change, have more human and financial resources. They can impose sanctions and penalties. They act under comparably higher pressure to provide legitimacy for their actions. They are the first responders in the event of extreme weather and other crises and they are observed more intensively by the public than government actors at other levels of the global multi-level governance system [21].

8. Conclusions

This article has shown that the acceleration of the diffusion of clean-energy technologies is a potentially strong option for climate policy (see also [22]). As Arthur stated long ago, mechanisms of reinforcement can be found in natural as well as in economic systems. Although his typology was abstract and theoretical, it shows many similarities with the empirical dynamics in climate governance that have been presented in this paper. It is, however, necessary to include the role of government in this analytical perspective to explain the special dynamics of climate-related governance.

Three mechanisms of reinforcement have been presented: (1) the dynamics of policy-driven domestic markets and innovation processes, that lead to a policy feedback, due to unexpected success but also due to the creation of a new interest base; (2) the dynamics of both the global climate policy arena and the global markets for clean-energy technologies: lead markets supported by country-to-country lesson-drawing; and (3) the dynamics of multi-level reinforcement, which is based on vertical and horizontal interaction and learning. It includes the up-scaling of decentralized innovations and the top-down implementation of climate policy measures on lower levels. In this paper we have focused on the horizontal dynamics of sub-national levels (such as networking, benchmarking, or competition between cities) being induced by top-down climate policy support. The three mechanisms are highly likely to support each other. In the EU they have sometimes (particularly after 2007) led to a syndrome of reinforcement. The list of possible accelerators may be longer than those presented. One additional likely mechanism of acceleration is the break-even point of simultaneously rising prices for fossil energy and the falling prices of renewable energies.

It seems that the accelerators discussed can best be understood within the system of global multi-level climate governance. The governance system of the EU can be regarded as the most advanced sub-system. Meanwhile the multi-level system of global climate governance seems to have achieved its own inherent logic. It can be characterized by typical horizontal and vertical dynamics as well as long-term stabilization mechanisms, based on institutional change, new economic interests, and policy feedback. This governance system has created opportunities for innovation and its diffusion. The broad variety of agents and possible interactions (Figure 5) could be seen as one of its main characteristics. The interaction between levels is another one; such vertical interactions are often connected with horizontal dynamics at different levels: pioneer activities and lesson-drawing, networking, and cooperation, as well as competition. They have become increasingly important, particularly at the sub-national level (Figures 6 and 7).

Several policy measures can be used to support this process and to stimulate acceleration, although a comprehensive strategy still needs to be developed. Thus far, these processes are mainly the result of an interactive learning-by-doing. The dynamics in most cases have been induced by competent practitioners. This means that they are not the result of scientific design; instead, they are most often unintended and unexpected.

The IPCC stated in its 5th Assessment Report that "…institutional and governance changes can accelerate a transition to low-carbon paths" [55]. However, there can be no doubt about the difficulty of translating the complex task of multi-level governance into a comprehensive strategy. There needs to be more research on best practices to draw better and more comprehensive conclusions for government strategies. The main general conclusions of this explorative analysis can be summarized as follows:

1) Translating climate policy objectives into the language of industrial policy and ecological modernization [21] is a strong option for climate policy (while it is not the only solution, since there are limits to technological approaches).

2) Ambitious goal-oriented climate policies can induce market growth and interactive technological learning (secondary innovation). Successful learning by doing, increased capacity, and the creation of new interests could lead to a policy feedback with even higher ambition.

3) Multi-level governance is a highly important institutional opportunity structure for the innovation and diffusion of clean-energies and their supporting policies. Best practice can arise and lesson-drawing can take place at quite different points of the system of multi-level climate governance.

4) Therefore base policy on existing best practices at different levels and provide channels for lesson-drawing and interactive learning. Apply ambitious but realistic targets and credible implementation programs. Raise ambitions and targets in cases of unexpected success.

5) Give targeted support to sustainable R&D initiatives; use the lead market mechanism where possible.

6) Support lower levels of government and stimulate horizontal dynamics through benchmarking, competition, lesson-drawing, cooperation, and networking.
7) National governments—both as single and collective actors—so far exhibited the strongest capacities and therefore should lead with ambitious climate policies.

Conflicts of Interest

The author declares no conflict of interest.

References

1. Arthur, B. Self-reinforcing Mechanisms in Economics. In *The Economy as an Evolving Complex System*; Anderson, P., Arrow, K.J., Pines, P., Eds.; Addison-Wesley: Reading, MA, USA, 1988.
2. Watanabe, C.; Wakabayashi, K.; Miyazawa, T. Industrial Dynamism and the Creation of a "Virtuous Cycle" between R&D, Market Growth and Price Reduction. The Case of Photovoltaic Power Generation (PV) Development in Japan. *Technovation* **2000**, *20*, 225–245.
3. Hekkert, M.P.; Suurs, R.A.A.; Negro, S.O.; Kuhlmann, S.; Smits, R.E.H.M. Functions of innovation systems: A new approach for analyzing technological change. *Technol. Forecast. Soc. Change* **2007**, *74*, 413–432.
4. Bergek, A.; Jacobsson, S.; Carlsson, B.; Lindmark, S.; Rickne, A. Analyzing the functional dynamics of technical innovation systems. A scheme of analysis. *Res. Policy* **2008**, *37*, 407–429.
5. Intergovernmental Panel on Climate Change (IPCC). *Special Report on Renewable Energy Resources and Climate Change Mitigation (SRREN)*; IPCC: Geneva, Switzerland, 2011.
6. Pierson, P. When Effect Becomes Cause-Policy Feedback and Political Change. *World Polit.* **1993**, *45*, 595–628.
7. Patashnik, E.M. *Reforms at Risk: What Happens after Major Changes are Enacted*; Princeton University Press: Princeton, NJ, USA, 2008.
8. Jänicke, M. Dynamic Governance of Clean-Energy Markets: How Technical Innovation Could Accelerate Climate Policies. *J. Clean. Prod.* **2012**, *22*, 50–59.
9. Rose, R. *Lesson-Drawing in Public Policy. A Guide to Learning across Time and Space*; CQ Press: Chatham, NJ, USA, 1993.
10. Chandler, J. Trendy solutions: Why do states adopt sustainable energy portfolio standards? *Energy Policy* **2009**, *37*, 3247–3281.
11. Héritier, A.; Mingers, S.; Knill, C.; Becka, M. *Die Veränderung der Staatlichkeit in Europa*; Leske + Budrich: Opladen, Germany, 1994.
12. Kingdon, J.W. *Agendas, Alternatives and Public Policies*, 2nd ed.; Harper & Collins: New York, NY, USA, 1995.
13. Zahariadis, N. Ambiguity, Time, and Multiple Streams. In *Theories of the Policy Process*; Sabatier, P.A., Ed.; Westview Press: Boulder, CO, USA, 1999; pp. 73–93.
14. Jänicke, M.; Jacob, K. (Eds.) *Environmental Governance in Global Perspective*, 2nd ed.; Forschungszentrum für Umweltpolitik, Freie Universität Berlin: Berlin, Germany, 2007.

15. Stiglitz, J.E.; Lin, J.Y. (Eds.) *The Industrial Policy Revolution I: The Role of Government Beyond Ideology*; Palgrave Macmillan: New York, NY, USA, 2013.
16. Hallegatte, S.; Fay, M.; Vogt-Schilb, A. *Green Industrial Policies—When and How*; The World Bank, Policy Research Working Paper 6677: Washington, DC, USA, 2013.
17. United Nations (UN). *Industrial Policy for the 21st Century: Sustainable Development Perspectives*; UN Department of Economic and Social Affairs: New York, NY, USA, 2007.
18. United Nations Environment Programme (UNEP). *Towards a Green Economy: Pathways to Sustainable Development and Poverty Eradication*; UNEP: Nairobi, Kenya, 2011.
19. Organisation for Economic Co-operation and Development (OECD). *Towards Green Growth*; OECD: Paris, France, 2011.
20. World Bank. *Inclusive Green Growth. The Pathway to Sustainable Development*; The World Bank: Washington, DC, USA, 2012.
21. Jänicke, M. *Megatrend Umweltinnovation*, 2nd ed.; Oekom: München, Germany, 2012.
22. Stern, N.; Bowen, A.; Whalley, J. (Eds.) *The Global Development of Policy Regimes to Combat Climate Change*; The Tricontal Series of Global Economic Issues; World Scientific Publishing Co Pte Ltd.: Singapore, 2014; Volume 4.
23. Ernst & Young. *Eco-Industry, Its Size, Employment, Perspectives and Barriers to Growth in an Enlarged European Union*; EU Commission, DG Enviroment: Brussels, Belgium, 2006.
24. Dierkes, M.; Antal, A.B.; Child, J.; Nonaka, I. (Eds.) *Handbook of Organisational Learning and Knowledge*; Oxford University Press: Oxford, UK, 2001.
25. Jänicke, M. Trend Setters in Environmental Policy: The Character and Role of Pioneer Countries. *Eur. Environ.* **2005**, *5*, 129–142.
26. Beise, M.; Blazejczak, J.; Edler, D.; Jacob, K.; Jänicke, M.; Loew, T.; Petschow, U.; Rennings, K. The Emergence of Lead Markets for Environmental Innovations. In *Nachhaltige Innovation. Rahmenbedingungen für Umweltinnovationen*; Horbach, J., Huber, J., Schulz, T., Eds.; Oekom: München, Germany, 2003; pp. 13–49.
27. Rennings, K.; Schmidt, W. A Lead Market Approach towards the Emergence and Diffusion of Coal-fired Power Plant Technology. *Polit. Econ.* **2010**, *27*, 301–327.
28. Lacerda, J.S.; van den Bergh, J.C.J.M. International Diffusion of Renewable Energy Innovations: Lessons from the Lead Markets for Wind Power in China, Germany and USA. *Energies* **2014**, *7*, 8236–8263.
29. Jänicke, M.; Joergens, H.; Tews, K. Zur Untersuchung der Diffusion umweltpolitischer Innovationen. In *Die Diffusion Umweltpolitischer Innovationen im Internationalen System*; Tews, K., Jänicke, M., Eds.; Westdeutscher Verlag: Wiesbaden, Germany, 2005.
30. Witt, U. "Lock-in" *vs.* "critical masses"—Industrial Change under Network Externalities. *Int. J. Ind. Organ.* **1997**, *15*, 753–773.
31. REN21. *Renewables 2013. Global Status Report*; REN21: Paris, France, 2013.
32. REN21. *Renewables 2014. Global Status Report*; REN21: Paris, France, 2014.
33. Agence de l'Environnement et de la Maîtrise de l'Énergie (ADEME). *Energy Efficiency in the World Report (Study Produced for the World Energy Council)*; Paris, France, 2013.

34. Busch, P.O.; Joergens, H.; Tews, K. The Global Diffusion of Regulatory Instruments. The Making of a New International Environmental Regime. *Ann. Am. Acad. Polit. Soc. Sci.* **2006**, *598*, 146–167.

35. Quitzow, R. *The Co-evolution of Policy, Market and Industry in the Solar Energy Sector*; FFU-Report O6–2013; Forschungszentrum für Umweltpolitik/Freie Universität Berlin: Berlin, Germany, 2013.

36. Tiwari, R.; Herstatt, C. *India—A Lead Market for Frugal Innovations?* Working Paper Technology Innovation Management No. 67; Hamburg University of Technology: Hamburg, Germany, 2012.

37. Jänicke, M. Frugale Technik. *Ökol. Wirtsch.* **2014**, *29*, 30–36.

38. Organisation for Economic Co-operation and Development (OECD). *Green Growth in Cities*; OECD: Paris, France, 2013.

39. Bache, I.; Flinders, M. *Multi-Level Governance*; Oxford University Press: Oxford, UK, 2004.

40. Stephenson, P. Twenty years of multi-level governance: Where does it come from? What is it? Where is it going? *J. Eur. Public Policy* **2013**, *20*, 817–837.

41. Geels, F.W. The multi-level perspective on sustainability transitions: Responses to seven criticisms. *Environ. Innov. Soc. Transit.* **2011**, *1*, 24–40.

42. Schreurs, M.; Tiberghien, Y. Multi-Level Reinforcement: Explaining European Union Leadership in Climate Change Mitigation. *Glob. Environ. Polit.* **2007**, *7*, 19–46.

43. Jordan, A.; van Asselt, H.; Berkhout, F.; Huitema, D.; Rayner, T. Understanding the Paradoxes of Multi-Level Governing: Climate Change Policy in the European Union. *Glob. Environ. Polit.* **2012**, *12*, 43–66.

44. Delmas, E.; Montes-Sancho, M.J. U.S. State Policies for Renewable Energy: Context and Effectiveness. *Energy Policy* **2011**, *39*, 2273–2288.

45. Wolfinger, B.; Steininger, K.W.; Damm, A.; Schleicher, S.; Tuerk, A.; Grossman, W.; Tatzber, F.; Steiner, D. Implementing Europe's Climate Targets at the Regional Level. *Clim. Policy* **2012**, *12*, 667–689.

46. Bulkeley, H.; Castán Broto, V. Government by Experiment? Global Cities and the Governing of Climate Change. *Trans. Inst. Br. Geogr.* **2012**, *38*, 361–375.

47. Kern, K.; Bulkeley, H. Cities, Europeanization and Multi-Level Governance: Governing Climate Change through Transnational Municipal Networks. *J. Common Mark. Stud.* **2009**, *47*, 309–332.

48. Zhou, N.; He, G.; Williams, C. *China's Development of Low-Carbon Eco-Cities and Associated Indicator Systems*; Ernest Orlando Lawrence Berkely National Laboratory: Berkeley, CA, USA, 2012.

49. EU Commission. A Lead Market Initiative for Europe-Explanatory Paper on the European Lead Market Approach: Methodology and Rationale. In *Commission Staff Working Document*; (COM(2007)) 860 Final, SEC(2007); Commission of the European Communities: Brussels, Belgium, 2007.

50. World Bank. *Golden Growth—Restoring the Lustre of the European Economic Model*; The World Bank: Washington, DC, USA, 2011.

51. Rayner, T.; Jordan, A. The United Kingdom: A Paradoxical Leader? In *The European Union as a Leader in International Climate Change Politics*; Wurzel, K.W., Conelly, J., Eds.; Routledge: London, UK; New York, NY, USA, 2011; pp. 95–111.

52. Covenant of Mayors. Available online: http://www.covenantofmayors.eu/index_en.html (accessed on 12 June 2015).

53. Ostrom, E. Beyond Markets and States: Polycentric Governance of Complex Economic Systems. *Am. Econ. Rev.* **2010**, *100*, 641–672.

54. Sovacool, B.K. An International Comparison of Four Polycentric Approaches to Climate and Energy Governance. *Energy Policy* **2011**, *39*, 3832–3844.

55. Intergovernmental Panel on Climate Change (IPCC). *Fifth Assessment Report III: Climate Change 2014: Mitigation of Climate Change*; Cambridge University Press: New York, NY, USA, 2014.

Chapter 5:
Juridical Framework

The Smart City and the Green Economy in Europe: A Critical Approach

Rosario Ferrara

Abstract: It is shown in this article that the current European legislation makes the future progress of smart cities critically dependent on the advancement of the green economy and consequently on the further development of energy efficiency and of renewable energy sources. However, the lack of a clear legal framework capable of transforming the current pledges into binding rules at national level can jeopardize the establishment of a more direct and profitable link between the extensive European legislation on energy and environment, and the harmonious and efficient development of smart cities in Europe.

Reprinted from *Energies.* Cite as: Ferrara, R. The Smart City and the Green Economy in Europe: A Critical Approach. *Energies* **2015**, *8*, 4724-4734.

1. Introduction

The concepts of "smart city" and "green economy", as delineated in European regulations and directives, may sometimes appear to be rather cursory and are sometimes regarded as little more than simple slogans born in the shadow of the economic crisis and destined to disappear, almost without a trace, by the time the crisis is over.

While it is extremely difficult to make predictions in this regard, there might certainly be some elements of truth in this analysis; however it is also evident that today these conceptual categories, defined as values and objectives to be pursued and achieved, are heavily employed in the context of the general policies of the European Union.

Indeed, there are many documents of the European Union that are along these lines. They are often in the form of soft law propositions with the simple aim of introducing medium-long term targets and are subsequently followed by more binding regulatory measures, with a view to providing stable provisions to the general principles previously enunciated.

One of them is the document *Smart Cities & Communities* [1], issued by the European Commission in 2011, containing the outline of an optimal, virtually perfect model of smart city, in which it is possible to reconcile and combine economy and ecology [2], and in which it is, therefore, possible to implement any necessary synergy between the protection of the environment and the development of new technologies, including policies of environmental sustainability and transformation processes of the urban land [3].

In accordance with the definition that clearly inspired the document of the European Commission quoted above: "*A smart city...uses digital technologies to enhance performance and well-being, to reduce costs and resource consumption and to engage more effectively and actively with the citizens. Key smart sectors include transport, energy, health care, water and waste. A smart city should be able to respond faster to urban and global challenges than one with a simple 'transactional' relationship with its citizens*".

In other words: "*Interest in smart cities is motivated by major challenges, including climate change, economic restructuring, retail and entertainment services moving online, ageing populations, and pressures on public finances*". The terms 'intelligent city' and 'digital city' are also used [4].

These simple and concise excerpts taken from the most commonly used definitions show that the issue of smart cities is functionally related to that of environmental policies, and especially of environmental policies at the local level: cities, communities, regions. Thus, structure and governance of smart cities (or of smart communities, if related to wider areas) are based on the institutions of the so-called green economy, which is a key issue.

Indeed, the fundamental project of the green economy is the integration and reconciliation of economic growth and environmental values, as foreseen by the Article 11 of the Treaty on the Functioning of the European Union (TFEU). Consequently, in the papers of the scientific community as well as in EU documents, the smart city is substantially identified with the sustainable city, its smartness coinciding with its sustainability. This is going to become the new paradigm for the most important cities in the world [5], especially in the Western Hemisphere.

In this article, I shall outline the links between the considerable amount of *acquis communautaire* in the form of regulations, directives and soft law tools related to energy and environment with the vision of a smart city. The necessity of a formal framework to transform the current pledges into binding rules will be examined in the conclusions.

2. Analysis and Discussion

The proper frame of reference for smart cities is provided by a number of fundamental EU documents that regard the intelligent and efficient use of energy as the necessary medium-long term objective for the attainment of sustainability. The smart and consequently sustainable city must be capable of limiting the impact of the environmental overload caused by undue expenditure of energy and by an energy mix with a low share of renewable energy content.

Smart cities have even been predicted to become the future "*industrial (or) manufacturing cities*" [6], a globalized business model where the renewable energies can play, in a "smart context", an important economic role.

2.1. Precursory Concepts of the Smart Cities

While originally focused on ICT, the smart city concept has grown beyond this limited scope to include a better use of resources and a lower amount of emissions, which means more efficient urban transport networks, better waste disposal facilities, and energy-saving ways to light and heat buildings.

Being focused on sustainable mobility in urban environments and increased use of renewable energies the three initiatives contained in the *Intelligent Energy Europe Programme* (examined more in detail in the next section) have clearly contributed to paving the way to the Launch Conference of the Smart Cities and Communities Initiative by the European Commission in 2011.

From the very beginning, smart cities have been related to sustainability. This has become their most relevant factor of identification. In particular, sustainability has been associated with a greater enjoyment of life—a formulation clearly influenced by important doctrines (especially of French origin) that assert the right to pursue happiness. Indeed, one of the indispensable prerequisites of this right to happiness is the claim to a rational and, in fact, happy organization of the city, capable of adequately combining economy and ecology [7]. In other words a veritable *droit à la ville* [8], namely the right to a city that considers sustainability the founding character of its identity. It is, therefore, no surprise that Paris is a candidate to become, by 2050, a smart city of European level, a model of city life that has in sustainability (and in the "beauty" that naturally derives from it) its fundamental strength and indispensable identification [9].

The sustainability of the smart city appears to be a direct consequence of the EU strategy "20-20-20", heralding a comprehensive legislation on energy efficiency and production of energy from renewable sources. Its legal basis (*i.e.*, the directives resulting from the common strategy "20-20-20") can be found in two important provisions of the TFEU: article 192, second paragraph, which allows derogations from the ordinary procedure of decision in case of *"provisions primarily of a fiscal nature"*, of measures affecting the environment and of *"measures significantly affecting the choice of a Member State between different energy sources and the general structure of its energy supply"*, and article 194, under which EU policies in the energy sector must take into account *"... the need to preserve and improve the environment..."* [10,11].

2.2. A Review of the Relevant European Legislation

The probably best known measure is the "challenge" launched by the European Union in March 2007 with the fielding of the strategy "20-20-20" (the so-called climate and energy package), albeit in the less binding form of soft law. The Union's aim for the year 2020 is three-fold: reducing greenhouse gases by 20% (or 30%, provided an international agreement is attained on this issue); reducing energy consumption by 20% thanks to the gradual increasing of energy efficiency; increasing the share of renewable energy to at least 20%.

The climate and energy package was building on the results of *Intelligent Energy Europe Programme*, originally adopted in 2003. It foresaw the establishment of a roadmap for the energy choices of the European Union and the Member States to the distant 2050. The programme was divided into three main areas:

1. Fostering energy efficiency and the rational use of energy resources (SAVE initiative);
2. Promoting new and renewable energy sources and support energy diversification (ALTENER initiative);
3. Promoting energy efficiency and the use of new and renewable energy sources in the field of the transport (STEER initiative).

Five directives, one decision and one regulation are the most significant legal acts of the European Union that should be considered. They will be listed in chronological sequence.

(a) Directive on the so-called renewable energy sources (2009/28/EC);

It is considered a fundamental directive because it provides a legislative framework for the EU targets for greenhouse gas emission savings, and the production and promotion of energy from renewable sources. Moreover, the aim of the directive is to encourage energy efficiency, energy consumption from renewable sources in line with the overall goal "20-20-20", as well as the improvement of energy supply and the economic stimulation of a dynamic sector in which the European Union is trying to set an example.

Member States must also establish national action plans, which in addition to setting the share of energy from renewable sources in transport, production of electricity and heating by 2020, must also take into account the impact of other energy efficiency measures on final energy consumption. The aim of these plans is also to establish special procedures for the reform of planning and pricing schemes and of the access to electricity networks from renewable sources.

Cooperation between the Member States is also considered by the directive. The Member States can in fact transfer energy from renewable sources using a statistical accounting system and set up joint projects concerning the production of electricity and heating from renewable sources. This kind of cooperation is also possible with third countries, provided some conditions are satisfied: the electricity must be consumed in the territory of the European Union; after the month of June 2009 the electricity must be produced by a newly constructed installation; the quantity of electricity produced and exported must not benefit from any other support.

In addition, the Member States must ensure access to and operation of the grids. This is an important focus point because it requires the Member States to build the infrastructures necessary for the use of energy from renewable sources in the transport sector. It is hardly necessary to stress the importance of this issue in the context of the policies aiming to promote the smart cities and smart communities.

From a technical point of view, the directive takes into account energy from biofuels and bioliquids, in order to contribute to a reduction of at least 35% of greenhouse gas emissions. In particular, from the first of January 2017 their share in emissions savings should be increased to 50%. Biofuels and bioliquids should not use raw materials (produced either outside the European Union or within it) from land with high biodiversity value or with high carbon stock. Consequently, to benefit from financial support biofuels and bioliquids must be qualified as "sustainable", in accordance to the criteria of the directive. Again, an important issue for smart communities.

(b) Directive (2009/29/EC) on emission trading;
(c) Directive (2009/30/EC) (Fuel Quality Directive), with a strong impact on sustainable mobility and consequently on the development of the smart city concept;
(d) Directive (2009/31/EC) on carbon capture and storage;
(e) Decision (2009/406/EC), the so-called Effort Sharing Decision concerning the reduction of greenhouse gas emissions on the basis of solidarity between Member States and of sustainable economic growth across the Union;
(f) Regulation (2009/443/EC) Regulation of CO_2 emissions from cars;

These measures were all issued in 2009, *i.e.*, in the middle of the systemic financial and real economy crisis, in the hope that the green economy may contribute to overcoming it, as maintained also by US President Barack Obama in the immediate aftermath of the global crisis.

(g) Directive (2012/27/EU), which amends Directives 2009/125/EC and 2010/30/EU and repeals Directives 2004/8/EC and 2006/32/EC. This Directive integrates and completes the "package" of European standards issued in 2009.

Indeed this directive provides a broad framework concerning not only energy efficiency *sensu stricto*, *i.e.*, efficiency in energy use, (heating and cooling, article 14) and in energy transformation, transmission and distribution (article 15, with an important "warning" for the national energy regulatory authorities), but also the compliance with the minimum energy performance requirements of buildings used by public bodies, the energy efficiency obligation schemes, the so-called energy audits and energy management systems, metering and billing information, and so on.

Even the premises are very significant: "Whereas…(3) *The conclusions of the European council of 17 June 2010 confirmed the energy efficiency target as one of the headline targets of the Union's new strategy for jobs and smart, sustainable and inclusive growth (Europe 2020 Strategy)*; in addition: "…(11) *This directive…also contributes to meeting the goals set out in the Roadmap for moving to a competitive low carbon economy in 2050, in particular by reducing greenhouse gas emissions from the energy sector, and to achieving zero emission electricity production by 2050*" … "(17) *The rate of building renovation needs to be increased, as the existing building stock represents the single biggest potential sector for energy savings. Moreover buildings are crucial to achieving the Union objective of reducing greenhouse gas emissions by 80-90% by 2050 compared to 1990…*" This is, without any doubt, the core of the directive. It has the declared objective to increase the use of renewable energies (especially in the cities) by encouraging the development of financing facilities (such as special contributions [12]).

Chapter IV of the Directive (articles 16–20) on the "Horizontal provisions" includes issues that are of particular importance for the development of smart communities: article 17 concerns "information and training", whereas article 20 is dedicated to "Energy efficiency national fund, financing and technical support", which acknowledges a key point: "…*Member States shall facilitate the establishment of financing facilities, or use of existing ones, for energy efficiency improvement measures to maximise the benefits of multiple streams of financing*" (article 20, first paragraph). In other words, and in line with the article 192 (second paragraph), of the Treaty on the Functioning of the European Union, renewable energies will be financially sustainable only if they are economically supported by the Member States and by the European Union as well.

The Directive establishes a close relationship between the smart city concept and the renewable energy issue, which is an important element of the new green economy. Furthermore, the development of renewable energies, together with a policy of building renovation (as foreseen by the Directive), will increase the number of well-paid jobs and of skilled employment. This does match the anticipated character of smart cities (and of smart communities, clusters and regions): dynamic, well-educated and open to innovation.

Moreover, the smart city will be the ideal urban space where participation and democracy are promoted by the sophisticated interplay of the social actors, including both individuals and collective bodies, like political parties, consumer and environment protection associations, and other stakeholders. This is clearly envisaged by the Directive (art. 17): *"Member States shall ensure that information on available efficiency mechanisms and financial and legal frameworks is transparent and widely disseminated to all relevant market actors, such as consumers, builders, architects, engineers, environmental and energy auditors, and installers of building elements."*

The Directives 2009/72/EC and 2009/73/EC, as well as the three Regulations 714/2009, 715/2009 and 1775/2005 contained in the Third Energy Package, have also had a beneficial (if indirect) effect on the implementation of the "Europe 2020 Strategy" through a "competitive and sustainable supply of energy to the economy and the society". Indeed, the separation of companies' generation operations from their transmission activities (Three different options (ownership unbundling, independent system operator (ISO) and independent transmission operators (ITO)) are available so that the needs of different national market structures can considered.) has contributed to reducing the market power of the largest energy companies.

The new 2030 framework for climate and energy policies has been agreed upon by the EU Heads of Government in the month of October of 2014, after the publication of a green book by the Commission. In addition to a Green House Gas (GHG) emissions reduction target of 40%, the Policy Framework has set a target of at least 27% for renewable energy and energy efficiency enhancement by 2030. These targets are consistent with the Energy Roadmap 2050 set out by the Commission in 2011. However, no legislative act has been issued by the Commission at the date of this publication.

2.3. An Example of National Implementation: The Italian Case

As an example of how the European legislation is implemented in the regulations of the Member States, the Italian system is considered. The following key pieces of legislation have been passed by the Italian Parliament in response to European legal acts:

1. Decree Law N. 28/2011, by which the before mentioned Directive 2009/28/EC, has been implemented. It has set the general rules, delegating the detailed norms to a number of Ministerial Decrees issued by the Minister of Economic Development;

   i) Ministerial Decree of September 10, 2010, containing the national guidelines for the authorization of plants powered by renewable sources, according to article 12 of Legislative Decree n. 387/2003, which was the general regulation of the matter until the enactment of the Legislative Decree n. 28/2011;

   ii) Ministerial Decree of March 15, 2012 about definition and quantification of regional goals concerning renewable energy sources and the definition of how to handle cases of non-achievement of the objectives by Regions and Autonomous Provinces (Burden

Sharing), in compliance with Article 37 of the quoted Legislative Decree n. 28/2011, (definition and quantification of the medium-long term results that Regions and Autonomous Provinces commit to abide by in line with national targets set by 2020);

2. The Presidential Decree of April 16, 2013 on the energy performance certificate, setting rules for the verification of the professional qualification and independence of experts and organizations;
3. Law n. 90 of August 3, 2013, concerning the transposition of the Directive 2010/31/EU on energy performance in buildings into Italian law;
4. Legislative Decree n. 102 of July 4, 2014, implementing the Directive 2012/27/EU on energy efficiency. It provides a framework for concrete measures aimed at the promotion and implementation of the results of efficiency energy as already predetermined by the previously quoted D.M. of 15th March 2012, with a view to achieving the level of energy efficiency foreseen by European Union law.

Moreover, it would be useful to include norms and regulations issued by each Italian Region. However, they are mostly expenditure norms, *i.e.*, provisions of financial support with the aim of encouraging the use of renewable energies by economic operators and consumers [10].

In the other European countries transpositions into national laws went through similar legislative processes in accordance with the national procedures and practices specific to each issue.

In the United Kingdom, for instance, energy is a reserved matter for the Central Government, but the deployment of actual mechanisms for increasing the levels of renewables is a matter for Devolved Administrations. Thus, the Directive 2009/28/EC was implemented through a range of statutory instruments, such as the Renewables Obligation Orders (England and Wales/Scotland/Northern Ireland) issued by Devolved Authorities and the Feed-in Tariffs Order issued by the Minister of State of the National Government. Similarly, the Directive 2010/31/EU was implemented through the revision of the Housing Act 2004 and of the Energy Performance of Building Regulation 2007, contained in statutory instruments (England and Wales/Scotland/Northern Ireland). The strategy for the renovation of buildings referred to in the Directive 2012/27/EU has been established by the Secretary of State of the Central Government, after consultation with the other competent authorities (Northern Ireland departments, Scottish Ministers, and Welsh Ministers) in the Energy Efficiency (Building Renovation and Reporting) Regulations 2014 (S.I. 2014 No. 952).

2.4. Looming Difficulties

No doubt there is an important, definite and stable relationship between the new energy sources and the smart city (and the smart community): indeed a key feature of smartness is the capability of reducing greenhouse gas emissions in the urban spaces (the so-called public spaces).

However, this point of view does not go completely unchallenged. Indeed, it is often claimed that there is not enough evidence underpinning this close identification. For one thing, the renewable energies can prevail over traditional energy sources such as oil and gas only thanks to

498

the generous economic contribution of the European Member States. This opinion is shared by the European Union, as is well reflected by the content of the directives examined above.

Indeed subsidies granted to foster the use of renewable energy sources and to reduce greenhouse gas emissions protect the environment, but they introduce a dangerous distortion of the market. A doped market can easily collapse if the subsidies are suddenly lifted as might be the case in the present period of economic crisis. Even a progressive reduction of subsidies to the solar branch has given rise to a serious market disruption of this sector in Germany. The development of shale oil and shale gas extraction technologies, the recent volatility of oil prices, the geopolitical tensions involving major oil producers make predictions all the more unreliable. Thus, it can be wondered if the traditional energies have indeed a short future and how this lapse of time can be actually evaluated.

In other words, there are many important questions waiting for sure and unequivocal answers before the transition to a sustainable development can be reliably predicted. Another matter of concern is the growing tensions within the very environmental movement.

Indeed, most environmental associations have supported from the beginning the choice of new, renewable energies with a view to increasing the protection of the environment. This pragmatic point of view is in accordance with the so-called "shallow ecology", whose utilitarian attitude is criticized by a select minority of environmental associations opposing the use of the new energies for environmental reasons, as they are supposed to give rise to damages to the landscape, to a reduction of global space for agriculture, to a negative view impact, and so on. This is to some extent the natural consequence of a more general ideology, known as "deep ecology". It emphasizes the inherent worth of nature regardless of its instrumental utility to humans [13]. How deep ecology thinking may affect the development of smart cities is as yet unclear. In principle, some of the tenets of deep ecology [14] are not necessarily in conflict with the fundamental characters of smart cities. The complexity/complication argument for instance seems to be well in accordance with the vision of a composite, but well-structured urban space. However, the deep ecology discourse as it is conducted in practice tends to privilege the non-human environment and to become intrinsically "anti-urban".

In addition to the uncertainties mentioned above, it is also necessary to point out some reservations occasionally made against an uncritical, extensive use of the concept of "smart city".

For one thing, the existence of a positive prejudice toward the buildup of smart cities may lead to ignoring alternative urban development strategies for the improvement of the quality of life or, conversely, to underestimate the strong negative effects and consequences that the diffusion of new technologies and of networked infrastructures can occasionally give rise to [15].

2.5. A Serious Shortcoming: The Absence of a Binding Legal Framework

The new renewable energies are certainly relevant from a legal perspective: there are European Strategies, as well as European legal acts, especially directives, whose current trends have been analysed in previous sections; and partly as a consequence of this *acquis communautaire* there are, of course, national norms and rules that regulate projects and objectives, and, first, of all, there are legal definitions of the single energy sources.

On the other hand, is it possible to find, as of the present time, a legal definition for a smart city? Indeed, there exist both European and national strategies, as well as documents concerning the smart cities (smart communities, smart regions), but there does not seem to be anything really relevant from the point of view of the law. On the other hand, it is only the law that can legitimate the establishment of compulsory rules for individuals as well as for private and public collective bodies.

In other words, the smart city (regarded as a logical category) is presently relevant and to some extent well-defined only from a sociological point of view.

According to most general reports [16,17], the defining features and factors of a smart city can be classified as follows:

1. *Smart economy*, which means competitiveness, and consequently entrepreneurial skill, economic image & trademarks, productivity, flexibility of labour market, ability to innovate, *etc.*;
2. *Smart people*, in other words social and human capital, with high levels of qualification, social and ethnic plurality, flexibility and creativity, participation in public life, *etc.*;
3. *Smart governance*, which implies participation in decision-making process, availability of public and social services, a transparent governance, good political strategies, and perspectives [18];
4. *Smart mobility*, with both local and national/international accessibility, for the promotion of sustainable, innovative and safe transports systems;
5. *Smart environment*, in other words good policies for pollution prevention and environmental protection with a view to increasing the attractiveness of natural conditions and to promoting a sustainable resource management;
6. *Smart living*, the so-called quality of life, with cultural facilities, good health conditions, housing quality, education facilities, social cohesion, *etc.*

Indeed, these are all sociological concepts, apt to provide general objectives and general guidelines for the development of good public policies in line with the most general aim to elevate the quality of life in urban spaces.

In other words, it is certainly possible to conceive and design a development path for the future of our cities a development path eventually leading to a scenario close to the smart model, well in accordance with the European roadmap "*Europa 2050*". However, it is extremely difficult (if not outright impossible) to find, at the same time, some legal trail by which to introduce compulsory tools for the implementation of a "smart" policy (or of several "smart" policies) capable of increasing the overall attractiveness of public urban spaces in Europe.

3. Conclusions

The evolution of the concept of "smart city" has gone hand in hand with the growing concern for the protection of the environment, the development of renewable energy sources and with the necessity of increasing the overall quality of urban landscapes including a dynamic social and cultural environment, capable of attracting a well-educated and skilled workforce.

500

Each single component of this scenario has been the object of an extensive European legislation, whose subsequent implementations in each Member State has given rise to a reasonable legal harmonization. However, no such strategy has been attempted for the definition of smart cities as a whole.

There may be some real hurdles in such an attempt: different physical landscapes, distinctive urban residential cultures, a divergent morphology of infrastructures, contrasting norms, traditions and habits at local level. However, the history of European integration has shown that tackling and overcoming difficult tasks has been the key to veritable breakthroughs in the process of Europeanization.

It's time for the scientific community to analyze the development of a suitable legal framework for the European smart cities of the future and for the policy makers at European and national level to adopt the necessary decisions.

Conflicts of Interest

The author declares no conflict of interest.

References

1. European Commission. *Integrated Sustainable Urban Development, Cohesion Policy 2014–2020*; European Commission: Brussels, Belgium, 2012.
2. Morand-Deviller, J. Le juste et l'utile en droit de l'environnement. In *Pour un Droit Commun de l'environnement. Mélanges en l'honneur de Michel Prieur*; Dalloz: Paris, France, 2007; pp. 263–295.
3. Papa, R. Smart cities: Research, projects and good practices for infrastructures. *J. Land Use Mobil. Environ.* **2013**, *6*, 291–292.
4. Su, K.; Li, J.; Fu, H. Smart city and the applications. In Proceedings of the IEEE International Conference on Electronics, Communications and Control (ICECC), Zhejiang, China, 9–11 September 2011; pp. 1028–1031.
5. Kunzmann, K.R. Smart cities: A new paradigm of urban development. *CRIOS* **2014**, *7*, 9–19.
6. Gargiulo, C.; Pinto, V.; Zucaro, F. EU smart city governance. *J. Land Use Mobil. Environ.* **2013**, *6*, 356–370.
7. Jepson, E.J.; Edwards, M.M. How possible is sustainable urban development? An analysis of planners' perceptions about new urbanism, smart growth and the ecological city. *Plan. Pract. Res.* **2010**, *25*, 417–437.
8. Auby, J.B. *Droit De La Ville*; Lexisnexis: Paris, France, 2013.
9. Vincent Callebaut Architectures. Available online: http://vincent.callebaut.org/page1-img-parissmartcity2050.html (accessed on 27 January 2015).
10. Vivani, C. Ambiente ed energia. In *Trattato di Diritto Dell'ambiente*; Ferrara, R., Sandulli, M.A., Eds.; Giuffré: Milano, Italy, 2014; p. 503.
11. Ferrara, R. *I Principi Comunitari Della Tutela Dell'ambiente*; Giappichelli: Torino, Italy, 2006.

12. Ferrara, R. La tutela dell'ambiente. In *Trattato Di Diritto Privato Dell'unione Europea*; Ajani, G., Benacchio, G.A., Eds.; Giappichelli: Torino, Italy, 2006.

13. Ferrara, R. Etica, ambiente e diritto: Il punto di vista del giurista. In *Trattato di Diritto Dell'ambiente*; Ferrara, R., Sandulli, M.A., Eds.; Giuffré: Milano, Italy, 2014.

14. Naess, A. The shallow and the deep, long-range ecology movement. A summary. *Inquiry* **1973**, *16*, 95–100.

15. Graham, S.; Marvin, S. *Telecommunications and the City: Electronic Space, Urban Place*; Routledge: London, UK, 1996.

16. Giffinger, R.; Fertner, C.; Kramar, H.; Kalasek, R.; Pichler-Milanovic, N.; Meijers, E. *Smart Cities: Ranking of European Medium-Sized Cities*; Vienna University of Technology: Vienna, Austria, 2007.

17. Schönert, M. Städteranking und Imagebildung: Die 20 größten Städte in Nachrichten-und Wirtschaftsmagazinen. *BAW Monatsbericht* **2003**, *2*, 1–8.

18. Paskaleva, K.A. Enabling the smart city: The progress of city e-governance in Europe. *Int. J. Innov. Reg. Dev.* **2009**, *1*, 405–422.

Residual Mix Calculation at the Heart of Reliable Electricity Disclosure in Europe—A Case Study on the Effect of the RE-DISS Project

Markus Klimscheffskij, Thierry Van Craenenbroeck, Marko Lehtovaara, Diane Lescot, Angela Tschernutter, Claudia Raimundo, Dominik Seebach, and Christof Timpe

Abstract: In the EU, electricity suppliers are obliged to disclose to their customers the energy origin and environmental impacts of sold electricity. To this end, guarantees of origin (GOs) are used to explicitly track electricity generation attributes to individual electricity consumers. When part of a reliable electricity disclosure system, GOs deliver an important means for consumers to participate in the support of renewable power. In order to be considered reliable, GOs require the support of an implicit disclosure system, a residual mix, which prevents once explicitly tracked attributes from being double counted in a default energy mix. This article outlines the key problems in implicit electricity disclosure: (1) uncorrected generation statistics used for implicit disclosure; (2) contract-based tracking; (3) uncoordinated calculation within Europe; (4) overlapping regions for implicit disclosure; (5) active GOs. The improvements achieved during the RE-DISS project (04/2010-10/2012) with regard to these problems have reduced the total implicit disclosure error by 168 TWh and double counting of renewable generation attributes by 70 TWh, in 16 selected countries. Quantitatively, largest individual improvements were achieved in Norway, Germany and Italy. Within the 16 countries, a total disclosure error of 75 TWh and double counting of renewable generation attributes of 36 TWh still reside after the end of the project on national level. Regarding the residual mix calculation methodology, the article justifies the implementation of a shifted transaction-based method instead of a production year-based method.

Reprinted from *Energies*. Cite as: Klimscheffskij, M.; Van Craenenbroeck, T.; Lehtovaara, M.; Lescot, D.; Tschernutter, A.; Raimundo, C.; Seebach, D.; Timpe, C. Residual Mix Calculation at the Heart of Reliable Electricity Disclosure in Europe—A Case Study on the Effect of the RE-DISS Project. *Energies* **2015**, *8*, 4667-4696.

1. Introduction

Electricity flows to our houses and businesses from a mix of sources: from all the power stations that are connected to our power system [1]. Hence, in the physical sense, the origin of the energy that lights up our living rooms is just about as traceable as the origin of the dust on our shoes.

Nonetheless, Directive 2009/72/EC, Article 3(9) [2], requires electricity suppliers to disclose in electricity bills and in promotional materials the energy origin of the electricity they are selling. This process is commonly referred to as *electricity disclosure*, which constitutes of both explicit tracking and implicit allocation of electricity generation attributes (Information about the characteristics of the power production which need to be tracked, most importantly the "energy source and technology used for power production, the related CO_2 emissions and radioactive waste produced" [3]) from production to consumption.

In the EU, guarantees of origin (GO) as defined by Article 15 of the Directive 2009/28/EC, are the main mechanism for explicit tracking of electricity generation attributes. As long as not all electricity consumption is explicitly tracked to certain generation attributes, explicit tracking mechanisms always require the support of an implicit disclosure system, a *residual mix*, in order to avoid double counting. This is important, because consumers expect the tracking system to be reliable and their willingness to purchase renewable power can be significantly reduced if the attributes are double counted [4]. Residual mix is the main topic of this article.

This article presents the improvements in the implicit electricity disclosure of 16 (10 were chosen in the analysis due to their participation in the RE-DISS Project (Austria, Belgium, Denmark, Finland, Italy, Luxembourg, The Netherlands, Norway, Sweden and Switzerland) and 6 due to their relevance in terms of the European GO and/or electricity market (France, Germany, Ireland, Portugal, Slovenia and Spain)) selected European countries, achieved during the RE-DISS (Reliable Disclosure Systems in Europe, Berlin, Germany) project (4/2010–10/2012). These improvements are two-fold. First, RE-DISS has helped competent bodies of electricity disclosure in setting up a reliable disclosure system in their countries through a series of workshops, bilateral meetings, tailor-made recommendations and commenting of legal texts. The effect of these actions, and actions taken independently by Competent Bodies during the project, is quantified in Section 4.1, after comparing implicit disclosure errors before (2010) and after (2012) the project in Section 3.2. Second, RE-DISS has generated significant enhancements in the residual mix calculation methodology as developed by the E-Track [5] project. These enhancements, and their effects in the avoidance of double counting, are elaborated in Section 4.2 and later discussed in Section 5.

The article is structured as follows. Section 2.1 explains the importance of reliable electricity disclosure information in today's electricity markets. Sections 2.2 and 2.3 aim at elaborating how the current explicit and implicit (respectively) tracking and disclosure systems work in Europe and what is their role in renewable energy support. Section 3.1 presents five types of implicit disclosure problems discovered in the 16 selected countries before the RE-DISS project, and Section 3.2 describes how these problems have been overcome during the RE-DISS project, highlighting the major advancements. Section 4.1 quantifies the implicit disclosure error avoided due to these improvements. In Section 4.2, the residual mix calculation methodology presented in Section 2.3 is further studied. Also, the major deficiency of the methodology, corrected during the RE-DISS project, is depicted and new improvements are proposed. Section 5, shows what remains to be done in the field of implicit disclosure in the 16 countries as well as discusses the options for the further improvements of the residual mix calculation methodology.

The scientific scope of this study is the reliable implementation of the guarantee of origin and electricity disclosure mechanisms in the 16 selected European countries. Therefore critical assessment of the effects of GOs and disclosure on the RE market is left out of the research; the focus being on how well the mechanisms are operating. Furthermore, the European countries besides the 16 were not included due to restrictions of available data and small relevance to the international GO system.

Since the article focuses on the implementation of the mechanism itself, it also doesn't consider the sustainability of different energy sources. GO is a mechanism of consumer choice and does not in itself promote sustainable energy production and therefore detailed analysis of various fuels (including their Life-Cycle Assessment) is left out of the study.

The accuracy of the analysis is also limited in how well it considers so-called non-Reliable Tracking Systems. The main problem of these systems is the availability of data and therefore some volumes require estimation. Where estimations are used, this is clearly elaborated in the text.

Finally an important limitation regarding the results is that they are calculated at a national level. It is highly likely that problems beyond this analysis exist at individual supplier level, but these could not be assessed in the scope of this study, because data collection was made on a country rather than individual supplier level.

2. Electricity Tracking and Disclosure in Modern Electricity Markets

2.1. The Role of Electricity Disclosure in the Modern Electricity Markets

The cornerstone of a well-functioning internal electricity market is a deliberate choice of a supplier by the consumers. This choice is founded on general commercial aspects such as price, quality and reliability of service, but it also relates to the generation characteristics of the electricity supplied.

As mentioned, through 2009/72/EC, Article 3(9) [2], customers have the right to know the generation attributes of the electricity supplied to them, meaning from which energy source(s) the electricity was generated and what impacts this production had on the environment (CO_2 and radioactive waste generated). A reliable and transparent electricity disclosure system leads to trust in the energy market [6]. Also references [7,8] point out the importance of reliable disclosure information not only for consumers, but for energy companies as well.

Over the past years, energy regulators have put customer empowerment at the heart of their working program [9]. This development is vital, because according to [10,11] customers' trust in the supplier and the disclosure information increases when a governmental body is involved in checking this information. Also the European Commission (EC) has established the Citizens' Energy Forum in 2007 as a new regulatory platform based on the experiences gained in the Florence and Madrid Forums [12]. "The aim of the Forum is the implementation of competitive, energy-efficient and fair retail energy markets for consumers" [12], and this starts from empowered customers.

Trustworthy and well-presented ([6] highlight that because the electricity bills already tend to be confusing and complex, a clear structured, easy to understand disclosure information is essential. The information is most effective when placed on the bill, whereas placed in an attachment or appendix turns out to be less effective [7,11]. However the placing and appearance of disclosure information is not the topic of this article, but rather the reliability of this information) electricity disclosure information empowers the consumer to opt for desired sources of energy production and leads to a shift in the relevance of parameters for a decision to contract a supplier. Based on their survey, [6] see a possible link between transparent disclosure information available and the switching of electricity suppliers by consumers, but stress that this will only happen if the disclosure system is considered reliable and worthy of paying a premium. Consequently, through purchase decisions, customers are in the position to influence the policies, and in particular the energy mix, of suppliers and therefore indirectly also stimulate sustainable electricity production: "The advent of electricity disclosure has the potential to bring in a new era of citizen involvement in determining the national electricity generation mix" [6].

To sum up, the objectives of electricity disclosure are related to an increased market transparency, the consumers' right for information for making informed choices, education of consumers and stimulation of sustainable electricity generation [6,11]. Most importantly regarding this article, it is emphasized that the disclosure information needs to be reliable in order for these objectives to be achieved.

2.2. Guarantees of Origin as the Instrument for Explicit Tracking in Building Reliable Electricity Disclosure

In order to provide consumers with electricity disclosure information, electricity suppliers must be able to track the energy origin of the electricity back to its production. Tracking generally means a methodology for the accounting of generation attributes (energy source and environmental impacts) in the electricity market and their allocation to final consumption of electricity [3]. This is much more difficult than it seems at first glance, because the electricity market itself is complex.

Theoretically, there are three options how tracking could be implemented:

- along the physical flows in the electricity grid;
- along the trading arrangements (contracts) in the electricity market; or
- in a separate accounting mechanism which is independent from the physical flows and from electricity contracts.

Typical criticism towards electricity tracking is driven by the argument that electrons cannot be traced, which is a well-known and acknowledged fact [11]. The electricity flows in the grid are a result of the demand patterns, the grid structure and the operation of power plants. A flow in a certain direction is usually caused by the balance of a large number of electricity contracts, and is influenced by technical constraints or incidents. The flows in the grid follow physical laws rather than market activities and thus can hardly be a good basis for the tracking of attributes [13].

Physical trading of electricity (which excludes purely financial instruments for price hedging) might be a more suitable basis for tracking of attributes, as it represents the market activity of generators, traders, suppliers and consumers of electricity. Such approach probably comes close to the intuitive expectations by end-consumers of how tracking should work. However, liberalized electricity markets usually encompass different types of bilateral trading, power exchanges with spot and futures markets as well as balancing power, and thus may be too complex to be used for tracking purposes [14]. Furthermore, if qualities of electricity, in terms of their generation attributes, would be added to the electricity market, creating differences in the wholesale market price of electricity from different energy sources, this would severely damage the liquidity of the electricity market [13].

Therefore, the tracking of attributes for purposes of disclosure should in principle be separated from physical flows and from electricity contracts [15]. This allows a tailor-made design of the tracking mechanism according to the needs of the disclosure scheme.

The most prominent instrument for tracking the origin of electricity is the guarantee of origin (GO), which was first introduced by the European Commission Directives for renewable energy (2001/77/EC) and cogeneration (2004/8/EC), and has been taken over recently in the new Directives 2009/28/EC [16] and 2012/27/EC [17], respectively. According to Article 15 of 2009/28/EC, the GO is an accurate, reliable and fraud-resistant electronic document, issued by nationally appointed competent body, representing the generation attributes of 1 MWh of electricity production from renewable energy sources. The directive goes on to define that a GO can be transferred independently from the trading arrangements and physical flows of the associated electricity and cancelled (*i.e.*, used) to verify the origin of 1 MWh of electricity consumption for the purpose of electricity disclosure. The GO is now a reasonably standardized instrument, at least for the tracking of electricity from renewable energy sources. It is important to note that a guarantee of origin is an objective accounting instrument which aims at providing objective information and can therefore be used to support consumer choice. It does not in itself guarantee to the consumer that the specific electricity is sustainable or complies with a specific definition of "greenness" or "environmental additionality".

An important part of today's GO system is the European Energy Certification System (EECS) maintained by the Association of Issuing Bodies (AIB) (The AIB is a not for profit umbrella organization of Issuing Bodies of different countries. The AIB aims to ensure "the reliable operation of international energy certificate systems" through a standardized system EECS—The European Energy Certificate System [18]). EECS enables technically harmonized procedures for issuing, transferring and cancelling GOs [19] as well as international transfers between the AIB member countries through the AIB Hub [20]. There are currently 16 European countries in the AIB [21].

A reliable disclosure scheme has to be based on GOs, which can in some cases be complemented by other Reliable Tracking Systems (RTS) (such as: Homogenous disclosure mixes determined by a competent body for consumers in non-competitive electricity markets, as long as such segments still exist.

- Renewable energy support systems (e.g., German feed-in tariff) which require a defined allocation of the attributes of supported generation to consumers for disclosure purposes, which cannot be implemented reasonably based on GOs. In this case a pro-rata allocation of the attributes to all consumers which are paying for the support system can be the adequate solution;
- Under certain conditions, the contracts concluded by market participants in the physical electricity market (also known as contract-based tracking). The conditions include the central collection and supervision of such tracking data by an appointed competent body.

Such other Reliable Tracking Systems should only be introduced, if they provide added value to the tracking system (in addition to GOs and the residual mix). Furthermore they should be implemented in a reliable and transparent way, thus not endangering the reliability of disclosure information provided to consumers).

GOs feature the traceability of electricity back to its origin without disturbing the electricity market whilst avoiding double counting of electricity generation attributes in disclosure [13]. The E-Track project found out that having a reliable tracking system in place is the general expectation of all involved parties [4]. Reliability is also emphasized by the European Commission Directive: "Member States shall ensure that the same unit of energy from renewable sources is taken into account only once." [22].

When part of a reliable disclosure system, manifold possible effects of GOs have been identified:

- Support for renewable energy production

Producers can financially benefit from selling GOs. The prices on the international market are not public, but according to an estimate by the authors of this article, the price for large hydro power GOs currently ranges between 0.05 and 0.5 Euro/MWh [22]. However prices can be much higher especially for small hydro power and new renewable technologies. The extra revenue, albeit small compared to e.g., governmental renewable energy support schemes, might attract new investments [15].

- Interaction with governmental renewable support schemes

GOs can alleviate the budgetary burden of existing governmental renewable energy support schemes by allocating more of the burden to those willing to pay more. Besides relieving direct monetary load from governments, even more importantly, GOs relieve political burden. If consumers consider that renewable energy is worth a premium, they are encouraging the government to retain its current targets, and perhaps even to increase them, e.g., [15,23]. GOs give a voice to the people in an area traditionally dominated by the government: renewable energy support. It is especially important to involve the public into supporting renewable energy when considering the current problems faced by governments in maintaining their support schemes, e.g., [24–27].

- Supplier differentiation and consumer awareness

The customers choose their supplier based on the energy mix and the willingness to switch is higher when transparency on the market is high [28–30]. Further, the influence on energy behavior and the willingness to save energy are major drivers for having a functioning GO system in place [11,31].

- Harmonized tracking system in the EU

GOs are a harmonized mechanism within the EU and can cope with changes in the physical and market structures of electricity [13]. The international orientation of suppliers is becoming more relevant especially since the implementation of the new Internal Electricity Market Directive 2009/72/EC [2]. The fact that a growing number of countries fulfill the criteria on GOs and have adopted the EECS system opens an international demand and exchange of GOs. EECS-GOs alone currently cover more than 300 TWh of European renewable electricity production (Figure 1), which is impressive also due to the fact that large volumes of renewable electricity production are not eligible for GOs due to renewable energy support mechanisms and not all countries have yet adopted the EECS system.

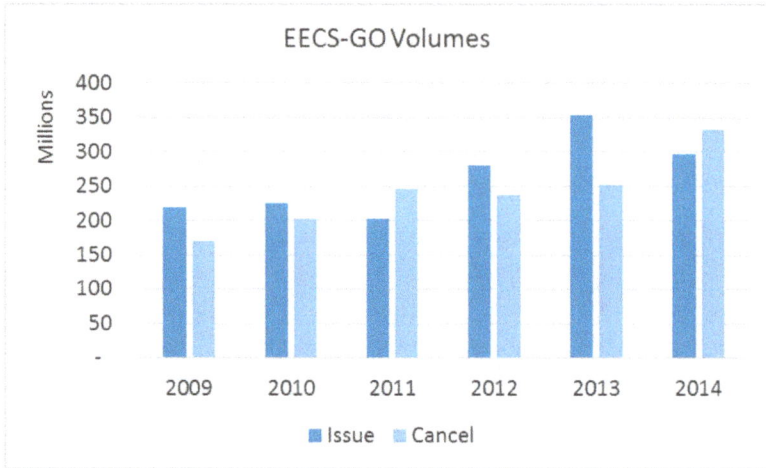

Figure 1. Millions of European Energy Certification System (EECS)-guarantees of origin (GOs) issued and cancelled (used) from 2009–2014 based on [32].

- Enabling operation of labels

According to [33], through the ability to track electricity, GOs enable the reliable operation of electricity labels. These labels use GOs to guarantee the origin of electricity, but add special criteria such as bird-, wildlife or nature conservation, CO_2 reductions, additionality (According to [15], in the context of renewable energy adoption, additionality indicates the additional renewable energy production compared to a Business As Usual case, due to a certain policy or action. For example GOs can deliver additionality, if they increase renewable energy production compared to the situation where no GOs would be issued) *etc.* into the electricity sold to a consumer. Labels can help customers in taking a decision in favor of a specific supplier, as labels are in general easy to understand and aim at a high recognition value.

To sum up, GOs are the mechanism of choice for explicit electricity tracking in the EU and have the potential to bring desired effects into the European electricity system. However, as outlined in Section 2.1, GOs will only be used if electricity disclosure is considered reliable by the consumers, *i.e.*, double counting of generation attributes is avoided [4]. The pronounced rules and guidelines for issuing, transferring and cancelling GOs, especially within the EECS system, efficiently hamper the possibilities for the origin to be tracked twice *explicitly*; but what about consumption not tracked with GOs?

The voluntary characteristic of using explicit tracking instruments is troublesome. Logically, if mostly renewable attributes are explicitly tracked, the remainder of consumption should be less rich in renewables than grid average. One of the most widespread reasons why double counting of renewable attributes still occurs is that competent bodies do not provide a *residual mix*, or do not require its usage for disclosure of consumption which is not explicitly tracked. This may lead to such consumption being disclosed with the production mix of the country where the renewable attributes presented by GOs (and RTSs) are included. The RE-DISS BPR recommend the explicit tracking of all electricity generation attributes, which supports reliable and more meaningful allocation of non-renewable attributes as well. If explicit tracking of electricity were extended to all types of generation attributes, the residual mix would not be required [34] (BPR number 11). (The RE-DISS project has recommended to expand the concept of GOs from electricity from renewable energy and high-efficient cogeneration to all energy sources, in order to support a reliable, explicit allocation of the attributes of any type of power generation [34]. This has already been done in e.g., Austria, Norway, Sweden and Switzerland [35])

The remainder of this article concentrates on the reliability of implicit disclosure, achievable through the correct implementation of the residual mix, because the aims of the GO system can only be achieved where the entire disclosure system is reliable. After all, who would be willing to pay for double counted renewable electricity?

2.3. Residual Mix as the Instrument of Reliable Implicit Electricity Disclosure

Apart from reports of E-TRACK [3,14], and later RE-DISS [36], electricity residual mixes are not well addressed in the academic literature. Raadal et al. [37] elaborate the need for a residual mix calculation in an electricity system with explicit tracking, with a focus on Nordic countries. [13] mention that "where default set of attributes is needed a residual mix should be used instead of uncorrected generation statistics in order to minimize multiple counting". This is also a key recommendation of the RE-DISS project [34]. Lastly, [15] has outlined problems in the residual mix calculation methodology, which will be further analyzed in Section 4.2.

Residual mix calculation prevents the double counting of explicitly tracked generation attributes by deducting them from the default mix, which is used for disclosure of non-explicitly tracked consumption. The residual mix is not to be used for any products which are differentiated with regard to the origin of electricity. Its use is restricted to bridging up the difference between the total electricity sales of a supplier and the attributes available from explicit tracking mechanisms where the supplier has not used explicit instruments for the disclosure of its entire sales [34].

The process of residual mix calculation might seem simple, but the international exchange of both electricity and GOs necessitates that also the calculation is coordinated among countries, which adds complexity. In a closed system where no connection to the external world exists, the consumption energy mix of a country equals, in volume and energy sources used, its production mix. But in real life, electricity as well as generation attributes (through GOs) are transferred across borders, which can significantly alter this equilibrium in a country.

The cross-border flow of explicit disclosure information (e.g., in the form of GOs) and of electricity can, at first sight, easily be described as a flow from one specific country to another specific country. One could therefore assume, that in an ideal case, each country should fill in the missing energy origin caused by net export of GOs with the residual mix of the country where the GOs were exported to (Also net physical import of electricity causes a deficit in available generation attributes of a country compared to its consumption, because part of the electricity is coming from abroad. In such case the missing attributes should ideally be derived from the residual mix of the exporting country. In practice however in the case of two European countries, the balancing is made with the European Attribute Mix (EAM), as with the balancing of attributes from GO transactions). In practice, such bilateral balancing would be highly complicated considering the international nature of the guarantee of origin markets (including also multiple cross-border transfers) and might likely not even be possible, because of the hen and egg problem: The residual mix of country A depends on that of B, which depends on that of C, which in turn depends on that of A. This might sound manageable with 3 countries, but the calculation needs to be Europe-wide.

One fundamental feature of the RE-DISS residual mix calculation methodology is the concept of a common attribute pool, generally known as the European Attribute Mix (EAM) (Which includes EU27, Norway, Iceland, Switzerland and Croatia but excludes Cyprus and Malta). Instead of different countries interacting with each other, they all interact with this common pool of attributes, which interconnects the domestic residual mixes the same way as the AIB Hub interconnects the explicit tracking of attributes (GOs). This approach does gain a "substitution mix" for the lack of attributes which does not to differentiate between the specific countries of net export of GOs. Although this obviously is a form of simplification of reality, it is consistent within the system borders of a group of countries which apply this approach, and corresponds to the concept of a pan-European internal market not only for electricity, but also for GOs.

The coordinated residual mix calculation for countries within the EAM area is divided into 4 phases (see Figure 2):

1. Data collection from each country:

   a. Net electricity production by tracking attribute including tracked externalities (CO_2 emissions and produced radioactive waste);
   b. Total electricity consumption;
   c. Data on explicitly tracked production attributes (imports, exports and cancellations of GOs and RTSs);
   d. Net electricity import and export *outside* the EAM area. In the case of net import, the residual mix (if exists; otherwise production mix) of the outside EAM country from where the electricity was imported.

2. Determination of the domestic residual mix and surplus/deficit of each country

   a. Determine *available attributes* by deducting exported and cancelled, and adding imported attributes to the generation mix (corrected with physical import or export to outside the EAM area). The mix of available attributes is the *domestic residual mix*;
   b. Determine *untracked consumption* by deducting cancelled attributes from total electricity consumption;
   c. Compare the volume of the available attributes with the volume of untracked consumption. If the amount of available attributes is greater/less than untracked consumption, the difference is surplus/deficit. The share of attributes in the surplus matches the domestic residual mix.

3. Determination of the European Attribute Mix including tracked externalities

   a. Combine the surpluses from all countries with a surplus.

4. Determination of the final residual mix of each country

a. In case of surplus, the final residual mix of the country is equal to the domestic residual mix minus the surplus;

b. Fill the domestic residual mix of deficit countries using the EAM until the volume of available attributes equals untracked consumption.

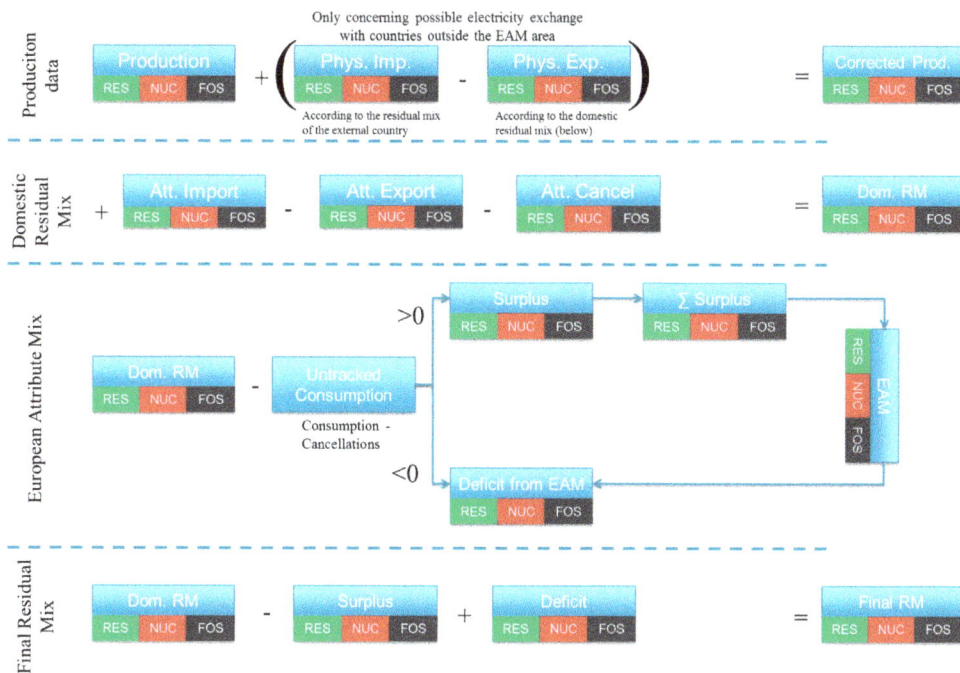

Figure 2. Residual mix calculation process based on [36].

A simplified numerical example of residual mix calculation is presented in Appendix section, Table A1.

Physically, electricity production and consumption in Europe equal each other in volume as long as electricity transfers to and from outside Europe are considered. International trading of GOs and electricity within Europe distort the equilibrium of available generation attributes and electricity consumption volumes on a national level, but on a European level the balance remains. The coordinated residual mix calculation, through EAM, returns this balance at the domestic level. Countries which have a surplus of generation attributes compared to their consumption (typically GO net importers and/or electricity net exporters), give attributes to the common pool and vice versa. Because of the physical balance, the total surplus equals in volume with the total deficit. The coordinated residual mix calculation is a simple yet powerful tool to allow international trading of generation attributes whilst avoiding double counting.

512

3. Analysis of Implicit Disclosure Problems before (2010) and after (2012) the RE-DISS Project

3.1. Implicit Disclosure Problems in Europe

Based on data collection of the disclosure practices of the 16 selected European countries, the RE-DISS project team found five different types of problems, which lead to double counting in implicit disclosure, due to deficiencies in the residual mix calculation:

- Uncorrected generation statistics used for implicit disclosure;
- Contract-based Tracking;
- Uncoordinated calculation within Europe;
- Overlapping Regions for Implicit Disclosure;
- Active GOs.

The following analysis of errors is based on qualitative and quantitative data collected for the 16 selected countries before (2010) and after (2012) the RE-DISS project.

3.1.1. Uncorrected Generation Statistics Used for Implicit Disclosure

Using uncorrected generation statistics for implicit disclosure appeared to be a straightforward approach followed by several countries, but it leads into the double counting of explicitly tracked attributes (often renewable), as explained in Sections 2.2 and 2.3. On the other hand, the relative shares of attributes which have not been explicitly tracked (particularly fossil and nuclear and their environmental impacts) are underestimated by such an uncorrected mix, because their share would be higher if the tracked attributes had been removed. The relevance of a proper residual mix calculation is best understood by comparing uncorrected generation statistics with residual mixes (Figure 3). The difference between the two is substantial.

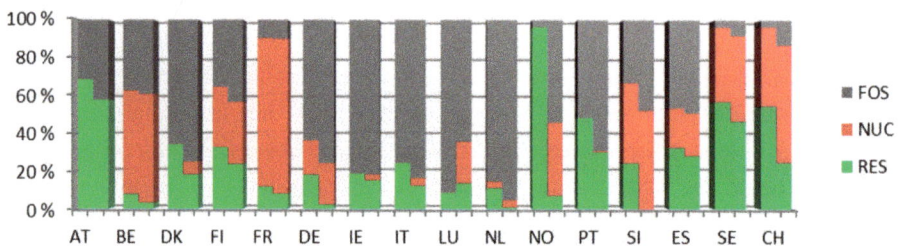

Figure 3. Generation (**left**) and residual (**right**) mixes of 2011 of selected European countries based on [38].

An alternative solution to the problem, without providing a residual mix, is utilized in Austria, Belgium, Germany, Luxembourg and The Netherlands, in which all renewable attributes are deducted from the generation statistics used for implicit disclosure [39]. This easy solution effectively excludes double counting of renewable attributes, but also leads to a disclosure error by neglecting possible remaining renewable shares in the residual mix and thus overestimating the fossil and nuclear shares including environmental impacts.

3.1.2. Contract-Based Tracking

Though in Section 2.2, it was explained that electricity can hardly be tracked along the trading arrangements in the electricity markets, contract-based tracking (CBT) is still allowed to be used for disclosure purposes in some countries. In CBT, the respective volumes which are physically delivered according to grid accounting rules to a supplier and his end-customers are, bilaterally between the buyer and seller, agreed to have been produced by a particular energy source or even a specific plant.

On top of the drawbacks already mentioned in Section 2.2, CBT is mostly not transparent for third parties and thus cannot be deducted from the residual mix of the country (Figure 4). This naturally results in double-counting of contract-based tracked attributes, which often include a higher than grid average share of renewable attributes. In order not to cause double counting, contract-based tracking has to fulfill the specifications for Reliable Tracking Systems as defined by the RE-DISS Best Practice Recommendations (BPR) by e.g., requiring central reporting of all tracked attributes to the national Competent Body, or that such contractual arrangements are accompanied with the transfer and use of GOs [34].

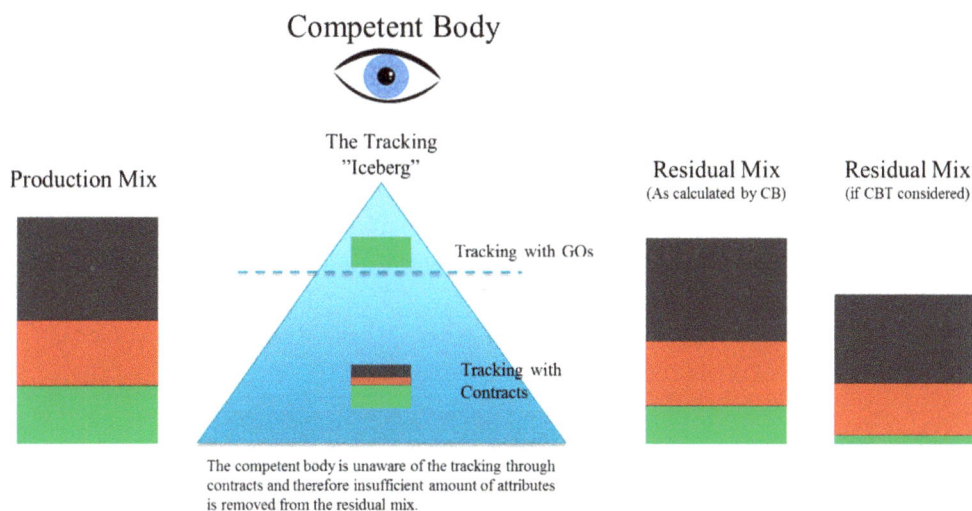

Figure 4. The problem of contract-based tracking as a reliable tracking system.

3.1.3. Uncoordinated Calculation of Residual Mixes within Europe

Uncoordinated calculation of residual mixes within Europe is a particularly relevant problem due to the international trade of both electricity and GOs. As explained in Section 2.3, such international trade means that a country might lack disclosure attributes. Without international coordination and the provision of the European Attribute Mix to balance surpluses and deficits, the deficit countries can either disclose the deficit as *unknown origin*, or expand the domestic residual mix to account for the entire untracked consumption.

For example, before adopting the EAM, Norway included in its residual mix a very high share of unknown origin, because most of the generation attributes were exported through GOs. Yet, given the production mix of Norway, it is likely that consumers presumed this unknown origin as renewable. Extrapolation of the domestic residual mix would have resulted in an even worse solution, because after accounting for the exported GOs the Norwegian domestic residual mix still contained a high share of renewable attributes, as practically all power production in Norway derives from renewable sources. Therefore the deficit would have been fulfilled by a much greener mix (expanding the Norwegian domestic residual mix) than the EAM, leading exported Norwegian GOs to be partially double counted. It should be highlighted that Norway has corrected the problem of unknown origin and currently uses the EAM for the filling up of its residual mix [39].

3.1.4. Overlapping Regions for Implicit Disclosure

The RE-DISS Best Practice Recommendations recommend that a residual mix should by default be calculated on a national level, but under a common decision, several countries can have a mutual residual mix if the electricity markets are closely integrated [34] (Recommendation 28). If some of the involved countries apply a common multi-country mix, and others only their national mix, the disclosure information of the latter countries will be overestimated. Although the problem has been acknowledged for several years, the Nordic countries (Denmark, Finland, Norway and Sweden) have yet to agree on a common solution. Sweden applies the common Nordic residual mix, while Norway and Denmark use their respective national mixes [39]. Finland has in the past unofficially used the common Nordic residual mix, but starting from 2013, the domestic residual mix will be officially applied [39]. This problem is especially relevant if the Nordic residual mix contains a higher renewable share than the Swedish mix, so that Sweden would benefit from such practice, while the other countries' renewable attributes would be double counted without them having a possibility to prevent this. But also in the opposite situation a disclosure error would occur due to loss of renewable attributes and double counting of others. A similar error also exists for countries which apply the overall ENTSO-E mix for all volumes of unknown origin (see Table 1).

3.1.5. Active GOs

Depending on how the residual mix is calculated, the problem of active GOs may occur if GOs issued for electricity production of year X are exported or cancelled after a defined disclosure deadline of the year X. The attributes presented by these GOs may already have been included in the residual mix of year X and thus the subsequent cancellation or export of the GOs would lead to double counting. This area was not well addressed in the residual mix calculation methodology developed by E-Track, but was taken on by RE-DISS and currently the methodology is robust against the active GO problem. The problem will be further discussed in Section 4.2.

3.2. Presence of Implicit Disclosure Problems in Europe before (2010) and after (2012) the RE-DISS Project

The RE-DISS project helped competent bodies of electricity disclosure to resolve problems described in Section 3.1 through consultation and dissemination actions, which led to the spreading of the RE-DISS Best Practice Recommendations for disclosure (BPR). During RE-DISS, seven workshops were carried out that counted with the participation of competent bodies from 16 countries. The first workshops focused on developing the BPR whereas towards the end of the project the emphasis was on supporting implementation. The BPR has been a living document throughout RE-DISS and the competent bodies have actively participated in its improvement.

The RE-DISS team also organized 16 (from Portugal, France, Greece, Latvia, Germany, Slovenia, Romania, Bulgaria, Czech Republic and Iceland) bilateral face-to-face meetings and a number of telephone conferences with competent bodies, performed consultancies of legal texts (for Germany, France, Poland and Iceland) and developed Country Profiles of 22 European countries that included an in-depth analysis of their electricity tracking system and provided tailor-made recommendations.

Through these actions, the RE-DISS project succeeded in triggering important enhancements in the field of electricity disclosure. Table 1 presents the status quo of the five problems in the 16 analyzed countries both before (1) and after (2) RE-DISS. For the sake of clarity it should be highlighted that the improvements listed in Table 1 are due to both efforts of RE-DISS and actions taken independently by the Competent Bodies.

In Table 1 marking X signifies that the problem is fully existing, (X) that it is partially existing and no marking that it is not existing in that country. Furthermore, if no residual mix is calculated in the country (problem 1), problems 3 and 5 are non-applicable. In such cases, the respective cell is marked grey.

4. Results

4.1. Effects from Corrections of Implicit Disclosure Problems during the RE-DISS Project

The effects of the improvements presented in Table 1 are quantified on an overall level in Figure 5 and by country in Figure 6. The results have been derived by simulating (By simulation it is meant that the residual mix of a specific country is calculated according to the national practices of that country. In other words practices, which are known to be unreliable in some cases are simulated and then compared to calculation results according to RE-DISS Best Practices) the implicit disclosure practices of the countries before 2010 (left column) and after 2012 RE-DISS (right column) according to the settings of Table 1. Columns in Figures 5 and 6 indicate volumes of erroneous disclosure compared to a situation where implicit disclosure is implemented according to RE-DISS BPR. Hence a positive column signifies double counting of the corresponding energy source, including externalities, and a negative column that the energy source is under-reflected.

Table 1. Improvements in implicit disclosure during RE-DISS [39].

Country	Problem					Description
	1	2	3	4	5	
Austria (1)	X			X		Before RE-DISS: No residual mix. ENTSO-e mix used for implicit disclosure.
Austria (2)	(X)			X		Improvements: All renewables filtered out of the ENTSO-mix before used for implicit disclosure.
Belgium (1)	(X)					Before RE-DISS: No residual mix. Production mix from which all RES filtered out used for implicit disclosure.
Belgium (2)	(X)					Improvements: No improvements.
Denmark (1) Denmark (2)	X	(X)	---		---	Before RE-DISS: No residual mix. CBT for nuclear and fossil. Improvements: Reliable and coordinated residual mix calculation. CBT of nuclear and fossil supervised.
Finland (1) Finland (2)	(X)	X (X)		X		Before RE-DISS: Residual mix of Finland based on the Nordic region. No legal status for residual mix: given as a recommendation by the Association of Energy Industries. Contract based tracking allowed. Improvements: Reliable and coordinated residual mix calculation set by legislation. CBT only for nuclear and fossil.
France (1)	X	X		X		Before RE-DISS: No residual mix. Mix of own production, contracts and ENTSO-e mix used for disclosure.
France (2)	X	X		X		Improvements: No improvements.
Germany (1)	X	X		X		Before RE-DISS: No residual mix. ENTSO-e mix as default value for disclosure. CBT, GOs, RECS and labels used for disclosure.

Table 1. *Cont.*

Country	Problem					Description
	1	**2**	**3**	**4**	**5**	
Germany (2)	(X)	(X)				Improvements National production mix, from which all renewables filtered out, used for implicit disclosure.
Ireland (1)			X			Before RE-DISS: Disclosure based on contracts and residual mix (residual mix accounts for contracts). Residual mix is not coordinated with other countries.
Ireland (2)						Improvements: Coordinated residual mix calculated.
Italy (1)	X	X	---		---	Before RE-DISS: No residual mix. Disclosure based on fuel mixes.
Italy (2)	X	X	X		(X)	Improvements: Residual mix calculated but not coordinated (deficit disclosed with Eurostat mix). It is not clear whether residual mix accounts for Active GOs.
Luxemburg (1)	---	---		---		Before RE-DISS: No disclosure.
Luxemburg (2)	(X)	X		X		Improvements: Disclosure system implemented. ENTSO-e mix from which all renewables filtered out used for implicit disclosure.
Netherlands (1)	(X)	X	X			Before RE-DISS: Residual mix calculated, but all renewables filtered out. Does not consider contracts and is not coordinated.
Netherland (2)	(X)	(X)	X			Improvements: Residual mix calculation considers contracts and is coordinated, but all renewables are filtered out.
Norway (1)			X	X		Before RE-DISS: Residual mix calculated, but not coordinated. Deficit attributes disclosed as unknown.
Norway (2)						Improvements: Deficit attributes replaced with the European Attribute Mix. Residual mix only accounts for year X certificates.
Portugal (1)	X	X				Before RE-DISS: No residual mix. Disclosure through contracts.
Portugal (2)	X	X				Improvements: Approach to a kind of residual mix.

518

Table 1. *Cont.*

Country	Problem 1	2	3	4	5	Description
Slovenia (1)	X	X		X		Before RE-DISS: No residual mix. Disclosure is based on contracts, GOs and ENTSO-e mix.
Slovenia (2)	X	X		X		Improvements: No improvements.
Spain (1)			X		(X)	Before RE-DISS: Residual mix is calculated, but not coordinated with other countries (domestic attributes expanded if needed). A problem with Active GOs might exist and GOs do not necessarily have to be cancelled in order to be used.
Spain (2)			X		(X)	Improvements: No improvements.
Sweden (1)				X		Before RE-DISS Residual mix based on the Nordic region. Contract based tracking allowed but accounted for. No legal status for residual mix: given as a recommendation by the
	(X)	(X)				Association of Energy Industries.
Sweden (2)				X		Improvements: Contract-based tracking not allowed (disclosure based on GOs or residual mix). Use of the residual mix obligated by law.
Switzerland (1)	X	X				Before RE-DISS No residual mix. Contract-based tracking allowed.
Switzerland (2)	---	(X)		---		Improvements: All electricity explicitly tracked with GOs (no residual mix needed).

The input data in all cases is that collected by RE-DISS for the 2011 residual mix calculation. This depicts that the volume to be disclosed remains constant throughout the analysis and hence that for each country the positive column equals in volume with the negative column, *i.e.*, if renewable attributes are double counted, nuclear and/or fossil attributes are automatically replaced by this amount in disclosure.

Other important settings and framework assumptions for the analysis are as follows:

- Effect of possible other problems besides the five listed have been neglected, e.g., problems relating to explicit tracking;
- in case CBT is allowed in the country, this has been assumed to cover:

  o 50% of untracked (not tracked with GOs or RTSs) domestic renewable production;
  o 20% of untracked domestic nuclear and fossil production;
  o for France (53% of RES and 56% of NUC and FOS) and Sweden (26% of RES, 17% NUC and 0% FOS), country specific estimates have been used.

- In case new legislation is currently being implemented and is scheduled to come into force in the near future, such progress has been taken into account for the after RE-DISS scenario. This is relevant for Germany, Sweden and Switzerland where the law or regulation is already ratified as well as for Finland and Italy where the ratification is in process.

Table 1 indicated that 12 (Austria, Denmark, Finland, Germany, Ireland, Italy, Luxembourg, The Netherlands, Norway, Portugal, Sweden, Switzerland) out of the 16 analysed countries made improvements regarding the 5 disclosure problems during the RE-DISS project. In 7 (Denmark, Finland, Germany, Italy, Luxembourg, Norway, Switzerland) of these countries, this significantly reduced the disclosure error as shown in Figure 6. Figure 5 illustrates that on a total level, double counting of renewable attributes decreased by 70 TWh (106−36) and the overall disclosure error by 168 TWh (243−75). Based on [40] these volumes represent 10% of electricity production from renewable energy sources and 7% of electricity consumption, respectively, of these 16 countries. Remaining double counting of renewable attributes after the project amounts to 36 TWh whereas the total disclosure error adds up to 75 TWh. However it should be noted that further errors may reside at individual supplier level as highlighted in Chapter 5.

No unknown origin was disclosed in the after RE-DISS case, compared to 102 TWh in the before case. This is also a significant improvement, because 95 TWh of the unknown origin was disclosed in Norway, where, given the production mix of the country, it is probable that consumers assume a green origin without better knowledge. The decreased amounts of renewable and unknown origin were correctly replaced by nuclear and fossil attributes, for which the negative disclosure error contracted by 58 TWh (76−8) and 110 TWh (134−24), respectively. The decreased disclosure error of fossil and nuclear attributes also increased the amount of disclosed externalities: CO_2 and radioactive waste, but these volumes were not covered by the analysis.

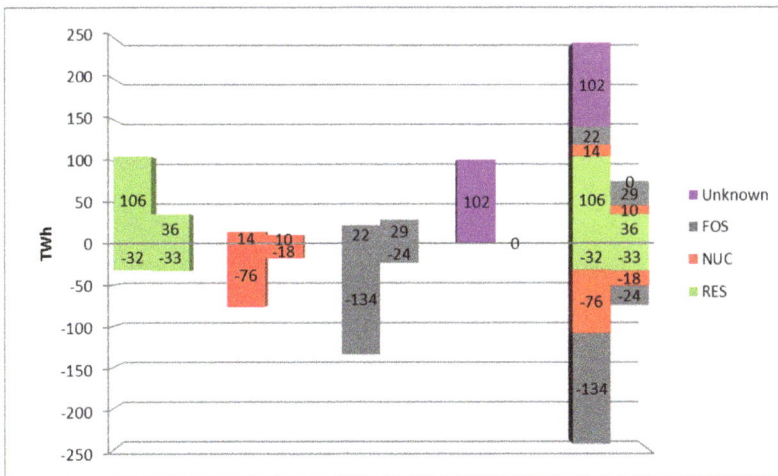

Figure 5. Total quantified implicit disclosure errors per energy source before 2010 (left-hand columns) and after 2012 (right-hand columns) RE-DISS.

520

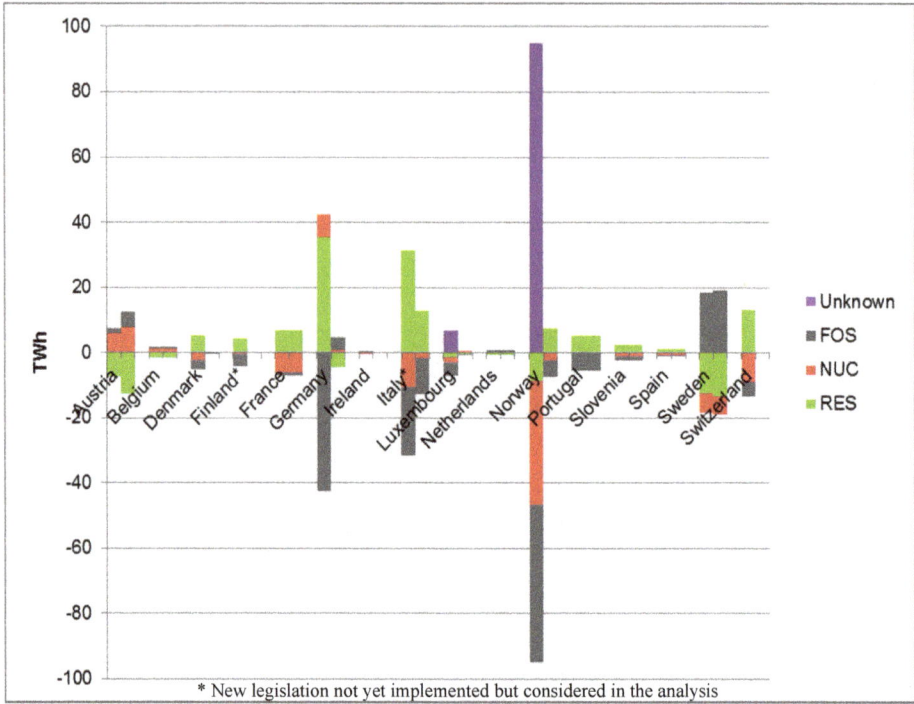

Figure 6. Quantified implicit disclosure errors per energy source and per country before 2010 (left-hand columns) and after 2012 (right-hand columns) RE-DISS.

Out of individual countries, Norway improved its disclosure the most, measured by volume of disclosure error avoided. This resulted from adopting the EAM for fulfilling deficit attributes (problem 3). Germany succeeded in making the largest reduction in double counting of renewable attributes, mostly by enhancements concerning problem 1. Instead of using uncorrected generation statistics, any renewable attributes are now eliminated from implicit disclosure, in Germany. Same amendment was made by Austria and Luxembourg, and was already used by Belgium and The Netherlands before RE-DISS. Although this approach does not eliminate the disclosure error altogether, it is effective against implicit double counting of renewable attributes. Also regarding problem 1, Italy made substantial progress by adopting a residual mix for implicit disclosure. The remainder of the implicit disclosure error in Italy would be avoided by using the EAM.

In Figure 6, the disclosure error is completely corrected in Denmark, Finland, Ireland and Switzerland. Switzerland overcame all problems by forcing all consumption to be explicitly tracked, which removes the need for implicit disclosure. Whereas the other three countries adopted a residual mix according to the RE-DISS BPR (although Finland still has a problem of unmonitored contract-based tracking of nuclear and fossil attributes).

Sweden was one of the top improvers on a qualitative level through the removal of contract-based tracking and adoption of mandatory use of the residual mix where GOs are not used. However in the quantitative analysis, these improvements were out shadowed by the persistence of problem 4: Sweden still uses the Nordic residual mix without a mutual decision between Nordic countries. In the residual mix of 2011, the Nordic residual mix contained a lower renewable share than the Swedish national mix, which meant that renewable attributes were not double counted according to the analysis, but rather under-reflected. This might change from year to year and hence the problem is important.

Throughout the analysis, improvements regarding problems 1 and 3 yielded highest results in the avoidance of disclosure errors and double counting of renewable attributes. However, the impact of developments regarding problem 2, contract-based tracking, was based on an estimated volume, and it is likely that these are in reality as important as those regarding problems 1 and 3. It should also be outlined that the analysis focused on measuring improvements per country, not per problem.

4.2. Improvements to the Residual Mix Calculation Methodology by RE-DISS

Based on data collection of the disclosure practices and quantitative tracking data of European countries, RE-DISS sought improvements to the residual mix calculation methodology. Though pioneering, the major deficiency of the E-TRACK's residual mix calculation model was the lack of detail regarding the time-frame and production year of GO transactions considered in the calculation [15]. Ideally, residual mix calculation of a certain year should reflect transactions of GOs issued for that year electricity production, which can be seen as the intention of the E-TRACK model. However, the 12 months' lifetime of GOs poses a problem for this *production year-based* approach, since not all GOs of production year X are cancelled or expired by the deadline for the disclosure of year X consumption (31st of March year X + 1 according to RE-DISS BPR number 5 [34]). Hence the respective attributes are included in the residual of year X. If in turn the calculation of year X + 1 only considers GOs of production year X + 1, the transactions of production year X GOs after the deadline fall out of the scope of both year calculations. As the case of 2010 depicts, it is crucial to account for the transactions of production year 2009 GOs also after the deadline:

When collecting residual mix calculation data for 2010, the RE-DISS project team observed that GO imports exceeded exports by 32 TWh. Comparison with the AIB statistics indicated that both exports and imports seemed to be missing, but exports in particular. Secondly, the total amount of GO cancellations was suspiciously low in the RM data (174 TWh) given the total cancellation volume during 2010 (211 TWh) [41]. Solving both of these issues was crucial, since according to the residual mix methodology described in Section 2.3, both lacking volume of exports and cancellations in the RM data, lead to production mixes not being corrected with a sufficient amount of explicitly tracked attributes, *i.e.*, implicit double counting.

The difference of imports and exports was a result of two different methods of collecting GO transaction data: Production year-based method (PYBM) and transaction-based method (TBM). The PYBM, used mainly for the Nordic countries, considered the transactions of only production year 2010 GOs that occurred before 31st of March 2011. This, as mentioned, was seen as the original intention of the calculation, but neglected the transactions of production year 2009 GOs which occurred after March 31, 2010. However, the registry systems of some countries were not capable of separating transaction statistics per production year of the GO. This signified that the transaction data for these countries consisted of all GO transactions that occurred during calendar year 2010 (TBM was used), without consideration of the production year of the GOs. The data was clearly inconsistent and neglected a significant volume of especially exports (Figure 7). This was due to the fact that PYBM was used for large exporters and TBM for significant importers and cancellers.

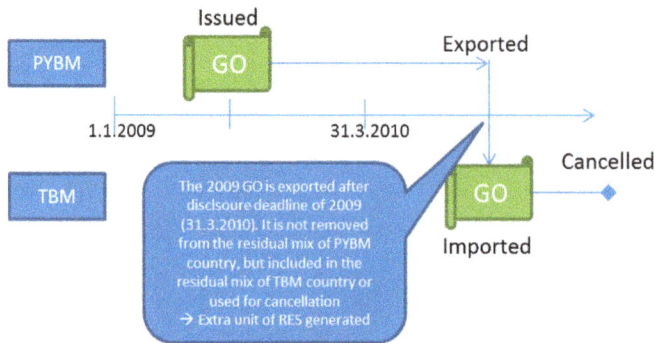

Figure 7. The problem of the production year-based method in residual mix calculation.

Based on detailed analysis of the data, the RE-DISS team was able to correct most of the problem for the 2010 calculation. The analysis also revealed that without corrections, explicit tracking of approximately 48 TWh of renewable and 21 TWh of nuclear attributes would have neglected in the residual mix calculation. For the 2011 calculation, the RE-DISS team set out to find a robust solution for the two major problems of the calculation:

- Transaction data being inconsistent between countries due to use of both PYBM and TBM;
- Active GOs from previous production years causing double counting.

As mentioned, the registry solutions of certain countries were, and still are, incapable of differentiating between production years in GO transaction data. Hence, the only applicable solution for the first problem was to apply TBM for all countries, and neglect the production year of GOs in the transaction data. Crucially, this also solved the problem of transactions of active GOs from previous production years, which were now considered automatically in the residual mix of 2011. In the 2011 calculation the double counting of Active GOs, prevented through TBM, was estimated by RE-DISS to be 40 TWh of renewable and 25 TWh of nuclear attributes.

As a consequence, the TBM, which was previously seen inferior to the PYBM, was lifted as the preferred method for residual mix calculation in RE-DISS. This was due to its ability to deliver a consistent methodology with the available data and to secure the avoidance of double counting in the existing legislative setting, which allows GOs to be usable for 12 months before expiry.

However the weakened accuracy of the residual mix in portraying explicit tracking of the correct production year attributes has stirred up discussions: "The residual mix is used to portray the generation attributes of a calendar year that are not used for explicit tracking... Thus, evidently, the theoretical argumentation to use PYB is better justified, because TB portrays the desired attribute flows poorer" [15]. The discussions have been further fueled by negative renewable volumes occurring in the domestic residual mixes of some countries. This may come about if the leftover of previous production year GOs is significant and if GOs are issued for a large share of the renewable production of the country. Negative renewable balances in final residual mixes of countries can be avoided through transferring the negativity to the European Attribute Mix, but this is not an optimal solution.

To enhance the accuracy of the calculation, the Shifted Transaction-Based Method (STBM) was introduced. STBM considers GO transactions between April 1st X and March 31st X + 1 for year X residual mix calculation, instead of the transactions of the calendar year X. This is logical since the time period to make cancellations for disclosing year X electricity consumption is accordingly shifted from the calendar year. Often most cancellation for year X disclosure occur in February and March of year X + 1, because it is the time suppliers close their electricity product portfolio for year X. The STBM considers this *cancellation peak* (see Figure 8) and the preceding transfer peak correctly in the residual mix of year X, whereas TBM would reflect it in the residual mix of year X + 1.

Figure 8. EECS-GO cancellations in 2012 [41].

A weakness of the Shifted Transaction-Based Method is that, though it decreases the chance for negative renewable balances, by increased accuracy, it does not eliminate the possibility entirely. Furthermore, a significant drawback of all methods that utilize import and export data, is their vulnerability to the GO market conditions. For example, GOs can be stocked in a country which has not yet implemented the expiry rule of GOs or which imposes little or no transaction costs. These patterns have been emphasized in the recent years, and can cause the renewable share in the residual mix to fluctuate significantly from year to year depending on whether the stock is built up or used that year. Even if the fluctuations are not powerful enough to turn the available renewable balance negative in a given year, they are hardly explainable and disturb confidence in the disclosure system. For example, the residual mix of Luxembourg contained a higher renewable share than its production mix in the 2011 calculation [38]. Robustness against market fluctuations is increasingly important due to the future emergence of exchanges in the GO market, e.g., [42].

The STBM was used for the residual mix calculation of 2012, though further solutions have also been considered to address the remaining weaknesses. To address these weaknesses, the Issuance-Based Method (IBM) might be a suitable approach, when certain preconditions are met. This method is presented in Section 5 in principle, but the detailed analysis is out of the scope of this paper.

5. Outlook and Discussion

As outlined, during the RE-DISS project (4/2010–10/2012), 16 selected European countries were able to reduce approximately 70 TWh of double counting of renewable attributes and 168 TWh of total implicit disclosure error. However, this only related to implicit disclosure errors on a national level and it is very likely that even greater errors reside in the individual supplier level. For example, if suppliers do not disclose product related mixes along with the total supplier mix as proposed in RE-DISS BPR number 39 [34], the resulting double counting can be even higher than that found in the analysis of Section 4.1. Therefore, in the future, emphasis should be put on the implementation and supervision of electricity disclosure at the supplier level.

But problems still persist on the national level as well. Out of the 16 countries in the analysis, only Denmark, Ireland and Switzerland had solved all five problems related to implicit disclosure (as well as Austria and Norway after the end of the project as explained in Table 2). Residual mix is still not calculated in France and Slovenia whereas Portugal has only taken initial steps towards a residual mix (problem 1). Likewise, Belgium, Germany and Luxembourg are not calculating a residual mix, but avoid double counting of renewable attributes (same applies partly for the Netherlands).

Contract-based tracking (problem 2) is still an issue in many countries as illustrated in Table 2, but it is better accounted for than in the beginning of the project. Spain and Italy have yet to coordinate their domestic calculation with other European countries (problem 3), whereas Sweden uses a regional approach for its residual mix calculation, which is not agreed upon with other countries in the Nordic region (problem 4). Table 2 presents the remaining steps to be made on a country level in correcting the five problems explained in Section 3.1.

Regarding improvements in the residual mix calculation methodology, RE-DISS solved the issue of double counting concerning Active GOs and introduced the Shifted Transaction-Based Method (STBM). The active GO problem was a result of emphasizing *accuracy* over *reliability* in the residual mix calculation. With RE-DISS coordinating efforts for Best Practices of electricity disclosure, such problems are unlikely in the future. However, it needs to be reminded that an explicit tracking system, requires a *reliable* implicit disclosure system as the backbone of operation. Hence reliability should always be emphasized over accuracy when the implicit system is developed. After all, the primary task of the residual mix is to secure full ownership of renewable attributes to consumers opting for green; not to perfectly portray attributes of correct production year to consumers buying regular electricity.

With this in mind, a future option for residual mix calculation, might be the Issuance-Based Method (IBM), which proposes a new perspective to the calculation. Instead of removing used production attributes (cancelled or net exported), the residual mix would be calculated by deducting attributes of issued guarantees of origin from the production mix and returning unused attributes (expired GOs) into it. The IBM is already correctly used by some competent bodies in the domestic residual mix calculation, but careful consideration is required in order to choose this as an approach for the determination of the EAM.

Table 2. Next steps in improving implicit disclosure on a national level based on [39].

Countries	Recommended Steps for Improving Implicit Disclosure
Austria	In Austria, explicit tracking of all electricity consumption will be required starting from 2013 electricity disclosure. Hence errors in implicit disclosure will be avoided in the future, but this was not covered by this analysis.
Belgium	To increase the accuracy of implicit disclosure, it is recommended that Belgium implement the RE-DISS residual mix, although the current method does not cause double counting of renewable attributes.
Denmark	Denmark has resolved the five problems related to implicit disclosure, presented in Section 3.1.
Finland	It is recommended that Finland ban unmonitored contract-based tracking of nuclear and fossil attributes.
Italy	It is recommended that Italy cooperate with other European countries in forming its residual mix to correctly account for international transfers of electricity and GOs in the residual mix and to prevent double counting of renewable attributes. Furthermore, it needs to be clarified whether the residual mix calculation accounts for previous production year GOs.
Netherlands	It is recommended that The Netherlands cease to filter out the possibly remaining renewable share from the residual mix. This would make the disclosure more accurate although the current form of implicit disclosure does not cause double counting of renewable attributes.
Sweden	It is recommended that Sweden use its domestic residual mix for implicit disclosure instead of the Nordic residual mix, unless Denmark, Finland and Norway also agree to use the Nordic residual mix.
Switzerland	Implicit disclosure is not used as all consumption is explicitly tracked. Hence errors in implicit disclosure are avoided.
Luxemburg	To obtain accurate implicit disclosure, it is recommended that Luxemburg implement the residual mix calculation according to the RE-DISS Best Practices, although the current method of Luxemburg does not cause double counting of renewable attributes.
Norway	Norway resolved the issue of active GOs after the end of the project (but this was not considered in the analysis) and has thus solved the five problems related to implicit disclosure, presented in Section 3.1.
Ireland	Ireland has resolved the five problems related to implicit disclosure, presented in Section 3.1.
France	It is recommended that France implement a residual mix calculation according to the RE-DISS BPR to eliminate implicit double counting of renewable attributes and to achieve more accurate disclosure.
Germany	To obtain accurate implicit disclosure, it is recommended that Germany implement the residual mix calculation according to the RE-DISS BPR, although the current method of Germany does not cause double counting of renewable attributes.
Portugal	It is recommended that Portugal implement a residual mix calculation according to RE-DISS BPR to eliminate implicit double counting of renewable attributes and to achieve more accurate disclosure.
Slovenia	It is recommended that Slovenia implement a residual mix calculation according to RE-DISS BPR to eliminate implicit double counting of renewable attributes and to achieve more accurate disclosure.
Spain	It is recommended that Spain cooperate with other European countries in forming its residual mix to correctly account for international transfer of electricity and GOs in the residual mix and to prevent double counting of renewable attributes. Furthermore, it needs to be clarified whether GOs need to be cancelled before they are used and whether the residual mix calculation accounts for previous production year GOs.

A clear benefit of IBM is that attributes presented by GOs are removed from the production mix of the year of production and country of issue. Hence, it is not affected by peculiarities in the GO market and cannot yield negative renewable balances, because, at maximum, renewable GOs can be issued for the volume of electricity production from renewable sources in the country. GOs of production year X, which are not cancelled by 31st March X + 1, are not without problems for the IBM either, because they disturb the balance between yearly production attributes and consumption to be disclosed. If a large part of GOs is unused before the disclosure deadline, the calculation removes more from the production than consumption side. This means that deficit exceeds surplus in the residual mix calculation and that the EAM should be expanded. Of course, if the volume of active GOs remains constant from year to year, the effect cancels itself out, but this can hardly be relied on.

Expanding of the EAM is not something that should be considered lightly, since the foundation of the coordinated calculation is on the balance of generation attributes and consumption. Therefore, from the perspective of calculating the EAM, the IBM might be a viable option only when implemented in connection with a so-called early expiry solution. The early expiry suggests that year X electricity consumption can only be disclosed with GOs of production year X and hence all remaining production year X GOs should expire at the disclosure deadline of year X. This should in theory not be contradictory to the lifetime of GOs as set by Article 15, paragraph 3 of Directive 2009/28/EC as it states that a GO should be used *within* 12 months of the associated electricity production. However, this is highly debated and countries interpreted the Directive differently in national legislation. Furthermore, to avoid potential arbitrage, all countries should use a harmonized disclosure deadline, which is not currently the case.

Linking the production year of the GO to the consumption year for which it can be used is a logical solution, since electricity consumed in a given year is effectively derived from the energy sources used for electricity production during that year. However, a change of this magnitude into existing procedures on a European level requires wide acceptance from national competent bodies of electricity disclosure and might not be possible without a revision of the EC Directive.

Acknowledgments

The authors of this article would like to thank all competent authorities who participated in the workshops and accepted to endorse the Best Practice Recommendations, which contain the RE-DISS methodology for the calculation of the Residual Mix. Without their involvement and commitment to change national regulations and practices, the RE-DISS work would not have led to the improvements described in this paper.

The authors would also like to acknowledge AIB's and RECS International's constructive and active support of the team in discussing and disseminating the findings of the project.

The authors are grateful to the European Commission for supporting the RE-DISS project through the IEE programme (Contract No. IEE/09/761/SI2.558253).

Appendix: Numerical Example of Simplified Residual Mix Calculation

In this chapter an example is given on how the residual mix is calculated based on the steps elaborated in Chapter 2.3. This example is to be used together with the process description in Chapter 2.3. For simplification, environmental indicators as well as exchange with countries outside of the European Attribute Mix are not included in the example.

Table A1. Simplified example of residual mix calculation process.

	Country A	Country B	Country C	Country D	SUM		Explanation
1. Data Collection							
1a. Production Mix							
RES	5	10	80	40	135	TWh	
NUC	30	0	40	0	70	TWh	
FOS	40	60	10	20	130	TWh	
1b. Consumption	100	40	140	55	335	TWh	
1c. Certificate Transactions							
(all RES)							
Imp.	60	30	5	0	95	TWh	
Exp	5	5	60	25	95	TWh	
Canc.	50	15	20	10	95	TWh	
2. Domestic RM							
2a. Available Attributes							
RES	10	20	5	5	40	TWh	1a.Prod+ 1c.Imp-1c.Exp-1c.Canc.
NUC	30	0	40	0	70	TWh	1a.Prod
FOS	40	60	10	20	130	TWh	1a.Prod
SUM	80	80	55	25	240	TWh	
2b. Untracked Consumption	50	25	120	45	240	TWh	1b.Cons-1c.Canc.
2c. Surplus/Deficit							
Surplus	30	55			85	TWh	IF 2a.SumAvailable Attributes >2b. Untracked Consumption—> Surplus = Sum 2a-2b
Deficit			65	20	85	TWh	IF 2a.SumAvailable Attributes <2b. Untracked Consumption—> Deficit = 2b-sum 2a
3. European Attribute Mix							
3a. Surplus to EAM							
RES	3.75	13.75	0	0	17.5	TWh	2c.Surplus*2a.Share of RES in Available Attributes
NUC	11,25	0	0	0	11.25	TWh	2c.Surplus*2a.Share of NUC in Available Attributes
FOS	15	41.25	0	0	56.25	TWh	2c.Surplus*2a.Share of FOS in Available Attributes
SUM	30	55	0	0	85	TWh	

528

Table A1. *Cont.*

	Country A	Country B	Country C	Country D	SUM	Explanation
4. Final Residual Mixes						
4a. RM of Surplus domain						
RES	6.25	6.25			TWh	2a.Av. Att RES—3a.Surplus RES
NUC	18.75	0			TWh	2a.Av. Att NUC—3a.Surplus NUC
FOS	25	18.75			TWh	2a.Av. Att FOS—3a.Surplus FOS
4b. RM of Deficit domain						
RES			18.38	9.12	TWh	2a.Av. Att. RES + 2c.Deficit*3a.ShareofRESinEAM
NUC			48.60	2.65	TWh	2a.Av. Att. NUC + 2c.Deficit*3a.ShareofNUCinEAM
FOS			53.01	33.24	TWh	2a.Av. Att. FOS + 2c.Deficit*3a.ShareofFOSinEAM
Final RM (%)						
RES	13%	25%	15%	20%		
NUC	38%	0%	41%	6%		
FOS	50%	75%	44%	74%		
Production Mix (comparison)						
RES	7%	14%	62%	67%		
NUC	40%	0%	31%	0%		
FOS	53%	86%	8%	33%		

Conflicts of Interest

The authors declare no conflict of interest.

References

1. Trevino, L. *Liberalization of the Electricity Market in Europe: An overview of the Electricity Technology and the Market Place. Chair Management of Network Industries—MIR*; College of Management of Technology, Ecole Polytechnique Federale de Lausanne: Lausanne, Switzerland, 2008.
2. 2009/72/EC. Directive 2012/27/EC of the European Parliament and of the Council of 25 October 2012 on Energy Efficiency, Amending Directives 2009/125/EC and 2010/30/EC and Repealing Directives 2004/8/EC and 2006/32/EC. Available online: http://eur-lex.europa.eu/LexUriServ/LexUriServ.do?uri=OJ:L:2009:211:0055:0093:EN:PDF (accessed on 15 April 2015).
3. Timpe, C.; Seebach, D. Best Practice for the Tracking of Electricity. Recommendations from the E-Track II Project. Deliverable 10 of the IEE Project "A European Tracking System for Electricity—Phase II (E-TRACK II)". Available online: http://www.e-track-project.org/docs/E-TRACK%20II_WP7_Recommendations.pdf (accessed on 15 April 2015).

4. Jansen, J.C.; Seebach, D. Requirements of Electricity End-Users on Tracking of Electricity Generation Attributes and Related Policies. D7 of WP 5 of the E-Track II Project (Final Report). A Report Prepared as Part of the EIE Project "A European Tracking System for Electricity—Phase II (E-Track II)". Available online: http://www.e-track-project.org/docs/final/WP5_D7_Consumer%20requirements%20report_V2.pdf (accessed on 15 April 2015).

5. The E-TRACK Project. Available online: http://www.e-track-project.org (accessed on 17 April 2015).

6. Boardman, B.; Palmer, J. Electricity disclosure: The troubled birth of a new policy. *Energy Policy* **2007**, *35*, 4947–4958.

7. Wilhite, H. An assessment of experiences in the U.S.A. with power and emission disclosure information for energy consumers. Available online: https://www.duo.uio.no/handle/10852/32691 (accessed on 15 April 2015).

8. Markard, J.; Holt, E. Disclosure of electricity products—Lessons learnt from consumer research as guidance for energy policy. *Energy Policy* **2003**, *31*, 1459–1474.

9. CEER 2012. Draft Work Programme 2013 of the Council of European Energy Regulators. Available online: http://www.energy-regulators.eu/portal/page/portal/EER_HOME/EER_CONSULT/CLOSED%20PUBLIC%20CONSULTATIONS/CROSSSECTORAL/2013%20Work%20Programme/CD/C12-WPDC-22–06_WP%202013-PC_18-Jun-12.pdf (accessed on 3 November 2012).

10. Miller, R.D.; Ford, J.M. *Shared Savings in the Residential Market: A Public/Private Partnership for Energy Conservation*; Energy Task Force, Urban Consortium for Technology Initiatives: Baltimore, MD, USA, 1985.

11. Aasen, M.; Westskog, H.; Wilhite, H.; Lindberg, M. The EU electricity disclosure from business perspective—A study from Norway. *Energy Policy* **2010**, *38*, 7921–7928.

12. EC 2013. Homepage of Citizens' Energy Forum of the European Council. Available online: http://ec.europa.eu/energy/gas_electricity/forum_citizen_energy_en.htm (accessed on 30 August 2012).

13. Lise, W.; Timpe, C.; Jansen, J.C.; ten Donkelaar, M. Tracking electricity generation attributes in Europe. *Energy Policy* **2007**, *35*, 5855–5864.

14. Timpe, C. A European Standard for the Tracking of Electricity. Available online: http://ec.europa.eu/energy/intelligent/projects/sites/iee-projects/files/projects/documents/e-track_summary_slides.pdf (accessed on 15 April 2015).

15. Klimscheffskij, M. Tracking of Electricity in the EU—From Directives to Practice. Master Thesis, School of Science, Aalto University, Espoo, Finland, 2011.

16. Directive 2009/28/EC of the European Parliament and of the Council of 23 April 2009 on the Promotion of the Use of Energy from Renewable Sources and Amending and Subsequently Repealing Directives 2001/77/EC and 2003/30/EC. Available online: http://eur-lex.europa.eu/legal-content/EN/TXT/PDF/?uri=CELEX:32009L0028&from=en (accessed on 15 April 2015).

17. 2012/27/EC. Directive 2009/72/EC of the European Parliament and of the Council of 13 July 2009 Concerning Common Rules for the Internal Market in Electricity and Repealing Directive 2003/54/EC. Available online: https://www.energy-community.org/pls/portal/docs/36275.PDF (accessed on 15 April 2015).

18. AIB 2011. Annual Report 2010. Association of Issuing Bodies. Available online: http://www. aib-net.org/portal/page/portal/AIB_HOME/NEWSEVENTS/Annual_reports (accessed on 15 April 2015).

19. AIB 2013a. EECS Rules. Association of Issuing Bodies. Available online: http://www. aib-net.org/portal/page/portal/AIB_HOME/EECS/EECS_Rules (accessed on 15 April 2015).

20. AIB 2013b. The AIB Hub. Association of Issuing Bodies. Available online: http://www. aib-net.org/portal/page/portal/AIB_HOME/FACTS/EECS%20Registries/AIB_Hub (accessed on 11 December 2012).

21. AIB 2013c. AIB Members. Association of Issuing Bodies. Available online: http://www.aib-net.org/portal/page/portal/AIB_HOME/FACTS/AIB%20Members/AIB%20Members (accessed on 1 August 2012).

22. EEX 2015. Guarantees of Origin—EEX Power Derivatives. Available online: https://www.eex.com/en/market-data/power/derivatives-market/guarantees-of-origin#!/2015/03/19 (accessed on 24 March 2015).

23. Raadal, H.; Dotzauer, E.; Hanssen, O.J.; Kildal, H.P. The interaction between electricity disclosure and tradable green certificates. *Energy Policy* **2012**, *42*, 419–428.

24. Iisd 2012. Trends in investor claims over feed-in tariffs for renewable energy. News article of International Institute for Sustainable Development. Available online: http://www.iisd.org/itn/2012/07/19/trends-in-investor-claims-over-feed-in-tariffs-for-renewable-energy/ (accessed on 1 December 2012).

25. Ref 2011. Turkey implements feed-in tariff, Spain cuts. News article of Renewable Energy Focus. Available online: http://www.renewableenergyfocus.com/view/14994/turkey-implements-feed-in-tariff-spain-cuts/ (accessed on 13 October 2012).

26. Reuters 2010. Czech government adopts a plan to rein in renewables. News article of Reuters. Available online: http://www.reuters.com/article/2010/08/25/us-czech-renewables-idUSTRE67O4DP20100825?utm_source=feedburner&utm_medium=feed&utm_campaign=Feed%3A+reuters%2FUSgreenbusinessNews+%28News+%2F+US+%2F+Green+Business%29&utm_content=Google+Reader (accessed on 20 September 2012).

27. Italy cuts FiTs in an effort to balance renewables growth. News article of Renewable Energy Focus. Available online: http://www.renewableenergyfocus.com/view/25145/italy-cuts-fits-in-an-effort-to-balance-renewables-growth/ (accessed on 13 October 2012).

28. Bird, L.; Wüstenhagen, R.; Aabakken, J. A review of international green power markets: Recent experience, trends, and market drivers. *Renew. Sustain. Energy Rev.* **2002**, *6*, 513–536.

29. Hansla, A.; Gamble, A.; Juliusson, A.; Gärling, T. Psychological determinants of attitude towards and willingness to pay for green electricity. *Energy Policy* **2007**, *36*, 768–774.

30. Statistics on switching customers in Austria 2013. Unpublished material.

31. Sogge, C.V.; Øen, J. Klima for Redusert Strøm for Bruk? En Focus Gruppe Studie I Spenningsfeltet Mellom Forbruk Og Miljø, Kunnskap Og Verdier. Master Thesis, University of Life Science, Ås, Norway, 2007.

32. AIB 2015. AIB statistics. Association of Issuing Bodies. Available online: http://www. aib-net.org/portal/page/portal/AIB_HOME/FACTS/Market%20Information/Statistics (accessed on 20 April 2015).

33. PWC 2009. Green Electricity Making a Difference: An International Survey of Renewable Electricity Labels. Available online: https://www.pwc.dk/da/publikationer/assets/green-electricity-making-a-difference.pdf (accessed on 11 April 2015).

34. RE-DISS 2012a. Best Practice Recommendations for the implementation of Guarantees of Origin and other tracking systems for disclosure in the electricity sector in Europe. Available online: http://reliable-disclosure.org/ (accessed on 15 April 2015).

35. RE-DISS 2012c. Country Profiles. Available online: http://www.reliable-disclosure.org/documents/ (accessed on 15 April 2015).

36. Klimscheffskij, M. Electricity Residual Mix Calculation According to the RE-DISS Project—Methodology of Residual Mix Calculation. Appendix to the RE-DISS Best Practice Recommendations; 2012. Available online: http://reliable-disclosure.org/ (accessed on 15 April 2015).

37. Raadal, H.L.; Nyland, C.A.; Hanssen, O.J. Calculation of residual mixes when accounting for the EECS (European Electricity Certificate System)—The need for a harmonised system. *Energies* **2009**, *2*, 477–489.

38. RE-DISS 2012b. European Residual Mixes 2011. Available online: http://www.reliable-disclosure.org/upload/147-RE-DISS_2011_Residual_Mix_Results_v1_1.pdf (accessed on 15 April 2015).

39. RE-DISS 2012d. *Country-specific Qualitative Data Sheets*; RE-DISS project; European Commission: Brussels, Belgium, 15 October 2012.

40. ENTSO-e 2012. Detailed Monthly Production Data for All Countries for a Specific Range of Time. Excel-Spreadsheet. Available online: https://www.entsoe.eu/data/data-portal/production/ (accessed on 1 May 2012).

41. AIB 2013d. EECS Market Statistics. Association of Issuing Bodies. Available online: http://www.aib-net.org/portal/page/portal/AIB_HOME/FACTS/Market%20Information/Statistics (accessed on 31 October 2012).

42. EEX 2013. Guarantees of origin—Overview. Available online: http://www.eex.com/en/goo (accessed on 17 November 2012).

Chapter 6:
Societal Issues

Current and Future Friends of the Earth: Assessing Cross-National Theories of Environmental Attitudes

Karen Stenner and Zim Nwokora

Abstract: Empirical studies of public opinion on environmental protection have typically been grounded in Inglehart's post-materialism thesis, proposing that societal affluence encourages materially-sated publics to look beyond their interests and value the environment. These studies are generally conducted within, or at best across, Western, democratic, industrialized countries. Absence of truly cross-cultural research means the theory's limitations have gone undetected. This article draws on an exceptionally broad dataset—pooling cross-sectional survey data from 80 countries, each sampled at up to three different points over 15 years—to investigate environmental attitudes. We find that post-materialism provides little account of pro-environment attitudes across diverse cultures, and a far from adequate explanation even in the affluent West. We suggest that unique domestic interests, more than broad value systems, are driving emerging global trends in environmental attitudes. The environment's future champions may be the far from 'post-material' citizens of those developing nations most at risk of real material harm from climate change and environmental degradation.

Reprinted from *Energies*. Cite as: Stenner, K.; Nwokora, Z. Current and Future Friends of the Earth: Assessing Cross-National Theories of Environmental Attitudes. *Energies* **2015**, *8*, 4899-4919.

1. Introduction

Climate change is a global 'tragedy of the commons' [1]. This means that solutions will require cooperation by developed and developing nations. An important, and perhaps necessary, condition for such cooperation is widespread concern among ordinary people about the potential effects of climate change. Given that environmental protection may be accompanied by real costs, policy action to combat climate change seems significantly more likely when political leaders face pressure from their citizens. As Scruggs and Benegal explain, 'while the earth's climate may not react to what people think about the climate, elected politicians often do' [2] (p. 515). And even non-elected political leaders must be somewhat sensitive to the sentiments of their societies in order to maintain the legitimacy of their regimes [3].

The translation of citizens' concerns into policy action is a complex process and there has been research investigating several key linkages in this process. One strand of research, and the one to which we aim to contribute, investigates the overarching sources of citizens' environmental attitudes. As we explain in detail in the next section, this research aims to identify the factors that account for variation in pro-environment attitudes across (and within) nations. Influential theories draw attention to individuals' social position—as determined by class, income and education—and individuals' values, namely the issues and concerns they choose to prioritize. While there are connections between these two approaches, they are usually treated as distinct perspectives on how individuals form their environmental attitudes. The first approach emphasizes the 'structural' determinants of attitudes while the second stresses that individuals, independent of their social background, can *choose* which values to prioritize.

There are a number of other research agendas likewise relevant for understanding the connection between pro-environment attitudes and policy action on climate change. One agenda investigates the relationship between citizens' attitudes and their patterns of behavior [4,5]. Motivating this research is the observation that attitudes do not consistently predict actions. For example, older people often behave in more environmentally friendly ways than younger people despite having weaker pro-environmental sentiments [6] (p. 770). Furthermore, there are numerous studies that examine how citizens' environmental attitudes become institutionalized in environmental movements [7–9]. These studies show that different political opportunity structures provide environmental groups with greater or lesser obstacles in their attempts to consolidate, grow, and influence policy-makers. Finally, there are studies that examine variation in the success of environmental movements, seeking to identify which circumstances enable them to secure a foothold in mainstream politics, either through an influential Green party or the adoption of pro-environment policies by older, established parties [10,11].

In this article, we analyze the determinants of citizens' environmental attitudes drawing on a much larger and far more comprehensive dataset than has been examined to date. In particular, we focus our investigation on postmaterialism theory, developed by Inglehart [12–14], which has been a highly influential framework for understanding the sources of 'green' sentiment. It posits that environmental concern depends upon satisfaction of basic material needs. It is thought that generations socialized in material affluence will lift their sights to appreciate and preserve the environment even when there are real costs to doing so. Our dataset enables a rigorous test of this thesis, as well as other general accounts of environmental attitudes, across 80 countries surveyed on up to three occasions over 15 years.

The remainder of this paper is organized as follows. In the next section we summarize the dominant theories of environmental attitudes. We then describe the sample, variables, and models that we employ to test these theories. We present our findings in two subsequent sections. The first presents a broad-brush overview of trends we observe in the data. In the second we use hierarchical linear modelling to assess the merits of the general theories against our cross-national data. Their limitations lead us to argue in favour of more context-specific theories of environmental concern. We conclude by discussing how this change of perspective alters our expectations of future global environmental politics.

2. The Determinants of Public Concern about the Environment: Existing Studies

Existing studies of environmental attitudes are rooted in two theoretical schools [15]. The 'social-structural' view presents pro-environment opinions as characteristic of certain groups defined by age, class, education, income and other cleavages. The main alternatives to 'social bases' explanations are theories highlighting 'value systems'. A consistent research finding is that pro-environment views are negatively associated with youth and positively associated with education and residence in urban areas. However, in general, most studies confirm the greater importance of values compared to social-structural factors. Thus, in their review of the scholarship, Oreg and

Katz-Gerro conclude that 'social-psychological constructs such as values, attitudes, and beliefs have been more successful in predicting pro-environmental behaviours' than structural theories [15] (p. 463).

Since the 1970s, Inglehart's postmaterialism thesis has been the most prominent theoretical framework for understanding the sources of 'green' sentiment [6,16,17]. It posits that the relative abundance of material (*i.e.*, economic and physical) security in advanced industrial countries has led to postmaterialist priorities (including environmental concern) among generations that came of age during periods of material security. The theory derives from Maslow's argument that human needs can be represented as a hierarchy. Concern for the environment is a high-level need that is generally only reached by individuals (and by implication, societies) once they have satisfied their lower-level needs for physical and economic security. Thus, it is thought that generations socialized in material affluence tend to lift their sights to appreciate and preserve the environment even when there are real costs to doing so. Because postmaterialism relies on gradual opinion change resulting from different socialization contexts the thesis differs from the simpler affluence hypothesis, which posits that environmentalism is a 'normal' good directly (and positively) related to income [18,19].

The theory was tested on Western citizens during a period of economic expansion when the most serious environmental policy problems were deferred for later resolution [20]. Additionally, postmaterialism was measured with valence questions that allowed people simply to express their general values rather than their willingness to protect the environment when faced with real trade-offs [21]. Despite these limitations postmaterialism quickly gained pre-eminence, perhaps because it predicted an increasingly wealthy world becoming more supportive of environmentalism, a much rosier future than other projections of global environmental politics.

The past two decades, however, have seen sustained challenges to the theory of postmaterialism on both methodological and practical grounds. Methodologically, critics claim Inglehart's postmaterialism measure does not capture the properties of the concept [22,23], with some arguing that postmaterialism is a proxy for education and affluence [24], or even that the materialism-postmaterialism dimension is non-existent [25,26]. At a practical level, the ineluctable rise of Western postmaterialists looks far from certain. Western economies stagnated just as the costs of environmental concern began to bite. And some research suggests that, in cases such as the United States, support for environmental action stagnated and perhaps even declined during the period of sustained economic growth in the late 1990s and early 2000s [27]. Globally, the current struggle to reach a workable agreement on greenhouse gas emissions reveals 'generally increasing antagonism between North and South' [28] (p. 137). In the North, even pro-environment voters are often 'strongly resistant to the reality of higher taxes or energy prices' [21] (p. 15).

Moreover, while postmaterialism theory predicts environmental concern will be greatest in materially-sated Western populations, the frequently negative environmental experiences in developing countries provide alternative, and potentially more powerful, stimuli for concern [29]. Environmentalism in developing countries appears often to be driven by local practicalities, and only weakly related to social structures or values. For example, pro-environment attitudes have been influenced by rising sea levels in Bangladesh [30]; soil degradation in China [31]; destruction of communal lands in Nigeria [32]; and health and safety concerns in South Korea [33]. Inglehart has responded by presenting postmaterialism as a partial theory applying only to rich countries, while environmentalism in poor countries is explainable by reference to 'objective problems' [34].

These challenges suggest limits to postmaterialism as a universal theory of environmental sentiment, but the attack is a scattered one. First, the methodological critiques tend to focus not on environmental issues but on testing the validity of the postmaterialism measure. Environmentalism is peripheral to this, with even the twelve-item index of postmaterialism including only one question on the environment [26] (p. 938). Second, comparative studies have focused disproportionately on developed nations [16,18,35–38]. Some research has used the International Social Survey Programme [19,39], but even the latest ISSP data (from the year 2000) extends to only 26 countries, most of which are Western. A handful of recent studies use the more expansive World Values Survey, but without drawing on more recent waves of data [16,40,41] which, notably, have deliberately expanded coverage of non-Western, developing countries. Our investigation has even broader scope than the approximately 60 countries examined in the most comprehensive of these studies to date [16]. Third, there has been no alternative theoretical framework developed to explain public opinion on the environment. It has proved easier to dismantle the postmaterialism thesis than to replace it.

Our paper addresses these three gaps. First, we focus directly on the utility of postmaterialism as a predictor of environmental attitudes. Second, we provide a more rigorous test of postmaterialism, as well as other values and social-structural theories, by drawing on pooled survey data collected from a broad cross-section of countries (covering all major regions) over 15 critical years in global environmental politics. And third, we ultimately argue that any adequate explanation of environmental attitudes must consider—in addition to values and social-structural variables—the confluence of local problems, issue salience, and calculations of domestic interest.

Both the values and social-structural explanations are cast as general, cross-national theories and can therefore be tested using large-n statistical techniques. To do so, we draw on a pooled cross-sectional dataset including nearly 150,000 survey respondents—drawn from 80 countries whose populations were randomly sampled on up to three occasions over a 15 year period—substantially exceeding the range of existing large-n studies. Observing, by these means, any evident cross-cultural differences or over-time changes in environmental attitudes, our investigation is able to offer insights into recent patterns and possible future trends.

We acknowledge at the outset that our empirical design is not without limitations. The 'objective problems' thesis cannot be *directly* tested here in the absence of data suitable for gauging the relationship between local environmental events and the dynamics of public opinion. In addition, we must emphasise that the dataset we analyze contains greater cross-national than longitudinal variation: for no country do we have more than three independent waves of survey data drawn within the 15-year period, and there is no panel of continuing respondents embedded in any of these surveys. Even so, our approach still offers the most systematic and comprehensive test to date of the leverage offered by postmaterialism (and other general theories) in accounting for environmental attitudes. The obvious alternative, a meta-analysis of existing studies, would immediately confront the problem that past empirical work focuses disproportionately on Western countries. In any case, we would certainly argue that a large-n study of a diverse cross-national dataset—pooling jointly-designed surveys sharing a great number of relevant measures (both of critical outcomes, and key explanatory factors)—is likely to produce more reliable and readily generalizable findings about the universals (or lack thereof) in the dynamics of environmental concern.

3. Data

3.1. Sample and Sub-Samples

We use data from the cumulative World Values Survey (1981–2008) (WVS) to test cross-national theories of environmental attitudes. The WVS is a 'worldwide investigation of socio-cultural and political change' [42] (p. 528). The first wave of surveys (1981) spanned 22 countries, and country coverage has been expanded significantly in the waves that have followed. At the time of this analysis, the WVS database consisted of 261 large-scale random sample surveys drawn from 99 countries across nearly three decades, incorporating 355,298 respondents. Norris observes that recent WVS data captures the attitudes of approximately 88 per cent of the world's population 'covering all six inhabited continents' [42] (p. 528). When the data were distilled to include only those surveys with two critical measures of support for environmental protection, plus all major explanatory variables (discussed below), our remaining total was 135 samples drawn from 80 countries, at up to three time points over 15 years (1994–2008), and ultimately incorporating 149,559 respondents (see Supplementary Materials, Table S1 for details). Some 'countries' are more accurately described as 'societies' as they are not uncontested, fully sovereign states: East and West Germany (WVS drew separate samples even after reunification); Hong Kong ('special administrative region' of People's Republic of China); Taiwan (sovereignty disputed by PRC); Puerto Rico (unincorporated US territory); Bosnia Federation and Republika Srpska (two entities comprising sovereign country of Bosnia and Herzegovina). However, given our focus on the values of distinct *societies*, there is a strong case for treating such entities as separate analytic units. Recent research using WVS data generally follows this approach [16].

Despite this trimming of the data, the sample's coverage greatly exceeds previous analyses based on WVS data (including Inglehart's). Because of the sample's breadth we are able to investigate environmental attitudes in six sub-samples (see Table S1): Western and non-Western countries (according to Huntington [43]); democracies and autocracies (following Pemstein et al. [44]); and developed and developing economies (using a median split of countries ranked by their log GDP per capita at the time of the WVS survey, with GDP data from Heston et al. [45]). Notice that these sub-samples overlap, so (for example) there are some non-Western / high-GDP countries (e.g., Japan, Argentina, Croatia) and some democratic nations that are neither Western nor high-GDP (e.g., India, Brazil, Turkey).

3.2. Dependent Variable: Support for Environmental Protection

The WVS over the years has measured a wide array of environment-related variables. But given that our key objective was detection (or refutation) of any universal influences on environmentalism across diverse cultures, we focused in on and restricted our analysis to just two relevant items, these being asked of enough WVS samples to allow for broad cross-national comparisons. Equally critical, these two items both tap willingness to bear economic costs to protect the environment, and reflect real and concrete choices. In our view, they present respondents with a more meaningful test of environmental concern than valence questions [27,37] and items requiring them simply to express their values in the abstract. In the end, these were the only two environmental questions on the cumulative WVS that fully satisfied both of these criteria (common question, presenting a real and explicit trade-off). The first question (*b002*) asks whether respondents agree with an 'increase in taxes if used to prevent environmental pollution', on a four-point scale ranging from 'strongly disagree' to 'strongly agree'. The second question (*b008*) has respondents choose between 'protecting the environment' and 'economic growth and creating jobs'. Here we arranged responses on an ordinal scale, with 'other answer' a neutral response between the two proffered options. Our overall measure of environmentalism (mean = 0.56, sd = 0.30, range = 0–1 across 11 points) averaged these two items, after first rescoring each to range from 0 to 1. Note that subsequent extensive testing of the robustness of our results to alternative constructions of the dependent variable confirmed that our central findings and the conclusions drawn from them do not alter significantly if we substitute one or the other of these environmental variables for the overall measure formed by combining the two.

At this point we should note that, here and throughout, our general practice is to re-score all variables (and any measures constructed from them) that have no 'natural metric' to range (in practice) from 0 to 1, which greatly simplifies the comparison and interpretation of estimated effects (coefficients). Variables that do have their own natural metric (in this case: age, time, and average GDP per capita in childhood) are of course left in those original units, with results interpreted accordingly.

3.3. Independent Variable: Values and Social-Structural Factors

3.3.1. Values

Foremost among potential explanatory variables is, of course, postmaterialism. Here we rely on Inglehart's 12-item index of postmaterialism furnished by the WVS [14] (pp. 415–416). This index is formed from three separate batteries where respondents indicate their first and second preferences from among four values (two postmaterialist and two materialist options). The postmaterialist values are: 'giving people more say in how things are done', 'making our cities and countryside more beautiful', 'giving people more say in government decisions', 'protecting free speech', 'progressing toward a less impersonal and more humane society', and 'a society where ideas count more than money'. The materialist stances are: 'maintaining high economic growth', 'ensuring strong defense forces', 'maintaining order', 'fighting rising prices', 'a stable economy', and 'fighting crime'. We rescored the index to range from 0 to 1.

We also included a standard measure of 'left-right' self-placement [46,47], where respondents placed their own 'views' in 'political matters' on a ten-point scale (rescored to range from 0 to 1), labelled 'left' and 'right' at either end. While the meaning of these labels varies between countries, pro-environment stances generally form part of the bundle of 'left-wing' attitudes in most countries covered by the WVS.

We also included a measure of authoritarianism [48–51]: an individual's preference for obedience and conformity over freedom and difference. This value dimension cuts across the traditional 'left/right' divide [47]. We measured authoritarianism through child-rearing values [49] with respondents choosing from a list of 'qualities that children can be encouraged to learn at home' those they 'consider to be especially important'. The choice of 'obedience' reflects authoritarianism, and 'independence' and 'imagination' its inverse. We also incorporated a separate item where respondents indicated whether it would be 'a good thing' if there was 'greater respect for authority' in the future. These four items were reversed as appropriate and equally weighted in the overall measure (rescored to range from 0 to 1). We expect that authoritarians, inclined to obedience and conformity, are likely to privilege national economic strength over protection of vulnerable ecosystems.

3.3.2. Social-Structural Factors

To test the thesis that being socialized in a climate of material security promotes environmentalism we included a variable directly measuring the national economic prosperity experienced in respondents' youth, specifically their country's log GDP per capita averaged across their 18 years of childhood (henceforth 'affluent socialization'). It is important to bear in mind that this variable, like all others in the analysis, remains an individual-level measure, although it is informed by aggregate country-level data. Respondents surveyed even at the same time in the same country will accrue a very wide array of societal affluence scores, since (for example) someone who has just come of age will have experienced in their childhood a different economic and cultural climate than that prevailing across the 18 years during which (say) some elderly counterpart being interviewed contemporaneously was socialized. A survey of around 1000 respondents will typically yield about 60 different 'affluent socialization' scores, just in that one country in a particular year, so the range is considerable. The levels will then vary enormously—with the ranges perhaps not even overlapping—across different societies.

In sum, this critical measure for our analysis takes the most widely accepted indicator of national economic prosperity, observes its level for each of the 18 years spanning a particular respondent's upbringing, and then averages across those 18 scores to reflect the *societal affluence* that that particular respondent experienced in their childhood. The measure reflects societal wealth (not household wealth) at the time one was growing up, because it is societal affluence (rather than the prosperity of one's own parents/household) that is purportedly critical for nurturing a postmaterial perspective.

We did test two other measures of material affluence that reflected the individual's current household prosperity, rather than that of the society in which they were raised. We included, first, a measure of family income decile within nation (rescored from 0–1); and, second, satisfaction with one's household financial situation (10 point scale, rescored 0–1). We then extended beyond *material* security to measure how satisfied respondents were with their 'life as a whole these days' (10 point scale, rescored 0–1), and whether 'taking all things together' they were 'very happy, quite happy, not very happy, or not at all happy' (4 point scale rescored 0–1).

We also tested the impact of several variables relating to social location. We used an 8-point variable reflecting highest level of education attained, from incomplete primary schooling (scored 0) to postgraduate qualification (scored 1). Age was simply measured in decades (*i.e.*, 65 years was scored 6.5), but the concept of generation presented difficulties given that the purportedly critical eras of socialization varied across countries. In the end, we covered the main cultural fault line common to many countries via a dummy variable scored '1' if the respondent was born since the 1980s ('Gen Y' in the West, and variously named elsewhere), and '0' otherwise.

3.4. Control Variables: Trust and Institutional Confidence; Sex, Religion, and Ethnicity

The perceived effectiveness of interventions may well rest on one's confidence in fellow citizens and social institutions. After all, environmental protection is a collective action problem and trust is often necessary for the solution of such problems [52,53]. Thus, an individuals' willingness to sacrifice economic growth for leap-of-faith efforts at environmental protection might be tempered by low inter-personal trust or lack of institutional confidence. Neither has been routinely included in prior research. We controlled for inter-personal trust using the item: 'generally speaking, would you say that most people can be trusted or that you need to be very careful in dealing with people?' (dummy variable 1/0). We gauged institutional confidence by ascertaining whether respondents had 'a great deal of confidence, quite a lot of confidence, not much confidence, or none at all' in each of the following: the armed forces, press, labour unions, police, parliament, civil services, television, government, political parties, justice system, and major companies (averaged and re-scored to range 0 to 1 overall; $\alpha = 0.87$).

Finally, we also controlled for sex (with men generally considered less sympathetic to environmental causes), as well as religious denomination and ethnicity. The latter were controlled in all analyses via an extensive set of dummy variables: 14 categories for denomination and 18 categories for ethnicity (full results are available from authors upon request).

4. Broad Patterns and Trends in Support for Environmental Protection

Before moving to the multivariate analyses, we first consider whether any patterns can be discerned by looking in broad overview at the earliest and latest data available for each of the 80 countries. Table 1 ranks these countries based on: (1) citizens' average scores (according to the latest WVS survey) on environmentalism, postmaterialism, and 'affluent socialization'; as well as (2) national scores (at the time of the latest WVS survey) for level of democracy and log GDP per capita.

Focusing first on countries with multiple waves of data, the immediate surprise is that there seems to have been no general rise in support for environmental protection, and this over a decade (1996–2006 is the typical span) ranging from adoption of the Kyoto Protocol through to its ratification by most signatories. We see only a waxing and waning of support far too idiosyncratic and country-specific to suggest a global, regional, or cultural shift in sentiment. If anything, there has been a general downturn in willingness to bear economic costs to protect the environment. Of the 41 countries with environmentalism measured on at least two occasions, 25 experienced a decline in public support between their earliest and latest WVS waves, averaging 7 percentage points (that is, −0.07 on the measure's 0–1 scale). Moreover, the downward trend is actually most marked in some of the more affluent countries, including South Korea (−0.25), Taiwan (−0.15), West (−0.16) and East Germany (−0.13).

Table 1. Countries ranked by environmentalism, postmaterialism and affluence.

Country (Years)	Environ. (earliest)	Environ. (latest)	Environ. Rank	Postmaterial. Rank	Affluent Socializ. Rank	logGDP/cap. Rank
El Salvador (1999)	.	0.78	1	23	48	52
Dominican Rep. (1996)	.	0.75	2	9	51	50
Andorra (2005)	.	0.71	3	1	14	22
Vietnam (2001–2006)	0.68	0.71	4	46	70	64
Norway (1996–2007)	0.64	0.69	5	4	4	1
Puerto Rico (1995–2001)	0.72	0.68	6	8	24	17
Tanzania (2001)	.	0.66	7	41	76	78
China (1995–2007)	0.67	0.66	8	77	69	45
Switzerland (1996–2007)	0.49	0.65	9	2	1	3
Canada (2000–2006)	0.61	0.65	10	7	7	5
Burkina Faso (2007)	.	0.63	11	45	75	76
Mexico (1996–2005)	0.57	0.63	12	16	28	27
Croatia (1996)	.	0.63	13	17	19	35
Sweden (1996–2006)	0.70	0.62	14	3	6	6
Peru (1996–2006)	0.54	0.62	15	20	40	48
Cyprus (2006)	.	0.62	16	35	27	21
Mali (2007)	.	0.62	17	55	78	74
Philippines (1996–2001)	0.61	0.62	18	36	66	66
India (1995–2006)	0.45	0.62	19	40	71	63
Australia (1995–2005)	0.62	0.61	20	13	8	4
Belarus (1996)	.	0.61	21	58	36	58
Argentina (1995–2006)	0.49	0.60	22	21	26	30
Bangladesh (1996–2002)	0.60	0.60	23	49	73	72
Turkey (1996–2007)	0.62	0.60	24	31	35	31
Russia (1995)	.	0.60	25	75	46	38
Chile (1996–2006)	0.59	0.60	26	15	37	28
Finland (1996–2005)	0.48	0.59	27	12	10	11
Rwanda (2007)	.	0.59	28	24	74	75
Guatemala (2004)	.	0.58	29	44	45	51
Italy (2005)	.	0.58	30	10	11	13
New Zealand (1998–2004)	0.53	0.58	31	19	9	16
Czech Republic (1998)	.	0.58	32	47	13	23
Venezuela (1996)	.	0.58	33	28	22	36
Brazil (1997–2006)	0.60	0.58	34	25	39	43
Moldova (1996–2006)	0.61	0.57	35	52	54	65
Trinidad & Tobago (2006)	.	0.56	36	42	23	18
Ghana (2007)	.	0.56	37	57	72	71
Bosnia Fed. (1998–2001)	0.51	0.56	38	66	60	53

Table 1. *Cont.*

Country (Years)	Environ. (earliest)	Environ. (latest)	Environ. Rank	Postmaterial. Rank	Affluent Socializ. Rank	logGDP/cap. Rank
Serbia (1996–2006)	0.58	0.56	39	64	47	42
Iran (2007)	.	0.56	40	43	30	32
Spain (1995–2007)	0.60	0.56	41	18	15	14
Azerbaijan (1997)	.	0.56	42	76	53	67
Japan (1995–2005)	0.57	0.56	43	22	18	12
Indonesia (2006)	.	0.55	44	56	65	61
Thailand (2007)	.	0.55	45	30	61	41
Macedonia (1998–2001)	0.57	0.54	46	74	55	49
Slovenia (1995–2005)	0.54	0.54	47	14	17	19
Albania (1998–2002)	0.50	0.54	48	80	56	60
Taiwan (1994–2006)	0.69	0.54	49	78	41	15
Rep. Srpska (1998–2001)	0.56	0.54	50	73	59	54
Malaysia (2006)	.	0.53	51	27	43	29
Georgia (1996–2009)	0.63	0.53	52	72	58	55
Latvia (1996)	.	0.53	53	48	25	47
Slovakia (1998)	.	0.52	54	65	20	25
Kyrgyzstan (2003)	.	0.52	55	54	52	69
Estonia (1996)	.	0.52	56	37	29	40
Morocco (2007)	.	0.52	57	60	63	62
USA (1995–2006)	0.55	0.52	58	32	2	2
Ukraine (1996–2006)	0.57	0.51	59	50	33	46
Jordan (2007)	.	0.50	60	62	50	59
Montenegro (1996–2001)	0.58	0.50	61	71	57	56
Bulgaria (1997–2006)	0.54	0.50	62	69	44	33
Armenia (1997)	.	0.50	63	67	62	68
Uruguay (1996–2006)	0.64	0.48	64	11	34	34
Nigeria (1995)	.	0.47	65	39	67	73
Hong Kong (2005)	.	0.47	66	61	21	8
South Korea (1996–2005)	0.71	0.46	67	38	42	20
Romania (1998–2005)	0.60	0.46	68	63	49	37
Poland (1997–2005)	0.51	0.45	69	29	31	24
Lithuania (1997)	.	0.44	70	68	32	39
Singapore (2002)	.	0.43	71	34	12	7
Ethiopia (2007)	.	0.43	72	26	79	79
Zambia (2007)	.	0.42	73	33	68	70
Zimbabwe (2001)	.	0.42	74	51	80	80
Uganda (2001)	.	0.42	75	59	77	77
Egypt (2008)	.	0.42	76	70	64	57
W. Germany (1997–2006)	0.57	0.41	77	5	3	9
South Africa (1996–2006)	0.38	0.40	78	53	38	44
Hungary (1998)	.	0.39	79	79	16	26
E. Germany (1997–2006)	0.47	0.34	80	6	5	10

Sources: For attitudes WVS; for GDP Heston, Summers and Aten (2011): purchasing power parity (int'l prices).

Considering now all 80 countries in Table 1 (including those with just one wave of data), there are clearly significant deviations from the expected pattern of environmentalism being nourished by a climate of postmaterialism and national affluence. In the top quartile of Table 1 there are conforming cases, such as Norway, Switzerland, Canada, Sweden and Australia. But there are at least as many cases that contradict the postmaterialism theory, including Vietnam, Tanzania, China, Burkina Faso, Mali, the Philippines and India. No country has witnessed a more explosive growth in environmentalism than India: a hefty 17 percentage point rise between 1995 and 2006 (from 0.45 to 0.62 on the 0–1 scale). But it remains a country with hardly a whiff of postmaterialist sentiment and where, despite recent economic growth, the average citizen remains impoverished. These unexpected 'greenies' in Asia and Africa seem willing to bear real costs to protect the environment, without the impetus of affluence (either in childhood or at present), or any disavowal of their pressing material concerns.

At the bottom of Table 1, Egypt, Uganda, Zimbabwe, Zambia and Ethiopia all manifest the expected intersection of impoverishment, fixation on materialism, and disregard for the environment. But again, there are departures from the expected pattern with East and, especially, West Germany the most obvious deviations. Although one of the first publics to support a 'Grün' party and elect green candidates to a national legislature, Germany has experienced a precipitous decline in environmental concern, despite long experience with affluence and still high levels of postmaterialism. For instance, in the 1997 WVS, both West and East Germans displayed a willingness to 'increase taxes to prevent environmental pollution' that essentially halved over the next couple of years, and remained around that level in 2006. It seems that Germans' continuing desire for self-expression and self-actualization no longer translates consistently into concern for the environment.

Although never notably inclined toward postmaterialism, the citizens of the US and, to a lesser extent, Singapore and Hong Kong also represent publics that have experienced considerable affluence without lifting their aspirations to environmental protection. And we should emphasize, again, that while South Koreans' disregard for the environment saw them hovering around the bottom of these rankings—scoring just 0.46 on environmentalism—they plummeted to that lowly position from an almost unparalleled score of 0.71 over a decade in which their levels of postmaterialism and affluence barely shifted.

5. Cross-National Determinants of Support for Environmental Protection

This broad overview indicates that postmaterialism provides, at best, an incomplete account of the variance in public opinion on environmental issues. The results of our hierarchical linear modelling (see Table 2) tend to confirm this and show, further, that other values and social-structural theories are likewise weak explanations of public support for environmental protection.

These 'fixed effects' (reported in upper panel, Table 2) are unstandardized multiple regression coefficients (with standard errors in parentheses). They are derived from hierarchical linear modelling of a two-level random intercept model, with the grouping structure of the data consisting of nested groups that vary according to WVS 'national wave' (a certain nation in a particular year). We see evidence of variation in the intercepts across these WVS national waves (see 'random effects', middle panel, Table 2). Comparing the fit of the random intercept model to a standard regression model yields a likelihood ratio test (*vs.* linear regression) of 5500 (p = 0.000). One can think of this 'random intercept' HLM as equivalent to a standard regression model with an exhaustive series of dummy variables entered into the analysis, to represent each of the WVS national waves. The lesser variance explained by the standard regression model relative to one allowing the intercepts to vary across WVS national waves (see 'variance explained', lower panel, Table 2) shows clearly that we can greatly improve (especially outside the West) our account of environmental concern if we accommodate cross-wave/cross-national differences in baseline attitudes toward environmental protection. On its own, this tends to support our central intuition that a good deal of the variation in environmental sentiment is linked to changeable, malleable, local conditions, and *not* so heavily influenced by purportedly universal ideological/psychological influences and enduring socio-structural factors.

As for those fixed effects, it is notable, first, that the coefficients attached to the passage of time are insignificant across samples and, with one exception, negative, suggesting that, if anything, environmentalism has marginally decreased over time. The slight exception to this downward drift was in the core of the West, where there may have been a trivial (and statistically insignificant) rise in willingness to bear the economic costs of environmental protection. Multiplying the unstandardized coefficient b by the full range of time—0.0008×14 years—yields just 0.0112: a rise that amounts to a mere one percentage point boost (across the 0–1 scale of the dependent variable, environmentalism) over the decade and a half following the Kyoto Protocol.

5.1. Postmaterialism

The multivariate model explains about 14 percent of the variance in Western attitudes toward environmental protection, but less than 7% outside the West: less than half the explanatory power. In short, the cross-national theories work better in the West but, even there, they leave much unexplained. Consistent with Inglehart's research, postmaterialism is the most important source of environmental sentiment in Western nations. Even so, it accounts for less than 3 per cent of the variance in attitudes toward environmental protection. As for its magnitude of impact, moving across the full range of the postmaterialism scale—from hardened materialists to ardent devotees of self-actualization—shifts us only a fifth of the way across the environmentalism measure, increasing support for environmental protection by just 20 percentage points. Specifically, the unstandardized coefficient in the Western subset ($b = 0.197$) indicates that a one-unit increase in postmaterialism (*i.e.*, the full range of the 0–1 measure) is expected to increase environmentalism by about 0.20 on its 0–1 scale, holding all else constant. Moreover, the influence and impact of postmaterialism are roughly halved outside the West. Across the entire dataset (merging 135 samples), it does remain the most important explanatory variable but the overall effect is still modest. Postmaterialism apparently explains just 1% of the variance in environmentalism around the globe, and boosts support for environmental protection by just 12 percentage points.

Table 2. Determinants of public support for environmental protection, across cultures.

				Sub-Samples			
	All	West	Non-West	Higher GDP	Lower GDP	Democracies	Autocracies
Fixed Effects	b(s.e.)	b(s.e.)	b(s.e.)	b(s.e)	b(s.e.)	b(s.e.)	b(s.e.)
time (year 0-14)	−0.0012(0.0015)	0.0008(0.003)	−0.0016(0.0017)	−0.0022(0.0019)	−0.0015(0.002)	−0.0006(0.0017)	−0.0013(0.0031)
post-materialism	0.12(0.003) *	0.197(0.008) *	0.089(0.004) *	0.159(0.005) *	0.071(0.005) *	0.136(0.004) *	0.072(0.007) *
left-right self-plcmnt	−0.032(0.003)*	−0.169(0.009) *	−0.005(0.004)	−0.074(0.005) *	0.002(0.004)	−0.04(0.004) *	0.014(0.007) *
authoritarianism	−0.021(0.003) *	−0.073(0.007) *	0.001(0.004)	−0.039(0.005) *	0.003(0.005)	−0.03(0.004) *	0.009(0.006)
GDPpc in R's youth	−0.0011(0.0029)	−0.0107(0.0094)	0.0056(0.0032)	−0.0053(0.0045)	0.0068(0.0044)	−0.0114(0.0038) *	0.0162(0.0048) *
income decile	0.013(0.003) *	0.014(0.007) *	0.016(0.004) *	0.019(0.005) *	0.005(0.005)	0.014(0.004) *	0.005(0.007)
financial satisfaction	0.033(0.003) *	0.039(0.008) *	0.035(0.004) *	0.03(0.005) *	0.039(0.005) *	0.037(0.004) *	0.028(0.007) *
life satisfaction	0.045(0.004) *	0.032(0.01) *	0.046(0.004) *	0.047(0.006) *	0.043(0.005) *	0.047(0.004) *	0.042(0.007) *
happiness	0.05(0.004) *	−0.01(0.01)	0.062(0.004) *	0.034(0.006) *	0.062(0.005) *	0.042(0.004) *	0.069(0.007) *
education level	0.071(0.003) *	0.102(0.006) *	0.062(0.003) *	0.083(0.004) *	0.058(0.004) *	0.081(0.003) *	0.042(0.005) *
age (in decades)	−0.0035(0.0007)*	−0.0071(0.0017)*	−0.0017(0.0008) *	−0.0041(0.0011) *	−0.002(0.001) *	−0.0047(0.0009) *	−0.0022(0.0013)
"gen Y"/born 1980s+	−0.009(0.003) *	−0.023(0.008) *	−0.005(0.003)	−0.015(0.004) *	−0.004(0.004)	−0.01(0.004) *	0.000(0.005)
male	−0.005(0.001) *	-0.015(0.003) *	−0.002(0.002)	−0.012(0.002) *	0.002(0.002)	−0.007(0.002) *	0.001(0.003)
confid in institutions	0.096(0.005) *	0.102(0.014) *	0.093(0.005) *	0.085(0.008) *	0.104(0.007) *	0.089(0.006) *	0.104(0.009) *
inter-personal trust	0.032(0.002) *	0.039(0.004) *	0.026(0.002) *	0.035(0.002) *	0.024(0.003) *	0.031(0.002) *	0.03(0.004) *
Cons	0.421(0.036) *	0.593(0.191) *	0.298(0.04) *	0.448(0.099) *	0.294(0.049) *	0.55(0.047) *	0.192(0.06) *
Random Effects							
sd(cons)	0.077(0.005) *	0.071(0.011) *	0.076(0.006) *	0.069(0.006) *	0.068(0.007) *	0.073(0.006) *	0.082(0.011) *
sd(residual)	0.286(0.000) *	0.277(0.001) *	0.287(00.001) *	0.281(00.001) *	0.289(00.001) *	0.284(00.001) *	0.289(0.001) *
Variance Explained							
-standard model	0.073	0.14	0.065	0.083	0.08	0.082	0.067
-varying intercepts	0.11	0.168	0.102	0.117	0.108	0.115	0.101
LR test vs. lin reg: chibar2(01)	5500	824	4409	2453	2165	3543	1446
Prob >= chibar2	0.000	0.000	0.000	0.000	0.000	0.000	0.000
# observations	149,559	28,080	121,479	71,183	78,376	106,371	43,188
# groups	135	27	108	69	66	99	36
average obs/group	1108	1040	1125	1032	1188	1075	1200

Note: Cell entries (see 'Fixed Effects', upper panel) are unstandardized multiple regression coefficients, and standard errors (* significant at p < 0.05 or better), derived from hierarchical linear modelling of two-level random intercept model, nested according to WVS 'national wave' (a certain nation in a particular year); *sd(cons)* = between-national-waves standard deviation; *sd(residual)*=between-individuals standard deviation; *LR test vs. lin reg: chibar2(01)* = likelihood ratio test for significant difference between the two models (random intercept vs. linear regression), which is distributed chi-square with 1 degree of freedom; *Prob ≥ chibar2* = p-value for chi-square test. Source: World Values Survey (1994–2008). See text for elaboration, measurement and scoring of variables.

5.2. Other Values

It is only in the West that environmental attitudes are at all ideologically structured. This is unsurprising given Western politicians' longer experience in marketing green issues as extensions of the traditional 'left-right' economic divide. Apart from the influence of postmaterialism, support for environmental protection across affluent Western democracies also depends on left-right self-placement and, to a lesser degree, authoritarianism. For example, Westerners who place themselves at the extreme 'right-wing' are around 17 percentage points less willing than ardent 'leftists' to tolerate taxes and restraints on economic growth to protect the environment.

5.3. Education

While scholars point to education as the primary means through which people are exposed to the values associated with environmental protection, our results indicate that non-Western education does not effectively transmit these values. The environmental returns to education are meager outside affluent Western democracies. Even there, it apparently takes a lot of education to boost green sentiment. Individuals with higher degrees are predicted to be only 8–10 percentage points more supportive of environmental protection than those not completing elementary school (an effect that drops to just 4–6 percentage points outside affluent Western democracies). Education explains just one percent of the variance in environmentalism even in the West.

5.4. Age and Generation

The effects of age are likewise culturally variant and surprisingly slight. Age generally proves to be a (statistically) significant dampener of enthusiasm for environmental protection. But the effects are very modest, especially outside of wealthy Western democracies, where the youngest and eldest differ in willingness to protect the environment by barely a percentage point. Even within affluent Western democracies, ageing many decades—say, maturing from 20 to 80—tends to diminish environmentalism by four percentage points at most; the youngest are barely more enthusiastic than the eldest. Moreover, those born since the 1980s ('Gen Y') are actually a couple of percentage points *less* concerned for the environment than earlier generations.

5.5. Affluent Socialization

Finally, contrary to theoretical expectations, our results suggest that a prosperous upbringing generally has no significant impact on environmental concern later in life, and if anything, may somewhat diminish environmentalism, especially in democratic societies (as may have been the case in Germany and the USA, for example). Conversely, impoverished childhoods—perhaps upbringings marred by famine or flood, or at least precariously dependent on careful management of natural resources—may build yearning for environmental protection, at least among democratic citizens with some tangible control over their collective fates (see penultimate column of Table 2). This pattern is evidenced in the democratic countries of Asia (e.g., Philippines, India, Bangladesh) and Africa (e.g., Mali). These effects are statistically significant but very modest. Generally speaking, across democracies, an increase of two standard deviation units in childhood experience of national prosperity tends to *depress* environmentalism by about 2 percentage points. This finding proves robust to repeated sampling, and to substituting in varied indicators of respondents' childhood experience of national well-being (e.g., alternately deploying Log GNI per capita, or the UN Human Development Index, again averaged across the 18 years of the respondent's childhood).

Notice this stands in sharp contrast to the only other significant result concerning societal affluence in childhood (see final column, Table 2): that upbringings in impoverished *non*-democratic societies—where people have little control of political outcomes (North/East Africa, for example)—do not generally provide fertile ground for nurturing environmental concern, a finding more in line with conventional theories.

6. Toward Context-Sensitive Theories of Environmental Concern

Our findings show the limited leverage to be gained from cross-national theories of environmental concern. We agree with Brechin's conclusion that 'environmentalism is most likely a complex social phenomenon, a mixture of social perceptions, local histories and environmental realities, international relationships and influences, and unique cultural and structural features of particular countries and regions' [14] (p. 807). Because of this complexity, both the levels and sources of environmentalism are likely to vary substantially both across and within nations. Accounts of environmental public opinion should, therefore, be grounded in the contextual features that impact a population's perception of the meaning, necessity, costs and benefits of environmental protection. Clearly these contextual features may take a variety of forms, and may occasionally include value systems such as postmaterialism, as well as social-structural variables. However, three factors that are wholly absent from the theories tested in this paper are likely to be important in context-sensitive accounts.

First, a population's experience of environmental events will likely impact on their attitudes toward environmental protection. The literature has shown this to be true for poor countries but it seems no less likely to be a key factor for wealthy populations. In his OPSV thesis, Inglehart still maintains that values account for environmental attitudes in rich countries, but our analyses here attest to their weak explanatory power. Inclusion of objective problems in our models would surely improve understanding of environmental sentiments in rich and poor countries alike.

Second, context-sensitive theories should be attentive to the salience of environmental issues in a particular time and place. Green sentiment is likely to be greater when environmental issues are highly salient [20]. However, salience is a variable, not a constant. While not ignoring issue salience, postmaterialism theory tends to view it in simplistic terms as a function of popular sentiments. Yet the reality is that motivated political leaders and mobilized activists, and specific events—such as a warmer-than-expected winter—can heighten issue salience, which may in turn influence public attitudes on the environment [20,54].

Third, to properly account for the opinions of a particular population, theories must give due weight to the (materialist) calculations of the costs and benefits of environmental protection. Controlling pollution can be expensive and the benefits from polluting activity may be substantial. In most situations, therefore, citizens are unlikely to take an absolutist stance, choosing instead to balance the benefits of a cleaner environment against the costs of environmental protection. Countries, and constituencies within countries, may disagree on the appropriate level of environmental protection, based on their perceptions of the accompanying material costs and benefits.

By their nature, postmaterialist accounts (and others in a similar vein)—even when explicitly cross-national—tend to discount multi-factor explanations of environmental concern, in general, and material considerations, in particular. But in their absence, it is difficult to account for some of the curious empirical results from our cross-national testing.

Certainly, the United States has witnessed a decline since its 1970s peak of environmental concern, presumably partly due to a creeping recognition of the costs of action, with the public coming to 'realize the immensity of the social and financial costs of cleaning up our air and water' [55] (p. 43). Consistent with theories of political responsiveness to citizens' opinions [56], the United States was a vanguard nation in the 1972 UN Conference on the Human Environment and in several environmental protection treaties that followed [57] (p. 336). But by the 1990s, the United States had abandoned its leadership role and had become a 'laggard and obstructionist' in the area of environmental politics [57] (p. 336). Since the 1980s, the EU nations have replaced America as leaders of the pro-environment agenda. However, we would expect the EU soon to follow the American pattern. Awareness and alarm should eventually be succeeded by a gradual waning of the issue, especially to the extent that pursuing the pro-environment agenda is perceived as costly, and most of the adverse consequences are borne by others (in developing countries).

Our empirical results support the notion that European countries are beginning to baulk at the costs of curbing their pollution. Initially these costs were masked by the global agreement to make 1990s emissions (rather than projected 'business as usual' emissions) the baseline for assessing country reductions post-Kyoto. This agreement made Europeans (and especially Germans) appear better 'green citizens' than they really were. Closing down East German industries created a post-reunification emissions reduction 'windfall' that enabled Germany to easily meet its obligations while leaving room for other EU states' emissions to grow, sometimes significantly [21,58]. As this dividend has now been consumed, further reductions will require economic sacrifices that are likely to be increasingly unpalatable to European mass publics. Our data show, for instance, that from 1997 to 2006, both East and West Germans plummeted roughly 15 percentage points in professed willingness to bear real economic costs to protect the environment.

Our results also show that environmental concern is high in countries like Vietnam, China, the Philippines and India, which have few of the purported preconditions for postmaterialist sentiments but suffer the adverse consequences of serious environmental events. Conversely, in Hong Kong and Singapore, the seemingly favourable preconditions for postmaterialism have yielded no abundance of postmaterialists or environmentalists. Both sets of observations contradict Inglehart's thesis, with environmental concern seeming to depend more on a palpable sense there is a problem to be solved. This condition is realized where there are numerous victims of environmental events but absent when a population is relatively insulated from substantial environmental costs.

7. Conclusion: Current and Future Friends of the Earth

Overall, our findings paint a less rosy picture of global mass support for environmental protection than that anticipated by popular cross-national theories. The postmaterialism thesis, in particular, envisages a world growing increasingly wealthy and industrialized (implicitly: modernized and westernized), which evolves predictably toward greater enthusiasm for environmental protection. Our analysis has exposed the weak foundations underlying this optimistic prediction.

It is only in affluent countries, and the Western world in particular, that attitudes toward environmental protection are driven by values such as postmaterialism, left-right ideology or authoritarianism. Even in the West, while environmental attitudes are *better* explained by values than by any other cross-national factor, they are not *well* explained by values. Without extensive longitudinal data we are unable to say whether Western attitudes are less structured by values than they *were* but certainly they are not heavily structured by values *now*. And outside of the affluent West, these attitudes are hardly structured at all. Little of the variance in public opinion outside the West can be explained using any of the universal, cross-national theories. The general pattern of results is consistent with mass opinion being determined largely by factors specific to a particular population, namely, the occurrence of environmental events, the salience of environmental issues on the domestic political agenda, and calculation of the costs and benefits accruing from environmental protection.

In contrast to the dominant cross-national accounts, what would a more context-specific explanation of variation in environmental concern imply for the future of global environmental politics? We think it suggests three major amendments to the 'Inglehart story'.

First, our confirmation of the poor explanatory power of cross-national accounts suggests that the future of environmental politics will be highly unpredictable. There are no psychological or socio-structural factors that consistently boost or diminish environmental sentiment across nations, regions and cultures. Pro-environment attitudes have diverse origins, and may therefore have varying future trajectories dependent on local contexts. Since experience, salience, and interests are not broadly shared, future environmental politics will be more uncertain and the bases of international cooperation unstable.

Second, the West may be the wrong place to look for leadership of the pro-environment agenda. Even in Western nations with a large proportion of postmaterialists, that orientation is not a strong predictor of pro-environment attitudes. Other distinctive features of Western nations, notably wealth and education, do not appreciably promote environmentalism. Of course, the environmental issue is sufficiently grave that it will not be soon or easily ignored, but its salience will decline and support for costly measures will dissipate. A useful boon to the movement in the West was provided by the Stern Review (2006), which determined that the costs of inaction would greatly exceed the costs of action to protect the environment. This logic may compel Western governments to do more to solve the world's emissions problems. But the actions that result from such calculations should not be interpreted as postmaterialist in origin. Rather, they are a consequence of straightforward materialist assessment of domestic costs and benefits.

The third implication is that calculations of costs and benefits may induce some major developing countries, especially China and India, to take the lead in the area of environmental protection. And again, these countries are likely to be motivated more by interests than values. In China, for instance, environmental protection efforts are increasingly driven by concern for public health and/or economic benefit. For example, policies intended to improve sanitation, reduce respiratory illness, ease congestion, or attract tourism can also mitigate emissions and increase interest in environmental protection [59,60]. Clearly, it is possible to care *for* the environment without caring *about* the environment, and around the globe environmental action may increasingly be driven by such divergent forces.

Finally, we can also imagine a new wave of environmentalism springing from *anti*-materialism instead of post-materialism: a reaction against industrialization, and in favour of a simpler, safer, 'traditional' life. Certainly this is conceivable for some parts of West Africa, where protective efforts are already afoot in Mali and Burkina Faso lauding traditional indigenous adaptations in agriculture, soil care, and land use management. But the impact of these efforts on global pollution is likely to be more limited because their proponents lack resources and influence beyond their domestic spheres.

To conclude, it may be that there are few universals regarding the sources of environmental sentiments. No psychological or socio-structural factors reliably predict public opinion on environmental protection across nations, regions and cultures. Instead, pro-environment attitudes appear to have diverse origins that depend largely on local peculiarities. This leaves us ill-equipped to make confident predictions about the likely success of any future environmental agreements, which must rest on the cooperation of representatives of hundreds of idiosyncratic publics. The most confident prediction we can make is that the values-based frameworks scholars have typically used to explain environmental attitudes, and the politics that result, are of declining utility for analyzing the future of environmentalism.

Supplementary Materials

Supplementary materials can be accessed at: http://www.mdpi.com/1996-1073/8/6/4899/s1.

Acknowledgments

The authors express their appreciation to the World Values Survey Association (www.worldvaluessurvey.org) for the World Values Survey data whose analysis constitutes the core of this paper. We would also like to thank this journal's anonymous reviewers for their comments, which provoked significant improvements in the paper.

Author Contributions

Both authors contributed substantially to this research article. Stenner conceived and designed the study, analysed the data, and reported and interpreted the results. Nwokora wrote the first draft of the paper, including the literature review. Stenner and Nwokora contributed equally to subsequent expansion and revision of the paper. Both authors have approved the submitted and accepted versions of the manuscript.

Conflicts of Interest

The authors declare no conflict of interest.

References

1. Hardin, G. The tragedy of the commons. *Science* **1968**, *162*, 1243–1248.
2. Scruggs, L.; Benegal, S. Declining public concern about climate change: Can we blame the great recession? *Global Environ. Change* **2012**, *22*, 505–515.
3. Miller, M. Elections, information, and policy responsiveness in autocratic regimes. *Comp. Polit. Stud.* **2015**, *48*, 691–727.

4. Ajzen, I. *Attitudes, Personality, and Behavior*; Open University Press: Milton Keynes, UK, 2003.

5. Diekmann, A.; Preisendörfer, P. Green and greenback: The behavioral effects of environmental attitudes in low-cost and high-cost situations. *Ration. Soc.* **2003**, *15*, 441–472.

6. Olofsson, A.; Öhman, S. General beliefs and environmental concern: Transatlantic comparisons. *Environ. Behav.* **2006**, *38*, 768–790.

7. Kitschelt, H. Political opportunity structures and political protest: Anti-nuclear movements in four democracies. *Br. J. Polit. Sci.* **1986**, *16*, 57–85.

8. Van der Heijden, H.-A. Political opportunity structure and the institutionalization of the environmental movement. *Environ. Polit.* **1997**, *6*, 25–50.

9. Van der Heijden, H.-A. Environmental movements, ecological modernisation and political opportunity structures. *Environ. Polit.* **1999**, *8*, 199–221.

10. Meguid, B. Competition between unequals: The role of mainstream party strategy in niche party success. *Am. Polit. Sci. Rev.* **2005**, *99*, 347–359.

11. Rohrschneider, R. New party versus old left realignments: Environmental attitudes, party policies, and partisan affiliations in four West European countries. *J. Polit.* **1993**, *55*, 682–701.

12. Inglehart, R. The silent revolution in Europe: Intergenerational change in post-industrial societies. *Am. Polit. Sci. Rev.* **1971**, *65*, 991–1017.

13. Inglehart, R. *The Silent Revolution: Changing Values and Political Styles among Western Publics*; Princeton University Press: Princeton, NJ, USA, 1977.

14. Inglehart, R. *Modernization and Post-Modernization: Cultural, Economic and Political Change in 43 Societies*; Princeton University Press: Princeton, NJ, USA, 1997.

15. Oreg, S.; Katz-Gerro, T. Predicting proenvironmental behavior cross-Nationally: Values, the theory of planned behavior, and value-belief-norm theory. *Environ. Behav.* **2006**, *38*, 462–483.

16. Dunlap, R.E.; York, R. The globalization of environmental concern and the limits of the postmaterialist values explanation: Evidence from Four Multinational Surveys. *Sociol. Quart.* **2008**, *49*, 529–563.

17. Steel, B.S.; Warner, R.L.; Lovrich, N.P.; Pierce, J.C. The Inglehart-Flanagan debate over postmaterialist values: Some evidence from a Canadian-American case study. *Polit. Psychol.* **1992**, *13*, 61–77.

18. Mostafa, M.M. Wealth, post-materialism and consumers' post-environmental intentions: a multilevel analysis across 25 nations. *Sustain. Develop.* **2011**, doi:10.1002/sd.517.

19. Franzen, A. Environmental attitudes in international comparison: An analysis of the ISSP surveys 1993 and 2000. *Soc. Sci. Quart.* **2003**, *84*, 297–308.

20 Pralle, S.B. Agenda-setting and climate change. *Environ. Polit.* **2009**, *18*, 781–799.

21. Harrison, K.; Sundstrom, L.M. The comparative politics of climate change. *Global Environ. Polit.* **2007**, *7*, 1–18.

22. Bean, C.; Papadakis, E. Polarized priorities or flexible alternatives? Dimensionality in Inglehart's materialism-postmaterialism scale. *Int. J. Public Opin. Res.* **1994**, *6*, 264–288.

23. Carlisle, J.; Smith, E.R.A.N. Postmaterialism *vs.* egalitarianism as predictors of energy-related attitudes. *Environ. Polit.* **2005**, *14*, 527–540.

554

24. Duch, R.M.; Taylor, M.A. Postmaterialism and the economic condition. *Am. J. Polit. Sci.* **1993**, *37*, 747–779.

25. Davis, D.W.; Davenport, C. Assessing the validity of the postmaterialism index. *Am. Polit. Sci. Rev.* **1999**, *93*, 649–664.

26. Davis, D.W.; Dowley, K.M.; Silver, B.D. Postmaterialism in world societies: Is it really a value dimension? *Am. J. Polit. Sci.* **1999**, *43*, 935–962.

27. Nisbet, M.C.; Myers, T. Twenty Years of Public Opinion about Global Warming. *Public Opin. Quart.* **2007**, *71*, 444–470.

28. Wagner, L.M. Identifying US preferences and a way forward in the ozone, climate and forests regimes. *Global Environ. Polit.* **2008**, *8*, 137–142.

29. Brechin, S.R. Objective problems, subjective values, and global environmentalism: evaluating the postmaterialist argument and challenging a new explanation. *Soc. Sci. Quart.* **1999**, *80*, 793–809.

30. McGranahan, G.; Balk, D.; Anderson, B. The rising tide: Assessing the risks of climate change and human settlements in low elevation coastal zones. *Environ. Urbaniz.* **2007**, *19*, 17–37.

31. Erda, L.; Wei, X.; Hui, J.; Hu, Y.; Li, Y.; Bai, L.; Xie, L. Climate change impacts on crop yield and quality with CO_2 fertilization in China. *Philosoph. Trans. Royal Soc. B* **2005**, *360*, 2149–2154.

32. Adeola, F.O. Environmental contamination, public hygiene, and human health concerns in the third world: The case of Nigerian environmentalism. *Environ. Behav.* **1996**, *28*, 614–646.

33. Kim, D.-S. Environmentalism in developing countries and the case of a large Korean city. *Soc. Sci. Quart.* **1999**, *80*, 810–829.

34. Inglehart, R. Public support for environmental protection: Objective problems and subjective values in 43 Societies. *Polit. Sci. Polit.* **1995**, *28*, 57–72.

35. Abramson, P.R.; Inglehart, R. Generational replacement and value change in eight West European societies. *Br. J. Polit. Sci.* **1992**, *22*, 183–228.

36. Franzen, A.; Vogl, D. Two decades of measuring environmental attitudes: A comparative analysis of 33 countries. *Global Environ. Change* **2013**, *23*, 1001–1008.

37. Ivanova, G.; Tranter, B. Paying for environmental protection in a cross-national perspective. *Aus. J. Polit. Sci.* **2008**, *43*, 169–188.

38. Rohrschneider, R. Citizens' attitudes toward environmental issues: Selfish or selfless? *Comp. Polit. Stud.* **1988**, *21*, 347–367.

39. Franzen, A.; Meyer, R. Environmental attitudes in cross-national perspective: A multilevel analysis of the ISSP 1993 and 2000. *Eur. Sociol. Rev.* **2010**, *26*, 219–234.

40. Gelissen, J. Explaining popular support for environmental protection: A multilevel analysis of 50 nations. *Environ. Behav.* **2007**, *39*, 392–415.

41. Knight, K.W.; Messer, B.L. Environmental concern in cross-national perspective: The effects of affluence, environmental degradation, and world society. *Soc. Sci. Quart.* **2012**, *93*, 521–537.

42. Norris, P. The globalization of comparative public opinion research. In *The Sage Handbook of Comparative Politics*; Robinson, N., Landman, T., Eds.; London, UK, 2009; pp. 522–540.

43. Huntington, S.P. The clash of civilizations? *Foreign Aff.* **1992**, *72*, 22–49.

44. Pemstein, D.; Meserve, S.A.; Melton, J. Democratic compromise: A latent variable analysis of ten measures of regime type. *Polit. Anal.* **2010**, *18*, 426–449.

45. Heston, A.; Summers, R.; Aten, B. Penn World Table Version 7.0 (PWT 7.0) 2011. Center for International Comparisons of Production, Income, and Prices, University of Pennsylvania. Available online: http://pwt.econ.upenn.edu/php_site/pwt_index.php (accessed on 1 July 2011).

46. Evans, G.; Heath, A. The measurement of Left-Right and libertarian-authoritarian values in the British electorate. *Qual. Quant.* **1995**, *29*, 191–206.

47. Stenner, K. Three kinds of "conservatism". *Psychol. Inq.* **2009**, *20*, 142–159.

48. Altemeyer, R. *The Authoritarian Specter*; Harvard University Press: Cambridge, MA, USA, 1996.

49. Feldman, S.; Stenner, K. Perceived threat and authoritarianism. *Polit. Psychol.* **1997**, *18*, 741–770.

50. Hetherington, M.J.; Weiler, J.D. *Authoritarianism and Polarization in American Politics*; Cambridge University Press: New York, NY, USA, 2009.

51. Stenner, K. *The Authoritarian Dynamic*. Cambridge University Press: New York, NY, USA, 2005.

52. Levi, M.; Stoker, L. Political trust and trustworthiness. *Ann. Rev. Polit. Sci.* **2000**, *3*, 475–507.

53. Lubell, M. Familiarity breeds trust: collective action in a policy domain. *J. Polit.* **2007**, *69*, 237–250.

54. McCright, A.M.; Dunlap, R.E. Anti-reflexivity: The American conservative movement's success in undermining climate science and policy. *Theor. Cult. Soc.* **2010**, *27*, 100–133.

55. Downs, A. Up and down with ecology—The 'issue-attention' cycle. *Public Interest* **1972**, *28*, 38–50.

56. Page, B.I.; Shapiro, Y. Effects of public opinion on policy. *Am. Polit. Sci. Rev.* **1983**, *77*, 175–190.

57. Kelemen, R.D. Globalizing European Union environmental policy. *J. Eur. Public Pol.* **2010**, *17*, 335–349.

58. Schreurs, M.A.; Tiberghien, Y. Multi-level reinforcement: Explaining European Union leadership in climate change mitigation. *Global Environ. Polit.* **2007**, *7*, 19–46.

59. Baer, H.; Singer, M. *Global Warming and the Political Ecology of Health: Emerging Crises and Systematic Solutions*; Left Coast Press: Walnut Creek, CA, USA, 2009.

60. Koehn, P.H. Underneath Kyoto: Emerging subnational government initiatives and incipient Issue-Bundling opportunities in China and the United States. *Global Environ. Polit.* **2008**, *8*, 53–77.

The Socio-Demographic and Psychological Predictors of Residential Energy Consumption: A Comprehensive Review

Elisha R. Frederiks, Karen Stenner and Elizabeth V. Hobman

Abstract: This article provides a comprehensive review of theory and research on the individual-level predictors of household energy usage. Drawing on literature from across the social sciences, we examine two broad categories of variables that have been identified as potentially important for explaining variability in energy consumption and conservation: socio-demographic factors (e.g., income, employment status, dwelling type/size, home ownership, household size, stage of family life cycle) and psychological factors (e.g., beliefs and attitudes, motives and intentions, perceived behavioral control, cost-benefit appraisals, personal and social norms). Despite an expanding literature, we find that empirical evidence of the impact of these variables has been far from consistent and conclusive to date. Such inconsistency poses challenges for drawing generalizable conclusions, and underscores the complexity of consumer behavior in this domain. In this article, we propose that a multitude of factors—whether directly, indirectly, or in interaction—influence how householders consume and conserve energy. Theory, research and practice can be greatly advanced by understanding what these factors are, and how, when, where, why and for whom they operate. We conclude by outlining some important practical implications for policymakers and directions for future research.

Reprinted from *Energies*. Cite as: Frederiks, E.R.; Stenner, K.; Hobman, E.V. The Socio-Demographic and Psychological Predictors of Residential Energy Consumption: A Comprehensive Review. *Energies* **2015**, *8*, 573-609.

1. Introduction

In recent years, a growing body of research has sought to identify the key factors underlying patterns of residential energy consumption and conservation. In particular, many studies have been conducted to investigate different types of energy consumer "profiles" in an effort to pinpoint precisely what factors are associated with energy-saving and energy-wasting behavior (e.g., [1–5]). A number of important determinants have been identified, ranging from situational factors in the external environment (e.g., contextual, structural and institutional factors) through to more person-specific attributes of consumers themselves (e.g., socio-demographic, psychological and motivational factors) [6–12]. Yet efforts to summarize, integrate and synthesize the key findings across studies have failed to keep pace. The current paper addresses this gap by conducting a comprehensive review of published research on the socio-demographic and psychological determinants of household energy consumption and conservation. In the literature, behaviors related to energy conservation are sometimes categorized into "curtailment" behaviors (*i.e.*, ongoing day-to-day actions to reduce consumption, such as setting thermostats, switching off lights, limiting use of heating/cooling and ventilation systems, *etc.*) and "efficiency" behaviors (*i.e.*, once-off actions to save energy, such as investing in home improvements like insulation, solar panels, energy-efficient appliances, new technology, *etc.*) [13,14]. In this article, we focus on both categories of energy usage behavior. By doing so, we aim to provide researchers, practitioners and policymakers with a deeper understanding of what person-specific factors might explain different patterns of household energy usage, and thereby provide valuable insights on when, where, how, why and for whom energy-efficient interventions might serve to promote and sustain new energy-conserving practices.

Advancing our understanding of the key factors shaping consumers' energy-related behavior is important for many reasons. Against a backdrop of global concerns over climate change and rising greenhouse gas emissions, renewable and sustainable energy use has become a key challenge and opportunity for improving the overall social-ecological resilience of communities worldwide. Globally, researchers and policymakers are investing significant resources in designing cost-effective solutions and new technology to increase household energy efficiency and conservation. Yet there is vast scope for improvement, as reflected by recent calls for greater integration of social and behavioral sciences in energy research [15]. Solving many of the world's energy-related problems requires not only technological advances, but also changes in human behavior—and successfully shifting the behavior of consumers in the desired direction (*i.e.*, toward more efficient and sustainable practices) is facilitated by first identifying potential causal and explanatory variables (predictors and "mediators") and various contingencies (interactions or "moderators") that might impact the nature, intensity, frequency and duration of behavior across time and contexts.

So why is it important to identify the correlates of energy consumption and conservation? The answer is simple: to know how to intervene, and with whom, where and when. It is necessary to understand what drives household energy consumption and conservation in order to determine how these behaviors can usefully be altered by consumer-focused interventions, technological solutions, public policy initiatives and other such strategies.

In an effort to integrate key insights from the literature, our paper begins by providing a theoretical overview of residential energy usage, with a focus on describing how the processes and predictors of energy consumption and conservation have been conceptualized to date. Drawing heavily on published work from the social and behavioral sciences, we then review research and empirical evidence on the individual-level predictors of household energy use in an effort to identify the key characteristics and variables that explain consumers' energy-related behavior. This includes a review of the major socio-demographic factors that have been touted as explaining individual differences in household energy consumption and conservation, as well as the psychological and motivational attributes of consumers that have also been hypothesized to play a role. We review publications that present both primary and secondary research, and studies that employ a range of designs and methodologies. In outlining our key findings and conclusions, we provide a brief summary of research in the body of the article itself, with a more detailed review of empirical evidence and citations appearing in the accompanying table. Finally, we draw out the implications of our key findings for theory, research and practice, with a focus on identifying some cost-effective behavioral solutions to influence household energy consumption and conservation.

2. Theoretical Background: Conceptualizing Energy Consumption and Conservation

Over the past few decades, the factors underpinning individual differences in pro-environmental attitudes and behavior have been examined from a range of different theoretical perspectives (for reviews, see [16–20]). Due to the complex and dynamic nature of behavior in this domain, a wide variety of conceptual models have been hypothesized and countless studies have been conducted to investigate the variables influencing environmentally significant decision-making and action. Some of the most influential and commonly cited perspectives, theories and models of pro-environmental behavior include: Hines *et al.*'s [21] model of responsible environmental behavior; Ajzen's theory of planned behavior [22,23]; Guagnano *et al.*'s [24] attitude-behavior-external conditions (ABC) model; Stern *et al.*'s [25] value-belief-norm (VBN) theory; Blake's [26] conceptualization of the barriers between environmental concern and action; Stern's [19] framework of environmentally significant behaviors and causal variables; and Kollmuss and Agyeman's [20] model of pro-environmental behavior.

This theoretical research from the broad domain of pro-environmental behavior has extended to the more specific area of residential energy conservation, with recent years witnessing an increased focus on identifying the specific factors that influence household energy usage (e.g., consumption) and changes in energy use over time (e.g., curtailment and efficiency behaviors) [6,13,14,27]. An exhaustive summary of all relevant theories, frameworks and conceptual models of household energy use is beyond the scope of this paper. However, some of the most influential and commonly cited approaches include: Van Raaij and Verhallen's [8] behavioral model of residential energy use; Costanzo *et al.*'s [10] socio-psychological model of energy conservation behavior; and Stern and Oskamp's [28] causal model of resource use. Some researchers have also applied Hägerstrand's [29,30] time-geographic approach to study household energy-related activities [31,32]; and Schatzki's [33] practice theory to study the unconscious habits and technological structures that influence residential energy consumption [34,35]. Rogers' [36,37] diffusion of innovations theory has also been used to explain consumers' decision-making and behavior in the context of residential energy consumption, specifically in terms of the adoption of energy-saving practices and products [38–42].

While various theoretical perspectives have emerged in the literature, there is no single conceptual framework or model that is universally accepted by scholars as providing an all-inclusive explanation of energy consumption and conservation, nor any single approach that precisely predicts individual differences in such behavior. Rather, the extant literature seems to indicate that the issue of what distinguishes above- and below-average energy users—or "energy-wasting" and "energy-saving" consumers—is so complex that it is difficult to capture in a single framework [20,21]. Further, while empirical evidence indicates that some variables may be better predictors of energy consumption than others, the findings have still been far from consistent across time, contexts, samples of participants, and studies. This inconsistency may be partly an artifact: due to energy-related "behavior" being conceptually and operationally defined in different ways—for example, it can be measured in terms of overall household energy consumption (e.g., kilowatts per hour usage), changes in specific everyday practices (e.g., curtailment actions), or adoption of certain energy-efficient technology (e.g., efficiency actions), among many others. And the role of different explanatory variables can appear to vary depending on exactly how a "behavior" is defined and measured, and the relationships specified (or "allowed") in the model. The inconsistency may also be due to the fact that very few studies have rigorously tested causal relationships using the appropriate scientific methodology (*i.e.*, randomized controlled trials), with many relying on non-experimental designs that can only explore correlations between variables. In the absence of well-designed, consistently specified, and rigorously conducted empirical research, it is impossible to draw firm conclusions regarding the precise causal impact of certain factors on energy consumption and conservation.

Nevertheless, several researchers have made progress with integrating different perspectives in a bid to advance the literature and resolve inconsistent findings [14,43–45]. This effort has yielded some clarity and there is now general agreement that several broad yet interrelated categories of variables may explain individual differences in household energy use. These explanatory variables include a range of socio-demographic factors (e.g., income, education, household size, dwelling type, stage of family life cycle), psychological factors (e.g., knowledge, values, attitudes, motivations, intentions, social norms) and external contextual and situational factors (e.g., socio-cultural, economic, political, legal, institutional forces), among others.

The literature now features models that better articulate the multiplicity of forces underpinning energy consumption and conservation. In early research, Costanzo *et al.* [10] proposed a social-psychological model of energy conservation consisting of two interacting sets of factors: psychological (*i.e.*, factors shaping consumers' information-processing and decision-making, such as perception, evaluation, understanding and memory) and positional/situational (*i.e.*, factors that facilitate or constrain consumers' actions, such as disposable income, home ownership, home repair skills, and own-home technology). More recently, Abrahamse *et al.* [14] proposed that both micro-level factors (e.g., preferences, attitudes, values, abilities, opportunities) and macro-level factors (e.g., availability of new technology, economic and population growth, government regulations and policies, socio-cultural change) can influence household energy consumption. Kollmuss and Agyeman [20] have also distinguished multiple influences of pro-environmental behavior, such as demographic factors (e.g., gender, years of education), external factors (e.g., social, cultural, economic, institutional), and internal factors (e.g., motivation, environmental knowledge, awareness, values, attitudes, emotion, locus of control, responsibilities, priorities). Similarly, Stern [19] has proposed that environmental behavior is shaped by a range of attitudinal variables (e.g., general environmentalist predisposition, behavior-specific norms and beliefs, perceived costs/benefits, non-environmental attitudes), personal capabilities (e.g., literacy, social status, financial resources, behavior-specific knowledge and skills), contextual factors (e.g., social norms and expectations, material costs/rewards, available technology, advertising, and laws, policies and regulations), and habits and routines.

Over the years, researchers have increasingly favored these integrative approaches, which view energy consumption and conservation as arising from an ongoing interaction of multiple factors (e.g., [10,14,19,20]). We follow this approach by conceptualizing household energy usage as a complex process with a range of predictors—including both individual and situational factors, and their interaction—that jointly influence the energy-related practices and behavior of households (Figure 1).

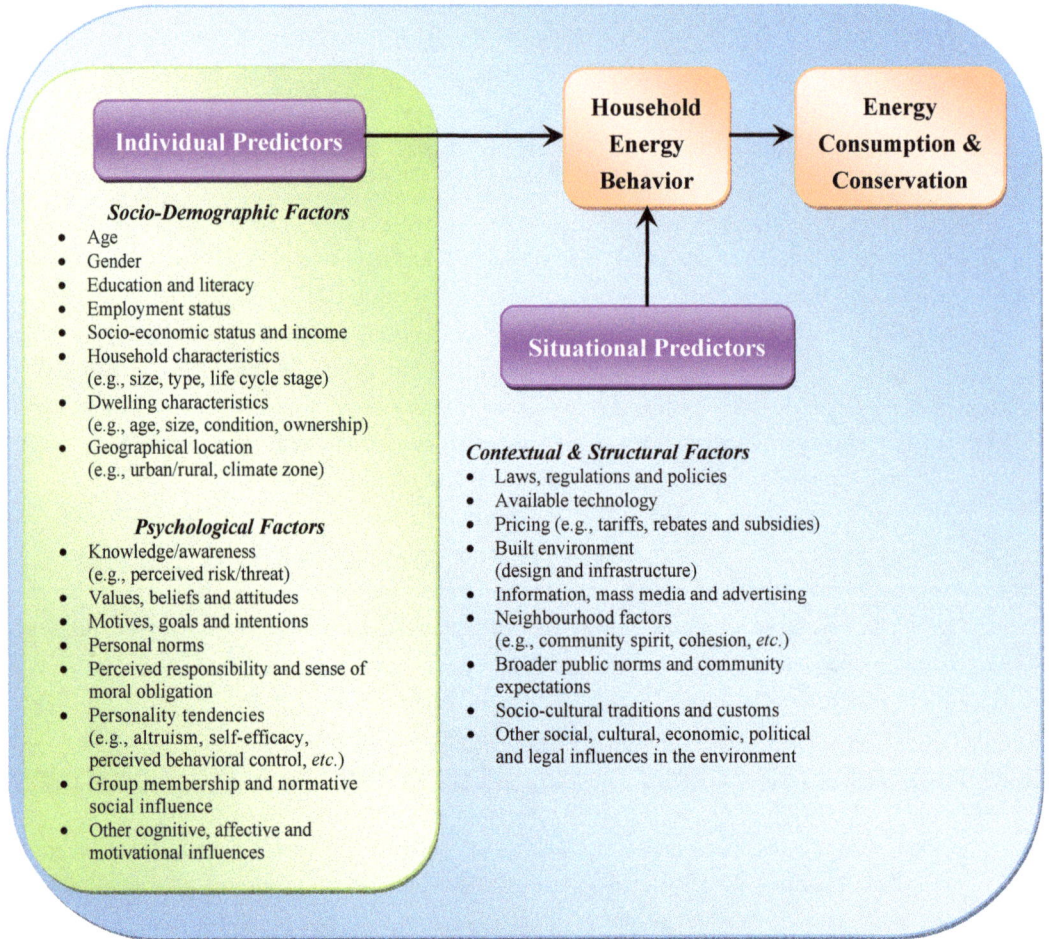

Figure 1. Integrative conceptualization of the various individual (socio-demographic and psychological) and situational (contextual and structural) factors that may influence household energy consumption and conservation.

While we use this integrative conceptualization as our overarching framework for understanding energy consumption and conservation, our review only centers on a subset of factors from the framework: the individual-level predictors of household energy use. This is in keeping with traditional psychological perspectives of pro-environmental behavior, which focus primarily on person-specific factors.

3. The Current Review

3.1. Focus and Scope of the Review

While not discounting the important role of contextual and situational factors, this paper will focus squarely on reviewing the most commonly-examined individual factors correlated with household energy consumption and conservation. Comparatively less emphasis will be placed on macro-level predictors in the broader environment, which are often social, technological and institutional constraints that prevent householders from acting (or enable householders to act) in a certain way regardless of their particular socio-demographic features, psychological attributes and other person-specific characteristics. For example, contextual forces such as government regulations, public policies and other aspects of the broader social, cultural, economic, and political environment (e.g., public infrastructure, electricity prices, government sensitivity to public and interest group pressures, mass media, advertising campaigns, financial markets) can influence patterns of household energy usage, often independent of any individual-level influences. These macro-level factors may also place constraints on policymakers, who are faced with making public policy decisions about the energy industry and consumers within relatively fixed societal and institutional boundaries. While it is important to recognize the potential impact of these contextual factors, they fall outside the scope of this review, whose focus is on first elucidating the individual (*i.e.*, human behavior) part of the equation. Nonetheless, many of the individual variables discussed herein inherently reflect the interface between people and their environment, to the extent that such factors are inextricably linked with one another (as in the case of normative social influence, for example).

3.2. Procedure for the Review

A rigorous process was followed to identify relevant literature for this review. First, a systematic search of the academic literature was undertaken using a number of bibliographic databases in the social, behavioral and environmental sciences (e.g., PsychINFO, ScienceDirect, SpingerLink, Wiley Online Library), as well as other internet search engines and online resources. Subject headings and keywords used in the search process included: residential energy consumption, residential energy conservation, household energy consumption, household energy conservation, and household energy use. Publications from the domains of energy, industry and the built environment were also examined. As part of this literature search process, the reference lists, bibliographies and citations of retrieved literature were also scanned for additional sources. We confined our search to studies conducted in Western countries, written in English, and published since the late 1970s. Evidence from non-western contexts was excluded due to potentially significant and consequential differences in the socio-demographic and psychological determinants of residential energy usage in less developed countries. Studies conducted before the 1970s were also excluded due to concerns about their applicability and generalizability to contemporary contexts (many patterns and predictors of household energy use may have changed over the past 40 years). Our overriding objective was to draw valid conclusions for current contexts and consumers;

up-to-date knowledge and insights are essential for devising cost-effective and readily scalable solutions to contemporary energy-related challenges.

In conducting our review, we examined both primary and secondary evidence, and included studies with a wide range of research designs and methodologies. To make our review as comprehensive as possible, we considered not just studies where socio-demographic and psychological factors were the focal variables of interest, but also those that shed light on such factors incidentally, as a by-product of the primary analyses, as well as publications that usefully synthesized earlier findings. While our primary objective was to review published research in the specific domain of household energy usage, we also considered key insights from the broader domains of pro-environmental behavior (e.g., conservation activities) and resource usage (e.g., consumption), as well as empirical findings regarding the psychology of human behavior more generally. Our systematic review of the literature ultimately identified a large number of journal articles, books and book chapters, working papers, conference proceedings and reports—which in totality formed the basis for the key findings and conclusions presented herein.

3.3. Structure of the Review

In the sections that follow, we summarize the key findings from our comprehensive review of the literature, first for socio-demographic predictors and then for psychological factors. While a range of variables have been hypothesized to explain variation in household energy consumption and conservation, we focus only on the individual-level factors most often touted as distinguishing "energy-wasting" from "energy-saving" consumers. Prior research suggests that residential energy use is more strongly related to socio-demographic variables, whereas changes in residential energy use over time are more dependent on psychological and motivational variables [27]. We summarize here the broad findings that have emerged to date for both categories of predictors. To maintain focus and clarity, the main body of this article presents only a concise summary of general findings, with the accompanying table furnishing a more comprehensive review of prior research—including sources of supporting evidence and full citations.

4. Overview of Key Findings

4.1. Socio-Demographic Predictors

As evident in the top portion of Table 1, household energy consumption and conservation are associated with a wide range of socio-demographic variables. The relatively unchanging opportunities and constraints that people confront when seeking to engage in certain activities may significantly influence how much energy a particular consumer or household can use at any moment in time. Socio-demographic factors such as household income, dwelling type and size, home ownership, family size and composition, and life cycle stage are just some of the many factors that may influence these opportunities and constraints, and thereby indelibly shape the amount, frequency and duration of a household's energy use. Our review of the literature reveals a good number of sources examining the effects of standard socio-demographic factors like age, gender, income, home ownership and household size, but comparatively little assessment of variables such as householders' technical expertise and ownership of home technology. Despite considerable uncertainty introduced by the complex interactions that can occur between multiple factors over time (see the diversity of results evident in Table 1), some general findings to emerge regarding the socio-demographic predictors of energy usage include the following:

- There is inconsistent empirical support for age and gender differences in energy consumption, with any effects tending to be rather small and/or statistically insignificant.
- Education tends to be associated with increased knowledge, awareness and concern regarding environmental issues (such as energy efficiency), however, higher levels of education generally do not lead certainly and directly to pro-environmental behavior (e.g., saving energy).
- Employment status of household occupants (e.g., full-time, part-time, retired or unemployed) may indirectly impact energy consumption, by influencing household income and socio-economic status, which in turn can constrain the household's financial capacity to invest in efficiency measures. Links between occupational status and acceptance of energy-saving strategies have also been examined, but there is limited and inconsistent evidence that this strongly influences energy consumption.
- Household income tends to be positively related to residential energy consumption, but may also enhance household capacity to invest in products and improvements that increase energy efficiency (e.g., to purchase new appliances and more energy-efficient technology).
- Household size (number of people per residence) tends to be positively associated with energy consumption, such that larger families generally consume more energy overall. However, energy usage per capita tends to be lower in larger households, presumably due to the sharing of energy services among multiple residents.
- Dwelling size (floor space, number of rooms/floors, *etc.*) appears to be positively related to household energy consumption, with larger dwellings typically using more energy. Additionally, people residing in detached dwellings (free-standing homes and townhouses) tend to consume more energy than those in multi-unit dwellings (apartments and units).
- Homeowners tend to make larger capital investments in energy conservation measures (e.g., household improvements to increase energy efficiency, purchase of new technology and energy-saving devices) than those living in rental housing.
- Stage of family life cycle appears to be an important predictor of household energy use, with energy consumption typically peaking during the child-rearing years, presumably due to associated changes in household work (e.g., cleaning, cooking, laundry), childcare, and family activities (e.g., in-home entertainment, recreation). The presence or absence of family members—including changes in family composition over time (e.g., the birth of a baby, an older child leaving home)—may also influence levels and patterns of household energy consumption.

Table 1. Socio-demographic and psychological factors associated with household energy consumption and conservation.

Category	Predictor	Impact on household energy consumption and conservation behavior
Socio-demographic factors	Age	• Overall, age does not consistently emerge as a statistically significant predictor of household energy use. Some research supports a positive association between age and energy consumption, such that energy usage increases as the household head grows older. This may be because older people are less likely to adopt energy efficiency measures (given more negative perceptions of the likely cost/benefit ratio and return on investment), possess less knowledge of energy problems and solutions, have lower income and/or poorer home conditions, or require more cooling/heating than younger people to be comfortable [6,46–51]. However, this positive association is far from consistent, with many studies failing to detect any statistically significant effects of age [7,21,27,52,53] or even suggesting that older people are actually more likely to be energy-savers and committed to sustainable energy use (e.g., [1,43,54]).
		• Some studies have even proposed a curvilinear relationship with age, in various forms, where energy consumption peaks either (a) during the middle stages of the life cycle, perhaps with the larger households typical of mid-life having higher energy requirements [50,55], or conversely (b) for younger and older households, perhaps because both tend to live in smaller households with higher per capita consumption, and take fewer energy-saving actions than those in middle-age [56].
	Gender	• The effects of gender on household energy usage seem to be inconsistent, minimal or statistically insignificant. Some research seems to indicate that women exhibit more pro-environmental attitudes and behavior than men [20,43,57,58], while others find no significant relationship [6,21,27,51,59,60].
		• It may be that gender differences in socio-economic conditions and lifestyles (e.g., exposure to poverty, child rearing responsibilities, *etc.*) sometimes constrain the ability of women to conserve energy [61,62]. However, gender-based differences tend to dissipate once one controls for the effects of confounding variables such as household size, income and even age [58].
	Education	• Some studies have reported significant effects of education on pro-environmental behavior and/or energy usage [7,46,63]. But increased education does not typically translate directly into more pro-environmental behavior per se [20]. Rather, across many domains of human behaviour there is often a "knowledge-action gap" [43,64–66], not only in terms of general pro-environmental behavior but also (more specifically) in regard to household energy consumption.
		• For example, several studies have found that education level has no significant impact on either the number of conservation activities [52,67] or household energy consumption [53,54]. While others have found more educated people are slightly more likely to display pro-environmental behavior, these effects are either statistically insignificant [21] or far weaker than the impact of socio-demographic, psychological and motivational factors that are more proximal to actual behavior [56].
	Employment status	• Type of employment (full-time, part-time, retired or unemployed) of household members—particularly the head of the household—may indirectly impact energy consumption by influencing the household's socio-economic status, confidence in income security and/or financial capability to invest in efficiency measures (e.g., new energy-saving technology).

Table 1. *Cont.*

Category	Predictor	Impact on household energy consumption and conservation behavior
	Employment status	• Consumers in full-time employment tend to have more disposable income to spend on day-to-day energy use and energy-intensive appliances, but also more money to invest in one-off energy-saving measures (e.g., solar panels, insulation, energy-efficient light bulbs). They also tend to spend fewer hours per day at home compared to part-time, retired or unemployed consumers, which may contribute toward less household consumption. Indeed, some research suggests that full-time employment is significantly related to making home improvements to conserve energy [47]. It has been proposed that compared to being retired, unemployed or even in part-time employment, full-time employment of the head-of-household may raise consumers' confidence in their capacity to undertake home improvements. • Some research has also found that people with higher-status occupations may be slightly more accepting of certain energy conservation strategies [60]; however, these effects have not been consistently observed (e.g., [67,68]) and as such, the influence of occupational status is unlikely to be strong and/or explain substantial variability in household energy consumption.
Socio-demographic factors	Income	• Household income appears to be one of the strongest socio-demographic predictors of residential energy use and conservation. Most studies have found positive associations between household income and residential energy consumption, suggesting that higher-income households tend to consume more energy than lower-income households [6,7,27,53,54,69–72]. At the same time, however, there is also evidence that higher-income households may be more willing and/or able to conserve energy because they can afford the financial costs of energy-saving investments, such as purchasing new efficient technology [59]. • The effects of socio-economic factors may differ for day-to-day energy usage as opposed to the one-off adoption of energy efficiency measures. Household income is closely linked with factors such as employment status, education and household size—all of which reflect situational characteristics that may facilitate or constrain energy-related behavior by providing the means by which people can perform everyday actions and take one-off steps to save energy. For example, higher-income households typically own and use more electrical appliances than lower-income households, which may lead to differences in energy use [53,71]. At the same time, a higher total income may also increase a household's capability to invest in one-off energy efficiency measures (e.g., solar panels, insulation, energy-saving devices) that serve to conserve energy. • While most research supports a positive relationship between income and household energy consumption, findings are variable across studies—for example, some studies report weak to insignificant effects [46]. Some researchers have even concluded that middle-income households may be the most likely income group to save energy because low-income consumers are unable to reduce their energy use, while high-income consumers are unwilling to reduce their energy use [52,73].

Table 1. *Cont.*

Category	Predictor	Impact on household energy consumption and conservation behavior
Socio-demographic factors	Household size	• Total household energy consumption is positively related to family or household size and composition (*i.e.*, number of persons per residence), such that larger families/households typically consume more energy compared to smaller families/households [6,27,53,69,74]. This may be because larger households generally: (a) possess and/or use more energy-intensive appliances; (b) have more disposable income to spend on energy; and (c) have greater energy demands and requirements (*i.e.*, more cooking, cleaning, washing, heating/cooling, *etc.*). The presence or absence of family members from a household, as well as changes in family composition over time (*i.e.*, new-born baby, older child leaving home, *etc.*) may also influence household energy consumption [8,68]. • However, the relationship between household size and energy use is not perfectly linear and may actually differ when energy consumption is measured per capita as opposed to per household. The sharing of energy services among multiple household members generally leads to lower energy use per capita in larger households, all else being equal [72,75]. That is, the association between household size and energy use is reversed when consumption is measured on a per capita basis. For example, while the highest consuming households tend to be larger families (*i.e.*, couples with children), single-person households actually use the highest amount of energy per capita followed by couple-only, single-parent and two-parent families, respectively [71]. One study has found that two-person households consume around 17% less energy per person for residential and transportation activities than single-person households, with three-person households using more than one-third less energy per person [72]. • The impact of household size may also differ for energy consumption as opposed to conservation. While larger households tend to consume more energy overall, they may also make greater investments in energy efficiency measures. Some research suggests that household size may even have a curvilinear relationship with energy conservation (see Curtain 1976, cited in [50]), with three to four-person households reporting a greater history of past conservation efforts and less expected difficulty in future conservation.
	Dwelling type and size	• Dwelling type (e.g., free-standing houses, townhouses/duplexes, residential units, apartments, *etc.*) appears to directly influence energy consumption because different dwellings vary in important characteristics such as the number of rooms/floors, amount of floor space, degree of insulation, sun and wind exposure, the attributes of energy-using equipment, the number and kind of household appliances, and other critical design features such as wall-cavity insulation, double glazing, energy-efficient heating, ventilation and cooling systems (e.g., [8,68,76]). Some early research has estimated that considering all factors, as much as half of total household energy use depends on the characteristics of existing equipment and the dwelling, with the remainder determined by the features and behavior of occupants [75].

Table 1. *Cont.*

Category	Predictor	Impact on household energy consumption and conservation behavior
	Dwelling type and size	• Households residing in larger dwellings, as indexed by number of rooms and floor space (e.g., freestanding homes) typically consume more energy than households residing in smaller dwellings (e.g., units or apartments) [54]. Some research has found that households living in detached houses, townhouses and semi-detached dwellings consumed 74% more electricity than those living in multi-unit dwellings [71]. At the same time, some research has also found that households residing in detached houses are actually more willing to engage in energy conservation activities than those residing in apartment blocks [59]. • Some evidence also indicates that dwelling characteristics can influence the behavior of household members themselves, thereby impacting energy use indirectly. For example, residing in a larger dwelling may signal to consumers that their household uses considerable electricity/gas and that energy-savings and home improvements are therefore more desirable or necessary [47].
Socio-demographic factors	Dwelling age	• The age of a house/dwelling is often expected to be positively associated with household energy consumption, primarily due to the lower energy efficiency standards of older dwellings (e.g., less efficient heating/cooling, poor insulation, energy wasting appliances, *etc.*). • However, some studies have failed to detect statistically significant effects of dwelling age on consumer participation in energy conservation activities such as home energy audits [56], with more recent research also suggesting that newer homes with more energy efficiency appliances do not necessarily reduce a household's overall energy expenditure. In fact, homeowners residing in older dwellings may even be more inclined to adopt energy-efficient measures than those residing in newer dwellings, particularly if older dwellings are in physically or aesthetically in poor condition and require the installation of new appliances or building components [46].
	Home ownership	• Home ownership (e.g., rented *vs.* owner-occupied) appears to indirectly impact energy consumption by way of influencing household investment in energy efficiency measures [10,77,78]. Sardianou [59] summarises key findings from a number of studies (e.g., [43,67,70,77,79,80]) to demonstrate why home ownership is often associated with greater availability of, and access to, energy efficiency measures. Compared to renters, homeowners are more likely to invest in energy efficiency measures because they tend to be wealthier and have greater financial security, hold longer tenure, and receive greater return on energy efficiency investments. Conversely, renters tend to be poorer, more transient, and less willing and/or capable of making home improvements (either due to limited financial resources, lack of control and/or fewer incentives), thereby leading to less financial investment in energy-efficient devices and new technology. • Some early research on the determinants of gasoline and home heating energy conservation found that home ownership was among the most powerful socio-demographic factors distinguishing conservers from non-conservers [1], with subsequent research revealing that it is also one of the most important factors explaining large capital investment in household energy-saving measures [77]. Barr *et al.* [43] have also recently suggested that home ownership may provide consumers with a sense of personal control and belonging that encourages them to focus more conscious attention toward saving energy.

568

Table 1. *Cont.*

Category	Predictor	Impact on household energy consumption and conservation behavior
	Home ownership	• However, other studies have found that homeowners consume more energy than tenants (for a list of studies, see [78]). For example, a recent Australian study found that households who rent their homes actually consumed less energy than homeowners [71].
Socio-demographic factors	Stage of family life cycle	• Many researchers have investigated how household life cycle relates to consumer's economic behavior (e.g., [81,82]). The stage of a family's life cycle—typically defined as a combination of criteria such as family members' age, marital status, and family size/type—appears to be one of the strongest predictors of household energy consumption (for reviews, see [50,55,75]). This is because family life cycle stages are linked with differences in household needs, priorities, and activities—all of which can help explain variability in household energy demands and usage levels. Families in different stages of the life cycle have vastly different attributes in terms of household work (e.g., cleaning, cooking, laundry), childcare, and in-home entertainment (e.g., TV/computer, visits of friends/family, hobbies and recreation, sleeping and resting/relaxation), all of which may influence patterns of energy consumption and conservation. • A reasonable body of research has supported a curvilinear pattern whereby household energy consumption peaks during the child-rearing years. For example, studies have found that families with children typically use more energy than families without children (at either earlier or later stages of the life cycle) or families with children who have left home [50]. According to Lutzenhiser [83], household life cycle (*i.e.*, family age and composition) differences have been reported across a range of criteria including areas such as heating, electricity use, housing needs, overall energy efficiency, building/appliance characteristics, and levels of carbon dioxide pollution. • These differences in household energy consumption across the lifespan may arise from concurrent changes in some of the socio-demographic predictors of energy use, particularly factors such as employment, income, house type/size, and household composition. Because household characteristics may influence the behavior, activities and lifestyles of occupants, any changes in these characteristics over time may subsequently change the energy usage of the household [75].
	Geographical location	• Regional differences in climate, temperature and geography are key determinants of household energy use, with studies in the northern hemisphere finding that households located in more southern regions (*i.e.*, warmer temperatures) tend to consume less energy than households in more northern regions (*i.e.*, colder temperatures) [6,8]. • Rural areas have also been found to have higher levels of energy use than urban areas [8], with these regional differences purportedly arising due to variability in types of houses (e.g., freestanding dwellings *vs.* apartments), life-style characteristics, and house orientation to sunlight and wind. Geographical location may also impact homeowners' attitudes and preferences toward energy conservation—for example, due to the effects of the local governments' actions to encourage and reward energy efficiency measures and behavior [46].

Table 1. *Cont.*

Category	Predictor	Impact on household energy consumption and conservation behavior
Socio-demographic factors	Ownership of home technology & technical expertise	• Some researchers have suggested that householders who possess "high-tech" consumer products (e.g., electrical goods, computers, *etc.*) may be attracted toward all types of technical innovation (including energy-saving devices), alongside having more disposable income to invest in such products (*i.e.*, greater financial capacity to purchase new technology); thus, ownership of general (non-energy) home technology may be associated with ownership of more specific energy-saving devices, systems and equipment (e.g., energy efficient appliances, solar power) that assist with energy conservation [10]. • Some research also suggests that the presence of a household member with technical knowledge and skills in home repairs (e.g., home appliance and automotive repairs)—that is, a "handyperson"—is positively related to energy conservation, presumably because such people may have a better understanding of new technology [10] and be more capable of performing installation and ongoing maintenance tasks for energy-saving technology [40,46]. However, "do-it-yourself" consumers may also be less inclined to purchase unfamiliar energy efficiency equipment and appliances if they perceive the installation and maintenance of these items to be complicated, burdensome or requiring expert skills [84]. Indeed, Kollmuss and Agyeman [20] have cited some research showing that very detailed technical knowledge does not inherently facilitate or increase pro-environmental behavior.
Psychological factors	Knowledge & problem awareness	• In the context of energy consumption, energy-related knowledge reflects one's level of knowledge, awareness and understanding of energy costs, energy-saving behavior, and the consequences of such behavior [8]. While greater knowledge, awareness and understanding of environmental issues such as energy conservation tend to be positively associated with pro-environmental behavior (e.g., saving energy) [70,85], greater knowledge and/or awareness does not directly and automatically lead to more pro-environmental behavior per se—that is, there is often a "knowledge-action gap" [14,43,64,65,86]. • Empirical evidence shows that only a small portion of pro-environmental behavior can be directly linked to environmental knowledge and awareness, with Kollmuss and Agyeman [20] citing some research to suggest that at least 80% of the motives for pro-environmental action are other internal and situational factors. While there are exceptions, most studies have failed to consistently detect statistically significant relationships between knowledge and problem awareness and pro-environmental behavior such as energy conservation [14,87,88]. Thus, while greater knowledge and problem awareness is generally positively related to energy savings (and negatively related to energy usage), this relationship is likely to be weak and/or insignificant.

Table 1. *Cont.*

Category	Predictor	Impact on household energy consumption and conservation behavior
		• Values reflect a global, abstract and relatively enduring set of beliefs, ideals and standards that serve as guiding principles in life (e.g., a person's general sense of right *vs.* wrong), whereas attitudes reflect more specific positive or negative evaluations of a particular idea, object, person, situation or activity [89–92]. Many scholars have examined the role of values, attitudes and beliefs in the context of pro-environmental behavior and, more specifically, residential energy usage [6,8,27,93–97]. Some early studies found support for the notion that holding more pro-environmental values, attitudes and beliefs will lead to more pro-environmental behavior [21,93,98]. For example, an early meta-analysis by Hines *et al.* [21] reported a positive relationship between attitudes and pro-environmental behavior, suggesting that people with more positive attitudes were more likely to report engaging in environmentally responsible behavior than those with less positive attitudes. In terms of the specific pro-environmental behavior of sustainable energy use, Becker *et al.* [93] found that householder's attitudes toward thermal comfort and convenience were the most powerful predictors of household energy use; and Seligman *et al.* [98] reported a relatively strong positive association between personal values and residential electricity consumption.
Psychological factors	Values, attitudes & beliefs	• Nevertheless, most empirical evidence indicates that the strength of these associations is often inconsistent, weak and/or insignificant [6,27,53,54,70,99], especially when compared to the effects of socio-demographic factors [7]. It appears that positive values, attitudes and beliefs toward the environment may encourage sustainable behavior (e.g., energy savings, adoption of efficiency measures, *etc.*), but they do not inherently lead to actual reductions in energy use per se [46,53,54,100]. This discrepancy has been referred to as a "value-action gap" and/or "attitude-action gap", and has been observed across many domains of human behavior [26,101–103]. Consistent with this notion, daily life illustrates many situations where people express strong beliefs about the negative consequences of environmental problems (e.g., global warming, climate change, reliance on fossil fuels), or positive evaluations of sustainability and "green" technologies (e.g., renewable energy sources), but fail to translate those beliefs, values and attitudes into practical actions to limit household energy use.
		• Several scholars have offered useful explanations for how and why environmental attitudes have varying, and typically very small, impacts on pro-environmental behavior (e.g., [8,20,68,102,104]). For example, people typically make choices and behave in ways that minimize costs and maximise benefits to themselves (in terms of time, effort, money, comfort, *etc.*) rather than based on what is "best" for others and the environment [93,94]. Moreover, while attitudes may lead to positive intentions to save energy, various intervening factors (e.g., lack of knowledge about effective actions, social norms, perceived personal responsibility, self-efficacy, anticipated cost-benefit trade-offs, situational and institutional factors, *etc.*) may block this intention from being realised into actual behavior [8,68].

Table 1. *Cont.*

Category	Predictor	Impact on household energy consumption and conservation behavior
Psychological factors	Motives, intentions & goals	• Motives are the driving forces or impulses that initiate, guide and maintain goal-directed behavior; that is, the specific reasons why a person acts in a certain way at any given time. Most modern theories define motivation as the process that shapes the intensity, direction and persistence of effort that a person allocates toward achieving a particular goal or desired end state [105–111]. A range of theoretical models have been proposed to explain the various motivations that underpin different types of pro-environmental behavior, including sustainable use of energy [17,44]. Some scholars have drawn on the theory of planned behavior [23] to argue that people make reasoned choices and behave in a way that yields "optimal" outcomes in terms of minimising costs and maximising benefits (in terms of time, effort, money, social approval, *etc.*). Others have focused more heavily on moral and normative concerns by arguing that pro-environmental behavior is positively associated with specific motives related to altruistic, biospheric, prosocial and self-transcendent value orientations (*i.e.*, values beyond one's immediate self-interests), environmental concern, and a sense of moral obligation (see [17,44,63,96,112–114]). • Distinctions have also been made between self-transcendent and self-enhancing goals (for more information, see [89,90]). Some research suggests that whereas self-transcendence goals (e.g., promoting the interests of others and the external world) are positively related to a range of pro-environmental behaviors, the relationship between self-enhancement goals (e.g., focusing on oneself and one's interests) and action is either negative or non-significant. Indeed, Schultz and Zelezny [97] have concluded that people who place a higher value in self-transcendent life goals typically express greater care and concern for environmental issues, as well as exhibiting more pro-environmental behavior. Conversely, those who place higher value in self-enhancing life goals typically have more egotistical concerns about environmental issues and display less pro-environmental behavior. • Intrinsic motives—that is, motivation that stems from personal interest, enjoyment or satisfaction in an activity itself, regardless of external pressures or rewards—have also been associated with pro-environmental behavior [115]. De Young [116] proposed four different intrinsic satisfactions and associated motives that may underpin environmental sustainability: satisfaction from striving for behavioral competence (e.g., enjoyment from solving problems and completing tasks); satisfaction from frugal, thoughtful consumption (e.g., enjoyment from survival based on careful management of finite resources); satisfaction from participating in the community (e.g., enjoyment from being involved in community activities); and satisfaction from luxuries (e.g., enjoyment from convenience and access to new/novel products). Research suggests that individuals who possess intrinsic motivation tend to engage in more durable, sustainable behavior [116–118]. • Researchers have also distinguished between primary and specific motives, with Kollmuss and Agyeman [20] suggesting that the larger primary motives that influence a wide range of behavior (*i.e.*, altruistic and social values around living a pro-environmental lifestyle) are often surpassed or overridden by more immediate, selective motives (*i.e.*, specific motives that influence particular actions and often evolve around one's own needs, such as being comfortable, saving money/time, reducing effort, *etc.*).

Table 1. *Cont.*

Category	Predictor	Impact on household energy consumption and conservation behavior
Psychological factors	Motives, intentions & goals	• Goal framing theory [119–121] has been offered as a framework for integrating diverse concepts from the above theoretical perspectives. This theory proposes that at any given moment, human behavior arises from multiple motivations, and goals guide or "frame" how people think, feel and act. Three main motives or "goal frames" have been identified as relevant for predicting pro-environmental behavior: gain goal frames (*i.e.*, a desire to protect and improve one's resources or possessions, such as to save money, protect financial security, *etc.*); normative goal frames (*i.e.*, a desire to act appropriately in line with social and moral standards, that is, to behave in the "right" way); and hedonic goal frames (*i.e.*, a desire to seek pleasure and avoid pain) (for a comprehensive review, see [122]). While motivations are rarely homogeneous and multiple goals can influence behavior, hedonic goal frames are assumed to exert the strongest effects. However, little empirical research exists to support this model, as it has yet to be scientifically tested in the environmental or residential energy usage domains. • In the general psychology literature, many studies also reveal a discrepancy between intentions and behavior, *i.e.*, an "intentions-action gap" [123,124]. For example, a meta-analysis by Sheeran [124] estimated that intentions explain only about 28% of the variance in future behavior, with a more recent mega-analysis finding weak support for the overall impact of changing behavioral intentions on subsequent change in behavior—in fact, it was concluded that a medium-to-large sized change in intention leads to only a small-to-medium change in behavior [125]. Thus, while people driven by certain intentions may be more inclined to engage in energy-saving behavior, simply possessing these intentions does not automatically translate to behavior.
	Personal norms	• It has been suggested that altruistic behavior is activated by personal norms, and acting in manner that is consistent with one's personal norms may lead to positive feelings of pride and self-satisfaction whereas acting in a manner inconsistent with personal norms may lead to negative feelings of guilt and regret. According to the norm activation model [126], pro-social behavior is influenced by moral or personal norms—*i.e.*, feelings of strong moral obligation to perform certain types of pro-social behavior, including pro-environmental actions such as energy conservation (for reviews, see [6,16,27]). For personal norms to be activated, however, a person must first be aware that their behavior has an impact on others and/or the environment (*i.e.*, there must be awareness of consequences), and also feel a sense of personal responsibility for such impacts (*i.e.*, termed "ascription of responsibility"). Consistent with this notion, Abrahamse and Steg [27] have suggested that consumers are likely to feel a stronger obligation to save energy if they believe that energy consumption negatively impacts the environment, and that they are personally responsible. • Evidence from several studies supports the role of moral norms and personal responsibility in explaining individual differences pro-environmental behavior [17,21,77,127,128]. However, the strength of this effect of personal norms is questionable, with some research suggesting that personal norms only guide behavior when they are focal [129]. Thus, while it may be assumed that personal norms will influence household energy usage, simply possessing more positive personal norms is unlikely to directly translate to changes in energy consumption or conservation.

Table 1. *Cont.*

Category	Predictor	Impact on household energy consumption and conservation behavior
Psychological factors	Perceived responsibility	• Perceived responsibility reflects the attribution of responsibility (*i.e.*, self-blame, accountability, liability, obligation, *etc.*) for energy conservation to oneself rather than away from oneself to other people, the government, industry bodies, environmental groups, or other external entities [8]. It is often argued that feeling personally responsible for environmental problems (e.g., accepting blame for ecological damage caused by excessive energy use) and for protecting the environment (e.g., feeling obligated to combat climate change by reducing carbon emissions) is positively associated with pro-environmental behavior. A number of researchers have proposed that people who feel personally responsible for environmental problems tend to feel a stronger obligation to help minimise or mitigate them, thereby activating personal norms (e.g., moral obligation to act) and increasing one's willingness to act pro-environmentally (e.g., [6,27]). Denying one's own responsibility, on the other hand, may diffuse blame to an external entity and indicate that there is no need to change one's behavior or lifestyle [8]. Accepting personal responsibility for sustainable energy use is therefore hypothesized to be a positive predictor of energy-saving behavior [43,94,95,130]. However, the strength of this relationship may be weak due to the same processes implicated in the aforementioned "value-action gap" [26,101–103].
		• In terms of empirical evidence, Hummel *et al.* [131] (cited in Van Raaij & Verhallen, 1983a) found that perceived self-blame of energy consumers was related to a greater willingness to save energy, whereas diffusing blame for the energy crisis was related to less willingness to conserve. In subsequent research, Hines *et al.* [21] proposed that personal responsibility may be expressed in various ways, for example in terms of the environment as a whole (e.g., personal responsibility to protect the environment, social responsibility to the community) and/or in terms of only one facet of the environment (e.g., personal responsibility felt for reducing energy). Based on a meta-analysis, these authors concluded that those who feel some degree of personal responsibility toward the environment are more likely to display responsible environmental behavior than those who do not feel personally responsible. Given that other research from the general psychological literature shows there is often a discrepancy between intentions and action, however, the strength of the relationship between perceived responsibility and the specific pro-environmental behavior of energy conservation may not always be consistent or reliable.
	Locus of control, self-efficacy, and perceived behavioral control	• Locus of control reflects a person's perception of whether they have the capability to enact change and/or control events that impact them. Individuals with a strong internal locus of control believe that they can exercise personal control over their own decisions, life circumstances and outcomes (*i.e.*, belief that events arise primarily from internal factors, such as one's own motivation and actions), whereas those with a strong external locus of control believe that decisions, life circumstances and outcomes are controlled by environmental factors outside their influence (*i.e.*, belief that events arise primarily from external factors, such as other people, the government, socio-economic influences, *etc.*). This factor is similar to perceived instrumentality, self-efficacy, perceived behavioral control [27,123,124].

574

Table 1. *Cont.*

Category	Predictor	Impact on household energy consumption and conservation behavior
Psychological factors	Locus of control, self-efficacy, and perceived behavioral control	• A large body of literature suggests that locus of control is associated with a person's values, attitudes and intentions to engage in pro-environmental behavior such as energy conservation (for a recent empirical analysis of the various linkages between perceived behavioural control and constructs such as attitudes, social and moral norms, feelings of guilt, and intentions to engage in pro-environmental behaviour, see [16]); however, the extent to which this then translates to action is highly questionable, with behavioral effects sometimes very weak and/or insignificant. • Compared to individuals with a strong internal locus of control, those with an external locus of control may be less likely to behave in a pro-environmental way because they perceive that such behavior is inefficacious and "doesn't make a difference" [20]. In the context of energy consumption, for example, householders may be more likely to conserve energy if they believe that reducing consumption will be effective in yielding valued outcomes, such as reducing costs and/or increasing benefits (*i.e.*, protecting the environment, saving money, *etc.*); but less likely to conserve energy if they believe that their personal contributions are marginal and ineffective. Consistent with this notion, an early meta-analysis by Hines *et al.* [21] revealed a strong positive association between locus of control and pro-environmental behavior, such that individuals with an internal locus of control were more likely to report engaging in environmentally responsible behavior than those with a more external locus of control. However, subsequent empirical studies have revealed mixed findings. Sheeran *et al.* [123,124] have reported that individuals with equivalent perceived behavioral control may oftentimes differ in their subsequent behavior.
	Perceived cost: benefit ratio	• People are often motivated by self-interest and try to select alternatives that yield the highest benefit for the lowest cost—where "benefits" and "costs" may include scarce or valued resources such as time, effort, money, social status/acceptance, convenience, comfort, and so forth. Both economic and behavioral cost-benefit tradeoffs may influence pro-environmental behavior such as household energy consumption and conservation (for further details, see [8,122]). For example, Midden and Ritsema [130] explored several categories of perceived advantages and disadvantages of energy conservation that may be important: personal disadvantages (e.g., beliefs regarding loss of comfort, coldness, unhealthiness, behavioral constraints, *etc.* imposed by an energy-saving lifestyle), societal advantages (e.g., beliefs regarding less environmental pollution, more energy for future generations, world energy supplies, *etc.*) and personal responsibility (e.g., beliefs regarding a sense of duty/responsibility).

Table 1. *Cont.*

Category	Predictor	Impact on household energy consumption and conservation behavior
	Perceived cost: benefit ratio	• From an economic perspective, financial costs (or benefits) include the monetary expenses (or potential savings) that households incur from consuming and/or conserving energy [43,76,95]. The high financial costs of adopting one-off efficiency measures may decrease the likelihood of engaging in conservation initiatives, with long-term monetary payoffs also playing an important role. While people may be keen to purchase new appliances and undertake house improvements to optimise energy efficiency (e.g., installing solar panels, insulation, low-energy appliances), the immediate financial costs incurred by such activities may constrain them from doing so, or otherwise act as a disincentive (particularly if there are no immediate benefits). At the same time, energy usage costs may impact homeowners' choice of efficiency measures in the opposite direction: that is, consumers who perceive the costs of consumption or inefficiency to be high might be more motivated to take extra steps to reduce consumption (and thus utility bill expenses), particularly if they believe that non-investment in energy-saving measures is unlikely to reduce costs, or potentially make existing costs even worse [46,77]. Indeed, a recent study by Nair *et al.* [46] found that individuals who perceived their household energy costs to be high were more likely to adopt investment measures compared to those who perceived their costs to be low, which suggest that increases in energy prices may actually encourage consumers to actively search for, and invest in measures that will yield energy savings. • The concept of time inconsistency—that is, the tendency for people to be very short-sighted when some costs or benefits are immediate, but more farsighted when all costs and benefits are in the future—is also relevant for understanding the potential impact of cost: benefit appraisals. In daily life, there are countless situations where people procrastinate, postpone decisions, or delay actions because they are viewed as costly in the short-term, despite offering long-term benefits. In the context of energy usage, consumers may be reluctant to outlay the high monetary costs of investing in one-off efficiency measures today even if it leads to substantial monetary savings on energy bills (and environmental benefits, such as reduced carbon emissions) in a few years. Considerable evidence supports this tendency for people to value immediate rewards (and dislike immediate costs) far more than they value future rewards (and dislike future costs) [132–134].
Psychological factors	Need for personal comfort	• Personal comfort, particularly the perceived loss of comfort that any energy-saving measure might impose, may have a sizeable impact on household energy consumption [43,76,94,95,130]. Any decrease in personal comfort, or perceived threat to lifestyle quality, may reduce the likelihood of engaging in conservation behavior. Empirical research has found that consumers' perceptions of comfort and health are related to energy consumption in both summer and winter seasons [55,93,98]. Some early research found that the combined effect of comfort and health was a significant predictor, accounting for 30% of the variability in a household's actual electric consumption. Results revealed that the more a household perceived energy-saving behavior as leading to discomfort and ill-health, the more energy that particular household consumed [98].

Table 1. *Cont.*

Category	Predictor	Impact on household energy consumption and conservation behavior
	Need for personal comfort	• More recently, Barr *et al.* [43] examined the level of comfort that people with different characteristics are willing to accept in relation to energy-saving behavior. Results revealed that while over 60% of "committed environmentalists" were willing to sacrifice some comfort in order to save energy, less than 25% of "non-environmentalists" were willing to do so. Furthermore, while less than 20% of "committed environmentalists" rated "feeling comfortable around the home" to be an important issue to them, this factor was considered important for almost 60% of "non-environmentalists".
Psychological factors		• It is well established that human beings make social comparisons, follow the behavior of other people, conform to social norms—*i.e.*, the explicit and/or implicit rules, guidelines or behavioral expectations within a group or society that guide what is considered normal and/or desirable [135–139]. Two distinct types of social influence can motivate human action to conform: injunctive norms, which raise a person's awareness of the attitudes and/or behavior that are typically approved or disapproved by a social group (*i.e.*, what people should think or do); and descriptive norms, which raise a person's awareness of the attitudes and/or behavior that are typically adopted, supported or performed by a social group (*i.e.*, what people actually think or do). Various factors can strengthen or weaken normative influence (e.g., group cohesion, group size, social support), but the final result—conformity—tends to be consistent and pervasive [140].
	Normative social influence	• Considerable research has identified group membership and normative social influences as having significant impacts on energy consumption and conservation [10,12,130]. In early work, Costanzo *et al.* [10] proposed a social-psychological model of energy conservation that highlighted the importance of social influence, diffusion and reference groups (*i.e.*, friends, family, other social networks) in promoting and maintaining energy conservation. This model proposes that information transmitted via social diffusion is more likely to influence behavior because it tends to be more easily perceived, favourably evaluated, and better understood and remembered than information transmitted via traditional means of education, marketing and advertising. As such, interpersonal sources of information may be more influential than media appeals in eliciting and sustaining reductions in energy use. Other researchers have made similar arguments, with Stern ([11], p. 1229) suggesting that the personal opinions and actions of one's friends may have a more powerful influence over household energy choices than expert advice, even if the latter is better informed.

577

Table 1. *Cont.*

Category	Predictor	Impact on household energy consumption and conservation behavior
Psychological factors	Normative social influence	• Extensive evidence from the behavioral economics and behavior change literatures supports the impact of normative information on residential energy usage (for an overview, see [141]). Many studies have examined the behavioral effects of providing consumers with information about descriptive norms—*i.e.*, personalised messages or communication containing details of one's energy consumption relative to a neighbourhood norm—with extensive evidence supporting the behavioral impact of this comparative information (e.g., [141–146]). For example, Nolan *et al.* [146] found that delivering a descriptive normative message (*i.e.*, information about the conservation behavior of one's neighbours) motivated consumers to save more energy than a control message or any other messages that included appeals traditionally accorded motivational power (e.g., protecting environment, saving money, being socially responsible). Ayres *et al.* [144] found similar support for the effectiveness of descriptive norms, with two experiments revealing that normative social information can lead to energy savings of between 1.2% and 2.1%. More recently, Allcott [141] found that providing descriptive normative information led to an average residential energy saving of 2.0%. However, these effects were heterogeneous, with above-average consumers saving far more energy than below-average users.

- Ownership of non-energy technology (e.g., "high-tech" products like computers and gadgets) is often related to greater use of energy-saving devices and systems (e.g., energy efficient appliances). The presence of "handy" household members with technical knowledge and skills in home repairs (e.g., home appliance and automotive repairs) has also been linked with energy conservation. However, very detailed technical knowledge does not consistently promote pro-environmental behavior.

- Regional differences in climate, temperature and geography are closely related to energy use, with households located in colder zones typically consuming more energy than households in warmer zones. Households in rural regions also tend to have higher levels of energy use than those in urban areas, other things being equal.

4.2. Psychological Factors Related to Household Energy Consumption

While socio-demographic factors clearly play an important, albeit complex role in household energy consumption and conservation, a range of person-specific psychological factors may also have powerful effects (e.g., [8,9,11,27,104]). As shown in the bottom portion of Table 1, some of the psychological factors most commonly associated with household energy usage include: knowledge and problem awareness (both of environmental and energy issues); beliefs, values and attitudes; motives, intentions and goals; subjective appraisals and perceptions (e.g., cost-benefit trade-offs; perceived behavioral control); personality tendencies (e.g., self-efficacy, locus of control); and personal and social norms. Our review of the literature shows considerable attention being paid to the influence of values, attitudes and beliefs, as well as motivational constructs such as goals and intentions, but relative less emphasis on investigating variables such as locus of control and self-efficacy. This variation in the attention paid to different psychological constructs is reflected in Table 1, and adds another dimension to our understanding of the extant literature and what we can rightly make of the evidence currently available.

Before reviewing the key findings regarding these psychological factors, we point out that over the years there has been some variation and even marked shifts in the definition of some of these constructs, particularly beliefs, values, attitudes and motives (for more detailed reviews, see [89–91,117,147–151]). There is considerable overlap among the latter factors, in particular, and ongoing debate over their precise definitions and degree of relatedness. Some scholars use the terms somewhat interchangeably while others argue that they represent conceptually and operationally distinct constructs. For example, some researchers have examined the value basis of environmental beliefs and behavior by distinguishing between egoistic, altruistic and biospheric values, value orientations and/or attitudes (e.g., [97,112,114,152]), whereas others have explored these same categories—egoistic, altruistic and biospheric—as applied to motives (e.g., [96,114,153,154]). We will avoid unnecessarily complicating the current review by simply adopting the most common conceptualizations and usages of each construct.

As shown in Table 1, some of the general findings to emerge from research exploring the specific psychological and motivational variables that influence patterns of household energy consumption and conservation include the following:

- Knowledge, awareness and understanding of environmental issues (e.g., energy-related problems) does not always lead directly and consistently to pro-environmental behavior such as energy conservation. Rather, there may often be a "knowledge-action gap" [65], such that increasing knowledge and awareness does not routinely translate into congruent behavioral change, perhaps due to the influence of various moderating factors that may constrain or facilitate energy-related behavior.

- Likewise, pro-environmental values, beliefs and attitudes do not reliably translate to congruent changes in energy consumption or conservation, with the relationship between values and behavior ultimately contingent upon various moderating factors, such as knowledge, problem awareness, household technology, socio-demographic constraints, and the like. In the end, there may often be a marked "value-action gap" and/or "attitude-action gap" [26,102,103].

- Likewise, we might reasonably expect that people who are driven by certain goals (e.g., self-transcendence *versus* self-enhancing goals; hedonic *versus* gain frames) and motives (e.g., pro-social, altruistic) will be inclined toward energy-saving behavior. But again, the relationship between "good intentions" and actual behavior depends ultimately on moderating factors. Again, we are often left with a marked "intention-action gap" [123,124], with possession of environmentally friendly goals and motives failing to translate—reliably and consistently—into environmentally friendly behavior, such as energy conservation.

- Personal norms (e.g., feeling a strong moral obligation to act in a pro-social, altruistic manner) tend to encourage pro-environmental behavior such as energy conservation. But this relationship may be contingent on awareness of the consequences of one's behavior and ascription of felt responsibility for these behavioral consequences.

- Perceived responsibility for environmental issues and problems tends to be positively associated with pro-environmental behavior and sustainable consumption, presumably because people who feel personally responsible for a particular problem also tend to feel a stronger obligation to help minimize and mitigate it, thereby activating personal norms (e.g., moral obligation to act). However, the precise strength of these associations depends on a range of other mediating and moderating factors.

- Perceived behavioral control (and the associated construct of self-efficacy) tends to be positively associated with pro-environmental behavior such as energy conservation, such that individuals with an internal locus of control are more likely to engage in pro-environmental behavior than those with a more external locus of control. Similar to personal norms and perceived responsibility, however, the strength of this association depends on a range of other mediators and moderators.

- Both economic and behavioral cost-benefit tradeoffs may influence energy consumption and conservation, with people tending (other things being equal) to select courses of action that yield the highest benefit for the lowest cost (in terms of time, effort, money, status/prestige, social approval, comfort, convenience, *etc.*). However, research in behavioral economics shows that people are also frequently prone to a range of cognitive biases, heuristics and other anomalies in their decision-making and behavioral choices—including around environmental protection, renewable and sustainable technologies, and energy consumption—which cause them to act in seemingly "irrational" ways that diverge markedly from traditional economic models of behavior [104,155–158].

- Personal comfort, particularly the perceived loss of comfort that energy-saving measures may entail, can have a powerful influence on household energy usage. Any decrease in personal comfort, or reduction in lifestyle quality, may reduce the likelihood of householders engaging in energy conservation behavior.

- Group membership and normative social influence (e.g., the perceived energy-related practices of one's peers or neighbors, and social pressure from family/friends to save energy) can significantly influence household energy use. Much research indicates that people tend to behave in ways similar to those around them (*i.e.*, people desire normalcy and often exhibit conformity). This is largely due to the effects of social norms—those explicit and implicit "rules" or expectations that guide what is deemed normal, common and/or desirable behavior in society. In terms of pro-environmental actions, injunctive norms (*i.e.*, perceptions of what attitudes and behavior are approved/desired by a social group with whom one associates or identifies) and descriptive norms (*i.e.*, perceptions of what attitudes and behavior are normal/common among this social group) can both exercise great influence over behavior.

4.3. Summary of Key Findings and Conclusions

Our comprehensive literature review has revealed that household energy consumption and conservation are associated with a number of socio-demographic and psychological variables, but that these associations are not always substantial, straightforward or consistent, making it difficult (and certainly more difficult than is typically assumed) to draw definitive conclusions across studies. Indeed, it is clear that most of the factors we have reviewed actually interact with other variables, often in rather complex ways, and that their impact is heavily contingent upon those "moderating" factors. It is not simply a matter of household energy use being shaped—in a direct and linear fashion—by just a few principal individual-level factors. Rather, there are a multitude of variables (predictors, mediators and moderators) that together influence the nature, intensity and duration of behavior around energy consumption and conservation [10,14,19,20]. This complexity and inconsistency pose some challenges for drawing firm conclusions about specific effects (e.g., the size and direction of a particular variable's impact on household energy use), and especially for generalizing findings more broadly. Accordingly, we strongly recommend that researchers and practitioners exercise due caution when drawing inferences regarding the effects of individual variables as reported herein, without taking careful account of the complex interplay among the various factors.

In terms of socio-demographic predictors, our review suggests that several factors (e.g., household income, dwelling type/size, home ownership, family size/composition) are strongly associated with household energy usage, but in some cases the effects are mixed. For example, while a few studies suggest curvilinear effects on energy consumption for certain socio-demographic factors (e.g., age, income, stage of family life cycle), this non-linear pattern does not always hold up in other studies. To illustrate, some research has suggested that middle-income households are actually most likely to save energy, with low-income households (already, by necessity, consuming little energy) simply unable, and high-income households unwilling, to reduce usage [52,73]. However, most research has observed a simple linear association [6,7,27,53,54,69–72]. Moreover, the relationship (whether curvilinear or otherwise) between income and energy use is expected also to be influenced by the greater capacity of higher income householders to invest in energy efficiency technologies and measures. If we take proper account of these nuances, it would be misleading simply to claim that higher income leads to greater household energy consumption. This pattern of results for income is just one example of the many complexities we identified in the literature. It is clear that the extent to which socio-demographic variables influence household energy usage depends on complex and dynamic interactions among different factors, sometimes simultaneous, and other times unfolding over time.

In terms of psychological predictors of energy usage, our review identified several factors that seem to play an important role, with normative social influence being especially powerful. But the results for many of the other psychological factors we reviewed were again far from consistent and conclusive across studies. For example, we identified a wealth of research investigating the impact on household energy usage of variables such as knowledge and awareness; beliefs, values and attitudes; goals, motives and intentions; and personal and social norms. Yet the available evidence indicates that environmentally friendly knowledge and values do not reliably predict environmentally friendly actions—there is often a sizable discrepancy between "good intentions" and actual behavior. Furthermore, the empirical evidence on balance suggests that the effects of many psychological factors (like values, attitudes and beliefs) on subsequent energy behavior tend to be small and/or weak [6,27,53,54,70,99]—often failing to attain statistical significance—especially compared to the effects of socio-demographic factors [7]. For instance, Poortinga et al. [7] found that while attitudinal variables explained a mere 2% of variation in home energy use, the variance explained increased to 15% after taking into account several socio-demographic variables. The relatively poor correspondence between psychological factors and actual energy use suggests that future energy-saving initiatives must direct considerable additional efforts toward helping people act in accordance with their underlying values, beliefs and attitudes, and ultimately, to translate their good intentions into tangible changes in energy consumption and conservation.

In summary then, while some general trends have emerged from the literature, it is clear that predicting and explaining household energy consumption and conservation is considerably more complex than often assumed. This complexity has previously been remarked not only for the specific domain of energy usage, but also for the broader domain of pro-environmental behavior (see Hines et al.'s [21] early meta-analysis). Similar to many other forms of environmentally significant behavior, household energy usage is a complex phenomenon, which is worked upon—directly and indirectly—by a great variety of factors. In the end, a multiplicity of forces interact to influence the nature, intensity and duration of household energy conservation [10,14,19,20]. If the researcher or practitioner seeks specific guidance—lessons applicable to a particular type of householder, context, or point in time—then we must caution them always to take care to undertake their own focused study, one that can reveal the complex interplay of forces bearing upon their specific problem and population of interest.

Nevertheless, the general trends and broad conclusions we have managed to draw out remain illuminating to the extent that they highlight how different types of consumers can have markedly different socio-demographic, psychological and behavioral profiles. When designing and implementing energy-saving interventions, it would be useful for policymakers to identify what unique household profiles exist in their target population. Different types of consumers and households are bound to have vastly different characteristics, needs, and living arrangements. The environments in which people live, and their ability and willingness to control energy use by taking certain actions, will vary widely. To take a simple example, it is likely that conventional energy-saving tips aimed at homeowners living in free-standing dwellings are far less applicable and persuasive for those living in master-metered apartments or subsidized accommodation [159]. It is also likely that strategies promoting financial investment in one-off efficiency measures (e.g., home improvements, such as installing energy-saving retrofits or purchasing new energy efficient technology) are better targeted at high-income households that can afford to outlay money for such measures. Low-income households may benefit more from inexpensive behavioral strategies that help them to recognize and modify certain key energy-wasting practices [51]. By understanding the unique profiles of customers, policymakers will be better placed to identify and target opportunities for effective behavior change, along with the messages and motivational strategies most likely to sustain that change in the specific population of interest.

5. Practical Implications and Directions for Future Research

The key findings and conclusions presented in this paper have important implications for future research and practice. Greater knowledge and understanding of precisely what drives energy consumption and conservation in households, alongside when, where, how, why and for whom this occurs, can make a valuable contribution toward the cost-effective design and delivery of consumer-focused behavioral interventions to promote energy efficiency. Developing innovative, evidence-based solutions to reduce energy consumption—particularly solutions that are cost-effective, mass-scalable and generalizable to broad sections of the community—is currently a major priority at local, national and international levels. Any viable long-term solution to curtailing rising residential energy usage relies on addressing the major determinants of consumer behavior. This naturally includes consideration of the various socio-demographic and psychological characteristics of individuals themselves, alongside immediate contextual factors (which are still bound to an individual's psychology via the automatic perception of, or deliberate appraisal of, their environment) that influence behavior. While promoting societal acceptance and uptake of new energy efficient technology and low-emission "green" energy sources can go some way toward solving the world's energy-related problems, longer-term behavior change in the day-to-day usage of such technology and the enactment of other everyday energy-consuming practices is also at the crux of achieving significant reductions in residential energy usage.

To date, a range of strategies have been developed to encourage pro-environmental behavior among consumers, including behavior change interventions to reduce residential energy consumption and/or improve efficiency [44,160–164]. Such interventions have typically targeted many of the individual-level factors (or the individual's immediate environment) reviewed in this paper. Interventions have ranged from so-called "antecedent strategies" aimed at changing the factors that precede consumer behavior—such as basic information provision and education; goal-setting and commitment strategies; and the use of social/group norms, peer influence and social modeling—through to more "consequence" strategies aimed at changing the outcomes of such behavior—such as self-monitoring; delivering feedback (on one's behavior or performance); and the use of rewards (intrinsic and extrinsic) and other incentives [14,165]. While the literature suggests that all of these strategies have the potential to motivate pro-environmental behavior, the effects have been far from robust and consistent across studies—certain strategies have been found to be effective in some contexts, for some people, and for some types of behavior, but not others (for an overview, see [166]).

Interestingly, the efficacy of different behavioral interventions appears to be highly domain-specific—that is, contingent on the specific type of pro-environmental behavior in question. In an extensive meta-analysis, Osbaldiston and Schott [161] found statistically significant variation in the effect sizes of treatments for different types of pro-environmental behavior (e.g., public recycling, public energy conservation, water conservation, gasoline conservation, curbside recycling, central location recycling, home energy conservation, home energy adoption, and other behaviors), such that no single treatment or intervention was highly effective across all of the behaviors. Rather, there was considerable variability across the different types of behavior in the extent to which certain interventions were (in)effective relative to others. In terms of home energy conservation, treatments that included social modeling, commitment and rewards were found to be most effective, with goal-setting, cognitive dissonance and feedback showing modest effects. In contrast, treatments involving instructions, justifications and prompts to save energy had comparatively weaker, if any, effects. While more empirical research is clearly needed, these results suggest that there may be value in examining the underlying socio-demographic and psychological correlates of specific energy-related practices when designing interventions, rather than simply focusing on the more general domains of consumption and conservation. It may well be that specific energy-related practices (*i.e.*, showering, laundering, space heating/cooling) have different underlying predictors, such that marked variation exists in the responsiveness of these specific practices to different treatments. In addition, practitioners and policymakers are strongly advised to undertake a comprehensive analysis of their specific target population of interest before designing and implementing their own interventions in the field. In particular, it is important to take into account the socio-demographic and psychological profiles of the target population, as well as the relevant contextual factors and experiences (social, cultural economic, political, environmental) that may influence this population.

In parallel, the overall success of any tailored intervention to motivate and sustain positive change in consumer behavior can be enhanced by gaining greater knowledge of the specific antecedents (*i.e.*, predictors) of such behavior, as well as by better understanding the underlying explanatory variables (*i.e.*, mediators) and factors that may influence the nature, intensity, frequency and duration of that behavior (*i.e.*, moderators). This review has highlighted that there are various socio-demographic and psychological factors that may predict (albeit to differing degrees) energy consumption and conservation. In terms of changing behavior, therefore, practitioners and policymakers would be well-placed to focus greatest attention toward those predictors that are most strongly and consistently related to energy usage, and most malleable and responsive to external influences. For example, compared to traditional information-intensive interventions such as educational campaigns that aim to increase knowledge and modify deep-seated beliefs and values, lower-cost strategies that capitalize on behavioral economics principles (e.g., message framing, choice architecture and incentives) to target psychological factors such as cost-benefit appraisals and social norms may prove more impactful [104]. At the same time, it is also imperative to consider the socio-demographic and psychological profiles of individual consumers and households, to ensure behavioral strategies are appropriately tailored and customized to the target population of interest. Finally, both before and after implementing any behavior change intervention, it is critically important for policymakers to consider cost-effectiveness and return-on-investment—not only compared to business-as-usual (*i.e.*, compared to not implementing the intervention at all), but equally importantly, compared to other strategies that may achieve similar results but in a far more/less expensive and mass-scalable manner.

Moving forward, there is still vast scope to extend our understanding of unique customer and household profiles by drawing on the key findings from our review. In particular, the literature could be advanced by developing and testing an evidence-based framework for consumer segmentation that incorporates many of the socio-demographic and psychological factors variables discussed in this paper, and that successfully and usefully distinguishes consumers with different energy-consuming patterns of use. A systematic and consistent framework—validated by empirical evidence—would enable researchers, policymakers and industry experts to better predict how different types of energy consumers are likely to behave in different contexts and at different points in time. Such insight would also enable the design and delivery of tailored intervention efforts that might ultimately be more cost-effective than alternative mass-market solutions.

6. Conclusions

In conclusion, this article has demonstrated that there are a number of individual-level predictors of household energy consumption and conservation. Based on a review of theory and evidence from the social and behavioral sciences, we have identified two broad categories of variables that are commonly proposed as explaining variability in energy usage: socio-demographic and psychological factors. While the influence of specific predictors within each of these categories has not always been consistent or conclusive across studies, we have sought to bring some clarity to the literature by summarizing some of the more robust, generalizable findings that have emerged to date. In doing so, we have highlighted the importance of taking multiple factors into account when aiming to design and deliver strategies that reduce consumption and increase conservation. By shedding more light on precisely what drives consumer behavior, this paper provides practitioners and policymakers with useful insights for developing cost-effective solutions that target and exploit these individual-level predictors of household energy consumption and conservation. We hope that the key findings from our review help to advance the design and delivery of behavior change interventions that will ultimately assist individual consumers, households and entire communities achieve greater sustainability in the use of energy, both now and in the future.

Author Contributions

All three authors were involved in conceiving the aims, objectives, scope and structure of the review. Elisha Frederiks was responsible for conducting the literature review and writing the manuscript. Karen Stenner and Elizabeth Hobman both reviewed and edited the manuscript drafts. All authors have therefore been involved in the preparation and have approved the submitted manuscript.

Conflicts of Interest

The authors declare no conflict of interest.

References

1. Painter, J.; Semenik, R.; Belk, R. Is there a generalized energy conservation ethic? A comparison of the determinants of gasoline and home heating energy conservation. *J. Econ. Psychol.* **1983**, *3*, 317–331.
2. Rosson, P.J.; Sweitzer, R.W. Home heating oil consumption: Profiling 'efficient' and 'inefficient' households. *Energy Policy* **1981**, *9*, 216–225.
3. Guerra Santin, O. Behavioural patterns and user profiles related to energy consumption for heating. *Energy Build.* **2011**, *43*, 2662–2672.
4. Gaspar, R.; Antunes, D. Energy efficiency and appliance purchases in europe: Consumer profiles and choice determinants. *Energy Policy* **2011**, *39*, 7335–7346.
5. Pedersen, M. Segmenting residential customers: energy and conservation behaviours. In Proceedings of the 2008 ACEEE Summer Study on Energy Efficiency in Buildings, Pacific Grove, CA, USA, 17–22 August 2008; Volume 7, pp. 229–241.
6. Abrahamse, W.; Steg, L. Factors related to household energy use and intention to reduce it: The role of psychological and socio-demographic variables. *Hum. Ecol. Rev.* **2011**, *18*, 30–40.
7. Poortinga, W.; Steg, L.; Vlek, C. Values, environmental concern and environmental behavior: A study into household energy use. *Environ. Behav.* **2004**, *36*, 70–93.
8. Van Raaij, W.F.; Verhallen, T.M.M. A behavioral model of residential energy usage. *J. Econ. Psychol.* **1983**, *3*, 39–63.
9. Wilson, C.; Dowlatabadi, H. Models of decision making and residential energy use. *Annu. Rev. Environ. Resour.* **2007**, *32*, 169–203.
10. Costanzo, M.; Archer, D.; Aronson, E.; Pettigrew, T. Energy conservation behavior: The difficult path from information to action. *Am. Psychol.* **1986**, *41*, 521–528.
11. Stern, P. What psychology knows about energy conservation. *Am. Psychol.* **1992**, *47*, 1224–1232.
12. Stern, P. Psychological dimensions of global environmental change. *Annu. Rev. Psychol.* **1992**, *43*, 269–302.

13. Gardner, G.; Stern, P. *Environmental Problems and Human Behavior*; Pearson: Boston, MA, USA, 2002.

14. Abrahamse, W.; Steg, L.; Vlek, C.; Rothengatter, T. A review of intervention studies aimed at household energy conservation. *J. Environ. Psychol.* **2005**, *25*, 273–291.

15. Sovacool, B.K. Diversity: Energy studies need social science. *Nature* **2014**, *511*, 529–530.

16. Bamberg, S.; Möser, G. Twenty years after hines, hungerford, and tomera: A new meta-analysis of psycho-social determinants of pro-environmental behaviour. *J. Environ. Psychol.* **2007**, *27*, 14–25.

17. Vining, J.; Ebreo, A. Emerging theoretical and methodological perspectives on conservation behavior. In *Handbook of Environmental Psychology*; Bechtel, R.B., Churchman, A., Eds.; Wiley: New York, NY, USA, 2002; pp. 541–558.

18. Wapner, S.; Demick, J.; Yamamoto, T.; Minami, H. *Theoretical Perspectives in Environment-Behavior Research: Underlying Assumptions, Research Problems, and Methodologies*; Kluwer Academic/Plenum Publishers: New York, NY, USA, 2000.

19. Stern, P. Toward a coherent theory of environmentally significant behavior. *J. Soc. Issues* **2000**, *56*, 407–424.

20. Kollmuss, A.; Agyeman, J. Mind the gap: Why do people act environmentally and what are the barriers to pro-environmental behavior? *Environ. Educ. Res.* **2002**, *8*, 239–260.

21. Hines, J.M.; Hungerford, H.R.; Tomera, A.N. Analysis and synthesis of research on responsible environmental behavior: A meta-analysis. *J. Environ. Educ.* **1987**, *18*, 1–8.

22. Ajzen, I. From intentions to actions: A theory of planned behavior. In *Action-Control: From Cognition to Behavior*, Kuhl, J., Beckmann, J., Eds.; Springer: Heidelberg, Germany, 1985; pp. 11–39.

23. Ajzen, I. The theory of planned behavior. *Organ. Behav. Hum. Decis. Process.* **1991**, *50*, 179–211.

24. Guagnano, G.A.; Stern, P.C.; Dietz, T. Influences on attitude-behavior relationships: A natural experiment with curbside recycling. *Environ. Behav.* **1995**, *27*, 699–718.

25. Stern, P.; Dietz, T.; Abel, T.; Guagnano, G.; Kalof, L. A value-belief-norm theory of support for social movements: The case of environmentalism. *Hum. Ecol. Rev.* **1999**, *6*, 81–97.

26. Blake, J. Overcoming the 'value-action gap' in environmental policy: Tensions between national policy and local experience. *Int. J. Justice Sustain.* **1999**, *4*, 257–278.

27. Abrahamse, W.; Steg, L. How do socio-demographic and psychological factors relate to households' direct and indirect energy use and savings? *J. Econ. Psychol.* **2009**, *30*, 711–720.

28. Stern, P.C.; Oskamp, S. Managing Scarce Environmental Resources. In *Handbook of Environmental Psychology*; Stokols, D., Altman, I., Eds.; John Wiley & Sons, Inc.: New York, NY, USA, 1987; pp. 1043–1088.

29. Hägerstrand, T. What about people in regional science? *Pap. Reg. Sci. Assoc.* **1970**, *24*, 6–21.

30. Hägerstrand, T. Geography and the study of interaction between nature and society. *Geoforum* **1976**, *7*, 329–334.

31. Palm, J.; Ellegård, K. Visualizing energy consumption activities as a tool for developing effective policy. *Int. J. Consum. Stud.* **2011**, *35*, 171–179.

32. Ellegård, K.; Palm, J. Visualizing energy consumption activities as a tool for making everyday life more sustainable. *Appl. Energy* **2011**, *88*, 1920–1926.

33. Schatzki, T.R. *Social Practices: A Wittgensteinian Approach to Human Activity and the Social*; Cambridge University Press: Cambridge, UK, 1996.

34. Gram-Hanssen, K. Residential heat comfort practices: Understanding users. *Buildi. Res. Inf.* **2010**, *38*, 175–186.

35. Gram-Hanssen, K. New needs for better understanding of household's energy consumption—behaviour, lifestyle or practices? *Archit. Eng. Des. Manag.* **2014**, *10*, 91–107.

36. Rogers, E.M. *Diffusion of Innovations*, 1st ed.; Free Press: New York, NY, USA, 1962.

37. Rogers, E.M. *Diffusion of Innovations*, 5th ed.; Free Press: New York, NY, USA, 2003.

38. Darley, J.M.; Beniger, J.R. Diffusion of energy-conserving innovations. *J. Soc. Issues* **1981**, *37*, 150–171.

39. Faiers, A.; Cook, M.; Neame, C. Towards a contemporary approach for understanding consumer behaviour in the context of domestic energy use. *Energy Policy* **2007**, *35*, 4381–4390.

40. Darley, J.M. Energy conservation techniques as innovations, and their diffusion. *Energy Build.* **1978**, *1*, 339–343.

41. Mahapatra, K.; Gustavsson, L. An adopter-centric approach to analyze the diffusion patterns of innovative residential heating systems in sweden. *Energy Policy* **2008**, *36*, 577–590.

42. Dieperink, C.; Brand, I.; Vermeulen, W. Diffusion of energy-saving innovations in industry and the built environment: Dutch studies as inputs for a more integrated analytical framework. *Energy Policy* **2004**, *32*, 773–784.

43. Barr, S.; Gilg, A.W.; Ford, N. The household energy gap: Examining the divide between habitual- and purchase-related conservation behaviours. *Energy Policy* **2005**, *33*, 1425–1444.

44. Steg, L.; Vlek, C. Encouraging pro-environmental behaviour: An integrative review and research agenda. *J. Environ. Psychol.* **2009**, *29*, 309–317.

45. Van den Bergh, J.C.J.M. Environmental regulation of households: An empirical review of economic and psychological factors. *Ecol. Econ.* **2008**, *66*, 559–574.

46. Nair, G.; Gustavsson, L.; Mahapatra, K. Factors influencing energy efficiency investments in existing swedish residential buildings. *Energy Policy* **2010**, *38*, 2956–2963.

47. Powers, T.L.; Swan, J.E.; Lee, S.-D. Identifying and understanding the energy conservation consumer: A macromarketing systems approach. *J. Macromarket.* **1992**, *12*, 5–15.

48. Hartman, R.S.; Doane, M.J. The estimation of the effects of utility-sponsored conservation programmes. *Appl. Econ.* **1986**, *18*, 1–25.

49. Curtin, R.T. Consumer adaptation to energy shortages. *J. Energy Dev.* **1976**, *2*, 38–59.

50. Frey, C.J.; LaBay, D.G. A comparative study of energy consumption and conservation across family life cycle. *Adv. Consum. Res.* **1983**, *10*, 641–646.

51. Poortinga, W.; Steg, L.; Vlek, C.; Wiersma, G. Household preferences for energy-saving measures: A conjoint analysis. *J. Econ. Psychol.* **2003**, *24*, 49–64.

52. Verhage, B.J. Stimulating energy conservation: Applying the business heritage of marketing. *Eur. J. Market.* **1980**, *14*, 167–179.

53. Gatersleben, B.; Steg, L.; Vlek, C. Measurement and determinants of environmentally significant consumer behavior. *Environ. Behavi.* **2002**, *34*, 335–362.

54. Ritchie, B.; McDougall, G.; Claxton, J. Complexities of household energy consumption and conservation. *J. Consum. Res.* **1981**, *8*, 233–242.

55. Fritzsche, D.J. An analysis of energy consumption patterns by stage of family life cycle. *J. Market. Res.* **1981**, *18*, 227–232.

56. Tonn, B.; Berry, L. Determinants of participation in home energy audit/loan programs: Discrete choice model results. *Energy* **1986**, *11*, 785–795.

57. Zelezny, L.C.; Chua, P.-P.; Aldrich, C. Elaborating on gender differences in environmentalism. *J. Soc. Issues* **2000**, *56*, 443–457.

58. Clark, C.F.; Kotchen, M.J.; Moore, M.R. Internal and external influences on pro-environmental behavior: Participation in a green electricity program. *J. Environ. Psychol.* **2003**, *23*, 237–246.

59. Sardianou, E. Estimating energy conservation patterns of greek households. *Energy Policy* **2007**, *35*, 3778–3791.

60. Olsen, M.E. Public acceptance of consumer energy conservation strategies. *J. Econ.Psychol.* **1983**, *4*, 183–196.

61. Clancy, J.; Roehr, U. Gender and energy: Is there a northern perspective? *Energy Sustain. Dev.* **2003**, *7*, 44–49.

62. Oparaocha, S.; Dutta, S. Gender and energy for sustainable development. *Curr. Opin. Environ. Sustain.* **2011**, *3*, 265–271.

63. Semenza, J.C.; Hall, D.E.; Wilson, D.J.; Bontempo, B.D.; Sailor, D.J.; George, L.A. Public perception of climate change: Voluntary mitigation and barriers to behavior change. *Am. J. Prev. Med.* **2008**, *35*, 479–487.

64. Kennedy, T.; Regehr, G.; Rosenfield, J.; Roberts, S.W.; Lingard, L. Exploring the gap between knowledge and behavior: A qualitative study of clinician action following an educational intervention. *Acad. Med.* **2004**, *79*, 386–393.

65. Courtenay-Hall, P.; Rogers, L. Gaps in mind: Problems in environmental knowledge-behaviour modelling research. *Environ. Educ. Res.* **2002**, *8*, 283–297.

66. Katzev, R.D.; Johnson, T.R. *Promoting Energy Conservation: An Analysis of Behavioral Research*; Westview Press: Boulder, CO, USA, 1987.

67. Curtis, F.A.; Simpson-Housley, P.; Drever, S. Communications on energy: Household energy conservation. *Energy Policy* **1984**, *12*, 452–456.

68. Van Raaij, W.F.; Verhallen, T.M.M. Patterns of residential energy behavior. *J. Econ. Psychol.* **1983**, *4*, 85–106.

69. Biesiot, W.; Noorman, K.J. Energy requirements of household consumption: A case study of The Netherlands. *Ecol. Econ.* **1999**, *28*, 367–383.

70. Brandon, G.; Lewis, A. Reducing household energy consumption: A qualitative and quantitative field study. *J. Environ. Psychol.* **1999**, *19*, 75–85.

71. Holloway, D.; Bunker, R. Planning, housing and energy use: A review, urban policy and research. *Urban Policy Res.* **2006**, *24*, 115–126.

72. O'Neill, B.C.; Chen, B.S. Demographic determinants of household energy use in the United States. *Popul. Dev. Rev.* **2002**, *28*, 53–88.

73. Cunningham, W.H.; Joseph, B. Energy conservation, price increases and payback periods. In *Advances in Consumer Research*; Hunt, H.K., Ed.; Association for Consumer Research: Ann Abor, MI, USA, 1978; Volume 5, pp. 201–205.

74. Benders, R.M.J.; Kok, R.; Moll, H.C.; Wiersma, G.; Noorman, K.J. New approaches for household energy conservation-in search of personal household energy budgets and energy reduction options. *Energy Policy* **2006**, *34*, 3612–3622.

75. Schipper, L.; Bartlett, S.; Hawk, D.; Vine, E. Linking life-styles and energy use: A matter of time? *Annu. Rev. Energy* **1989**, *14*, 273–320.

76. Verhallen, M.M.; van Raaij, W.F. Household behavior and the use of natural gas for home heating. *J. Consum. Res.* **1981**, *8*, 253–257.

77. Black, J.S.; Stern, P.C.; Elworth, J.T. Personal and contextual influences on household energy adaptations. *J. Appl. Psychol.* **1985**, *70*, 3–21.

78. Rehdanz, K. Determinants of residential space heating expenditures in Germany. *Energy Econ.* **2007**, *29*, 167–182.

79. Stern, P.; Gardner, G. Psychological research and energy policy. *Am. Psychol.* **1981**, *36*, 329–342.

80. Walsh, M. Energy tax credits and housing improvement. *Energy Econ.* **1989**, *11*, 275–284.

81. Murphy, P.E.; Staples, W.A. A modernized family life cycle. *J. Consum. Res.* **1979**, *6*, 12–22.

82. Wells, W.; Gubar, G. Life cycle concept in marketing research. *J. Market. Res.* **1966**, *3*, 355–363.

83. Lutzenhiser, L. Social and behavioral aspects of energy use. *Annu. Rev. Energy Environ.* **1993**, *18*, 247–289.

84. Mayer, P.C. Do-it-yourself and energy conservation. *Contemp. Econ. Policy* **1996**, *14*, 116–118.

85. Herberlein, T.A.; Warriner, G.K. The influence of price and attitude on shifting residential electricity consumption from on- to off-peak periods. *J. Econ. Psychol.* **1983**, *4*, 107–130.

86. Sligo, F.X.; Jameson, A.M. The knowledge-behavior gap in use of health information. *J. Am. Soc. Inf. Sci.* **2000**, *51*, 858–869.

87. Staats, H.J.; Wit, A.P.; Midden, C.Y.H. Communicating the greenhouse effect to the public: Evaluation of a mass media campaign from a social dilemma perspective. *J. Environ. Manag.* **1996**, *46*, 189–203.

88. Geller, E.S. Evaluating energy conservation programs: Is verbal report enough? *J. Consum. Res.* **1981**, *8*, 331–335.

89. Schwartz, S.H. *Universals in the Content and Structure of Values: Theoretical Advances and Empirical Tests in 20 Countries*; Academic Press: Orlando, FL, USA, 1992; Volume 25.

90. Schwartz, S.H. Are there universal aspects in the structure and contents of human values? *J. Soc. Issues* **1994**, *50*, 19–45.

91. Schwartz, S.H.; Bilsky, W. Toward a universal psychological structure of human values. *J. Personal. Soc. Psychol.* **1987**, *53*, 550–562.

92. Eagly, A.H.; Chaiken, S. Attitude structure and function. In *Handbook of Social Psychology*; Gilbert, D.T., Fiske, S.T., Lindzey, G., Eds.; McGraw-Hill: New York, NY, USA, 1998; pp. 269–322.

93. Becker, L.J.; Seligman, C.; Fazio, R.H.; Darley, J.M. Relating attitudes to residential energy use. *Environ. Behav.* **1981**, *13*, 590–609.

94. Samuelson, C.D.; Biek, M. Attitudes toward energy conservation: A confirmatory factor analysis. *J. Appl. Soc. Psychol.* **1991**, *21*, 549–568.

95. Seligman, C.; Kriss, M.; Darley, J.M.; Fazio, R.H.; Becker, L.J.; Pryor, J.B. Predicting summer energy consumption from homeowners' attitudes. *J. Appl. Soc. Psychol.* **1979**, *9*, 70–90.

96. Schultz, P.W. New environmental theories: Empathizing with nature: The effects of perspective taking on concern for environmental issues. *J. Soc. Issues* **2000**, *56*, 391–406.

97. Schultz, P.W.; Zelezny, L.C. Reframing environmental messages to be congruent with American values. *Hum. Ecol. Rev.* **2003**, *10*, 126–136.

98. Seligman, C.; Darley, J.M.; Becker, L.J. Behavioral approaches to residential energy conservation. *Energy Build.* **1978**, *1*, 325–337.

99. Cook, S.W.; Berrenberg, J.L. Approaches to encouraging conservation behavior: A review and conceptual framework. *J. Soc. Issues* **1981**, *37*, 73–107.

100. Anker-Nilssen, P. Household energy use and the environment—A conflicting issue. *Appl. Energy* **2003**, *76*, 189–196.

101. Boulstridge, E.; Carrigan, M. Do consumers really care about corporate responsibility? Highlighting the attitude-behaviour gap. *J. Commun. Manag.* **2000**, *4*, 355–368.

102. Flynn, R.; Bellaby, P.; Ricci, M. The 'value-action gap' in public attitudes towards sustainable energy: The case of hydrogen energy. *Sociol. Rev.* **2010**, *57*, 159–180.

103. Huddart-Kennedy, E.; Beckley, T.M.; McFarlane, B.L.; Nadeau, S. Why we don't "walk the talk": Understanding the environmental values/behaviour gap in Canada. *Hum. Ecology Rev.* **2009**, *16*, 151–160.

104. Frederiks, E.R.; Stenner, K.; Hobman, E.V. Household energy use: Applying behavioural economics to understand consumer decision-making and behaviour. *Renew. Sustain. Energy Rev.* **2015**, *41*, 1385–1394.

105. Baumeister, R.F.; Vohs, K.D. *Handbook of Self-Regulation: Research, Theory, and Applications*; Guilford Press: New York, NY, USA, 2004.

106. Carver, C.S.; Scheier, M.F. *On the Self-Regulation of Behavior*; Cambridge University Press: New York, NY, USA, 2001.

107. Latham, G.P.; Pinder, C.C. Work motivation theory and research at the dawn of the twenty-first century. *Annu. Rev. Psychol.* **2005**, *56*, 485–516.

108. Locke, E.A.; Latham, G.P. New directions in goal-setting theory. *Curr. Dir. Psychol. Sci.* **2006**, *15*, 265–268.

109. Locke, E.A.; Latham, G.P. Building a practically useful theory of goal setting and task motivation: A 35-year odyssey. *Am. Psychol.* **2002**, *57*, 705–717.

110. Locke, E.A.; Latham, G.P. What should we do about motivation theory? Six recommendations for the twenty first century. *Acad. Manag. Rev.* **2004**, *29*, 388–403.

111. Steel, P.; König, C. Integrating theories of motivation. *Acad. Manag. Rev.* **2006**, *31*, 889–913.

112. Stern, P.; Dietz, T. The value basis of environmental concern. *J. Soc. Issues* **1994**, *50*, 65–84.

113. Stern, P.; Dietz, T.; Kalof, L. Value orientations, gender, and environmental concern. *Environ. Behav.* **1993**, *25*, 322–348.

114. Schultz, P.W. The structure of environmental concern: Concern for self, other people, and the biosphere. *J. Environ. Psychol.* **2001**, *21*, 327–339.

115. Deci, E.L.; Ryan, R.M. *Intrinsic Motivation and Self-Determination in Human Behavior*; Plenum Press: New York, NY, USA, 1985.

116. De Young, R. Expanding and evaluating motives for envrionmentally responsible behavior. *J. Soc. Issues* **2000**, *56*, 509–526.

117. Pelletier, L.G. A Motivational analysis of self-determination for pro-environmental behaviors. In *Handbook of Self-Determination Research*; Deci, E.L., Ryan, R.M., Eds.; University of Rochester Press: Rochester, NY, USA, 2002; pp. 205–232.

118. Pelletier, L.G.; Sharp, E. Persuasive communication and proenvironmental behaviours: How message tailoring and message framing can improve the integration of behaviours through self-determined motivation. *Can. Psychol.* **2008**, *49*, 210–217.

119. Lindenberg, S. Intrinsic motivation in a new light. *Kyklos* **2001**, *54*, 317–342.

120. Lindenberg, S. Social rationality *versus* rational egoism. In *Handbook of Sociological Theory*; Turner, J.H., Ed.; Kluwer Academic/Plenum Publishers: New York, NY, USA, 2001; pp. 635–668.

121. Lindenberg, S. Prosocial Behavior, Solidarity and Goal-Framing Processes. In *Solidarity and Prosocial Behavior*; Fetchenhauer, D., Flache, A., Buunk, B., Lindenber, S., Eds.; Kluwer: Amsterdam, The Netherlands, 2006; pp. 23–44.

122. Lindenberg, S.; Steg, L. Normative, gain and hedonic goal frames guiding environmental behavior. *J. Soc. Issues* **2007**, *63*, 117–137.

123. Sheeran, P.; Abraham, C. Mediator of moderators: Temporal stability of intention and the intention-behavior relation. *Pers. Soc. Psychol. Bull.* **2003**, *29*, 205–215.

124. Sheeran, P. Intention-behavior relations: A conceptual and empirical review. *Eur. Rev. Soc. Psychol.* **2002**, *12*, 1–36.

125. Webb, T.L.; Sheeran, P. Does changing behavioral intentions engender behavior change? A meta-analysis of the experimental evidence. *Psychol. Bull.* **2006**, *132*, 249–268.

126. Schwartz, S.H. Normative influences on altruism. In *Advances in Experimental Social Psychology*; Berkowitz, L., Ed.; Academic Press: New York, NY, USA, 1977; Volume 10, pp. 221–279.

127. Wiidegren, O. The new environmental paradigm and personal norms. *Environ. Behav.* **1998**, *30*, 75–100.

128. Harland, P.; Staats, H.; Wilke, H.A.M. Explaining proenvironmental intention and behavior by personal norms and the theory of planned behavior. *J. Appl. Soc. Psychol.* **1999**, *29*, 2505–2528.

129. Kallgren, C.A.; Reno, R.R.; Cialdini, R.B. Focus theory of normative conduct: When norms do and do not affect behavior. *Pers. Soc. Psychol. Bull.* **2000**, *26*, 1002–1012.

130. Midden, C.J.H.; Ritsema, B.S.M. The meaning of normative processes for energy conservation. *J. Econ. Psychol.* **1983**, *4*, 37–55.

131. Hummel, C.F.; Levitt, L.; Loomis, R.J. Perceptions of the energy crisis who is blamed and how do citizens react to environment-lifestyle trade-offs? *Environ. Behav.* **1978**, *10*, 37–88.

132. Loewenstein, G.; Thaler, R.H. Anomalies: Intertemporal choice. *J. Econ. Perspect.* **1989**, *3*, 181–193.

133. Thaler, R.H. Some empirical evidence on dynamic inconsistency. *Econ. Lett.* **1981**, *8*, 201–207.

134. Thaler, R.H. Toward a positive theory of consumer choice. *J. Econ. Behav. Organ.* **1980**, *1*, 39–60.

135. Cialdini, R.B.; Trost, M.R. Social influence: Social norms, conformity and compliance. In *The Handbook of Social Psychology*, 4th ed.; Gilbert, D.T., Fiske, S.T., Lindzey, G., Eds.; McGraw-Hill: New York, NY, USA, 1998; Volumes 1 and 2, pp. 151–192.

136. Feldman, D.C. The development and enforcement of group norms. *Acad. Manag. Rev.* **1984**, *9*, 47–55.

137. Cialdini, R.B. *Influence: Science and Practice*, 5th ed.; Pearson Educational Incorporated: Upper Saddke River, NJ, USA, 2003.

138. Cialdini, R.B.; Kallgren, C.A.; Reno, R.R. A focus theory of normative conduct: A theoretical refinement and reevaluation of the role of norms in human behavior. *Adv. Exp. Soc. Psychol.* **1991**, *24*, 201–234.

139. Turner, J.C. *Social Influence*; Brooks/Cole: Pacific Grove, CA, USA, 1991.

140. Cialdini, R.B. Crafting normative messages to protect the environment. *Curr. Dir. Psychol. Sci.* **2003**, *12*, 105–109.

141. Allcott, H. Social norms and energy conservation. *J. Public Econ.* **2011**, *95*, 1082–1095.

142. Allcott, H.; Mullainathan, S. *Behavioral Science and Energy Policy*; Ideas42: Cambridge, MA, USA, 2010.

143. Allcott, H.; Mullainathan, S. *External Validity and Partner Selection Bias*; NBER Working Paper No. 18373; National Bureau of Economic Research (NBER): Cambridge, MA, USA, 2012.

144. Ayres, I.; Raseman, S.; Shih, A. *Evidence from Two Large Field Experiments that Peer Comparison Feedback Can Reduce Residential Energy Usage*; NBER Working Paper No. 15386; National Bureau of Economic Research (NBER): Cambridge, MA, USA, 2009.

145. Costa, D.L.; Kahn, M.E. *Energy Conservation "Nudges" and Environmentalist Ideology: Evidence from a Randomized Residential Electricity Field Experiment*; NBER Working Paper No. 15939; National Bureau of Economic Research (NBER): Cambridge, MA, USA, 2010.

146. Nolan, J.M.; Schultz, P.W.; Cialdini, R.B.; Goldstein, N.J.; Griskevicius, V. Normative social influence is underdetected. *Pers. Soc. Psychol. Bull.* **2008**, *34*, 913–923.

147. Kahneman, D.; Tversky, A. Choices, values and frames. *Am. Psychol.* **1984**, *39*, 341–350.

148. Eccles, J.S.; Wigfield, A. Motivational beliefs, values, and goals. *Annu. Rev. Psychol.* **2002**, *53*, 109–132.

149. Latham, G.P. *Work Motivation: History, Theory, Research, and Practice*; Sage Publications, Inc.: Thousand Oaks, CA, USA, 2007.

150. Weiner, B. *Human Motivation: Metaphors, Theories, and Research*. Sage: Newbury Park, CA, USA, 1992.

151. Fishbein, M.; Ajzen, I. *Belief, Attitude, Intention, and Behavior: An Introduction to Theory and Research*; Addison-Wesley: Reading, MA, USA, 1975.

152. De Groot, J.I.M. Value orientations to explain beliefs related to environmental significant behavior: How to measure egoistic, altruistic, and biospheric value orientations. *Environ. Behav.* **2007**, *40*, 330–354.

153. Bruni, C.M.; Schultz, P.W. Implicit beliefs about self and nature: Evidence from an IAT game. *J. Environ. Psychol.* **2010**, *30*, 95–102.

154. Splash, C. Non-economic motivation for contingent values: Rights and attitudinal beliefs in the willingness to pay for environmental improvements. *Land Econ.* **2006**, *82*, 602–622.

155. Camerer, C.F.; Loewenstein, G.; Rabin, M. *Advances in Behavioral Economics*; Princeton University Press: Princeton, NJ, USA, 2004.

156. Kahneman, D. Maps of bounded rationality: Psychology for behavioral economics. *Am. Econ. Rev.* **2003**, *93*, 1449–1475.

157. Pesendorfer, W. Behavioral economics comes of age: A review essay on advances in behavioral economics. *J. Econ. Lit.* **2006**, *44*, 712–721.

158. Pollitt, M.G.; Shaorshadze, I. The role of behavioural economics in energy and climate policy. In *Handbook on Energy and Climate Change*; Fouquet, R., Ed.; Edward Elgar: Cheltenham, UK, 2013; pp. 523–546.

159. McMakin, A.H.; Malone, E.L.; Lundgren, R.E. Motivating residents to conserve energy without financial incentives. *Environ. Behav.* **2002**, *34*, 848–863.

160. Geller, E.S. The challenge of increasing proenvironmental behavior. In *Handbook of Environmental Psychology*; Bechtel, R.B., Churchman, A., Eds.; Wiley: New York, NY, USA, 2002; pp. 525–540.

161. Osbaldiston, R.; Schott, J.P. Environmental sustainability and behavioral science: A meta-analysis of proenvironmental behavior experiments. *Environ. Behav.* **2012**, *44*, 257–299.

162. Stern, P. Information, incentives, and proenvironmental consumer behavior. *J. Consum. Policy* **1999**, *22*, 461–478.

163. Winkler, R.C.; Winett, R.A. Behavioral interventions in resource conservation: A systems approach based on behavioral economics. *Am. Psychol.* **1982**, *37*, 421–435.

164. Yates, S.; Aronson, E. A social psychological perspective on energy conservation in residential buildings. *Am. Psychol.* **1983**, *38*, 435–444.

165. Geller, E.S.; Berry, T.D.; Ludwig, T.D.; Evans, R.E.; Gilmore, M.R.; Clarke, S.W. A conceptual framework for developing and evaluating behavior change interventions for injury control. *Health Educ. Res. Theory Pract.* **1990**, *5*, 125–137.

166. Schultz, P.W. Strategies for promoting proenvironmental behavior: Lots of tools but few instructions. *Eur. Psychol.* **2014**, *19*, 107–117.

Can the BestGrid Process Improve Stakeholder Involvement in Electricity Transmission Projects?

Nadejda Komendantova, Marco Vocciante and Antonella Battaglini

Abstract: The European Union has set ambitious targets for deployment of renewable energy sources to reach goals of climate change mitigation and energy security policies. However, the current state of electricity transmission infrastructure is a major bottleneck for further scaling up of renewable energy in the EU. Several thousands of kilometers of new lines have to be constructed and upgraded to accommodate growing volumes of intermittent renewable electricity. In many countries, construction of electricity transmission projects has been delayed for several years due to concerns of local stakeholders. The innovative BESTGRID approach, reported here, brings together transmission system operators (TSOs) and non-governmental organizations (NGOs) to discuss and understand the nature of stakeholder concerns. This paper has three objectives: (1) to understand stakeholder concerns about the deployment of electricity transmission grids in four pilot projects according to five guiding principles: need, transparency, engagement, environment, and impacts on human health as well as benefits; (2) to understand how these principles can be addressed to provide a basis for better decision-making outcomes; and (3) to evaluate the BESTGRID process based on feedback received from stakeholders and the level of participation achieved according to the ladder of Arnstein. This paper goes beyond a discussion of "measures to mitigate opposition" to understand how dialogue between TSOs and the public—represented mainly by NGOs and policy-makers—might lead to a better decision-making process and more sustainable electricity transmission infrastructure deployment.

Reprinted from *Energies*. Cite as: Komendantova, N.; Vocciante, M.; Battaglini, A. Can the BestGrid Process Improve Stakeholder Involvement in Electricity Transmission Projects? *Energies* **2015**, *8*, 9407-9433.

1. Introduction

Goals of Climate Change Mitigation and Energy Security Policies

Climate security goals require a significant reduction of the level of CO_2 and other greenhouse gas (GHG) emissions to mitigate the potentially catastrophic risks of climate change [1]. However, anthropogenic GHG emissions continued to increase in the period from 2000 to 2010 with CO_2 emissions from fossil fuel combustion and industrial processes contributing about 78% of total GHG emissions. Energy supply contributes to 47% of this increase [2]. To achieve climate security goals, emissions should be cut by 50% globally, and by 80% in industrialized countries by 2050 [3]. To achieve the mitigation scenario of 450 ppm CO_2 by 2100, reductions of over 90% of CO_2 emissions from energy supply—compared to 2010 levels—are needed in the period between 2040 and 2070 [2]. In 2011, the European Commission published a roadmap to achieve a reduction of GHGs by at least 80% by 2050 [4]. The roadmap foresees five alternative pathways; for all of these, renewable energy generation plays a significantly stronger role than it does today.

Furthermore, the goals of European energy security policy require that energy generation is restructured toward a greater share of renewable and low carbon sources [5,6]. In October 2014, EU leaders agreed on the 2030 policy framework for climate and energy, which set a GHG reduction target of 40% compared to 1990, as well as an increase in the share of renewable energy to at least 27% of the EU's energy consumption by 2030. Security of energy supply is a high priority in the European policy-making process and the electrical power transmission system is an essential element to achieving this goal. The 2030 framework sets policy targets and different aspects and dimensions of energy security are discussed, from "availability of sufficient supplies at affordable price" [7] to "uninterrupted physical availability on the market of energy products at a price which is affordable to consumers" [8]. According to the International Energy Agency, energy supply is secure if it is adequate, affordable, and reliable [9]. According to [10] energy security has several dimensions, including acceptability and social and environmental stewardship. This aspect of energy security is often associated with unacceptable environmental and social impacts and a significant number of existing scholarly articles on energy security include environmental performance of energy systems in their definition of energy security [11].

Today, the grid is considered to be one of the major barriers to a further increase in renewable energy sources. There are also reasons to believe that to guarantee a reliable and secure electricity supply, construction of a new electricity transmission infrastructure is needed, as well as upgrading the existing one. In particular, cross-country interconnectors are required as well as the deployment of smart grid technologies to manage and optimize different energy supply options. According to the Ten-Year Network Development Plan [12], which was developed by the power transmission system operating body ENTSO-E, the construction of roughly extra 52,300 km of high voltage power lines across Europe is required. The extension of electricity grids by only 1.3% per year will allow for an increase of 3% of generation capacity and integration of 125 GW of renewable energy sources [13].

Currently, several options to provide the required energy transition are under discussion. These include key elements such as geographically dispersed units, integration of different renewable sources, storage, integration with demand (including adapted demand), and decentralized generation in micro-grids. Additional research about the need for large-scale transmission projects is required, especially in light of the growing importance of distributed or decentralized energy generation (DG) and its consequences for the development of the grid and for high voltage transmission. The term DG applies to electric power generation without distribution networks or on the customer side of the network. Even though DG might have significant benefits, such as reduction of network losses, it could also have additional costs; for example, the protection system may need to be redesigned [14]. Therefore both options, centralized and decentralized energy generation, should be considered.

Currently the process of constructing, extending, and upgrading electricity grids in Europe is extremely slow, and stakeholders question the need for infrastructure projects as well as the assumptions used to determine this need. Claims that new developments are needed to deliver the energy future they favor are also met with skepticism [15–19]. Failing to reach an agreement with local stakeholders on the deployment and siting of projects can cause lengthy and costly delays to the planning process and can even jeopardize the project altogether [20,21]. However, despite substantial efforts in recent years (TEN-E Legislation and national legislation in several EU countries requesting participatory activities) for addressing stakeholders' concerns, scientific evidence of the efficiency of action is still missing.

Taking into account the reality of existing conditions, we examine participatory governance in climate change mitigation and energy security, using existing evidence regarding stakeholders concerns. We formulate and address the following research questions:

- What are the main stakeholder concerns about deployment of electricity transmission grids in Europe?
- What are successful actions to address these concerns?
- What level of participation can be achieved in electricity transmission infrastructure project siting in Europe?

2. Background

2.1. Stakeholders Concerns beyond Not-in-My-Backyard Behavior

Today, stakeholders' concerns about new transmission line infrastructure are different, compared to when the existing architecture was built in the early 20th century. At that time, infrastructure projects were viewed by many as representing technological progress, providing jobs and contributing to increased levels of social wellbeing [22]. Now transmission line infrastructure is viewed by many as a blight on the landscape and a threat to biodiversity, with negative impacts on property values [23]. Furthermore, in past decades a series of technological accidents and environmental disasters negatively influenced stakeholder perceptions and acceptance of new infrastructure [24]. Environmental NGOs have contributed to this discourse by explaining and highlighting the impact that technologies and infrastructure projects have on nature and human health. Today, local citizen groups and NGOs are able to quickly mobilize and articulate opposition towards or support of infrastructure projects [24]. Such a change in views can be partially explained by the distinction proposed by Inglehart (1995) on materialist and post-materialist values; he argued that value priorities shifted significantly in the last decades of the 20th century, from concerns over economic and physical security towards quality of life and freedom of self-expression [25].

It is, therefore, important to understand stakeholders' concerns [20], which are often connected with externalities caused by grid transmission projects, such as the visual, health, and environmental effects as well as perceived loss of property value [26]. These concerns are generally defined as not-in-my-backyard (NIMBY) behavior [27]. The classical understanding of when NIMBYism will arise is when the advantages of the renewable energy sources (RES) projects are perceived to be only at global or national levels, while the impacts of such projects mainly affect only the local population and environment [28].

Pronouncing a local response to new development as NIMBYism can itself be biased, and it is necessary to bear in mind that the NIMBY concept has been frequently questioned, for example by Wolfsink (2012) [29]. In a larger sociopolitical context, NIMBY terminology is frequently used to explain local opposition as selfishness, irrationality, and ignorance. If a researcher applies this concept, they also seem to classify local opposition as such [27]. Often concerns go beyond NIMBY and are about procedural justice, such as the lack of opportunity to express views; this feeling that "voices are not heard" can lead to local opposition [30].

There is also a great difference between accepting deployment of RES in general, and acceptance of the reality of RES generation, an electricity transmission project, or a company on the ground [31]. Often the attitudes of inhabitants towards RES projects can be influenced by factors such as trust in the company, the perceived need for the project, or impacts during and after the construction phase. Opposition to the project can be generated by inhabitants' skepticism towards the company or the authority that wants to develop the project. There is evidence that many protesting inhabitants do so because of their opposition to a company's energy policy or because of the way the project was planned and implemented, rather than specifically opposing infrastructure itself [32]. Inhabitants' skepticism may also be considered as a place-protective action, caused by a reaction to developments that might disrupt existing emotional attachments and threaten place-related identity processes [33]. Public opposition to a project also often focuses on environmental impacts during the installation and construction phases of the project [28].

However, the most frequent concern is local stakeholders questioning the need for a project. In particular, local communities often question the necessity of large-scale infrastructure to address a global problem such as climate change [34]. In the case of electricity transmission projects, stakeholders often want to understand the purpose and possible alternatives—for example, decentralized energy generation [29], which foresees deployment of energy generation near the point of use [35], therefore potentially matching supply with demand. The results of a survey conducted with stakeholders at local and national levels do not support the assumption about a gap in responses but rather show the value of adoption of the place-based approach [36].

2.2. Growing Importance of Public Participation

According to the Universal Declaration of Human Rights, people have the right to participate in decision-making that affects their lives [37]. There has been no clear evidence of the efficacy of participatory processes in addressing stakeholders' concerns. More specifically, there are multiple ways to design and run a participatory process, but no clear rules to guide stakeholders in choosing the most effective strategies. The traditional view was that decisions regarding technical issues should be concentrated in the hands of experts and scientists [38]. For example, some scholars suggested that stakeholders' involvement in complex decisions may be limited with regards to capacity and knowledge; for example, the ability of stakeholders to understand concepts of "uncertainty" and "variability" might be limited, therefore leaving the leading role in the decision-making process to scientists would be best [39]. However, the need of local stakeholders' participation is being increasingly recognized, as expert knowledge can also be limited, particularly in relation to local knowledge on the ground. Moreover, in some cases, experts disagree among themselves [40] and local knowledge can become decisive for conflict resolution.

Evidence shows that the conceptual approach towards "social acceptance," which dominates in the literature, can be biased and misleading if it is understood as passive agreement to something that one cannot change [41]. In the past, project developers frequently used this understanding of acceptance to assess the efficacy of the participatory process and therefore provide stronger arguments in favor of the project. However, decision-making in participatory governance processes can even result in the cancellation of the project or deliver new alternatives, like construction of alternative infrastructure or use of a different geographic location [22].

The terminology "acceptance" is multi-layered and multi-dimensional. In relation to energy innovation, acceptance was studied by Wüstenhagen *et al.* [34] and includes a number of potential attitudes towards energy innovation (including RES), such as apathy, passive acceptance, approval, and active support. Such acceptance can take place in policy decision-making, markets, or communities. The definition of "acceptance" by Wüstenhagen is currently used by most researchers to introduce the institutional character of acceptance. It also includes sociopolitical acceptance of new institutional frameworks, for example about how to re-structure decision-making processes towards necessary involvement and participation, as well as institutional shifts in the energy sector away from top-down, centralized, large-scale, fossil fuel-based energy generation, towards inclusive decision-making.

The term "social acceptance" is a multifaceted concept and refers to acceptance of all relevant stakeholders. Otherwise, we refer to acceptance of a certain group as "acceptance by the local public," "acceptance by energy companies," or "acceptance by local governments". When the term "acceptance" applies to laypeople, such as inhabitants of affected communities or citizen of certain countries, we speak about "public opposition" and "public involvement." It should be noted that social acceptance referring to stakeholders in favor of infrastructure has been studied much less frequently [26]. The position of the proponents is not often questioned; those opposing the development are seen as the issue.

It remains a challenge to understand how the broad concept of energy security, the related need for grid infrastructure, and the complexity of the entire system can be embedded and based on participative democratic practices leading to better outcomes. Research exists on the factors that influence local communities' acceptance of new power lines in light of a transition to low carbon energy generation [41]. However, available evidence is limited, and generally linked to "willingness to pay" research.

Indeed, from the policy perspective, the assessment of the need for large-scale electricity transmission infrastructure is included in scenario frameworks and national power grid plans, which should be consistent with European grid plans and scenarios developed by the grid operator. However, only in a few European countries are these assessments of need the subject of consultation with stakeholders. Furthermore, evidence has shown [41] that research and policy processes should abandon the prescriptive assumption that we know *a priori* who is "right" about the need for a project. For example, a meta-analysis on existing research about public attitudes to wind power showed that the entire discourse about wind infrastructure siting was dominated by five key assumptions: the majority of the public supports wind power, opposition to wind power is deviant, opponents are ignorant and misinformed, the reason for understanding opposition is to overcome it, and trust is a key. This research showed that trust is indeed a key issue but that a greater trust must also be given to stakeholders and to their knowledge during the decision-making process [42].

In view of the evidence mentioned above, and the need to address increasing public opposition, a number of European countries have developed measures to address citizens' concerns and translated them into legal requirements aimed at improving and defining regulation for stakeholder engagement. Stakeholders' feedback has led to a more careful evaluation of the standards for minimum distances to settlements, clear and robust rules for nature protection and mitigation measures, more transparent planning processes, and a greater ability to influence them [16]. These are all important elements contributing to the legitimacy of the infrastructure siting process and the acceptability of its outcomes.

In order to understand the outcomes of the participatory process, it is necessary to evaluate the process itself. Usually, such a process incorporates certain characteristics that might have an influence on its effectiveness [43]. Most of these elements discussed in the literature are procedural rather than substantive [44], therefore it is easier to evaluate how effective the entire process is, rather than the efficiency of single measures.

One of the most well-known methodologies used to evaluate the participatory process was developed by Arnstein (1969) [45] and Rau (2012) [46]. According to Arnstein's ladder, there are eight elements of participation, classified into three levels (Figure 1). These levels represent the range of intensity in citizen engagement.

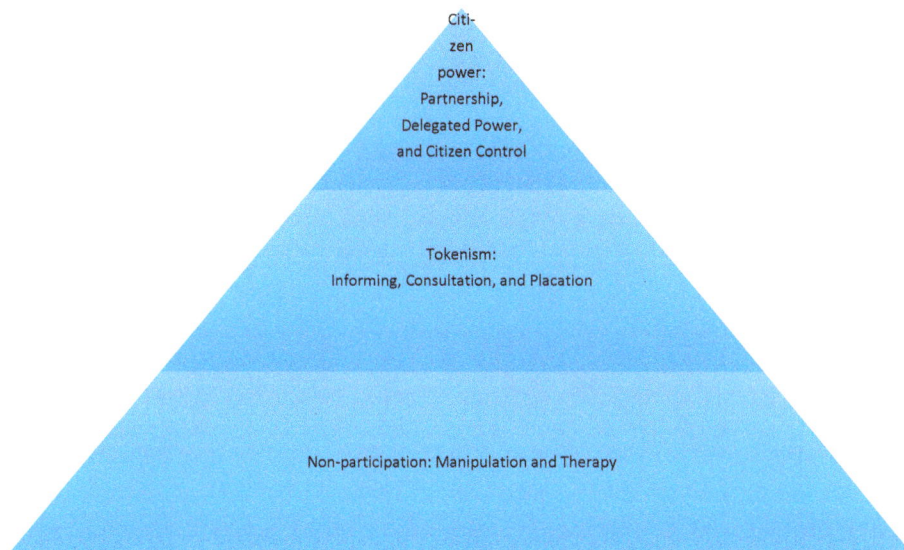

Figure 1. Ladder of participation [45,46].

At the bottom of the ladder, the *"non-participation"* level includes two very different elements: manipulation and therapy. Manipulation occurs when powerful actors use less powerful and/or disadvantaged population groups to achieve their goals. For example, energy companies in previous centuries sited copper mining infrastructure in developing countries [47]. The therapy element is instead applied to cure a problem or to educate the stakeholders involved. This is a sort of "Decide–Argue–Defend" model that achieves public support largely through public relations for a proposed infrastructure development plan.

The second level, *"tokenism,"* includes informing, consultation, and placation. Informing is one of the steps necessary to legitimize the process. However, this level only allows for a one-way flow of information and participants are not provided with channels for feedback. Public participation is excluded as there is no opportunity for the public to influence the decision-making process. The public is only informed about decisions that have already been made. Unlike in *therapy*, the intention here is to share information, providing a certain level of transparency about the project and decision-making processes.

At the same level, *"consultation"* represents the opportunity for the public to provide input during the decision-making process. Local residents are asked about their concerns, and these concerns should be considered by the project developers. Each input also receives a response explaining how it was considered in the final decision. In this phase, a set of tools to collect stakeholders' concerns is suggested and includes data collection via surveys, neighborhood meetings, and public enquiries.

Placation is the next step on the ladder, in which special cooperation, mainly with representatives of stakeholder organizations, is put in place by, for example, setting up advisory roles. In this case, project developers retain the right to judge the legitimacy or feasibility of the advice. At this level stakeholders are invited to participate in the process before any decision is taken and are provided with multiple opportunities to influence the decision-making process.

The third level is about citizen power and includes partnership, delegated power, and citizen control. In *partnerships*, the power is redistributed through negotiation between citizens and power holders who share planning and decision-making responsibilities. The have-not, poor, or powerless citizen can negotiate and engage in trade-offs with power holders through, for example, joint committees. *Citizen control* occurs when citizens handle the entire process of planning, policymaking, and management of the project with no intermediaries. At this level citizens also have the majority of the decision-making seats in the committees, or full managerial power.

2.3. BESTGRID Approach

This innovative approach was developed in the framework of the BESTGRID project (www.bestgrid.eu). Its goal is to deploy electricity grids with the minimum possible impact on the environment. The approach also stresses the need to involve stakeholders in the decision-making process as early as possible. This approach was also formulated in the European Grid Declaration (EGD) on Electricity Network Development and Nature Conservation in Europe [48], which defines a set of principles on how to build power lines without harming nature. The Declaration was developed under the guidance of the Renewables Grid Initiative (RGI) secretariat and signed by more than 30 large institutions across Europe.

In 2013 RGI was granted funds within the European Commission's "Intelligent Energy Europe" program to implement the verbal commitments contained in the EGD in real projects on the ground. This resulted in the project BESTGRID—testing better practices. The consortium consists of five TSOs, Elia, TenneT, 50Hertz, National Grid, and Terna; national NGOs Germanwatch and BirdLife International; several local NGOs, such as Bond Beter Leefmilieu Vlaanderen vzw (BBL), Natagora, Fédération Inter-Environnement Wallonie (IEW), DUN, NABU; scientific research institute IIASA; and RGI as coordinator. The BESTGRID project is based on the involvement of stakeholders, such as NGOs, in joint development and implementation of action plans to deal with issues of public acceptance and environmental protection by siting electricity transmission infrastructure. It suggests that the joint efforts of four TSOs, as well as national and local NGOs—backed by academia—could contribute to a higher level of stakeholder engagement and a better understanding of stakeholders' concerns, using five pilot grid development projects as case studies. NGOs and academic institutions help TSOs improve their stakeholder engagement and environment protection by jointly developing action plans. However, as all pilot projects covered by BESTGRID are still ongoing, it is uncertain to what extent the feedback from stakeholders will influence the outcomes of the projects.

All the BESTGRID pilot projects are high-voltage transmission lines. These are usually operated by one national transmission system operator. Germany is an exception, with four transmission systems operators. There are currently no high-voltage lines operated by third parties in Europe. The only exception is merchant lines—commercial investments that are not regulated assets and not part of remuneration schemes, which foresee a fixed rate of return. Usually merchant lines are leased to grid operators or private actors for specific purposes, so their numbers are very limited. In this paper we discuss high voltage lines in AC and DC up to a voltage level of 380 KW.

Superconductor high voltage (SCHV) technologies are potential alternatives and these innovations are already in operation in a number of places, such as Korea and China. Even though this innovation is not widely applied, it could be scaled up soon. Indeed, this is an issue of sociopolitical acceptance relevant for high-voltage (HV) transmission, as SCHV might substantially reduce environmental impact, particularly in terms of visibility and electro-magnetic fields. However, the existing regulatory framework in Europe is not yet ready for this innovation, and therefore we do not discuss it in this paper.

Here, we examine the concerns of stakeholders in light of the following five subject areas or guiding principles: need, engagement, transparency, environment, and benefit. These principles were identified in the scientific literature as being crucial to understanding existing support of or opposition to the siting of infrastructure projects [26,35,42].

We have already discussed "need"; the principle *"engagement"* refers to the need to engage civil society, in particular local communities as well as other interested stakeholders. Engagement is a two-way communication process, which foresees not only the provision of information but also the collection of views, perceptions, and concerns on a particular project and its relation to the broader achievement of the energy transition. Engagement goes beyond simple consultation, which is also a two-way communication process, and implies not only that information is exchanged constantly between public and project developers but also that there is dialogue and negotiation between all parties. This allows for changing of opinions and influences decision-making [39]. Evidence shows that the engagement process should start as early as possible [49]. However, the optimum time for engagement is still unclear because the concept of "as early as possible" can result in a process that starts "too early" to be meaningful to stakeholders, whereas delaying the process could result in stakeholders feeling that they were consulted "too late." Hänlein (2015) described the risk of missing the optimum time for engagement as the "participatory paradox" [50].

The principle "*transparency*" states that information about the project and decision-making should be transparent at each stage of the process, and the potential for stakeholders to influence the outcomes should be made clear. Transparent participatory processes should be built on discussions and inputs from local stakeholders and involve views and concerns regarding possible alternatives, technologies, impacts, costs, compensations, and benefit-sharing options.

The principle "*environment*" is based on the fact that, in Europe, environmental protection law is well developed and rather strong. However, the quality and accuracy of implementing environmental protection laws vary substantially from country to country. The value and benefits of fully implementing existing legislations are not fully appreciated across Europe. This principle stresses the need to understand the views and concerns of different stakeholders regarding environmental protection and to address these concerns by planning and constructing new infrastructure. This planning process should also involve feedback from local stakeholders to help avoid harming the environment. Strategic environmental planning also provides an opportunity to consider local stakeholders' concerns related to the possible impacts on human health—such as electro-magnetic fields (EMF)—and to design routes that avoid protected areas and human settlements.

The principle "*benefit*" investigates the possibilities of providing benefits or compensations to local communities [51] as well as creating additional co-benefits for local communities.

The BESTGRID project provides an opportunity to involve a select group of targeted stakeholders in each pilot project. The process of selecting stakeholders was very different across all pilot projects. In the case of Stevin and Waterloo-Braine's Alleud connections of Elia, the local NGOs BBL and Fédération Inter-Environnement Wallonie (IEW) firstly conducted stakeholder mapping to understand who the active stakeholders were. Then the NGOs invited these stakeholders for workshops and round table discussions. In pilot projects by TenneT and 50Hertz, stakeholder events were organized in cooperation with local NGOs and invitations to participate were sent out to a broader range of stakeholders. In the pilot project of National Grid, invitations were sent to a number of stakeholders who already had experience in cooperating with the TSO. These were mainly civil society organizations' representatives and residents of affected communities, the private sector, and authorities (Table 1). BESTGRID also provided funds to enable national and local NGOs, such as Germanwatch and BirdLife, to be directly involved in designing the TSOs' activities on engagement and the environment by providing comments on action plans and guidance on how to address environmental protection and engagement concerns.

Table 1. Realization of the BESTGRID approach in four pilot projects.

	Elia	TenneT	50Hertz	National Grid
Need (understanding needs and concerns of stakeholders)	Stakeholder mapping. Interviews to collect concerns.	Stakeholder mapping. Survey to collect concerns.	Stakeholder mapping. Conducted by public relations agency.	Stakeholder mapping. Interviews to collect concerns.
Transparency	Workshop with local residents to provide information on planning.	Public information events, media campaign.	Workshops in two communities to provide information on planning.	Mini-workshops in two communities with civil society organizations.
Engagement	NGOs providing comments on action plans. Round table discussions.	NGOs providing comments on action plans. Round table discussions.	NGOs providing comments on action plans. Round table discussions.	NGOs providing comments on action plans. Mini-workshops.
Environment	Round table discussion between TSO and NGOs to discuss environmental issues.	Information markets.	Round table discussion between TSO and NGOs to discuss environmental issues. Media campaign on EMF, mobile information office with EMF measurements.	Mini-workshop with NGOs.
Benefit or compensation	Not foreseen at the current stage.	Not foreseen at the current stage.	Not foreseen at the current stage.	Not foreseen at the current stage.

Questions about the "need" for the project were addressed in a series of workshops with stakeholders, where local residents and representatives of organizations could express their concerns and ask TSOs directly for additional information. Round table discussions were mainly used to address the issues related to the "environment" and "engagement"; environmental NGOs were also able to provide their input and advice on environmental protection during the planning phase of the projects. The TSO 50Hertz also addressed the issue of the EMF and its impact on human health.

The process also included several project meetings where the BESTGRID team members discussed the progress of the action plan's implementation and provided their comments and advice. For instance, NGOs advised TSOs on how to approach local residents. The decisions on actions were taken and realized jointly by TSOs and NGOs.

3. Methodology

3.1. Data on Stakeholders' Concerns and Feedback Regarding the BESTGRID Approach

Data were collected by the joint efforts of the entire BESTGRID team, the International Institute for Applied Systems Analysis (IIASA), and the Renewables Grid Initiative (RGI), as well as national and local NGOs such as Germanwatch, BirdLife, IEW, BBL, DUN, and NABU. IIASA developed data collection protocols, provided input and methodological guidance on different data collection methods—questionnaires, surveys, and feedback forms—and analyzed the results. IIASA, Germanwatch, and RGI provided records of public information events and round table discussions for all pilot projects. Locally involved NGOs such as NABU, DUN, IEW, and BBL provided feedback forms about the BESTGRID project.

In Belgium, IEW and BBL, in cooperation with IIASA, conducted extensive in-depth interviews with local stakeholders. The interviews were based on an open-ended question survey intended to identify concerns of key stakeholders, such as impacts on the environment and human health, as well as land use and energy supply planning issues, for both overhead lines and underground cables. Interviews also provided an opportunity to collect feedback about the BESTGRID project. Interviews lasted approximately two hours and were recorded. Scripts from all the interviews were provided to IIASA in French for analysis of recorded concerns.

As the major goal of the research was to understand stakeholders' concerns, we mainly collected views, feedback, and concerns from organized stakeholders representing civil society, so that we could address the issues of social acceptance or opposition to the projects. All interviews, focus group discussions, and observations were conducted with consideration of ethical issues, such as informed consent. Participants were informed of the goals of the BESTGRID project, and were assured that their concerns, views, and opinions would be the subject of scientific research. Chatham House rules were applied to guarantee the anonymity of participants.

We applied content analysis and data mining to understand which concerns were expressed most frequently. Content analysis is often used for interpretation of documents, which are used for communication processes on infrastructure siting. This method is used to identify evidence from texts such as the frequency of most used words [52]. The material for analysis came from reports about public information events produced by the TSOs and NGOs, as well as from interviews conducted by the NGOs and observations on the site of public information events.

In order to map the interactions between NGOs and TSOs in the pilots of the BESTGRID project, we applied the framework on participation development, as proposed by Arnstein and Rau, to map different levels of stakeholders' participation. We applied this framework to understand how cooperation between TSOs and NGOs could foster public acceptance by minimizing conflicts and, therefore, enhancing the efficiency of the process of deployment of electricity transmission grids.

3.2. Case Study Method

The multiple cases study was used as our research design [53]. The case study allows holistic analysis of events, decisions, projects, policies, institutions, or other systems [54]. This method has been applied in a range of disciplines such as psychology, sociology, economics, political science, geography, and medical science. It has gained popularity recently in testing hypotheses and researching questions that are difficult to answer with existing statistical methods. Focusing on large-scale infrastructure projects, such as the Channel tunnel between Great Britain and France, the case study method helped to identify biases in the decision-making process regarding large-scale infrastructure projects [55]. Other strengths of the method include in-depth analysis, high conceptual validity, an understanding of the context and process as well as the causes of a phenomenon, linking causes and outcomes, and fostering new hypotheses and new research questions [56].

We developed our research protocol according to the methodology of Yin, which includes multiple cases and components of research design such as questions, assumptions, units of analysis within case studies, data, criteria for interpretation of data, and validation. The case-research protocol creates replication logic providing opportunity for generalization and is described in Komendantova *et al.* [57]. This protocol allowed the various actions by the BESTGRID partners to be transformed into a common framework of actions and guiding principles. This common methodological framework, which categorized actions taken in pilot projects, helps us to understand differences and similarities across the projects.

There are two types of questions related to data collection and stakeholder mapping. The first type relates to all pilots and is therefore called "common to all pilots" questions. The second type relates only to specific actions within certain pilot projects and workshops with local governments (Figure 2). The protocol for data collection is described in Komendantova *et al.* [58].

We conducted a case study analysis of the following pilot projects: Waterloo–Braine-l'Alleud and Stevin connections by Elia, the Bertikow–Pasewalk connection by 50Hertz, NEMO by National Grid, and SUED.LINK by TenneT. All these projects are designed to serve large-scale, centralized energy generation. It is outside the scope of this paper to discuss alternatives like decentralized generation because researchers seldom have the opportunity to observe the participatory process in "real-world" pilot projects. As these projects did not include alternatives such as decentralized generation, the comparison of received primary data from this research with secondary existing data on alternatives would not be scientifically robust.

Until June 2014, *Elia* planned to construct a 150 kV onshore underground cable between the two regions of Braine-l'Alleud and Waterloo in the South of Brussels in Wallonia. This project was needed to guarantee a reliable power supply in view of the projected increasing demand from Waterloo. The project should start from the year 2018 and integrate electricity generated by renewable energies. In 2013, the line between Braine-l'Alleud and Waterloo was selected as a pilot project for BESTGRID. The infrastructure project was expected to cover both urban areas and green zones, thus making it an interesting complementary project to other pilots selected to be part of BESTGRID. In June 2014 all actions of Elia were cancelled because the project was postponed for several years. One of the reasons for this was the publication of new data and forecasts from the distribution system operator regarding the Waterloo zone, which showed that the growth of electricity consumption in the region was stabilizing, and that a number of requests for connections for wind energy were rejected.

The remaining pilot activities were therefore moved to another ongoing project by Elia, the Stevin project, the biggest grid project in Belgium for many years. It will involve the construction of 380 kV electricity lines between Zeebrugge and Zomergem, in the vicinity of Ghent. Out of the 47 km of transmission lines, 10 km should be underground. The project will integrate electricity from offshore wind farms, will connect to the NEMO project—the interconnector between the United Kingdom and Belgium—and also contribute to on-shore decentralized energy and harbor development. The project will cross eight communities and two provinces, traversing very densely

606

populated areas as well as natural and landscape protected areas. The fact that it is planned in an area with very dispersed single-family houses and urban settlements is a major challenge.

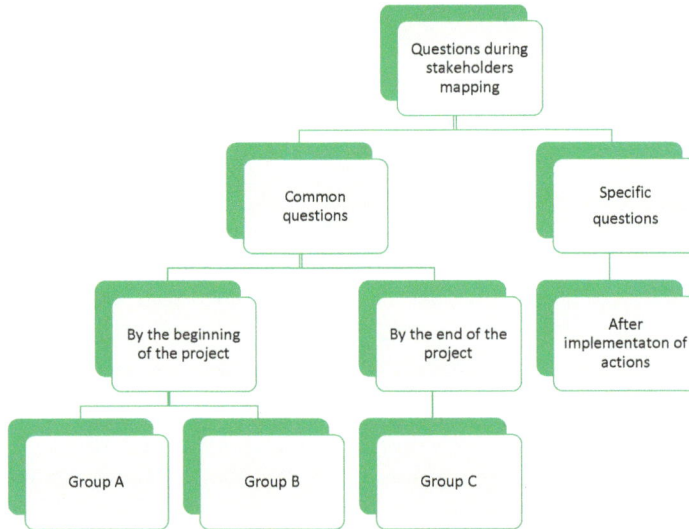

Figure 2. Data collection protocol [58].

TenneT, in cooperation with the German TSO TransnetBW, is currently developing the largest energy transmission infrastructure project in Germany, the SUED.LINK. This power transmission line is around 800 km, with a transmission capacity of 4 GW, and will deliver electricity generated from wind energy in the North Sea to consumers in central and southern Germany, providing a connection to an area between Wilster in Schleswig-Holstein and Grafenrheinfeld in Bavaria. SUED.LINK is the first line to be constructed using high-voltage direct current (HVDC) lines, which enable large volumes of electricity to be transported over long distances. The 800 km line is planned to go through dozens of densely populated regions, thus requiring information to be delivered and discussed with thousands of citizens.

50Hertz is planning a 380 kV overhead line of approximately 29 km between two communities: Bertikow in Brandenburg and Pasewalk in Mecklenburg-Vorpommern. The project will replace the existing 220 kV line to integrate the growing volumes of electricity from renewable energy sources generated in the Uckermark region and Western Pomerania. Currently, wind turbines in Bertikow are generating 330 MW, and this is expected to increase to up to 800 MW by 2020. There are currently 13 alternatives for the corridor, each approximately 1 km wide, adding up to a total of 11 km. Current plans are that the new lines will be constructed along the existing 220 kV and 110 kV lines from 1950, which will be partly replaced or coupled with the new 380 kV line. The need for the new line was legally set by the German authorities in the year 2013 and it is expected to become operational in 2019.

Under the *National Grid* pilot, a retrospective evaluation of the consultation and permitting phases for the project is planned. The chosen pilot is the NemoLink project, a joint venture between National Grid and the Elia group constituting a 130 km interconnection between the United

Kingdom and Belgium. The subsea cable will run from Pegwell Bay in the United Kingdom to Zeebrugge in Belgium and will pass through English, French, and Belgian waters.

4. Results

The review of actions that were planned or carried out in the pilots showed that different participatory elements were used. We classify these elements according to Arnstein's ladder (Table 2). We also classify the concerns of stakeholders according to five subject areas or guiding principles (need, engagement, environment, transparency, and benefit), which are described in the background section.

Table 1. Actions within four pilot projects, according to Arnstein's Ladder.

	Need	Transparency	Engagement	Environment	Benefit
Therapy	TenneT media campaign; Elia, TenneT, 50Hertz, and NG stakeholder mapping	NG: mini-workshops on planning procedures	NG: mini-workshops where stakeholders who had experience of cooperation with the TSO were invited	50Hertz: video about EMF	No
Information	Elia and 50Hertz: workshops with local residents on the need of the project	Elia and 50Hertz: workshops with local residents on the details of planning; TenneT: public information campaign	TenneT, Elia, and 50Hertz: public information events providing information to organized stakeholders about engagement opportunities	NG: mini-workshops on the impacts on the environment; 50Hertz: mobile office information tour on impacts of EMFs	Elia, 50Hertz, TenneT, and NG: workshops and round table discussions on the benefits of the project

Table 2. *Cont.*

	Need	Transparency	Engagement	Environment	Benefit
Consultation	Elia: interviews to collect stakeholder concerns; TenneT: surveys to collect concerns and feedback on public information events	TenneT: public information markets	Elia, 50Hertz, NG, and TenneT: partnership with Germanwatch on stakeholder engagement	50Hertz and Elia: round table discussions on impacts on the environment	No
Placation	No	50Hertz and NG: partnership with Germanwatch on action plans to increase transparency	No	50Hertz and NG: partnership with BirdLife on environmental protection	No
Partnership	No	Elia and TenneT: partnership with Germanwatch on action plans to increase transparency	No	Elia and TenneT: partnership with BirdLife on environmental protection	No
Delegation	No	No	No	No	No
Control	No	No	No	No	No

This table shows that almost all actions foreseen by TSOs fall in the middle of Arnstein's ladder, varying from therapy to information and consultation. No manipulation was found. Even though selection of the participating group of stakeholders by the project developers is defined in the literature as manipulation [55], we still consider the mini-workshops of the National Grid as a higher level of participation because national NGOs had a chance to comment on the process of organization of these workshops. Most of the TSOs' actions are at the levels of information and consultation. In our view, the info markets introduced by TenneT can be considered to be a mixture of information and consultation. They allow project developers to introduce themselves to stakeholders and to encourage an open dialogue. Round table discussions and mini-workshops allow a smaller number of participants to discuss specific issues. Independent observers recorded stakeholders' feedback and the TSOs replied to concerns raised by stakeholders.

Table 2 also shows the level of participation needed to address each guiding principle. For instance, therapy would entail providing information to address the perception of risks, such as impacts on human health, while the information level could address all five guiding principles. The higher we climb the ladder, the lower the number of guiding principles addressed by BESTGRID. For instance, at the partnership level, only the environment and transparency principles were addressed through cooperation between national and local NGOs and TSOs.

During the planning procedures, round table discussions, and workshops, NGOs provided input into the planning of infrastructure. This feedback corresponds to a level of consultation and placation where NGOs can play an advisory role but TSOs retain the right to judge the legitimacy or feasibility of the advice received. We attribute the efforts to settle action plans in cooperation between NGOs and TSOs at the level of placation for the pilots of National Grid and 50Hertz because decisions were frequently taken without involving NGOs or informing them about the decisions already taken. The cooperation between NGOs and two other TSOs, Elia and TenneT, allowed for more extensive feedback and a higher level of participation, reaching the level of partnership between TSOs and NGOs by jointly developing the action plans. However, this consultation was also restricted as the options and issues to be raised were defined by the developers and authorities. Finally, a key factor was that the need for the projects was identified by the developers and authorities, who were not prepared to discuss alternatives.

Existing evidence suggests that the middle level of Arnstein's ladder is the most frequent level of stakeholders' participation in developed economies. However, only consultation, involvement, or collaboration with stakeholders occurs at this level. This does not equate to full participation, but rather suggests tokenism as stakeholders cannot influence decisions.

The BESTGRID approach goes beyond this level, as it allows for a stronger and systematic collaboration between TSOs and NGOs. This can be considered an innovative approach because the two groups are involved in the same project, their collaboration is voluntary, based on the recognition of common interests and respect of each other priorities, and it is guided by an independent organization, in our case, the Renewables Grid Initiative. The NGOs Germanwatch and BirdLife actively commented on plans and provided the TSOs with guidance on the design and implementation of actions relating to public acceptance and environmental protection. These recommendations stimulated TSOs to implement some extra measures not requested by legislation. For instance, the following recommended participation and consultation tools were implemented: info-markets, mobile bus tours, and roundtable discussions. The feedback from participants showed that info-markets and information events were successful in demonstrating the need for the project to the public and organized stakeholders. The roundtable discussions helped to collect feedback and new suggestions on actions to protect the environment.

Existing evidence regarding the process of siting electricity transmission projects in other countries corresponds with our results. For example, in its review of ongoing infrastructure projects, the U.S. Environmental Protection Agency came to the conclusion that public participation may occur at multiple levels and at different stages of the project, and that different stakeholders may choose to engage at different levels. Generally, a higher level of participation requires more effort from both project developers and stakeholders, and therefore attracts fewer stakeholders. In light of this, we mapped stakeholders and their level of engagement across the four pilot projects as well as the different levels of Arnstein's ladder (Table 3).

Table 2. Stakeholders in four pilot projects.

	Therapy	Information	Consultation	Cooperation
Elia		Local residents	Environmental stakeholders	IEW, BBL, Germanwatch, BirdLife
50Hertz	Local residents	Local residents, NGOs	Environmental and critical stakeholders	DUH, NABU, Natagora, Germanwatch, BirdLife
TenneT	Local residents	Local residents	Environmental and critical stakeholders	Germanwatch, BirdLife
National Grid	Established group of stakeholders	Local residents	Environmental stakeholders	Germanwatch, BirdLife

Content analysis of reports, protocols, and interviews containing the concerns of both organized stakeholders and laypeople, showed that the guiding principle "need" was one of the most questioned principles in all three pilots and was mentioned 235 times. "Environment" was also discussed often (mentioned 149 times) as well as "transparency" (mentioned 125 times). The guiding principles "benefit" and "engagement" were raised significantly less frequently (45 and 53 times, respectively). Figure 3 shows the patterns of concerns regarding guiding principles.

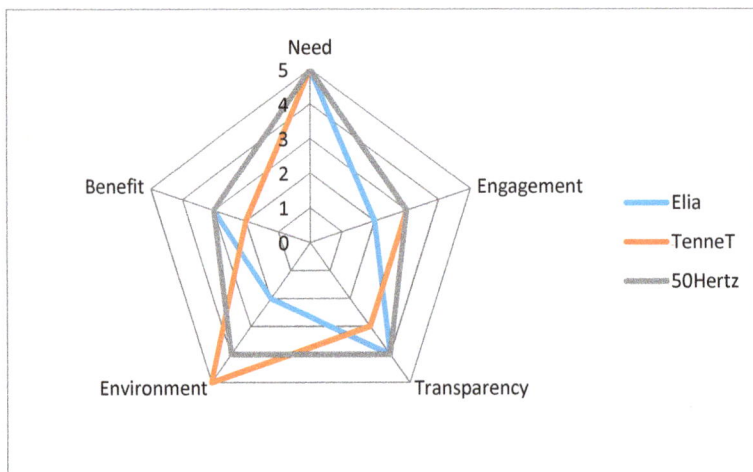

Figure 3. Concerns according to five guiding principles.

The need for the projects was often unclear to local inhabitants. Generally speaking, they questioned why the projects were needed, in light of available alternatives such as decentralized energy generation and efficiency measures. They also requested clarification regarding factors influencing discussion about the need of the project, such as future energy demand in the region, population dynamics, migration of industries, energy efficiency measures, and options for decentralized energy generation.

"Environment" was the second most heatedly discussed principle and it raised many very specific concerns. In the case of the Elia project, concerns were limited and mainly pertained to the impacts during the construction period. This is largely due to the location and the relative size of the Elia project. In the TenneT pilot, the impacts on the environment, including those on nature and ecosystems, were fiercely debated, as well as the visibility impacts on the landscape and impacts of EMF on human health. In this case, stakeholders perceived the direct involvement of environmental NGOs very positively, but concerns about landscape impacts and EMFs were only partly addressed.

The BESTGRID actions allowed stakeholder concerns to be heard and understood. BESTGRID also provided the opportunity for NGOs to advise on how to deal with the concerns raised, including the provision of detailed information on the impacts of EMF, conducting EMF measurements, and involving independent experts to talk about the risks of EMFs. These recommendations were partly implemented by 50Hertz during the mobile information tour and round table discussions. The recommendations on visibility effects included the possibilities of merging old and new lines or applying new design options to the construction of pylons in consultation with stakeholders using 3D visualization tools. Additional research is needed in many cases to better understand whether these measures can reduce social opposition to infrastructure projects.

The principle "transparency" was frequently discussed. Stakeholders requested clear information as they had doubts regarding the level of transparency of information and, in particular, questioned the sources of electricity to be fed into the grid. In all pilots, stakeholders requested more information about the preferred corridor, the criteria used for its selection, and the details of the planning process.

Transparency of information is often connected with the issue of trust in the companies, officials, or sources of information. During the public information events, concerns about the trustworthiness of information were raised several times, especially with respect to published information or experts engaged by TSOs to report about contested issues, such as the impacts of EMF. During the focus group discussions, the issue of trust also appeared several times, namely regarding the sources of information. Public authorities and academics were regarded as the most trustworthy source of information about the risks or the need for the project.

The guiding principle "benefits" was least frequently discussed. The major concerns were co-benefits, such as the modernization of roads in the region during construction, and compensation to separate groups of stakeholders such as landowners or to the entire community.

The principle "engagement" did not attract much attention among stakeholders. Generally, they did not believe that they would have an opportunity to engage, and therefore questioned the value of engagement as a result of previous experience. Stakeholders reported their doubts that their voices would be considered and complained that information regarding the different types of possible participation was limited (Table 4).

The BESTGRID approach has not led to a higher level of public participation such as delegation or citizen control. Evidence shows that achieving these levels is possible; for example, citizen control proved to be successful in the implementation of some wind projects where public resistance turned into public support [59].

However, the approach developed and deployed in BESTGRID can be regarded as an improvement in the quality of engagement and an important learning experience for all participants. Despite the fact that the information provided by TSOs on the need for specific electricity grids, projects for supporting the energy transition, and energy security significantly contributed to open discussions about the need for the electricity transmission projects, it remains questionable whether TSOs should have to provide such information or whether this should be the responsibility of other stakeholders, such as regulators, policy-makers, or civil society itself.

Table 3. Concerns according guiding principles.

Guiding Principle	Elia (Underground Cable)	TenneT	50Hertz
Need	Unclear need as it is not certain if energy consumption in the region will be growing	Unclear need because of decentralized generation options	Unclear need for the project
Engagement	Optimum time for engagement, perception that voices will not be heard	Place of public information events, where anybody could pass by, not only already informed stakeholders	Information about who will be involved in the discussions on the project
Transparency	Planned corridor, sources of electricity	Criteria of selecting priority corridor, sources of electricity	Planning procedures, source of electricity, EMFs
Environment	Impacts of construction, visibility effects	Visibility impacts, security of transmission system, impacts from EMFs	Impacts on environment, visibility impacts, impacts from EMFs
Benefit	Modernization of routes during construction period, jobs, and impulses for socioeconomic development		Compensation to land-owners, compensation to environment

For the Elia and 50Hertz pilot projects, we also collected feedback on the usefulness of the approach and the activities that ran as part of BESTGRID. The responses generally indicated a high level of interest and satisfaction from local stakeholders. Organized stakeholders in particular, rather than laypeople, were positive about the participatory process. Some mentioned that such participatory processes would require more time but could indeed provide stronger arguments in favor of projects, and therefore save time at a later stage of the permitting procedures. One interesting recommendation from both experts and laypeople was that ideas and propositions for possible corridor alternatives should be systematically collected.

In the case of Elia, stakeholder organizations, such as public authorities and NGOs, declared that the project helped to raise awareness and increase knowledge levels (87% of respondents). For the majority of stakeholders (82%), their interest in BESTGRID and the project increased after the public information events. Respondents appreciated the content discussed at the events, with 45% of participants judging it to be very good and 55% good. However, generally speaking, it seems that additional information is desired. Overall, 35% declared their satisfaction with the information provided during the workshop, but 65% of participants indicated their wish for more information. It is important to note that, according to stakeholders, more information is required from local authorities about the details of the planned corridor and its alternatives, the possibilities for participation and engagement, the planning process, and the expected benefits generated by the project for affected communities.

In the case of 50Hertz, evaluation of the feedback from stakeholders showed that 85% of all participants considered the round table discussions interesting and that they provided stakeholders with important information and suggestions. A total of 80% of participants thought that the local stakeholders and experts from the scientific field who participated in discussions had sufficient expertise and provided valuable input, and 75% of all participants will definitely consider participating in follow-up events. However, participants noticed that the value of round table discussions is limited because it is more difficult to reach a broad range of interested stakeholders through such a forum.

5. Discussion

Our results showed that concerns about the need for a project, issues of visibility, and impacts on human health as well as transparency of information about planning procedures are more relevant than concerns about compensations or benefits from the projects. These results correspond with existing evidence. For example, Porsius et al. (2014) [60] suggested that perceptions about the impacts from EMF on human health are among the greatest concerns regarding the deployment of electricity transmission grids. The results of their study showed an increase in non-specific health complaints and attribution of these complaints to a new power line in a residential area. Tempesta et al. (2014) [61] found that overhead transmission lines have major impacts on how people perceive landscape aesthetics; a pylon occupying even a minor fraction of the view can spoil the look of the landscape. Their study showed that the social benefits of burying high voltage power lines would exceed the costs in some parts of the country, such as the mountains, where the willingness of population to pay, per kilometer of power line eliminated, is higher. Elliott and Wadley (2012), who studied homeowners' perceptions of high-voltage overhead transmission lines, also identified pylon design, health effects, and visual and noise impacts as being among the major concerns [62].

In the current study, many stakeholders raised concerns about the need for large-scale transmission projects in light of available alternatives. Several alternatives were discussed such as decentralized and distributed energy generation, energy efficiency measures, and alternative routings for the transmission lines. A more detailed discussion about centralized versus decentralized or distributed energy generation was out of the scope of this research; however, this would be a very interesting topic for follow-up research. This is a very complex issue with many pros and cons for both options and several existing studies try to understand the advantages and disadvantages of both. Such research would also involve the very complex question of to what extent stakeholders' feedback should be included in a discussion about energy security issues. Furthermore, existing governance systems should be studied in more detail to understand how participatory governance works in different countries and how differing degrees of centralization or decentralization of decision-making as well as different regulatory and institutional frameworks shape the process of stakeholder involvement in discussions about energy security issues.

Regarding the guiding principle "transparency," it seems that it is not only connected with the availability of information and its clarity but also with the level of trust in the source of this information. This result also corresponds to the available evidence. When researching public opposition to new energy infrastructure, Devine-Wright (2012) investigated the role of place attachments as well as the project-related constructs such as perceived impacts, trust, and procedural justice [23]. Based on surveys from inhabitants in southwest England, he found that perceptions of positive and negative impacts of the project, followed by trust in developer and procedural justice, are identified as major factors influencing public opposition. Cotton and Devine-Wright (2011) studied the discourse among stakeholders and the local community in response to a new power line [63]. They revealed three discourses about procedures for public participation, trust in the transmission system operators, and the need for the project in light of the possibility of centralized or decentralized energy generation. In a comparative study on public beliefs about high-voltage power lines in Norway, Sweden, and the United Kingdom, Aas *et al.* (2014) used correlation analysis to reveal that the strongest correlation is between acceptance or support of the project and trust in TSO's activities [64]. Trust, therefore, seems to be one of the major factors influencing social support.

Currently there is no common agreement on how to measure either the efficiency of stakeholders' participation methods or the degree of stakeholders' participation that should be reached to make the project successful. A general lack of empirical consideration regarding the quality of participation methods arises from confusion caused by the lack of an appropriate benchmark for the evaluation and comparison of results regarding participatory processes [65]. It can be difficult to evaluate the quality of participation exercises; the evaluation often focuses more on the quality of the process itself rather than on its outcomes. During the last decade, there have been a number of attempts to specify criteria against which the effectiveness of the participatory process can be assessed. However, there are certain limitations to these attempts, and the developed criteria were not assessed in terms of how they could be applied in practice [66].

The evaluation of procedural justice and the organization of the participatory process in BESTGRID clearly show that full engagement with stakeholders has not been achieved. However, we believe that the heterogeneity of public stakeholders and the complexity of the issues impede comprehensive and deep engagement. Despite this, the value of the BESTGRID process has been fully acknowledged by the participating consortium partners as well as by the interviewed stakeholders. The value is mainly driven by the cooperation with organized stakeholders who represent different public interests and contribute different worldviews. Moreover, cooperation with national and local NGOs stimulates project developers to go beyond standard practices and motivates them to make regular improvements to the infrastructure planning process. These reasons, as well as the observations made during the course of the BESTGRID project, lead us to consider that the cooperation between TSOs and project developers more generally needs to be encouraged and facilitated as it has the potential to deliver better projects and more legitimate outcomes.

It is, therefore, essential for policy makers and energy regulators to consider inclusion of participatory and consultation activities in the decision-making process, as well as to provide measures of how these activities can be realized. In particular, we refer to the fact that capacity among NGOs is very limited and often subject to the availability of funds. In the case of BESTGRID, NGOs received funds from the Intelligent Energy Europe program to carry out the tasks allocated to them in the BESTGRID project. Once the project is terminated, these funds will no longer be available and their expertise and time will need to be moved to different issues, where funds can be raised. While some TSOs could be willing to invest resources into an NGO's capacity scheme because of the value they bring to the project, this option is considered problematic for a number of reasons. First of all, NGOs need to remain free to follow their own motivations and principles. If funded by TSOs, they could lose credibility and weaken their position in society. Generally, energy regulators will not allow TSOs to recover the costs for financing capacities within NGOs. The ability of securing capacity among key NGO actors will be important for the realization of the desired grid expansion, because organized groups usually form for reasons of opposition and only very seldom to support the infrastructure siting process.

Today public participation is strongly institutionalized and it is assumed that citizens have an opportunity to express their concerns with the help of organized groups, such as NGOs [63]. However, because of current new institutional arrangements, such as governance beyond the state, in reality citizens do not have more opportunities to express their concerns [67]. In several cases of infrastructure siting it is still the same groups of stakeholders, such as experts and decision-makers, with the power and ability to influence the processes according to their interests. If BESTGRID is assessed by using the ladder of Arnstein, it still mainly includes degrees of tokenism and still no trace of citizen power, which is a sign of unequal distribution of power between different groups of stakeholders.

What we have also learned from the BESTGRID process is that there are basic steps that should to be fully addressed in any consultation process. In the case of transmission grids, the "need" is a fundamental discussion point that, complemented with high levels of transparency and information provision, requires attention in the very early stages. The "need" for a new project is always the result of a set of assumptions and visions of the future. These have to be disclosed and efforts should be made to meet the desires of stakeholders to be engaged in the process of shaping assumptions and future scenarios. We have also learned that a stable and robust environmental protection governance framework benefits project developers because it addresses stakeholders' concerns and creates the basis for cooperation with civil society groups. However, in the BESTGRID project, we did not achieve any insight as to how to address those concerns related to the impacts on the landscape caused by infrastructure as well as visibility effects. Addressing these questions will require additional research.

Moreover, use of the term "acceptance" of a project, whose need was identified by policy-makers in cooperation with experts, should be applied with caution. As Batel *et al.* (2013) [41] pointed out, "acceptance" is often connected with the toleration of something that is inevitable. Other authors [43,66] also argue that this terminology implies a top-down, normative perspective when acceptance from inhabitants is needed only to legitimize a decision or to allow infrastructure construction without protests. Focusing only on acceptance can divert attention away from other aspects of human reasoning such as support, resistance, apathy, *etc.* BESTGRID showed that sociopolitical acceptance would also involve institutional changes that indeed would open up the possibility for inhabitants to express their opinion about the need for transmission and generation projects. Currently this is carried out in a top-down process and research on social acceptance shows this [29]. This approach hinders the establishment of an innovative renewables-based power supply.

Many different methods for stakeholder participation—going beyond standard practices and legal requirements—were applied during the duration of BESTGRID. However, the most appropriate techniques for public participation are likely to be hybrids of different methods [43]. The BESTGRID approach is the first project to have been constructed around the active cooperation between NGOs and TSOs. It can be considered a good method to address the many challenges that the infrastructure siting process faces today and could be a source of inspiration for future work.

Acknowledgments

The work described in this paper was supported by the European Commission in the frame of the Intelligent Energy for Europe program. The paper reflects the results generated by the BESTGRID project and we would like to thank the members of the BESTGRID project, especially Joanne Linnerooth-Bayer from IIASA, Anthony Patt from ETH Zurich, Antina Sanders and Theresa Schneider from the Renewables Grid Initiative, and Rotraud Hänlein from Germanwatch for their feedback, comments, and input. We would like to thank our project partners, the transmission system operators Elia, TenneT, 50Hertz, and National Grid as well as NGOs such as Germanwatch, BirdLife International, IEW, DUH, BBL, NABU, and Natagora. We are also very grateful to several experts and to the inhabitants of communities affected by the pilot projects who found time to participate in stakeholders' interviews as well as the focus group and were generous with their information and recommendations. Any remaining errors of fact or interpretation are those of the authors.

Author Contributions

Nadejda Komendantova, Marco Vocciante and Antonella Battaglini designed the data collection protocol, collected data and evaluated the results, conducted literature analysis and drafted the article jointly.

Conflicts of Interest

The authors declare no conflict of interest.

References

1. Kunreuther, H.; Gupta, S.; Bosetti, V.; Cooke, R.; Dutt V.; Ha-Duong, M.; Held, H.; Llanes-Regueiro, J.; Patt, A.; Shittu, E.; *et al*. Integrated Risk and Uncertainty Assessment of Climate Change Response Policies. In *Climate Change 2014: Mitigation of Climate Change. Contribution of Working Group III to the Fifth Assessment Report of the Intergovernmental Panel on Climate Change*; Edenhofer, O., Pichs-Madruga, R., Sokona, Y., Farahani, E., Kadner, S., Seyboth, K., Adler, A., Baum, I., Brunner, S., Eickemeier, P., *et al*., Eds.; Cambridge University Press: Cambridge, UK; New York, NY, USA, 2014.
2. IPCC. Summary for Policymakers. In *Climate Change 2014, Mitigation of Climate Change. Contribution of Working Group III to the Fifth Assessment Report of the Intergovernmental Panel on Climate Change*; Edenhofer, O., Pichs-Madruga, R., Sokona, Y., Farahani, E., Kadner, S., Seyboth, K., Adler, A., Baum, I., Brunner, S., Eickemeier, P., *et al*., Eds.; Cambridge University Press: Cambridge, UK; New York, NY, USA, 2014.
3. Riahi, K.F.; Dentener, D.; Gielen, A.; Grubler, J.; Jewell, Z.; Klimont, V.; Krey, D.; McCollum, S.; Pachauri, S.; Rao, B.; *et al*. Energy Pathways for Sustainable Development. In *GEA (2012): Global Energy Assessment—Toward a Sustainable Future*; Cambridge University Press: Cambridge, UK; New York, NY, USA, 2012; The International Institute for Applied Systems Analysis: Laxenburg, Austria, 2012.
4. European Commission. A Roadmap for moving to a competitive low carbon economy in 2050. Communication from the Commission to the European Parliament, the Council and the European Economic and Social Committee and the Committee of the regions; European Commission: Brussels, Belgium, 2011.
5. Battaglini, A.; Lilliestam, J.; Haas, A.; Patt, A.G. Development of SuperSmart Grids for a more efficient utilization of electricity from renewable resources. *J. Clean. Prod.* **2009**, *17*, 911–918.
6. German Aerospace Center. *Trans-Mediterranean Interconnection for Concentrating Solar Power*; German Aerospace Center (DLR) Institute of Technical Thermodynamics, Section Systems Analysis and Technology Assessment: Stuttgart, Germany, 2006.
7. Yergin, D. Ensuring Energy Security. *Foreign Aff.* **2006**, *85*, 69–82.
8. Ocaña C.; Hariton A. *Security of Supply in Electricity Markets. Evidence and Policy Issues*; International Energy Agency, OECD/IEA: Paris, France 2002.
9. Ölz, S.; Sims, R.; Kirchner, N. *Contribution of Renewables to Energy Security*; IEA Information Paper; OECD/IEA: Paris, France, 2007.
10. Kruyt, B.; van Vuuren, D.; de Vries, H.J.M.; Groenenberg, H. Indicators for energy security. *Energy Policy* **2009**, *37*, 2166–2181
11. Sovacool, B.; Brown, M. Competing Dimensions of Energy Security: An International Perspective. *Annu. Rev. Environ. Resour.* **2010**, *35*, 77–108.

12. TYNDP. *10-Year Network Development Plan 2012*; European Network of Transmission Systems Operators for Electricity: Brussels, Belgium, 2012.

13. ENTSO-E. *Ten-Year Network Development Plan*; European Network of Transmission System Operators for Electricity (ENTSO-E): Brussels, Belgium, 2012.

14. Ackermann, T.; Göran, A.; Söder, L. Distributed generation: A definition. *Electr. Power Syst. Res.* **2001**, *57*, 195–204.

15. European Transmission System Operators (ETSO). *Overview of the Administrative Procedures for Constructing 110 kV to 400 kV Overhead Lines*; ETSO: Brussels, Belgium, 2006.

16. Battaglini, A.; Komendantova, N.; Brtnik, P.; Patt, A. Perception of barriers for expansion of electricity grids in the European Union. *Energy Policy* **2012**, *47*, 254–259.

17. Sander, A. From "Decide, Announce, Defend" to "Announce, Discuss, Decide? Suggestions on how to Improve Public Acceptance and Legitimacy for Germany's 380 kV Grid Extension. Master's Thesis, Lund University, Lund, Sweden, September 2011.

18. Schneider, T.; Battaglini, A. Efficiency and Public Acceptance of European Grid Expansion Projects: Lessons Learned across Europe. *Renew. Energy Law Policy Rev.* **2013**, *1/2013*, 42–51.

19. Schweizer-Ries, P. *Umweltpsychologische Untersuchung der Akzeptanz von Maßnahmen zur Netzintegration Erneuerbarer Energien in der Region Wahle-Mecklar (Niedersachsen und Hessen);* Forschungsgruppe Umwelt Psychologie: Magdeburg, Germany, 2010.

20. Kunreuther, H.; Linnerooth-Bayer, J.; Fitzgerald, K. *Siting Hazardous Facilities: Lessons from Europe and America*; Wharton Risk Management and Decision Processes Center: Philadelphia, PA, USA, 1994.

21. Renewable Grid Initiative. *Beyond Public Opposition to Grid Expansion: Achieving Public Acceptance: Transparency, Participation, Benefit Sharing*; Renewable Grid Initiative: Brussels, Belgium, 2011.

22. Cowell, R., Owens, S. Governing space: planning reform and the politics of sustainability. *Government Policy* **2006**, *24*, 403–421.

23. Devine-Wright, P. Explaining "NIMBY" Objections to a Power Line: The Role of Personal, Place Attachment and Project-Related Factors. *Environ. Behav.* **2012**, *45*, 761–781.

24. Beierle, T.; Cayford, J. *Democracy in Practice: Public Participation in Environmental Decisions*; RFF Press: Washington, DC, USA, 2002.

25. Inglehart, R. *Modernization and Postmodernization: Cultural, Economic and Political Change in 43 Societies*; Princeton University Press: Princeton, NJ, USA, 1997.

26. Cohen, J.J.; Reichl, J.; Schmidthaler, M. Re-focusing research efforts on the public acceptance of energy infrastructure: A critical review. *Energy* **2014**, *76*, 4–9.

27. Burningham, K.; Barnett, J.; Thrush, D. The Limitations of the NIMBY Concept for Understanding Public Engagement with Renewable Energy Technologies: A Literature Review, Beyond Nimbyism research project Working Paper 1.3. Available online: http://geography.exeter.ac.uk/beyond_nimbyism/deliverables/bn_wp1_3.pdf (accessed on 27 November 2006).

28. Kaldellis, J., Kapsali, M., Kaldelli, E., Katsanou, E. Comparing recent views of public attitude on wind energy, photovoltaic and small hydro applications. *Renewable Energy* **2013**, *52*, 197–208.

29. Wolfsink, M. The research agenda on social acceptance of distributed generation in smart grids: Renewable as common pool resources. *Renew. Sustain. Energy Rev.* **2012**, *16*, 822–835.

30. Bell, D.; Gray, T.; Haggett, C. The "Social Gap" in Widn Farm Policy Siting Decisions: Explanations and Policy Responses. *Environ. Politics* **2005**, *14*, 460–477.

31. Krohn, S.; Damborg, S. On public attitudes towards wind power. *Renewable Energy* **1999**, *16*, 954–960.

32. Ek, C. Public acceptance and private attitudes towards "green" electricity: the case of Swedish wind power. *Energy Policy* **2005**, *33*, 1677–1689.

33. Devine-Wright, P. Beyond NIMBYism: Towards an integrated framework for understanding public perceptions of wind energy. *Wind Energy* **2005**, *8*, 125–139.

34. Wüstenhagen, R.; Wolsink, M.; Bürer, M.J. Social acceptance of renewable energy innovation: An introduction to the concept. *Energy Policy* **2007**, *35*, 2683–2691.

35. Barthes, Y.; Mays, C. *High Profile and Deep Strategy: Communication and Information Practices in France's Underground Laboratory Siting Process*; Technical Note SEGR/98, 18; Institute De Protection Et De Surete Nucleaire: Paris, France, 1998.

36. Batel, S.; Devine-Wright, P. A critical and empirical analysis of the national-local "gap" in public responses to large-scale energy infrastructures. *J. Environ. Plan. Manag.* **2015**, *58*, 2015.

37. *Media Effects: Advances in Theory and Research*, 2nd; Bryant, J., Zillmann, D., Eds.; Lawrence Erlbaum Associates, Inc.: Mahwah, NJ, USA, 2002.

38. Perhac, R. Comparative Risk Assessment: Where Does the Public Fit In? *Sci. Technol. Hum. Value* **1998**, *23*, 221–241.

39. Rowe, G.; Frewer, L. Public participation methods: A framework for evaluation. *Sci. Technol. Hum. Values Winter* **2000**, *25*, 3–29.

40. Jasanoff, S. The political science of risk perception. *Reliab. Eng. Syst. Saf.* **1998**, *59*, 91–99.

41. Batel, S.; Devine-Wright, P.; Tangeland, T. Social acceptance of low carbon energy and associated infrastructures: A critical discussion. *Energy Policy* **2013**, *58*, 1–5.

42. Aitken, M. Why we still don't understand the social aspects of wind power: A critique of key assumptions within the literature. *Energy Policy* **2010**, *38*, 1834–1841.

43. Smith, P.; McDonough, M. Beyond Public Participation: Fairness in Natural Resource Decision Making. *Soci. Nat. Resour. Int. J.* **2001**, *14*, 239–249.

44. Middendorf, G.; Busch, L. Inquiry for the public good: Democratic participation in agriculture research. *Agric. Hum. Values* **1997**, *14*, 45–57.

45. Arnstein, S.R. A Ladder of Citizen Participation. *J. Am. Plan. Assoc.* **1969**, *35*, 216–224.

46. Rau, I.; Schweizer-Ries, P.; Hildebrandt, J. The Silver Bullet for the Acceptance of Renewable Energies? In *Vulnerability, Risks, and Complexity: Impact of Global Change on Human Habitats*; Kabisch, S., Kunath, A., Schweizer-Ries, P., Steinführer, A., Eds.; Gottingen: Hogrefe, Germany, 2012; pp. 177–191.

47. McCallum, David B. and Santos, Susan L. Comparative Risk Analysis for Priority Setting. *Human Ecological Risk Assessment J.* **1997**, *3*, 1215–1234.

48. Renewable Grid Initiative. *European Grid Declaration on Electricity Network Development and Nature Conservation in Europe*; Renewable Grid Initiative: Berlin, Germany, 2011.

49. Rottmann, K. *Recommendations on Transparency and Public Participation in the Context of Electricity Transmission Lines*; Position Paper; Germanwatch: Bonn, Germany, 2014.

50. Hänlein, R. *Public Participation and Transparency in Power Grid Planning. Recommendations from the BESTGRID Project*; Germanwatch: Bonn, Germany, 2015.

51. Schneider, T.; Sander, A. *European Grid Report: Beyond Public Opposition, Lessons Learned Across Europe*; Renewable Grid Initiative: Berlin, Germany, 2012.

52. Shaw, S., Elston, J., Abbott, S. Comparative analysis of health policy implementation: The use of documentary analysis. *Policy Studies* **2004**, *25*, 259–266.

53. Yin, R.K. *Case Study Research: Design and Methods*, 3rd ed.; Sage: Thousand Oaks, CA, USA, 2003.

54. Thomas, G. A typology for the case study in social science following a review of definition, discourse and structure. *Qual. Inq.* **2011**, *17*, 511–521.

55. Flyvberg, B.; Holm, M.; Buhl, S. How common and how large are cost overruns in transport infrastructure projects? *Transp. Rev.* **2003**, *23*, 71–88.

56. Flyvberg, B. Five Misunderstandings about Case-Study Research. *Qual. Inq.* **2006**, *12*, 219.

57. Komendantova, N.; Linnerooth-Bayer, J.; Patt, A. Methodological and theoretical framework mapping commonalities and differences of separate pilot study action plans onto a common framework of actions and guiding principles. Deliverable 2.2, BESTGRID Project, Brussels, 2014. Available online: http://www.bestgrid.eu/uploads/media/D2.2_Common_Methodology_Framework.pdf (accessed on 31 March 2015).

58. Komendantova, N.; Linnerooth-Bayer, J.; Patt, A. Common Protocol for Data Collection and Recording to Ensure Comparability across Pilot Projects of the Quantified Indicators. Deliverable 2.3, BESTGRID Project, Brussels, 2014. Available online: http://www.bestgrid.eu/uploads/media/D2.3_Data_Collection_Protocol.pdf (accessed on 24 October 2013).

59. Xavier, R.; Komendantova, N.; Jarbandhan, V.; Nell, D. Participatory Governance in the Transformation of the South African Energy Sector: Critical Success Factors for Environmental Leadership. *Clean. Prod. J.* **2014**, to be submitted.

60. Porsius, J.; Claassen, L.; Smid, T.; Woudebberg, F.; Timmermans, D. Health responses to a new high-voltage power line route: Design of a quasi-experimental prospective field study in the Netherlands. *BMC Public Health* **2014**, *14*, 237.

61. Tempesta, T.; Vecchiato, D.; Girardi, P. The landscape benefits of the burial of high voltage power lines: A study in rural areas of Italy. *Landsc. Urban Plan.* **2014**, *126*, 53–64.

62. Elliott, P.; Wadley, D. Coming in Terms with Power Lines. *Int. Plan. Stud.* **2012**, *17*, 197–210.

63. Cotton, M.; Divine-Wright, P. Discourses of energy infrastructure development: A Q-method study of electricity transmission line siting in the UK. *Environ. Plan.* **2011**, *43*, 942–960.

64. Aas, O.; Devine-Wright, P.; Tangeland, T.; Batel, S. Public beliefs about high-voltage powerlines in Norway, Sweden and the United Kingdom: A comparative survey. *Energy Res. Soc. Sci.* **2014**, *2*, 30–37.

65. Lowndes, V.; Pratchett, L.; Stoker, G. Trends in Public Participation: Part 1—Local Government Perspectives. *Public Adm.* **2001**, *79*, 205–222.

66. Webler, T. "Right" Discourse on Citizen Participation. An Evaluative Yardstick. In *Competence and Fairness in Citizen Participation. Evaluating Models for Environmental Discourse*; Renn, O., Webler, T., Wiedermann, P., Eds.; Springer: Dordrecht, The Netherlands; Boston, MA, USA, 1995.

67. Swyngedouw, E. Governance Innovation and the Citizen: The Janus Face of Governance-beyond-the-State. *Urban Studies* **2005**, *42*, 1991–2006.

MDPI AG
Klybeckstrasse 64
4057 Basel, Switzerland
Tel. +41 61 683 77 34
Fax +41 61 302 89 18
http://www.mdpi.com/

Energies Editorial Office
E-mail: energies@mdpi.com
http://www.mdpi.com/journal/energies

www.ingramcontent.com/pod-product-compliance
Lightning Source LLC
Chambersburg PA
CBHW051927190326
41458CB00026B/6429